Nanomaterials and Nanochemistry

C. Bréchignac P. Houdy M. Lahmani
(Eds.)

Nanomaterials and Nanochemistry

With 461 Figures and 26 Tables

E·MRS
EUROPEAN MATERIALS
RESEARCH SOCIETY

Springer

Catherine Bréchignac, PhD
Member of l'Académie des sciences (French Academy of Sciences)
President of the CNRS
Centre universitaire Paris-Sud, Laboratoire Aimé Cotton
Bâtiment 505, 91405 Orsay Cedex, France
E-mail: catherine.brechignac@lac.u-psud.fr

Philippe Houdy, PhD
Université d'Évry
Boulevard François Mitterrand, 91025 Évry Cédex, France
E-mail: philippe.houdy@univ-evry.fr

Marcel Lahmani, PhD
Club Nano-Micro-Technologie de Paris
Boulevard François Mitterrand, 91025 Évry Cédex, France
E-mail: marcel.lahmani@univ-evry.fr

Translation from the French language edition of
"Les nanosciences – Nanomatériaux et nanochimie"

Library of Congress Control Number: 2007928828

ISBN 978-3-540-72992-1 Springer Berlin Heidelberg New York

Springer is a part of Springer Science+Business Media.

springer.com

Typesetting: Data prepared S. Lyle and by SPi using a Springer LaTeX macro package
Cover design: WMX Design GmbH, Heidelberg, using a figure from the Hanbücken–Neddermeyer collaboration, Appl. Surf. Sci. **234**, 307 (2004)

Printed on acid-free paper SPIN 11743941 57/3180/SPi 5 4 3 2 1 0

Foreword to the French Edition

Nanomaterials constitute an important branch in the burgeoning field of nanoscience. Size reduction can lead to a whole range of new physicochemical properties and a wealth of potential applications. However, access to these nanostructured entities requires the development of suitable methods for their elaboration.

This book, aimed at MSc or PhD students and young engineers, research scientists and teachers, provides a complete review of all relevant aspects from the fabrication of nanomaterials able to carry out new functions to the self-assembly of complex structures.

Part I provides a theoretical description of the basic principles and fundamental properties of nanomaterials, whilst Part II treats the physical and chemical properties of nanoscale structures. Methods for designing and fabricating such structures are then discussed in Parts III and IV.

In Part V, a great many industrial applications, some still under development, are used to demonstrate the significant economic potential of these new structures and their consequences in various areas of everyday life.

Supramolecular chemistry can provide novel ways of moving forward in this domain. Indeed, molecular recognition phenomena, based on molecular information, can be used to form supramolecular materials in a spontaneous but controlled manner, by self-organisation starting from their components. Self-organisation processes thus represent a powerful method for building functional nanomaterials, which may provide a way of avoiding ever more delicate and costly nanofabrication and nanomanipulation processes.

It seems fair to hope that the meeting of supramolecular chemistry with materials science will soon open up new lines of development in nanoscience and nanotechnology. The present work lays the foundations on which these prospects may be pursued.

Collège de France, May 2006 *Jean-Marie Lehn*

Preface to the French Edition

The present book *Nanoscience II – Nanomaterials and Nanochemistry* has been designed as the natural continuation of *Nanoscience I – Nanotechnology and Nanophysics*. It seemed to us to provide an essential complement, considering the significant developments and economic potential of nanomaterials. Many applications of nanomaterials will undoubtedly use current technology, with a few modifications. However, as work proceeds in this area, there is every reason to think that the new properties they give rise to will also lead to major industrial developments.

The chapters of the book are grouped into five main parts:

- The fundamental physicochemical principles and the basic features of matter on the nanoscale.
- The basic properties relevant to this state of matter.
- Methods for designing nanomaterials and nanoparticles.
- Fabrication processes for nanostructured bulk materials and nanoporous materials.
- A selection of current and future industrial applications.

As a guide to the layout of the book, let us recall a few general ideas.

First of all, what is meant by the term 'nanomaterial'? From an etymological standpoint, it would not appear to be very explicit. Indeed, the prefix 'nano' used in scales of physical units means one billionth, or 10^{-9}, of the relevant unit. In the present case it refers to the nanometer, or one billionth of a meter. When we use the term nanomaterial, we are thus specifying an order of magnitude of a geometric dimension. But then what is it in nanomaterials that is of nanometric dimensions?

To answer this question, we must now consider the second part of the term, viz., 'material'. A material is matter that has been transformed or adapted to be able to fulfill some particular function. One can say that this matter has been functionalised. Many materials we use and which appear to the naked eye to be of a perfectly continuous constitution are in fact made up of grains of crystallised matter with dimensions often of the order of the micron (one millionth of a meter, or 10^{-6} m). This is true in particular for most metals and ceramics in common use, but it is not the case for glasses and so-called plastics, which are amorphous, or can be considered as such for the purposes of the present discussion. These micrometric grains are of course very small compared with the dimensions of the objects generally made with such materials. However, they are very large compared with the dimensions of the atoms that make them up. Indeed, atoms have diameters ten thousand times smaller than these grains. Consequently, there are some $(10^4)^3 = 10^{12}$ or a thousand billion iron atoms in a grain of steel of diameter 1 micron.

Forty years ago, it was realised that the properties of certain materials could be modified, improved or adapted in specific ways if, during the fabrication process, the grains making them up could be made much smaller. The first 'nanomaterials' were born. They can be found today in many and varied fields of application, from cosmetics, through magnetic and electronic recording devices to precision cutting tools. Further research and new developments are under way to invent or improve novel nanomaterials, exploiting the way their properties depend on grain sizes.

More recently, over the past twenty years or so, the term 'nanomaterials' has also sometimes been used to refer to matter in which the atoms make up assemblages with dimensions of the order of a few nanometers. A priori, these assemblages, known as clusters, have nothing in common with nanomaterials as they were previously defined. By their very nature, these new materials, unlike their predecessors, can only be conceived on the nanometric scale. However, they too can exhibit quite exceptional properties and are currently the subject of much scientific interest both on the level of fundamental research and for their prospective applications. The elaboration of memory cells on a quasi-molecular scale can be cited as one of the most exciting of these prospects.

To get a clearer idea of the distinction between these two families of nanomaterials, let us take the example of solid architectures made from carbon atoms:

- In the solid state, carbon is known to occur in two crystal forms: graphite and diamond. Both can be produced in the form of very small grains, a few nanometers in size. Carbon can therefore be produced at least in the form of a powder, comprising nanograins of graphite or diamond. One thus seeks to establish how the properties of graphite or diamond will vary with the grain dimensions.
- Furthermore, it has now been known for around twenty years how to make a type of molecule known as a fullerene, the most familiar being C_{60}, which comprises 60 carbon atoms. We have also discovered, even more recently, how to create another special kind of architecture from carbon atoms, namely carbon nanotubes. C_{60} like the nanotubes is neither graphite nor diamond reduced to the nanometric length scale. They are both entirely novel entities, totally different from the traditional forms of solid carbon.

Conceptually, therefore, there seem to be two large families of nanomaterials and hence two communities of research scientists which have evolved independently of one another. These two communities can be distinguished in the following ways:

- by the nature and spirit of the fundamental research they carry out,
- by the applications, which are conventional for the first community because they generally seek to improve or optimise the performance of a material that is already known and used in the same field, e.g., greater data or energy storage capacity, increased hardness or greater aptitude for plastic deformation, etc. In contrast, the prospective applications are completely novel in the second family of nanomaterials, e.g., carbon nanotube memories, implying basic computer processing units on the molecular scale!

However, this distinction cannot be so clearly made in the case of metals. After all, is there a fundamental distinction between a cluster of silver atoms and a nanometric silver grain? Can we not consider a silver nanograin containing $10 \times 10 \times 10 = 10^3$ atoms as a rather large silver cluster? Is this not an artificial distinction between the two communities and the two concepts of what constitutes a nanomaterial?

From a historical perspective, the distinction between these two communities and the two concepts would appear to be justified. One community, using the so-called bottom-up approach, started with the atom and built up nano-objects from there, while the other, adopting a top-down approach set out from standard bulk materials to design and produce the same materials but made up from nanometric grains.

Likewise, the development of processes and products based on advanced knowledge of the chemistry of molecular or particle synthesis, or supramolecular chemistry, will lead to a wide range of objects with novel properties as regards strength, optics, electronics, magnetism, biology, and so on.

In the end we should therefore arrive at a single physicochemistry of nano-objects, a multiscale physicochemistry that will take into account the organisational state and properties of nanograins as a function of their size or the number of atoms making them up.

Université de Bourgogne, Dijon, May 2006 *Jean-Claude Nièpce*

Acknowledgements

We would like to thank all members of the French nanoscience community (CNRS, CEA, universities, Grandes Ecoles, industry) who gave a very favourable welcome to the writing of these pedagogical introductions to nanotechnology and nanophysics, nanomaterials and nanochemistry (presented here), and nanobiotechnology and nanobiology (to be published soon), and without which they would have been impossible. Special thanks go, of course, to all those who contributed to these books.

We would also like to thank the late Hubert Curien of the Academy of Sciences (Paris) and Jean-Marie Lehn (Nobel Prize for Chemistry) for contributing the forewords to volumes I and II of this series, and also Patrice Hesto who gave invaluable advice when the project first began.

We warmly acknowledge the material and financial support of the French Ministry of Research, orchestrated by Jean-Louis Robert of the Department of Physics, Chemistry, and Engineering Sciences, and Michel Lanoo, Director of the Department of Physical Sciences and Mathematics at the CNRS.

Likewise, our warmest thanks go to Claude Puech, President of the Club NanoMicroTechnologie, everyone at the LMN (Laboratoire d'étude des Milieux Nanométriques at the University of Evry, France) and the GIFO (Groupement des Industries Françaises de l'Optique) for their administrative and logistical support.

Finally, we would like to thank Henri Van Damme and Dominique Givord for their continued scientific support, especially during copy-editing sessions, and Paul Siffert of the European Materials Research Society for supporting the English edition of the book.

Marcel Lahmani, Catherine Bréchignac and Philippe Houdy

Contents

Part IV Fabrication of Nanostructured Bulk Materials and Nanoporous Materials

Part V Applications of Nanomaterials

List of Contributors

Pascal Aubert
Laboratoire d'étude
des milieux nanométriques
Université d'Evry
Bld F. Mitterrand
91025 EVRY Cedex, France
pascal.aubert@univ-evry.fr

David Babonneau
Laboratoire de Métallurgie Physique
Université de Poitiers SP2MI
Bd. M. et P. Curie BP 30179
86962 Futuroscope Cedex
Chasseneuil, France
david.babonneau@univ-poitiers.fr

Frédéric Bernard
Institut Carnot de Bourgogne
Département Nanosciences
Université de Bourgogne
9 avenue Alain Savary
BP 47870
21078 Dijon, France
fbernard@u-bourgogne.fr

Serge Berthier
Université Denis Diderot
Institut des Nanosciences
de Paris, France
berthier@ccr.jussieu.fr

Jean-Claude Bertolini
Institut de Recherches
sur la Catalyse
Université Claude Bernard Lyon I
2 avenue Albert Einstein
69626 Villeurbanne Cedex, France
jean-claude.bertolini@catalyse.cnrs.fr

Laurent Bonneviot
Laboratoire de chimie
Ecole Normale Supérieure de Lyon
46 allée d'Italie
69364 Lyon Cedex 07, France
laurent.bonneviot@ens-lyon.fr

Catherine Bréchignac
President of the CNRS
France

Francisco José Cadete Santos Aires
Institut de Recherches
sur la Catalyse
Université Claude Bernard Lyon I
2 avenue Albert Einstein
69626 Villeurbanne Cedex, France
francisco.aires@catalyse.cnrs.fr

Florent Calvo
Laboratoire de Physique Quantique
Institut de Recherche
sur les Systèmes Atomiques
et Moléculaire Complexes
Université Paul Sabatier
118 route de Narbonne
31062 Toulouse, France
florent.calvo@irsamc.
ups-tlse.fr

Matthieu Chagny
DGTec
178 rue de Mayoussard
38430 Moirans, France
matthieu.chagny@dgtec.fr

Yannick Champion
Centre d'Etudes de Chimie
Métallurgique
15 rue Georges Urbain
94407 Vitry-sur-Seine, France
champion@glvt-cnrs.fr

Elisabeth Charlaix
Université Lyon 1
Laboratoire de Physique
de la Matière Condensée
et Nanostructures
43, boulevard du 11 novembre
69622 Villeurbanne Cedex, France
elisabeth.charlaix@lpmcn.
univ-lyon1.fr

Alexa Courty
Laboratoire des matériaux
mésoscopiques et nanométriques
Université Pierre et Marie Curie
4, place Jussieu
75005 Paris, France
courty@sri.jussieu.fr

Patrick Couvreur
UMR CNRS Physico-chimie,
Pharmacotechnie et Biopharmacie
Université de Paris-Sud,
5 rue Jean-Baptiste Clément
92296 Châtenay-Malabry, France
patrick.couvreur@cep.u-psud.fr

Jean-Alain Dalmon
Institut de Recherches
sur la Catalyse
Université Claude Bernard Lyon I
2 avenue Albert Einstein
69626 Villeurbanne Cedex
jean-alain.dalmon@catalyse.
cnrs.fr

Renaud Denoyel
MADIREL
Centre de Jérôme
Université de Provence,
13397 Marseille Cedex 20, France
renaud.denoyel@up.univ-mrs.fr

Francesco Di Renzo
Ecole normale supérieure
de chimie de Montpellier
8 rue Ecole Normale
34296 Montpellier, France
direnzo@enscm.fr

Véronique Dufaud
Laboratoire de chimie
Ecole Normale Supérieure de Lyon
46 allée d'Italie
69364 Lyon Cedex 07, France
vdufaud@ens-lyon.fr

Zhi-Qiang Feng
Laboratoire de mécanique
et énergétique d'Evry
Université d'Evry-Val d'Essonne
40, rue du Pelvoux
91020 Evry Cedex, France
feng@iup.univ-evry.fr

Marc Fivel
Génie physique et mécanique
des matériaux
ENSPG BP 46
38402 Saint-Martin d'Hères
France
marc.fivel@inpg.fr

Eric Gaffet
Nanomaterials Research Group
Université de technologie
de Belfort-Montbéliard
90010 Belfort Cedex, France
eric.gaffet@utbm.fr

Dominique Givord
Laboratoire Louis Néel
CNRS BP 166
38042 Grenoble, France
givord@grenoble.cnrs.fr

Claude Henry
CRMCN-CNRS, Campus de Luminy
Case 913, 13288 Marseille
Cedex 09, France
henry@crmcn.univ-mrs.fr

Philippe Houdy
Laboratoire d'étude
des milieux nanométriques
Université d'Evry
Bld F. Mitterrand
91025 EVRY Cedex, France
philippe.houdy@univ-evry.fr

Dorothée Ingert
Laboratoire des matériaux
mésoscopiques et nanométriques
Université Pierre et Marie Curie
4, place Jussieu
75005 Paris, France
ingert@sri.jussieu.fr

Pierre-Paul Jobert
DGTec
178 rue de Mayoussard
38430 Moirans, France
pierre-paul.jobert@dgtec.fr

Pierre Joli
Laboratoire systèmes complexes
Université d'Evry-Val d'Essonne
40, rue du Pelvoux
91020 Evry Cedex, France
pierre.joli@iup.univ-evry.fr

Pierre Labastie
Collisions agrégats réactivité
Institut de Recherche
sur les Systèmes Atomiques
et Moléculaire Complexes
Université Paul Sabatier
118 route de Narbonne
31062 Toulouse, France
pierre.labastie@irsamc.
ups-tlse.fr

Sid Labdi
Laboratoire d'étude
des milieux nanométriques
Université d'Evry
Bld F. Mitterrand
91025 EVRY Cedex, France
sid.labdi@univ-evry.fr

Marcel Lahmani
Laboratoire d'étude
des milieux nanométriques
Université d'Evry
Bld F. Mitterrand
91025 EVRY Cedex, France
lahmani@bp.univ-evry.fr

Christophe Laurent
CIRIMAT
Université Paul Sabatier
118 route de Narbonne
31062 Toulouse Cedex 09, France
laurent@chimie.ups-tlse.fr

Gérard Le Caër
GMCN Université de Rennes I
263, avenue du Général Leclerc
Campus de Beaulieu
35042 Rennes Cedex, France
gerard.le-caer@univ-rennes1.fr

Florence Lecouturier
Laboratoire National des Champs Magnétiques Pulsés
Université Paul Sabatier
143, avenue de Rangueil BP 14245
31432 Toulouse Cedex 4, France
lecouturier@lncmp.org

Jacques Livage
Collège de France
Laboratoire de la Matière Condensée
Université Pierre et Marie Curie
4, place Jussieu
75252 Paris Cedex 05, France
livage@ccr.jussieu.fr

Patrick Maestro
Rhodia Recherches et Technologies,
Aubervilliers, France
patrick.maestro@eu.rhodia.com

Alain Mocellin
Institut National Polytechnique de Lorraine
Ecole des Mines de Nancy
Parc de Saurupt
54042 Nancy Cedex, France
alain.mocellin@mines.inpl-nancy.fr

Jean-Claude Nièpce
Institut Carnot de Bourgogne
Département Nanosciences
Université de Bourgogne
9 avenue Alain Savary
BP 47870
21078 Dijon, France
jcniepce@u-bourgogne.fr

Jean-François Nierengarten
Groupe de Chimie des Fullèrenes et des Systèmes Conjugués
Laboratoire de Chimie de Coordination du CNRS
205 route de Narbonne
31077 Toulouse Cedex 4, France
jfnierengarten@lcc-toulouse.fr

Joël Patarin
Laboratoire de matériaux à porosité controlée
Ecole Nationale Supérieure de Chimie de Mulhouse
3, rue A. Werner
68093 Mulhouse Cedex, France
joel.patarin@uha.fr

Alberto Pimpinelli
LASMEA
Université Clermont 2
24, avenue des Landais
63177 Aubières Cedex, France
Visiting professor
University of Maryland, USA
alpimpin@univ-bpclermont.fr

Laurent Pizzagalli
Laboratoire de Métallurgie Physique
Université de Poitiers SP2MI
Bd. M. et P. Curie BP 30179
86962 Futuroscope Cedex
Chasseneuil, France
laurent.pizzagalli@univ-poitiers.fr

Emmanuelle Reynaud
LARMAUR
Université de Rennes I
35042 Rennes, France
emmanuelle.reynaud@univ-rennes1.fr

Johannes Richardi
Laboratoire des matériaux
mésoscopiques et nanométriques
Université Pierre et Marie Curie
4, place Jussieu
75005 Paris, France
richardi@ccr.jussieu.fr

Jean-Paul Rivière
Laboratoire de Métallurgie Physique
Université de Poitiers SP2MI
Bd. M. et P. Curie BP 30179
86962 Futuroscope Cedex
Chasseneuil, France
jean.paul.riviere@univ-poitiers.fr

Jean-Luc Rousset
Institut de Recherches
sur la Catalyse
Université Claude Bernard Lyon I
2 avenue Albert Einstein
69626 Villeurbanne Cedex
jean-luc.rousset@catalyse.cnrs.fr

Didier Roux
Saint-Gobain, La Défense
Paris, France
didier.roux@saint-gobain.com

Tanguy Rouxel
LARMAUR
Université de Rennes I
35042 Rennes, France
tanguy.rouxel@univ-rennes1.fr

Nathalie Solladié
Groupe de Synthèse
de Systèmes Porphyriniques
Laboratoire de Chimie
de Coordination du CNRS
205 route de Narbonne
31077 Toulouse Cedex 4
France
solladie@lcc-toulouse.fr

Michel Soulard
Laboratoire de matériaux
à porosité controlée
Ecole Nationale Supérieure
de Chimie de Mulhouse
3, rue A. Werner
68093 Mulhouse Cedex, France
michel.soulard@uha.fr

Olivier Spalla
Département de recherche
sur l'état condensé,
les atomes et les molécules
CEA Saclay
91191 Gif-sur-Yvette Cedex
France
spalla@drecam.saclay.cea.fr

Fernand Spiegelman
Laboratoire de Physique Quantique
Institut de Recherche
sur les Systèmes Atomiques
et Moléculaire Complexes
Université Paul Sabatier
118 route de Narbonne
31062 Toulouse, France
fernand.spiegelman@irsamc.ups-tlse.fr

Abdelhafed Taleb
Laboratoire d'électrochimie
et de chimie analytique
Université Pierre et Marie Curie
4, place Jussieu
75005 Paris, France
ataleb@ccr.jussieu.fr

Ludovic Thilly
Laboratoire de Métallurgie Physique
Université de Poitiers SP2MI
Bd. M. et P. Curie BP 30179
86962 Futuroscope Cedex
Chasseneuil, France
ludovic.thilly@univ-poitiers.fr

Christophe Tromas
Laboratoire de Métallurgie Physique
Université de Poitiers SP2MI
Bd. M. et P. Curie BP 30179
86962 Futuroscope Cedex
Chasseneuil, France
christophe.tromas@univ-poitiers.fr

Fabrice Vallée
LASIM, University of Lyon I
43 boulevard du 11 novembre 1918
69622 Villeurbanne Cedex, France
f.vallee@lasim.univ-lyon1.fr

Henri Van Damme
PCPMD
Ecole supérieure de physique
et de chimie industrielles
10 rue Vauquelin
75231 Paris Cedex 05, France
henri.vandamme@espci.fr

Marc Verdier
Laboratoire de Thermodynamique
et Physicochimie Métallurgique
Domaine Universitaire
1130, rue de la Piscine B.P. 75
38402 Saint-Martin d'Hères
France
m.verdier@ltpcm.inpg.fr

Maria Zei
Laboratoire d'étude
des milieux nanométriques
Université d'Evry
Bld F. Mitterrand
91025 EVRY Cedex, France
mzei@univ-evry.fr

Part I

Basic Principles and Fundamental Properties

1

Size Effects on Structure and Morphology of Free or Supported Nanoparticles

C. Henry

1.1 Size and Confinement Effects

1.1.1 Introduction

There are two ways of approaching the properties of nanoscale objects: the bottom-up approach and the top-down approach. In the first, one assembles atoms and molecules into objects whose properties vary discretely with the number of constituent entities, and then increases the size of the object until this discretisation gives way in the limit to continuous variation. The relevant parameter becomes the size rather than the exact number of atoms contained in the object.

In the second case, one considers the evolution of the properties of a sample as its size is whittled down from macroscopic toward nanometric lengths. It is this approach that we shall examine here, whilst mentioning zones of overlap and exclusion between the two approaches.

1.1.2 Fraction of Surface Atoms

Consider a homogeneous solid material of compact shape (let us say spherical) and macroscopic dimensions (let us say millimetric). Most of its properties will be related to its chemical composition and crystal structure. This is what is traditionally studied in the physics and chemistry of solids. For an object of this size, the surface atoms comprise a negligible proportion of the total number of atoms and will therefore play a negligible role in the bulk properties of the material. Note, however, that surface atoms will nevertheless play a predominant role in properties involving exchanges at the interface between the object and the surrounding medium. This is the case, for example, when we consider chemical reactivity (and catalysis) and crystal growth, which are discussed later in the book.

It can be seen from Fig. 1.1 that, when the size of the object is reduced to the nanometric range, i.e., $< 10\,\mathrm{nm}$, the proportion of surface atoms is

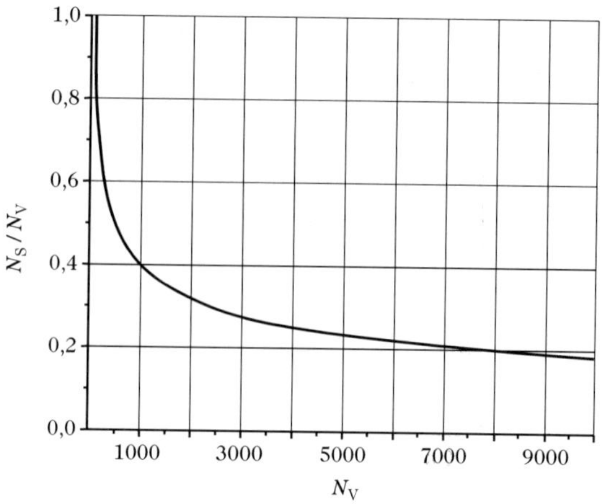

Fig. 1.1. Proportion of surface atoms for a spherical particles comprising N_v atoms with N_s at the surface

no longer negligible. Hence, at 5 nm (around 8,000 atoms), this proportion is about 20%, whilst at 2 nm (around 500 atoms), it stands at 50%. This proportion can be estimated for the transition metals by the relation

$$\frac{N_s}{N_v} \approx \frac{1}{2R} , \tag{1.1}$$

where R is the radius in nm. This empirical law gives a proportion of surface atoms of 100% for a size of 1 nm. Of course, (1.1) is no longer valid for smaller dimensions. We shall see that the fact that a large fraction of the atoms are located at the surface of the object will modify its properties. To tackle this question, we shall need to review certain physical quantities associated with surfaces, namely the specific surface energy and the surface stress.

1.1.3 Specific Surface Energy and Surface Stress

The specific surface energy γ (J/m^2) can be represented as the energy produced by cleaving a crystal divided by the surface area thereby created. More generally, the specific surface energy can be defined as follows. In order to increase the surface area of an object by an amount dA, e.g., by changing the shape of the object, the work required to do this will be

$$\mathrm{d}W = \gamma \mathrm{d}A . \tag{1.2}$$

γ is the specific surface energy. In this case, the area of the object has increased by displacing atoms from the bulk to the surface. However, one could also increase the area by stretching it, i.e., keeping the number of surface atoms constant. The work required to do this will then be

$$\mathrm{d}W = g_{ij}\mathrm{d}A \,, \tag{1.3}$$

where g_{ij} is the surface stress in J/m^2. This is a tensorial quantity because it depends on the crystallographic axes. The surface stress is related to the elastic stresses resulting from deformation of the surface (strain). It is related to the specific surface energy by

$$g_{ij} = \delta_{ij}\gamma + \frac{\partial \gamma}{\partial u_{ij}} \,, \tag{1.4}$$

where u_{ij} is the strain tensor and δ_{ij} the Kronecker symbol. Note that for a liquid there is no strain tensor and $g_{ij} = \gamma$. Indeed, if one tries to increase the surface area of a liquid, the bulk atoms will move to the surface to keep the density constant. The surface stress reduces to the specific surface energy.

1.1.4 Effect on the Lattice Parameter

Let us now consider the effects of the increase in the surface-to-volume ratio as the object size decreases. To do this, we consider first the very simple case of a liquid sphere of diameter $2R$. Due to the curvature of the surface, a pressure is generated toward the inside of the sphere. The excess pressure ΔP inside the sphere, in the purely hydrostatic case, is given by the Laplace equation

$$\Delta P \mathrm{d}V = \gamma \mathrm{d}A \,, \tag{1.5}$$

where $\mathrm{d}V$ is the volume change corresponding to a change $\mathrm{d}A$ in the area of the droplet. In the case of a sphere, (1.5) takes the form

$$\Delta P = \frac{2\gamma}{R} \,. \tag{1.6}$$

For a spherical solid, the specific surface energy must be replaced by the surface stress tensor g_{ij}. To simplify the problem, consider the case of a solid with simple cubic structure. In this case, γ is isotropic and we have

$$g = \gamma + A\frac{\mathrm{d}\gamma}{\mathrm{d}A} \,. \tag{1.7}$$

Moreover, we may recall the definition of the compressibility, viz.,

$$\chi = -\frac{\Delta V}{v\Delta P} \,, \tag{1.8}$$

where v is the atomic volume of the solid, which can also be defined as a^3, where a is the lattice parameter. Combining (1.5) and (1.8), We obtain the relative variation of the lattice parameter:

$$\frac{\Delta a}{a} = -\frac{2}{3}\chi\frac{g}{R} \,. \tag{1.9}$$

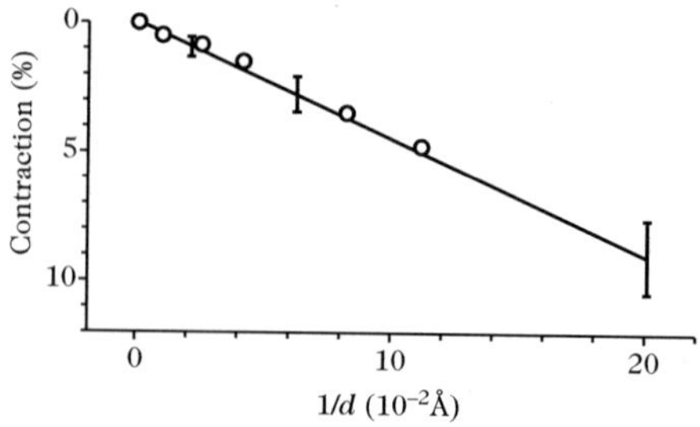

Fig. 1.2. Contraction of the lattice parameter of copper clusters as a function of the reciprocal of their diameter. *Circles* correspond to measurements of electron energy loss near an ionisation threshold (SEELFS). Taken from De Crescenzi et al. [1]. The *straight line* shows measurements of X-ray absorption (EXAFS). Taken from Apai et al. [2]

We thus find that there is a contraction of the crystal lattice due to the pressure exerted toward the interior of the particle. This contraction is proportional to the surface stress and inversely proportional to the particle size. The lattice contraction in nanometric particles has been observed on many occasions. Figure 1.2 shows the change in lattice parameter for copper clusters, measured by the electron energy loss technique known as SEELFS and also by X-ray absorption near an ionisation threshold (see below).

Determination of Local Order in a Material. EXAFS and SEELFS

These two techniques exploit the absorption of energy by an atom under the impact of a beam of X-ray photons or high-energy electrons.

The first of these, also the oldest, goes by the name of extended X-ray absorption fine structure (EXAFS). It refers to fine structure spectroscopy in the vicinity of an X-ray absorption threshold. An X-ray photon is absorbed by a given atom in the material, exciting an electron from an inner electron shell to an unoccupied state above the Fermi level, which corresponds to a well-defined energy for each type of atom, whence the chemical sensitivity of the method. The excited atom relaxes by emitting an electron whose wave function interacts with neighbouring atoms. If the atoms are in a crystal lattice, interference will occur between the wave function of the photoelectron and the wave functions of neighbouring atoms. This will cause a change in the absorption of X rays by the target atom (and neighbouring atoms), which will be detected in the form of low amplitude oscillations in the X-ray absorption spectrum near the chosen atomic absorption threshold. A full analysis of this technique can be found in [3].

In practice, carrying out a Fourier transform of these oscillations, one obtains the radial distribution of atoms in the vicinity of the target atom (up to a phase factor). For example, if we consider a solid with NaCl-type structure and examine a sodium threshold, we find a first peak in the radial distribution corresponding to the Na–Cl separation, followed by a second peak corresponding to the Na–Na separation, and a third peak corresponding to the second Na–Cl separation. The intensities of the various peaks are proportional to the number of atoms in the considered coordination sphere. This technique can thus be used to measure the local order in monatomic or multiatomic materials. In contrast to X-ray diffraction, it can be used on objects with no long-range order or on small clusters. In the latter case, it is extremely useful for determining the lattice parameter (see Fig. 1.2).

The technique known as surface extended electron energy loss fine structure (SEELFS) is analogous to EXAFS, except that the atom is excited by an electron of well-defined energy (usually in the range 3–10 keV) and the energy loss spectrum is measured near an ionisation threshold of the relevant target atom. Since the electrons do not penetrate very far into the material, this technique can only be used to study surfaces or thin films. Spectra are analysed in an analogous way to those produced by EXAFS, but for a quantitative analysis, one must take into account the fact that the excitation is obtained by electrons. The reader is referred to [4] for more details.

As can be seen from Fig. 1.2, the contraction varies linearly with the reciprocal of the particle size. For a diameter of 2 nm, it is 2%. According to (1.9), the gradient of the straight line yields the value of the surface stress as 3.35 J/m^2. The pressure exerted on the crystal lattice is then 6.7 GPa, which is extremely high.

It is reasonable to ask how far (1.9) remains valid. Put another way, can one still appeal to quantities like the specific surface energy and the surface stress, quantities defined in the context of macroscopic thermodynamics, when dealing with nanoscale systems? To address these questions, one may turn to numerical simulation. Indeed, good (semi-empirical) interatomic potentials are available for describing metals, i.e., n-body potentials in which each bond depends on the local atomic environment, in contrast to the so-called pairwise potentials [5–7].

Figure 1.3 shows the change in the lattice parameter as a function of the reciprocal of the radius of spherical particles, obtained by numerical simulation using EAM-type (embedded atom method) semi-empirical potentials [8]. The relationship is linear down to a size of about $4a_0$, where a_0 is the lattice parameter. This corresponds to a diameter of 2.5–3 nm. One might expect to find that at smaller sizes the relationship expressed by (1.9) would no longer be valid. However, it seems that this discrepancy is rather due to the fact that the specific surface energy and surface stress are no longer constant. Indeed, with the same kind of simulation, these two quantities have been calculated for different (spherical) particle sizes and the results do indeed show that they are no longer constant below a diameter of about 2–3 nm. In fact, they increase as the size continues to decrease. One might think that these deviations are due to the constraint, imposed in the calculation, of a spherical particle shape which, as we shall see below, does not correspond to the equilibrium shape of the crystal particles. In fact, this is not the case for, as we shall show in Sect. 1.2.2, even for nanoparticles having their equilibrium shape, the surface energy and surface stress nevertheless increase as the size decreases.

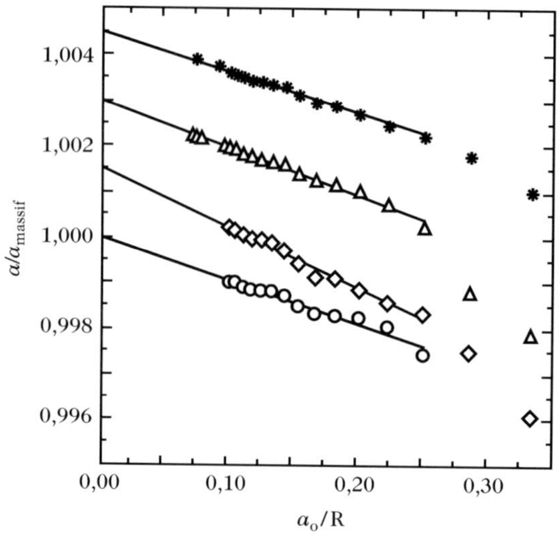

Fig. 1.3. Change in lattice parameter, relative to the bulk solid, as a function of the reciprocal of the radius for spherical clusters (with fcc structure) of Ag (*stars*), Au (*triangles*), Cu (*diamonds*), and Pt (*circles*). Numerical simulations at 0 K. Taken from Swaminarayan et al. [8]

1.1.5 Effect on the Phonon Density of States

Another effect of size reduction can be seen in the lattice dynamics of nanometric particles. When the surface-to-volume ratio reaches a certain value, the phonon spectrum broadens [9, 10]. Phonons are quasi-particles representing the vibrational modes of atoms in the lattice. On the low frequency side, the broadening is due to the contribution of surface atoms which have softer modes. The broadening toward higher frequencies is due to the lattice contraction which corresponds to increased rigidity in the system, itself the consequence of increased interatomic forces. These changes in the distribution of the phonon spectrum also affect the thermodynamic properties of the system. Note in particular that there is an increase in the vibrational entropy and that the specific heat deviates from a T^3 dependence at low temperatures.

The increase in the surface-to-volume ratio when the size decreases also has a significant effect on the melting temperature of nano-objects. This point will be dealt with in the chapter on phase transitions.

1.2 Nanoparticle Morphology

1.2.1 Equilibrium Shape of a Macroscopic Crystal

The shape of a crystal generally depends on growth conditions, which are usually very far from equilibrium, and for this reason it is not unique. However,

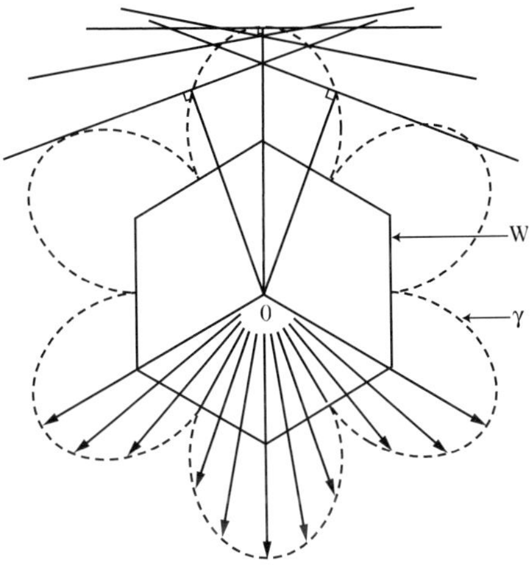

Fig. 1.4. Wulff construction of the equilibrium shape of a crystal from the γ-graph (*dashed curve*). O is the center of the crystal. The hexagon represents a projection of the equilibrium shape of the crystal (Wulff polyhedron). From [11]

under conditions of thermodynamic equilibrium, the shape of a crystal is unique. This last result was first obtained by Wulff over a century ago [12]. The solution to this problem consists in minimising the total surface energy E_s. For a liquid the result is immediate: one obtains a sphere. For a crystal, the specific surface energy γ depends on the orientation of the crystal face. One must therefore minimise

$$E_\mathrm{s} = \sum_i \gamma_i A_i \,, \tag{1.10}$$

where the index i represents the different facets with areas A_i and specific surface energy γ_i. Wulff showed that the minimal energy is obtained for a polyhedron in which the central distances h_i to the faces are proportional to their surface energies γ_i. This is the well-known Wulff theorem:

$$\frac{\gamma_i}{h_i} = \text{constant} \,. \tag{1.11}$$

Using this theorem, if we know the dependence of the surface energy on the orientation, we can easily construct the equilibrium shape of the crystal (see below).

Wulff Construction for the Equilibrium Shape of a Crystal

The Wulff construction provides an easy way of determining the equilibrium shape of a crystal if one knows the dependence of the specific surface energy on the crystal orientation (γ-graph). Consider a projection of the γ-graph along an axis of symmetry of the crystal, as shown in Fig. 1.4. Starting from the center of symmetry O, draw the radial vectors out to each point of the γ-graph and then draw straight lines normal to the radial vectors at these points. The inner envelope obtained from the set of all these normals represents the projection of the equilibrium shape of the crystal along the chosen crystal axis (a hexagon in the case illustrated). It is clear from the figure that the facets of the equilibrium shape correspond to the cusps of the γ-graph at the minima of the surface energy. When the temperature of the crystal comes close to the melting temperature, the cusps will be less and less deep (the anisotropy of the surface energy decreases) and the equilibrium shape tends to spherical.

At 0 K, the equilibrium shape contains only a few different faces with the lowest surface energies. For metals with face centered cubic (fcc) structure, the equilibrium shape is a truncated octahedron exposing the faces (111) and (100) (see Fig. 1.5). For a metal with body centered cubic (bcc) structure, the shape is a dodecahedron. For ionic crystals which have a high degree of surface energy anisotropy, a single face shows up and the equilibrium shape of crystals like NaCl or MgO is a cube.

1.2.2 Equilibrium Shape of Nanometric Crystals

The Wulff theorem can be rigorously proven for a macroscopic crystal. But what happens for a nanometric crystal? From the theoretical point of view, this problem can be handled using numerical simulation. As we have seen, we now possess realistic interatomic potentials for most metals. Molecular dynamics can then be used to seek the shape corresponding to minimal energy for constructions of different sizes. Many calculations of this kind have been carried out over the past few years (see for example [13]). Figure 1.5 shows the four main shapes that arise for fcc metals, namely, the truncated octahedron, the cubo-octahedron, the icosahedron, and the truncated decahedron (also known as the Marks decahedron).

These calculations show that for very small sizes the icosahedron is the most stable shape. This result is easy to understand, because it is the most compact shape: it maximises the binding energy by having a shape very close to spherical and it exposes only (111) facets which have the lowest surface energy. However, this structure is not a crystal structure. It cannot therefore extend to macroscopic crystals. In an icosahedron, the central atoms are subject to a very high degree of compression, whilst the surface atoms are relaxed. If an icosahedron is made to grow, the energy related to these stresses becomes too large compared with the gain in compactness and the structure tends to the normal one, i.e., fcc in this case. Figure 1.6 shows the stability diagrams of copper, silver, gold, nickel, palladium and platinum clusters up to 50,000 atoms [13].

It can be seen that, beyond a certain size, the equilibrium shape changes from an icosahedron to a truncated octahedron. The cubo-octahedron (not

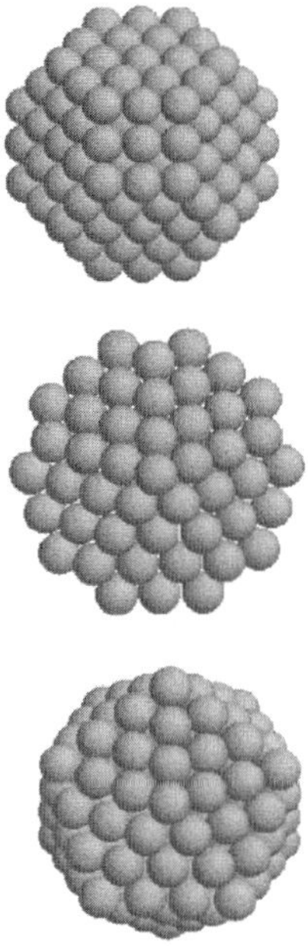

Fig. 1.5. Morphology of nanoparticles. (**a**) Truncated octahedron with 201 atoms. (**b**) Cubo-octahedron with 147 atoms. (**c**) Icosahedron with 147 atoms. (**d**) Truncated decahedron with 146 atoms

Table 1.1. Magic numbers for different clusters: icosahedron, cubo-octahedron, truncated (Marks) decahedron, and truncated octahedron (Wulff polyhedron for an fcc crystal)

Icosahedron, cubo-octahedron	13	55	147	309	561	923	1 415	2 057
Marks decahedron	75	100	146	192	238	247	268	318
Truncated octahedron	38	116	201	225	314	405	807	1 289

shown in Fig. 1.6) is always less stable than the truncated octahedron for a metal with fcc structure. Note, however, that there is an intermediate structure, the truncated decahedron, which is slightly more stable than the truncated octahedron over a certain size range. It is remarkable that the transition from the non-crystalline structure (icosahedron, truncated decahedron) to the fcc structure occurs at very different sizes depending on the metal: below 200 atoms for gold and around 30,000 atoms for copper. Note also that the calculated sizes correspond to closed shell polyhedra, i.e., with no vacancies or adatoms at the surface. These closed shells are obtained for very precise numbers of atoms called magic numbers (see Table 1.1). The magic numbers correspond to clusters with higher stability than clusters with a neighbouring number of atoms.

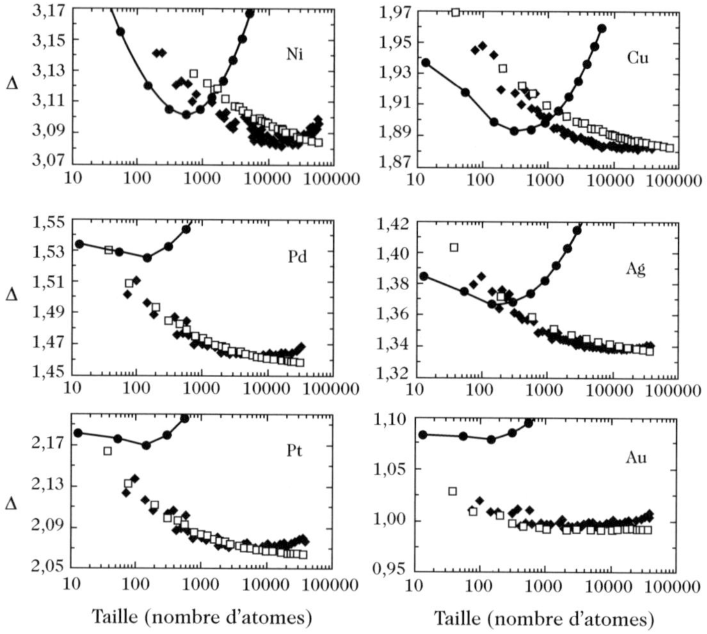

Fig. 1.6. Stability of different structures of Cu, Ag, Au, Ni, Pd and Pt clusters as a function of the number of atoms N. *Black circle*: icosahedron. *Black diamond*: truncated decahedron. *White square*: truncated octahedron. Calculations made using molecular dynamics simulations with an N-body potential. Taken from Mottet et al. [13]

For clusters with a number of atoms intermediate between two consecutive closed shells, the shape can be different and it may even oscillate between shapes with fivefold symmetry (icosahedron, decahedron) and shapes corresponding to an fcc structure [14]. In any case, it should be noted that the energy difference between the various structures for very small clusters is actually very low, so that in practice, at finite temperatures, a range of shapes is observed. In situ electron microscope observations show that the shape of small metallic particles fluctuates incessantly between different structures, passing through disordered structures. Figure 1.7 shows a series of high-resolution electron microscope images of the same gold particle containing about 450 atoms. With the electron microscope, we visualise the projections of columns of atoms. It can be seen that the particle alternates between an fcc structure and a structure with fivefold symmetry.

This phenomenon of shape and structure fluctations, known as quasimelting, has been widely studied. It was first thought that it was an artifact of electron microscopy due to the large amount of energy transferred to the sample by the electron beam (100–300 keV). However, with the development of new microscopes operating with very low currents, it was shown that, although

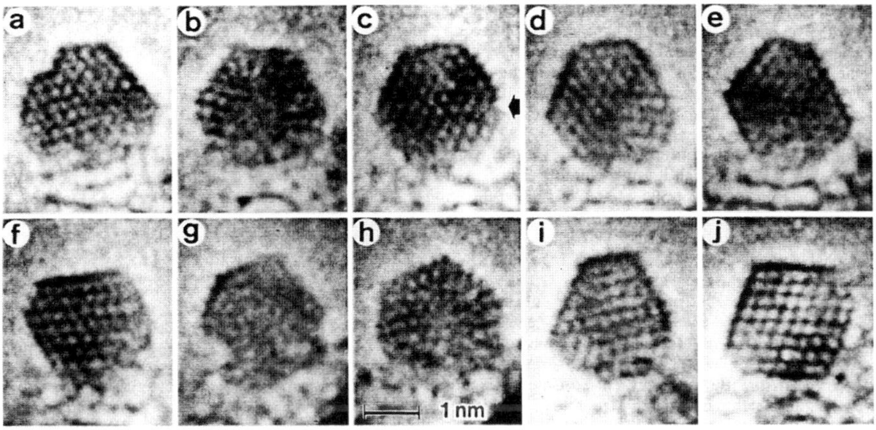

Fig. 1.7. High-resolution electron microscope images of a 2-nm gold cluster comprising 459 atoms. The structure fluctuates during the observation period. The particle changes between an fcc truncated octahedral shape [(**e**), (**f**), and (**j**)], a polyhedron with fcc structure and a twinned structure [(**a**), (**d**), and (**i**)], and a multitwinned icosahedral structure [(**b**) and (**h**)]. Taken from Iijima and Ichihashi [15] with kind permission of the American Physical Society ©1986

in some cases the electron beam can accelerate the process, the quasimelting phenomenon is an intrinsic feature of the very small size of the particles. Ajayan et Marks [16] calculated the free energy of metal particles with different structures and morphologies using a continuous model based on macroscopic thermodynamic quantities (surface energy, surface and bulk elastic energy, entropy). Using their results, the authors were able to plot the phase diagram shown in Fig. 1.8.

This shows once again that at small sizes the icosahedron is the most stable, followed by the decahedron and finally the Wulff polyhedron (crystal structure). At the smallest sizes, there is a new phase known as quasimelting. This corresponds to objects in which the energy difference between the various structures is low enough to allow them, at finite temperatures, to fluctuate between these structures (see Fig. 1.7). Doraiswamy and Marks tried to check these theoretical predictions by making quantitative measurements of the appearance of the different structures, observing gold particles with high-resolution electron microscopy [17]. They showed that, for sizes in the range 2–8 nm, observation frequencies agree with theoretical predictions. Icosahedral particles are more frequently observed at very small sizes, whilst above a size of 3–4 nm, monocrystalline particles predominate. The frequency of occurrence of decahedral particles increases with the size up to about 7 nm, exceeding the frequency of icosahedra at about 4 nm.

We have seen that for very small clusters the crystal structure is not the most stable. For larger particles, the thermodynamically most stable structure is the Wulff polyhedron, e.g., the truncated octahedron for a metal with

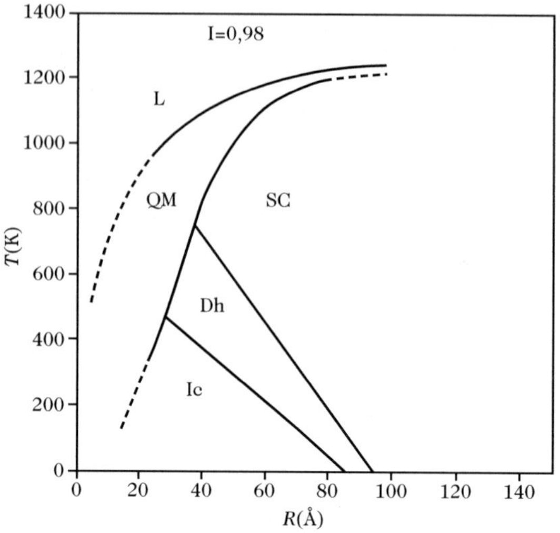

Fig. 1.8. Theoretical structural phase diagram for gold particles. Stability of different structures as a function of size and temperature: Wulff polyhedron (SC), decahedron (Dh), icosahedron (Ic), quasimelting (QM), liquid (L). Taken from Ajayan and Marks [16]

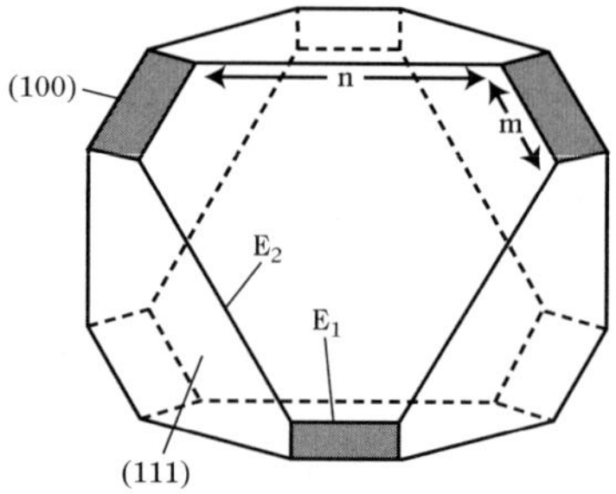

Fig. 1.9. Wulff polyhedron for an fcc crystal (truncated octahedron). The edges between (100) and (111) faces and between pairs of (111) faces have lengths m and n and energies E_1 and E_2, respectively. Taken from Marks [18]

fcc structure which is the equilibrium shape of a macroscopic crystal. The proportions of the different facets making up the polyhedron is defined by the Wulff theorem and the associated construction (see p. 9). However, it is interesting to ask for what sizes these results are valid. In fact, there are two reasons for raising this question:

- The values obtained for the surface energy of macroscopic surfaces may no longer be valid.
- The edges separating the different facets are composed of atoms with lower coordination than the atoms in the facets themselves, and the specific

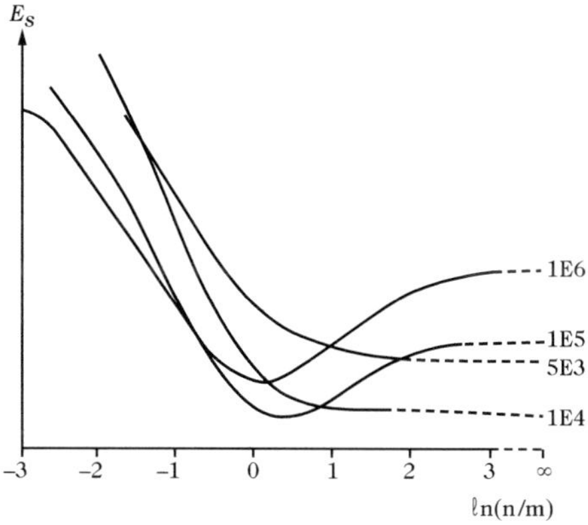

Fig. 1.10. Total surface energy for a truncated octahedron as a function of the relative extent of the (111) facets, given by the parameter n/m (see Fig. 1.9), calculated for clusters of 5,000, 10,000, 100,000 and one million atoms. Taken from Marks [18]

surface energies (extending the notion to 1D here) are thus higher and should perhaps be included in the total surface energy.

Marks tackled this question in 1985 [18]. Marks considered the case of an equilibrium shape corresponding to a truncated octahedron like the one in Fig. 1.9, whose faces are all (111) or (100) facets.

Edges separating two (111) faces have length n atoms whilst those separating a (111) face and a (100) face contain m atoms. The total surface energy is calculated for clusters comprising a variable number N of atoms and it is then minimised with respect to n and m, which are coupled variables. For this calculation, a simple 'broken bond' model was used. Figure 1.10 shows the results obtained for several cluster sizes [18].

The values of the specific surface energy are respectively 3 and 4 for the (111) and (100) faces, 5 for the edges, and 6 for the vertices. They correspond to the number of broken bonds as compared with an atom in the bulk. For large clusters, the minimum surface energy occurs for a value of the ratio n/m between 1 and 2, and it increases as the cluster size is decreased. However, below a size of around 50,000 atoms, i.e., a diameter of about 10 nm, there is no longer any minimum. The surface energy is minimised for n/m infinite, i.e., the (100) facets must have disappeared and the minimal energy shape is therefore an octahedron. This result is interesting because it shows that the equilibrium shape of an fcc cluster can be different from the equilibrium shape of a macroscopic crystal. Of course, the size at which this transition occurs depends on the values attributed to the various surface energies. We shall see in Sect. 1.2.3 on supported clusters that the disappearance of the (100) facets can be observed experimentally.

To find out what effect the edges have on the specific surface energies of clusters, we have used the same model, taking an equilibrium shape

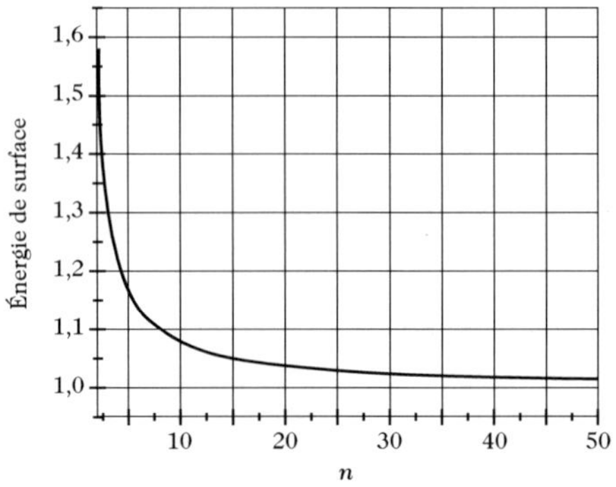

Fig. 1.11. Average specific surface energy as a function of the number of atoms n in the edges for a Wulff polyhedron ($n = m$). The average specific surface energy of an infinite crystal is normalised to unity. (Its actual value for the energy parameters used here is 3.2)

corresponding to $n = m$. Figure 1.11 shows the dependence of the average specific surface energy (total surface surface energy divided by the total number of surface atoms) on the cluster size.

We see that the specific surface energy does vary with the cluster size. The variations are large below about $n = 20$, which corresponds to a size of around 10 nm. For the smallest cluster, which contains 38 atoms, the surface energy is about 56% greater than the surface energy of a macroscopic crystal, which is not a negligible difference. Naturally, the results depend on the choice of energy parameters, but qualitatively, the effects remain the same, i.e., an increase in the specific surface energy of the cluster as the size decreases and the disappearance of (100) facets at small sizes. These effects are due to the presence of the atoms from the edges, a presence that is no longer negligible at small sizes. Another consequence of edge effects is that, if we calculate the anisotropy $\gamma_{(100)}/\gamma_{(111)}$ of the surface energy using the Wulff theorem as expressed by (1.11), we find that it increases as the size decreases. For the broken bond model, it is equal to

$$\sqrt{3}\frac{n+m}{n+2m} .$$

As we move toward large sizes, n/m tends to 1 and it converges to $2/\sqrt{3} \approx 1.15$, whereas for small sizes, n/m tends to infinity and it is equal to $\sqrt{3}$.

Note that this model is only valid if the crystal structure remains because, as we have seen, at very small sizes, a non-crystalline structure with fivefold symmetry can occur. These morphological and structural changes depend on the nature of the material.

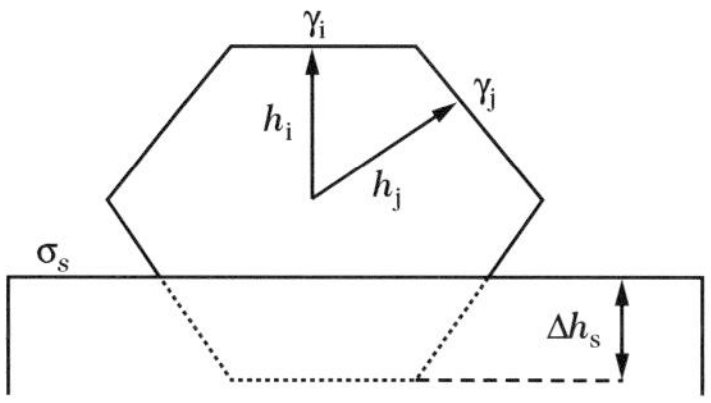

Fig. 1.12. Schematic representation of the equilibrium shape of a supported crystal. In equilibrium, the crystal interacting with the substrate assumes the form of the Wulff polyhedron (of the free crystal), truncated at the interface to a height of Δh_s. Taken from Henry [20]

1.2.3 Morphology of Supported Particles

Wulff–Kaichew Theorem

Having described what happens for free particles, i.e., not fixed to any supporting surface, let us now consider the equilibrium shape of a supported crystal, i.e., interacting with some supporting surface. This problem was solved by Kaichew [19], who showed that the equilibrium shape of a supported crystal is the Wulff polyhedron (obtained for a free crystal), truncated at the interface due to the existence of an interaction energy between the (crystal) particle and the substrate (see Fig. 1.12). The amplitude Δh_s of the truncation is given by the Wulff–Kaichew relation

$$\frac{\Delta h_s}{h_i} = \frac{E_{adh}}{\gamma_i} , \qquad (1.12)$$

where h_i is the central distance to the facet parallel to the interface, which has specific surface energy γ_i, and E_{adh} is the adhesion energy. The latter is defined as the energy required to separate to infinite distance two crystals made respectively from the deposited material and the substrate material, initially in contact over a unit area. Two extreme cases should be considered: the case of zero adhesion energy, so that $\Delta h_s = 0$, which amounts to the case of a free crystal; and the case where $E_{adh} \geq 2\gamma_i$, which corresponds to a truncation equal to $2h_i$, and a situation in which the crystal perfectly wets the substrate, i.e., it forms a continuous plane film.

The size effects discussed for a free crystal will also be relevant for supported crystals and must therefore be taken into account.

Stress Effects

The Wulff–Kaichew theorem implicitly assumes that the crystal and substrate have the same crystal lattice, i.e., the same structure and the same lattice parameter. In practice, this is rarely the case. The crystal may than be distorted to fit onto the substrate lattice, thereby creating interface stresses. The stored

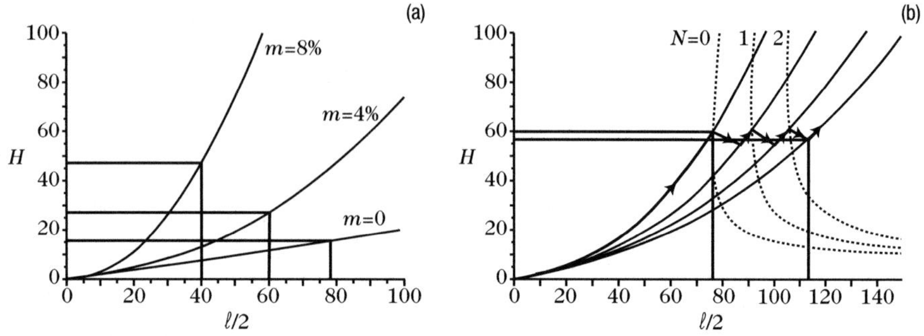

Fig. 1.13. Equilibrium shapes of stressed crystals. The profile of the crystals is represented for different values of the lattice mismatch m. *Continuous curves* represent the locus of intersection of the upper edge with the plane of the figure. (**a**) Stressed crystals. (**b**) Above a critical size, the crystals relax plastically through the successive appearance of dislocations. *Dotted curves* represent the thermodynamic criterion for the appearance of dislocations of order N $(= 0, 1, 2, \ldots)$. Taken from Müller and Kern [22]

elastic energy must then be taken into account when calculating the equilibrium shape of the supported crystal. This problem has only been tackled recently [21,22]. Müller and Kern [22] gave an analytic solution in a simplified case, using the laws of elasticity. They considered a parallelipiped equilibrium shape with base l and height H. Figure 1.13 shows half the crystal in profile. The continuous curves represent the loci of the upper edge as a function of the size for different values of the lattice mismatch m (Fig. 1.13a). The lattice mismatch is equal to the difference between the values of the lattice parameters divided by the value of the lattice parameter for the substrate.

For perfect lattice matching ($m = 0$), we obtain a straight line (see Fig. 1.13a), which indicates that the aspect ratio (the ratio H/l of the height to the size) is constant, i.e., the equilibrium shape is independent of the size. We then recover the Wulff (or Wulff–Kaichew) theorem. For nonzero lattice mismatch, the aspect ratio is no longer constant. In fact it grows with the size of the crystal, i.e., the equilibrium shape does now depend on the crystal size. Moreover, the greater the lattice mismatch, the more the aspect ratio will increase. This result can be understood quite simply from the fact that the crystal tries to reduce the stresses, localised near the interface, by reducing the area of that interface. However, when the size of the crystal increases, the stored elastic energy increases and, at some point, it becomes more economical energywise to undergo plastic relaxation by creating dislocations.

Figure 1.13b shows that, beyond a certain size, the crystal relaxes by emitting a dislocation. The upper edge then moves toward a curve corresponding to a lower lattice mismatch, before a second dislocation is introduced and the process starts over again. It can be seen that the height of the crystal oscillates around a certain value. This process will continue until a regular network

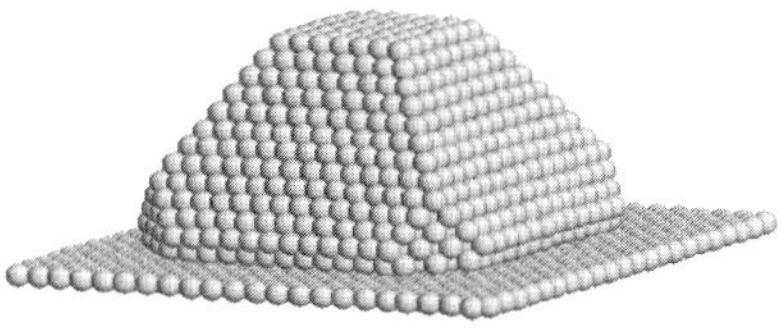

Fig. 1.14. Molecular dynamics simulation for Pd/MgO (100). Equilibrium shape of a 5-nm cluster. Taken from Vervisch et al. [23]

of interface dislocations has been created which relaxes the crystal toward its normal lattice parameter. Very recently, this problem of the equilibrium shape of stressed crystals has been studied for a specific system, Pd/MgO (100), using molecular dynamics simulation [21]. For Pd, the authors used a semi-empirical potential based on the tight-binding method in the second moment approximation, where the electron density of states is characterised only by the average bandwidth, which provides a good representation of the transition metals [6].

To simulate the Pd–MgO interaction, potential curves had to be constructed using ab initio energy calculations. Using these potentials, the authors were able to simulate Pd crystals of different sizes, treating the support as fixed. Figure 1.14 shows the equilibrium shape obtained for a 5-nm Pd particle. The equilibrium shape is indeed that of the truncated Wulff polyhedron. The calculation gives the adhesion energy directly. This energy is graphed as a function of the size of the Pd cluster in Fig. 1.15a.

As the size of the Pd particle increases, the adhesion energy drops (in absolute value), starting from a value of about $1\,\mathrm{J/m^2}$ (this energy being graduated in eV per atom in Fig. 1.15a), down to a roughly constant value of $0.85\,\mathrm{J/m^2}$. Below a size of about 7 nm, this trend is no longer followed monotonically. For sizes of about 3.5 and 7 nm, there are minima as indicated by the arrows in the figure. Figure 1.15d shows the occupation percentage of oxygen sites on the substrate, these being the most favorable absorption sites for isolated atoms or a monolayer pseudomorphic to the substrate. The drop in (the absolute value of the) adhesion energy corresponds to a reduction in occupation of the oxygen sites. This is easily understood from the fact that the metal–substrate binding energy is maximal at oxygen sites, so that the total energy of metal–substrate bonds, which represents most of the adhesion energy, is then reduced (see Fig. 1.15b).

But how do we explain these strange minima in the adhesion energy? We can obtain a better understanding of what is happening if we examine Fig. 1.16, which shows variations of the lattice parameter in the first Pd layer, at the interface, in the directions parallel and perpendicular to the substrate. Parallel to the substrate, the Pd lattice is dilated, which one would have expected, since Pd has an fcc structure like MgO, but with an 8% smaller lattice parameter. The contraction in the direction perpendicular to the interface

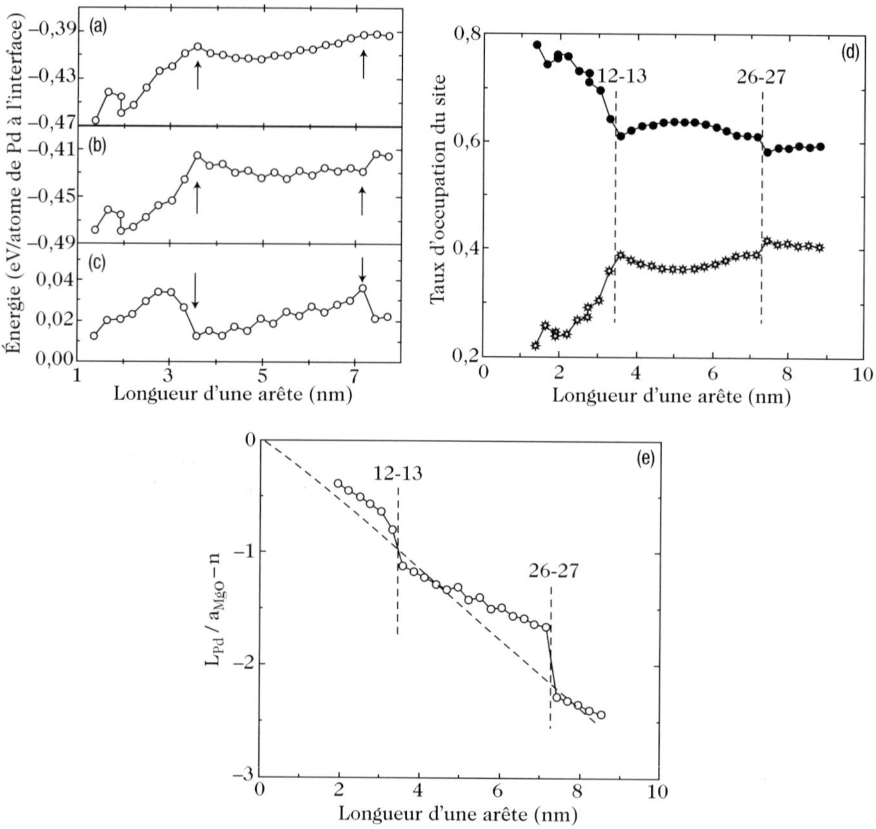

Fig. 1.15. Adhesion energy (expressed by negative values in the calculation) and occupation of oxygen sites on the surface as a function of the size of the Pd cluster supported on MgO (100). The adhesion energy (**a**) has two parts: the contribution from Pd–MgO bonds (**b**) and the contribution from Pd–Pd bonds at the interface (**c**). The occupation percentage of oxygen sites (*black circles*) and magnesium sites (*stars*) is shown in (**d**). The variation of the epitaxy parameter $L_{Pd}/a_{MgO} - n$ between the Pd and MgO lattices at the interfaces is shown in (**e**). It is equal to zero if all the Pd atoms are above oxygen sites (coherent epitaxy) and decreases linearly (*dotted curve*) for an incommensurable interface. Taken from Vervisch et al. [23]

respects the tetragonal deformation predicted by the theory of elasticity. Note, however, that these deformations are not uniform throughout the Pd particle. Indeed, in the direction parallel to the substrate, the first layer is considerably dilated, whereas the center of the cluster is slightly compressed.

The minima of the adhesion energy at 3.5 and 7 nm (see Fig. 1.15a) correspond in fact to minima in the distortion of the crystal lattice due to the appearance of dislocations in the Pd crystal lattice. When dislocations occur, this reduces the elastic stress and the total energy of the system with it, but

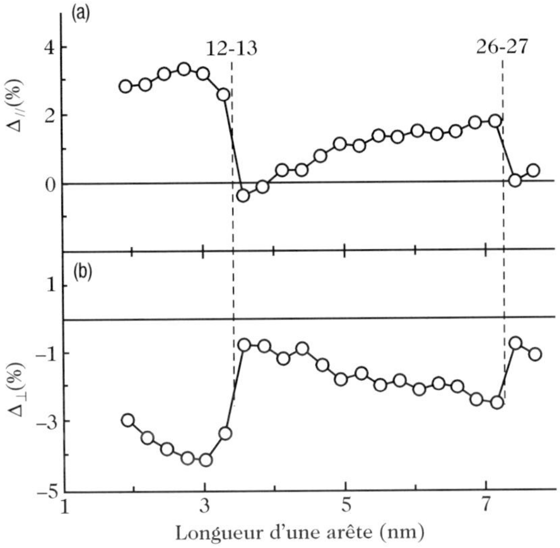

Fig. 1.16. Average strain in the first Pd layer in contact with the MgO surface (**a**) parallel and (**b**) perpendicular to the MgO (100) substrate. Taken from Vervisch et al. [23]

also the adhesion energy, because the (partial) relaxation of the Pd crystal lattice at the interface causes a drop in the number of Pd–O bonds.

Some Experimental Results

We shall now discuss some examples to see how the structure and morphology of supported clusters can be studied experimentally, relating the experimental results to the above theoretical predictions.

Morphology and Internal Structure

The oldest and most highly developed technique for studying the structure and morphology of supported clusters is transmission electron microscopy (TEM), and especially a variant of this known as high-resolution transmission electron microscopy (HRTEM). Let us note to begin with that the most delicate stage in these techniques is the preparation of the sample (see below).

Sample Preparation for TEM Observation

In order to observe a sample under the transmission electron microscope, it must be thin enough (of the order of a few tens of nanometers) to be transparent to electrons. If the supporting substrate is thick, the sample can be mechanically and/or chemically thinned, before ion milling. The last step can damage the crystal structure. One then has recourse to the carbon replica method. In this case, a thin film of carbon

Fig. 1.17. Electron microscope image of cubic MgO nanocrystals, exposing the (100) faces, obtained by burning a strip of Mg in the presence of oxygen in a vacuum chamber. The surface of the MgO nanocrystals is partially coated with gold clusters obtained by growth in ultrahigh vacuum. Taken from Giorgio et al. [24], with kind permission of Elsevier

is deposited (10–20 nm) on the sample surface by vacuum vapour deposition. The substrate is then dissolved in some solvent and the carbon membrane is transferred to the electron microscope with the metallic particles emprisoned within it. When layered crystals such as mica, molybdenite or graphite are used as substrate, it is then easy to cleave the sample to obtain thin enough substrates. Another method uses thin samples transparent to electrons as support for particle growth. These may be thin films prepared by vacuum vapour deposition or nanocrystals. MgO nanocrystals are often used because they are easy to produce by burning a magnesium strip. In addition, they are cubic in shape (see Fig. 1.17). One may then observe the particles in two different directions at right-angles to one another, thereby reconstructing their 3D shape. Figure 1.17 shows gold clusters obtained by growth on MgO nanocrystals. The atomic planes in both the substrate and the deposited particles can be visualised using high-resolution electron microscopy.

Figure 1.18 shows the morphology of Pd particles obtained by growth at high temperature (450°C) on an MgO (100) surface [25]. These samples have been annealed in ultrahigh vacuum at the same temperature for 5 hours to ensure that the shape is indeed at thermodynamic equilibrium. Note that, at this temperature, the vapour pressure of palladium is negligible and, for this reason, equilibrium can only be reached by surface diffusion of atoms in the particle.

Figure 1.18a shows a side view of a particle obtained by electron microscope imaging. Clearly visible are the incurving angles at the interface, undetectable in a top view (normal observation mode), as attested by Fig. 1.18b. The latter image, obtained by high-resolution TEM, shows the atomic planes within the particle, whence the crystal orientation of the Pd on the MgO substrate can be determined: Pd particles grow along a (100) plane on the MgO surface. By measuring the direction of the projections of the edges in the two

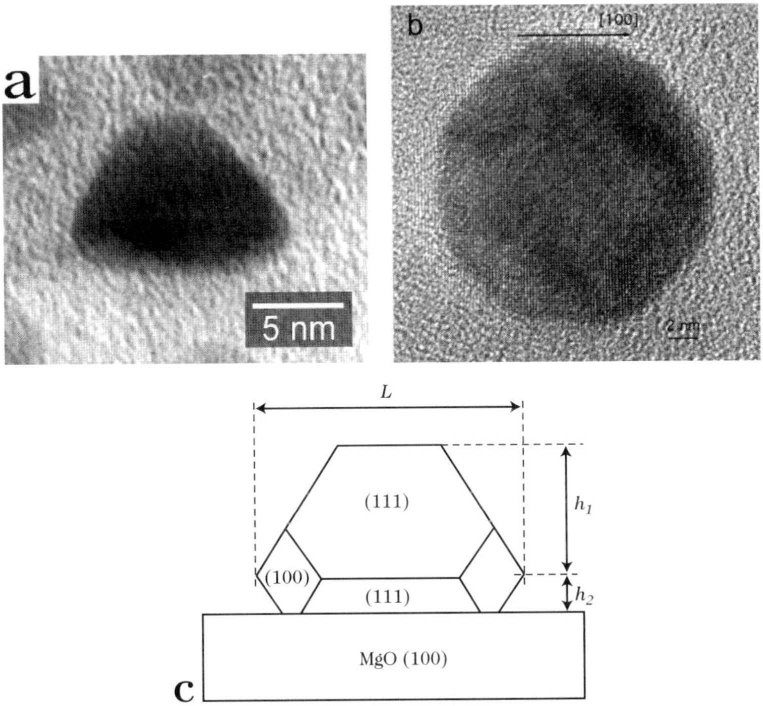

Fig. 1.18. Equilibrium morphology of Pd clusters grown epitaxially on MgO (100). (**a**) Side view of a 10-nm cluster in a ⟨110⟩ direction, obtained by TEM on a folded carbon replica. (**b**) HRTEM image of a 17-nm Pd particle, visualising {200} atomic planes separated by 0.2 nm. (**c**) Equilibrium shape (schematic) of Pd particles with a size of at least 10 nm, grown epitaxially on MgO (100). The observation direction is ⟨100⟩. Taken from Graoui et al. [25] with kind permission of Elsevier, and Prévot et al. [26]

perpendicular views of Figs. 1.18a and 1.18b, it is easy to reconstruct the 3D shape of the Pd particles, as shown schematically in Fig. 1.18c. The equilibrium shape of Pd particles on MgO is a truncated octahedron, as predicted by the Wulff–Kaichew theorem. According to this theorem, viz., (1.12), the adhesion energy is equal to the amplitude of the truncation of the equilibrium shape at the interface, i.e., $\Delta h_s = h_1 - h_2$, divided by the central distance h_1 of the upper facet and multiplied by the specific surface energy γ of this facet. h_1 and h_2 are obtained by a simple geometric construction, as shown in Fig. 1.18c, which applies to a side view in the ⟨110⟩ direction (Fig. 1.18a). The upper facet, parallel to the MgO surface, is a (100) plane because the Pd particles are grown in (100) epitaxy. Given that the surface energy of Pd (100) is $1.64\,\mathrm{J/m^2}$ [27] and that $\Delta h_s/h_1 = 0.55$, the adhesion energy of Pd on MgO (100) is equal to $0.90\,\mathrm{J/m^2}$. This value should be compared with

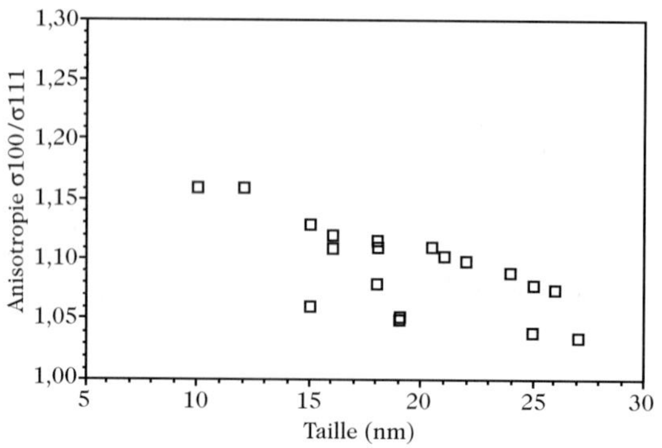

Fig. 1.19. Dependence of the surface energy anisotropy on the size of the Pd particles supported on MgO (100). Taken from Graoui et al. [28]

the one obtained by molecular dynamics simulations (see p. 17) which give $E_{\mathrm{adh}} = 0.85\,\mathrm{J/m^2}$.

The anisotropy $\gamma_{(100)}/\gamma_{(111)}$ of the surface energy is obtained from the Wulff theorem (1.11) by measuring the central distance to the corresponding facets on the profile images in the direction $\langle 110 \rangle$. Figure 1.19 shows that the surface energy anisotropy falls slightly as the particle size increases from 10 to 30 nm [28]. At 10 nm, it is equal to 1.15, close to the value predicted for a face centered cubic crystal at 0 K. However, we know that, when the temperature increases, the surface energy anisotropy falls (see Sect. 1.2.2) [29]. It tends to 1 as one approaches the melting temperature for which the shape becomes spherical. The value measured for large sizes, i.e., 1.05, is close to the values measured for metal clusters with sizes from one to a few microns [30].

The equilibrium shape for particles with sizes greater than or equal to 10 nm is thus the same as the one predicted by thermodynamics for a macroscopic crystal. For smaller particles, we have seen in the theoretical discussion that changes are predicted in the equilibrium shape for two main reasons: the influence of edge and vertex atoms which change the surface energies, and in the case of supported particles, the existence of stresses due to the difference in lattice parameter between crystal and substrate. Figure 1.20a shows a high-resolution electron microscope image in profile of a 5-nm Pd cluster supported on MgO (100). The cluster no longer has incurving angles at the interface with the substrate, signalling a greater wetting between the particle and the MgO substrate. However, as we shall see, it cannot necessarily be concluded that the adhesion energy is greater than for 10-nm particles. Using high-resolution electron microscope images (Fig. 1.20a), one can measure the average value of the interatomic distance (in fact, the distance between two columns of atoms) in each atomic layer parallel to the interface [31]. Figure 1.20b shows that the first plane of Pd atoms at the interface is accommodated to the MgO lattice, indicating that the Pd lattice is dilated by 8%. In subsequent layers, the lattice parameter decreases to reach the value in bulk Pd after just 3 atomic

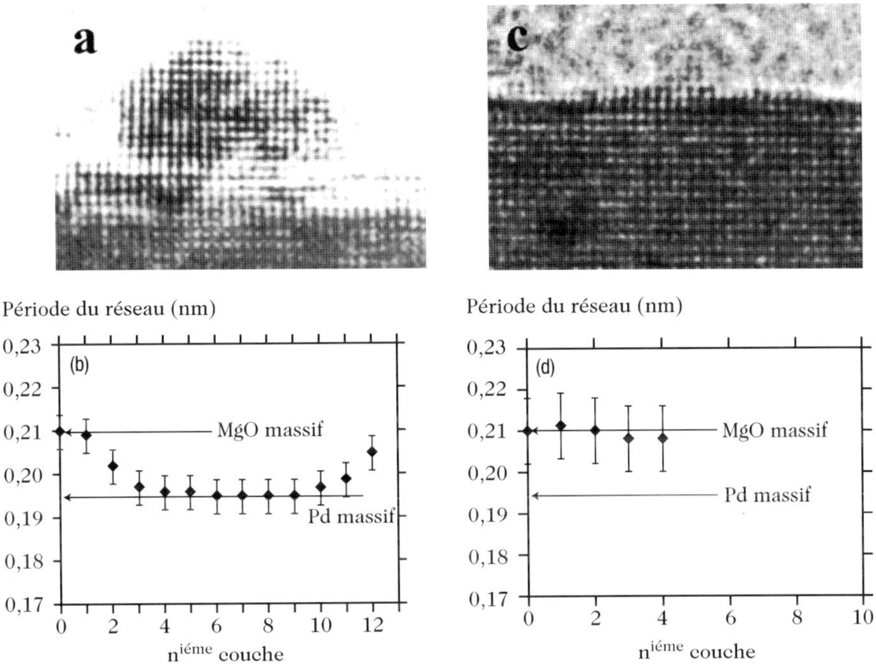

Fig. 1.20. Equilibrium shape of small Pd nanoparticles, grown epitaxially on MgO (100), observed in profile by high-resolution electron microscopy in the direction ⟨100⟩. (**a**) 5-nm particle. (**c**) 1.2 nm particle. (**b**) and (**d**) Variation of the lattice parameter in planes parallel to the interface, measured for particles of 5–6 nm and less than or equal to 2 nm, respectively. Taken from Giorgio et al. [31]

layers. In contrast, larger particles (10 nm or more) are completely relaxed. In this case the accommodation between the two lattices is achieved via a network of interface dislocations, this maintaining a perfect epitaxy [32]. The dilation observed at the top of the particle is an artifact of the high-resolution imaging [31].

The presence of stresses at the interface stores elastic energy in the particle and the equilibrium shape of the particle is no longer given by the Wulff–Kaichew theorem. The adhesion energy cannot therefore be calculated in this case. Very small particles (≤ 2 nm) have pyramidal shape (half an octahedron) and the (100) facets disappear (see Fig. 1.20c). The surface energy anisotropy is then greater than or equal to $\sqrt{3}$, agreeing with theoretical predictions. Figure 1.20d shows that these very small Pd particles have their crystal lattice dilated by 8% throughout. They are completely accommodated to the MgO lattice. As before, the Wulff–Kaichew theorem cannot be applied, because of the elastic energy due to stresses.

Electron microscopy is a very powerful technique for studying the structure and morphology of supported nanostructures. However, it is handicapped by the fact that it is generally an ex situ technique. In principle, it cannot therefore be used to monitor nanoparticle growth in situ. Note, however, that there are some electron microscopes operating in ultrahigh vacuum that have been used to study in situ growth. These microscopes are usually normal microscopes operating in a medium vacuum, but transformed to create an ultrahigh vacuum in the vicinity of the sample [33]. Around the world, there are also a few high-resolution transmission electron microscopes of commercial manufacture operating in ultrahigh vacuum [34]. At the present time, a new generation of transmission electron microscopes is being developed. These so-called environmental TEM can carry out in situ studies under pressures up to 10–50 mbar. They are more specifically destined for catalysis studies [35].

The scattering of X rays at small angles has proved to be a choice technique for studying supported nanoparticles. The grazing incidence version, known in full as grazing incidence small angle X-ray scattering (GISAXS), is a very powerful technique for in situ study of nanoparticle growth.

Grazing Incidence Small Angle X-Ray Scattering (GISAXS)

The working principle of this technique is illustrated schematically in Fig. 1.21. A beam of X-rays intercepts the surface at grazing incidence (angle α_i), close to the critical angle. The angles α_f and 2θ $(= 2\theta_\mathrm{i} + 2\theta_\mathrm{f})$ determine the directions along which the X rays are scattered. These three angles are related by the conservation of momentum q_x and q_y parallel or q_z perpendicular to the sample surface:

$$q_x = |k_\mathrm{i}|(\cos 2\theta \cos \alpha_\mathrm{f} - \cos \alpha_\mathrm{i}) \ , \qquad q_y = |k_\mathrm{i}| \sin 2\theta \cos \alpha_\mathrm{f} \ ,$$

$$q_z = |k_\mathrm{i}|(\sin \alpha_\mathrm{f} + \sin \alpha_\mathrm{i}) \ .$$

The sample can rotate around the normal to the surface through an angle ω which thereby defines the orientation of the incident X-ray beam relative to the crystal planes in the sample. To a first approximation, the average height h of the particles, their average size d, and their average separation D can be obtained from the reciprocals of the height, width and separation, respectively, of the scattering peaks.

The GISAXS technique has been used at the European Synchrotron Research Facility (ESRF) to study, for the first time in situ, in ultrahigh vacuum and in real time, the growth of metal nanoparticles on an MgO (100) surface [37].

Figure 1.22a shows a series of GISAXS images in a direction $\langle 110 \rangle$ of MgO, obtained during the growth of Pd nanoparticles on MgO (100). From about 0.2 nm, a scattering peak appears at 54.7° from the normal, corresponding to (111) facets on the Pd particles. Quantitative analysis of the images yields the average separation D between particles, their average height h, and their average size d (see Fig. 1.22b). Up to an equivalent thickness of about 1.5 nm, the aspect ratio h/d is constant and equal to 0.62. The images can be modelled quantitatively by truncated octahedra which correspond to the equilibrium shape of Pd particles observed ex situ by electron microscope (see Fig. 1.18). The observed aspect ratio can be used to obtain the adhesion energy, which

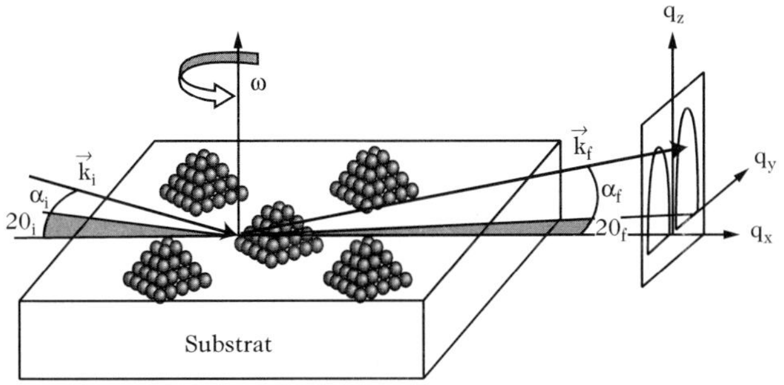

Fig. 1.21. Schematic of GISAXS experiment on supported clusters. Angles α_i, α_f, and 2θ are defined in the text. Taken from Renaud et al. [37]

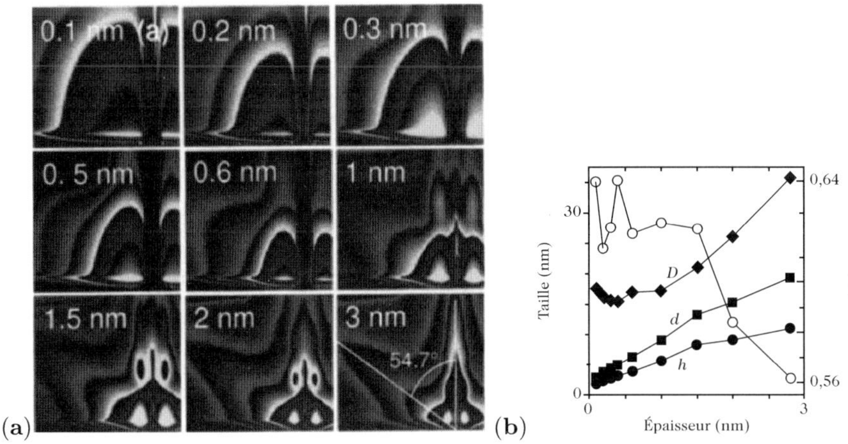

Fig. 1.22. Grazing incidence small angle X-ray scattering (GISAXS) in situ during growth of Pd nanoparticles on MgO (100), in ultrahigh vacuum. (**a**) Sequence of GISAXS images in the [110] direction of MgO as a function of the equivalent thickness of Pd deposited. The vertical direction is perpendicular to the surface and the horizontal direction is parallel to it. The intensity of scattered X-rays is on a logarithmic scale, each colour level corresponding to a factor of ten. See also the colour plate. (**b**) Average values of the distance D between particles (*black diamonds*), their size d (*black squares*), their height h (*black circles*), and their aspect ratio h/d (*white circles*) as a function of the deposited thickness, extracted from the GISAXS images. Taken from Renaud et al. [37], AAAS © 2003

is found to equal $1.1\,\mathrm{J/m^2}$, in fairly good agreement with the value obtained by electron microscope ($0.90\,\mathrm{J/m^2}$). The reduction in the aspect ratio above an equivalent deposited thickness of 1.5 nm is due to the coalescence of particles by growth. Once they have coalesced, the particles cannot spontaneously recover the equilibrium shape for kinetic reasons. The GISAXS technique is very promising because it can be used to monitor growth kinetics in real time

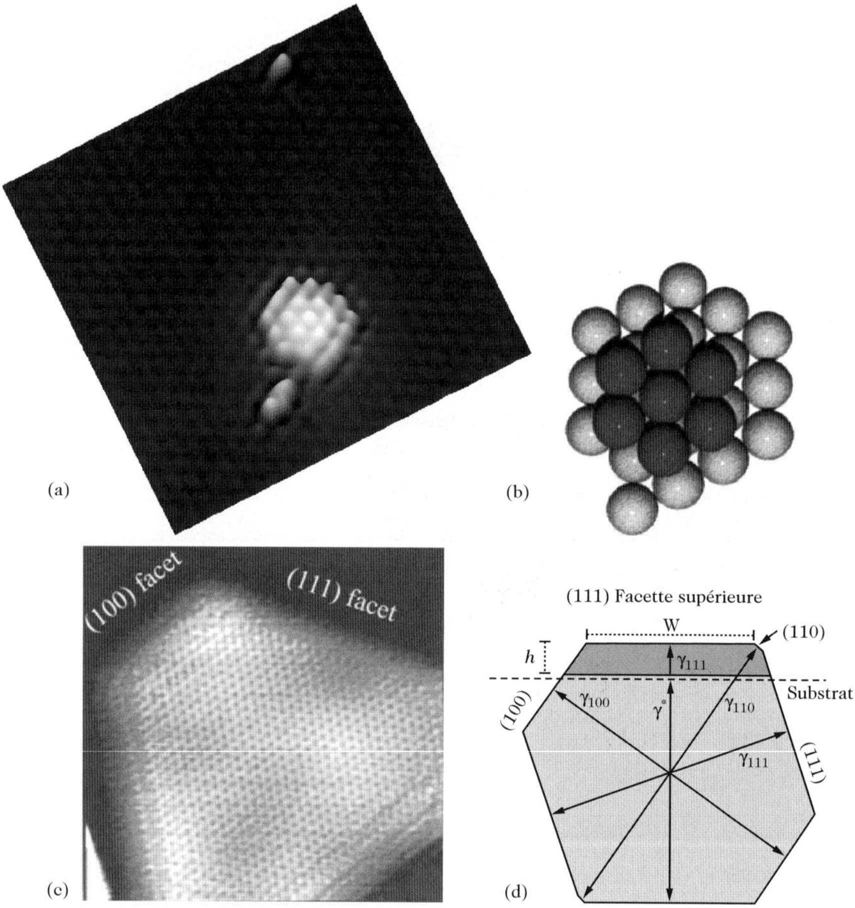

Fig. 1.23. STM images of supported Pd clusters. (**a**) Pd cluster containing 27 atoms on an MoS_2 (0001) surface. Taken from Piednoir et al. [40], with kind permission of Elsevier. (**b**) Ball model of the cluster in (**a**). (**c**) 5-nm Pd particle on an ultrathin film of alumina on an NiAl (110) surface. Taken from Hansen et al. [41]. (**d**) Equilibrium shape of Pd clusters on alumina. Taken from Hansen et al. [41], American Physical Society ©1999

with a time resolution better than one second, and it can in principle be used in a gaseous environment. It would then be possible to monitor in situ the change in equilibrium shape of nanoparticles in the presence of a gas. This technique complements electron microscopy, since it provides statistical data for a macroscopic population of clusters.

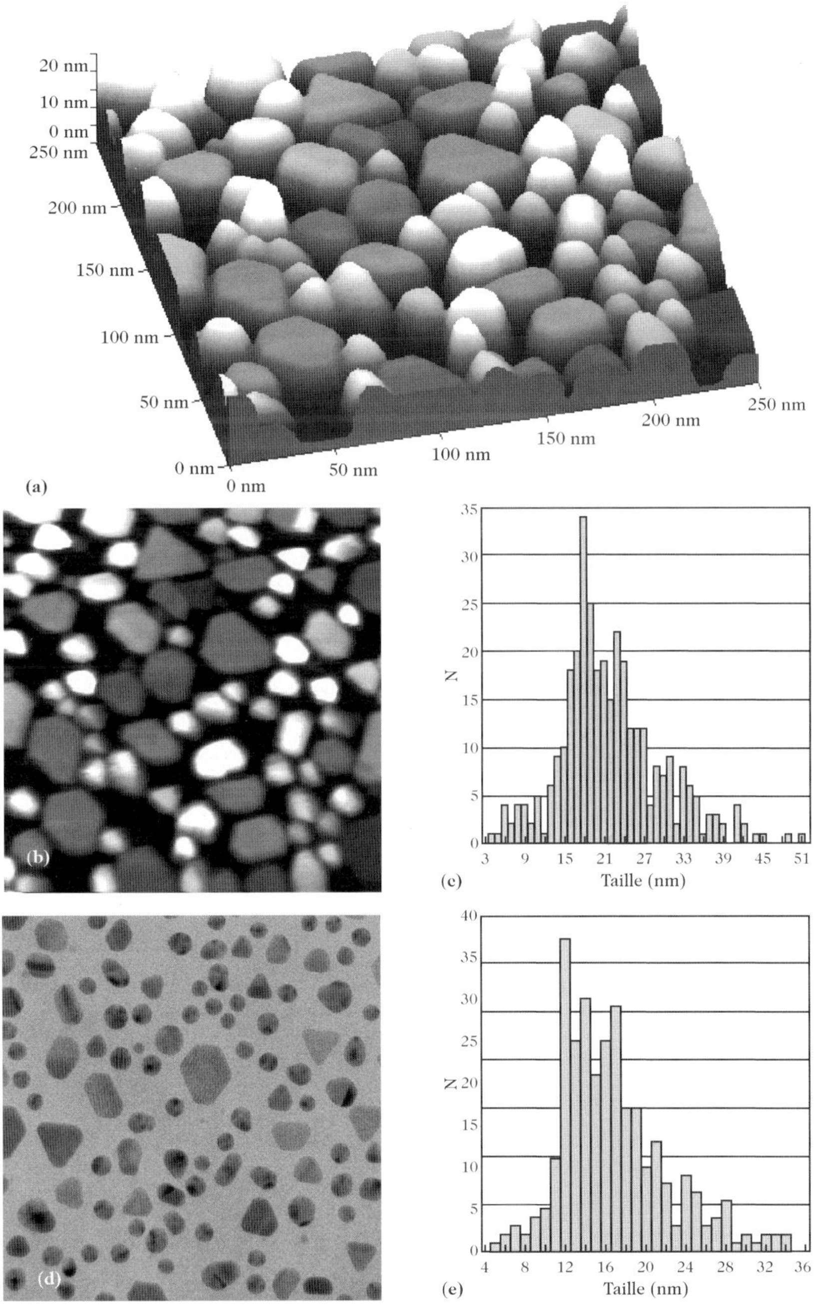

Fig. 1.24. Gold nanoparticles grown epitaxially on mica (100), visualised in situ in ultrahigh vacuum by AFM and ex situ by electron microscope. (**a**) 3D AFM image. (**b**) 2D AFM image. See also the colour plate. (**c**) Size histogram obtained from AFM images. (**d**) TEM image. (**e**) Size histogram obtained from TEM images. Taken from Ferrero [42]

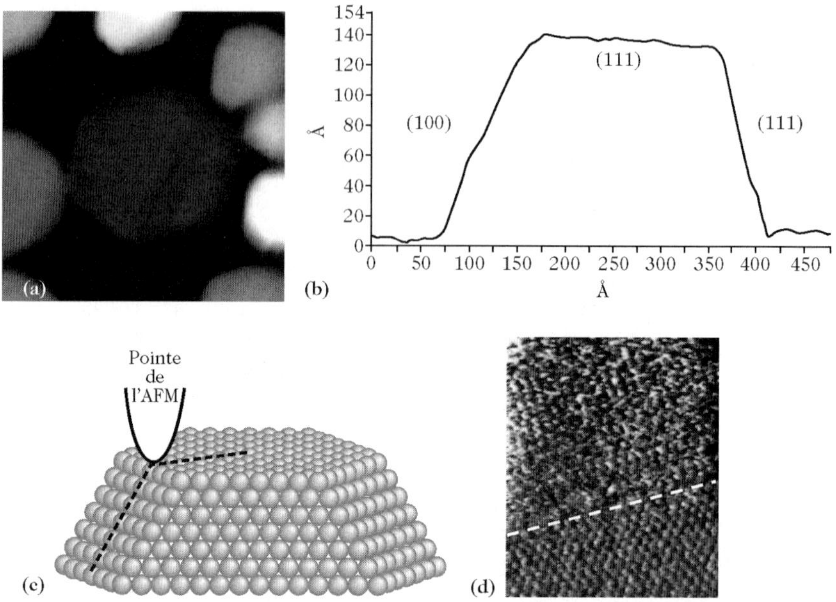

Fig. 1.25. Determining the morphology of gold nanoparticles supported on mica (100) by (contact mode) AFM. (**a**) 2D AFM image of a 27-nm hexagonal particle. (**b**) Profile of the particle in (**a**) in a direction perpendicular to the lateral facets, corrected for the convolution with the AFM tip. (**c**) Morphology of gold particles in (111) epitaxy. The upper facet is a (111) plane and the opposing lateral facets are (111) and (100) planes. (**d**) High-resolution image of the upper edge of the gold particle. The AFM tip scans the particle from the upper (111) facet toward a lateral (100) facet (see Fig. 1.25c). The *dashed line* indicates the edge separating the two types of facet. Taken from Ferrero et al. [42, 43]

Morphology and Surface Structure

The morphology of supported nanoparticles can also be obtained in situ by near-field microscopy techniques such as scanning tunneling microscopy (STM) or atomic force microscopy (AFM) (see Chaps. 3 and 4 in Vol. I of this series). These techniques can observe surfaces in 3D with resolution better than 0.1 nm. They are usually used to observe surfaces of monocrystals, revealing their atomic structure and defects such as steps, point defects, and so on. We shall see that in some circumstances, they can be applied to the study of supported nanoparticles. Scanning tunneling microscopy, which predates atomic force microscopy, was used for atomic scale observation of 2D gold and silver clusters on graphite substrates [38], and then a little later, 3D Pd clusters on graphite [39]. Figure 1.23a shows an STM image of a Pd cluster on an MoS_2 (0001) surface. The arrangement of the sulfur atoms can be seen on the substrate. The palladium cluster at the center is resolved on the atomic scale. It is two layers high and contains 20 atoms in the first layer and

7 in the second, arranged in (111) planes (see the ball model in Fig. 1.23b). The epitaxial relations can then be determined directly. Scanning tunneling microscopy is limited to the use of metallic or semiconducting samples, and cannot be used with insulating substrates like MgO or alumina. However, one can get round this difficulty by using ultrathin layers of oxides (thinner than 2 nm), obtained by epitaxial growth on metals. Figure 1.23c shows a 5-nm Pd cluster on an ultrathin alumina layer grown epitaxially on an NiAl (110) surface. The upper facet is resolved on the atomic scale and displays a (111) plane. The 3D shape can be determined. It corresponds to a truncated octahedron with aspect ratio 0.15 (see Fig. 1.23d). Assuming that the equilibrium shape has been reached, an adhesion energy of $2.8\,\mathrm{J/m^2}$ is determined. This is about three times as great as for MgO, which shows that the interaction of Pd with alumina is very strong.

Unlike STM, atomic force microscopy (AFM) can be used on any substrate and in particular on insulating substrates. Figure 1.24 shows AFM images of gold nanoparticles on a mica substrate. One of the advantages of near-field microscopy is that 3D images can be obtained directly (see Fig. 1.24a). Height measurements are extremely accurate (better than 0.1 nm). However, size measurements are subject to systematic error due to the convolution of the object shape with the tip shape (see below). This is particularly clear in Fig. 1.24.

Figures 1.24b and d show two views of the same sample obtained by AFM and electron microscopy, respectively. It is clear that the particles appear bigger in the AFM images. The size distributions obtained by the two techniques (Figs. 1.24c and 1.24e) show that by AFM the particles have an average size of 22 nm, whereas by electron microscopy the average size is only 18 nm. If the tip shape is known, the true particle size can be recovered (see below). After correcting for the convolution effect, the particle profile can be obtained very precisely. Figure 1.25b shows the profile of a gold particle of diameter 27 nm (Fig. 1.25a), obtained by AFM and corrected for the convolution with the AFM tip. Electron diffraction shows that the particles are in (111) epitaxy. In the (corrected) AFM profile, it can be seen that the lateral facets have slopes of 54° and 70°, which does indeed correspond to the angles between a (111) plane parallel to the substrate and (100) and (111) facets. The particle morphology is shown schematically in Fig. 1.25c. By AFM, the atomic lattice can be imaged on the facets of 30-nm particles [43]. Figure 1.25d shows an AFM image of the region around the upper edge of a particle. At the bottom of the image, which corresponds to the upper facet, we see the atomic lattice of a (111) plane of gold on part of the upper facet. Passing over the edge joining the lateral face, atomic resolution is lost and, beyond the edge, moving down a lateral facet, fringes can be seen perpendicular to the edge. These are in fact the [110] rows of a (100) plane. These observations provide sound confirmation of the particle morphology (Fig. 1.25c) deduced from the particle profiles.

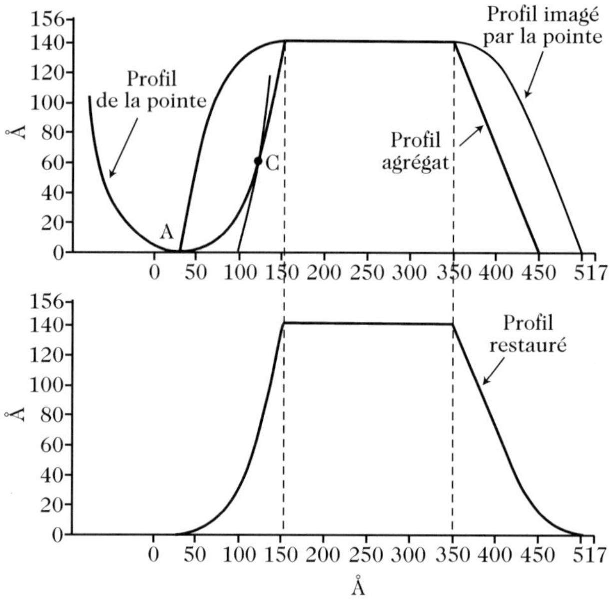

Fig. 1.26. Distortion of the image of a 3D particle due to the shape of the tip in near-field microscopy. Taken from Ferrero [42]

Convolution Between the Tip Shape and the Shape of a 3D Object Imaged by Near-Field Microscopy

Figure 1.26 (top) represents a line scanned by the tip of the microscope as it passes over a particle with fcc structure in (111) epitaxy, displaying lateral (111) and (100) facets. The projections of two of these facets (perpendicular to the plane of the figure) form straight lines at angles of 70.53° and 54.74° to the horizontal. The tip profile, assumed parabolic, is also illustrated. The line imaged is the locus of points of contact between the tip and the surface of the particle. It is clear that the height is correctly determined by the measurement, but the edges are rounded off and the size of the particle is increased. The convolution effect (not a convolution in the mathematical sense) increases as the radius of curvature of the tip increases and as the slopes of the facets increase. If the shape of the tip is known, the images can be corrected, as shown in the lower part of Fig. 1.26. On the corrected profile, the mid-height size of the particles is now correct, but information has been lost at the base of the particle, up to a height equal to the radius of curvature of the tip.

References

1. M. De Crescenzi, M. Diociaiuti, L. Lozzi, P. Picozzi, S. Santucci: Phys. Rev. B **35**, 5997 (1987)
2. G. Apai, J.F. Hamilton, J. Stöhr, A. Thomson: Phys. Rev. Lett. **43**, 165 (1979)

3. B.K. Teo: *EXAFS: Basic Principles, Data Analysis* (Springer, Berlin Heidelberg New York 1986)
4. M. De Crescenzi: Surf. Sci. Rep. **21**, 89 (1995)
5. S.M. Foiles, M.I. Baskes, M.D. Daw: Phys. Rev. B **33**, 7983 (1986)
6. V. Rosato, M. Guillopé, B. Legrand: Phil. Mag. A **59**, 321 (1989)
7. K.W. Jacobsen, J.K. Nørskov, M.J. Puska: Phys. Rev. B **35**, 7423 (1987)
8. S. Swaminarayan, R. Najafabadi, D.J. Solowitz: Surf. Sci. **306**, 367 (1994)
9. A. Kara, T.S. Rahman: Phys. Rev. Lett. **81**, 1453 (1998)
10. R. Meyer, S. Prakash, P. Entel: Phase Transitions, **75**, 51 (2002)
11. M.C. Desjonquères, D. Spanjaard: *Concepts in Surface Physics*, Springer-Verlag, Berlin, 1993
12. G. Wulff: Z. Krist. **34**, 449 (1901)
13. C. Mottet, J. Goniakowski, F. Baletto, R. Ferrando, G. Tréglia: Phase Transitions **77**, 101 (2004)
14. F. Baletto, C. Mottet, R. Ferrando: Phys. Rev. B **63**, 155408, (2001)
15. S. Iijima, T. Ichihashi: Phys. Rev. Lett. **56**, 616 (1986)
16. P.M. Ajayan, L.D. Marks: Phys. Rev. Lett. **60**, 585 (1988)
17. N. Doraiswamy, L.D. Marks: Phil. Mag. **71**, 291 (1995)
18. L.D. Marks: Surf. Sci. **150**, 358 (1985)
19. R. Kaichew: *Arbeitstagung Festkörper Physik*, Dresden, 1952, p. 81.
20. C.R. Henry: Cryst. Res. Technol. **33**, 1119 (1998)
21. J. Tersoff, F.K. Le Goues: Phys. Rev. Lett. **72**, 3570 (1994)
22. P. Müller, R. Kern: Surf. Sci. **457**, 229 (2000)
23. W. Vervisch, C. Mottet, J. Goniakowski: Phys. Rev. B **65**, 245411 (2002)
24. S. Giorgio, C.R. Henry, C. Chapon, G. Nihoul, J.M. Penisson: Ultramicroscopy **38**, 1 (1991)
25. H. Graoui, S. Giorgio, C.R. Henry: Surf. Sci. **417**, 350 (1998)
26. G. Prévot, O. Meerson, L. Piccolo, C.R. Henry: J. Phys.: Condens. Matter. **14**, 4251 (2002)
27. C. Liu, J.M. Cohen, J.B. Adams, A.F. Voter: Surf. Sci. **253**, 334 (1991)
28. H. Graoui, S. Giorgio, C.R. Henry: Phil. Mag. B **81**, 1649 (2001)
29. J.C. Heyraud, J.J. Métois: Surf. Sci. **128**, 334 (1983)
30. J.C. Heyraud, J.J. Métois: Acta. Metal. **28**, 1789 (1980)
31. S. Giorgio, C. Chapon, C.R. Henry: Phil. Mag. B **67**, 773 (1993)
32. G. Renaud: Surf. Sci. Rep. **32**, 1 (1998)
33. K. Heinemann, T. Osaka, H. Poppa: Ultramicroscopy **12**, 9 (1983)
34. Y. Kondo, K. Takayanagi: Science **289**, 606 (2001)
35. P.L. Hansen, J.B. Wagner, S. Helveg, J.R. Rostrup-Nielsen, B.S. Clausen, H. Topsoe: Science **295**, 2053 (2002)
36. J.R. Levine, J.B. Cohen, Y.W. Chung, P. Georgopoulos: J. Appl. Cryst. **22**, 528 (1989)
37. G. Renaud, R. Lazzari, C. Revenant, A. Barbier, M. Noblet, O. Ullrich, F. Leroy, Y. Borensztein, J. Jupille, C.R. Henry, J.P. Deville, F. Scheurer, J. Mane-Mane, O. Fruchart: Science **300**, 1416 (2003)
38. E. Ganz, K. Sattler, J. Clarke: Phys. Rev. Lett. **60**, 1856 (1988)
39. A. Humbert, M. Dayez, S. Granjeaud, P. Ricci, C. Chapon, C.R. Henry: J. Vac. Sci. Technol. B **9**, 804 (1991)
40. A. Piednoir, E. Perrot, S. Granjeaud, A. Humbert, C. Chapon, C.R. Henry: Surf. Sci. **391**, 19 (1997)

41. K.H. Hansen, T. Worren, S. Stempel, E. Laegsgaard, M. Bäumer, H.J. Freund, F. Besenbacher, I. Stensgaard: Phys. Rev. Lett. **83**, 4120 (1999)
42. S. Ferrero: Doctoral thesis, University of Aix, Marseille II, 2002
43. S. Ferrero, A. Piednoir, C.R. Henry: Nanoletters **1**, 227 (2001)

2

Structure and Phase Transitions in Nanocrystals

J.-C. Nièpce and L. Pizzagalli

2.1 Introduction

As shown in Chap. 1, the physical properties of conventional materials may change with the size of the grains making them up, even becoming totally different from what is observed in the bulk solid system. One then speaks of grain size dependence. This dependence can be put down to two more or less related effects:

- A size effect, or confinement effect. The nanograin behaves like a kind of box, within which the property may or may not exist [1]. Below a certain critical size, characteristics of the property depend on the grain size. This is the size or confinement effect. The way these characteristics change as a function of size is often non-monotonic and can exhibit extrema.
- A surface or interface effect. In the nanograin, the contribution from layers close to the surface occupies a more and more important place in the overall behaviour of the material as the grain size decreases [1]. The surface energy gradually becomes the dominating contribution to the total energy of the material. Such a property will evolve monotonically with size and can be treated within the framework of thermodynamics.

Barium Titanate

Barium titanate $BaTiO_3$ will be referred to often in the context of experimental results presented in this chapter. Thanks to its crystal structure and dielectric properties, it is a material with a wealth of applications, especially with regard to passive components in electronics and electrotechnics (see Chap. 28). Its crystal structure derives from the crystal structure known as perovskite (see Fig. 2.1). However, depending on the temperature and in normal atmospheric pressure, barium titanate can occur in four different crystal states, all derived from the perovskite structure (see Fig. 2.2).

Above about 120°C, $BaTiO_3$ adopts the ideal perovskite structure (space group $Pm3m$, centrosymmetric cubic, paraelectric), whereas below this temperature,

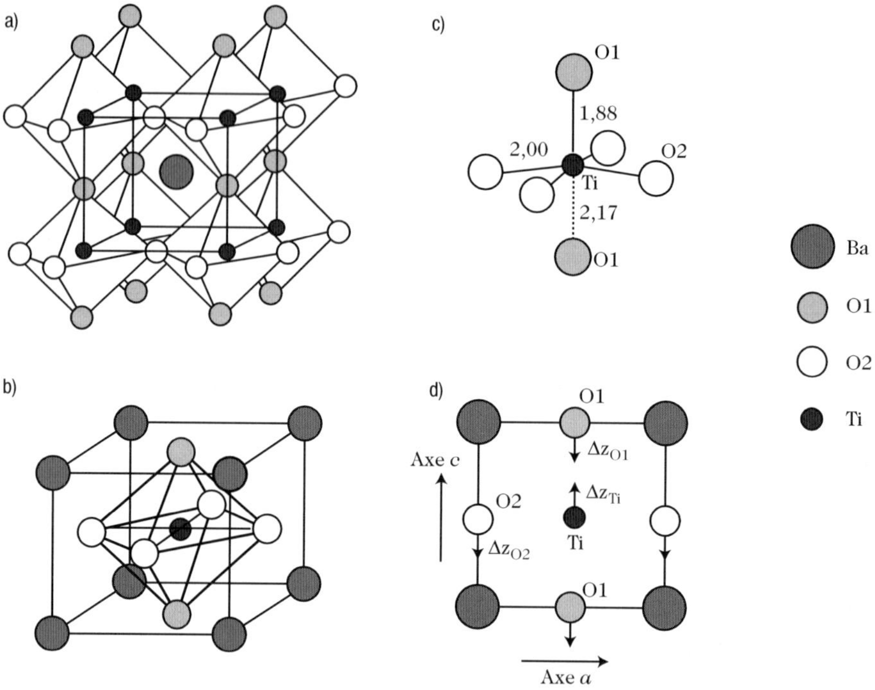

Fig. 2.1. Undistorted, i.e., cubic, perovskite crystal structure of $BaTiO_3$ above the Curie temperature. (**a**) Origin at a Ba^{2+} ion. (**b**) Origin at a Ti^{4+} ion. (**c**) and (**d**) Schematic representation of the tetragonal distortion of the latter below the Curie temperature

called the Curie temperature, the three crystal states of $BaTiO_3$ are no longer centrosymmetric. Owing to this feature in particular, $BaTiO_3$ then displays the interesting property of being ferroelectric. This fact and the consequences with regard to the dielectric properties of the transition between the ferroelectric and paraelectric states near the Curie temperature are the source of many applications for $BaTiO_3$. In particular, during the transition, the average dielectric constant reaches very high values, whence its use in ceramic capacitors (see Fig. 2.3).

Several properties, like magnetism for example, are grain size dependent. But what about the structure of the nanograins itself? In other words, is the reduction of grain size in the material accompanied by any modification, or even a complete transformation, of this structure? Consider the example of barium titanate $BaTiO_3$ in powder form, hence made up of nanograins of various sizes. Figure 2.4 shows the crystal lattice parameters at constant temperature $T = 25°C$ and pressure $P = 1$ atm as a function of the dimensions of these nanocrystals [2]. It is clear from the figure that there is a critical diameter $\Phi_c = 80$ nm for the elementary nanocrystals in the powder:

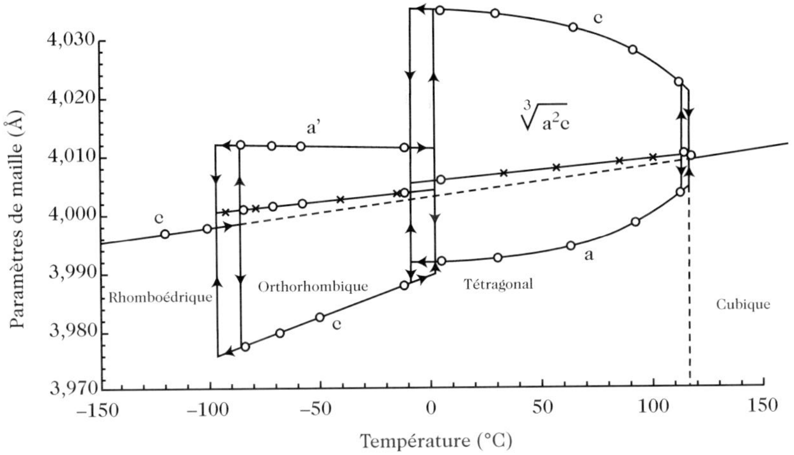

Fig. 2.2. Temperature dependence of the lattice parameters of a $BaTiO_3$ single crystal at standard atmospheric pressure, revealing the three phase transitions between the four most common crystal states of barium titanate

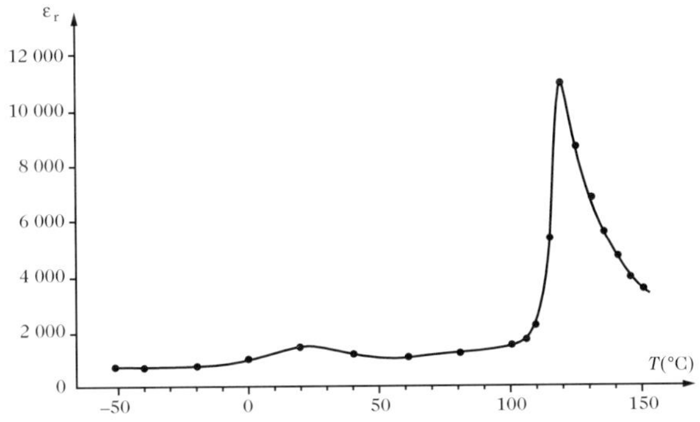

Fig. 2.3. Temperature dependence of the average dielectric constant in a pure $BaTiO_3$ ceramic, revealing the spectacular increase as it transits from the ferroelectric to the paraelectric state

- If $\Phi > \Phi_c$, $BaTiO_3$ crystallises with a tetragonal perovskite lattice, which is the same crystal structure as in the bulk solid. The tetragonal aspect ratio c/a gets smaller as the size of the nanocrystals decreases. However, it is a remarkable fact that the volume $v = a^2c$ of the unit cell remains constant [3].
- If $\Phi < \Phi_c$, on the other hand, $BaTiO_3$ crystallises with a cubic perovskite lattice, i.e., $c = a$. Here, in contrast to the tetragonal state, a increases, and so therefore does the volume $v = a^3$ of the unit cell, when the size of the nanocrystals gets smaller.

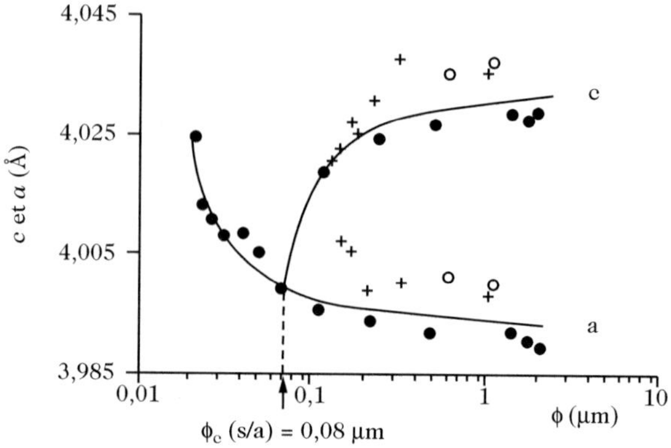

Fig. 2.4. Size dependence of the crystallographic lattice parameters a and c for $BaTiO_3$, at 25°C and standard atmospheric pressure in the powdered state [2]

In the light of this example, it is thus clear that the structure of nanomaterials depends on their characteristic dimensions. Phase transitions and variations in lattice parameters can be observed.

> The dimensions of nanocrystals thus seem to play an important role in their structure in the broadest possible sense:
>
> - with regard to the nature of the 'stable' phase, causing phase transitions at constant temperature and pressure,
> - with regard to the geometry and volume of the unit cell, for a given crystal symmetry.

There are two further pieces of data characterising nanocrystals as compared with the bulk solid: the dimensions and also the state of the outer surface or interface. Transitions in nanomaterials are studied by varying these parameters. Hence one can either reduce the size of the nanocrystals whilst keeping the surface in the same state, or one can modify the state of the surface or interface whilst conserving the size of the nanocrystal. The studies described here as examples use one or other of these methods. Note that, in the case of a size reduction, only the chemical composition of the surface or interface can be conserved. Indeed, the energy of the surface or interface will necessarily be modified owing to the reduction in the radius of curvature. We may say that, in this case, size reduction effects and surface effects are closely related.

Grain size dependence is a general property and we shall see below that it is relevant to a wide range of materials such as oxides, ceramics, semiconductors, and also metals. The size reduction of nanograins will also have a significant influence on the difficulty with which phase transitions can be brought about, by modifying the transition energies. Finally, structural states that do not exist in bulk solids, such as liquid–solid equilibria with segregation, have been

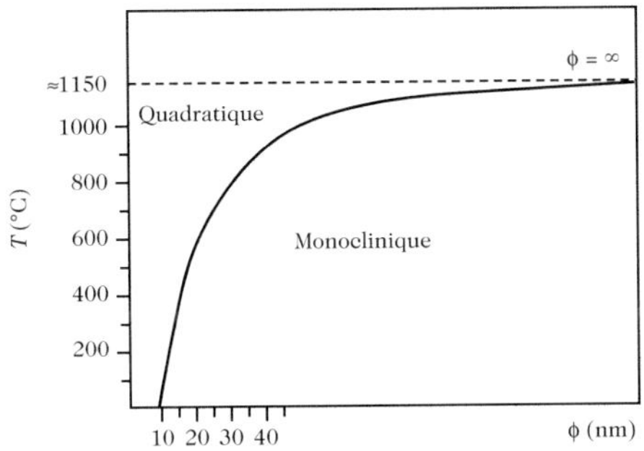

Fig. 2.5. Temperature of the phase transition in zirconia ZrO_2 as a function of the average diameter of nanocrystals in powders at standard atmospheric pressure [4,5]

observed for nanoparticles. In the rest of this chapter, we shall return to these various points and illustrate them with examples.

2.2 Crystalline Phase Transitions in Nanocrystals

2.2.1 Phase Transitions and Grain Size Dependence

There are many examples of phase transitions occurring whenever the characteristic dimensions of a material go beyond a certain critical value. More interestingly, such transitions are not restricted to any particular type of material. In the following, we shall give some examples, chosen among ceramics, metals, and also semiconductors.

Ceramics

The grain size dependence of a phase transition in a ceramic can be illustrated by the monoclinic–tetragonal transition occurring in zirconia. This property has been known and exploited for a long time now [4]. At standard atmospheric pressure and room temperature, zirconia crystallises in a monoclinic (low temperature) form, whereas at high temperature, above 1100–1150°C, its crystal structure is tetragonal. If ZrO_2 is in the form of 'crystals' of the order of 10 nm, it is the tetragonal form that is stable at room temperature. The transition temperature, somewhere between 1100 and 1150°C for micrometric crystals becomes lower as the dimensions of the nanocrystals decrease (see Fig. 2.5) [5]. In fact it has been shown recently that, for temperatures below those illustrated in Fig. 2.5, the tetragonal–monoclinic transition is still present and occurs for grain sizes of the order of 6.9 nm (at 175 K) [6].

Another situation where there is grain size dependence has been established in the case of the Verwey transition, first observed in 1939 by the discontinuous change in conductivity it causes. This transition occurs in magnetite Fe_3O_4, with the so-called spinel crystal structure. Although there is a wealth of literature on the subject, the mechanisms have not yet been clearly understood. However, a study for nanometric grains has been able to throw some light on the matter [7]. Experiments have also demonstrated a spectacular difference between the temperature predicted for large grains and that observed for nanometric grains (a difference of more than 70 K). It would thus appear that the Verwey transition constitutes another example of grain size dependent phase transition.

Metals

An example of a grain size dependent phase transition in metals is provided by nickel. In the bulk solid, this metal has fcc structure. However, recent experiments have brought to light a grain size dependent phase transition for very small Ni nanoparticles, chemically synthesised in solution [8]. Indeed for diameters less than 4 nm, the structure is still crystalline, but it becomes hexagonal close packed (hcp) rather than fcc. In the bulk material, an hcp structure is metastable, although it can be obtained in thin films in certain conditions. It should also be noted that the melting temperature of various metals is well known to depend on grain size [9].

Semiconductors

As far as semiconductors are concerned, it is likely that a crystal phase transition depending on nanoparticle size can occur in certain cases. However, the systems investigated as yet have not produced conclusive results. The case of CdS nanoparticles is worth mentioning in this context. These are obtained with a wurtzite-type structure (as in the bulk solid) for diameters greater than 5 nm, whereas for very small sizes (3 nm), a zinc blende structure is observed [10]. However, the latter is apparently metastable and seems to be obtained for purely kinetic reasons during formation.

> Grain size dependent crystalline phase transitions exist for a wide range of materials. This is therefore an intrinsic property of nanomaterials.

2.2.2 Elementary Thermodynamics of the Grain Size Dependence of Phase Transitions

Theory

The various examples described above show that the phase transitions, and more generally the phase diagrams of a material should be considered in a temperature–pressure–grain size space (or rather, reciprocal radius of curvature R for the latter degree of freedom). However, a general theory has not yet been put together. In this context, it is interesting to see what standard thermodynamics has to say about the matter [11]:

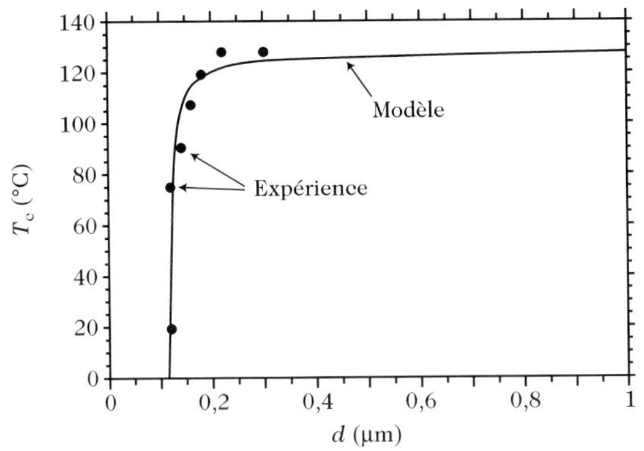

Fig. 2.6. Temperature of the tetragonal–cubic phase transition in $BaTiO_3$ as a function of the grain size. Calculated result from [11] and experimental result from [12]

- The stable states of the system are no longer governed by the free enthalpy G at constant T and P. This role is fulfilled by a generalised free enthalpy function G^* given by

$$G^* = G - 2\gamma V/R \, .$$

- G^* is no longer a state function of the system. γ is the surface energy, V the molar volume of the material, and R the radius of the nanocrystal, assumed spherical.
- The equilibrium state of a system made from nanometric grains is no longer obtained by the condition $\mathrm{d}G = 0$, but rather by $\mathrm{d}G^* = 0$.
- There is therefore equilibrium between the phase α and the phase β of the same body if $G^*_\alpha = G^*_\beta$.
- For given temperature and pressure, there is a critical radius R_c at which the phase transition occurs in nanometric grains.
- So the temperature T_c of the phase transition is, in particular, a function of the radius R of the nanocrystals:

$$T_c = T(P, R, \gamma_\alpha, \gamma_\beta) \, .$$

This relation has been clearly demonstrated for the tetragonal–cubic transition in $BaTiO_3$ (see Fig. 2.6).

Instability of the Crystal State

It has long been known that, beyond a certain size, the crystal becomes unstable, favouring an amorphous, hence disordered structure. The critical size obviously depends on the material. For example, in the case of silicon, a theoretical study has shown that this crystalline–amorphous transition occurs for sizes of the order of 3 nm [13]. This same study stresses that this phase transition is discontinuous. It is therefore quite appropriate here to speak of a grain size dependent phase transition.

The size of a nanocrystal thus plays the role of a kind of intensive quantity imposing the crystalline or amorphous state in the same way as the temperature or the pressure.

2.2.3 Influence of the Surface or Interface on Nanocrystals

Changes can thus be observed in the phase diagram of a material by varying only the size of the constituent grains. In the same way, one may expect to obtain similar effects by altering the surface state of these nanograins whilst keeping their dimensions constant. The modification of this surface or interface energy can be achieved in different ways: either by adsorption of various chemical species or molecules in the case of powdered systems or systems in solution (solid–gas or solid–liquid interface), or by compacting the nanograins, or embedding them within a matrix (solid–solid interface).

A Simple Example

Let us return to $BaTiO_3$. Quite recently, the role of the outer interface on transition temperatures in barium titanate has been investigated [14] (Fig. 2.7).

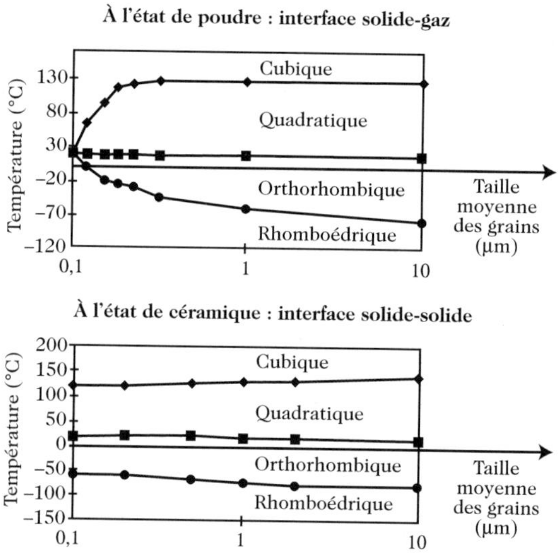

Fig. 2.7. Transition temperatures for $BaTiO_3$ as a function of grain size [14]. *Upper*: In the powdered state (solid–gas interface). *Lower*: In the ceramic state (solid–solid interface)

In the first case, $BaTiO_3$ is in powdered form and hence has an outer surface which is a solid–gas interface. In the second case, the material is in the form of a microstructure composed of agglomerated grains, i.e., a ceramic with varying degrees of density. In this case, there is a solid–solid interface between grains.

The two phase diagrams are clearly different. Thus, for a given grain size and temperature, different phases can be stabilised depending on the state of the interface.

Still in the second case, for a solid–solid interface, a very slight variation is observed in the transition temperatures as a function of grain size. Nanograins in a bulk solid material would therefore appear to behave like large grains. The surface effect would thus seem to disappear if the nanocrystal is bounded by an outer interface with nanocrystals that are identical to it.

A More Complex Example

Phase transitions on the nanoscale and in ternary systems have rarely been studied, despite the fact that quite spectacular modifications are to be expected in equilibrium diagrams established for large grains. Indeed, since the energy contribution of the interface between the phases increases when the grain size goes down, the phase separations observed in micrometric crystals should disappear at the nanoscale, leading to a single phase. New materials can thus be expected.

An interesting example is provided by nanometric titanium ferrites, which have the formula $(Fe_{3-x}Ti_x)_{1-\delta}O_4$, where δ (a parameter related to the average valency of cations) represents the deviation from the oxygen stoichiometry of the material [15]. Figure 2.8 shows the phase diagrams of the Fe–Ti–O system obtained for monocrystals that are at least micrometric (Fig. 2.8a), and for nanometric elementary crystals with solid–solid or solid–gas interfaces (Figs. 2.8b and c, respectively). In the latter case, the stability region of the spinel phase extends from $\delta = 0$ to $\delta = \delta_{\mathrm{max}}$ for titanium compositions with x in the range 0.25–0.75. For other compositions, only metastable spinel phases can be synthesised over the whole range of δ owing to their much larger grain sizes ($> 30\,\mathrm{nm}$), and for which the spinel phase is not stable. Because of the grain size, but also due to the particular interface, a unique face-centered cubic phase is thus stabilised over a wide range of phase diagrams, in contrast to the mixture of phases that occurs for micrometric crystals. Indeed, for the latter, an orthorhombic phase and a rhombohedral phase, or a rhombohedral phase and a face-centered cubic phase coexist over a large part of the diagram, depending on the titanium composition and the deviation from oxygen stoichiometry (see Fig. 2.8a). This phenomenon is explained by the fact that the surface energy that would be created by the fine grains, if these two phases were to coexist, would be too great, so that the system prefers to crystallise in a single phase in the case of nanometric grains.

Reducing the size of a nanocrystal increases the relative importance of the surface or interface between grains. Consequently, the state of the surface or interface also predetermines the properties of the nanomaterial.

2.2.4 Modification of Transition Barriers

In some cases, exotic crystalline phases have been observed unexpectedly in nanoparticles. For the main part, these phases are not the most stable structures, but are obtained during the formation or synthesis of the nanoparticles, probably for kinetic reasons. It is nevertheless remarkable that it is generally easier to produce these phases in nanocrystals than in the bulk solid. These phases are also generally more stable. For example, Ge nanoparticles formed via a vapour deposition technique possess a certain crystal structure known as ST12 whenever the particle size is less than 4 nm [16], whereas it is the diamond-type cubic structure of the bulk solid that is observed for larger nanoparticles. This ST12 structure also exists for the bulk material, but to obtain it, a considerable pressure must first be applied (around 10 Gpa), this causing the formation of a β-Sn type structure, before returning suddenly to atmospheric pressure. This ST12 structure is also metastable in the bulk solid and one only needs to anneal the system in order to recover the cubic diamond structure. What is interesting here is that this annealing stage need only be carried out at 200°C in the bulk solid, whereas one must go to temperatures above 800°C for nanoparticles! We may conclude that the reduction in size of nanocrystals is accompanied by a modification of the transition barriers between the phases.

The effect of size on phase transition barriers can be simply explained using the elementary thermodynamic model discussed above. We assume to begin with that the barrier height is directly proportional to the free energy difference ΔG^* between the two phases. Assuming that the molar volume of the two phases is the same, we obtain

$$\Delta G^* = \Delta G - 2\Delta\gamma\frac{V}{R} .$$

The variation of ΔG^* as a function of R thus depends only on the difference of surface energy $\Delta\gamma$ between the two phases. A reduction in the barrier height, and hence of the transition temperature, may thus occur when the size of the nanocrystals decreases. But it may also increase, as happens for the rhombohedral–orthorhombic transition in $BaTiO_3$ powders.

It has also been possible to explain modifications in transition barriers using criteria related to crystal defects in nanoparticles. In the bulk solid, phase transitions are helped by the presence of defects in the crystal structure. These facilitate nucleation of the new phase. In contrast, the smaller the nanocrystal, the less likely it is to contain any such defects. It may thus be assumed that size reduction will hinder the transition. However, for low or zero defect

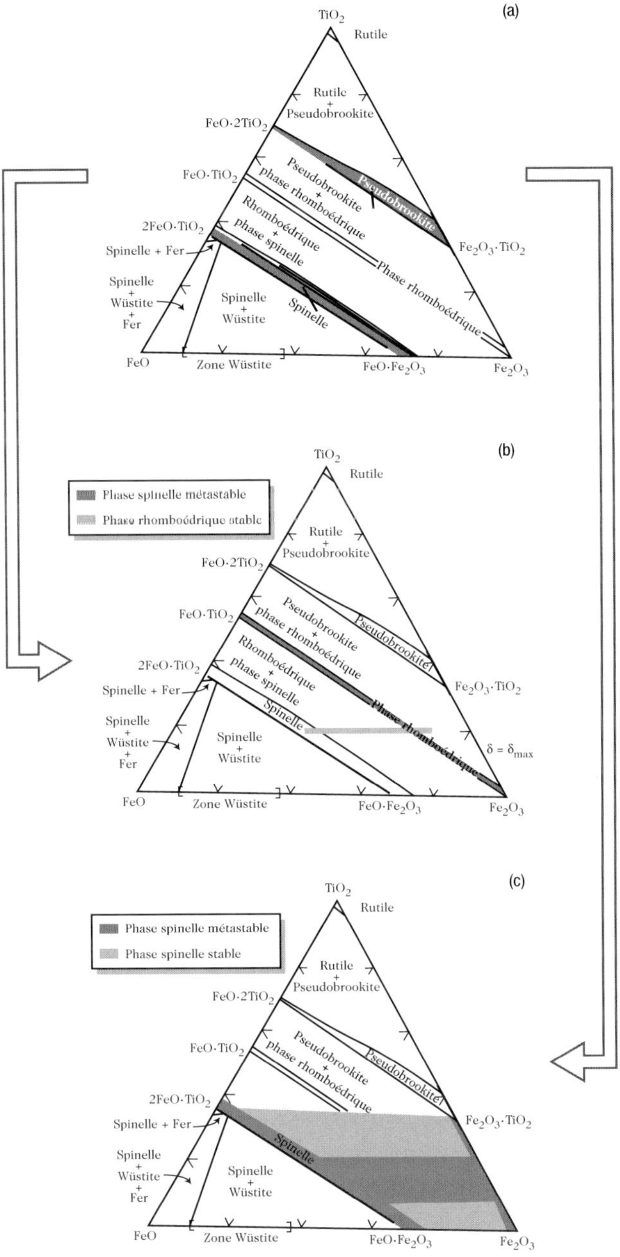

Fig. 2.8. (**a**) High temperature phase diagram for the Fe–Ti–O system in the form of micrometric crystals (pseudobrookite, an orthorhombic phase). (**b**) and (**c**) Phase diagrams for the Fe–Ti–O system, corrected in the case of nanomaterials. (**b**) Spinel and rhombohedral phases formed by mechanosynthesis, where only the titanium composition $x = 0.5$ has been studied [15], and where the interfaces are mainly of solid–solid type. (**c**) Stability regions of the spinel phase formed by soft chemistry when the surface energy is of solid–gas type

densities, recent work has shown that an increase in the size of Si or CdSe nanoparticles is accompanied by an increase in the pressure required to bring about the phase transition [17, 18]. There are therefore several mechanisms here and the variation of the transition barrier may be somewhat complex.

Phase transition barriers depend on the size of the nanocrystal, but in a complex way. This variation depends on the state of the surface or interface and the presence or otherwise of crystal defects.

2.3 Geometric Evolution of the Lattice in Nanocrystals

2.3.1 Grain Size Dependence

Reducing the size of nanocrystals changes the stability of the various crystal phases, so it is reasonable to ask whether this reduction is not also accompanied by some change in the lattice parameters for a given phase. For the moment, this question has received less attention than the phase transitions themselves, largely because experimental determination of lattice parameters is difficult in nanocrystals and requires high accuracy. This means that few data are yet available.

Lattice Geometry

Demonstrating Grain Size Dependence

Let us return to the example of $BaTiO_3$, already illustrated in Fig. 2.4 on p. 38. It has been observed that, in the case of the tetragonal perovskite lattice obtained for nanocrystal diameters $\Phi > 80\,\mathrm{nm}$, the tetragonal aspect ratio given by c/a falls off as Φ decreases. However, the volume $V = a^2c$ of the unit cell remains constant [3]. This is therefore a grain size dependence in the geometry of the crystal structure. This phenomenon was first discovered about fifty years ago [19] and then investigated further [3, 20].

Demonstrating the Absence of Grain Size Dependence

Despite the spectacular and entirely repeatable nature of this observation, it is not universal. Indeed, in the case of zinc oxide, for example, which crystallises according to a hexagonal system, no variation of the ratio c/a has been observed here when the dimensions of the elementary crystals are reduced from a few microns to a few nanometers [21].

This difference of behaviour is not yet understood. An explanation should probably be sought in the structural differences which induce major differences in physical properties, especially with regard to ferroelectricity. Indeed, the $BaTiO_3$ lattice is polar, thereby contributing to the ferroelectric nature of tetragonal $BaTiO_3$, whereas the ZnO lattice is not polar. It would seem that, in ferroelectric materials, size effects are relevant even for relatively large sizes and, in these materials, one must take into account changes in the polarisation at the interface between the solid and the surrounding medium. For small crystals of $BaTiO_3$ and considering only the effects on the crystal state, it is thus observed that the tetragonality of the lattice is already affected for sizes of the order of 1 μm.

Lattice Parameter

Beyond the critical diameter of 80 nm, $BaTiO_3$ stabilises in a cubic structure. Figure 2.4 shows an increase in the lattice parameter when the average grain size goes down. It would thus seem that there is a grain size dependence in the volume of the unit cell of the crystal lattice. In fact, as we shall see later, this phenomenon is not, at least not for the main part, related to the reduction in size of the nanocrystals, but is due rather to the action of certain heat treatments used on their surfaces to control their size.

> In some cases, a grain size dependence is observed in the parameters specifying the crystal structure. However, the few available examples are insufficient to conclude that this is a general effect, and a fortiori, cannot yet provide a full understanding.

2.3.2 Theory

Grain size dependence due to the surface or interface can be described using a thermodynamic approach, by considering the surface or interface as a whole, or using a microscopic approach, describing in detail the relaxation and reconstruction effects. For very small nanocrystals, there are also quantum confinement effects.

Thermodynamic Approach

Laplace's law is often used to describe the dependence of the lattice parameter on grain size. This law relates the pressure P_{int} inside the grains to the pressure P_{ext} outside, the surface energy γ, and the grain size ϕ:

$$P_{\text{int}} = P_{\text{ext}} + 4\gamma/\phi \, .$$

The term $4\gamma/\phi$ is positive, since γ is an excess energy at the surface of the solid and hence necessarily positive. Hence, when the grain size ϕ goes down, the term $4\gamma/\phi$ grows larger, leading to an increased pressure inside the grains. This would mean that the lattice parameter should decrease in every case. This is indeed what is observed for the ferrites of Co and Mn [22]. However, in both $BaTiO_3$ [3] and γ-Fe_2O_3 [23], the lattice parameter actually increases, in direct contradiction with the above law!

A more careful analysis can explain this apparent paradox, showing that it is the surface stress Γ that should appear in the last equation, rather than the surface energy [24]. So there is no longer any contradiction between reducing the grain size and observing an increase in the lattice parameter, since the pressure within the grain can be diminished if the stress Γ is negative. Note that the energy, which depends on the square of the stress, does remain positive. The thermodynamic approach shows that, depending on the state of the surface, one may see either a contraction or an expansion of the atomic bonds within the nanocrystal.

Microscopic Approach

When a surface is created, the atomic planes near the surface are usually displaced. The direction and magnitude of this relaxation depends on the type of material and also the orientation of the surface. In rare gas crystals or ionic crystals, for example, the bonds between the atoms are relatively long-range. In the bulk, the relative position of the atoms is determined by the competition between the mutual repulsion of the nuclei and long-range forces (Coulombic in the case of ionic crystals). In this case, the atomic planes at the surface expand [25,26]. In the case of face- or body-centered cubic metals, the situation is generally different: for close-packed planes without defects, a very small surface relaxation is generally observed. On the other hand, if a close-packed surface has a certain degree of roughness due to the presence of defects such as steps or islands, or if the surface is not close-packed, the outermost layer will relax considerably towards the core of the material, whilst certain layers further in will move toward the surface. As an example, Fig. 2.9 shows the various motions of the (210) surface planes in platinum [27]. These different relaxations compensate for the fact that the atoms at the surface have fewer near neighbours, whence the electron density is lower there. Relaxation is thus a way of compensating for changes in the electron density in each plane, in such a way as to make it as homogeneous as possible right out to the outermost surface.

Consider now a nanocrystal with approximately spherical shape. Relaxation of the first atomic planes will lead in this case to a pressure exerted by the surface on the core of the nanocrystal, and hence to an increase or a decrease in the lattice parameter. An effect is therefore induced not only by the adsorption state of the surface, but also by the surface itself. This effect is even more marked in semiconductors, where surfaces are generally reconstructed, in order to minimise the number of dangling bonds, which generates large stresses in the first layers. In a nanocrystal, these stresses will also contribute to the surface pressure and to the contraction or expansion of atomic bonds. For example, simulations of Ge nanocrystals with surface composed of

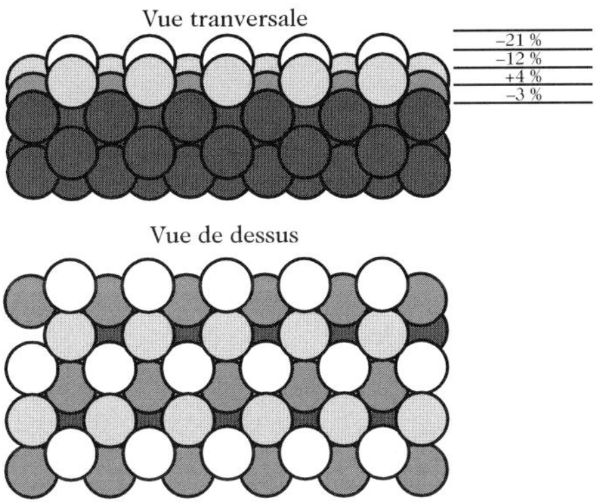

Fig. 2.9. Schematic description of surface relaxation effects in the case of a Pt (210) surface. These relaxations are deduced from models based on LEED (low energy electron diffraction) studies [26]

reconstructed layers have indicated a reduction in the structure parameter of up to 6% for 2-nm nanoparticles [28].

We have limited the discussion here to clean surfaces. In the more complicated case where there is adsorption on nanocrystal surfaces, it is not possible a priori to guess whether one will observe a contraction, or rather an expansion of the surface planes, i.e., a reduction or an increase in the lattice parameter. In the present state of our understanding, only a detailed analysis in each particular case can provide an answer to this question.

Quantum Confinement

A priori, there is no reason to expect any particular variation in the structure parameters due simply to a reduction in the volume of the nanocrystal. However, for very small dimensions, of the order of a few nanometers, a quantum confinement effect occurs, leading among other things to a change in the electronic structure. This change is accompanied by a shortening of the atomic bonds within the nanocrystal. This effect is relatively small and difficult to demonstrate experimentally. However, although this variation is small, it has been obtained in simulations for Ge nanoparticles (see Fig. 2.10) [28, 29]. The surfaces, passivated by hydrogen, have little influence in this case, and the bond contraction observed does indeed arise as a consequence of quantum confinement. The effect nevertheless remains very slight, even for very small nanocrystals with diameters of the order of 2 nm.

The variation of the parameters describing crystal structure can be understood in relation to the nanocrystal surface using either thermodynamic or microscopic approaches.

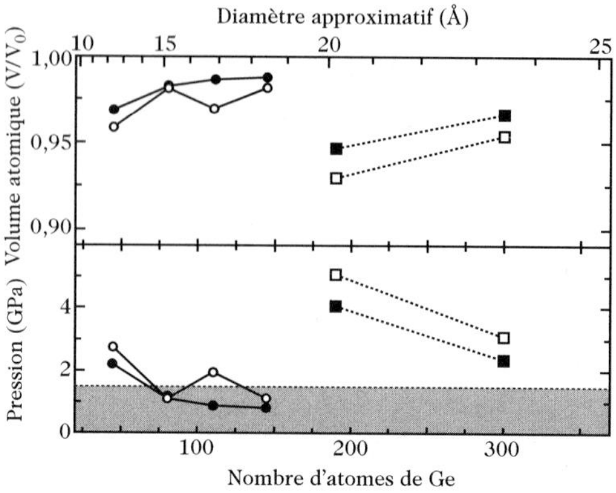

Fig. 2.10. Calculated variation of the atomic volume and the associated pressure as a function of the size of Ge nanocrystals. Two types of structure are considered: diamond cubic (*black symbols*) and ST12 (*white symbols*); and two surface states: hydrogen passivated (*circles*) and with reconstructed layers (*squares*)

2.3.3 Influence of the Nanocrystal Surface or Interface on the Lattice Parameter

Several experiments have shown a relationship between the adsorption state of nanocrystal surfaces and the change in the lattice parameter. Hence, for nanocrystals in ceramic powders like $BaTiO_3$ [30] or $SrTiO_3$ [31], the presence of water molecules and OH^- ions on the surface causes an increase in the lattice parameter. After desorption of the molecules by a suitable heat treatment, the values for the bulk solid are recovered.

Figure 2.11 shows how one may adjust the lattice parameter in γ-Fe_2O_3 by varying the surface energy of the nanocrystal. There are two clearly distinct regimes: first a contraction of the nanocrystal, and then an expansion. The first regime corresponds to a phase in which OH^- and H_2O are chemisorbed on the surface, and an initial water monolayer is formed, with water vapour pressures below a certain critical value. The second regime, on the other hand, corresponds to the formation of water multilayers by physisorption, and the relaxation of the oxide by strengthening of the bonds between the water layers [32]. The existence of two distinct regimes has also been demonstrated for iron nanoparticles coated with a thin layer of γ-Fe_2O_3 [33], and also when oxygen is adsorbed on carbon nanotubes [34].

The state of the surface or interface of a nanocrystal can affect the crystal structure parameters. However, it is still difficult to predict this effect or estimate its importance.

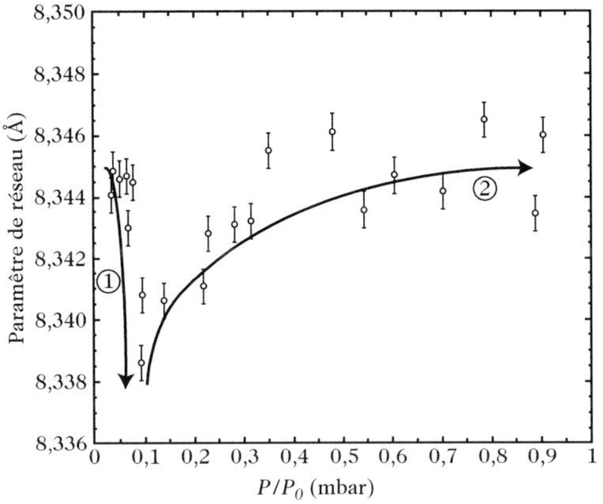

Fig. 2.11. Dependence of the lattice parameter for a nanometric γ-Fe_2O_3 powder on the state of adsorption of water vapour on the powder. For water vapour pressures below a critical value (around $P/P_0 = 0.1$), the formation of a monolayer appears to compress the oxide, whilst for high pressures, the formation of multilayers of water would appear to relax the oxide by strengthening the bonds between water layers [32]. (1) 630 kJ/mol: chemisorption of OH^{2-} and physisorption of H_2O (monolayer on γ-Fe_2O_3). (2) 45 kJ/mol: physisorption of H_2O (multilayer on H_2O)

2.3.4 Is There a Continuous Variation of the Crystal State Within Nanocrystals?

Experimental results using X-ray diffraction have shown that there is probably, in $BaTiO_3$, a gradient in the crystal organisation as one moves from the surface toward the core of a grain. The average state of the solid is all the more affected by this gradient as the grain size goes down [20]. However, some doubt has been raised concerning this result regarding the evolution of the microstructure in ferroelectric domains, because this too is influenced by grain size and this too shows up in the X-ray diffraction diagram [35, 36].

Simple Theoretical Approach

The simplistic idea of obtaining a nanometric grain by 'cutting' a nanometric chunk out of an infinite crystal leads one to distinguish the crystalline organisational states in the core of the grain and in those layers influenced by the proximity of the surface (see Fig. 2.12). Following the argument discussed above with regard to surface relaxation, it is easy to imagine that the lattice parameters may vary continuously as one moves away from the surface, inside the crystal.

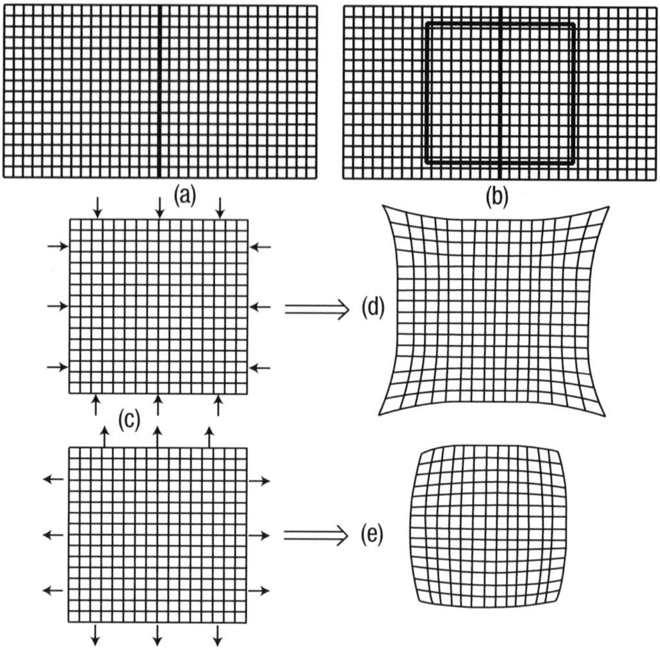

Fig. 2.12. Schematic representation of the distortion of the crystal lattices in layers close to the surface of a nanometric grain. (**a**) Infinite undistorted 2D lattice. (**b**) A chunk is cut from the undistorted 2D lattice. (**c**) Isolated chunk, held in the shape of the undistorted lattice by virtue of fictitious stresses equal to those that were exerted by the rest of the lattice on the same chunk before it was cut away. (**d**) Isolated chunk after removal of stresses exerted by the rest of the lattice: the case of an expansion after freeing from stresses. (**e**) Isolated chunk after removal of stresses exerted by the rest of the lattice: the case of a contraction after freeing from stresses

Simulation

The possibility of bond length variations within nanocrystals as a function of the distance of the bond from the surface has been investigated recently using ab initio simulations for Si and Ge [37]. Figure 2.13 shows the way the bond length changes as one moves away from the center of the nanograin. It turns out that, whatever the size of the nanocrystal considered, the crystal structure undergoes a slight dilation at the center which gradually diminishes to become a compression as one approaches the surface. This effect is more important here for Ge than for Si.

In certain cases, simulation predicts an inhomogeneous distribution of the structure parameters within nanocrystals. However, experimental determination remains difficult.

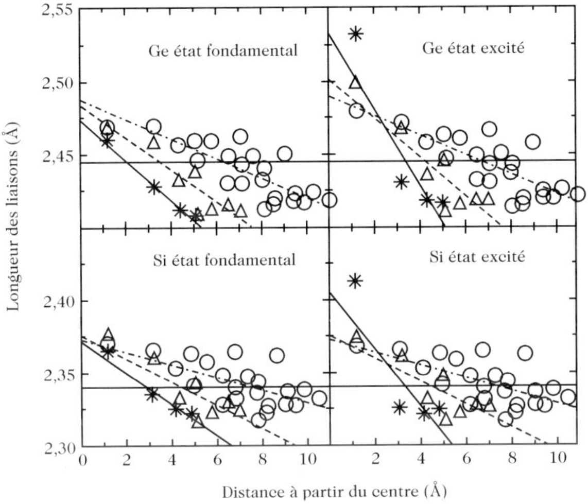

Fig. 2.13. Change in bond length as a function of the distance to the center of the nanocrystal for Ge (*top*) and Si (*bottom*) [37]

References

1. R.W. Siegel: Pour la Science **232** (February 1997) pp. 58–63
2. G. Caboche, F. Chaput, J.P. Boilot, J.C. Niepce: Silic. Indus. **58**, 103 (1993)
3. M. Hoang, J.C. Mutin, J.C. Niepce, O. Olivo: C.R. Acad. Sc. Paris **297**, series II, 899–902 (1983)
4. R.C. Garvie: J. Phys. Chem. **69**, 1238 (1965)
5. A. Methivier: Doctoral Thesis, Ecole des Mines, Saint-Etienne, France (1992)
6. P. Bouvier, E. Djurado, C. Ritter, A.J. Dianoux, G. Lucazeau: International J. Inorg. Mat. **3**, 647–654 (2001)
7. N. Guigue-Millot, N. Keller, P. Perriat: Phys. Rev. B **64**, 012402 (2001)
8. S. Illy, O. Tillement, F. Machizaud, J.M. Dubois, F. Massicot, Y. Fort, J. Ghanbaja: Phil. Mag. A **79**, 1021 (1999)
9. J.P. Borel: Surf. Sci. **106**, 1 (1981)
10. C. Ricolleau, L. Audinet, M. Gandais, T. Gacoin: Euro. Phys. J. D **9**, 565 (1999)
11. P. Perriat, J.C. Niepce, G. Caboche: J. Thermal Anal. **41**, 635–649 (1994)
12. K. Uchino, E. Sadanaga, T. Hirose: J. Amer. Ceram. Soc. **72**, 1555 (1989)
13. S. Veprek, Z. Iqbal, F.-A. Sarott: Phil. Mag. B **45**, 137 (1982)
14. N. Bernaben, A. Leriche, B. Thierry, J.C. Niepce, R. Waser: Fourth Euro Ceramics **5**, 203–210 (1995)
15. N. Millot, D. Aymes, F. Bernard, J.C. Niepce, A. Traverse, F. Bourée, B.L. Cheng, P. Perriat: J. Phys. Chem. B **5740** (2003)
16. S. Sato, S. Nozaki, H. Morisaki: Appl. Phys. Lett. **72**, 2460 (1998)
17. S.H. Tolbert, A.P. Alivisatos: Science **265**, 373 (1994)
18. L.E. Brus, J.A.W. Harkless, F.H. Stillinger: J. Am. Chem. Soc. **118**, 4834 (1996)
19. M. Anliker, H.R. Brugger, W. Känzig: Helv. Phys. Acta. **27**, 99 (1954)

20. S. Malbe, J.C. Mutin, J.C. Niepce: J. Chim. Phys. **89**, 825 (1992)
21. D. Louër, J.P. Auffredic, J.I. Langford, D. Ciosmak, J.C. Niepce: J. Appl. Cryst. **16**, 183–191 (1983)
22. R.E. Vandenberghe, R. Vanleerberghe, G.G. Robbrecht: Sol. State Chem.: Studies in Inorganic Chemistry **3**, 395 (1983)
23. P. Ayyub, M. Multani, M. Barma, V.R. Palkar, R. Vijayaraghavan: J. Phys. C: Solid State Phys. **21**, 2229 (1988)
24. P. Perriat, J.C. Niepce: High Temp. Chem. Processes **3**, 585–600 (1994)
25. B.J. Alder, J.R. Vaisnys, G. Jura: Phys. Chem. Solids **11**, 182 (1959)
26. P.W. Tasker: J. Phys. C **12**, 4977 (1979)
27. X.G. Zhang, M.A. Van Hove, G.A. Somorjai, P.J. Rous, D. Tobin, A. Gonis, J.M. Maclare, K. Heinz, M. Michl, H. Lindner, K. Muller, M.E. Hasi, J.H. Block: Phys. Rev. Lett. **67**, 1298 (1991)
28. L. Pizzagalli, G. Galli, J.E. Klepeis, F. Gygi: Phys. Rev. B **63**, 165324 (2001)
29. L. Pizzagalli, G. Galli: Mat. Sci. Eng. B **96**, 86 (2002)
30. P. Sarrazin, F. Bernard, G. Calvarin, J.C. Niepce, B. Thierry: J. Phys. IV **8**, 85–89 (1998)
31. F. Perrot-Sipple: Maîtrise de la taille de nanograins d'oxydes de structure pérovskite pour applications électro céramiques: Synthèse par chimie douce, broyage par attrition, Doctoral Thesis, Université de Bourgogne, France (1999)
32. T. Belin, N. Millot, F. Villieros, O. Bertrand, J.P. Bellat: J. Phys. Chem. B **108**, 5333 (2004)
33. T. Belin: Interface externe et propriétés structurales d'un solide nanométrique, doctoral thesis, Université de Bourgogne, France (2002)
34. A Fujiwara, K. Ishii, H. Suematsu, H. Kataura, Y. Maniwa, S. Susuki, Y. Achiba: Chem. Phys. Lett. **336**, 205–211 (2001)
35. N. Floquet, C.M. Valot, M.T. Mesnier, J.C. Niepce, L. Normand, A. Thorel, R. Kilaas: J. Phys. III **7**, 1105–1128 (1997)
36. A.I. Ustinov, L. Olikhovska, J.C. Niepce, F. Bernard: J. Prog. Phys. Met. **2**, 51–84 (2001)
37. H.-Ch. Weissker, J. Furthmuller, F. Bechstedt: Phys. Rev. B **67**, 245304 (2003)

3

Thermodynamics and Solid–Liquid Transitions

P. Labastie and F. Calvo

The thermodynamics of nanosystems differs significantly from the thermodynamics of macroscopic systems, because a certain number of variables such as the energy, entropy, etc., are only extensive in the thermodynamic limit, i.e., when the number of particles making up the system tends to infinity. Likewise, the equivalence of the Gibbs ensembles (microcanonical, canonical, and grand canonical) is only valid in this same limit. We shall see later (see Sect. 3.2) how thermodynamics can be applied to small systems.

In the framework of statistical thermodynamics, Gibbs defined three so-called ensembles, corresponding a priori to different physical situations [1]. However, these ensembles provide completely equivalent descriptions of macroscopic systems in the thermodynamic limit. Let us recall their main features:

- *Microcanonical Ensemble.* The system is assumed isolated so that its energy is constant. All the microscopic states corresponding to a given macroscopic state are equiprobable.
- *Canonical Ensemble.* The system has fixed composition, but it can exchange energy with a thermostat at temperature T. The probability of occupation of an energy state E is proportional to the Boltzmann factor $\exp(-E/k_\mathrm{B}T)$.
- *Grand Canonical Ensemble.* The system can exchange energy with a thermostat of temperature T and components with a reservoir of particles having chemical potential μ. The occupation probability of an N-particle state with energy E is proportional to $\exp\left[-(E-\mu N)/k_\mathrm{B}T\right]$.

Even if size effects such as structural changes or the influence of the surface can be quantitatively important, they do not lead to a qualitative change in the thermodynamic properties of single phase systems. On the other hand, phase transitions can be considerably altered:

- reduction of the transition temperature,
- reduction of the latent heat,
- broadening of the temperature range over which phases can coexist.

Moreover, phase transitions are a typically polyatomic effect. To characterise a phase, a certain minimum number of constituents is required. It makes no sense, for example, to ask whether a diatomic molecule is liquid or solid. But then, is there a size or size range for which phase transitions disappear?

The theme of this chapter on thermodynamics is thus the solid–liquid transition, which typifies the idea of phase transition. Beginning with macroscopic systems, we shall investigate their evolution with decreasing dimensions. We shall then see how thermodynamic concepts must be modified in order to understand what happens at very small scales. We shall find that a hypothesis is required regarding the stability of nanoparticles, and that this hypothesis is not always satisfied. We shall then treat the question of unimolecular evaporation, and use the results to specify the time span over which it is safe to assume that these particles are stable in the condensed state.

3.1 Size Dependence of the Solid–Liquid Transition

3.1.1 From the Macroscopic to the Nanometric

The Macroscopic Solid–Liquid Transition. Generalities

Given its importance in everyday life as well as in industrial processes, it is understandable that the solid–liquid transition should have been the subject of many studies, in both macroscopic and nanometric systems. However, there is no satisfactory theory for this transition. A good introduction to the phenomenology and empirical theories is given in [2, Chap. 3]. In the present context, we shall limit the discussion to the main ideas required to understand this chapter.

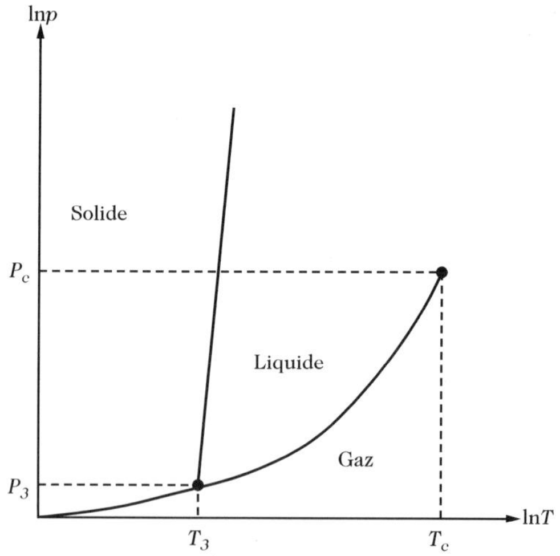

Fig. 3.1. Schematic phase diagram of a simple body

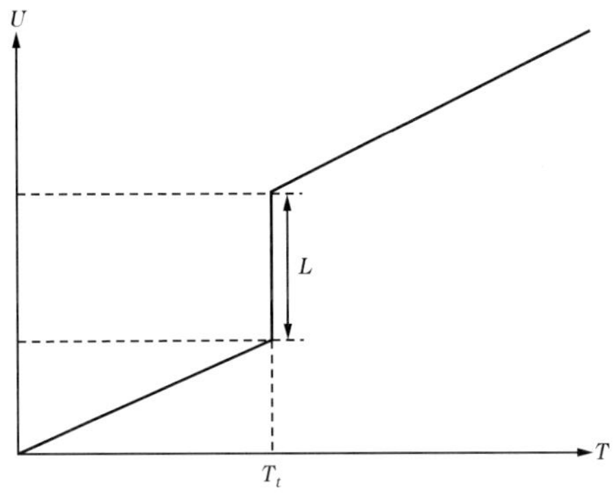

Fig. 3.2. Schematic caloric curve for a simple body at constant volume. Low temperature quantum effects are not represented

The phase diagram of a simple body is shown schematically in Fig. 3.1. Three phases are apparent:

- *Solid.* Below the melting temperature corresponding to the solid–liquid equilibrium and above the sublimation pressure corresponding to the solid–gas equilibrium curve.
- *Liquid.* Below the melting temperature and above the vaporisation pressure corresponding to the liquid–gas equilibrium curve.
- *Gas.* At low pressure, below the two previous curves. Above the critical point (P_c, T_c), there is no difference between liquid and gas.

The region that concerns us here is of course the liquid–solid equilibrium curve, or more precisely, its lower end, the so-called triple point (P_3, T_3), corresponding to the coexistence of three phases. Indeed, as we shall see shortly, an isolated nanoparticle is closer to a constant volume system than a constant pressure system. We thus begin by examining what happens when a solid macroscopic particle is heated at constant volume. We first follow the sublimation curve, a certain amount of matter being vaporised in order to maintain the low saturated vapour pressure in the container. If the volume of the container is not too large, so that not too much matter is vaporised, we end up at the triple point. At this point, although heat is still being supplied to the system, the temperature and pressure remain constant as long as the three phases coexist. When all the solid has been transformed into a liquid, we then follow the vaporisation curve, leaving it at a point depending on the volume chosen and on the size of the particle. A calorimeter is used to measure the energy supplied to the system. We then have a curve giving the internal energy as a function of the temperature, of the kind shown in Fig. 3.2, and called the caloric curve or heat curve.

The discontinuity in the energy at the transition is called the latent heat L. With the transition temperature T_t, the parameter L characterises the macroscopic transition in the context of thermodynamics.

Microscopic Theory of the Solid–Liquid Transition

At the present time, there is no satisfactory microscopic theory of the solid–liquid transition. Although it is not a recent discovery, the Lindemann criterion provides a simple and fairly reliable estimate of the melting temperatures of monatomic solids such as metals and noble gases. The key idea is that the amplitude of vibration of atoms in the crystal about their equilibrium position will increase with temperature. When this amplitude reaches a certain fraction f of the distance between nearest neighbours in the crystal lattice, the solid can no longer maintain its crystal structure and it subsequently melts. Experimentally, it is observed that for all monatomic solids $f \simeq 0.07$ [2]. In order to apply this to nanoparticles, we shall state this criterion in the form: a monatomic solid melts when the fluctuations in interatomic distances are of the order of 14%. Indeed, bond length fluctuations are equal to twice the vibrational amplitude of atoms located at each end of the bond. These fluctuations are given by the relative squared deviation δ defined by

$$\delta = \frac{2}{N(N-1)} \sum_{i<j} \frac{\sqrt{\langle r_{ij}^2 \rangle - \langle r_{ij} \rangle^2}}{\langle r_{ij} \rangle} , \tag{3.1}$$

where $\langle \ \rangle$ stands for a time average and r_{ij} is the distance between atoms i and j. δ is in fact also an average over all bonds of the relative fluctuation of each bond.

The solid–liquid transition is characterised by two thermodynamic parameters: the transition temperature T_t and the latent heat L. The Lindemann criterion provides a phenomenological prediction of the melting temperature, viz., the solid melts when the relative fluctuation in the interatomic distances reaches about 14%.

Size Effects on the Solid–Liquid Transition

The distinction between solid and liquid is obvious on our scale, in the sense that we can see immediately whether a glass of water is frozen or not. However, if we think about this for a moment, it is not so easy to find measurable criteria capable of distinguishing between the two states. An elementary school book gives the following definition: a liquid adopts the shape of the container in which it is held. But it would be hard to transpose this idea to the nanoscale! This is why, even though the reduction in the transition temperature of small particles was predicted as early as 1909 [3], experimental techniques have only been developed rather recently. Two types of method are currently available:

- observation of the structure of deposited particles by electron diffraction [4],
- calorimetric measurements on deposited [5] or free particles [6].

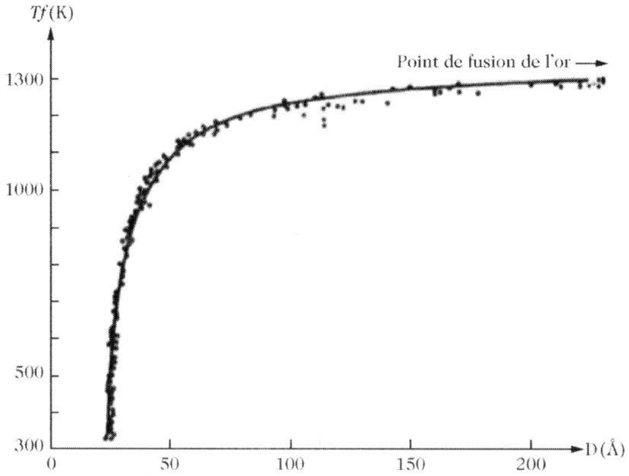

Fig. 3.3. Dependence of the solid–liquid transition temperature of gold on particle diameter. Taken from [4]. The transition is detected by the disappearance of Bragg peaks in the electron diffraction pattern. *Dots* represent experimental values. The *continuous curve* is fitted to (3.3) by the least squares method

We shall begin with deposited particles, containing at least a few thousand atoms. Measurements are usually made for a size distribution centered on a given radius.

Figure 3.3 shows the variation in the temperature of the solid–liquid transition for gold particles with sizes in the range 2–25 nm. The drop in the melting point is striking, from 1336 K (melting temperature of solid gold) to around 300 K! As the transition is directly detected by a change in the electron diffraction pattern, it is not possible to measure the latent heat. This has been done recently, however, by Lai and coworkers [5], thanks to progress in nanocalorimetry.

Figure 3.4 shows the measured caloric curves for tin particles of different radii. To show the energy discontinuity at the transition more clearly, the authors have subtracted the linear evolution due to the heat capacity of the system as a whole. The graph has a similar shape to the one in Fig. 3.2, except that the energy is no longer discontinuous and the transition stretches over a certain temperature interval ΔT. In fact, one reason for this in the experiment cited is that, since each size corresponds to a specific transition temperature, a size distribution automatically leads to a distribution of transition temperatures. We shall see below that there are also intrinsic reasons for the transition to broaden, although these concern much smaller sizes, i.e., a few hundred atoms (see Sect. 3.1.2).

Using the curves in Fig. 3.4, the thermodynamic parameters T_t and L can be extracted as a function of size. These are shown in Fig. 3.5. As for gold, there is a considerable reduction in the melting point at small sizes. Moreover, the latent heat also drops, by almost 70%. Now the classical thermodynamic models used up to recently to describe the change in melting point

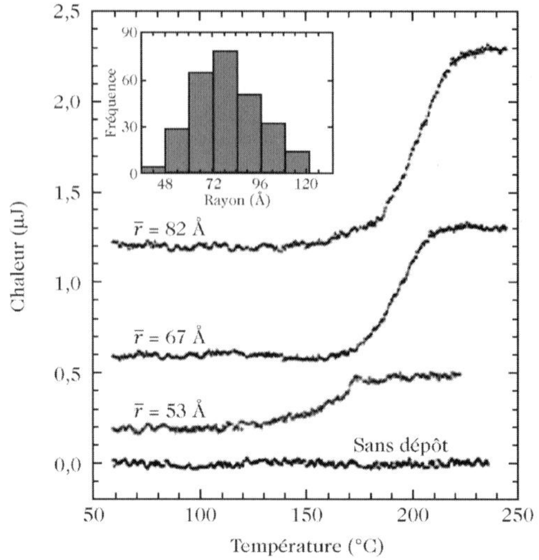

Fig. 3.4. Amounts of heat measured by calorimetry during the melting of deposited tin nanoparticles with average radii 5.3, 6.7 and 8.2 nm. Taken from [5]. The heat due to the heat capacity of the system as a whole has been subtracted so that only the contribution from the latent heat of transition of the clusters appears on the graphs. Curves are translated vertically for clarity

with particle size assumed that the latent heat was constant. This experiment demonstrates that such an assumption is unjustified, as had also been predicted by numerical calculations [7]. We shall nevertheless present these classical models in the next section, since they provide a simple interpretation of observed phenomena.

Classical Models

In order to model a small particle, surface effects cannot be ignored. The first effect due to the presence of a surface is to add a surface tension term to the thermodynamic potentials. If this correction is taken into account, assuming furthermore that the thermodynamic parameters do not depend on the size, the so-called classical models for the melting of small particles are obtained. All these theories are based on the Gibbs–Duhem equation, expressing the chemical potentials of the solid and the liquid, and the Laplace equation for the surface, taking surface tension into account. A first order expansion gives an expression of the form

$$1 - \frac{T_{\mathrm{m}}(r)}{T_{\mathrm{m}}(\infty)} = \frac{2\alpha}{\rho_{\mathrm{s}} L(\infty) r} , \tag{3.2}$$

where $T_{\mathrm{m}}(r)$ is the melting temperature of the particle of radius r, $T_{\mathrm{m}}(\infty)$ is the melting temperature of the bulk solid, ρ_{s} is the density of the solid,

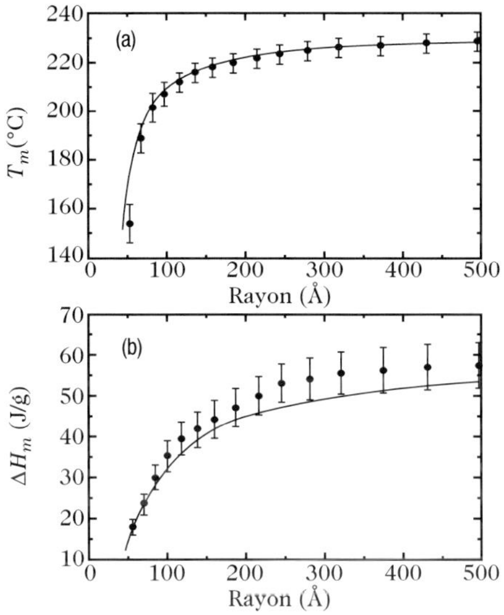

Fig. 3.5. Dependence of the thermodynamic melting parameters of deposited tin nanoparticles on the particle radius. Taken from [5]. (**a**) Solid–liquid transition temperature. *Points* represent experimental values. The *continuous curve* is fitted to (3.4) by the least squares method. (**b**) Latent heat per unit mass. *Points* represent experimental values. The *continuous curve* is fitted to (3.5) by the least squares method

$L(\infty)$ is the latent heat of fusion of the bulk solid, and α is a function of the solid–liquid surface tension γ_{sl}, the solid–vapour surface tension γ_{s}, and the liquid–vapour surface tension γ_{l}.

To obtain a more precise expression for α, one must choose a model for the solid–liquid equilibrium at the transition. There are two main choices here:

- *First Model.* This assumes a chemical-type equilibrium between entirely solid particles, entirely liquid particles, and the vapour. One then has

$$1-\frac{T_{\mathrm{m}}(r)}{T_{\mathrm{m}}(\infty)}=\frac{2}{\rho_{\mathrm{s}}Lr_{\mathrm{s}}}\left[\gamma_{\mathrm{s}}-\gamma_{\mathrm{l}}\left(\frac{\rho_{\mathrm{s}}}{\rho_{\mathrm{l}}}\right)^{2/3}\right], \tag{3.3}$$

  where ρ_{l} is the density of the liquid and r_{s} the radius of the solid particles. This expression has been used in [4] to interpret the curve in Fig. 3.3.
- *Second Model.* This assumes that, at the transition temperature, each particle, still in equilibrium with its vapour, contains a solid core surrounded by a layer of liquid of thickness δ. One then has

$$1-\frac{T_{\mathrm{m}}(r)}{T_{\mathrm{m}}(\infty)}=\frac{2}{\rho_{\mathrm{s}}L}\left[\frac{\gamma_{\mathrm{sl}}}{r-\delta}+\frac{\gamma_{\mathrm{l}}}{r}\left(1-\frac{\rho_{\mathrm{s}}}{\rho_{\mathrm{l}}}\right)\right]. \tag{3.4}$$

  Note that, even if δ is zero, this equation does not reduce to (3.3). We do have $\gamma_{\mathrm{s}}=\gamma_{\mathrm{sl}}+\gamma_{\mathrm{l}}$, but the ratio of the liquid and solid densities is not at the same power. This correction is nevertheless negligible compared with the many approximations made in these models.

The same type of model involving surface tension energies can be used to understand the change in the latent heat of the solid–liquid transition for particles of different radii r. In the framework of the first model (equilibrium between entirely solid and entirely liquid particles), we obtain

$$L(\infty) - L(r) = \frac{3}{r}\left(\frac{\gamma_\mathrm{s}}{\rho_\mathrm{s}} - \frac{\gamma_\mathrm{l}}{\rho_\mathrm{l}}\right) + \int_{T_\mathrm{m}(r)}^{T_\mathrm{m}(\infty)} \left[c_{p,\,\mathrm{liq}}(T) - c_{p,\,\mathrm{sol}}(T)\right]\mathrm{d}T \,. \tag{3.5}$$

Quite generally, the parameters determining the transition vary steadily with the size. The dominating term in these variations goes as $1/r$, i.e., as $N^{-1/3}$ if we refer to the number of particles N. This is why the graphs showing these variations are often plotted as a function of $N^{-1/3}$ rather than N.

Uncertainty in the Solid–Liquid Transition Temperature

As mentioned above, the transition seems less abrupt for small particles than in the bulk solid. Although the size dispersion of the sample, leading to a distribution of width ΔT in the transition temperatures, can in part explain this broadening effect, Imry [8] has demonstrated another more fundamental reason why the transition should broaden, which arises even if the particle sizes are perfectly selected.

Consider a particle with N constituents and neglect variations in the transition temperature and latent heat due to size differences. The probability $p(E)$ of an energy state E at temperature T is given by the Einstein formula

$$p(E) = \frac{\exp\left[S(E)/k_\mathrm{B} - E/k_\mathrm{B}T\right]}{Z(T)} \,, \tag{3.6}$$

where k_B is the Boltzmann constant and $Z(T)$ is a normalisation factor called the partition function. Now, if s_s and e_s are the entropy and energy per molecule in the solid, the energy per molecule in the liquid is given by $e_\mathrm{l} = e_\mathrm{s} + L$ and the entropy by $s_\mathrm{l} = s_\mathrm{s} + L/T_\mathrm{m}$, where L is the latent heat per molecule. We can now calculate the ratio of the probabilities of having a liquid particle and a solid particle:

$$\frac{p_\mathrm{l}}{p_\mathrm{s}} = \exp\left[\frac{NL}{k_\mathrm{B}}\left(\frac{1}{T_\mathrm{m}} - \frac{1}{T}\right)\right] \,. \tag{3.7}$$

It can thus be seen that this ratio goes from 0 to a very large value over an interval ΔT such that $\Delta(1/T)NL/k_\mathrm{B} \simeq 1$. We may deduce that

$$\Delta T = \frac{k_\mathrm{B}T_\mathrm{m}^2}{NL} \,. \tag{3.8}$$

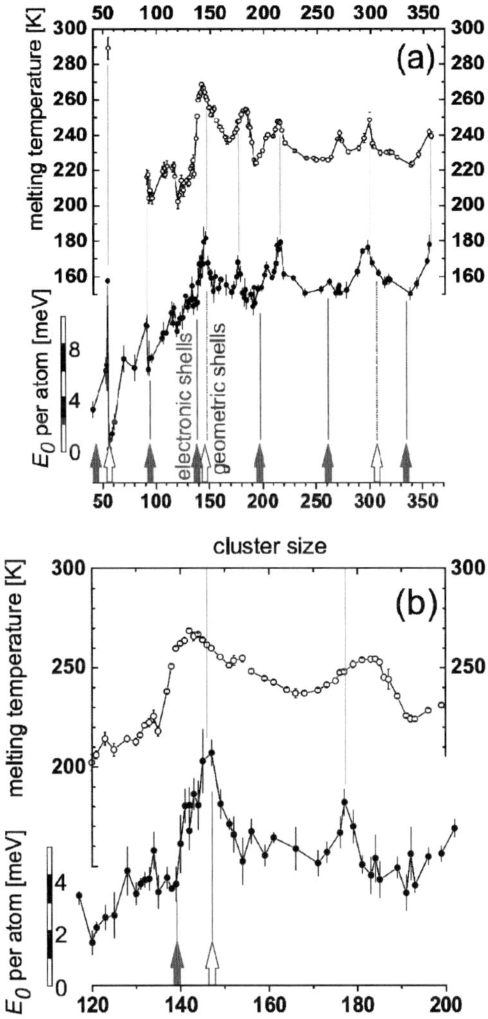

Fig. 3.6. Solid–liquid transition temperatures and latent heats (E_0) for sodium clusters, measured by Haberland and coworkers [6]. (**b**) is a magnified view of (**a**). Given the broadening of the transition, no exact value can be attributed to the latent heat. However, the experiment provides the differences from one size to another in a precise manner. The E_0 scale is thus a relative scale

For tin particles with diameter 5.3 nm, Fig. 3.5 implies that the latent heat per atom is about 4×10^{-21} J for particles of around 20,000 atoms. We may deduce that the width of the transition is of the order of 0.03 K, which is impossible to observe. However, for very small particle sizes, this width can become significant. For example, provided that the latent heat and the transition temperature do not vary too much, the width of the transition for 20 atoms would be of the order of 30 K. It remains to check that some meaning can be attached to the transition for such small sizes.

3.1.2 From Nanoparticles to Molecules

We shall now discuss experiments carried out on very small particles, containing only a few tens to a few hundreds of atoms, with size selection. There are very few methods available to do this and experimental results are scarce. This explains why numerical simulation experiments have been used to supplement the few available methods. The behaviour of a system is modelled and calculated by computer. Without wishing to exaggerate the scope of this kind of experiment, they have nevertheless proved extremely useful for studying universal phenomena like melting. Indeed, even though the model may be rather far removed from a real system, one may still hope that similar mechanisms and qualitative features will be generated.

Experiments

One of the great achievements of the last decade has been the possibility of carrying out nanocalorimetry experiments on size-selected clusters of a hundred or so atoms, by Haberland and coworkers in Freiburg (Germany). Figure 3.6 shows the latest results from this group [6]. In contrast to what is observed for larger particles, the evolution of the thermodynamic parameters now exhibits no regular trend with changing particle size. In fact, in this size range, more than half the atoms are at the surface. There is therefore no hope of basing a theory on any simple extrapolation of macroscopic parameters. Hence, certain sizes corresponding to some particular geometrical shape may be more stable than others. This happens for instance with the cluster of 147 atoms, which corresponds to a regular icosahedron. Moreover, the quantum confinement of electrons can have significant consequences, leading to stabilisation at certain sizes due to the closing of electron shells. This happens for the cluster with 138 valence electrons, for example. But in any case, the transition temperatures lie well below those for the bulk material (370.95 K), as do the latent heats (27 meV/atom for the bulk material). It would be interesting to see whether (3.3)–(3.5) give the right orders of magnitude here.

In a second recent experiment [9], the transition temperatures of tin clusters of a hundred or so atoms have been determined by gas phase chromatography. The basic idea is that the transition from solid to liquid changes the overall shape of the particle, which becomes spherical, whereupon it has a different mobility within a gas. Figure 3.7 compares this temperature with the temperature for bigger particles measured by various methods. The remarkable thing in this case is that the transition temperature is higher than in the macroscopic material. This result is attributed to a change in the solid structure in small tin particles, which is not directly revealed by the experiment.

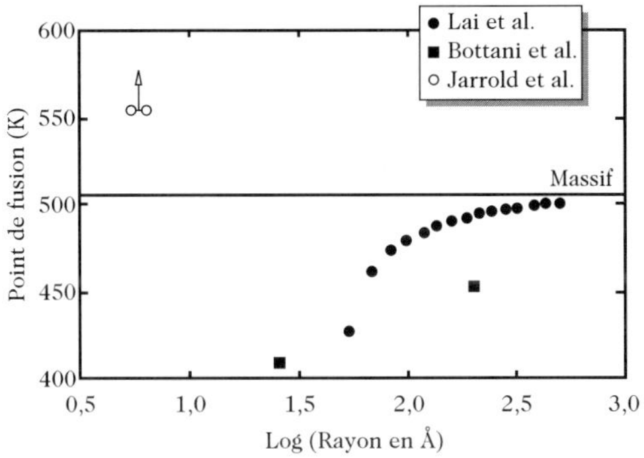

Fig. 3.7. Transition temperatures of tin clusters. *White circles* are measurements by Jarrold et al. [9]. *Black circles* are measurements by Lai et al. [5]. *Black squares* are measurements made by Kofman et al. using optical methods [10]

Numerical Results

When discussing the results of numerical experiments, it no longer suffices to simply cite the name of the element under investigation. The model must also be described. Here we present a model which has been so often studied that it has been given its own name: Lennard-Jonesium! It is a simple but realistic model, first used to simulate condensed phases, then applied to clusters since the beginning of the 1970s [11]. It was Berry et al. [12] who first used it to try to understand phase transitions in finite systems. In this model, material points interact in pairs through the Lennard-Jones (LJ) potential

$$v_{\mathrm{LJ}}(r) = e_0 \left[\left(\frac{r_0}{r} \right)^{12} - 2 \left(\frac{r_0}{r} \right)^{6} \right] . \tag{3.9}$$

This potential is zero at infinite distances, has a minimum with a value of $-e_0$ for $r = r_0$ and grows very fast as $r \to 0$. In simulations, $e_0 = 1$ and $r_0 = 1$, but well chosen values for these parameters provide an excellent model for the noble gases. A cluster of n atoms interacting by the Lennard-Jones potential will be denoted by LJ_n in the following. Many other models, realistic to varying degrees, have been used to calculate the properties of clusters of atoms or molecules. They are all more complicated than the LJ model and lead therefore to much heavier numerical calculations. For the qualitative investigation of melting, they provide no great advantage.

There are two main types of simulation method:

- *Molecular Dynamics.* The differential equations of motion are solved, generally in Hamiltonian form.
- *Monte Carlo.* Statistical averages are calculated directly.

To avoid evaporation during the simulation, a confinement term $v_{\mathrm{c}}(r)$ which becomes infinite when the atoms wander too far away is generally added to the potential. One thus ends up with a Hamiltonian of the form

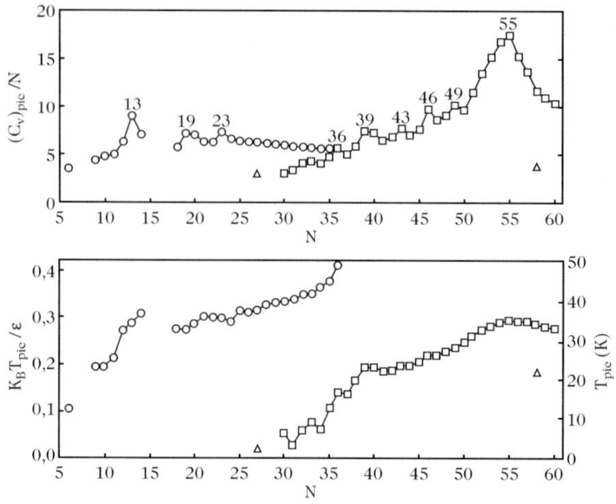

Fig. 3.8. Solid–liquid transition temperatures and maximum values of the peaks in the heat capacity for Lennard-Jones clusters. Taken from Frantz [13]

$$H = \sum_{i=1}^{N} \frac{p_i^2}{2m} + \sum_{i<j} v_{\mathrm{LJ}}(r_{ij}) + \sum_{i=1}^{N} v_{\mathrm{c}}(r_i) , \tag{3.10}$$

where r_{ij} is the distance between atoms i and j, and r_i is the distance between atom i and the origin. The atoms are thereby enclosed in a spherical cavity.

Figure 3.8 shows the calculated melting points of Lennard-Jones clusters containing a few tens of atoms. As in the Haberland experiment, the results reveal no regular trend for changing cluster size. Of course, certain types of structure can be related to the change in transition temperature from one size to another, but the diversity of these structures means that no general law can be given. Figure 3.8 also shows the maximum value of the heat capacity. This can only be related to the heat capacity if one also knows the transition width, as given approximately by (3.8). The small size of the clusters here makes it difficult to extract the latter parameter accurately.

The smallest cluster for which Frantz found a peak in the heat capacity and to which he attributes a transition temperature contains only 6 atoms! For such sizes, simulations are fast and reliable. One should still ask what this peak represents. Can we still speak of a phase transition? But if not, what criterion should we adopt? This is where theory can provide some answers, and this is why we shall devote the next section to some theoretical considerations concerning the thermodynamics of small systems.

3.2 Thermodynamics of Very Small Systems

3.2.1 General Considerations

Consider a simple thermodynamic system with volume V and containing N particles. This system is characterised by a function called the density of states, given in classical mechanics by an integral over the phase space, viz.,

$$\Omega(E,V,N) = \frac{\mathrm{d}}{\mathrm{d}E}\int_{H(\boldsymbol{q},\boldsymbol{p})\leq E,\, q\in V} \mathrm{d}\boldsymbol{q}\mathrm{d}\boldsymbol{p}\ , \tag{3.11}$$

where $\boldsymbol{q}$ stands for all position coordinates, $\boldsymbol{p}$ stands for the generalised momenta, and E is the total energy of the system. In quantum mechanics, the density of states is defined rather differently, but raises no particular difficulties in this context. We consider the following Gibbs ensembles:

- *Microcanonical Ensemble.* Variables (E, V, N). Entropy and temperature given by

$$S_\mu(E,V,N) = k_\mathrm{B}\ln\left[\Omega(E)\right]\ , \tag{3.12}$$

$$T_\mu(E,V,N) = \left(\frac{\partial S}{\partial E}\right)^{-1}\ . \tag{3.13}$$

- *Canonical Ensemble.* Variables (T, V, N). Setting $\beta = 1/k_\mathrm{B}T$, the partition function $Z(\beta,V,N)$ is defined by

$$Z(\beta,V,N) = \int \Omega(E,V,N)\mathrm{e}^{-\beta E}\mathrm{d}E\ . \tag{3.14}$$

  The entropy S and internal energy U are given by

$$U(\beta,V,N) = \frac{\displaystyle\int E\Omega(E,V,N)\mathrm{e}^{-\beta E}\mathrm{d}E}{Z(\beta,V,N)}\ , \tag{3.15}$$

$$S(\beta,V,N) = \frac{U(\beta,V,N)}{T} + k_\mathrm{B}\ln\left[Z(\beta,V,N)\right]\ . \tag{3.16}$$

- *Grand Canonical Ensemble.* Particles have variable size N. Since we wish to obtain the behaviour of a cluster of given size, we shall not consider this ensemble here.

Note that all the physics of the system is contained in the density of states $\Omega(E,V,N)$, whatever ensemble is considered. So by looking at the energy distribution in the canonical ensemble, one obtains the Boltzmann distribution, and hence the density of states, whereupon one may trace back to the microcanonical effects.

In the thermodynamic limit, i.e., when $N \to \infty$ with $U/N \to u$ and $V/N \to v$ (u and v finite), the microcanonical and canonical ensembles lead to the same predictions. This is what is meant by the equivalence of the ensembles. Moreover, there is a function $s(u, v)$ such that, as $N \to \infty$, the entropy per molecule $S/N \to s$. This is why the entropy is treated as an extensive function. However, for finite N, neither of these two properties is satisfied. The non-extensive nature of the system is already extant in the classical models mentioned above, since one adds a surface tension (energy proportional to $N^{2/3}$) to the bulk energy (proportional to N). In small clusters, the thermodynamic properties can vary enormously from one size to another, and there is really no place for extensivity here.

3.2.2 Non-Equivalence of the Gibbs Ensembles

One of the most spectacular effects of the non-equivalence of ensembles for finite systems is manifested in the convexity properties of the entropy. Indeed, in the thermodynamic limit, it can be shown that the entropy is a concave function of the energy (negative second derivative), and this is what explains the stability of the system. Moreover, it is easy to see that this concavity holds in all cases for the canonical ensemble, even in small systems.

The demonstration is straightforward. We begin by relating the heat capacity C_v to this second derivative. We have

$$\left.\frac{\partial^2 S}{\partial U^2}\right|_{V,N} = \frac{\partial}{\partial U}\frac{1}{T} = -\frac{1}{T^2}\left.\frac{\partial T}{\partial U}\right|_{V,N} = -\frac{1}{T^2 C_v} . \tag{3.17}$$

We must therefore show that $C_v \geq 0$. Now,

$$\begin{aligned} C_v(\beta, V, N) &= \left.\frac{\partial U}{\partial T}\right|_{V,N} = -k_{\mathrm{B}}\beta^2 \left.\frac{\partial U}{\partial \beta}\right|_{V,N} \\ &= k_{\mathrm{B}}\beta^2 \left[\frac{1}{Z(\beta, V, N)}\int E^2 \Omega(E, V, N)\mathrm{e}^{-\beta E}\mathrm{d}E - U^2\right] \\ &= k_{\mathrm{B}}\beta^2 \left[\frac{1}{Z(\beta, V, N)}\int (E-U)^2 \Omega(E, V, N)\mathrm{e}^{-\beta E}\mathrm{d}E\right] . \end{aligned} \tag{3.18}$$

The last term in square brackets represents the mean squared deviation of the Boltzmann energy distribution, with the probability distribution $\Omega(E, V, N) \times \mathrm{e}^{-\beta E}/Z(\beta, V, N)$. It is thus always positive.

In the microcanonical ensemble, C_v can no longer be expressed as a squared deviation and we shall see below that the entropy can contain convex regions in some cases. This implies a negative heat capacity in those regions. In other words, the temperature of the cluster can then be reduced by heating it! This counterintuitive effect occurs in other branches of statistical physics, e.g., self-gravitating systems [14] and nuclear physics [15]. We shall discuss the relationship between this effect and phase transitions below.

In fact, the absence of convex portions in the entropy in the canonical case arises due to the different nature of this ensemble. The microcanonical ensemble represents a single isolated system. In contrast, the canonical ensemble can be viewed in two different ways, either as describing a single system in contact with a thermostat, or as a statistical ensemble of isolated systems with Boltzmann energy distribution. In both cases, it is the result of a distribution imposed by an external agent.

These differences provide a useful guide in choosing the most suitable ensemble for experimental conditions. Hence, for deposited clusters or clusters embedded in a matrix, the canonical ensemble is the natural choice. Likewise, when observing a trapped cluster, one must use the microcanonical ensemble. Needless to say, however, some experiments can be more difficult to interpret.

As an example, consider a molecular beam of clusters. Each cluster is isolated and hence accurately described by a microcanonical ensemble. However, single clusters are rarely observed and experimental results are generally expressed as averages over a large number of individuals. Now the way the clusters are prepared influences their energy distribution. If the latter comprises a narrow peak around some given energy, we will be close to the microcanonical ensemble. On the other hand, if this distribution is more spread out, it may be more difficult to draw any conclusion. In practice, it is always possible to pass the beam through a thermalisation cell, where the distribution is rendered canonical. As mentioned above, microcanonical effects can be ascertained by looking at the energy distribution.

3.2.3 Dynamically Coexisting Phases

Observations

The aim here is to characterise phase transitions in nanoparticles. We begin with a numerical experiment using molecular dynamics to simulate a simple calorimetry experiment. Starting from the solid state (very low energy), a particle of size N is heated in a stepwise manner to higher energies, and the instantaneous kinetic temperature defined by

$$T_{\mathrm{kin}}(t) = \frac{2K(t)}{(3N-6)k_{\mathrm{B}}} \tag{3.19}$$

is observed, where $K(t)$ is the kinetic energy of the cluster at time t. For constant total energy E, the time average of $T_{\mathrm{kin}}(t)$ gives a value very close to the microcanonical thermodynamic temperature $T(E)$. Figure 3.9 shows the result for a Lennard-Jones cluster of 55 atoms. At low energies, it is clear that the kinetic temperature fluctuates around an average value corresponding to the temperature of the plateau. Successive plateaus are clearly visible. When the total energy increases, the average temperature and its fluctuations increase proportionally, until the plateaus disappear. Moreover, when the total energy reaches a certain threshold, obtained here after about 250 ps, the temperature drops sharply, then begins to grow again. This drop in temperature is accompanied by a sharp increase in the relative squared deviation of the bond lengths, δ given by (3.1), which goes from small values (< 0.05) to

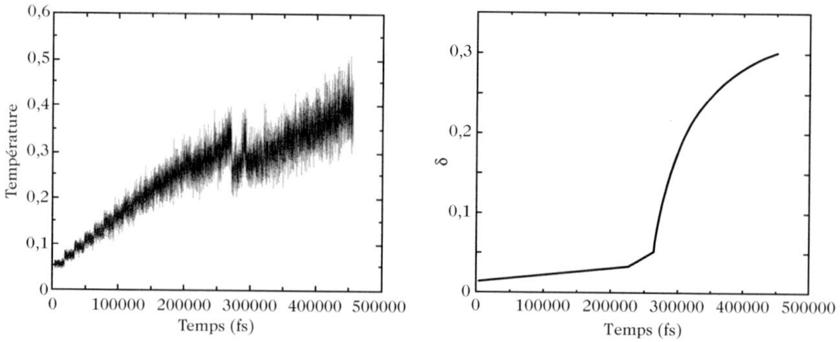

Fig. 3.9. *Left*: Instantaneous kinetic temperature of an LJ_{55} cluster heated in stages. *Right*: Mean squared deviation δ of the bond length calculated for the accumulated time averages

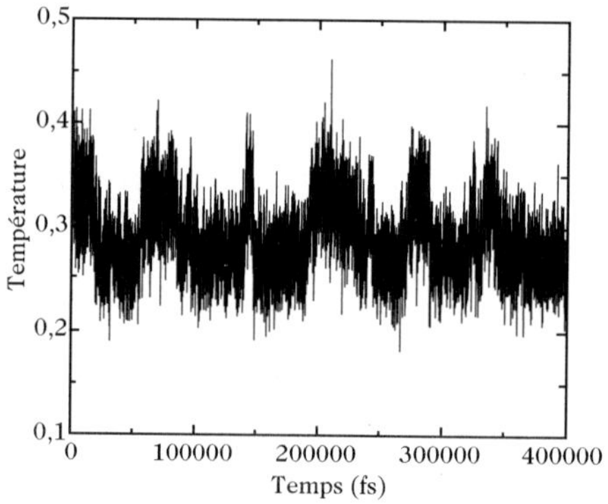

Fig. 3.10. Instantaneous kinetic temperature of an LJ_{55} cluster held at constant energy near the solid–liquid transition

more than 0.15. The Lindemann criterion mentioned above indicates that the system then becomes liquid. We are thus led to the paradoxical situation in which the liquid is colder than the solid at the energy at which the transition occurs.

When the simulation is left to evolve over a very long time at this energy (see Fig. 3.10), it turns out to be impossible to reach a stable regime. The temperature alternates at random between a 'hot' state (fluctuations about a high kinetic energy) and a 'cold' state. The cluster thus spends a certain time in each of these states, which we may identify with the solid and liquid phases. This phenomenon is called dynamic phase coexistence.

Furthermore, the sudden increase in δ is preceded by a more linear growth over a rather short time span. This increase is due to the greater mobility of the outer atoms of the cluster, an effect known as surface melting.

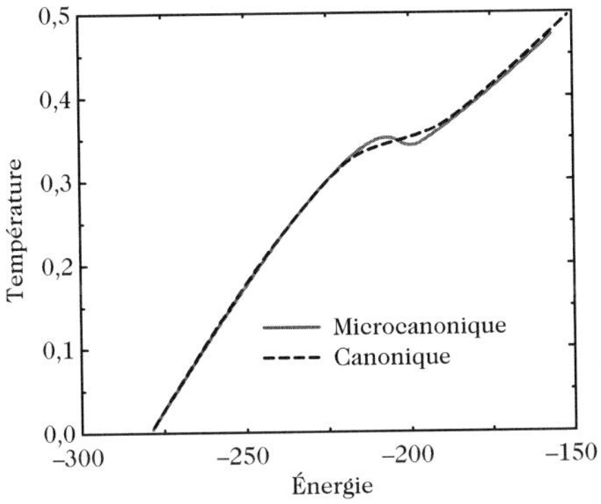

Fig. 3.11. Canonical and microcanonical heat curves for the LJ_{55} cluster

Interpretation

Heat curves can be plotted by making a detailed statistical analysis of the above simulation results. Figure 3.11 shows two curves obtained with microcanonical and canonical ensembles for the 55-atom cluster denoted by LJ_{55}. The first shows the behaviour of the isolated system, whilst the second is associated with a Boltzmann distribution of clusters. For the microcanonical curve, we observe the drop in temperature between two values of the internal energy. We deduce that, in this interval, the heat capacity is negative. For its part, the canonical curve is monotonic, in agreement with the fact that the heat capacity is positive in this ensemble, as given by (3.18).

As we saw above, a negative heat capacity is associated with a convex region of the entropy. Figures 3.12a and c summarise the situation. Once the entropy has been obtained, the probability distribution of the energy in the canonical ensemble can be constructed. Indeed, the logarithm of the probability $p(E,T)$ at temperature T is, up to an additive constant, equal to the difference between the entropy and the straight line of slope $1/T$. Each time the derivative of the entropy is equal to $1/T$, the probability $p(E,T)$ is extremal, as can be checked by working out the derivative of $\ln p$:

$$\frac{\partial \ln p}{\partial E} = \frac{\partial (S - E/T)}{\partial k_\mathrm{B} E} = \frac{1}{k_\mathrm{B}} \left[\frac{1}{T_\mu(E)} - \frac{1}{T} \right] , \qquad (3.20)$$

where $T_\mu(E)$ is the microcanonical temperature. In general, the probability only has one maximum, unless T is in the interval ΔT of Fig. 3.12a. The above derivative then vanishes at three points, and it is easy to see that these points correspond to two maxima on either side of one minimum. This is referred to as bimodality. The distribution shown in Fig. 3.12d illustrates this property when the slope $1/T$ is such that the straight line is tangent to the entropy curve at two points, in which case the two maxima have the same height.

Bimodality reflects the coexistence of phases in the canonical ensemble. This is closely connected to the phenomenon of dynamic coexistence discussed

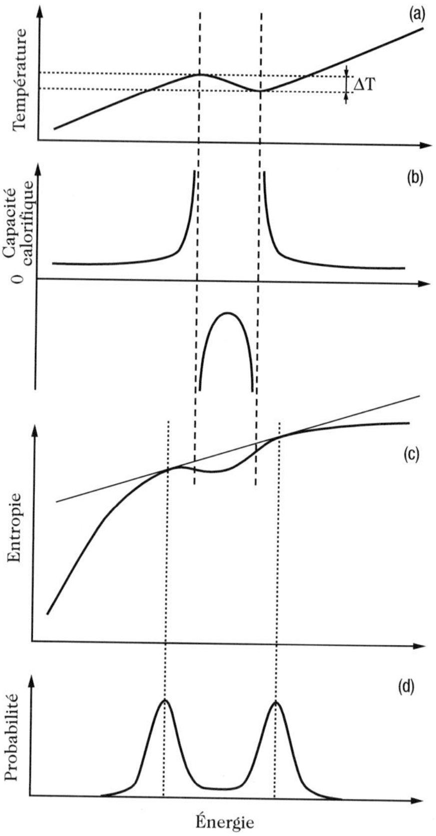

Fig. 3.12. Typical microcanonical properties as a function of the energy. (**a**) Temperature. (**b**) Heat capacity. (**c**) Entropy. The *straight line* tangent to the entropy curve at two points shows the construction of the canonical probability distribution represented in (**d**)

above. However, it is of a rather different nature, since what we have now is a statistical coexistence within a population of clusters, similar to a chemical equilibrium. This observation means we can relate each maximum of the distribution to a phase. Since the temperature is constant here, it is the high-energy (respectively, low-energy) phase which is associated with the liquid (respectively, solid). Moreover, the interval ΔT in Fig. 3.12a measures the width of the phase transition. It is effectively nonzero, as predicted from macroscopic considerations [see (3.8)].

According to what has been said so far, the following properties are equivalent:

- The microcanonical entropy is convex over a certain energy interval.
- The microcanonical heat capacity is negative in this interval.

- The canonical probability distribution is bimodal in a certain temperature interval.

These properties imply a coexistence between two states that can be called phases and hence provide a way of defining a phase transition in a finite system. Since they are all based on an intrinsic property of the system (the density of states), they fulfill the need for a criterion expressed at the end of Sect. 3.1.2. However, dynamic coexistence can sometimes be observed even when the entropy is never convex [16]. But this is a borderline case as far as phase transitions are concerned. The above properties have the advantage of being simple and can be used to identify phase transitions in astonishingly small systems where they remain valid.

3.2.4 Stability of an Isolated Particle. Thermodynamic Equilibrium

In the above discussion, we did not raise the question as to whether a thermodynamic treatment of nanoparticles remains valid. Such a treatment is based on the notion of a stable, hence enduring, equilibrium. Now it is easy to see that no finite canonical system in an infinite space is stable with respect to evaporation. A first difficulty in an infinite space arises from the fact that the motion of the center of mass is not restricted, so that the integral (3.11) includes a term proportional to the volume V, which will affect the partition function in (3.14). This problem is simply overcome by restricting to the center of mass frame. However, contributions from situations where the atoms are widely separated from one another each include once again a term proportional to V. They thus totally dominate in the limit $V \to \infty$, whatever the temperature (provided it is not zero). The system thus has zero probability of remaining bound together. In other words, the saturated vapour pressure of our particle is never strictly zero. It thus seems unjustified to base a study of very small particles on a thermodynamic equilibrium that does not exist.

This argument is not restricted to the canonical ensemble. It can be applied to the microcanonical ensemble, provided that the total energy is greater, no matter how slightly, than the dissociation energy. Indeed, in this case, dissociated situations make an infinite contribution to the density of states and are thus infinitely more probable than bound situations. In practice, the dissociation energy is usually much smaller than the heat energy. For example, for a particle containing $N = 100$ atoms, a dissociation energy E_0 of the order of 0.8 eV (value for sodium clusters) corresponds to a temperature of the order of $E_0/3Nk_B \simeq 30$ K. Such a low temperature is difficult to achieve in a beam of metal clusters.

Some caution would thus seem to be in order when applying the results of equilibrium thermodynamics to nanoparticles, since these particles can only exist for a finite lapse of time. However, this lapse of time can in practice be longer than most realistic experiments. In many cases, the cluster can be considered to be in a metastable equilibrium. One must then ensure that the duration of this state is long enough to allow the particle to reach an internal equilibrium, and that it has time to thermalise. The thermodynamic theory is thus valid whenever the observation time lies between the thermalisation time and the evaporation time. In the next section, we shall discuss several theories that can be used to estimate the lifetime of a nanoparticle.

3.3 Evaporation: Consequences and Observations

A diatomic molecule will dissociate whenever its excitation energy exceeds the binding energy during a single vibrational period. The situation is rather different on the macroscopic scale. Indeed, even if a material such as a gram of metal at room temperature contains a huge amount of thermal energy compared with the binding energy of a single atom, the typical dissociation time is extremely long due to the statistical distribution of this energy over all degrees of freedom of the system.

It is thus understandable that the time taken by an atomic or molecular cluster to dissociate will depend strongly on its internal energy, and also on its size. As in condensed matter, the phenomenon of evaporation will therefore be considered as a rare event for clusters.

Dissociation and evaporation are fundamental aspects of the study of small clusters, since they give access to one of the key properties of these systems, namely the binding energy. Binding energies and more generally the degree of stability of nanosystems are deduced using statistical theories able to relate them to certain experimental observables, such as the dissociation rate and the distribution of kinetic energy released.

3.3.1 Statistical Theories of Evaporation

The Rice–Ramsperger–Kassel Theory

The theory proposed by Rice and Ramsperger [17], then by Kassel [18], called the RRK theory, provides a simplified treatment of the dissociation of molecules containing an energy excess. It assumes that the molecule can be described as an ensemble of s harmonic oscillators that can exchange energy between them, in such a way that dissociation occurs whenever one of these harmonic modes has enough energy, i.e., more than the dissociation threshold E_0. The value of s depends on the entities making up the cluster:

- For a system of N atoms, $s = 3N - 6$.
- For a system of N linear (rigid) molecules, $s = 5N - 6$.
- For a system of N nonlinear (rigid) molecules, $s = 6N - 6$.

The problem then reduces to estimating the number of ways of distributing the total energy over the system, with an amount greater than E_0 in just one of the modes. In this framework, the reaction rate one seeks is this number normalised by the number of ways of distributing the total energy, whatever energy is deposited in the dissociating mode. Now the number of ways of distributing the energy E over s oscillators is given by $[E^{s-1}/(s-1)!]$. The dissociation rate is therefore given by

$$k(E) \propto \frac{(E-E_0)^{s-1}/(s-1)!}{E^{s-1}/(s-1)!} = \left(\frac{E-E_0}{E}\right)^{s-1} . \tag{3.21}$$

An improvement suggested by Kassel replaces the classical description of the oscillators by a quantum description. If ν is the frequency of vibration of the oscillators, then $p = E/h\nu$ and $q = E_0/h\nu$ represent the number of quanta contained in the total energy and the dissociation energy of the cluster, respectively. The number of ways of distributing j quanta among s oscillators is given by $(j+s-1)!/j!(s-1)!$. The dissociation rate is therefore given in the quantum formulation by

$$k(E) = \nu_0 \frac{(p-q+s-1)!/(p-q)!(s-1)!}{(p+s-1)!/p!(s_1)!} = \nu_0 \frac{p!(p-q+s-1)!}{(p+s-1)!(p-q)!} , \tag{3.22}$$

where the constant of proportionality ν_0 is usually treated as an adjustable vibration frequency.

It is easy to check that the last equation gives back the classical formula (3.21) when h tends to zero, or equivalently, when p and q tend simultaneously to infinity.

Unfortunately, the RRK theory often compares badly with experimental data, even in the quantum formulation, because it underestimates the dissociation rates by several orders of magnitude. A basic criticism that is often aimed at the RRK theory concerns the harmonic hypothesis for the oscillators. Anharmonicity is in fact crucial if an atom is to dissociate, and it is also likely to significantly alter the number of ways of distributing the energy among the various modes.

These difficulties led Marcus [19] to extend the predictions of the RRK theory and introduce the concept of transition state. When an atom moves away from the rest of the system, its potential energy may exhibit a non-monotonic profile with distance, including for example a so-called activation barrier relative to the equilibrium point of the reactants. The region in which this barrier is located, which characterises the transition state, is shown schematically in Fig. 3.13.

The Marcus theory, known as RRKM, assumes that all states available to the energy E are accessible and equiprobable (ergodicity). It also makes the hypothesis that dissociation is slower than the redistribution of energy over the ensemble of all modes in the system. With these approximations, Marcus was able to show that the reaction rate can be expressed in terms of the properties of the reactants and the transition state:

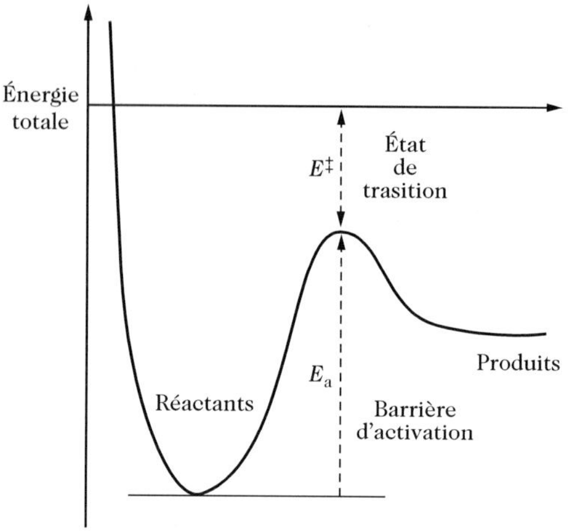

Fig. 3.13. Transition state along a dissociation path

$$k(E) = \frac{N(E^‡)}{h\Omega(E)} , \tag{3.23}$$

where $N(E^‡)$ is the sum of the vibrational states at the transition state of energy $E_a = E - E^‡$ and $\Omega(E)$ is the density of vibrational states of the reactants. The sum over states is the integral of the density of states with respect to the energy.

The first RRK theory is in fact a special case of the framework established by Marcus, in which the densities and sum of vibrational states are taken in the harmonic approximation, assimilating the transition state with the dissociation products. This last hypothesis is not generally justified, if only because the reactants have nonzero angular momentum, which induces a centrifugal barrier during dissociation.

Phase Space Theory

While the RRK and RRKM theories consider only the reactants undergoing dissociation, the phase space theory is concerned with the products and uses the principle of microreversibility. The foundations of this theory were laid by Weisskopf [20] in nuclear physics, before it was extended to physicochemistry by Light et al. [21], Klots [22], and Chesnavich and Bowers [23].

Consider the dissociation reaction

$$\mathrm{A} \rightleftharpoons \mathrm{B} + \mathrm{C} . \tag{3.24}$$

At equilibrium, the relative quantities [A], [B] and [C] of reactants and products satisfy the equation of microreversibility, viz.,

$$k_1[\mathrm{A}] = k_2[\mathrm{B}] \times [\mathrm{C}] , \tag{3.25}$$

where k_1 and k_2 are constants associated with the above dissociation reaction (3.24).

Let E and E_0 be the total available energy and the difference in binding energies of A and B+C, respectively, and let J be the total angular momentum of A. By the principle of microscopic reversibility, the total rate of reaction from A to B+C is the same as the total rate of the reverse reaction, from B+C to A:

$$R_{\mathrm{A}\to\mathrm{B+C}}(E,J) = R_{\mathrm{B+C}\to\mathrm{A}}(E-E_0,J) \,. \tag{3.26}$$

The left-hand side can be written as the product of the dissociation rate of A with the density of vibrational states Ω at energy E, not forgetting a factor to account for the rotational degeneracy of A:

$$R_{\mathrm{A}\to\mathrm{B+C}}(E,J) = k_{\mathrm{A}}(E,J)\Omega(E-E_{\mathrm{r}}^{\mathrm{A}})S_{\mathrm{r}}^{\mathrm{A}} \,, \tag{3.27}$$

where $E_{\mathrm{r}}^{\mathrm{A}}$ is the rotational energy of A and $S_{\mathrm{r}}^{\mathrm{A}}$ its rotational degeneracy factor, which depends on symmetry considerations. For example, if the molecule A can be modelled as a spinning top, then $S_{\mathrm{r}}^{\mathrm{A}} = (2J+1)^2$.

For the reverse reaction $\mathrm{B}+\mathrm{C}\to\mathrm{A}$, the relative energy of the products must also be taken into account. It can be shown that the total reaction rate takes the form

$$\begin{aligned} &R_{\mathrm{B+C}\to\mathrm{A}}(E-E_0,J) \\ &\quad= \frac{\sigma_{\mathrm{A}}}{\sigma_{\mathrm{B}}\sigma_{\mathrm{C}}}S_{\mathrm{r}}'\iint k_{\mathrm{BC}}(E_{\mathrm{r}},E_{\mathrm{t}},J)\rho_{\mathrm{t}}(E_{\mathrm{t}})\mathrm{d}E_{\mathrm{t}}\times\Omega'(E-E_0-E_{\mathrm{tr}})\mathrm{d}E_{\mathrm{tr}} \,. \end{aligned} \tag{3.28}$$

Here $k_{\mathrm{BC}}(E_{\mathrm{r}},E_{\mathrm{t}},J)$ is the partial evaporation rate at the relative translational energy E_{t} and at the rotational energy E_{r} for the formation of a collision complex with angular momentum J. $\rho_{\mathrm{t}}(E_{\mathrm{t}})$ is the density of translational states per unit volume, and $\Omega'(E-E_0-E_{\mathrm{tr}})$ is the density of vibrational states of the products B+C at the energy $E-E_0-E_{\mathrm{tr}}$, with $E_{\mathrm{tr}} = E_{\mathrm{t}}+E_{\mathrm{r}}$. E_0 is the energy difference between reactants and products and S_{r}' is the rotational degeneracy factor of the products. Finally, $\sigma_{\mathrm{A}}/\sigma_{\mathrm{B}}\sigma_{\mathrm{C}}$ is the ratio of the symmetry factors for each species.

After eliminating $\rho_{\mathrm{t}}(E_{\mathrm{t}})$ (analytic) and introducing the sum of rotational and orbital states $\Gamma_{\mathrm{ro}}(E_{\mathrm{tr}},J)$, a few further manipulations lead to the following expression for the dissociation rate:

$$k_{\mathrm{A}}(E,J) = \frac{\sigma_{\mathrm{A}}}{\sigma_{\mathrm{B}}\sigma_{\mathrm{C}}}\frac{S_{\mathrm{r}}'}{S_{\mathrm{r}}^{\mathrm{A}}}\frac{\displaystyle\int_{E_{\mathrm{tr}}^{\dagger}}^{E-E_0}\Omega'(E-E_0-E_{\mathrm{tr}})\Gamma_{\mathrm{ro}}(E_{\mathrm{tr}},J)\mathrm{d}E_{\mathrm{tr}}}{h\Omega(E-E_{\mathrm{r}}^{\mathrm{A}})} \,, \tag{3.29}$$

where $E_{\mathrm{tr}}^{\dagger}$ is the minimal accessible value of E_{tr} for fixed E and J, and the quantity Γ_{ro} is calculated by treating the products as solids with a specific symmetry, taking care to consider mechanical constraints related to conservation of total energy and angular momentum.

The Marcus theory is the same as the phase space theory if the transition state is taken at the maximum of the centrifugal barrier[1] and conservation of angular momentum is strictly taken into account. However, the distinguishing feature and the usefulness of the phase space theory come from the fact that its predictions are based purely on the properties of the reactants and products, rather than on the transition state. The latter is often difficult to localise and study, both in theory and in experiment.

Kinetic Energy Released During Dissociation

Apart from the dissociation rate, another very important observable, particularly from the experimental point of view, is the kinetic energy carried away (or released) during dissociation, and its statistical distribution. For example, in the phase space theory, the probability that a dissociation releases a kinetic energy in the interval from E_{tr} to $E_{\mathrm{tr}} + \mathrm{d}E_{\mathrm{tr}}$ is given by

$$P(E_{\mathrm{tr}}, E, J)\mathrm{d}E_{\mathrm{tr}} = \frac{\Omega'(E - E_0 - E_{\mathrm{tr}})\Gamma_{\mathrm{ro}}(E_{\mathrm{tr}}, J)\mathrm{d}E_{\mathrm{tr}}}{\displaystyle\int_{E_{\mathrm{tr}}^{\dagger}}^{E-E_0} \Omega'(E - E_0 - E_{\mathrm{tr}})\Gamma_{\mathrm{ro}}(E_{\mathrm{tr}}, J)\mathrm{d}E_{\mathrm{tr}}} . \tag{3.30}$$

This quantity, and the average

$$\langle E_{\mathrm{tr}} \rangle = \int E_{\mathrm{tr}} P(E_{\mathrm{tr}}, E, J)\mathrm{d}E_{\mathrm{tr}} ,$$

thus depends only on the properties of the products, and not those of the reactants.

Canonical Formulations

The expressions given in the last few sections are valid for a microcanonical system with fixed constant total energy and angular momentum. When the energy has a Boltzmann distribution, a canonical reaction rate $k(T)$ can be defined by the probability of finding the state of energy E at temperature T. Now the latter is nothing other than $\Omega(E) \times \exp(-E/k_{\mathrm{B}}T)/Z(T)$, where $Z(T)$ is the normalising partition function. In the framework of the RRK theory, $\Omega(E) \propto E^{s-1}$ and one obtains an Arrhenius-type formula, viz., $k(T) = \nu_0 \exp(-E_0/k_{\mathrm{B}}T)$. In the RRKM theory, one must define the partition

[1]The effective potential of the interaction between two bodies includes a repulsive term $J^2/2\mu r^2$ which dominates at long range. At shorter ranges, the potential becomes attractive again. The maximum of the effective potential then defines the centrifugal barrier.

function $Q(T)$ of the reactants and the partition function $Q^\dagger(T)$ restricted to the transition state:

$$k(T)_{\mathrm{RRKM}} = \frac{Q^\dagger(T)}{hQ(T)} . \tag{3.31}$$

The 'simple' statistical theories of evaporation (RRK) consider the problem from the point of view of the reactant and are not concerned with the final state of the products. They generally contain adjustable parameters.

The phase space theories treat the reactants and products on a par by introducing the microreversibility hypothesis. In principle, all parameters can be calculated a priori. However, various approximations lead to the Weisskopf, Engelking and Klots theories.

3.3.2 Link with the Solid–Liquid Transition. Numerical Results

The anharmonic statistical theories presented above show that the evaporation rate and the kinetic energy released depend explicitly on the vibrational density of states. Now we found in Sect. 3.2 that the density of states characterised the thermodynamic behaviour, in particular, during a phase transition in the system. Some connection is therefore to be expected between the solid–liquid transition and the observables related to dissociation.

In particular, in the framework of the phase space theory applied to evaporation, the study of the kinetic energy released during evaporation involves only the properties of the product cluster. If the latter exhibits the phenomenon of dynamical coexistence associated with the solid–liquid transition, then the large variations in the density of states are likely to have consequences for the variations in $\langle E_{\mathrm{tr}} \rangle$.

In fact, just such a connection was shown for the first time in a simulation by Weerasinghe and Amar [24] for argon clusters. On the basis of a considerable amount of theoretical work and molecular dynamics simulations, it was subsequently concluded that the phase space theory seems to be quantitatively appropriate for describing unimolecular evaporation of arbitrary atomic and molecular clusters.

Figure 3.14 shows the predictions of this theory, compared with numerical data for atomic evaporation of the $KrXe_{13}$ cluster. This heterogeneous system can dissociate along two channels, either by emitting the krypton atom, or by emitting one of the xenon atoms. The distributions of released kinetic energy differ in the two cases. However, the difference is not only quantitative, but also qualitative. Indeed, when a xenon atom is emitted, the $KrXe_{12}$ cluster produced has richer thermodynamic behaviour than Xe_{13}, due to the fact that a krypton atom can migrate from the core of the icosahedral cluster to its surface. This causes a slight inflection in the density of states that is more clearly visible in the microcanonical temperature.

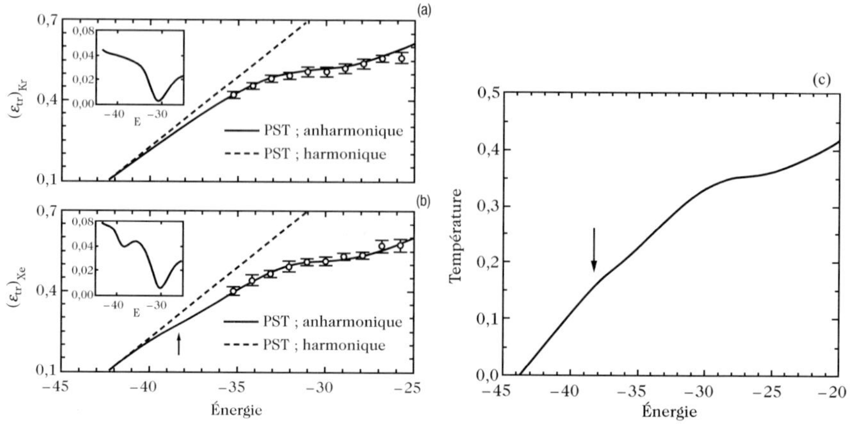

Fig. 3.14. (**a**) Average kinetic energy released during the dissociation $KrXe_{13} \longrightarrow Kr+Xe_{13}$ calculated using molecular dynamics (*circles*) and by applying the phase space theory (PST) in the harmonic and anharmonic approximations. *Insert*: Derivative of this quantity. (**b**) Average kinetic energy released during the dissociation $KrXe_{13} \longrightarrow Xe+KrXe_{12}$ calculated using molecular dynamics (*circles*) and by applying the phase space theory (PST) in the harmonic and anharmonic approximations. *Insert*: Derivative of this quantity. (**c**) Microcanonical temperature of the $KrXe_{12}$ cluster. In (**b**) and (**c**), *arrows* indicate core/surface isomerisation of the krypton atom in the $KrXe_{12}$ cluster

3.3.3 Experimental Investigation of Evaporation

Mass spectrometry is an ideal tool for studying free clusters. In particular, at the beginning of the 1980s, it revealed the existence of the so-called magic numbers characterising certain sizes of enhanced stability. In fact, as mentioned above, an isolated cluster or a cluster in a molecular beam is no longer perfectly stable once its internal energy exceeds the dissociation energy. Even if the cluster is large, it is condemned to evaporate an atom, even though this may take a very long time compared with the vibrational period. By evaporating an atom, the cluster loses energy and cools slightly. A system that is initially in a highly excited energy state will be able to evaporate on several occasions (sequential evaporation), cooling down at each stage in the process. Between two evaporations, cooling has the effect of reducing the dissociation rate. In practice, it is often considered that consecutive evaporation times increase at each step by an order of magnitude. This situation led Klots [25] to suggest the hypothesis that only the last evaporation before the observation of a cluster will have taken a nonzero time to occur. Klots also introduced the idea of an evaporative ensemble to describe the situation arising in time-of-flight spectrometry. In this ensemble, clusters with high energy dissociate before reaching the spectrometer exit. Below another limit, the evaporation rate becomes too low and clusters remain stable. It can be shown that the

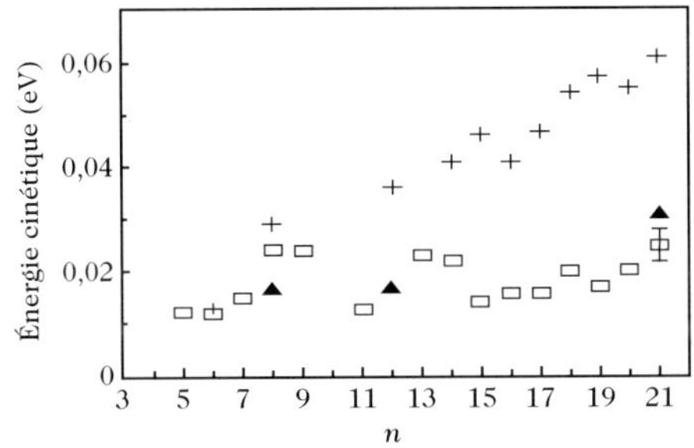

Fig. 3.15. Translational (□) and total (+) kinetic energies released during evaporation of Na_n^+ clusters. The anharmonic PST calculation is shown by *triangles*. Taken from [26]

energy distribution of the evaporative ensemble is significantly different from the initial canonical distribution and that it depends only on the time of flight from one end of the spectrometer to the other.

The Klots approximation also allows one to relate the relative abundance $\mathcal{I}(N)$ of a size measured by spectrometry to the separation energy $D(N) = E(N-1) - E(N)$, i.e., the energy required to remove an atom from the cluster of N atoms. This relation can be written

$$\mathcal{I}(N) = D(N+1) - \frac{3N}{G}\big[D(N+1) - D(N)\big] , \tag{3.32}$$

where G is the Gspann factor, equal to about 23.5 for monatomic clusters. By analysing the spectrometer signals, one can also estimate the total and translational kinetic energy distributions resulting from the last fragmentation. These distributions and their averages can then be compared with the theoretical predictions of the statistical models discussed earlier.

Figure 3.15, taken from [26], shows an example of the way the average translational and total (i.e., including the rotational component) kinetic energies of evaporation vary depending on the size of an Na_n^+ cluster, together with the results as calculated from the phase space theory.

Many very refined experiments, especially those of Märk et al. [28], seek to carry out accurate measurements of the distributions of kinetic energy released during the dissociation of molecules or clusters by combining spectrometry with analyses based on statistical models. One may thus reasonably expect to obtain experimental evidence for the solid–liquid transition in very small systems using observables relating to evaporation.

3.3.4 Beyond Unimolecular Evaporation

The dissociation phenomenon is omnipresent in the study of free clusters and is in no way restricted to evaporation alone. We shall limit ourselves here to listing certain situations involving very different mechanisms to those that underlie thermal evaporation.

In general, several fragmentation channels will be simultaneously available to an energetically excited cluster. For example, charged systems may prefer to emit an ion rather than a neutral atom. The investigation of these channels is particularly important in metals and van der Waals clusters. For the latter, measurements of the energy released provide indirect information about the nature and intensity of the chemical bonding in the core of the cluster. In metals, it has been suggested that dissociation may be possible by fission rather than evaporation for certain sizes. In particular, the case of symmetric fission $M_{2n}^{++} \longrightarrow M_n^+ + M_n^+$ has inspired a great many theoretical studies, but experimental evidence remains elusive.

As illustrated by the numerical example of $KrXe_{13}$, it should be possible to probe the structure of mixed clusters through their dissociative behaviour. However, competition between the main fragmentation channels raises a problem for statistical theories, because they cannot be applied without knowing the vibrational and rotational densities of states for a wide variety of subsystems.

Finally, certain metal clusters have been characterised as being able to evaporate not only atoms, but also electrons. This phenomenon is similar to thermionic emission and has been observed in tungsten [29].

Apart from 'natural' thermal fragmentation, much work has sought to describe and understand dissociation induced by short pulses, such as photoabsorption or photoionisation, or collision with an electron, an atom or a surface. The time scales for redistribution of energy are often longer than the exciting phenomenon itself and the hypotheses underlying conventional statistical theories are no longer justified.

3.3.5 Toward the Liquid–Gas Transition

Strontium Clusters

The temperature of clusters can be estimated from the released energy distribution and statistical models. When the quantity of energy E deposited in the cluster is controlled, spectrometric analysis reveals a maximal cluster temperature as a function of E, whose variations with E can be interpreted as those of a microcanonical heat curve. This is precisely what was achieved by Bréchignac et al. using small strontium clusters, as shown in Fig. 3.16.

Although the full interpretation of such an experiment remains delicate, data obtained for the larger clusters show that there is a plateau in the temperature variation, compatible with the existence of a vaporisation transition. Owing to the fact that they are nevertheless very small, it is not easy to characterise the initial condensed phase and the change of state detected here may in fact correspond to sublimation. It is interesting to observe in this context that other very different systems from a chemical point of view, viz., molecular

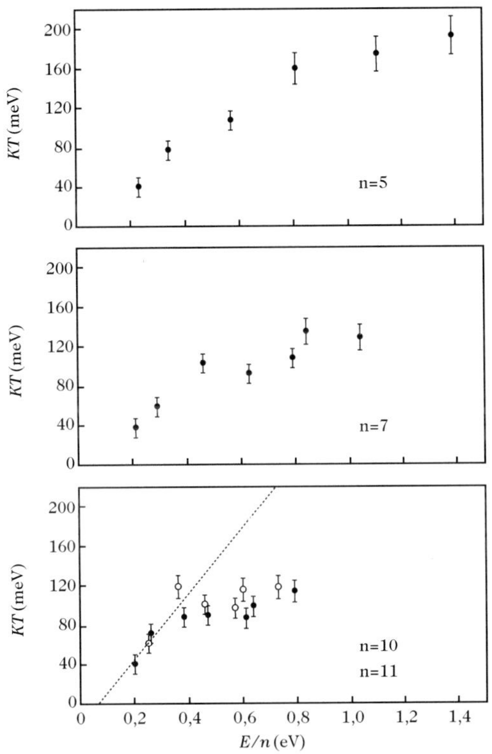

Fig. 3.16. Heat curves of Sr_n^+ determined from kinetic energies released by evaporation of strontium atoms. Taken from [27]

clusters of C_{60}, may tend to sublimate molecules rather than melt [30]. This would be consistent with the known instability of the liquid phase of fullerite.

Sodium Clusters

The experiment carried out by Haberland and coworkers to measure the heat curves of Na_n^+ clusters has been extended beyond the melting point [31]. The idea here complements the one used by Bréchignac et al. [26], in the sense that it is the measured temperature that reaches a limiting value when the system approaches the liquid–gas transition, owing to evaporative cooling. The vaporisation point is thereby identified by considering the average temperature as a function of the excitation energy (see Fig. 3.17).

The main problem relating to liquid–gas and solid–gas transitions in finite systems lies in the importance of the time factor: a free finite system is always condemned to vaporise, even if the time scales become ever longer as sequential evaporation proceeds. The stability of the condensed form can therefore only be defined relative to some previously specified observation time. In addition

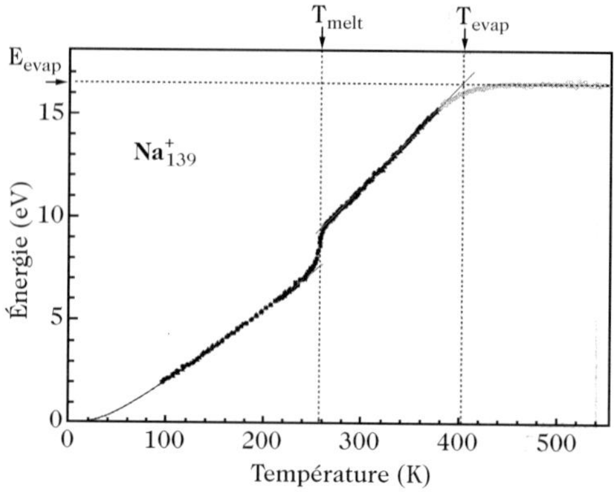

Fig. 3.17. Experimental heat curve for the Na_{139}^{+} cluster. Taken from [31]

to this restriction, it is difficult to define the pressure (and the volume) in the vicinity of a free system, although these two problems are in fact related [31].

Multifragmentation and the Theory of Nucleation

Evaporation and the liquid–gas transition are mainly distinguished by the magnitude of the initial excitation. These two phenomena correspond to two facets of a single situation, insofar as vaporisation can be described as a series of sequential evaporations, in which case the evaporative phenomenon may be slow.

In reality, when a cluster is strongly excited vibrationally, the excess energy can be rapidly dissipated via almost simultaneous dissociations. One then speaks of multifragmentation. It is worth trying to evaluate the subsequent evolution of such a cluster and to characterise its final state. As for evaporation, the observation time is a decisive parameter. It is possible to adopt a rather different point of view, doubtless more suitable for clusters in contact with a heat bath, describing an equilibrium situation in a given volume or pressure. This point of view is the one adopted in nucleation theories, outlined below.

Multifragmentation and nucleation theories do not seek to characterise the fine details of energy distributions or reaction rates associated with each elementary stage in the process, but are concerned rather with describing the matter distribution, given the initial (or total) size and the general conditions of excitation, i.e., energy or temperature, total volume, etc. In contrast to studies of unimolecular reactions, these theories are based on the concepts of macroscopic thermodynamics. It should be said, however, that many atomistic simulations have confirmed the conclusions whilst correcting certain hypotheses [32].

In the classical nucleation theory (CNT) [33], the system is shared into n_l atoms forming a liquid nucleus in equilibrium with n_g atoms forming a vapour around it. The liquid nucleus is treated as spherical and the interface with the vapour is assumed rigid. With these approximations, the free energy of the system is given by

$$G = n_{\mathrm{l}}\mu_{\mathrm{l}} + n_{\mathrm{g}}\mu_{\mathrm{g}} + 4\pi r^2\gamma \,, \tag{3.33}$$

where r is the radius of the liquid drop, γ is the surface tension, and μ_{l} and μ_{g} denote the chemical potentials of the liquid and vapour, respectively.

The energy of formation of the droplet from the vapour is

$$\Delta G = G - \mu_{\mathrm{g}}(n_{\mathrm{l}} + n_{\mathrm{g}}) = -\left(\frac{\mu_{\mathrm{g}} - \mu_{\mathrm{l}}}{v_{\mathrm{l}}}\right)\frac{4\pi}{3}r^3 + 4\pi r^2\gamma \,, \tag{3.34}$$

with

$$v_{\mathrm{l}} = \frac{4}{3}\pi r^3/n_{\mathrm{l}}$$

the volume of the droplet. Formation of the droplet requires $\mu_{\mathrm{l}} < \mu_{\mathrm{g}}$, allowing the free energy to reach a maximum at the critical radius r^* given by

$$r^* = \frac{2\gamma v_{\mathrm{l}}}{\mu_{\mathrm{g}} - \mu_{\mathrm{l}}} \,. \tag{3.35}$$

The value of ΔG at $r = r^*$ is called the nucleation barrier. The saturated vapour pressure p_{sat} in the presence of a plane interface must be corrected when the interface is spherical. It can be shown that its value is given by the Kelvin relation

$$\ln\frac{p}{p_{\mathrm{sat}}} = \frac{2\gamma v_{\mathrm{l}}}{k_{\mathrm{B}}Tr^*} \,. \tag{3.36}$$

Given the ratio $S = p/p_{\mathrm{sat}}$, the above relation can be used to find the critical radius r^*. Let us now determine the mass distribution $\{N_i\}$ at equilibrium. For reasonable hypotheses, the chemical activity a_i of a cluster of i atoms can be taken as the relative amount of these clusters with respect to the monomers, i.e., $a_i/a_1 \approx N_i/N_1$. If μ_i is the chemical potential of the cluster of i atoms and $\mu_i - \mu_1 = \Delta G_i$, we obtain an Arrhenius law for N_i:

$$N_i = N_1\exp(-\Delta G_i/k_{\mathrm{B}}T) \,. \tag{3.37}$$

Now ΔG_i is also given by

$$\Delta G_i = \mu_i - \mu_{\mathrm{g}}(n_{\mathrm{g}} + n_{\mathrm{l}}) = 4\pi r^2\gamma - (\mu_{\mathrm{g}} - \mu_{\mathrm{l}})n_{\mathrm{l}} \,.$$

Moreover $\mu_{\mathrm{g}} - \mu_{\mathrm{l}}$ is the change in the chemical potential when an atom moves from the vapour phase at pressure p to the bulk liquid in equilibrium with its saturated vapour. This quantity is thus also equal to $k_{\mathrm{B}}T\ln S$. If the droplet is considered homogeneous, we then have $n_{\mathrm{l}} = i$ and $r^2 = \alpha i^{2/3}$, where α is a positive constant. We thus obtain the equilibrium distribution

$$N_i = N_1 \exp\left(i \ln S - \alpha \frac{i^{2/3}}{k_B T}\right) . \qquad (3.38)$$

It is also possible to determine a nucleation rate which includes both sticking and evaporation phenomena at equilibrium [33]:

$$J = \frac{N_1^2}{\rho}\left(\frac{2\gamma}{\pi m}\right)^{1/2} \exp\left[-\frac{16\pi\gamma^3}{3\rho^2 (k_B T)^3 (\ln S)^2}\right] . \qquad (3.39)$$

However, the predictions of the classical nucleation theory do not generally compare well with experimental measurements. Various corrections have been proposed to improve agreement:

- The correction for the law of mass action considers the mechanical conservation of the position of the center of mass of the droplet, replacing the term $-ik_B T \ln S$ by $(1-i)k_B T \ln S$.
- Restricted self-consistency refers to the fact that the free energy of the liquid droplet should vanish when it contains only one monomer, replacing $\alpha i^{2/3}$ by $\alpha(i^{2/3} - 1)$.
- Taking into account the internal translational degrees of freedom of the liquid drop leads one to correct the prefactor of the exponential in (3.39) by $1/\Lambda_i^3$, where Λ_i is the de Broglie thermal wavelength.

Despite these corrections, the theory of nucleation remains rather unreliable. Deeper criticisms, due in particular to Dillmann and Meier [34], have led to a generalised expression for ΔG_i using an expansion in powers of $i^{1/3}$, along the lines of the liquid drop model in nuclear physics. The increased number of parameters is used to obtain quantitative agreement for the nucleation rates.

Another criticism concerns the rigidity of the liquid/vapour interface. It is sometimes important to take into account a finite thickness for this interface, especially in very small systems. The diffuse interface theory due to Gránásy [35] and theories based on the classical density functional for liquids [36] have been more successful, confirmed by atomistic simulations, allowing them to be extended to the case of solid nucleation.

References

1. B. Diu, C. Guthmann, D. Lederer, B. Roulet: *Physique statistique*, Hermann, Paris (1989)
2. P. Papon, J. Leblond, P.H.E. Meijer: *Physique des transitions de phases*, Dunod, Paris (1999)
3. P. Pawlow: Z. Phys. Chem. **65**, 545 (1909)
4. Ph. Buffat, J.-P. Borel: Phys. Rev. A **13**, 2287 (1976)
5. S.L. Lai, J.Y. Guo, V. Petrova, G. Ramanath, L.H. Allen: Phys. Rev. Lett. **77**, 99 (1996); M. Zhang, M. Yu. Efremov, F. Schiettekatte, E.A. Olson, A.T. Kwan, S.L. Lai, T. Wisdeler, J.E. Greene, L.H. Allen: Phys. Rev. B **62**, 10548 (2000)

6. M. Schmidt, H. Haberland: C.R. Acad. Sci. IV, **3**, 327 (2002); M. Schmidt, J. Donges, Th. Hippler, H. Haberland: Phys. Rev. Lett. **90**, 103401 (2003)
7. F. Ercolessi, W. Andreoni, E. Tosatti: Phys. Rev. Lett. **66**, 911 (1991)
8. Y. Imry: Phys. Rev. B **21**, 2042 (1980)
9. A.A. Shvartsburg, M.F. Jarrold: Phys. Rev. Lett. **85**, 2530 (2000)
10. C.E. Bottani, A. Li Bassi, B.K. Tanner, A. Stella, P. Tognini, P. Cheyssac, R. Kofman: Phys. Rev. B **59**, R15601 (1999)
11. D.J. McGinty: J. Chem. Phys. **55**, 580 (1971); J.J. Burton: ibid. **56**, 3123 (1972)
12. G. Natanson, F. Amar, R.S. Berry: J. Chem. Phys. **78**, 399 (1983); H.L. Davis, J. Jellinek, R.S. Berry: ibid. **86**, 6456 (1987); T.L. Beck, R.S. Berry: ibid. **88**, 3910 (1988)
13. D.D. Frantz: J. Chem. Phys. **115**, 6136 (2001)
14. D. Lynden-Bell: Physica A **263**, 293 (1999)
15. M. D'Agostino, F. Gulminelli, Ph. Chomaz, M. Bruno, F. Cannata, R. Bougault, F. Gramegna, I. Iori, N. Le Neindre, G.V. Margagliotti, A. Moroni, G. Vannini: Phys. Lett. B **473**, 219 (2000)
16. J.P.K. Doye, D.J. Wales: J. Chem. Phys. **102**, 9673 (1995)
17. O.K. Rice, H.C. Ramsperger: J. Am. Chem. Soc. **50**, 617 (1928)
18. L.S. Kassel: J. Phys. Chem. **32**, 225 (1928)
19. R.A. Marcus, O.K. Rice: J. Phys. Colloid Chem. **55**, 894 (1951); R.A. Marcus: J. Chem. Phys. **20**, 359 (1952)
20. V. Weisskopf: Phys. Rev. **52**, 295 (1937)
21. P. Pechukas, J.C. Light: J. Chem. Phys. **42**, 3281 (1965)
22. C.E. Klots: J. Phys. Chem. **75**, 1526 (1971)
23. W.J. Chesnavich, M.T. Bowers: J. Chem. Phys. **66**, 2306 (1977)
24. S. Weerasingh, F.G. Amar: J. Chem. Phys. **98**, 4967 (1993)
25. C.E. Klots: J. Chem. Phys. **83**, 5854 (1985); Nature **327**, 222 (1987)
26. C. Bréchignac, Ph. Cahuzac, B. Concina, J. Leygnier, B. Villard, P. Parneix, Ph. Bréchignac: Chem. Phys. Lett. **335**, 34 (2001)
27. C. Bréchignac et al.: Phys. Rev. Lett. **89**, 203401 (2002)
28. See for example S. Matt-Leubner, A. Stamatovic, R. Parajuli, P. Scheier, T.D. Märk, O. Echt, C. Lifshitz: Int. J. of Mass Spectr. **222**, 213 (2003)
29. T. Leisner, K. Athanassenas, O. Echt, O. Kandler, D. Kreisle, E. Recknagel: Z. Phys. D **20**, 127 (1991)
30. F. Calvo: J. Phys. Chem. B **105**, 2183 (2001)
31. M. Schmidt, T. Hippler, J. Donges, W. Kronmüller, B. von Issendorff, H. Haberland, P. Labastie: Phys. Rev. Lett. **87**, 203402 (2002)
32. P. Schaaf, B. Senger, J.-C. Voegel, R.K. Bowles, H. Reiss: J. Chem. Phys. **114**, 8091 (2001)
33. J. Frenkel: *Kinetic Theory of Liquids*, Dover (1955)
34. A. Dillmann, G.E.A. Meier: Chem. Phys. Lett. **160**, 71 (1989)
35. L. Gránásy: Europhys. Lett. **24**, 121 (1993)
36. X.C. Zeng, D.W. Oxtoby: J. Chem. Phys. **94**, 4472 (1991)

4

Modelling and Simulating the Dynamics of Nano-Objects

A. Pimpinelli

4.1 Introduction

The shape of an object is one of its most fundamental attributes. In our macroscopic world, shape is closely related to the very function of an object. It is difficult to imagine a fork having a very different shape, or for that matter, a function very different to the one it is usually used for.

But can the same be said in the nanoworld? More precisely, is there any correlation between the shape of a nano-object and its physical properties? We do not yet have a definitive answer to this question. In some cases, it is clear that such a correlation does exist. For example, it has been discovered that roughly spherical nanocrystals of cadmium selenide (a semiconducting material) can emit light of different colours depending on their size. More recently, we have learnt how to make rod-shaped nanocrystals (Fig. 4.1). These nanocrystals emit light just like their spherical relatives, but this time it is polarised along the rod axis. The energy interval between emission and absorption is also modified by altering the shape of the nanocrystal.

But what determines the shape of nano-objects? And is this shape always stable? Indeed, statistical physics teaches us that all physical characteristics of an object containing N atoms will fluctuate around its average value, with the relative fluctuations of extensive or additive quantities being proportional to $\sqrt{N}/N$ (see below). If $N = 1\,000$, the fluctuations are of the order of 3%, and if $N = 100$, they are more like 10%! So is it justified to speak of shape for objects in which all physical quantities are affected by thermal fluctuations, often having an amplitude almost as great as the quantity itself?

Fluctuations in Statistical Physics

The so-called additive quantities play a special role in the physics of macroscopic systems. A quantity A is said to be additive if, decomposing the system into N subsystems, the quantity A can be expressed as the sum of the values $A_1, A_2, \ldots, A_N$ taken by A in the corresponding subsystems, i.e.,

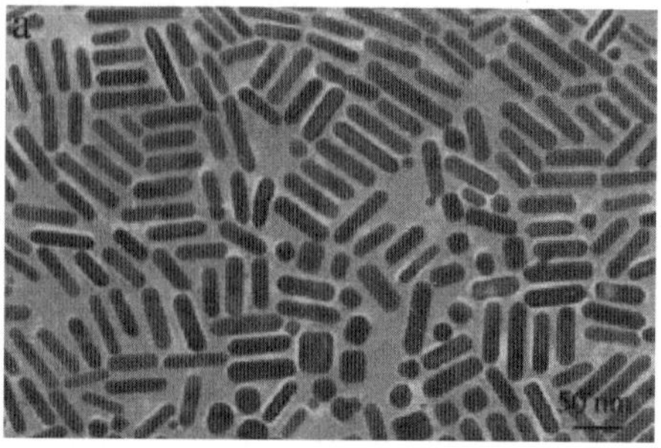

Fig. 4.1. Rod-shaped gold nanocrystals obtained by an electrolytic method whereby the shape of the crystallite can be controlled. From Wang et al. [6] with kind permission from Elsevier. Semiconducting nanocrystals of this shape have also been fabricated

$$A = A_1 + A_2 + \cdots + A_N .$$

Since $A_1, A_2, \ldots, A_N$ are random variables, the central limit theorem stipulates that their sum be a random variable with Gaussian distribution, and that its squared fluctuations increase linearly with the number N of subsystems. The fluctuations are therefore proportional to the square root of N. Since the average values increase as N, the relative fluctuations, i.e., fluctuations divided by average values, will decrease as $1/\sqrt{N}$.

To attempt to answer these questions, we shall discuss two cases: first, a free cluster, and then a pyramidal island on a crystal surface.

4.2 Free Clusters of Atoms. Molecular Dynamics Simulations

Clusters, particularly metal clusters, can be fabricated by collisions of atoms in a gas of neutral atoms. Once the seed has formed, other atoms will stick onto it and the cluster will grow. This cluster has a perfectly well ordered structure, but not always crystalline. Indeed, clusters of silver, which has an fcc crystal structure in the bulk, are often icosahedral or decahedral on nanometric scales. It is only at large (sic) sizes, around 9 nm in diameter ($N \approx 27\,000$), that nanoclusters assume an fcc structure. This is due to the fact that the equilibrium shape of an atomic cluster is determined by its surface.

At equal volumes, the shape that actually occurs is the one minimising the free energy. A crystal structure automatically minimises the bulk free energy, so it remains only to minimise the surface free energy. This is the Wulff theorem.

Molecular Dynamics Simulations

One method used to simulate the behaviour of a system containing up to a few hundred atoms is molecular dynamics (MD). This method is nothing other than the solution of Newton's dynamical equations describing the motion of the atoms making up the system. One needs to know the interaction potentials between atoms. The equations are then solved numerically by computer to obtain the positions and velocities of the particles at all times.

Newton's equations are integrated by taking discrete time steps. Typical durations for the time steps are of femtosecond order (10^{-15} s) and the total lapse of time covered by a simulation is anything between a few picoseconds and a few nanoseconds. In the simulations shown in Fig. 4.2, an atom was deposited on the cluster at a certain time. All the atoms were then free to move for 7 ns, at which point another atom was added to the cluster, and so on. In this way, cluster growth rates are obtained that are close to experimental values.

The advantage of an MD simulation compared with kinetic Monte Carlo simulations, for example (see p. 94), is that the cluster structure can be obtained in a realistic manner, this structure being chosen by the atomic dynamics rather than simply imposed at the outset. Metastable structures stabilised by the kinetics of the system can then emerge spontaneously during the simulation. However, today's computers can only simulate a few thousand atoms and only problems with a very high growth rate can be tackled.

However, when clusters have nanometric dimensions, the surface may sometimes play a more important role than the bulk. Indeed, an icosahedron, which gives rise to a perfectly ordered but non-crystalline structure, is bounded by facets of type $\{111\}$, with minimum energy. The latter are thus able to minimise the free surface energy very efficiently, at the expense of a large internal distortion. Since the surface energy dominates over the bulk energy, the icosahedron is the shape of lowest free energy for small clusters.

The next shape that turns up is the decahedron, with lesser internal deformation, followed by face-centered cubic nanocrystals (in fact, truncated octahedra with broad $\{100\}$ facets) at larger sizes. (For the latter, crystalline now, there is no internal deformation.) For example, for nickel, it has been calculated that the icosahedral structures are stable for $N < 2\,300$ and the decahedra for $2\,300 < N < 17\,000$, whilst the fcc polyhedra minimise the energy for higher values of N. However, during fabrication of metal nanoclusters, icosahedral particles with diameters greater than 10 nm are often found, and these are sizes for which such structures are not in principle stable, and hence observable.

These are therefore likely to be metastable structures, brought about by the growth kinetics of the cluster during its fabrication. Indeed, when an atom adheres to the growing cluster, it must diffuse across its surface to find a suitable low-energy site and incorporate itself into the system. Since surface diffusion is a thermally activated phenomenon, at low enough temperatures, it will only be efficient over a short range. Small clusters will thus be able to reach their minimal energy shape, but the larger ones will not. Instead, they

Fig. 4.2. Silver clusters obtained by numerical simulation of molecular dynamics type

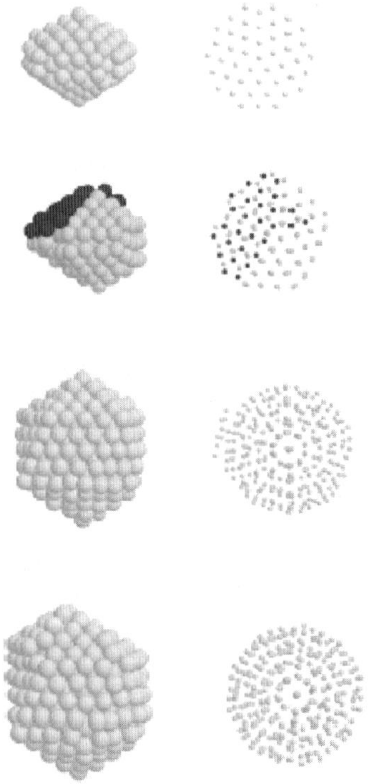

Fig. 4.3. Simulated growth of silver clusters using molecular dynamics at $T = 450\,\mathrm{K}$. The clusters shown contain $N = 146$, 181, 309 and 324 atoms, from top to bottom. An icosahedral stack (*black atoms*) can be seen to form on a facet of the $N = 181$ cluster. From this point on, the cluster, which started as a metastable decahedron, transforms into a metastable icosahedron

will be frozen into the structure, e.g., an icosahedral structure, produced when they were much smaller. Numerical calculations of molecular dynamics type (see above) confirm this scenario, showing that large silver icosahedra form more easily than octahedra, in suitable conditions of temperature and cluster growth rate (see Figs. 4.2 and 4.3).

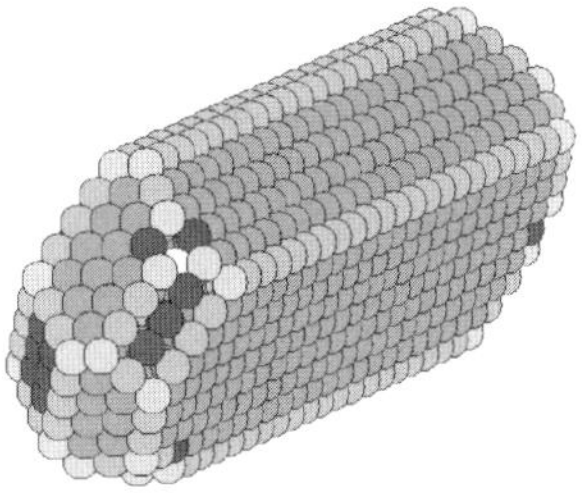

Fig. 4.4. Rod-shaped cluster of $N = 6\,000$ atoms. Kinetic Monte Carlo simulations have been used to investigate the relaxation of this nanocrystal toward its equilibrium shape

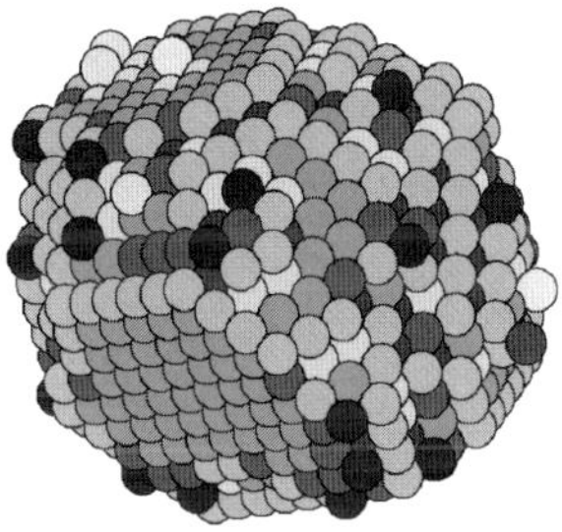

Fig. 4.5. The same cluster as in Fig. 4.3, but at higher temperature. Relaxation occurs by diffusion of atoms between non-facetted parts and is therefore easier than at low temperatures

The nucleation of metastable phases or shapes should come as no surprise. Even in the eighteenth century, Ostwald had already formulated his 'rule' stating that the phase which forms first is not necessarily the one which minimises the free energy, but is rather the one with the smallest nucleation barrier. The transition to the stable phase or shape is once again determined by a free energy barrier, which means that this transition can be difficult to accomplish.

In fact, when clusters have facets, as they do at low temperatures (Fig. 4.4), it is very difficult for them to change shape. Indeed, in order for the nanocluster to change from one shape to another, the facets must be destroyed and new ones created. The problem consists in nucleating the new facet in the right place, which requires the system to overcome a free energy barrier, the nucleation barrier. Consequently, the time required for a change in shape to occur will depend on the nucleation barrier, whether it be the time for a fluctuation to damp down or to appear, or for metastable shapes to evolve toward the equilibrium shape (see Fig. 4.5).

4.3 Evolution of Free and Supported Nanoclusters Toward Equilibrium. Kinetic Monte Carlo Simulations

The thermally activated nature of nucleation means that the time required for a large enough fluctuation to destroy a metastable shape becomes astronomically long if the temperature is low enough. For this reason, the rod-shaped nanocrystals mentioned above can exist with aspect ratios as high as 35. Dynamical studies over the very long time scales (from an atomic point of view!)

required for a nano-object to relax toward equilibrium mean that one must adopt another tack.

Kinetic Monte Carlo Simulations

Another numerical method, an alternative to molecular dynamics simulations, which provides a way of studying certain features in the evolution of a nano-object, is kinetic Monte Carlo simulation. The Monte Carlo method was invented to calculate the equilibrium properties of statistical systems such as the Ising model for magnetic systems. Local variations in the state of the system are made at random by drawing random numbers (hence the name 'Monte Carlo'), thereby sampling the whole phase space of the system and calculating averages for physical quantities such as the magnetisation. Indeed, the method allows one to weight each explored configuration with its statistical weight $\exp(-E/k_\mathrm{B}T)$, where E is the energy of the relevant configuration.

Kinetic Monte Carlo associates a time scale with each transition from one configuration to another. These transitions thus occur at a frequency (probability per unit time) ν proportional to the Arrhenius exponential of the free energy barrier E which separates the initial and final configurations, i.e., $\nu \propto \exp(-E/k_\mathrm{B}T)$. Once the frequency of each possible event has been fixed, the Monte Carlo simulation can calculate the temporal evolution of the system and estimate the value of any relevant statistical quantity as time goes by.

Less 'microscopic' than molecular dynamics, in the sense that atomic motions are not calculated using Newton's equations, but rather according to an effective dynamics, kinetic Monte Carlo can handle very large systems containing as many as several million atoms, not to mention coexisting phenomena characterised by very different time scales. Kinetic Monte Carlo can thus simulate the epitaxial growth of a thin film which may actually take several minutes or even hours (the deposition rate is typically one monolayer per minute for molecular beam epitaxy), whilst considering motions of atomic diffusion that may only last for a few tenths of a thousandth of a second.

Molecular dynamics calculations are totally unsuited to this type of problem, where one is concerned with the statistical properties of the objects as a function of time. One then turns to the so-called kinetic Monte Carlo simulations (see above).

These simulations make it possible to study morphological changes in a nanocluster which does not have the equilibrium shape and which evolves toward that shape if the temperature allows it (see Figs. 4.4 and 4.5). Controlling the morphology is also a major issue when patterning a crystal surface on the nanoscale, e.g., by etching, or by manipulating surface atoms one by one using a scanning tunneling microscope. Imagine stacking atoms over a region about 10 nm across and to a height of a few nanometers on a silicon surface (see Fig. 4.6).

At room temperature, the nanocluster is stable and its shape, whatever it may be, will barely change as time goes by. At higher temperatures, the structure survives but adopts its equilibrium shape, which is a sort of pyramid

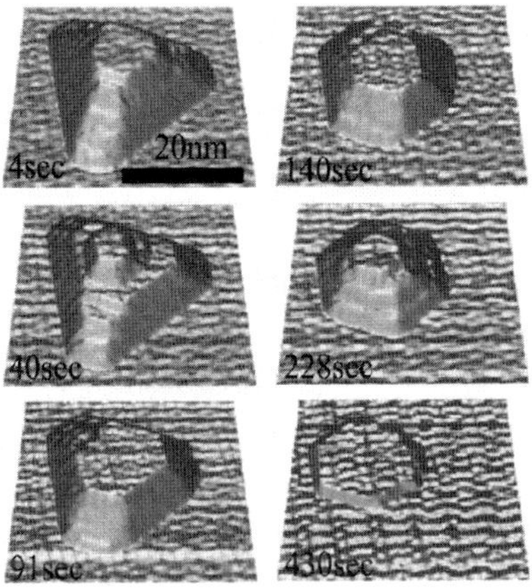

Fig. 4.6. Nanopyramid comprising roughly 12 levels on a (111) silicon surface. Heated to $T = 465°\mathrm{C}$, the pyramid evaporates atom by atom and level by level. After about 7 min, only the bottom level remains. Taken from Ichimiya et al. [4], with kind permission from Elsevier

whose precise shape is dictated by the symmetry of the crystalline substrate that serves to support it. If the substrate temperature is raised further, the pyramid begins to evaporate. At a temperature of $T = 440°\mathrm{C}$, the pyramid disappears at a rate of 3 atoms per second, then twice as fast at $T = 465°\mathrm{C}$ (see Fig. 4.6), and twice as fast again at $T = 485°\mathrm{C}$. Indeed, at these temperatures, the atoms can detach themselves from the pyramid and diffuse far away from it, allowing the surface to get rid of this rough feature and return to its equilibrium state, i.e., approximately smooth.

The island is not stable at any temperature, but as long as the heat energy remains low compared with the interatomic binding energy, the time required for it to evolve to any appreciable extent is very long. Paradoxically, the lower the temperature, the less stable the island becomes (at $T = 0\,\mathrm{K}$, the surface wants to be perfectly smooth), and the longer its lifetime becomes!

An engineer would ask whether these characteristics might be compatible with the operating conditions of a component. The physicist, on the other hand, seeks to establish how these characteristics arise from the individual behaviour of the atoms, and seeks also to quantify this relationship. It turns out that a very powerful tool is provided by the so-called scaling laws generated by the kinetics. For example, it has been found that the characteristic time τ required for one level of the silicon pyramid to disappear varies with the number N of atoms initially contained in the level according to a power law:

$$\tau(N) \sim N^{\beta} .$$

Experimentally, one obtains $\beta \approx 3/2$.

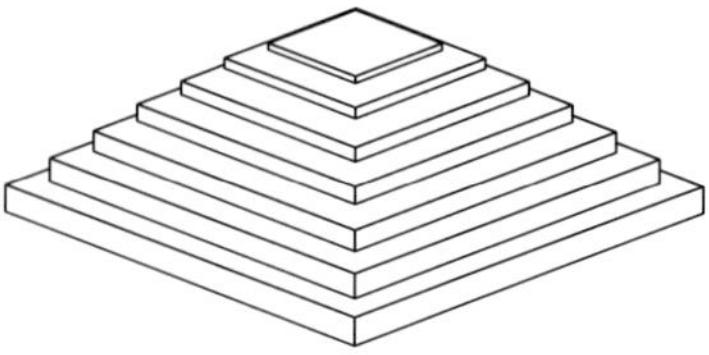

Fig. 4.7. Square-based pyramid. The top level has side L

This relation implies that an expansion of the size N of the system by a factor of b will change the characteristic time $\tau(N)$ by a known factor of b^β. If a level contains twice as many atoms as the one above, its lifetime will be about $2^{3/2} \approx 3$ times as long (at least), provided that one remains within the range of validity of the relation itself.

Apart from any practical use, it is very important to understand this type of scaling law, because such laws are directly related to the microscopic kinetic mechanisms that determine the physical phenomena they describe. For example, consider a pyramid comprising square levels of decreasing area stacked one on top of the other (shown schematically in Fig. 4.7). Let L be the length of the side of the top level, which is the first to start evaporating.

If the atoms evaporate mainly from the corners of the square, which are at distance L from one another, the evaporation rate of the square is then proportional to $1/L$. (The bigger the square, the fewer atoms are contained, proportionally, in the corners.) Since $L \sim N^{1/2}$, the rate of change of the number of atoms in the square level is therefore

$$\frac{\mathrm{d}N}{\mathrm{d}t} \approx 1/N^{1/2} \approx -N/\tau \ ,$$

which immediately implies that $\tau(N) \sim N^{3/2}$. However, if the atoms evaporate uniformly all along the perimeter of the square level, we will have instead

$$\frac{\mathrm{d}N}{\mathrm{d}t} \approx -N^{1/2} \approx -N/\tau \ ,$$

and hence $\tau(N) \sim N^{1/2}$. The observation and understanding of a kinetic scaling law thus provides a way of deducing the atomic behaviour which determines the stability (or lack of it) of a nanostructure, and this in turn may provide a way of controlling it.

We began with the shape of nano-objects and we have come to speak of the kinetics underlying their formation. Other factors are also involved in determining the shape of a nano-object. For example, the existence of magic values for the number of atoms making up a cluster. These are sizes at which the cluster is a perfect polyhedron, in which all atoms have maximal coordination number and the energy is therefore minimal. Another factor is the presence of surface reconstructions, or atomic rearrangements in nanocrystals, or again, the presence of impurities which affect the value of the surface energy.

Understanding and accounting for all these factors (often present at the same time) is one of the challenges that nanotechnology must face up to in order to achieve a perfect mastery of nanoclusters, the elementary building blocks of the nanoworld.

References

1. A. Pimpinelli, J. Villain: *Physics of Crystal Growth*, Cambridge University Press (1998)
2. F. Baletto, C. Mottet, R. Ferrando: Non-crystalline structures in the growth of silver nanoclusters, Euro. Phys. J. D **16**, 25 (2001)
3. N. Combe, P. Jensen, A. Pimpinelli: Changing shapes in the nanoworld, Phys. Rev. Lett. **85**, 110 (2000)
4. A. Ichimiya, K. Hayashi, E.D. Williams, T.L. Einstein, M. Uwaha, K. Watanabe: Decay of silicon mounds: Scaling laws and description with continuum step parameters, Appl. Surf. Sci. **175**, 33 (2001)
5. F. Nita, A. Pimpinelli: Scaling and crossovers in nano-island decay: A kinetic Monte Carlo study, Surf. Sci. **551**, 31 (2004)
6. Z.L. Wang, M.B. Mohamed, S. Link, M.A. El-Sayed: Crystallographic facets and shapes of gold nanorods of different aspect ratios, Surf. Sci. **440**, L809 (1999)

Part II

Physical and Chemical Properties on the Nanoscale

5

Magnetism in Nanomaterials

D. Givord

5.1 Introduction

This chapter is devoted to magnetic nanomaterials, i.e., systems with sizes between 1 nm and 100 nm. Such dimensions are of the same order of magnitude as the range of the various interactions that define magnetism in matter. This is the main reason why the properties observed in nanomaterials differ from those in the bulk state, and new magnetic effects can be created. Extremely interesting behaviour can be produced in functional materials, with potential applications in information technology (magnetic recording), telecommunications (non-reciprocal systems), energy transformation (high-performance soft materials and magnets), biology (functionalised magnetic particles), and medical engineering (mini- and microsensors or actuators).

Magnetic nanomaterials fall into two categories:

- systems with nanometric dimensions,
- systems with macroscopic dimensions, but made up of crystallites with nanometric dimensions.

We shall use the terms 'cluster' or 'nanoparticle' to refer to the first category, and 'nanostructured' or 'nanopatterned materials' to refer to the second.

In the first two sections, we review the main elements required to analyse magnetism in matter, without special reference to nanometric dimensions. The discussion of the properties of nanomaterials in Sects. 5.4–5.6 is then organised around the three main categories of properties by which these differ from bulk materials:

- Intrinsic properties of isolated clusters and nanoparticles (magnetisation, magnetic temperature, magnetic anisotropy) are described in Sect. 5.4.
- Coercivity and remanent magnetisation, which govern the process of magnetisation reversal and are referred to as extrinsic because they depend on the microstructure or nanostructure, are described in Sect. 5.5.

- The specific properties of nanostructured materials, resulting from the coupling between constitutive nanocrystallites, are the subject of Sect. 5.6.

Research on this type of material makes use of recent and remarkable experimental developments. However, it is not the aim of this chapter to describe these things, although we shall occasionally mention a new technique for elaborating or characterising nanomaterials if it is important for displaying some particular effect. Neither shall we discuss the applications of nanomaterials in any detail. The diversity of applications for these systems is illustrated in Sect. 28.2, which describes magnetic recording, and also in Sect. 31.3, which deals with applications in biology.

5.2 Magnetism in Matter

In this section, we review the three main phenomena that characterise magnetism in matter:

- the formation of the magnetic moment on the atomic scale,
- the occurrence of magnetic order, resulting from the strong interactions existing between atomic moments,
- the alignment of moments along some favoured crystallographic axis, leading to the phenomenon of magnetic anisotropy.

Iron, cobalt and nickel belong to the first series of magnetic elements, the $3d$ series, whilst the second series comprises the 14 rare earth elements. In the following sections, we describe the properties of materials composed of elements belonging to one or the other of these two series. Behaviour specific to each category of material is related to different characteristic properties of the $3d$ and $4f$ electronic shells.

5.2.1 Magnetic Moment

Magnetic Moment of an Atom

A moving electric charge is the source of a magnetic field and a magnetic moment can be associated with it. On the atomic scale, magnetism results from electron motion. There are two contributions to the magnetic moment m of an electron:

- The orbital moment m_l, due to the motion of the electron in its orbit. This has the form $m_l = -\mu_B l$, where l is the orbital angular momentum and μ_B is the Bohr magneton. The values $l = 0$, 1, 2 and 3 are attached to electrons in the s, p, d and f shells, respectively.
- The spin moment $m_s = -2\mu_B s$, where s is the spin, of purely quantum origin. It can be thought of as a spinning motion of the electron about its own axis. The spin is characterised by the value of s, which can only take two possible values, $+1/2$ and $-1/2$.

A different set of electrons rotating about the nucleus is associated with each element of the periodic table. In a given electron shell, there are $2l+1$ states available for each spin state, making a total of $2(2l+1)$ states (10 for a d shell, 14 for an f shell).

The distribution of the electrons over the available orbits of the given shell aims to minimise the energy associated with their mutual electrostatic repulsion. This energy contains the so-called exchange term, which depends on the spin and for this reason is responsible for magnetism in the matter. (The mechanism underlying exchange interactions is explained in Sect. 5.2.2.) The electron shells fill up progressively according to the Hund rules. The first of these rules says that the total spin S associated with all the electrons in the same electron shell is maximal, under the constraint that Pauli's exclusion principle be taken into account. This rule thus expresses the fact that all the electron spins tend to be parallel to one another. The second Hund rule states that the total orbital angular momentum L is maximal, with the restriction that the first rule takes precedence.

Viewed by the electrons, the motion of the nucleus creates a magnetic field acting on the spin moment. This field is the source of the spin–orbit coupling between the orbital and spin moments. The energy of the spin–orbit coupling is given by

$$E_{\text{spin–orbit}} = \lambda LS \,, \tag{5.1}$$

where λ is the spin–orbit coupling constant.

The orbital and spin contributions associated with all the electrons in a full (closed) electron shell tend to balance one another so that the resulting magnetic moment is zero. Magnetism is characterised only by partially filled electron shells.

Magnetic Moments in Matter. 4*f* and 3*d* Moments

In the solid state, the mixing of orbitals that results from covalence or the formation of energy bands often leads to the disappearance of the atomic magnetic moment. There are only two series of elements in which magnetism remains in the solid state. It is the internal structure of an unfilled electron shell which preserves magnetic effects in the solid state.

- The series of elements known as the rare earths, which goes from cerium to lutetium, corresponds to the progressive filling of the $4f$ shell. The electrons, localised in their atomic orbits, are essentially subject to the same exchange interactions as in an isolated atom. The magnetic moment is defined by the Hund rules described above (see Fig. 5.1a).
- The second series of magnetic elements, containing iron, cobalt, and nickel, corresponds to the filling up of the $3d$ shell. In insulating systems such as oxides, the $3d$ electrons are localised and the magnetic moment is still defined by the Hund rules. In metals, the $3d$ electrons are said to be itinerant,

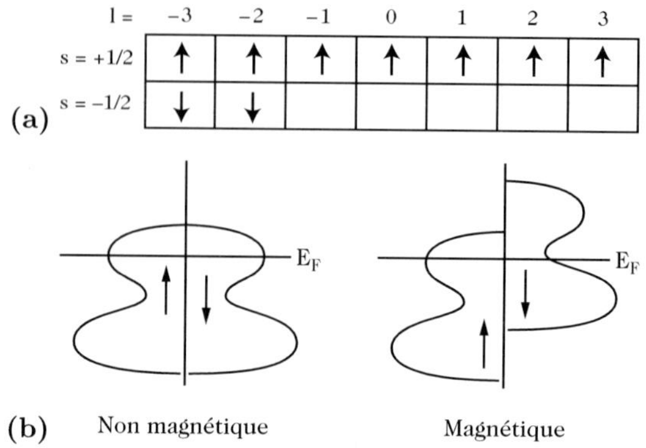

Fig. 5.1. Magnetic moment. (**a**) Localised system of electrons (dysprosium Dy). The $4f$ shell of dysprosium contains 9 electrons. The way they are distributed, as explained in the diagram, is governed by the Hund rules. S $(= 5/2)$ is maximum in agreement with the Pauli exclusion principle and L $(= 5)$ is maximum in agreement with the S maximum rule. (**b**) Itinerant electron system. The electrons are distributed in energy bands. The diagrams show the energy of states as a function of the density of states, i.e., the number of states per energy interval. Schematically, bands consist of two lobes, associated with bonding states at the bottom of the band and anti-bonding states at the top of the band. Electrons occupy available states up to the Fermi level. *Left*: There are as many spin-up (↑) electrons as spin-down (↓) electrons and the system is non-magnetic. *Right*: There are more spin-up (↑) electrons than spin-down (↓) electrons and the system is magnetic

forming an energy band with width of the order of 5 eV (about 50 000 K). The magnetic moment is no longer produced strictly on the atomic scale. Stoner showed that an alternative approach can be used to describe most of the observed behaviour. The whole set of electrons is considered, distributed over two half-bands, each containing $5N$ states, where N is the total number of atoms (see Fig. 5.1b), and characterised by the value of the electron spins $+1/2$ (↑) and $-1/2$ (↓). In the absence of exchange interactions, the minimal energy corresponds to equal filling of the two half-bands and the system is non-magnetic. Under the effect of exchange interactions, however, one half-band will be favoured because of the tendency for the electron spins to line up. The state of the system is defined by competition between these two terms. In the five elements Cr, Mn, Fe, Co and Ni, the minimum energy configuration is magnetic.

Despite the itinerant nature of the $3d$ electrons, they nevertheless remain essentially localised at each atomic site and at the end of the day many properties of $3d$ metals can be described by treating the magnetic moments as atomic. However, the distribution of moments considered at two different times will not be strictly the same, since the electrons can hop from one atom to another. The non-integral values of the magnetic moments per atom

($2.2\mu_B$ for Fe, $1.74\mu_B$ for Co, $0.60\mu_B$ for Ni) basically reflect the fact that they correspond to an average value taken over all the atoms making up the given system.

5.2.2 Magnetic Order

Exchange Interactions and Magnetic Order

The main interactions between electrons in matter are due to electrostatic repulsion. In quantum mechanics, the Hamiltonian describing them contains the so-called exchange term, resulting from the indistinguishability of electrons. As Heisenberg showed, magnetism then arises naturally as soon as the Pauli exclusion principle is taken into account. This states that two electrons cannot occupy the same quantum state defined by the space and spin variables, and thus requires the wave function for the electron ensemble to be antisymmetric in those variables. This in turn means that the interaction energy between electrons depends on their spin states. As mentioned above, it is the exchange interactions that cause the atomic magnetic moment. Then, insofar as the electrons in different atomic sites affect one another by exchange interactions, a magnetic coupling exists between the atomic moments of different atoms, and this is the source of magnetic order in a material.

The exact nature of the coupling between moments depends a great deal on the elements that are present, but also on the crystallographic arrangement of the atoms. For this reason, a whole range of what one might refer to as magnetic structures, i.e., arrangements of magnetic moments, can occur. When all the moments are parallel, the structure is said to be ferromagnetic (see Fig. 5.2a). In this case, the moment per unit volume of matter is called the spontaneous magnetisation and denoted by M_s.

In another case known as an antiferromagnetic arrangement, the moments are organised into two groups. Within the same group (called a sublattice), the moments are all parallel to one another. However, the moments of the two sublattices couple in an antiparallel manner and the resulting magnetisation is zero (see Fig. 5.2b).

In yet other systems, called ferrimagnetic systems, the numbers of atoms or the value of the magnetic moments are not the same in each sublattice. This time, cancellation is not complete and there remains some spontaneous magnetisation, as in the ferromagnetic materials (see Fig. 5.2c).

The exchange energy can be expressed phenomenologically by

$$E_{\text{exch}} = -\frac{1}{2}\sum_{i,j\neq i} J_{ij}S_iS_j \;, \tag{5.2}$$

where J_{ij} is the exchange integral, representing the coupling force between spins, and S_i and S_j are the spins carried by atoms i and j. The exchange interactions are assumed to be very short range. If all the atoms are identical

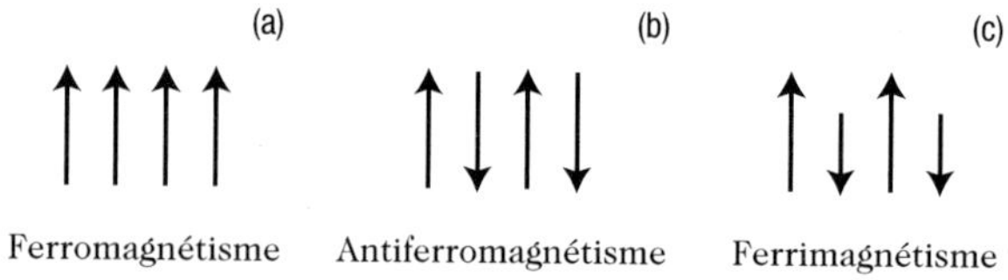

Fig. 5.2. Magnetic structures. (**a**) Ferromagnetic structure. All moments are parallel, with the same orientation. (**b**) Antiferromagnetic structure. Moments are arranged into two sublattices. The orientations of the moments alternate from one sublattice to the next. (**c**) Ferrimagnetic structure. Moments are arranged into two sublattices as in an antiferromagnetic structure, but the magnetisations in each sublattice are no longer equal and do not cancel one another completely

and considering only nearest neighbours, J_{ij} can be replaced by J, and S_i and S_j by S. The exchange energy per atom then becomes

$$e_{\text{exch}} = -\frac{1}{2} z J S^2 \,, \tag{5.3}$$

where z is the number of nearest neighbours of a given atom.

When J is positive, (5.3) implies that the minimal energy configuration corresponds to parallel coupling between all spins. This gives rise to ferromagnetism. More generally, the exchange interactions J_{ij} can vary for one atom i depending on its neighbours j, and from one atom i to another. When there are negative terms J_{ij}, they favour antiparallel coupling between moments i and j. Such terms are required to explain non-ferromagnetic structures, such as occur in antiferromagnetic or ferrimagnetic materials.

Molecular Field Model. Properties at Nonzero Temperature. Spin Waves

Equations (5.2) and (5.3) implicitly assumed a temperature of 0 K. At nonzero temperatures, thermal vibrations tend to destroy magnetic order. At a certain temperature (the Curie temperature in ferromagnetic and ferrimagnetic materials and the Néel temperature in antiferromagnetic materials) a transition occurs from the ordered magnetic state to the paramagnetic state, in which the magnetic moments are randomly oriented.

To first order, the effects of thermal vibrations can be described using the so-called molecular field model. Consider a ferromagnetic material. Since the spin is in fact proportional to the magnetic moment, (5.3), valid at 0 K, can be rewritten in the form

$$e_{\text{exch}} = -\frac{1}{2} \mu_{\text{at}} H_{\text{m},0} \,, \tag{5.4}$$

where μ_{at} is the atomic magnetic moment, proportional to the spin, and $H_{\text{m},0}$ is a fictitious magnetic field representing the exchange interactions and called the molecular field, first introduced by Weiss:

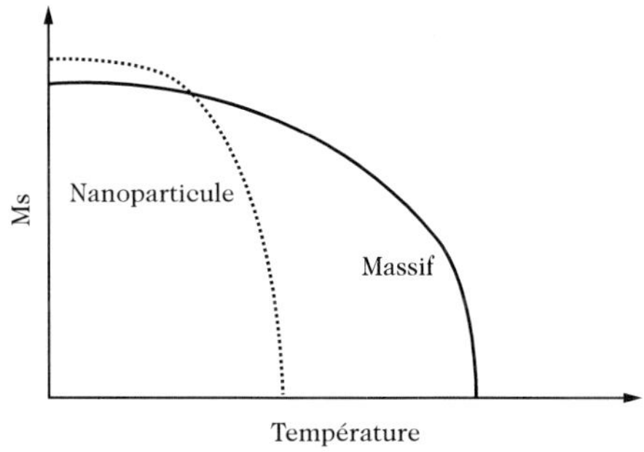

Fig. 5.3. Temperature dependence of spontaneous magnetisation M_s. In a nanoparticle, the magnetisation at low temperature is often greater than in the bulk solid. It varies more slowly at low temperatures than in the bulk, but more quickly at higher temperatures, since the the Curie temperature is lower than the Curie temperature in the bulk material

$$H_{m,0} = nM_{s,0} . \tag{5.5}$$

In this expression, n is the molecular field coefficient and $M_{s,0}$ the spontaneous magnetisation at 0 K, called the magnetisation at absolute saturation.

A central assumption in the molecular field model is that, when the temperature increases, the properties of the matter remain strictly homogeneous. The molecular field at temperature T can be expressed as a function of the spontaneous magnetisation at this temperature:

$$H_m = nM_s . \tag{5.6}$$

It can then be shown that the spontaneous magnetisation decreases steadily as the temperature increases (see Fig. 5.3) and vanishes at the Curie temperature T_C, which is given by

$$T_C = \frac{zJS^2}{3k} = \frac{n\mu_{at}^2}{3k} , \tag{5.7}$$

where k is the Boltzmann constant. The value of the Curie temperature provides a measure of the strength of the exchange interactions.

As was originally shown by Néel, the molecular field model can be extended to describe the behaviour of antiferromagnetic and ferrimagnetic materials. To go beyond this model, the disorder of the moments is considered to be produced at finite temperatures by collective excitations called spin waves. As far as magnetic moments are concerned, these are perfectly analogous to the collective vibrations of atoms known as phonons. For the same entropy as in the molecular field model, the loss of exchange energy is lower. When the wavelength λ of a spin wave increases, the angle between consecutive magnetic moments within the wave is reduced. The loss of exchange energy is therefore reduced. In a ferromagnetic material, the energy E_{SW} of the spin waves can be written in the form

$$E_{\mathrm{SW}} = Dq^2 , \tag{5.8}$$

where D is the rigidity constant of the spin waves and $q = 2\pi/\lambda$ is the wave vector.

Magnetic Order in Different Categories of Material

Localised $4f$ electrons are located in an electronic shell that is well protected from the effects of the environment by the presence of the outer $5s$, $5d$ and $6s$ shells. Interactions between $4f$ electrons located on different atoms are therefore negligible. The kind of interactions we appealed to above to explain the exchange coupling mechanism cannot be invoked here to account for the magnetic structures observed in metallic systems composed of rare earth elements. In fact, an indirect form of exchange coupling does exist in these systems, involving electrons in the outer $5d$ and $5s$ shells as intermediaries. This coupling, known as RKKY (Ruderman–Kittel–Kasuya–Yoshida) coupling, is relatively weak and is said to be long range. Indeed, it can extend as far as 1–2 nm, whereas in other exchange coupling mechanisms, the range of the interactions never exceeds one or two times the interatomic distance. The magnetic order temperatures are always lower than room temperature. RKKY interactions tend to stabilise the various types of magnetic order, i.e., ferromagnetic, antiferromagnetic, and also helimagnetic, characterised by a spiral arrangement of moments.

In $3d$ transition metal systems, the exchange coupling is intimately related to the itinerant nature of the electrons. When they hop from atom 1 to atom 2, the electrons conserve their spin. They then interact with electrons present on atom 2, their spin tending to lie parallel with these, according to the first Hund rule (see Sect. 5.2.1). The resulting ferromagnetic coupling force is explained by the intra-atomic nature of the exchange terms coming into play. In iron, cobalt and nickel, the Curie temperature is much higher than room temperature, reaching 1 380 K in cobalt.

In insulating transition metal compounds, the exchange mechanism involves a mixture of the $3d$ wave functions and the p wave functions of the anions, such as oxygen. The hybridisation of the wave functions depends significantly on the type of crystallographic background. Hence, the mechanism known as superexchange gives rise to coupling of different signs and the magnetic arrangements are very often antiferromagnetic (MnO, CoO, NiO) or ferrimagnetic (γ-Fe_2O_3, Fe_3O_4). In some cases, several coupling mechanisms compete and non-aligned magnetic arrangements of moments can exist (manganates).

5.2.3 Magnetocrystalline Anisotropy

Origins of Magnetocrystalline Anisotropy

Electrons in a magnetic shell are often subject to interactions with charges in other atoms in the neighbourhood, which constitute the crystalline field.

The corresponding electrostatic interaction depends on the orientations of the orbits of the magnetic electrons in this crystalline field. To the favoured orientation of the orbit there corresponds a favoured orientation of the magnetic moment, since magnetic moment and electron orbit are closely related. The difference in energy of the crystalline field between orientations of the magnetic moments constitutes the energy of magnetocrystalline anisotropy between the two associated crystallographic axes.

In a so-called uniaxial system, there is just one easy direction of magnetisation. To lowest order, the uniaxial anisotropy energy per unit volume can be written in the form

$$E_{\mathrm{A}} = K \sin^2 \theta \,, \tag{5.9}$$

where K is the magnetocrystalline anisotropy constant and θ is the angle between the moments and the easy axis of magnetisation. The anisotropy field H_{A} is defined by

$$H_{\mathrm{A}} = \frac{2K}{M_{\mathrm{s}}} \,. \tag{5.10}$$

It represents the field strength required to align the moments along a field applied perpendicularly to the easy axis of magnetisation.

Magnetocrystalline Anisotropy of Transition Metals and Rare Earth Metals

The magnetic anisotropy can be one or two orders of magnitude stronger in materials composed of $4f$ elements than in those made from $3d$ elements. To discuss this phenomenon, two important facts need to be considered. On the one hand, the $4f$ shell, further in than the $3d$ shell, is much less sensitive to the background crystalline field. On the other, the spin–orbit coupling force, approximately proportional to the square of the atomic number, is much stronger in rare earth elements than in $3d$ transition metals.

In order to understand a physical phenomenon, one must consider the competing interactions in decreasing order of importance. In the present case, the exchange interactions are taken to determine the maximal value of the spin moment, and the magnetocrystalline anisotropy results from competition between the spin–orbit coupling and the crystalline field. In rare earth materials, the spin–orbit coupling dominates the crystalline field. It favours a maximal value of the orbital moment, aligned with the spin moment. Now a strong orbital moment indicates a highly asymmetric electron orbit. (There is an analogy with a current loop, whose magnetic moment increases in proportion to its area.) The energy of an orbit placed in a given environment depends all the more on its orientation as it becomes more asymmetric. This is basically why rare earth compounds and alloys often exhibit strong magnetocrystalline anisotropy.

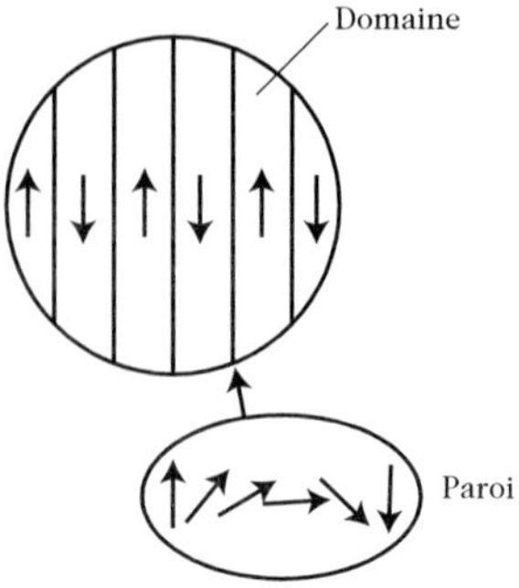

Fig. 5.4. Magnetic domains and walls. In order to minimise the energy of the demagnetising field, matter tends to divide up into domains which differ in the orientation of the magnetisation. On the boundary between two such domains, a wall is formed, within which the magnetisation gradually rotates to provide continuity

For materials made from transition elements, the crystalline field must be considered before the spin–orbit coupling. Electron orbits adopt shapes that follow from the symmetry of their surroundings and the associated orbital moment is often small, e.g., zero for cubic symmetry. The spin–orbit coupling tends to induce an orbital moment in the direction of the spin moment, but the associated distortion of the electron orbit is not favourable for the crystalline field. The orbital moment induced by this mechanism is thus small, e.g., a few percent of the spin moment in the metals Fe, Co, and Ni, and the magnetic anisotropy, which is roughly proportional to it, is likewise small.

5.3 Magnetisation Process and Magnetic Materials

On the basis of the ideas introduced so far, the moments in a ferromagnetic material should all turn out to be parallel in order to minimise the exchange energy, lined up along the easy axis of magnetisation in order to minimise the anisotropy energy. In reality, matter is observed to divide up into so-called magnetic domains (see Sect. 5.3.1), within which the moments do indeed adopt the expected configurations, but such that, from one domain to another, the direction of the magnetic moments actually alternates (see Fig. 5.4). This division into domains allows the system to minimise an energy term that we have neglected up to now, namely the demagnetising field energy which results from the action of the magnetised matter on itself.

When the magnetic domains are formed, magnetic domain walls also come into being between them (see Sect. 5.3.1). The formation of a wall has a certain cost in anisotropy and exchange energy. The magnetisation process (see Sect. 5.3.2) describes the action of a magnetic field on magnetised matter. What is observed results from complex competition effects between all the energy terms so far described. Various families of magnetic materials can be distinguished, depending on their response to an applied magnetic field. In soft materials, the magnetisation follows the field easily, whereas in harder materials, it tends to resist the effects of the field. In materials used for magnetic recording, the saturated magnetisation state constitutes one bit of information which can be reversed from state 1 to state 0, or the opposite.

5.3.1 Energy of the Demagnetising Field. Domains and Walls

Division into Domains. Demagnetising Factors

The energy of the demagnetising field is given by

$$E_{\mathrm{D}} = -\frac{1}{2}\mu_0 \int_V \boldsymbol{M} \cdot \boldsymbol{H}_{\mathrm{D}} \mathrm{d}V \,, \tag{5.11}$$

where $\boldsymbol{M}$ is the local magnetisation, $\boldsymbol{H}_{\mathrm{D}}$ is the demagnetising field at the relevant point and V is the volume of magnetic matter. It can be shown that E_{D} can also be written in the form

$$E_{\mathrm{D}} = \frac{1}{2}\mu_0 \int_{\mathbb{R}^3} (\boldsymbol{H}_{\mathrm{D}})^2 \mathrm{d}^3 R \,, \tag{5.12}$$

where the symbol $\int_{\mathbb{R}^3}$ indicates that the integral is taken over the whole of space. Equation (5.12) shows that E_{D} is always positive. The configuration of moments that minimises E_{D} is such that $H_{\mathrm{D}} = 0$.

The energy of the demagnetising field, although less than the exchange energy, is not involved in establishing magnetic order. However, the associated interactions are long range and impose the division of the matter into domains with dimensions of micrometric order (see Fig. 5.4). On the scale of a material element big enough to contain several domains, the magnetisation is practically zero and the demagnetising field likewise, being proportional to it.

The energy density of the demagnetising field in a uniformly magnetised object can be expressed in the form

$$E_{\mathrm{D}} = \frac{1}{2}\mu_0 N M^2 V \,, \tag{5.13}$$

where M is the magnetisation and V the volume. In the general case, it is no simple matter to calculate the coefficient N. For an object with ellipsoidal shape, one may introduce a tensor $\boldsymbol{N}$. Let x, y, and z be the principal axes of the ellipsoid. Then $\boldsymbol{N}$ has three components, N_x, N_y, and N_z, known as the demagnetising factors, which satisfy $N_x + N_y + N_z = 1$. For a sphere, for example, we have $N_x = N_y = N_z = 1/3$. The magnetic demagnetising field is uniform and given by

$$\boldsymbol{H}_{\mathrm{D}} = -\boldsymbol{N}\boldsymbol{M} \,. \tag{5.14}$$

In an object with arbitrary shape, it is often justifiable to treat the demagnetising field as homogeneous to the first order of approximation.

Domain Walls

The region of transition between two neighbouring domains is called a domain wall, or Bloch wall. In this region, the moments gradually rotate round from an initial position in which they are parallel to the direction of the moments in one domain, to a final position in which they are parallel to the direction of the magnetisation in the adjacent domain. To a first approximation, the domain walls form planes running right across the sample. It can be shown that the energy of the demagnetising field resulting from the formation of these walls is actually very small. However, the moments within the walls are neither strictly parallel to one another, nor aligned with any easy axis of magnetisation. The energy lost in wall formation, called the wall energy, is thus determined purely by competition between anisotropy energy and exchange energy. For a uniaxial system with anisotropy constant K, the wall energy expressed per unit area of the wall is equal to $\gamma = 4\sqrt{AK}$, and the thickness of the wall is given by $\delta = \pi\sqrt{A/K}$. The constant A in the expressions for γ and δ is the exchange constant, a parameter representing exchange interactions and proportional to the Curie temperature: $A \propto Nk\xi^2 T_{\mathrm{C}}$, where N is the number of atoms per unit volume and ξ is the distance between nearest neighbour atoms. In iron, $A \approx 18 \times 10^{-11}$ J/m and $K \approx 50\,\mathrm{kJ/m^3}$, whence $\gamma \approx 3 \times 10^{-3}\,\mathrm{J/m^2}$ and $\delta \approx 100\,\mathrm{nm}$. The Curie temperatures of all magnetic materials used in applications are higher than room temperature, generally lying in the range 500–1 000 K. The value of A is therefore always of the order of 10^{-11} J/m. In contrast, K may vary over several orders of magnitude, so that $K \approx 10^3\,\mathrm{J/m^3}$ in ultrasoft materials, whereas $K \approx 10^7\,\mathrm{J/m^3}$ in hard materials. The wall energy thus has values between one tenth and ten times the value for iron, with wall thicknesses between 300 nm in soft materials and 5 nm in hard materials.

Note that the energy of a wall increases rapidly when its thickness cannot take the equilibrium value δ. The most important case is that of a confined wall. The increase in energy e of the wall is due to the loss of exchange energy resulting from the fact that the magnetic moments must rotate rapidly from one atomic plane to the next in order to complete the 180° rotation over the available distance.

5.3.2 The Magnetisation Process

Displacement of Walls

When a magnetic field H_{app} is applied to a ferromagnetic material, the Zeeman energy of coupling with the field is given per unit volume by

$$E_{\mathrm{Z}} = -\mu_0 M H_{\mathrm{app}} . \tag{5.15}$$

Equation (5.15) states that magnetisation tends to occur along the field direction, starting from the initial situation resulting from the division into

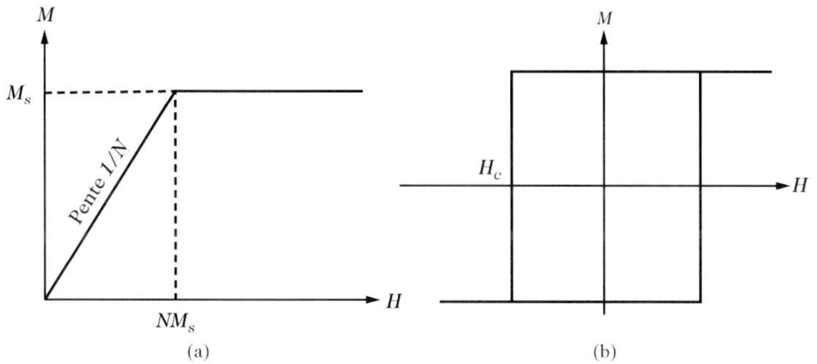

Fig. 5.5. The magnetisation process. (**a**) Variation of the magnetisation due to wall displacement. The magnetisation varies linearly with the field. The slope of this variation is $1/N$, where N is the demagnetising factor. It is independent of temperature. (**b**) Magnetisation reversal in a coercive system. The magnetisation resists the applied field, until this field reaches the value of the coercive field H_c

domains and characterised by $M = 0$. The magnetisation increases by displacement of the walls in such a way that domains with magnetisation along the field grow larger to the detriment of the others. In a homogeneous material and assuming that the energy of the demagnetising field is correctly given by (5.13), the total energy density can be written (note the omission of the index on the factor N)

$$E_T = \frac{1}{2}NM^2 - \mu_0 M H_{app} . \qquad (5.16)$$

The wall energy does not depend on the position of the wall and this is why it does not enter into (5.16). Minimising, one deduces that the magnetisation has the linear dependence

$$M = \frac{1}{N}H . \qquad (5.17)$$

The slope of the functional dependence here, given by $1/N$, is known as the demagnetising gradient. It depends only on the shape of the sample and is independent of temperature.

Wall displacement proceeds until a state is reached in which all moments are aligned with the field. This happens for $H_{app} = NM_s$. In stronger fields, the magnetisation remains constant and is said to be saturated (see Fig. 5.5a).

Coercivity

Suppose a new field is applied to a sample in the saturated magnetisation state, but in the opposite direction to the initially applied external field. According to (5.17), the magnetisation should grow in the opposite direction in order to minimise the Zeeman energy. However, the initial nucleation of

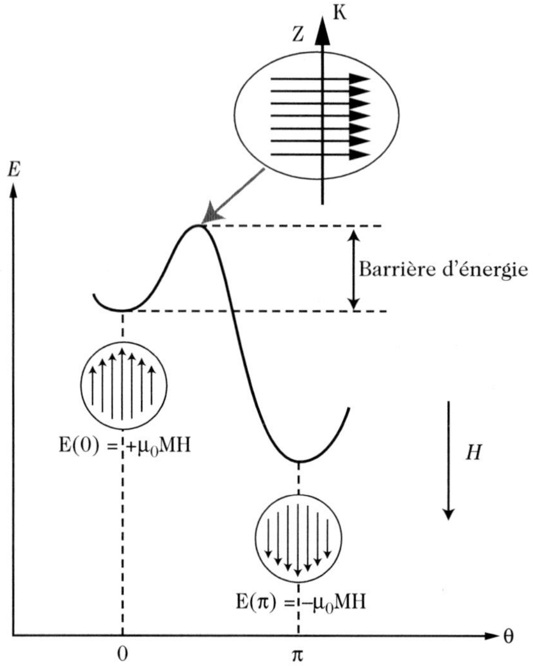

Fig. 5.6. Coercivity. In a saturated material, magnetisation reversal, assumed to proceed by coherent rotation of all the moments at the same time, involves transit through an unfavourable intermediate state where the moments are actually perpendicular to the easy axis of magnetisation

a new domain in the saturated state requires the system to overcome an energy barrier. Indeed, a moment taking part in the magnetisation reversal must necessarily go through a situation in which it lies perpendicular to the easy axis of magnetisation. In this unfavourable situation, the anisotropy energy is maximal (see Fig. 5.6).

The coercivity of a material is its capacity to resist the effect of an applied field. When the applied field is exactly antiparallel to the initial magnetisation direction z, the coercive field H_c, at which the magnetisation swings into line with the field (see Fig. 5.5b), is equal to the anisotropy field:

$$H_c = 2K/M_s = H_A . \tag{5.18}$$

As soon as nucleation occurs, all the moments reverse in phase in the process known as coherent rotation, first described by Stoner and Wohlfarth. More generally, the coercive field of such a system is a function of the angle θ between the applied field and the z axis (with the convention that $\theta = 0$ corresponds to a field antiparallel with the initial direction of the moments):

$$H_c(\theta) = H_{SW}(\theta) = \frac{H_A}{(\sin^{2/3}\theta + \cos^{2/3}\theta)^{3/2}} . \tag{5.19}$$

Equations (5.18) and (5.19) express the fact that it is the magnetic anisotropy that underlies coercivity. However, in the vast majority of real systems, the coercive field is much weaker than the anisotropy field. This discrepancy with the coherent rotation model is due to the presence of structural defects. Since the anisotropy is reduced at defect sites, nucleation is thereby facilitated, and nucleation is accompanied by the formation of a wall separating the nucleus from the rest of the material with opposite magnetisation. Complete reversal of the magnetisation proceeds by the propagation of the wall through the material. In this case, the angular variation of the coercive field, very different from what would correspond to coherent rotation, is given rather by an expression of the form

$$H_c(\theta) \approx \frac{H_c(0)}{\cos\theta} , \tag{5.20}$$

where $H_c(0)$ is the coercive field obtained for $\theta = 0$. Equation (5.20) means that only the projection of the applied field in the initial magnetisation direction plays an active role in the reversal process. Since the coercive field is much weaker than the anisotropy field, the reversible rotation of the moments toward the applied field can be neglected.

5.3.3 Magnetic Materials

Magnetic materials based on iron, cobalt, or nickel (see Sect. 5.2.3) have the behaviour of soft magnetic materials owing to their weak anisotropy. The coercivity is low and the variation of the magnetisation with the applied field is dominated by the wall displacement process described by (5.17). The magnetisation induced by a weak excitation field is in turn the source of a magnetic field or flux. When the applied field is time dependent, the induced magnetisation is also time dependent. The voltage thereby generated provides the signal exploited in transformers or sensors.

The magnetocrystalline anisotropy characteristic of the $4f$ electrons is much stronger in rare earth materials. However, the magnetic properties disappear below room temperature (see Sect. 5.2.2). In rare earth–transition metal compounds, there are strong $3d$–$4f$ interactions. They result from hybridisation between $5d$ electrons in the rare earth atoms and $3d$ electrons in the transition atoms. By this mechanism, the magnetism of the rare earths is preserved at high temperature. In systems where the magnetocrystalline anisotropy is uniaxial, there may be a strong coercivity. This kind of hard magnetic material is used in high-performance permanent magnets. The best known are the compounds $SmCo_5$, with hexagonal symmetry, and $Nd_2Fe_{14}B$, with quadratic symmetry. Potential energy can be stored in an applied field antiparallel to the magnetisation, weaker in strength than the coercive field. It is transformed into mechanical energy in motors and actuators.

In materials used for magnetic recording, the coercivity must be great enough to ensure the stability of the stored data, but it must not be so strong that the magnetisation cannot be reversed, as required during the writing process, when a field of moderate strength is applied. Most of the materials used here are cobalt alloys, this being the element most likely to exhibit coercivity in the $3d$ series.

5.4 Magnetism in Small Systems

The various interactions causing magnetism in matter and discussed in the previous sections are all very short range interactions:

- Exchange interactions at atomic sites, leading to the Hund rules and the formation of the atomic moment.
- Exchange interactions between electrons on different atoms, which lead to magnetic order. The range of these interactions depends on the system, going from roughly the interatomic distance in transition metal oxides, to a few interatomic distances in transition metals, and up to 1–2 nm in the rare earth metals.
- The spin–orbit coupling, an atomic phenomenon, and the crystalline field, dominated by the interactions of electrons at one site with the charges on nearest neighbour atoms, which together underly the phenomenon of magnetic anisotropy.

In this section, we shall discuss the properties of clusters and nanoparticles, using the term 'cluster' to refer to very small objects, up to about 1 nm, and nanoparticles for larger nanometric dimensions. Their properties may differ from those in the bulk solid state, owing to the fact that their dimensions are of the same order of magnitude as the range of the relevant interactions. In clusters, studied in flight, only the magnetic moment can be measured. Deposition on a substrate produces an agglomeration and hence an increase in the volume of the objects. Magnetic order and anisotropy can then be examined.

5.4.1 Magnetic Moments in Clusters

Enhancement of the Magnetic Moment in Clusters

In localised electron systems, the atomic magnetic moment obtained by applying the Hund rules is determined purely by electrons localised on the same atom (see Sect. 5.2.1 and Fig. 5.1a). Due to the strictly local character of the interactions, there is no reason why the value of the magnetic moment should depend on the size of the objects.

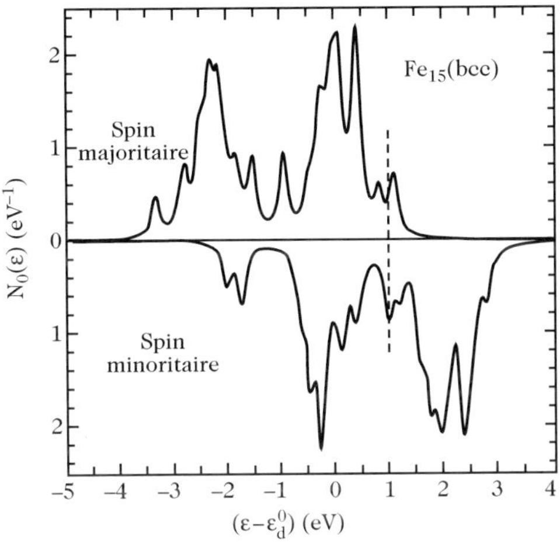

Fig. 5.7. Theoretical band structure of a cluster comprising 15 iron atoms (from [3]). The two lobes associated with bonding and anti-bonding states are narrower than in the bulk (see the diagram in Fig. 5.1b). This phenomenon leads to an increase in the magnetic moment carried by the iron atoms

The first measurements of the magnetic moments of free clusters of $3d$ ferromagnetic metals, such as iron and cobalt, were made on clusters of 100–500 atoms (diameter 1–2 nm) [1, 2]. These clusters are obtained by ablation of a target under an intense laser beam and subsequent condensation of vapourised atoms in an atmosphere of neutral atoms such as helium. The magnetic moments of the clusters are deduced from the extent to which they are deflected when passed through a region containing a magnetic field gradient. (This is the same principle as the Stern–Gerlach experiment to measure the spins of silver atoms.) In order to analyse the results, one must take into account the fact that the clusters are superparamagnetic (see Sect. 5.5.2). For iron, the results suggest an average atomic moment of $2.2\mu_B$, equal to the value in the bulk solid. For cobalt, the experiments, more accurate here, lead to a value of $2.08\mu_B$, to be compared with $1.72\mu_B$ in the bulk.

The theoretical band structure of a cluster comprising 15 iron atoms, separated into the successive contributions of the various atomic layers or shells making it up, has been calculated by Pastor et al. (see Fig. 5.7) [3]. Qualitatively, the energy distribution of the states in each half-band has two lobes, as in the bulk.

The lower energy lobe is associated with delocalised bonding states, whilst the higher energy lobe comes from localised anti-bonding states. However, the band width is less than in the cluster. Indeed, the band states are formed from linear combinations of atomic states. The more states are involved, the broader is the energy distribution of the states. The small number of atoms within a cluster thus explains the relative reduction in band width compared with the characteristic value for the bulk solid.

When a magnetic moment is formed, a certain number of electrons is transferred from one half-band to another. From the point of view of the non-magnetic terms, the energy loss is smaller in the case of a narrow band. The narrowing of the *d* band in clusters accounts for the observed tendency for the 3*d* moment to increase.

Note that the value of the magnetic moment of atoms in a cluster is affected by further, secondary mechanisms [3]. There is a reduction in interatomic distances which leads to an increase in the itinerancy of the electrons, and hence to an increased mixing of orbitals and in the end, a reduction of the atomic magnetic moment. Going the other way, the reduced symmetry of the environment of the atoms causes a reduction in band width that therefore favours an increased magnetic moment.

Ferromagnetism of Certain Metals That Are Paramagnetic in the Bulk

Measurements similar to those made on Fe and Co clusters have been carried out on small rhodium clusters containing up to 32 atoms. These show that there is a magnetic moment of the order of $1\mu_B$/atom [4]. This is a quite spectacular result since in the bulk state rhodium is not magnetic at all. The stabilisation of the ferromagnetism of 4*d* electrons in clusters can be attributed to the narrowing of the *d* band just mentioned to explain the increased moments of Fe or Co.

Spin and Orbital Moments in Clusters

Measurements of circular dichroism using X-rays, developed over the last few years with the advent of reliable X-ray synchrotron sources, make it possible to ascertain the spin and orbital moments separately. Apart from the enhancement of the spin moment already explained, a similar enhancement of the orbital moment has been detected for small iron clusters [5].

In Sect. 5.2.3, we explained the connection between the orbital moment and magnetic anisotropy. In thin films, Bruno [6] has shown that the enhancement of the orbital moment and the increase in magnetocrystalline anisotropy (see Sect. 5.4.3) are both related to the local reduction of symmetry at the surface. However, in clusters, all moments are coupled in parallel by exchange. The value of the orbital moment depends on the orientation of the moments relative to the local easy axis of magnetisation. The latter is different for each surface considered and, assuming the clusters to be spherical, the increase in orbital moment should vanish by symmetry. Two phenomena then explain the existence of strong orbital moments revealed experimentally. Firstly, there is an anisotropic relaxation which breaks the spherical symmetry of the clusters, and secondly, the cluster atoms are arranged into successive shells and the cancelling effects only come fully into effect for systems comprising filled (closed) atomic shells.

5.4.2 Magnetic Order in Nanoparticles

The Nature of Magnetic Order

The ferromagnetic nature of the bulk solid is conserved in clusters of Fe, Co, or Ni. This result can be related to the fact that, in the $3d$ metals, the interaction mechanism between magnetic moments is insensitive to slight changes in the atomic environment (see Sect. 5.2.2).

In nanoparticles of transition metal oxides, measurements reveal a reduction in the average magnetisation. This happens for maghemite nanoparticles (γ-Fe_2O_3), a collinear ferrimagnetic material in the bulk. For an insulating system, such a reduction in magnetisation cannot be attributed to a lower iron moment, since the latter does not depend on the size of the system. In fact, Mössbauer spectroscopy on the Fe^{57} nucleus reveals non-collinear arrangements of the moments [7]. Such arrangements are characteristic of atoms located in a low-symmetry environment and subject to magnetic interactions of various signs. Numerical simulation confirms the highly non-collinear nature of the calculated arrangements whenever the reduced symmetry of the environment of surface atoms is taken into account.

Non-collinear arrangements of the same origin also occur in nanoparticles of systems that are antiferromagnetic in the bulk state, such as NiO [8]. There is no reason why the moments should cancel one another exactly. The result is that antiferromagnetic nanoparticles carry a small magnetic moment.

Order Temperature and Thermal Variations of the Spontaneous Magnetisation

Few studies have been concerned with determining the magnetic order temperature of very small particles. This is due to the experimental difficulties involved in such measurements. But despite the lack of experimental data, a reduction in the order temperature has been noted compared with the bulk state. This can simply be attributed to a reduction of the molecular field resulting from the lower coordination of surface atoms [see (5.3)].

Furthermore, the thermal variation of the magnetisation at low temperatures should be influenced by the discrete nature of the energy levels of spin waves, in systems of the smallest dimensions. For a cluster comprising n atoms, there are n possible spin wave energies, distributed over an energy range up to kT_c, which is the value corresponding to spin waves of highest energy. Due to the discrete number of possible spin waves, there is a discontinuity between the energy level of the ground state and the energy of the lowest energy spin wave [9]. The physical picture associated with this phenomenon is that it is not possible to excite spin waves with wavelength greater than the diameter of the nanoparticles. At low temperatures, as long as it is not possible to excite even the lowest energy spin waves, the magnetisation must stay at its absolute saturation value. Hence in a certain temperature range, the thermal variation of the magnetisation of a ferromagnetic nanoparticle must be less than it is in the bulk state. Of course, at higher temperatures, it must become greater, due to the reduced value of the Curie temperature. The expected thermal variation of the spontaneous magnetisation is shown schematically in Fig. 5.3. These theoretically predicted effects have not yet been corroborated experimentally.

5.4.3 Magnetic Anisotropy in Clusters and Nanoparticles

Surface Anisotropy

As early as 1954, Néel noted that the reduction in symmetry at the surface of a thin magnetic layer can lead to a very significant increase in magnetocrystalline anisotropy [10]. Assuming that the electrostatic charges acting on the electrons in the magnetic shell of atom i are centered at the site of the nearest neighbour atoms j in its environment, a contribution to the anisotropy denoted by w_{ij} can be associated with each atom. Since an anisotropy term tends to orient the magnetic moments in some favoured direction, it is natural to express w_{ij} to lowest order in the form

$$w_{ij} = k(\cos^2 \Phi_j - 1/3) , \tag{5.21}$$

where Φ_j is the angle between the direction of the magnetic moment of atom i and the vector joining this atom to atom j. The total surface anisotropy energy E_{SA} is obtained by summing over all atoms in the environment:

$$E_{\mathrm{SA}} = \sum_j k \cos^2 \Phi_j . \tag{5.22}$$

For cubic symmetry, the various terms in (5.22) cancel one another exactly and the order 2 anisotropy is zero. This explains the weak anisotropy generally found in materials that crystallise with cubic symmetry. However, some atoms are missing in the neighbourhood of surface atoms and this means that the various terms in (5.22) no longer completely cancel. The resulting anisotropy is often greater by one or two orders of magnitude than would occur in the bulk state.

Anisotropy in Nanoparticles

In Sect. 5.5.2 we shall show that the value of the magnetic anisotropy for nanoparticles can be deduced from the so-called blocking temperature. In cobalt nanoparticles, the anisotropy obtained is equal to $3 \times 10^7 \mathrm{J/m^3}$, which is almost two orders of magnitude greater than in the bulk [11].

In spherical symmetry, even if the magnetic anisotropy of a surface atom is high, the total anisotropy should vanish for reasons of symmetry, according to the argument given in Sect. 5.4.1 for the orbital moment. In the present case, as in the last, the strong anisotropy in nanoparticles must be associated with anisotropic relaxation and the fact that the outer atomic shell of the nanoparticles is not filled.

5.5 Magnetostatics and Magnetisation Processes in Nanoparticles

In this section, we discuss the way magnetisation occurs in nanoparticles. Below a certain critical size, the formation of domain walls costs energy and the single-domain state becomes the most stable state. The particles are also sensitive to thermal activation. Above a certain temperature, called the blocking temperature, their magnetisation fluctuates over the energy barrier caused by anisotropy and they become superparamagnetic.

On the macroscopic scale, magnetisation reversal occurs by nucleation and propagation of walls (see Sect. 5.3.2). On the nanoscale, coherent rotation has been demonstrated for the first time. It is through an understanding of the coherent rotation process that it has been possible to analyse the effects of thermal activation quantitatively and show that in some cases magnetisation reversal can occur by a macroscopic manifestation of a quantum phenomenon known as the tunnel effect.

5.5.1 Single-Domain Magnetic Particles

The way matter divides into domains, as described in Sect. 5.3.1, is determined by competitition between the wall energy that is thereby lost and the energy of the demagnetising field that is gained. The walls are quasi-2D objects and the wall energy is therefore a surface term. When the particle volume is reduced, this term becomes greater than the energy of the demagnetising field, itself a bulk term. For very small dimensions, the most stable magnetisation state is one in which there is just one domain. To find the corresponding critical volume, Kittel [12] considered a spherical particle of radius R, characterised by a uniaxial anisotropy constant K. Its energy E_1 in the single-domain state contains only one term, due to the demagnetising field:

$$E_1 = \frac{1}{6}\mu_0 M^2 \frac{4}{3}\pi R^3 . \tag{5.23}$$

When this same particle is divided into two equal domains, Kittel assumes that the energy of the demagnetising field is divided by two. The corresponding energy E_2 is

$$E_2 = 4\sqrt{AK}\pi R^2 + \frac{1}{12}\mu_0 M^2 \frac{4}{3}\pi R^3 , \tag{5.24}$$

where the term in R^2 represents the wall energy (see Sect. 5.3.1). The critical radius R_c is obtained for $E_1 = E_2$, viz.,

$$R_c = \frac{36\sqrt{AK}}{\mu_0 M_s^2} . \tag{5.25}$$

In metallic iron, the critical radius R_c obtained in this way is of the order of 7 nm. Below this size, the nanoparticles are spontaneously magnetised. Such particles are typical objects used in magnetic recording.

5.5.2 Thermal Activation and Superparamagnetism

In this section, we consider a magnetic particle with uniaxial anisotropy and radius well below the critical value R_c. In zero field, the two orientations of the magnetisation along the easy axis, denoted $\uparrow$ and $\downarrow$, have equal energy. An energy barrier of height $E = KV$ separates these two states (the same diagram as in Fig. 5.6, if we had $H = 0$). At low temperatures, the magnetisation of each particle remains blocked in one of the two directions, imposed either by chance as the particle cooled from higher temperatures, or by application of some magnetic field stronger than the coercive field (see Sect. 5.3.2).

As the temperature increases, thermal activation must be taken into account. The characteristic time required to overcome a barrier of height E is given by the Arrhenius law

$$\tau = \tau_0 \mathrm{e}^{E/kT} , \tag{5.26}$$

where τ_0 is the intrinsic magnetisation reversal time, of the order of 10^{-9} s. On macroscopic scales, $E/kT \gg 1$ and τ is extremely long. Thermal activation will have no effect on the magnetisation state of the particle. But on very small scales, the energy barrier E, with height proportional to the particle volume, is much lower. Above the temperature T_B known as the blocking temperature, the particle magnetisation fluctuates between its two possible orientations. When a physical measurement is made at a temperature above T_B, the particle behaves as though it is no longer ferromagnetic, although in reality the atomic magnetic moments are still rigidly coupled together. This is the phenomenon known as superparamagnetism.

There is a characteristic acquisition time t_c associated with each measurement technique. For example, $t_c \approx 1$ s for magnetisation measurements, and $t_c \approx 10^{-8}$ s for Mössbauer spectroscopy. For an intermediate value of τ, the particle is superparamagnetic as far as magnetisation measurements are concerned, but blocked for Mössbauer spectroscopy. More precisely, the value obtained for T_B depends on the experimental technique used according to the expression deduced from (5.26):

$$E = KV \approx kT_B \ln \frac{t_c}{\tau_0} . \tag{5.27}$$

The superparamagnetism phenomenon characterises objects of very small dimensions. As far as magnetisation measurements are concerned, the blocking temperature of a 3-nm cobalt nanoparticle is of the order of 30 K, but it approaches room temperature for a diameter of 6 nm.

For nanoparticles of known volume (deduced, for example, from transmission electron microscope observations), the magnetic anisotropy can be deduced from the value of the blocking temperature using (5.27) (see Sect. 5.4.3).

The remanent magnetisation of a particle is zero above the blocking temperature. Such a particle cannot therefore be used in a magnetic recording medium. The superparamagnetism phenomenon specifies a physical lower limit on the size of particles that can be used for magnetic recording. The corresponding ultimate limit on the recording density is of the order of 50 Gbits/cm^2, close to the recording densities achieved today. A significant increase in the recording density achievable with a magnetic medium will probably involve some way of overcoming or circumventing the superparamagnetic limit.

5.5.3 Coherent Rotation in Nanoparticles

Consider a spherical particle of radius R with saturated magnetisation in the direction $\uparrow$, subject to a magnetic field applied in the direction $\downarrow$. On the macroscopic scale, magnetisation reversal involves the nucleation of a wall on a defect, followed by propagation of this wall (see Sect. 5.3.2). After nucleation, the wall surface increases until it reaches the maximal value πR^2. An energy barrier thus comes into play in the propagation mechanism. This barrier is given by

$$E_{\mathrm{p}} = \gamma \pi R^2 . \tag{5.28}$$

By a similar argument to the one used in the last section, this barrier can be compared with the energy barrier $E_{\mathrm{SW}} = KV$ characterising the process of coherent rotation. For $R < 3\gamma/4K$, $E_{\mathrm{p}} > E_{\mathrm{SW}}$ and nucleation at defects becomes inoperative due to the propagation barrier. Magnetisation reversal must then occur via coherent rotation. In cobalt, $\gamma \approx 10^{-2}\,\mathrm{J/m^2}$ and $K \approx 5 \times 10^5\,\mathrm{J/m^3}$. The associated critical radius is $R_{\mathrm{cr}} = 15\,\mathrm{nm}$.

Although the coherent rotation theory of magnetisation reversal dates from 1948, it could not be confirmed experimentally until it became possible to measure the individual magnetic properties of very small objects, in which the nucleation–propagation scenario is impossible. Magnetisation reversal has been studied in 4-nm cobalt nanoparticles using a high-sensitivity magnetometry technique based on micro-SQUIDs [13]. The nature of the reversal process was deduced by measuring the angular dependence of the coercive field, which corresponds exactly to the prediction of the Stoner–Wohlfarth theory given by (5.19) (see Fig. 5.8).

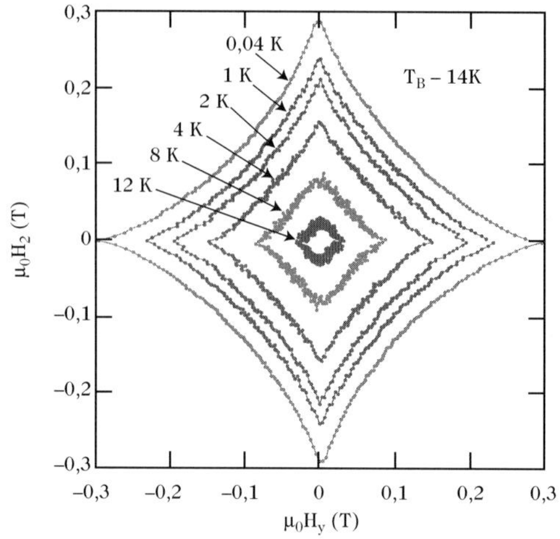

Fig. 5.8. Angular dependence of the coercive field for a single Co nanoparticle of diameter 3 nm, measured at different temperatures. Taken from [14]. In this polar diagram, the amplitude of the coercive field at angle θ is represented by the length of the vector making angle θ with the horizontal axis and joining the origin to the experimental curve

5.5.4 From Thermal Activation to the Macroscopic Tunnel Effect

Experimental confirmation of coherent rotation opened the way to investigation of both thermal and quantum activation effects in the reversal process. When such effects come into play, magnetisation reversal is not induced by the applied field alone, but may occur before the energy barrier vanishes. We define the parameter ε by [14]

$$\varepsilon = 1 - \frac{H_{\mathrm{app}}}{H_{\mathrm{A}}} . \tag{5.29}$$

In the general case, the energy barrier ΔE is given by the approximate relation

$$\Delta E = 4KV \left(\frac{2}{3}\right)^{3/2} \frac{|\cos\theta|^{1/3}}{1 + |\cos\theta|^{2/3}} \varepsilon^{3/2} = E_0 \varepsilon^{3/2} . \tag{5.30}$$

The probability $P(t)$ that the magnetisation has not been reversed by time t is given by

$$P(t) = \mathrm{e}^{-t/\tau} , \tag{5.31}$$

where τ is the characteristic reversal time. In a thermal activation process in independent entities, τ is given by (5.26) and we deduce

$$kT \ln \frac{\tau}{\tau_0}^{2/3} = E_0^{2/3} \left(1 - \frac{H_{\mathrm{app}}}{H_{\mathrm{SW}}}\right) . \tag{5.32}$$

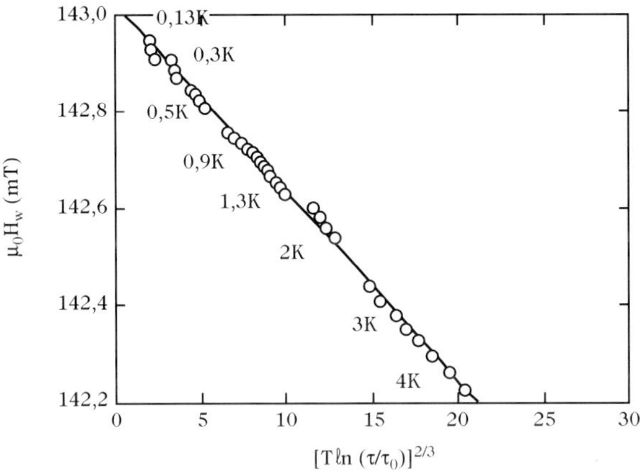

Fig. 5.9. Dependence of the reversal field H_{SW} on the variable $[T\ln(\tau/\tau_0)]^{2/3}$. Such a scaling law indicates that the relevant activation phenomena are purely those of thermal activation. Taken from [14]

From an experimental standpoint, we seek the characteristic reversal time τ at each temperature T and for a series of values of the applied field. The applied field associated with a given τ is called the waiting field H_{w}. For cobalt nanoparticles, H_{w} is plotted in Fig. 5.9 as a function of $[T\ln(\tau/\tau_0)]^{2/3}$. The experimental data lie on a universal curve obtained for the value $\tau_0 = 3\times10^{-9}$ s and the particle volume is deduced to be $(25\,\mathrm{nm})^3$, in good agreement with electron microscope measurements. In this case, the observed activation phenomena are therefore of a purely thermal nature.

Insofar as the magnetic states of a particle on either side of the energy barrier are not strictly orthogonal, magnetisation reversal becomes feasible by macroscopic tunnel effect through the barrier. The first confirmation of such effects was obtained for molecular clusters comprising around ten atoms [15]. In an assemblage of nanoparticles, it is difficult to establish the occurrence of reversal by tunnel effect in an unambiguous manner. Let us once again consider a single particle. The experimental approach used to demonstrate the occurrence of a reversal of magnetisation in a single particle via tunnelling exploits the fact that there is greater mixing of the states on either side of the barrier when the angle between the applied field and the easy axis is increased. The theoretical angular dependence of the temperature at which quantum effects become equal to thermal effects is shown in Fig. 5.10. Barium ferrite particles of diameter 10 nm, comprising some 10^5 atoms, follow the theoretical curve very closely [16].

In metallic particles, experiments carried out so far have never been able to establish the occurrence of reversal by the macroscopic tunnel effect. The absence of quantum effects is attributed to coupling between the particles and

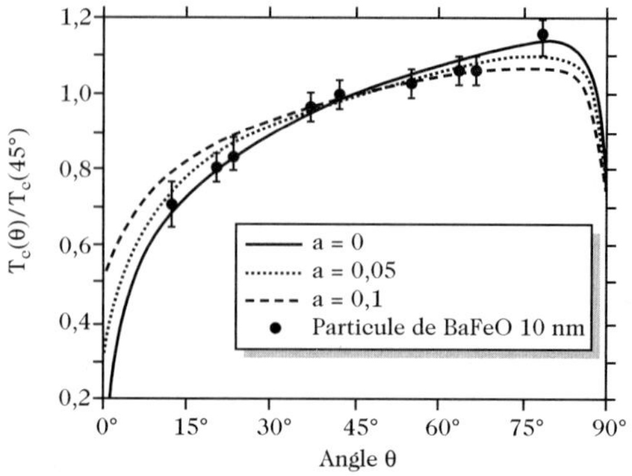

Fig. 5.10. Experimental demonstration of reversal by macroscopic tunnel effect in BaFeO nanoparticles. The experimental angular dependence of the temperature at which quantum effects become equal to thermal effects is in close agreement with the theoretical prediction. Taken from [14]

the matrix, in which process conduction electrons serve as intermediaries, and the effective size of the objects is then too great.

5.6 Magnetism in Coupled Nanosystems

In this section we shall be concerned with the third category of magnetic nanomaterials, i.e., those made up of nanostructured systems, with macroscopic dimensions, themselves made up of exchange-coupled magnetic nanoparticles. Coupling between ferromagnetic nanoparticles leads to a whole range of novel behaviour: ultrasoft magnetism, enhanced remanence, and the so-called exchange spring effect. The exchange bias effect, or unidirectional anisotropy, is characteristic of ferromagnetic systems coupled with an antiferromagnetic matrix.

5.6.1 Exchange-Coupled Nanocrystals. Ultrasoft Materials and Enhanced Remanence

To investigate the behaviour of coupled nanoparticles, we consider an assemblage of magnetic nanocrystals. These are assumed to be identical, randomly oriented, and characterised by the anisotropy constant K. Further, the same constant $A > 0$ is assumed to represent the exchange interactions within each nanocrystal and between nanocrystals. We seek magnetic configurations defined on small length scales and we can therefore neglect dipole interactions. Exchange interactions favour parallel coupling of all moments in the system, while magnetic anisotropy favours their alignment along the easy magnetisation axis of each nanocrystal. Since the exchange interactions are much stronger than those of the crystalline field, one might think that they would dominate here. However, the nanocrystals are coupled together by their surfaces alone, so that the exchange energy is involved as a surface term, whereas the anisotropy energy is a bulk term as usual. For a volume V containing N nanoparticles, the average anisotropy $\overline{K}$, defined statistically, is given by

$$\overline{K} = \frac{K}{\sqrt{N}} = \frac{Kv^{1/2}}{V^{1/2}} , \tag{5.33}$$

where v is the volume of one nanoparticle. The characteristic length for magnetisation reversal for a given $\overline{K}$ can be considered equal to the thickness of the associated wall, i.e., $\delta_{\overline{K}} = \pi\sqrt{A/\overline{K}}$ (see Sect. 5.3.1). Hence, $\overline{K}$ depends on V through N, while V depends on $\overline{K}$ through $\delta_{\overline{K}}$. The correlation volume V_{corr} is obtained self-consistently by writing $V_{\mathrm{corr}} = \delta^3_{\overline{K}}$. This yields [17]

$$V_{\mathrm{corr}} = \frac{A^6}{K^6 v^3} , \tag{5.34}$$

and the average anisotropy $\overline{K}$ is a function of the ratio D/δ between the particle diameter D and the wall thickness δ:

$$\overline{K} = K\frac{D^6}{\delta^6} . \tag{5.35}$$

For a material with Curie temperature of the order of 1 000 K, we have $A \approx 1$–2×10^{-11} J/m. In transition metal magnetic alloys such as Fe–Si or Fe–Ni, we have $K \approx 10^4\ \mathrm{J/m^3}$. For $v = 10^3\ \mathrm{nm^3}$, the correlation volume deduced from (5.34) is macroscopic, of the order of $1\ \mathrm{cm^3}$. The average anisotropy deduced from (5.35) is of the order of $10^2\ \mathrm{J/cm^3}$. This is extremely small, because averaged over a very large number of particles.

For iron alloys, the dependence of the coercivity on the particle size is shown in Fig. 5.11 [18]. At micrometric scales, a reduction in particle size is accompanied by an increase in average coercivity. This is a classic metallurgical phenomenon which is not our concern here. However, an extremely fast drop in coercivity occurs at nanometric dimensions ($D \approx 10$ nm). Since the coercivity is basically due to anisotropy (see Sect. 5.3.2), this result may be considered as the signature of a significant drop in effective anisotropy. Indeed, the experimentally observed decrease in coercivity obeys a D^6 dependence, agreeing with (5.35).

The residual coercivity of this type of system is due to an effect that we have neglected up to now, namely, magnetostriction. The material distorts to minimise the interactions of the crystalline field. This deformation is then itself a source of anisotropy and a secondary coercivity. Optimised nanocrystallised alloys such as FeSiBNbCu or FeZrB are actually made up of magnetic nanocrystals within an amorphous matrix which is itself magnetic. Magnetostriction is positive in nanocrystals but negative in the amorphous phase, so that the two compensate one another.

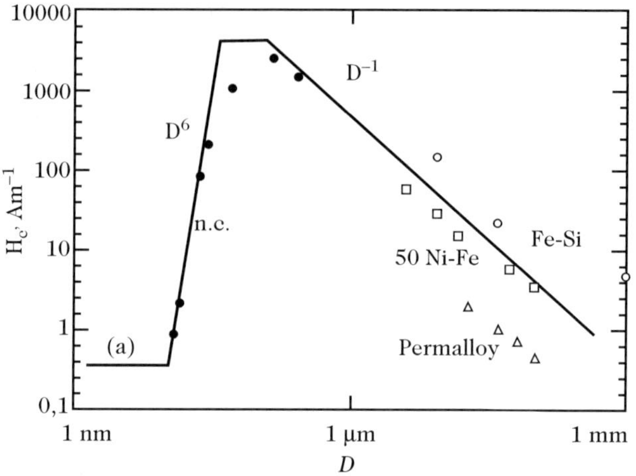

Fig. 5.11. Dependence of the coercivity of iron alloys on the size of particles of which they are composed. Note the significant decrease in coercivity at small dimensions, reflecting the fact that the magnetic moments are exchange-coupled over distances well above the particle size. Taken from [18]

In intermetallic rare earth–transition metal compounds, the magnetocrystalline anisotropy can reach values of the order of $K \approx 10^7\,\mathrm{J/m^3}$. The correlation volume deduced from (5.34) is of the order of $10^3\,\mathrm{nm}^3$, similar to the size of the individual nanocrystals. The moments within each crystal are thus aligned along the local easy axis of magnetisation. The lowering of the anisotropy resulting from parallel alignment of all the moments over a volume much greater than the volume of the nanoparticles is no longer effective. The characteristic magnetocrystalline anisotropy remains high in such systems and very large coercivities are obtained. These materials are used to make high-performance permanent magnets.

To minimise the exchange interactions at the interface between nanocrystals, magnetic moments close to the particle surface tend to adopt an intermediate alignment between the local easy axes of magnetisation. For this reason, the remanent magnetisation M_r, corresponding to saturation of all particles in the hemisphere favoured by the field, is greater than the value $M_\mathrm{s}/2$ expected for full alignment along the easy axes of magnetisation. This is the phenomenon known as enhanced remanence, used to improve the properties of magnets, in general proportional to M_r^2 (see Fig. 5.12) [19].

5.6.2 Coercivity in Nanocomposites

Some nanocrystallised alloys are heterogeneous, associating magnetically soft particles with other hard particles. When the size of the nanocrystals is small enough, these materials exhibit a coercivity $\mu_0 H_\mathrm{c}$ of the order of 0.5 T (see, for

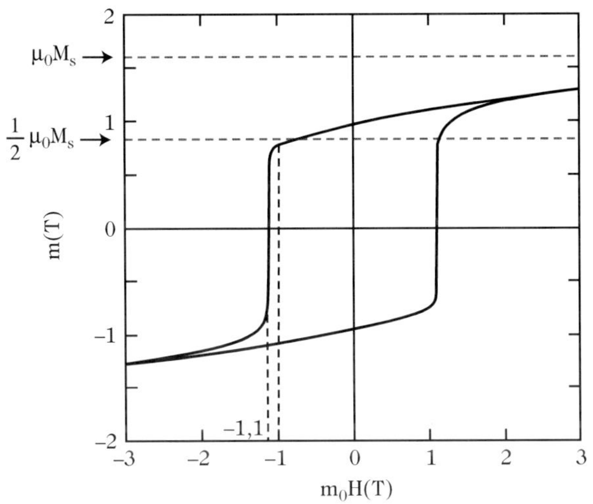

Fig. 5.12. Enhanced remanence in NdFeB-type nanostructured alloys. The remanent magnetisation is greater than the value $0.5M_s$, indicating that moments close to the interfaces between particles tend to share a common average direction

example, the NdFeB system in Fig. 5.13 [20]). To describe this phenomenon, we assume that the applied field is weaker than the coercive field of the hard particles, so that the magnetisation of these particles remains frozen. Due to exchange coupling between soft and hard particles at the interface, the reversal of magnetisation in the soft particles requires a magnetisation configuration characterised by progressive rotation of the moments in these particles, very similar to what happens in a domain wall. However, this rotation of the moments can only occur over the size of these particles, of the order of 10–20 nm, well below the equilibrium thickness of a domain wall. As already explained in Sect. 5.3.1, the energy of a confined wall is greater than the energy of a wall with the equilibrium thickness. The soft phase therefore resists the formation of a confined wall and hence becomes coercive. The experimental value of the coercive field is in qualitative agreement with the theoretical value [21].

When this type of material was discovered, it was hoped that new families of hard magnetic materials might be produced, having a strong remanent magnetisation by virtue of the contribution of the soft iron-rich phases and exhibiting a satisfactory coercivity. The properties so far obtained have not lived up to expectations. Materials in which the easy magnetisation axes of hard particles are assumed to be oriented in some common direction appear to exhibit better properties than those so far achieved. Recent preparations of this type of material are quite promising [22].

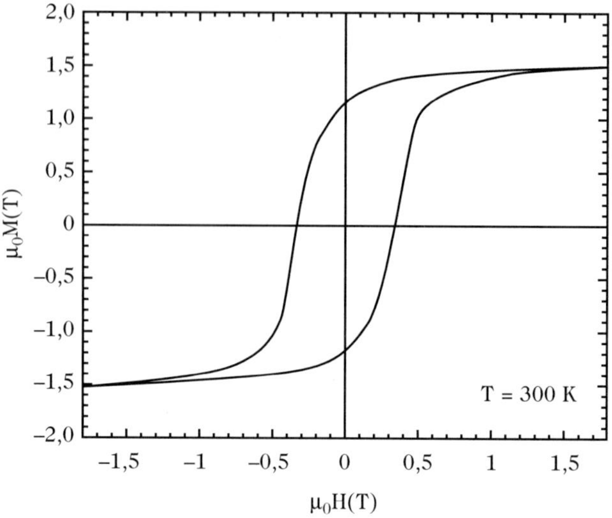

Fig. 5.13. Hysteresis cycles of an NdFeB nanocomposite alloy. This material is made from a mixture on the nanometric scale of a hard phase $Nd_2Fe_{14}B$ with two soft phases, Fe and Fe_3B. There is no room for a magnetic domain wall to develop within the soft material, which thus has a very high coercivity

5.6.3 Exchange Bias in Systems of Ferromagnetic Nanoparticles Coupled with an Antiferromagnetic Matrix

In order to achieve independent recording of data on each bit, materials used for magnetic recording are made from decoupled nanoparticles. It is not known how to control the orientations of the particles, and the direction of the moments fluctuates from one nanoparticle to another. To reduce read noise, one bit of information comprises some 1 000 particles, and the average magnetisation direction is then much better defined than that of the individual particles. At today's densities of 20 Gbits/cm^2, the bit size is around $0.2 \times 0.1\,\mu m^2$ and the volume of the particles is around 100 nm^3. The very small size of the particles raises the question of stability for the recorded data. For a characteristic fluctuation time of 100 yr and assuming that $\tau_0 \approx 10^{-9}$ s, the barrier height must be of the order of 40 kT [see (5.26)] and it follows that $K = 10^6\,J/m^3$, a very high level of anisotropy.

There are various ways of pushing back the superparamagnetic limit. One of these is to find a way of preparing materials in which the magnetisation directions of all nanocrystals are aligned. A single crystal could then play the role of one data bit.

Another way of increasing the blocking temperature is illustrated by the behaviour of cobalt nanoparticles with diameters of 3–4 nm [23]. Embedded in a non-magnetic carbon or aluminium oxide matrix, the nanoparticles exhibit their usual behaviour, becoming superparamagnetic as soon as the

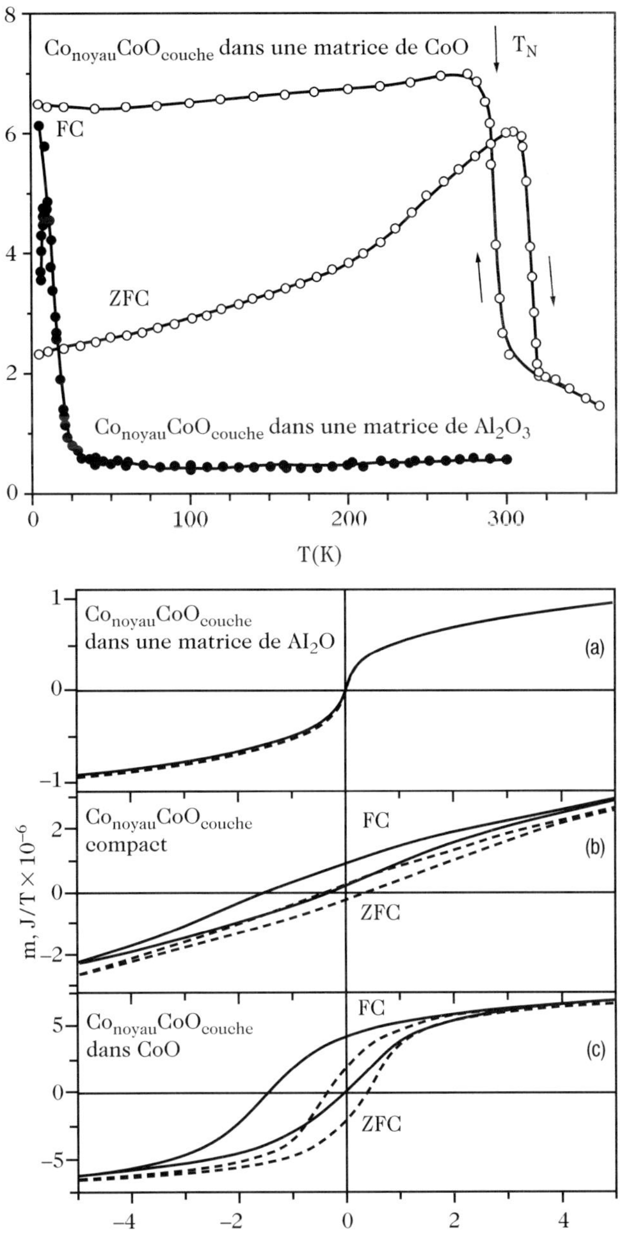

Fig. 5.14. Exchange bias in Co nanoparticles interacting with CoO. Taken from [23]. Suppression of superparamagnetism for ferromagnetic cobalt nanoparticles within an antiferromagnetic cobalt matrix. *Top*: Temperature dependence of remanent magnetisation. *Bottom*: Hysteresis cycles. FC = field cooling, ZFC = zero field cooling

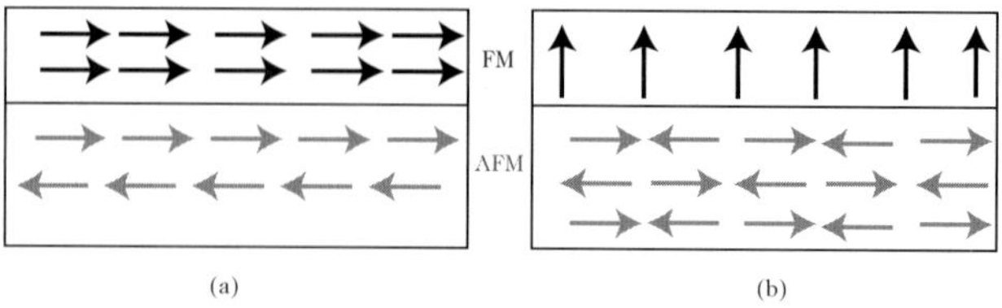

Fig. 5.15. Exchange-bias coupling mechanism. (**a**) Uncompensated interface. (**b**) Compensated interface. FM = ferromagnetic, AFM = antiferromagnetic

temperature goes above 30 K. The same nanoparticles no longer become superparamagnetic when they are embedded in an antiferromagnetic matrix (see Fig. 5.14a). At the same time, it is observed that the hysteresis cycle is shifted along the axis of the fields (see Fig. 5.14b).

It has long been known that there is coupling at the interface between a ferromagnetic material and an antiferromagnetic material [24]. This is the so-called exchange bias mechanism. The picture usually given is shown schematically in Fig. 5.15a. The moments of a single sublattice of the antiferromagnetic material are located at the interface and this results in a fundamental asymmetry in the coupling. A particular orientation of the ferromagnetic moments is favoured (towards the right in Fig. 5.15a). The mechanism explains the shift in the hysteresis cycles, as observed in Fig. 5.14b. However, the coupling amplitude calculated in this model is much greater than what is actually observed experimentally. In the general case, there is no reason why the number of moments in each sublattice of the antiferromagnetic material should not be roughly the same. We can therefore propose an opposite extreme hypothesis, in which we assume that the interface is 'compensated'. In this case, coupling occurs by distortion of the antiferromagnetic structure at the interface, as shown in Fig. 5.15b. There is then an extra energy term and it provides a very effective barrier against fluctuations in the magnetic moment. In a schematic way, the coupling is the source of a very significant enhancement of the anisotropy of the ferromagnetic particles. The calculated coercive field is close to what is observed experimentally. The use of antiferromagnetic matrices may thus provide a way of producing recording densities greater than the upper limits that can be reached by current methods.

References

1. W.A. de Heer, P. Milani, A. Châtelain: Spin relaxation in small free iron clusters, Phys. Rev. Lett. **65**, 488 (1990)
2. J.P. Bucher, D.C. Douglas, L.A. Bloomfield: Phys. Rev. Lett. **66**, 3052 (1991); S.N. Khanna, S. Linderoth: Magnetic behavior of clusters of ferromagnetic transition metals, Phys. Rev. Lett. **67**, 743 (1991)

3. G.M. Pastor, J. Dorantes-Davila, K.H. Bennemann: Size and structural dependence of the magnetic properties of small 3d-transition-metal clusters, Phys. Rev. Lett. **40**, 7642 (1989)
4. A.J. Cox, J.G. Louderback, L.A. Bloomfield: Experimental observation of magnetism in rhodium clusters, Phys. Rev. Lett. **71,** 924 (1993)
5. K.W. Edmonds, C. Binns, S.H. Baker, S.C. Thornton, C. Norris, J.B. Goedkoop, M. Finazzi, N.B. Brookes: Phys. Rev. **60**, 472 (1999)
6. P. Bruno: Physical origins and theoretical models of magnetic anisotropy, in: *Magnetismus von Festkörpern und grenzflächen*, published by P.H. Dederichs, P. Grünberg, and W. Zinn, 24 IFF-Ferienkurs, Forschungszentrum Jülich P, 24-.& -24.28 (1993)
7. J.M.D. Coey: Phys. Rev. Lett. **27**, 1140 (1971)
8. R.H. Kodama, S.A. Makhlouf, A.E. Berkowitz: Finite size effects in antiferromagnetic NiO nanoparticles, Phys. Rev. Lett. **79**, 1393 (1997)
9. P.V. Hendriksen, S. Linderoth, P.A. Lindgärd: Finite-size modifications of the magnetic properties of clusters, Phys. Rev. **48**, 259 (1993)
10. L. Néel: Anisotropie magnétique superficielle et surstructures d'orientation, J. Phys. Rad. **15**, 225 (1954)
11. B.J. Hickey, M.A. Howson, D. Greig, N. Wiser: Enhanced magnetic anisotropy energy density for superparamagnetic particles of cobalt, Phys. Rev. **53**, 32 (1996)
12. C. Kittel: Rev. Mod. Phys. **21**, 541 (1949)
13. W. Wernsdorfer, E. Bonet Orozco, K. Hasselbach, A. Benoit B. Barbara, N. Demoncy, A. Loiseau, D. Boivin, H. Pascard, D. Mailly: Phys. Rev. Lett. **78**, 1791 (1997)
14. W. Wernsdorfer: Classical and quantum magnetization reversal studied in nanometer-sized particles and clusters, Adv. Chem. Phys. **118**, 99 (2001)
15. J.R. Friedman, M.P. Sarachik, J. Tejada, R. Ziolo: Phys. Rev. Lett. **76**, 3830 (1996); L. Thomas, F. Lionti, R. Ballou, D. Gatteschi, R. Sessoli, B. Barbara: Nature **383**, 145 (1996)
16. W. Wernsdorfer, E. Bonet Orozco, K. Hasselbach, A. Benoit, D. Mailly, O. Kubo, H. Nakano, B. Barbara: Phys. Rev. Lett. **79**, 4014 (1997)
17. G. Herzer: IEEE Trans. Mag. **25**, 3327 (1989)
18. R. Schäfer, A. Hubert, G. Herzer: J. Appl. Phys. **69**, 5325 (1991)
19. R.W. Mc Callum, A.M. Kadin, G.B. Clemente, J.E. Keen: J. Appl. Phys. **61**, 3577 (1987)
20. R. Coehoorn, D.B. de Mooij, J.P.W.B. Duchateau, K.H.J. Buschow: J. Phys. (Paris), Colloq. **49**, C8-669 (1988); R. Coehoorn, D.B. de Mooij, D. de Waard: J. Magn. Magn. Mat. **80**, 101 (1989)
21. E. Kneller, R. Hawig: IEEE Trans. Mag. **27**, 3588 (1991)
22. D. Lee, J.S. Hilton, S. Liu, Y. Zhang, G.C. Hadjipanayis, C.H. Chen: Hot-pressed and hot-deformed nanocomposite $(Nd,Pr,Dy)_2Fe_{14}B/\gamma$-Fe based magnets, IEEE Trans. Mag. **39**, 2947 (2003)
23. V. Skumryev, S. Stoyanov, Yong Zhang, G.C. Hadjipanayis, D. Givord, J. Nogués: Beating the superparamagnetic limit with exchange bias, Nature **423**, 850 (2003)
24. W.H. Meiklejohn, C.P. Bean: New magnetic anisotropy, Phys. Rev. **102**, 1413 (1956)

Books and Review Articles

25. E. de Lacheisserie (Ed.): *Magnétisme* (2 volumes), Collection Grenoble Sciences, EDP Sciences (2000)
26. P. Bruno: Physical origins and theoretical models of magnetic anisotropy, in: *Magnetismus von Festkörpern und grenzflächen*, published by P.H. Dederichs, P. Grünberg, W. Zinn, 24 IFF-Ferienkurs, Forschungszentrum Jülich P, 24-.& -24.28 (1993)
27. X. Batlle, A. Labarta: Finite-size effects in fine particles: Magnetic and transport properties, J. Phys. D: Appl. Phys. **35**, R15–R42 (2002)
28. W. Wernsdorfer: Classical and quantum magnetization reversal studied in nanometer-sized particles and clusters, Adv. Chem. Phys. **118**, 99–190 (2001)
29. M.E. McHenry, D.E. Laughlin: Nanoscale materials development for future magnetic applications, Acta Mater. **48**, 223–238 (2000)

6

Electronic Structure in Clusters and Nanoparticles

F. Spiegelman

6.1 Introduction

The physics of clusters or nanoparticles occupies an intermediate position between atomic and molecular physics and condensed matter physics. The study of nanoparticles is in part motivated by the various types of application, ranging from the development of novel nanoscale materials for electronics, photonics or magnetism, to applications in chemistry (e.g., catalysis), atmospheric physics (e.g., droplets, carbon-bearing particles), or astrophysics (e.g., silicate- or carbon-bearing grains). However, a deeper understanding of the fundamental properties of matter remains one of the key issues. Indeed, nanoparticles bridge the gap between the properties of isolated atoms or small molecules and the properties of condensed matter, wherein their behaviour can be followed as a function of the number of constituents, from a few units up to a hundred thousand atoms. A whole range of properties are relevant to this fundamental quest, including questions of structure, spectroscopy, magnetism, dynamics, thermodynamics, and mechanics. In this context, electronic structure is the heart of the matter because, although it is not always made explicit, it directly or indirectly determines all these properties through the complex interaction resulting from electrostatic and sometimes electromagnetic forces between atomic nuclei and electrons. Hence, electronic structure gives rise to the various types of binding and ensures the stability of the resulting edifice. We may classify the different types rather summarily in the following groups: covalent binding, metallic binding, ionic binding, van der Waals binding, and hydrogen binding. Although well understood in molecules and in bulk solids, these can assume novel forms in clusters and nanoparticles.

These properties do not evolve monotonically with size. Nanoparticles are obviously governed by the same interactions as periodic systems but display their own specific features. As the size decreases from the macroscopic limit towards nanoparticles and clusters, properties first exhibit a continuous evolution related to the growing importance of the surface and possibly the shape of the nanoparticle (close-packed, cubic, facetted) without qualitative change in the electronic structure. In this size range, properties can be described using scaling laws. Of course, the electronic structure at the surface depends on the different coordinations of the the atoms there, but also on hybridisation and electrostatic conditions that differ from those in the bulk.

Below a critical size, properties become more likely to change in the bulk itself, thereby affecting the cluster as a whole. On the structural level, the icosahedral transition is well known. Size-dependent transitions also occur for the electronic structure. This can be exemplified by the metal–insulator transition in clusters of divalent metals, induced by the separation of the s and p energy bands at small sizes. When the number of atoms or constituents is decreased further, one ends up in a regime where properties are reconstructed at each new size. At this point, a discrete form of behaviour is superposed on the scaling laws. This is the size range characterised by finite size effects. These are due partly to the discrete nature of the atomic structure, which must now be taken into account explicitly, and partly to the quantum nature of the electrons participating in the chemical binding. In covalent systems, binding is ensured by electron delocalisation among several atoms. This delocalisation may sometimes be more extensive, as in the π systems, even extending to the whole cluster, as in the case of delocalised electrons in metals. At small sizes, electronic and geometrical properties can become highly interdependent. Ionisation or addition of an electron is likely to have repercussions for the structure. An external perturbation in the form of an electric or magnetic field, or some specific type of environment, e.g., solvent, matrix, can also strongly perturb or even determine the properties of such a system. Another feature here is that classical concepts as simple as volume or area can no longer be clearly defined owing to the spatial extension of the electron wave function in quantum mechanics. This spatial extension also leads to the tunnel effect, so important in near-field microscopy and manipulation, techniques which have underscored the rise of nanoscience. In addition to all this, the energy levels, which group into bands in periodic solids, become discrete in confined systems. Hence, small metallic clusters always exhibit a finite gap between the highest occupied molecular orbital (HOMO) and the lowest unoccupied molecular orbital (LUMO).

In the size range where each size becomes a case in its own right, interpolation and extrapolation of physical properties are of little use. At certain specific sizes, entities may display extreme properties or even adopt geometric and electronic configurations which just do not exist in molecular matter or bulk solids. The remarkable stability of the fullerene C_{60} is a case in point. Discovered by Smalley [136], this molecule has a cage structure associated with a very high degree of spatial symmetry, and reveals a particular disposition for π bonds. By analysing the intensities in mass spectrometry, it has been shown that certain sizes, the so-called magic sizes, are more abundant than others. For alkali metal clusters, a decisive example was brought to light by Knight et al. [39]. The increased abundances for the magic numbers $n = 2$, 8, 20, 40, 58, etc., are shown in Fig. 6.1. This sequence, associated with increased

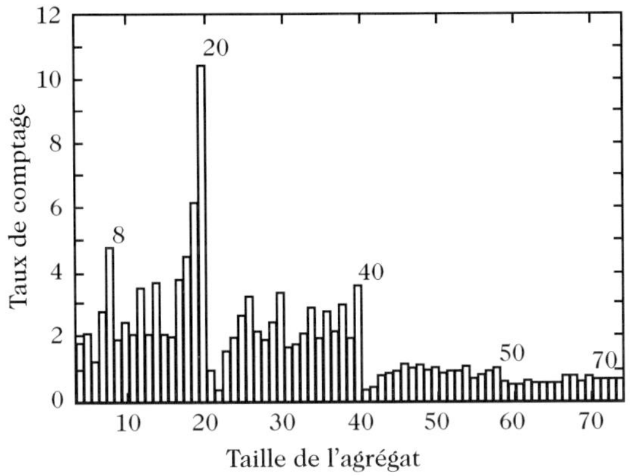

Fig. 6.1. Mass spectrum of sodium clusters in a molecular beam. Taken from Knight et al. [39]

stability of clusters involving certain well defined numbers of electrons, indicates a structuring of the electron shells and establishes an analogy with the properties of superatoms or the shell structure of atomic nuclei. In this context, electronic properties are determined by the quantum nature of a system of fermions confined within a spherical potential well. This novel electronic structure plays a key role in the stability, fragmentation, optical properties, or reactivity of metallic particles.

The present chapter aims to initiate the reader in the description of electronic structure in clusters and nanoparticles. We shall begin by presenting the classical liquid-drop model, and introducing the scaling laws. The classical model provides a reasonable description of the stability of simple metallic clusters in which electrons are essentially delocalised with relatively uniform density, and interactions between atoms are screened and largely non-directional. This model can also be used to investigate charge stability. As already mentioned, the correlation between experimental data and electronic or structural properties is not always simple or one-to-one, in particular when finite size effects become relevant. This is all the more true given that microclusters can adopt a variety of structural configurations, often very different from those of the crystal, and which must be redetermined for each new size. Clusters can also exhibit a rather different kind of electronic structure to those encountered in saturated molecules typical of chemical binding. For example, one finds hypervalent situations, but also the opposite case of frustated, unsaturated or dangling bonds at the surface. This non-standard electronic structure may be responsible for finite size effects, leading sometimes to unexpected geometrical configurations and/or close-run energy competition between a great many possible isomers.

In this context, theory and simulation are of the utmost importance, especially when it comes to electronic structure. Joint progress in computing,

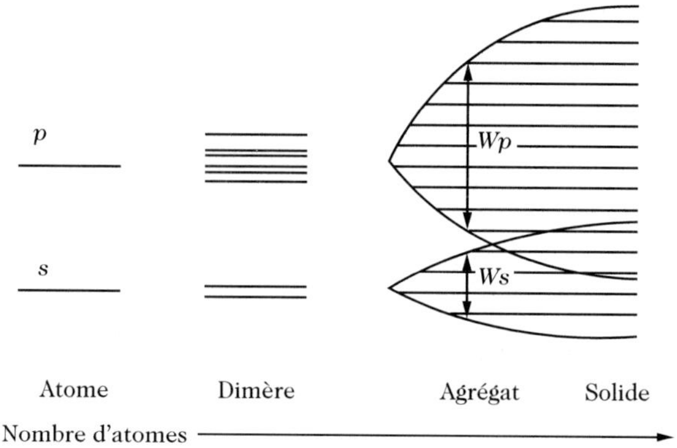

Fig. 6.2. Schematic variation of the structure of the s and p electronic shells from an atom to the bulk solid, exemplified by the alkali and alkaline earth metals. W_s and W_p are the band widths

theory and calculation techniques mean that electronic structure can be determined ab initio, i.e., from first principles, for an ever wider range of increasingly complex systems, by directly solving the equations of quantum mechanics without introducing empirical parameters. This approach has been the source of many computational codes for electronic structure, developed by solid-state physicists and quantum chemists. Although not specifically designed to deal with nanoparticles, certain codes can now be used to tackle non-periodic systems containing more than a hundred atoms (at least for calculating the electronic structure in a given geometric configuration). Certain versions are also integrated into dynamical codes. The advantage with ab initio calculations is not only quantitative. When used to analyse data concerning small or periodic systems, they often provide a way of building appropriate approximate models which make calculation and simulation possible in size ranges that cannot be tackled directly. In the field of electronic structure and molecular simulation, numerical experimentation has become an indispensable complement to laboratory experimentation using real systems.

Section 6.3 discusses the concepts and the main methods used to determine electronic structure, with an introduction to the explicit methods used in quantum chemistry and a brief presentation of the density functional theory which has become the favoured technique for exploring structural and electronic properties in nanosystems. Section 6.3.5 describes some approximate methods that prove useful for exploring nanosystems containing a large number of atoms or for carrying out intensive sampling in molecular dynamics simulations, going beyond the current possibilities of ab initio methods. The aim in Sect. 6.4 is to describe the electronic structure of typical species: metallic nanoparticles, van der Waals clusters, ionic clusters, and carbon nanotubes and nanoparticles. Section 6.5 is devoted to valence transitions, i.e., qualitative changes in electronic structure as a function of size (see Fig. 6.2). Finally,

Sect. 6.7 reviews some current subjects of interest and possibilities for future development.

A selection of reference books which discuss in greater detail the electronic structure of clusters, nanoparticles, and nanosystems is given at the beginning of the reference list at the end of the chapter [1–8].

6.2 Liquid-Drop Model

The simplest approach to describing variations in the physical properties of a finite system is the liquid-drop model [31,32,35–38], which takes into account the different classical energy contributions. This type of model, used to provide a simple description of the stability of the nucleus in nuclear physics, has been transposed to metallic clusters, reflecting the analogies exhibited by the properties of matter when confined to a finite volume, electrons within clusters, or nucleons in the nucleus. This analogy has a quantum extension in the jellium model, used to describe the properties of finite fermion systems. The classical liquid-drop model accounts for average and continuous variations of certain properties as the system size decreases.

The classical liquid-drop model extrapolates from the macroscopic properties of the condensed phase and those of surfaces using the laws of proportionality. Let a_{v} be the energy per atom in the macroscopic solid, i.e., minus the binding energy ε_{b} per atom:

$$a_{\mathrm{v}} = -\varepsilon_{\mathrm{b}} = \lim_{n \to \infty} \frac{E(n) - nE(1)}{n} .$$

The energy per unit volume of a cluster of n atoms is thus

$$E_{\mathrm{v}}(n) = a_{\mathrm{v}} n .$$

This contribution to the energy is negative and the stability of the solid for large n implies that $E(n) < nE(1)$. Making the approximation that the system is spherical, a reasonable assumption for simple metal clusters of fairly large dimensions, and assuming also that each atom occupies the same volume in the cluster as in the solid, the radius of the cluster is given by

$$R_0 = r_0 n^{1/3} .$$

In alkali metal clusters, for example, r_0 is the Wigner–Seitz radius, corresponding to the average volume occupied by a valence electron in the solid. However, a spherical particle has a surface area of $4\pi R_0^2 = 4\pi r_0^2 n^{2/3}$. The binding energy of each surface atom is lower, having lower coordination than atoms in the bulk. Inserting the data for a plane surface with surface tension τ_{s}, an energy deficiency given by

$$E_{\mathrm{s}} = \tau_{\mathrm{s}} 4\pi r_0^2 n^{2/3} = a_{\mathrm{s}} n^{2/3}$$

can thus be associated with surface atoms, where a_s is the binding energy deficiency of a surface atom. This contributes positively to the energy. One must also take into account the curvature $1/R_0$ of the cluster surface, which further reduces the coordination of surface atoms and hence their binding energy. An energy deficiency proportional to the curvature and the number of surface atoms is also associated with this contribution:

$$E_c = a_c n^{1/3} ,$$

where a_c is also a positive coefficient. It is clear that this droplet model cannot be transposed to highly non-uniform nanoparticles, nor to typical 1D or 2D systems, such as very small carbon clusters, nor indeed to the fullerenes, which are hollow clusters. However, in many cases, other scaling laws can be formulated with the help of parameters such as the coordination number, the distance to nearest neighbours, or the widths of bands of electronic levels.

Let $E(n, Q)$ be the energy of a system of size n and charge Q. For the liquid drop, the energy $E(n, 0)$ of a neutral system can then be expanded in powers of $n^{1/3}$, viz.,

$$E(n, 0) = a_v n + a_s n^{2/3} + a_c n^{1/3} ,$$

from which the binding energy is immediately deduced to be

$$e_b(n) = -\frac{E(n, 0) - nE(1, 0)}{n} .$$

Furthermore, the dissociation energy D_n^p corresponding to fragmentation into two subunits of arbitrary sizes p and $n - p$ is then given by

$$D_n^p = -E(n, 0) + E(n - p, 0) + E(p, 0) .$$

Note that, in this model, the binding energy per atom $e_b(n)$ and also the dissociation energy D_n^1 required for the evaporation of one atom both increase monotonically with the size.

This type of expansion can be extended to the electrostatic energy in the case of charged metallic nanoparticles, treated as classical conductors, i.e., with surface charge. If Q electrons are removed from a solid surface, the energy required is QW, where W is the work function of the solid. However, in the case of a spherical nanoparticle, one must add to this work function the electrostatic contribution corresponding to the classical Coulomb energy of a charged conducting sphere of radius R, viz.,

$$E_{\text{Coul}} = \frac{1}{2}\frac{Q^2}{R_0} = \frac{1}{2}\frac{Q^2}{r_0} n^{1/3} .$$

The total energy can then be expanded as a series in $R_0 = r_0 n^{1/3}$. Taking the energy of the dissociated atoms as zero,

$$E(n,q) = a_{\rm v}n + a_{\rm s}n^{2/3} + a_{\rm c}n^{1/3} + QW + a_q Q^2 r_0^{-1} n^{-1/3} .$$

Naturally, the coefficients depend on the elements. The classical value of a_q is 1/2. A further term $c_q Q r_0^{-1} n^{-1/3}$ arises owing to the quantum spatial extension of the electron density beyond the classical surface of the droplet ($c_q \approx -0.08$ for sodium) [33, 34].

The classical spherical droplet approximation is only appropriate in weakly anisotropic compact systems. However, deformations can be considered [34, 52], especially in the study of fragmentation and fission in charged systems.

The electrostatic part also gives an approximation to the ionisation potentials $I(n,Q)$ and the electron affinities $A(n,Q)$:

$$I(n,Q) = E(n,Q+1) - E(n,Q) = W + \left[a_q(2Q+1) - c_q\right] r_0^{-1} n^{-1/3} ,$$

$$A(n,Q) = E(n,Q) - E(n,Q-1) = W + \left[a_q(2Q-1) + c_q\right] r_0^{-1} n^{-1/3} .$$

6.3 Methods for Calculating Electronic Structure

The quantum description begins by solving the Schrödinger equation. In this context, the nanoparticle is treated, like a molecule, as a quantum system comprising n nuclei labelled by a and N electrons labelled by i. The nuclei are characterised by their mass M_a, their charge $z_a e$, and their coordinates $\boldsymbol{R}_a$ ($R_{1a} = x_a, R_{2a} = y_a, R_{3a} = z_a$), and the electrons by their mass m, their charge $-e$, and their coordinates $\boldsymbol{r}_i$ ($r_{1i} = x_i, r_{2i} = y_i, r_{3i} = z_i$). We shall also use the generic notation $\boldsymbol{R} = \{\boldsymbol{R}_a\}$ for the nuclei and $\boldsymbol{r} = \{\boldsymbol{r}_i\}$ for the electrons. In the following, we shall basically be presenting the time-independent treatment of the electron problem.

Unless otherwise indicated, we use the atomic system of units in which $m = 1$, $e = 1$, $\hbar = 1$, and $4\pi\epsilon_0 = 1$. The unit of length is the Bohr radius $a_0 = 0.529\,167$ Å, the unit of energy or hartree is twice the ionisation potential of the hydrogen atom, i.e., 1 hartree = 2 rydberg = 27.211 395 7 eV, or in wave number, 219 474.625 cm^{-1}.

The total Hamiltonian of the system is

$$H(\boldsymbol{R},\boldsymbol{r}) = -\sum_{a=1}^{n} \frac{1}{2M_a}\Delta_a + H_{\rm el}(\boldsymbol{R},\boldsymbol{r}) + \sum_{a<b}^{N} \frac{Z_a Z_b}{|\boldsymbol{R}_a - \boldsymbol{R}_b|} .$$

It includes the kinetic energy $T_{\rm n}$ of the nuclei, expressed in quantum mechanics by means of Laplacian operators Δ_a with respect to nuclear coordinates $\boldsymbol{R}_a$, the electron Hamiltonian $H_{\rm el}$, and the Coulomb interaction energy between nuclei, denoted by $V_{\rm nn}$. The electron Hamiltonian

$$H_{\rm el}(\boldsymbol{R},\boldsymbol{r}) = \sum_{i=1}^{N}\left(-\frac{1}{2}\Delta_i - \sum_{a=1}^{n}\frac{Z_a}{|\boldsymbol{r}_i - \boldsymbol{R}_a|}\right) + \sum_{i<j}^{N}\frac{1}{|\boldsymbol{r}_i - \boldsymbol{r}_j|} = T_{\rm e} + V_{\rm en} + V_{\rm ee}$$

includes the kinetic energy of the electrons

$$T_{\mathrm{e}} = -\sum \frac{1}{2}\Delta_i \, ,$$

the Coulomb interaction V_{en} between electrons and nuclei, and the electron–electron interaction V_{ee}.

6.3.1 Born–Oppenheimer Approximation. Surface Potential

The Born–Oppenheimer approximation takes advantage of the mass difference $M_a \gg m$ to decouple the motion of the nuclei with coordinates $\boldsymbol{R} = \{\boldsymbol{R}_a\}$ and the motion of the electrons with space and spin coordinates $\boldsymbol{x} = \{\boldsymbol{x}_i\} = \{\boldsymbol{r}_i, \sigma_i\}$. It thus assumes that the nuclei, moving more slowly, can be treated as fixed on the time scale of the electron dynamics. This approximation is generally appropriate for the ground state and low energy excitations, provided that the nuclei remain slow-moving.

The first step in this decoupling process consists in solving the time-independent Schrödinger equation for the electrons when the nuclei have fixed positions $\boldsymbol{R}$ which are treated at this stage as parameters:

$$(H_{\mathrm{el}} + V_{\mathrm{nn}})\Psi_m(\boldsymbol{x}, \boldsymbol{R}) = E_{\mathrm{el}}^m(\boldsymbol{R})\Psi_m(\boldsymbol{x}, \boldsymbol{R}) \, .$$

It is only in the second stage of the process that we treat the motion of the nuclei. In a quantum treatment of the nuclei, the Born–Oppenheimer approximation consists in writing the global wave function in the form of a product $\chi_v(\boldsymbol{R})\Psi_m(\boldsymbol{x}, \boldsymbol{R})$ of a nuclear wave function and an electron wave function. We then obtain a Schrödinger equation for the motion of the nuclei:

$$\left[-\sum_{a=1}^{n} \frac{1}{2M_a}\Delta_a + E_{\mathrm{el}}^m(\boldsymbol{R}) \right] \chi_v(\boldsymbol{R}) = E_v^m(\boldsymbol{R})\chi_v(\boldsymbol{R}) \, .$$

This is also called the adiabatic approximation in the sense that the electron wave function adapts strictly to each geometric configuration of the atoms.

The functions $E_{\mathrm{el}}^m(\boldsymbol{R})$ represent the potential energy of the forces exerted on the nuclei in the system, either in the electron ground state Ψ_0, or in the excited electron states Ψ_m $(m > 0)$. These functions, which depend parametrically on the geometrical variables of the system (structural configuration), are the potential energy surfaces. In each geometric configuration $\boldsymbol{R}$, the forces are obtained by differentiating with respect to the nuclear coordinates, i.e.,

$$\boldsymbol{F}_a^m = -\boldsymbol{\nabla}_a E_{\mathrm{el}}^m(\boldsymbol{R}) \, .$$

The electron potential energy surfaces (see Fig. 6.3) thus determine the structural and dynamic properties in the Born–Oppenheimer approximation.

In classical dynamics simulations, the motions of the nuclei (associated with the momenta $\boldsymbol{P}_a$) are governed by Newton's equations

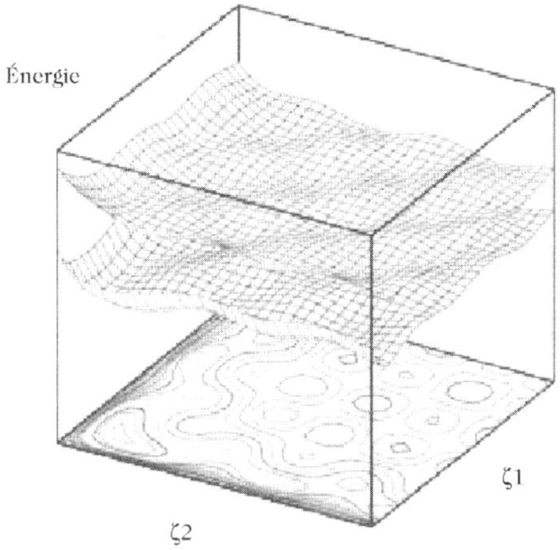

Fig. 6.3. Schematic view of electron potential energy surfaces as a function of two internal coordinates ζ_1 and ζ_2

$$\frac{\mathrm{d}\boldsymbol{P}_a}{\mathrm{d}t} = -\boldsymbol{\nabla}_a E_{\mathrm{el}}^m(\boldsymbol{R}) \,.$$

The minima of the potential energy surfaces $E_{\mathrm{el}}^m(\boldsymbol{R})$ correspond to stable equilibrium positions. For a 3D system, these are the geometric configurations in which the derivatives $\partial E_{\mathrm{el}}^m(\boldsymbol{R})/\partial R_{\mu a}$ of the potential energy with respect to the coordinates $R_{\mu a}$ vanish, and $3n-6$ eigenvalues of the matrix of second derivatives $\partial^2 E_{\mathrm{el}}^m(\boldsymbol{R})/\partial R_{\mu a}\partial R_{\nu b}$ are positive (not counting the translation and rotation modes). It is sometimes convenient to use non-Cartesian coordinate systems ζ_λ, e.g., $3n-6$ internal coordinates, or coordinates picking out certain global deformation modes, such as the average radius of the cluster, or quadrupole deformation modes.

Saddle points on the surfaces, with $3n-7$ positive second derivatives and one negative, are used to define reaction paths between stable isomers. The topology of the potential energy surface clearly plays a key role in questions of stability, isomerisation, dynamics, and thermodynamics of molecular systems, clusters, and nanoparticles (see Chap. 3).

When the system is finite rather than periodic, some differences arise because the number of electrons may play an important role. Indeed, the potential energy surface of the electronic states varies with the charge and the level of excitation. It follows that the stable geometrical structures of clusters possessing different numbers of electrons can vary significantly. The equilibrium geometries of cations ($\boldsymbol{R}_{\mathrm{eq}}^{+}$), anions ($\boldsymbol{R}_{\mathrm{eq}}^{-}$), or the neutral system ($\boldsymbol{R}_{\mathrm{eq}}$) are not generally the same. If we neglect the quantum vibrational energy of the nuclei, or zero point energy (ZPE), the ionisation potential in the so-called adiabatic limit is defined as the minimal energy required to take the system from the most stable neutral isomer to the most stable ionised isomer, i.e., the energy difference between the absolute minima of the ground state adiabatic

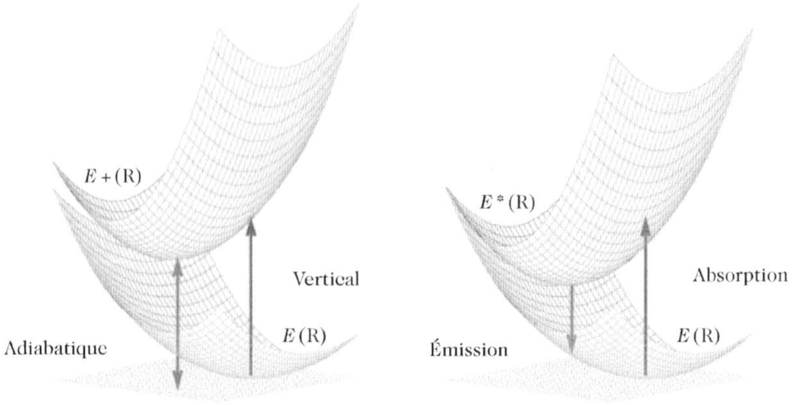

Fig. 6.4. *Left*: Definition of adiabatic and vertical ionisation potentials. *Right*: Absorption and emission processes after relaxation on the excited potential energy surface

potential surfaces $E_0^+(\boldsymbol{R}_{\mathrm{eq}})$ and $E_0(\boldsymbol{R}_{\mathrm{eq}})$ of the charged and neutral systems, respectively:

$$PI = PI_{\mathrm{A}} = E_0^+(\boldsymbol{R}_{\mathrm{eq}}^+) - E_0(\boldsymbol{R}_{\mathrm{eq}}) \; .$$

However, the vertical ionisation potential PI_{v} is usually used, corresponding to frozen geometric relaxation:

$$PI_{\mathrm{V}} = E_0^+(\boldsymbol{R}_{\mathrm{eq}}) - E_0(\boldsymbol{R}_{\mathrm{eq}}) \; .$$

This definition corresponds to the classical Franck–Condon approximation, in which the nuclei do not have time to reorganise themselves during an electron process, ionisation, or attachment of an electron. Likewise, one can define the vertical detachment energy E_{VD} from a negative ion, different from the adiabatic electron affinity. This double definition reflects the interference between electronic structure and geometric structure in finite systems, also encountered in the dynamics of excited states (see Fig. 6.4). Geometric relaxation of the system on a potential surface excited by absorption of a photon can thus lead to a different emission wavelength to the one inducing the absorption process, even in the absence of non-adiabatic processes involving a change in the electron surface. These concepts no longer necessarily apply to ultra-intense processes whose description may require a direct treatment of the electron dynamics in a non-adiabatic formalism (time-dependent treatment of the electrons).

6.3.2 Ab Initio Calculation of Electronic Structure

The ab initio approach to electronic structure in molecular systems, nanoparticles and nanostructures consists in solving the electron Schrödinger equation from first principles without introducing empirical parameters. There are two methods: explicit solution of the quantum N-body problem, approximating observables by calculating the many-particle wave function [9–13], and the density functional theory [18, 19].

If we associate with each electron, labelled by i, a single-particle space (or orbital) wave function $\varphi_i(\boldsymbol{r})$ and a spin wave function σ, we can define a spinorbital wave function $\phi_i(\boldsymbol{x})$, where $\boldsymbol{x}$ combines space and spin coordinates. For a set of electrons, the product of the spinorbitals occupied by each electron can be used to generate a many-electron wave function. However, owing to the fact that electrons are fermions, the total wave function must be antisymmetric under permutations of these indistinguishable electrons, so that they satisfy the Pauli exclusion principle. A simple product would not satisfy this requirement. The result is therefore antisymmetrised in the form of a Slater determinant:

$$\Phi = \frac{1}{\sqrt{N!}} \sum_P (-1)^p P\left[\phi_1(\boldsymbol{x}_1)\phi_2(\boldsymbol{x}_2)\dots\phi_N(\boldsymbol{x}_N)\right] ,$$

an antisymmetric linear combination of the products, summed over all possible permutations P of the N electrons in the N occupied spinorbitals ϕ_i. The exponent p indicates the sign of the permutation P. Variational solution of the many-electron Schrödinger equation with a wave function of this form leads, by minimising the energy with respect to the orbitals, to an integro-differential equation for the eigenvalues. The solution spinorbitals are eigenfunctions of a single-electron operator, the Fock operator, which describes each electron in the Coulomb field of the ionic nuclei and in the mean field of the Coulomb and exchange contributions due to the other electrons:

$$\left(-\frac{1}{2}\Delta - \sum_a \frac{Z_a}{|\boldsymbol{r}-\boldsymbol{R}_a|}\right)\phi_i(\boldsymbol{x}) + \sum_j \int \frac{\phi_j^*(\boldsymbol{x}')\phi_j(\boldsymbol{x}')}{|\boldsymbol{r}-\boldsymbol{r}'|}\phi_i(\boldsymbol{x})\mathrm{d}\boldsymbol{x}' \\ - \sum_j \int \frac{\phi_j^*(\boldsymbol{x}')\phi_i(\boldsymbol{x}')}{|\boldsymbol{r}-\boldsymbol{r}'|}\phi_j(\boldsymbol{x})\mathrm{d}\boldsymbol{x}' = \epsilon_i\phi_i(\boldsymbol{x}) .$$

This is the Hartree–Fock (HF) approximation. The HF equations can be solved iteratively in such a way as to obtain the identity between the occupied orbitals which generate the mean field part of the Fock operator and its eigenfunctions. This is the known as the self-consistent field (SCF) technique.

However, the most general many-electron wave function is a linear combination of such Slater determinants, i.e.,

$$\Psi = C_0\Phi_0 + \sum_{ik} C_i^k \Phi_i^k + \sum_{ijkl} C_{ij}^{kl}\phi_{ij}^{kl} + \cdots = \sum_K C_K \Phi_K .$$

The determinants Φ_K (where K is the generic excitation index) are obtained from the reference determinant Φ_0, generally the HF determinant, by replacing certain occupied single-electron wave functions ϕ_i, ϕ_j, etc., in the reference by unoccupied wave functions, sometimes called virtual wave functions, ϕ_k^*, ϕ_l^*, etc. One then speaks of configuration interactions (CI). These post-Hartree–Fock methods allow one to take into account electron correlations and so go beyond the mean field approximation. The wave function Ψ thus accounts for fluctuations in the occupation numbers of orbitals via the excited determinants in the configuration interaction expansion. For the ground state, the variational solution involves minimising the electron energy with respect to the coefficients C_K:

$$E_0 = \min\langle\Psi_0|H_{\rm el}|\Psi_0\rangle \ .$$

In practice, minimising the energy amounts to diagonalising the Hamiltonian matrix when it is expressed in terms of a basis of Slater determinants, assumed orthonormal. Diagonalisation of $H_{\rm el}$ also provides other solutions corresponding to excited states. However, the size of the CI matrix increases combinatorically with the number of electrons and the dimension of the single-electron wave function basis $\{\phi_i\}$. For $N = N_\alpha + N_\beta$ electrons α and β distributed over M orbitals and without considering simplifications due to space or spin symmetries, the CI matrix relative to the determinant basis has dimension $C_M^{N_\alpha} \times C_M^{N_\beta}$. Although it is now possible to diagonalise very large matrices (of the order of 10^6–10^9), the full CI methods remain limited to a few electrons. They are therefore subject to approximation in practice. For example, the second or fourth order perturbation techniques known as Möller–Plesset MP2/MP4 treat the determinants generated from the single-determinant Hartree–Fock solution to second or fourth order of perturbation, respectively. Other techniques involve truncating the CI matrix in order to restrict to the occupied and virtual (active space) orbitals likely to be involved in the excitation processes treated variationally. This is the complete active space configuration interaction (CASCI) technique. The multiconfigurational self-consistent field (MCSCF) or complete active space self-consistent field (CASSCF) techniques simultaneously optimise the orbitals and the configuration interaction coefficients. The multi-reference configuration interaction (MRCI) methods generate singly and doubly excited determinants which interact with an arbitrary space of reference determinants, itself multiconfigurational. Perturbation techniques can be generalised to the case where the generating space is of multi-reference type, as in the CAS-PT2 method, for example, able to treat the excited states. Finally, the coupled cluster (CC) method uses an exponential of the excitation operator X to generate the multiconfigurational wave function $\Psi = (\exp X)\Phi_0$.

If the aim is to monitor size dependence, it is important that the energy approximations should satisfy the requirements of additivity and extensivity. The MP2 and MP4 methods and the coupled cluster methods do satisfy this criterion, but are at present mainly adapted to the ground state. The MCSCF, MRCI and CAS-PT2 methods can handle the excited states, but they do not necessarily respect these criteria.

In practice, in quantum chemistry programs, orbitals are expressed in the form of linear combinations of Gaussian functions centered on the atoms. This is known as linear combination of atomic orbitals or LCAO. Pseudopotentials [167] or effective core potentials (ECP) may provide a significant technical advantage when it comes to reducing the number of electrons to be taken into account. Indeed, electrons in the core of the atoms can be replaced by effective operators. Only the valence electrons then come into the calculation, the bare atomic ions being replaced by ionic atomic cores with effective charge $Z^* = Z - N_c$, where N_c is the number of core electrons of the atom carrying the pseudopotential.

Although the CI methods are in principle exact, they converge slowly with the Gaussian basis and/or with the dimension of the CI matrix, and the computation time is often too long to allow direct real-time integration of the energy in the dynamical codes, except for very small systems.

6.3.3 Density Functional Theory

In the last section, we discussed methods for explicit construction of the wave function. The density functional theory (DFT) provides an alternative that concentrates on the density in the calculation of the energy and electron properties [15, 17, 19, 127, 128]. The many-electron wave function is a complex function on $\mathbb{R}^{3N}$. This representation can be reduced by introducing the one-particle electron density

$$\rho(\boldsymbol{r}) = \sum_{\sigma_1\sigma_2\ldots\sigma_N} \int_{\boldsymbol{r}_2\ldots\boldsymbol{r}_N} \Psi(\boldsymbol{r}, \boldsymbol{x}_2, \ldots \boldsymbol{x}_N) \mathrm{d}\boldsymbol{\sigma}_1 \mathrm{d}\boldsymbol{\sigma}_2 \ldots \mathrm{d}\boldsymbol{\sigma}_N \mathrm{d}\boldsymbol{r}_2 \ldots \mathrm{d}\boldsymbol{r}_N ,$$

where the integral is carried out over all spin variables and $N-1$ space variables. The density, normalised to the number of electrons so that $\int_{\boldsymbol{r}} \rho(\boldsymbol{r})\mathrm{d}\boldsymbol{r} = N$, is a function on $\mathbb{R}^3$ with simpler structure than the many-electron wave function.

According to the Hohenberg–Kohn theorem [15], generalised by Lévy and Lieb [16] to the degenerate case, the ground state energy of an electronic system can be expressed as a functional of the electron density $\rho(\boldsymbol{r})$, i.e.,

$$E_{\mathrm{el}}[\rho] = F[\rho] + \int V_{\mathrm{ext}}(\boldsymbol{r})\rho(\boldsymbol{r})\mathrm{d}\boldsymbol{r} .$$

In this equation, $F[\rho]$ is a functional which represents the energy contributions intrinsic to the electron system, i.e., the kinetic energy, Coulomb energy, exchange energy, and correlation energy. The value of this function thus depends on the electron density distribution in $\mathbb{R}^3$. The function is universal in the sense that it is the same for all electron systems and does not depend on the environment of the electrons, represented by the potential V_{ext}, which describes the Coulomb forces exerted on the electrons by the nuclei and any external fields imposed on the system. For each particular system, there is therefore a one-to-one correspondence between the potential V_{ext} and the density. The ground state energy is obtained by minimising the functional with respect to the density. Unfortunately, the functional $F[\rho]$ is not known exactly and only approximate functionals are available in practice.

In the most widely used and fully developed version, the operator character of the theory results from use of the Kohn–Sham (KS) equations [17]. These authors had the idea of decomposing the functional $E_{\text{el}}[\rho]$ by explicitly bringing out the known kinetic contribution T_0 of a fictitious system of non-interacting electrons in the potential V_{ext}, whose wave function can thus be reduced to a single Slater determinant and whose density

$$\rho(\boldsymbol{r}) = \sum_i^{\text{occ}} f_i |\phi_i(\boldsymbol{r})|^2$$

is identical to that of the real system. The coefficients f_i are the occupation numbers of the orbitals ϕ_i ($f_i = 2$ for occupied orbitals in the simple case of closed shell systems). The energy then decomposes according to

$$E_{\text{el}}[\rho] = T_0[\rho] + \frac{1}{2}\iint \frac{\rho(\boldsymbol{r})\rho(\boldsymbol{r}')}{|\boldsymbol{r}-\boldsymbol{r}'|}\mathrm{d}\boldsymbol{r}\mathrm{d}\boldsymbol{r}' + E_{\text{xc}}[\rho] + \int V_{\text{ext}}(\boldsymbol{r})\rho(\boldsymbol{r})\mathrm{d}\boldsymbol{r} \;,$$

in which the second term represents the classical Coulomb contribution of an electron system with density ρ, whilst E_{xc} contains the exchange and correlation contributions, but other corrections as well, in particular, the correction to the kinetic energy that results from the interaction between electrons $T[\rho] - T_0[\rho]$.

By minimising the total energy with respect to the density defined by the occupied orbitals ϕ_i and imposing orthogonality, i.e., $\int \phi_i^*(\boldsymbol{r})\phi_j(\boldsymbol{r})\mathrm{d}\boldsymbol{r} = \delta_{ij}$, one obtains the Kohn–Sham (KS) equations

$$\left[-\frac{1}{2}\Delta + \int \frac{\rho(\boldsymbol{r}')}{|\boldsymbol{r}-\boldsymbol{r}'|}\mathrm{d}\boldsymbol{r}' + \frac{\delta E_{\text{xc}}[\rho]}{\delta\rho(\boldsymbol{r})} + V_{\text{ext}}(\boldsymbol{r})\right]\phi_i(\boldsymbol{r}) = \epsilon_i\phi_i(\boldsymbol{r}) \;.$$

The KS equations have similar structure to the Hartree–Fock equations. They describe the motion of each electron in a self-consistent effective potential V_{eff}.

The many-electron DFT–KS wave function in the form of a single determinant should not be confused with the actual wave function, but considered as a way of expressing the density that is then appropriate for calculating the energy via the functional.

There are many approximations for the unknown functional or its functional derivative $V_{\text{xc}} = \delta E/\delta\rho(\boldsymbol{r})$. The simplest, often used for nanosystems, is the so-called local density approximation (LDA), in which the exchange–correlation energy is written in the form

$$E_{\text{xc}}[\rho] = \int V_{\text{xc}}^{\text{LDA}}[\rho(\boldsymbol{r})]\mathrm{d}\boldsymbol{r} \;.$$

Here $V_{\text{xc}}^{\text{LDA}}[\rho(\boldsymbol{r})]$ is the exchange–correlation energy per particle in a uniform electron gas of density $\rho(\boldsymbol{r})$. The variant known as the local spin density approximation (LSDA) allows electron densities $\rho_\alpha(\boldsymbol{r}) \neq \rho_\beta(\boldsymbol{r})$ and hence spin-polarised spinorbitals. However, other more effective functionals have been developed [19–24], taking into account the gradient of the density (generalised gradient approximation or GGA), or combining a functional approximation to exchange with the exact Hartree–Fock exchange (BLYP, B3LYP, BPW91). Note, however, that these functionals do not correctly represent dispersion forces.

In most applications, the Kohn–Sham orbitals are expanded with respect to either Gaussian LCAO bases, or plane wave bases, possibly extended by localised functions (augmented plane wave basis or APW). KS orbitals are also directly expressed on a real space lattice in some applications. Pseudopotential techniques can also be used to reduce the calculation to the valence electrons alone.

6.3.4 Charge Analysis

Ab initio methods can be used to calculate the electron density $\rho(\boldsymbol{r})$. However, it is often instructive to try to analyse the distribution of this density over the atoms. But the analysis of atomic charges via an ab initio description immediately encounters a fundamental difficulty: in the context of quantum mechanics, the atomic charge in a crystalline or molecular system is not an observable, only the charge density being measurable. This is clearly connected to the quantum and non-separable nature of the electron wave function and the overlap between atomic wave functions localised on different atoms, which is indeed the origin of the chemical bond. The concept of local charge is therefore ambiguous and several definitions can be used. Mulliken defines the atomic electron populations ρ_a in an LCAO formalism by considering the weight

$$p_{aa}^i = f_i \sum_\mu |c_{a\mu}^i|^2$$

of the occupied molecular orbitals ϕ_i over the basis functions $\chi_{a\mu}$ of an atom a and sharing the so-called overlap contributions

$$p_{ab}^i = \frac{f_i}{2} \sum_{\mu \in a, \nu \in b} (c_{a\mu}^{*i} c_{b\nu}^i + c_{a\mu}^i c_{b\nu}^{*i}) S_{a\mu, b\nu}$$

between atoms a and b ($b \neq a$). This provides the Mulliken estimate for the atomic charges q_a:

$$q_a = Z_a - \sum_i^{\text{occ}} \left(p_{aa}^i + \sum_{b \neq a} p_{ab}^i \right) .$$

One disadvantage of the Mulliken analysis is that the result is not invariant under choice of LCAO basis and can become quite unreliable when the basis contains atomic functions that are too diffuse. Other definitions have been given. For example, charges can be defined from an analysis of the wave function in terms of natural bond orbitals (NBO) [116, 117]. In another approach, Bader has introduced an analysis based on the topology of the density $\rho(\boldsymbol{r})$ [114,115]. This method divides up the density into disjoint topological regions $\boldsymbol{\Omega}_a$ of $\mathbb{R}^3$ bounded by surfaces through which there is no flux of density, i.e.,

$$\boldsymbol{\nabla}\rho\cdot\boldsymbol{n} = 0 ,$$

where $\boldsymbol{n}$ is a unit vector normal to the surface. These density regions, centered on the atoms, define the Bader topological charges. Integrating the density over each region around an atom, an electron population can then be attributed to it:

$$\rho_a = \int_{\boldsymbol{\Omega}_a} \rho(\boldsymbol{r})\mathrm{d}^3\boldsymbol{r} .$$

Extending the analysis of the electron density, the description of the bond in terms of electron localisation functions (ELF) can then be used to visualise the contributions of electron pairs to the bonds simply from knowledge of the density.

6.3.5 Approximate and Semi-Empirical Descriptions

DFT-type ab initio methods are currently applied to the study of systems containing tens or even hundreds of atoms [67, 68], depending on the nature of the elements and the type of function used, in calculations with fixed nuclear geometry or in constrained relaxation processes. Approximate models can push back the size limit, but more importantly, they make it possible to implement less time-consuming energy calculations which can therefore be used in intensive simulations like Monte Carlo or molecular dynamics. Different models, sometimes historically independent, can be formally integrated into the density functional framework.

DFT can in principle operate directly from the density without involving the KS orbitals. This requires knowledge of the kinetic energy functional $T[\rho]$. Semi-classical approximations can be obtained from an expansion in powers of $\hbar$. The simplest functional describes a homogeneous electron gas:

$$T_{\mathrm{TF}}[\rho] = \int \frac{3}{10}(3\pi^2)^{2/3}\rho^{5/3}\mathrm{d}\boldsymbol{r} ,$$

which defines the Thomas–Fermi model [165]. It can be improved to the extended Thomas–Fermi model (ETF) by including inhomogeneneity corrections involving the gradient or the Laplacian of the density [166, 170]. This type of functional is sometimes used in formulations of DFT without orbitals. Orbital effects are then ignored, but quantum corrections can in some cases be included via the Strutinsky method [168, 178], wherein the density ρ is expanded about its semi-classical value ρ_{ETF}, viz., $\rho = \rho_{\mathrm{ETF}} + \delta\rho$, and the correction due to the quantum orbital structure is calculated in a perturbative formulation. The semi-classical methods have been widely applied to metallic nanoparticles and clusters [40, 154, 178].

The tight-binding method, also known as the Hückel model in chemistry, begins by expressing the single-electron Hamiltonian h in an LCAO basis reduced to the valence atomic orbitals $\chi_{a\mu}$, and then parametrising the diagonal (atomic levels) and non-diagonal (hopping integrals $t_{a\mu,b\nu}$) matrix elements [26, 106]

$$h_{a\mu,b\nu} = \langle \chi_{a\mu} | h | \chi_{b\nu} \rangle \,.$$

The total energy can be expressed in the general form

$$E[\rho] = \sum_i f_i \epsilon_i + G[\rho] \,,$$

as a function of the eigenvalues of the Hamiltonian ϵ_i. The first term represents the energy of delocalisation over the sites, or band energy. The functional $G[\rho]$ contains the energy of interaction between ions, and any residual exchange–correlation energy not included in the one-electron contributions, as well as corrections for double counting in the sum. The electron density can be discretised over the atomic sites. If the basis is considered to be orthogonal, the density at each site is then

$$\rho_a = \sum_i f_i \left(\sum_\mu |c^i_{a\mu}|^2 \right) \,, \qquad \phi_i = \sum_{a\mu} c^i_{a\mu} \chi_{a\mu} \,.$$

The diagonal elements $h_{a\mu,a\mu}$ are usually taken as constant and equal to the energies $\epsilon_{a\mu}$ of the atomic levels, while the hopping terms are approximated by analytic expressions, and the term G is reduced to a simple additive potential describing pairwise repulsive interactions between ions. Parameters are generally provided directly by experimental data or ab initio calculations (molecules and small clusters, or the condensed phase). A crucial point is transferability to the wide range of situations encountered in nanoparticles, where one must describe both electrons in the bulk and electrons at the surface. In particular, hybridisation of bonds can differ from what happens in the solid. To improve flexibility and take the environment into account, the diagonal elements can be expressed as functions of the variations in the effective electron density $\delta\rho_a$ associated with the atomic environment:

$$h_{a\mu,a\mu} = \epsilon_{a\mu} + \Phi(\delta\rho_a) \,, \qquad \delta\rho_a = \sum_b \gamma(R_{ab}) \,.$$

This expression, with $\delta\rho_a$ pairwise additive here, is inspired by classical energy approximations often used in molecular dynamics studies of transition metal nanoparticles, in particular when the potentials depend on the environment (embedded atom model or EAM potentials) [57, 60, 61, 63–65]:

$$E\{\boldsymbol{R}_a\} = \sum_a \Phi(\delta\rho_a) + \sum_{a<b} V(\boldsymbol{R}_a - \boldsymbol{R}_b) ,$$

where $V(R)$ is a repulsive pairwise potential and $\Phi(\delta\rho_a)$ expresses the variation in the site energy associated with atom a as a function of the environment of this atom. The function Φ allows one to include many-body contributions analytically and represents electron contributions (band energy in the bulk solid). In the context of the EAM philosophy, it is also worth mentioning the effective medium theory, which attributes to each atom an energy contribution equal to that of the same atom in an equivalent homogeneous medium, sometimes corrected by terms describing local perturbations.

Finally, when charge transfer is high, e.g., ionic clusters, heterogeneous systems, it may also be important via a self-consistent calculation to include electrostatic corrections in the diagonal elements of the tight-binding model, viz.,

$$h_{a\mu,a\mu} = \epsilon_{a\mu} - U_{aa}q_a - \sum_{b\neq a} q_b U_{ab} ,$$

these being induced by on-site charges and charges on neighbouring sites, with $q_b = Z_b - \rho_b$. The quantities U_{ab} are single-centre ($a = b$) or double-centre Coulomb interactions.

6.3.6 Energy Bands and Densities of States

In periodic solids, bands of electron levels are characterised by a density of states $g(\epsilon)$. If l denotes the degeneracy of an atomic level, n the number of atoms, and $M = nl$ the total number of states,

$$g(\epsilon) = \frac{1}{nl}\sum_i \delta(\epsilon - \epsilon_i) .$$

By projection, local site densities $g_a(\epsilon)$ are defined by

$$g_a(\epsilon) = \frac{1}{l}\sum_{\mu=1}^{l} g_{a\mu}(\epsilon) = \frac{1}{l}\sum_{\mu=1}^{l}\left[\sum_{i=1}^{M} c_{a\mu}^{*i} c_{a\mu}^{i}\delta(\epsilon - \epsilon_i)\right] ,$$

normalised so that

$$\int g_{a\mu}(\epsilon)\mathrm{d}\epsilon = \int g_a(\epsilon)\mathrm{d}\epsilon = 1 .$$

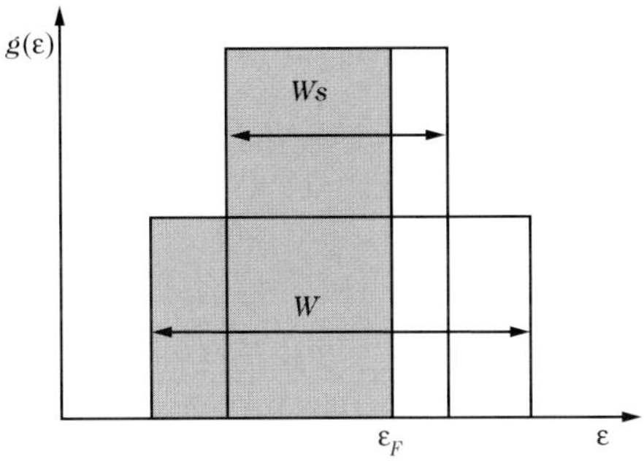

Fig. 6.5. Schematic variation of the local electron density of states in the bulk and at the surface. W_s and W are the band widths associated with surface and bulk sites, respectively. *Grey*: Occupied states below the Fermi level ϵ_F

The distributions of the densities of states can be characterised by their moments of order p. For a band generated by a degenerate atomic level of energy ϵ_0, the first few moments are:

$$\mu_0 = \frac{1}{ln}\sum_{a\mu}\langle a\mu|h^0|a\mu\rangle = 1 , \qquad \mu_1 = \frac{1}{ln}\sum_{a\mu}\langle a\mu|h|a\mu\rangle = \epsilon_0 ,$$

$$\mu_2 = \frac{1}{ln}\sum_{a\mu}\langle a\mu|h^2|a\mu\rangle = \epsilon_0^2 + \frac{1}{ln}\sum_{a,b\neq a,\mu,\nu}(t_{a\mu,b\nu})^2 .$$

The centered second moment is proportional to the number of bonds and hence to the atomic coordination z. This plays a particularly important role in nanoparticles, where a great many atoms are located at the surface [29,106]. Now the density of states is narrower at the surface than in the bulk, since the coordination number of the atoms is smaller there (see Fig. 6.5). Hence for a band with uniform rectangular density and width W_a at site a,

$$W_a = 12\sqrt{\sum_{b\mu\nu}(t_{a\mu,b\nu})^2} ,$$

and we deduce that

$$g_a(\epsilon) = \frac{1}{\sqrt{12\sum_{b\mu\nu}(t_{a\mu,b\nu})^2}} .$$

From the relation giving the number of electrons per site, i.e.,

$$N_a = l\int_{-\infty}^{E_F} g_a(\epsilon)\mathrm{d}\epsilon ,$$

it is easy to calculate the Fermi level:

$$E_F = \left(\frac{N_a}{l} - \frac{1}{2}\right) W_a ,$$

and also the electron contribution to the cohesive energy of the site:

$$E_a = l \int_{-\infty}^{E_F} (\epsilon - \epsilon_a) g_a(\epsilon) d\epsilon = -N_a \left(1 - \frac{N_a}{l}\right) \frac{W_a}{2} ,$$

proportional to the site band width. The energy maximum is obtained when the band is half filled, i.e., $N_a/l = 1/2$. In fact, a filled band $N_a/l = 1$ does not contribute to the energy. For the noble metals, the contribution of the d electrons to the band energy thus occurs via s–d hybridisation, which partly depopulates the d band.

From the expression for the band width, it is easy to deduce an approximate analytical form for an EAM-type potential. Using an exponential parametrisation for the hopping integral as a function of the interatomic distance R_{ab} and then short-range repulsion, treated in the form of pairwise contributions, the following expression is obtained:

$$E = \sum_a \left\{ \gamma \sum_b \exp\left[-p\left(\frac{R_{ab}}{\sigma_0} - 1\right)\right] - \sqrt{\zeta^2 \sum_b \exp\left[-q\left(\frac{R_{ab}}{\sigma_0} - 1\right)\right]} \right\} .$$

Hence the band energy, as the square root of a sum, explictly introduces N-body terms. With σ_0, the parameters γ and p determine the form of the repulsion term V, while ζ and q determine the band energy Φ. In practice, the parameters are often adjusted to reproduce certain properties of the periodic solid, e.g., lattice parameter, cohesive energy, elastic properties, thermodynamic data.

6.4 Applications to Some Typical Examples

6.4.1 Metallic Nanoparticles

In simple metals (especially the alkali metals, but also the noble metals), electrons in the inner shells retain their atomic character while delocalisation of the valence electrons significantly screens the ionic structure. The result is a single-electron effective potential V_{eff} that is practically constant throughout the bulk. It is therefore essentially the size and shape that determine the structure of the discrete electron levels in a finite metallic particle. One of the most remarkable discoveries concerning clusters of simple metals is that they have an electronic shell structure [39]. The shell structure is a consequence of the confinement of almost free valence electrons in a finite volume of roughly spherical shape. There is an obvious analogy with other confined systems of fermions, such as electrons in an atom, or nucleons in atomic nuclei. Even for small sizes ($n \geq 8$), the shape of free metallic clusters is close to that of a compact droplet. The simplest quantum extension of the liquid-drop model is the uniform spherical jellium model, which ignores the discrete structure of the ions and represents their charge by a continuous positive distribution or jellium whose density ρ_0 can be identified with the charge densities of the same ions in an infinite metal. The volume of the cluster is taken to be that of a metallic sphere of radius R_0:

$$\rho_+(r) = \begin{cases} \rho_0 & r \le R_0 , \\ 0 & r > R_0 , \end{cases}$$

and the total charge of the jellium distribution is $Z = N = 4\pi R_0^3 \rho_0/3$. It is convenient to introduce the radius r_0 of the average spherical volume $v_0 = 4\pi r_0^3/3$ occupied by an electron of the metal, or Wigner–Seitz radius.. For monovalent metals, this volume corresponds to an ionic density equal to the electron density, i.e., $3/4\pi r_0^3$. As in the classical droplet, it is convenient to express the cluster radius as a function of the number of electrons, viz., $R_0 = r_0 N^{1/3}$. In the electrostatic field of the spherical jellium, the electrons are subject to a potential energy

$$V_\mathrm{J}(r) = \begin{cases} \dfrac{R_0^2}{2r_0^3}\left(\dfrac{r^2}{R_0^2} - 3\right) & r \le R_0 , \\ -\dfrac{R_0^3}{r_0^3}\dfrac{1}{r} & r > R_0 , \end{cases}$$

parabolic inside the cluster and Coulomb outside. The nanoparticle has thus been simplified to a many-electron system in a central field (see Fig. 6.6).

In a self-consistent single-electron description (HF or DFT), each electron is subject to an effective potential $V_\mathrm{eff}^\mathrm{SCF}$, a sum of the positive jellium potential $V_\mathrm{ext} = V_\mathrm{J}$, the average Coulomb potential, and the exchange potential (and correlation potential in DFT), which screen the jellium. The spherical symmetry allows one to factorise into a product of a radial function and a spherical harmonic $Y_{lm}(\theta, \phi)$, viz.,

$$\phi_{klm}(r) = R_{kl}(r) Y_{lm}(\theta, \phi) ,$$

thereby reducing the Schrödinger equation to a 1D problem. Here k is the principal quantum number, i.e., the number of nodes in the radial function. Figure 6.7 illustrates the self-consistent potential [43] for the cluster Na_{20}, which has electronic structure $1s^2 1p^6 1d^{10} 2s^2$ in the spherical jellium model, with the $1f$ LUMO unoccupied.

The radial potential V_eff can be approximated by model shapes, e.g., a square well potential, or a harmonic potential of frequency $k_0 = \Omega_0^2$, viz.,

$$V_\mathrm{eff}(r) = -V_0 + \frac{1}{2}\Omega_0^2 r^2 ,$$

or again, a Woods–Saxon function [46]:

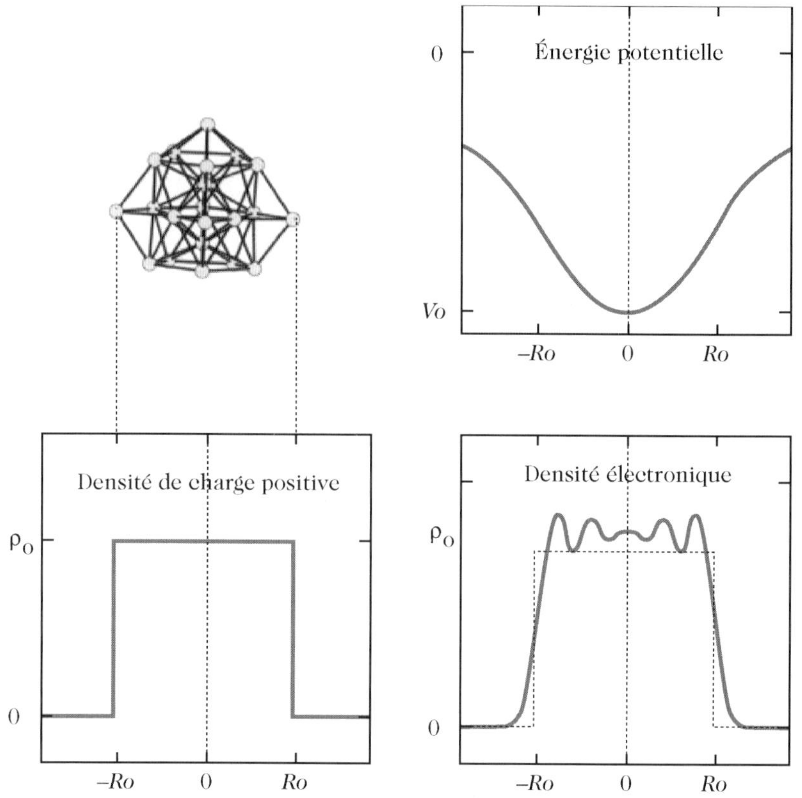

Fig. 6.6. Metallic cluster approximated by a uniform spherical jellium. *Left*: Nuclear structure modelled by a constant radial distribution of positive charge in a sphere. *Right*: Potential energy V_J of the jellium (*top*) and total electron density (*bottom*). R_0 is the radius of the distribution of positive charges and ρ_0 its density

$$V_{\mathrm{eff}}(r) = -\frac{V_0}{1+\exp\left[\alpha(r-r_0)\right]} .$$

The use of any of these potentials (or the 'exact' potential $V_{\mathrm{eff}}^{\mathrm{SCF}}$) may lead to certain differences in the details of the electronic levels, but the shell structure is essentially related to their degeneracy.

The electronic levels ϵ_{kl} are thus characterised by their radial quantum number k, and their orbital angular momentum l. Taking spin into account, the degeneracy of each level is $2(2l+1)$. The degenerate levels define the shells associated with the quantum numbers k and l, with $1s$, $1p$, $1d$, $2s$, $1f$, $2p$, $1g$, $2d$, $1h$, $3s$, $2f$, $1i$, $3p$, $1j$, etc., in increasing order (the exact order of the levels varying with the potential V_{eff} considered). The fundamental electronic configuration is obtained by filling up the levels in order of increasing energy. One consequence is that, as in atoms, clusters with closed (filled) outer shells have greater stability compared with neighbouring sizes. Indeed, the effective

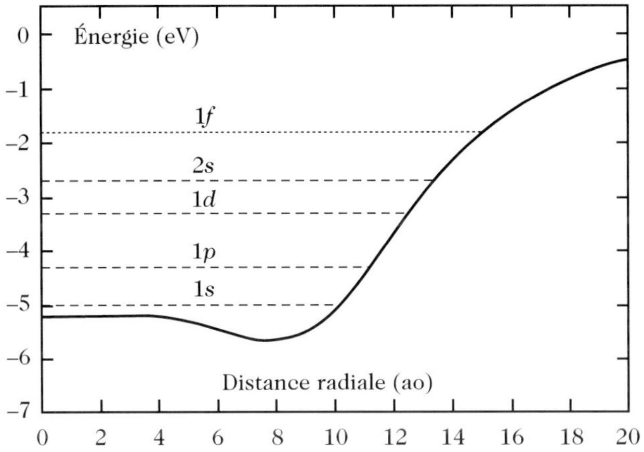

Fig. 6.7. Self-consistent potential $V_{\text{eff}}^{\text{SCF}}$ and single-electron levels $\epsilon_{kl}^{\text{SCF}}$ for Na_{20} in the spherical jellium model, from a DFT/KS/LDA calculation by Eckardt [43]. The $1s$, $1p$, $1d$ and $2s$ levels are occupied, while the $1f$ level is unoccupied

potential acting on an electron in a shell is stabilised as that shell fills up: the attractive force due to the increased charge is not completely balanced by the increased screening of the electrons. This is no longer true for the first electron in a new shell, which for one thing occupies a higher level in the effective potential, and for another, fully experiences the screening by electrons in the shell immediately below. The spherical jellium model explains the abundances observed in molecular beam experiments for clusters of specific sizes known as magic sizes, having particular numbers of electrons, i.e., $N = 2$, 8, 18, 20, 34, 40, 58, 70, 92, 138, and so on, magic numbers associated with closed electron shells, as opposed to geometric magic numbers related to the stacking of atoms, which characterise the abundances of other types of cluster. There may also be quasi-degeneracies and accidental degeneracies. This happens in the harmonic oscillator model. Since $\epsilon_{k,l} = \epsilon_{k+1,l-2}$, the $1d$ and $2s$ levels turn out to be degenerate (see Fig. 6.8).

In neutral alkali clusters made from monovalent atoms, the magic sizes n are the same as the electronic magic numbers, i.e., $n = N$. The shell structure is also reflected by the way the ionisation potentials and electron affinities vary with the cluster size. Indeed in alkali clusters, ionisation potentials have maxima for the electron magic sizes, followed by a sudden drop for the next size up. This stability induced by filling of electronic shells and hence by the number of electrons is found shifted by one unit for the magic sizes associated with cations, i.e., $n_+ = N-1$, or one unit in the other direction for anions, i.e., $n_- = N+1$, relative to the values for neutral clusters. Hence, in simple metallic systems, finite-size quantum effects induce an electronic shell structure that is superposed on the classical continuous evolution toward the properties of the bulk solid predicted by the liquid-drop model. Trivalent metals, e.g.,

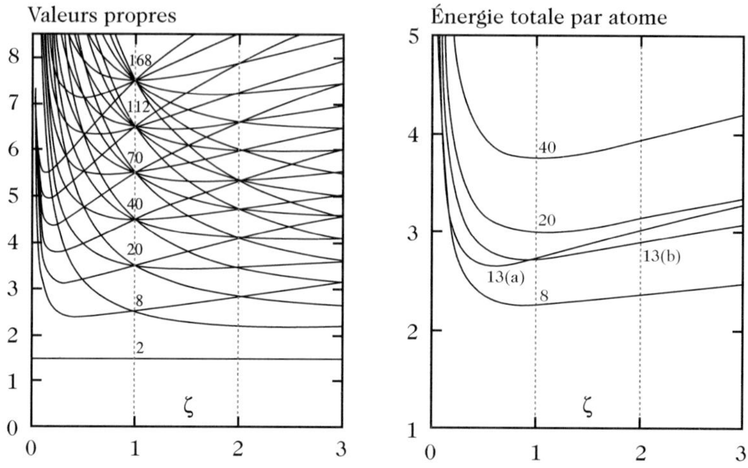

Fig. 6.8. Single-electron spectrum (*left*) and total electron energy per atom (*right*) in the independent-electron spheroidal harmonic jellium model. Energies are in units of Ω_0. The reference level for the single-electron energies is $-V_0$ and for the total energies is $-NV_0$. (In a self-consistent calculation, V_0 would be a function of N.) The energy of the positive jellium itself is not taken into account. *Right*: Na_{13} with spin (**a**) $M_s = 1/2$ and (**b**) $M_s = 5/2$

aluminium or gallium, also display electronic shell structure [197, 199, 200], governed by the relation between the number of electrons and the number of atoms ($n = 3N$). The uniform spherical jellium model is widely used to describe the optical properties of spherical nanoparticles embedded in insulating matrices [205]. Indeed, it has recently been shown that the reactivity of silver clusters with oxygen is also strongly influenced by the electron magic numbers [59].

In free clusters, non-spherical deformations of the quantum droplet must be taken into account. One cause of such deformation results from the fact that, for incomplete filling of the outer shell (cluster with open outer shell), several many-electron configurations may turn out to be degenerate. For example, the jellium configuration of Na_{13} is $1s^2 1p^6 1d^5$. The highest spin state ($M_s = 5/2$) is non-degenerate and, in an ab initio calculation, would be stabilised by the exchange terms, via a mechanism analogous to the Hund rules in atoms. However, this state is not the ground state because the exchange integrals are smaller than in the atoms owing to the delocalisation of the wave functions in a greater volume. The possibility of distributing the electrons in the open outer half-filled $1d$ shell generates a set of degenerate determinants for $M_s = 1/2$. The degeneracy is reduced if biaxial or triaxial deformations of the spherical jellium are allowed. Removal of the sphericity constraint stabilises certain levels to the detriment of others. For each spin state, the cluster adopts the most energetically favourable deformation. In the high spin state of Na_{13}, all the levels are occupied by one electron and the effects cancel. This no longer happens for the low spin states, for which the energy-dependent filling rule leads to only the lowest levels being occupied. This is an instance of the Jahn–Teller effect [45], a very general quantum effect which also induces the deformation of certain atomic nuclei, or a reduction of molecular symmetry. In Na_{13}, it is the low spin state, thereby stabilised by deformation, which is the ground state.

Triaxial (ellipsoidal) deformations can be treated using the anisotropic harmonic oscillator model, corresponding to an effective potential

$$V_{\text{eff}} = -V_0 + \frac{1}{2}\Omega_0^2\left(\frac{x^2}{a^2}+\frac{y^2}{b^2}+\frac{z^2}{c^2}\right) ,$$

where a, b, and c are the ellipsoidal deformation coefficients of the jellium charge density. (The isotropic harmonic oscillator of the spherical model with frequency Ω_0 is retrieved by setting $a = b = c = 1$.) For a uniform positive charge density, charge conservation requires the jellium volume to be conserved too, and hence $abc = 1$. In a (biaxial) spheroidal model, two axes are equivalent, i.e., $b = c \neq a$, and the cluster has cylindrical symmetry, so that the deformation leads to flattened ($c < a = b$) or oblong ($c > a = b$) shapes. The energy levels of the spheroidal oscillator are given by

$$\epsilon(\nu_x,\nu_y,\nu_z) = -V_0 + \frac{\Omega_0}{a}\left(\nu_x+\frac{1}{2}\right) + \frac{\Omega_0}{b}\left(\nu_y+\frac{1}{2}\right) + \frac{\Omega_0}{c}\left(\nu_z+\frac{1}{2}\right) ,$$

and can then be expressed as a function of the deformation parameter $\zeta = c/a$, viz.,

$$\epsilon(\nu_x,\nu_y,\nu_z,\zeta) = -V_0 + \Omega_0(\nu_x+\nu_y+1)\zeta^{1/3} + \Omega_0\left(\nu_z+\frac{1}{2}\right)\zeta^{-2/3} ,$$

corresponding to a total electron energy

$$E = \sum \epsilon(\nu_x,\nu_y,\nu_z,\zeta) .$$

This expression does not take into account variations in the electrostatic energy of the distorted positive jellium itself. The jellium does not yield a good approximation of the classical Coulomb energy of ions which have a discrete character. Some authors have suggested determining the discrete positions of the nuclei in a self-consistent (iterative) way in an electronic potential with spherical symmetry (spherically averaged pseudopotentials, or SAPS), or cylindrical symmetry (cylindrically averaged pseudopotentials, or CAPS) [78] resulting from the spherical or cylindrical average of the potentials of the ions. At each iterative step, the discrete positions of the nuclei are redetermined in this electronic potential with constrained symmetry.

Self-consistent solution of the spheroidal jellium model has been carried out by Eckardt et Penzar in a DFT calculation [77]. Triaxial deformations

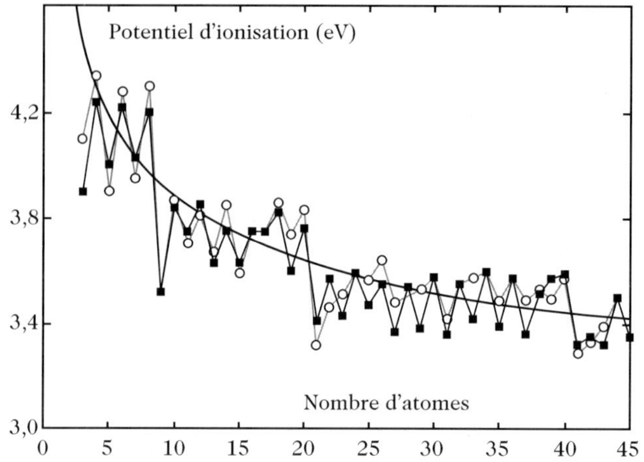

Fig. 6.9. Ionisation potentials of neutral sodium clusters Na_n. *Black squares*: Measurements by Homer [173]. *White circles*: Values calculated using the triaxial jellium model by Yannouleas et al. [178]. *Continuous curve*: Continuous decrease when shell effects are disregarded

have been studied systematically by Yannouleas and Landman [178] in a semiclassical ETF treatment corrected for electronic shell effects. Figure 6.9 shows the ionisation potentials thereby calculated. Even in the simple jellium model, almost quantitative results can be obtained by taking into account ellipsoidal deformations (for sodium), illustrating the importance of the coupling between geometric structure and electronic properties. In fact, shell effects remain in the non-spherical jellium, but are lessened by the effects of distortion. Note also that there is a very marked odd/even alternation in the ionisation potentials, resulting as for electronic shells from the stabilisation of clusters whose HOMO contains two paired electrons in the same orbital, compared with those where it contains only one electron.

Various studies, generally empirical, have led to more sophisticated versions of the jellium model. One point is that the cluster surface cannot be considered as abrupt. Jellium distributions decreasing continuously in the radial direction have thus been investigated. Another point is that the surface always has a certain roughness due to the discrete structure of the nuclei. This can be corrected by an angular modulation function [201]. Corrections relating to the field of the nuclei, even in the bulk, have been introduced more systematically using crystalline field methods or pseudopotentials. Finally, some authors have considered uniform jellium models bounded by various more realistic facetted polyhedra to describe, for example, cold aluminium clusters which conserve an fcc crystalline structure [164].

Alkali clusters have served as prototypes for theoretical and experimental studies. Beyond the jellium model, their structural and electronic properties have been investigated in many ab initio studies, using both configuration interaction methods [3, 71, 73, 108, 110] and the density functional theory [51, 74, 161]. For larger clusters, the CAPS approach [79] has been used up to several tens of atoms. It is also worth mentioning the tight-binding models, used in global and unconstrained search algorithms for low-energy isomers, or in simulations requiring large numbers ($\approx 10^6$) of energy calculations [75, 76, 156]. Calculations explicitly taking into account the discrete structure of the nuclei generally confirm the qualitative trends predicted for the electronic structure in the distorted jellium model. Explicit introduction of discrete structure is nevertheless necessary, even for the alkali metals, in order to obtain fully quantitative results, in particular with regard to electronic excitations. It is obviously unavoidable in the investigation of dynamic and thermodynamic effects.

The study of electronic excitations has provided a choice tool for testing the electronic structure of metallic clusters and especially collective excitations, or plasmons. The Mie theory [206] gives a phenomenological description of the electrons in a metallic cluster subject to an electromagnetic field. In the infinite metal, plasmons are collective oscillations of electrons, i.e., the electrons oscillate in phase. Each electron treated classically here is subject to three forces: a restoring force $-\omega_\mathrm{p}^2 r$ exerted by the positive ions, which derives from a harmonic potential $\omega_\mathrm{p} r^2/2$, a damping force $\Gamma \mathrm{d}\boldsymbol{r}/\mathrm{d}t$, proportional and opposite to the velocity and which describes the dissipation resulting from collisions, and the force induced by the external electric field $\boldsymbol{E}_0\mathrm{e}^{-\mathrm{i}\omega t}$. In a 1D model, the motion of each electron is given by

$$\frac{\mathrm{d}^2\boldsymbol{r}}{\mathrm{d}t^2} + \Gamma\frac{\mathrm{d}\boldsymbol{r}}{\mathrm{d}t} + \omega_\mathrm{p}^2\boldsymbol{r} = -\boldsymbol{E}_0\mathrm{e}^{-\mathrm{i}\omega t} ,$$

and the induced dipole moment associated with the collective motion of N electrons oscillating in phase in the field is

$$\boldsymbol{D}(\omega) = \alpha(\omega)\boldsymbol{E}(\omega) = N\boldsymbol{r} = \frac{N}{\omega_\mathrm{p}^2 - \omega^2 - \mathrm{i}\omega\Gamma}\boldsymbol{n} ,$$

where $\boldsymbol{n}$ is the unit vector in the direction of the field. In the limit of constant electric field, i.e., setting $\omega = 0$, we recover the static dipole polarisability $\alpha(0) = N/\omega_0^2$. For a monovalent metal, the plasmon frequency is $\omega_\mathrm{p} = \sqrt{3/R\pi r_0^3}$. The photoabsorption cross-section is proportional to the square of the modulus of the oscillating electronic dipole $\boldsymbol{D}$:

$$\sigma(\omega) \propto \frac{N\omega^2}{(\omega_\mathrm{p}^2 - \omega^2)^2 + \omega^2\Gamma^2} .$$

This is the classical plasmon formula describing the collective oscillation of the electrons induced by the electric field in the restoring potential of the ions. If charge neutrality is respected in the bulk, the motion must be essentially an oscillation of the surface charge of the cluster parallel to the electric field, at the resonance frequency ω_0, broadened by collisions. Spherical metallic droplets exhibit collective oscillations of plasmon type, also referred to as giant resonances in nuclear physics. The oscillation of the spherical electronic and ionic charge densities also induces surface charges polarised in the direction of the inducing field. The oscillation frequency ω_0 is shifted from the value for the plasmon in the infinite solid $\omega_0 = \omega_p/\sqrt{3}$ due to the spherical shape.

For an ellipsoidal metallic droplet, the anisotropy can give rise to different resonance frequencies ω_0/a, ω_0/b and ω_0/c in the three directions. This provides a qualitative way of relating the plasmon resonances to the shape anisotropy of the cluster. It should be mentioned, however, that coupling between one-particle excitations can also significantly affect the position and widths of the plasmon resonances, even splitting the absorption lines. Hence, the relation between Mie resonances and shape is not always one-to-one. Truly quantitative predictions without ad hoc parameters can only be obtained by including the discrete ionic structure, e.g., using pseudopotentials [79]. Furthermore, the widths of the giant resonances can only be determined by a suitable treatment of the electron–vibration coupling.

6.4.2 Molecular Clusters

Molecular clusters are generally made up of highly stable monomers, i.e., atoms with filled outer shells or molecules with relatively high ionisation potentials and excitation energies, above 10 eV for rare gas atoms, and from 5 to 7 eV for molecular monomers [132–135].

Interactions between monomers do not lead to significant electron delocalisation and intermolecular binding energies generally remain weak for neutral systems in the absence of charge transfer. Gas phase molecular clusters are often considered as precursors for studying elementary processes and microscopic aspects of solvation in the liquid phase. Water clusters and nanoparticles are also relevant in atmospheric physics. Clusters of polycyclic aromatic molecules are hypothesised by astrophysicists to interpret certain infrared bands observed in the interstellar medium. Other applications such as conduction properties in low-dimensional assemblies of columnar structures may be relevant in nanotechnology.

The description of these clusters in the electronic ground state is a matter of intermolecular interactions [187,188]. Short-range interactions are governed by the overlap between the electron wave functions of the fragments and the Pauli exclusion principle. The associated energy is repulsive and increases exponentially as the distance between monomers decreases. Long-range interactions correspond to configurations in which the monomer functions have little or no overlap. They are governed by electrostatic interactions and dispersion. Weaker than covalent or metallic binding (except for charged constituents), the long-range interactions determine the asymptotic shapes of the potential energy surfaces. When the interacting systems are neutral and bound together by no other type of binding, these intermolecular interactions constitute the only attractive contribution. They generally stabilise the overall system at interfragment distances that are longer than those for chemical or metallic binding. Rare gas clusters are the prototype, although certain features are simplified: pointlike constituents, zero permanent dipole moments, isotropic polarisabilities, and atomic internal electronic structure.

In the following, we shall denote atomic or molecular monomers by A and their charge by Q_A. The presence of charged monomers ($Q_A \neq 0$) may contribute to more strongly bound structures and even to charge delocalisation, as we shall see below. Intermolecular contributions to the energy are derived by treating the interactions W_{AB} between monomers as a perturbation to their Hamiltonian H_0 when they are infinitely far apart:

$$H = H_0 + W = \sum_A H_A + \sum_{A<B} W_{AB} \, .$$

The zero order wave function is the product of the isolated monomer wave functions $\{\Phi_0^A\}$:

$$\Phi_0 = \mathcal{A}\left[\Phi_0^A \Phi_0^B \ldots\right] ,$$

where $\mathcal{A}$ is an intermonomer antisymmetrisation operator. Antisymmetrisation leads to exchange interactions at medium and short range, but will be neglected in the following where we are more concerned with long-range interactions. In the perturbative theory of intermolecular interactions, the perturbation W is expressed as a function of the multipole moment operators of the isolated fragments. The first contributions are the total charge Q_A, the dipole moment operator $\boldsymbol{D}_A$, the quadrupole moment tensor $\Theta_A^{\mu\nu}$, and so on.

To the first order of perturbation, the interaction energy is the expected value of W for the unperturbed function. Neglecting exchange and overlap,

$$\begin{aligned}\Delta E^{(1)} &= \sum_{A<B} \frac{Q_A Q_B}{R_{AB}} + \sum_{A,B\neq A} Q_B \frac{\boldsymbol{R}_{AB}}{R_{AB}^3} \cdot \langle \boldsymbol{D}_A \rangle \\ &+ \sum_{A<B} \left[\frac{\langle \boldsymbol{D}_A \rangle \cdot \langle \boldsymbol{D}_B \rangle}{R_{AB}^3} - 3 \frac{(\langle \boldsymbol{D}_A \rangle \cdot \boldsymbol{R}_{AB})(\langle \boldsymbol{D}_B \rangle \cdot \boldsymbol{R}_{AB})}{R_{AB}^5} \right] + \cdots \, .\end{aligned}$$

This expansion represents the interactions between the average values of the multipole moments associated with the charge distributions of the unperturbed wave functions of the fragments. This yields successively the charge–charge interaction (going as $1/R$), the permanent dipole–charge interaction (going as $1/R^2$), the permanent dipole–permanent dipole interaction (going as $1/R^3$), the permanent quadrupole–charge interaction (going as $1/R^4$), and so on.

To first order, the wave function responds to the intermolecular perturbation by acquiring components over the whole set of excited configurations. The latter may be generated by single excitations, double excitations in a single monomer, and double excitations resulting from simultaneous single excitations in two monomers. With this analysis in terms of local excitations, the second order interaction energy is separated into three main contributions:

$$\Delta E^{(2)} = \sum_A \Delta E_A^{(2)} + \sum_A E_A^{\mathrm{pol}} + \sum_{A<B} E_{AB}^{\mathrm{disp}} \,.$$

The first term represents the variation of the intrafragment correlation (contribution from double excitations in the same fragment). The second is the polarisation of the fragment A by the charge distributions of the other fragments (induction contribution described by single excitations in each fragment). The last term is the dispersion or van der Waals interaction. It has no classical counterpart and represents the coupling between simultaneous single excitations in two fragments A and B. Other contributions to $\Delta E^{(2)}$ would arise if exchange terms were considered.

The excitation energies $E_A^m - E_A^0$ relative to the ground state are all positive, while the three contributions to the energy are negative and generate an attractive interaction. Expressing the matrix elements $\langle \Phi_0 | W | \Phi_m \rangle$ as a function of the dipole moments of transitions in the monomers, we obtain the main contributions to the intermolecular energy. The induction energy is

$$E_A^{\mathrm{pol}} = - \sum_{B \neq A, m>0} \frac{\left(\langle \Phi_A^0 | \boldsymbol{D}_A | \Phi_A^m \rangle \dfrac{Q_B \boldsymbol{R}_{AB}}{R_{AB}^3} + \cdots \right)^2}{E_A^m - E_A^0} \,,$$

while the dispersion energy is

$$E_{AB}^{\mathrm{disp}} = - \sum_{m,l>0} \frac{1}{E_B^l + E_A^m - E_A^0 - E_B^0} \left(\frac{\langle \Phi_A^0 | \boldsymbol{D}_A | \Phi_A^m \rangle \cdot \langle \Phi_B^0 | \boldsymbol{D}_B | \Phi_B^l \rangle}{R_{AB}^3} \right.$$
$$\left. - \frac{3 (\langle \Phi_A^0 | \boldsymbol{D}_A | \Phi_A^m \rangle \cdot \boldsymbol{R}_{AB}) ((\langle \Phi_B^0 | \boldsymbol{D}_B | \Phi_B^l \rangle \cdot \boldsymbol{R}_{AB})}{R_{AB}^5} + \cdots \right)^2 .$$

The above expressions show that, for a dimer ($R_{AB} = R$), the main contribution, the dipole polarisation energy, goes as $C_4 R^{-4}$, whereas the dispersion energy goes as $C_6 R^{-6}$. Higher order multipole contributions generate terms in reciprocal even powers of R, viz., $C_8 R^{-8}$, $C_{10} R^{-10}$, and so on.

Introducing the electric dipole polarisability tensors $\alpha_A^{\mu\nu}$ ($\mu = x, y, z$) of the monomers, expressed in second order perturbation theory by

$$\alpha_A^{\mu\nu} = 2 \sum_{m>0} \frac{\langle \Phi_A^0 | D_A^\mu | \Phi_A^m \rangle \langle \Phi_A^m | D_A^\mu | \Phi_A^0 \rangle}{E_A^m - E_A^0} \,,$$

the dipole response of each monomer A to the total electric field $\boldsymbol{f}_A$ created at A by the charge distributions of the other monomers can be written

$$E_A^{\mathrm{pol}} = -\frac{1}{2}\sum_{B\neq A}\sum_{\mu\nu}\alpha_A^{\mu\nu}\frac{Q_B R_{AB\mu}}{R_{AB}^3}\frac{Q_B R_{AB\nu}}{R_{AB}^3} = -\frac{1}{2}\sum_{A}\sum_{\mu\nu}\alpha_A^{\mu\nu}\boldsymbol{f}_{A\mu}f_{A\nu} ,$$

where $R_{AB\mu}$ and $f_{A\mu}$ are the components of the vectors $\boldsymbol{R}_{AB}$ and $\boldsymbol{f}_A$. This is the form of the classical induced dipole polarisation.

Another way of analysing the energy is to write it as an expansion in many-body contributions:

$$E = \sum_A E_A^{(1)} + \sum_{A<B} V_{AB}^{(2)} + \sum_{A<B<C} V_{ABC}^{(3)} + \cdots .$$

Each m-body contribution $V^{(m)}$ is defined as the difference between the energy of the system of m monomers and the total lower order contribution. The simplest approximation consists in truncating this expansion at its two-body terms and hence taking it to have a pairwise additive form. However, higher order terms must often be included. In particular, we have seen that the polarisation forces are not additive whenever the system has several localised charges. The electric fields then created at each entity by the other charges of the system involve many-body terms. To second order in perturbation theory, the dispersion forces are additive, but many-body effects, generally weak, do occur at higher orders of perturbation. Overlap also generates many-body terms in the short-range repulsive forces. Finally, pairwise potentials are generally poorly suited to describing systems with highly delocalised or diffuse electrons.

There are several types of pairwise potential. The best known is undoubtedly the Lennard-Jones (LJ) potential [187, 188]

$$V(R_{AB}) = \epsilon\left(\frac{\sigma^{12}}{R_{AB}^{12}} - \frac{\sigma^6}{R_{AB}^6}\right) ,$$

for which the dissociation energy and equilibrium distance are $D_{\mathrm{e}} = \epsilon$ and $R_{\mathrm{e}} = \sigma^6\sqrt{2}$, respectively. The LJ potential affords a reasonable description of the repulsion–dispersion forces in atomic van der Waals clusters, and also in molecular clusters, provided that orientational anisotropy is taken into account. However, more precise empirical atom–atom potentials are often proposed, in particular with extra dispersion terms associated with the coefficients C_8 and C_{10}.

When considering interactions between large, highly non-spherical molecules, multipole expansions have the disadvantage of involving high order multipole distributions. It is also important to handle steric hindrance effects correctly. It is then better to return to point models, representing the charge distributions of the monomers by equivalent pointlike charges, e.g., centered on the atoms, including localised multipoles if necessary, to imitate the actual electrostatic potential. The repulsion–exchange–dispersion term can be approximated in certain generic forms [190, 191]:

$$E_{\text{rep/exch/disp}} = \sum_{a \in A, n \in B} k_a k_b G_{ab} V(R_{ab}/R^0_{ab}) ,$$

$$V(\zeta) = (A \exp^{-\alpha\zeta} + B \exp^{-\beta\zeta}) - \left(\frac{c_6}{\zeta^6} + \frac{c_8}{\zeta^8} + \frac{c_{10}}{\zeta^{10}} \right) ,$$

where the parameters R^0_{ab} characterise the range of the interactions, the coefficients k_a depend on the atoms, and the coefficients G_{ab} depend only on the types of the pairs of atoms a and b. $\zeta = R_{ab}/R^0_{ab}$ and $V(\zeta)$ is a reduced interaction. The coefficients c_6, c_8 and c_{10} are the reduced coefficients of the dispersion interactions associated with the reciprocal powers of ζ.

Molecular clusters in their neutral ground state are weakly bound because the electrons of the monomers do not form true chemical bonds. The same is not true when they are ionised [100, 102–104] or excited [162, 163]. We begin by distinguishing non-resonant and resonant cases.

In heterogeneous charged systems, the charge tends to be localised on the monomer with the lowest ionisation potential. This situation can be described by a single wave function $\Phi^+_{A\mu}$ in which the orbital $\phi_{A\mu}$ of monomer A (usually the HOMO) has been ionised. This localisation produces a polarisation of neighbouring neutral monomers corresponding to the solvation of an ion [134]: in addition to the ionisation energy $I_{A\mu}$ of the monomer, the dispersion contributions from pairs AB are replaced by more attractive polarisation contributions A^+B. This means that the charged monomer tends to be localised at the center of the cluster in such a way as to maximise the number of favourable interactions by efficiently polarising all the neutral molecules organised thereby into solvation layers. This is no longer necessarily true for highly anisotropic systems where steric issues may prove predominant. The mutual interactions between dipoles induced on neutral monomers can also play a significant role. Finally, for small sizes, relaxation of the intramolecular geometrical degrees of freedom of charged monomers can also have considerable influence.

In homogeneous systems, the ionisation potentials $I_{A\mu}$ of isolated monomers are equivalent and there is a possibility of charge being delocalised over all the monomers. This is charge resonance. One then seeks a wave function [104, 193] in the form of a linear combination of locally ionised configurations $\Phi^+_{A\mu}$, viz.,

$$\Psi^+_m = \sum_{A\mu} c_{A\mu} \Phi^+_{A\mu} .$$

A charge resonance Hamiltonian then describes hole transfers. Using hole creation operators $a^\dagger_{A\mu}$ and annihilation operators $a_{A\mu}$ in the HOMO $\phi_{A\mu}$ of a given monomer, the resonance Hamiltonian has the simple second quantised form

$$h = \sum_{A} (I_{A\mu} + V^+_{A\mu}) a^\dagger_{A\mu} a_{A\mu} + \sum_{A<B} t_{A\mu,B\nu} a^\dagger_{A\mu} a_{B\nu} ,$$

where $V^+_{A\mu}$ is the interaction energy of the system with a hole in the orbital $\phi_{A\mu}$. This energy is determined as in the non-resonant case. The term $t_{A\mu,B\nu}$ is a hopping integral, describing the transfer of a hole from one entity to the other. It is a function of the overlap between the orbitals $\phi_{A\mu}$ and $\phi_{B\nu}$ of the relevant monomers. Note that there may be significant delocalisation even in heterogeneous systems. Delocalisation between inequivalent species A and B then depends on the values of the hopping integrals $t_{A\mu,B\nu}$ compared with the difference $|I_{A\mu} + V^+_{A\mu} - I_{B\nu} - V^+_{B\nu}|$ between the 'dressed' ionisation potentials of the two species. It may also be necessary to consider several ionisation configurations $\Phi_{A\mu}$ and $\Phi_{A\nu}$ for the same monomer in the case of degeneracy or near-degeneracy of the internal orbital structure of each monomer below the Fermi level.

Let us examine the situation for rare gas dimers, starting with He_2, which has the simplest electronic structure. The ground state Φ_0 of He_2, very weakly bound, is well described by a single determinant. Neglecting overlap between the atomic functions, one can use either a delocalised description in terms of molecular orbitals $\Phi_0 = |\sigma_g^2\sigma_u^2|$, or an equivalent localised description in terms of atomic orbitals $\Phi_0 = |s_a^2 s_b^2|$. This arises because, in a filled valence shell, the localised and delocalised occupied orbitals are related to one another by a unitary transformation. The ground state of the molecular ion He_2^+ is stabilised by a bond that results from the fact that the highest antibonding molecular valence orbital σ_u contains only one electron, thus leading to an overall state of bonding that is reinforced by attractive polarisation forces (weak in the case of helium, but stronger in the heavier rare gases). Hence, the dissociation energies of homogeneous molecular ions are of the order of one electronvolt. One way of obtaining the valence states of the He_2^+ ion is to use a delocalised description by removing one electron from each of the σ_g (bonding) or σ_u (antibonding) orbitals, thereby generating two single-determinant configurations $\Phi_u^+ = |\sigma_g^2\sigma_u|$ and $\Phi_g^+ = |\sigma_g\sigma_u^2|$. The equivalent possibility involves expressing the Hamiltonian relative to the basis of valence-bond-type charge resonance configurations $\Phi_a^+ = |s_a s_b^2|$ and $\Phi_b^+ = |s_a^2 s_b|$. In the latter basis, the Hamiltonian has matrix

$$h = \begin{pmatrix} I_a^+ + V_a^+ & t_{ab} \\ t_{ab} & I_a^+ + V_a^+ \end{pmatrix} ,$$

where $V_a^+ = V_b^+$ is the energy of interaction of the configurations He^+He and $HeHe^+$, corresponding to an ionised atom interacting with a neutral atom at finite distance. The non-diagonal term t_{ab}^+ (< 0) is the hopping integral of the hole between the two resonant configurations. The eigenstates of the Hamiltonian are the molecular configurations Φ_u^+ and Φ_g^+. The state ${}^2\Sigma_u^+$ is thus stabilised by the hopping integral, whereas the state ${}^2\Sigma_g^+$ is destabilised.

In heavier rare gas ions, the valence shell np generates four molecular valence orbitals, viz., σ_g and π_u bonding and π_g and σ_u antibonding. Since the hopping integral between atomic np_σ orbitals (pointing along the axis) is greater than that between np_π orbitals (perpendicular to the axis), the energy gap between the σ orbitals is larger than the one between the π orbitals, thus determining the energy hierarchy $\sigma_g < \pi_u < \pi_g < \sigma_u$. Molecular ion configurations are characterised by a reversed energy hierarchy with respect to that of the orbitals, i.e., ${}^2\Sigma_u^+ <{}^2 \Pi_g <{}^2 \Pi_u <{}^2 \Sigma_g^+$. In the ground state ${}^2\Sigma_u^+$, the equilibrium distance is significantly shortened compared with that of the neutral van der Waals pair. Owing to the greater overlap between the np orbitals when they are aligned with the interatomic axis (np_σ) than when they are perpendicular to it (np_π), the delocalisation of a σ hole may be spread out axially, thereby greatly stabilising the linear geometries in small $n = 3$ and $n = 4$ ionised clusters. Hence, charge delocalisation over the symmetric linear trimer Ar_3^+ corresponds to a charge close to 0.5 on the central atom and 0.25 on each of the end atoms. For other sizes or species, the missing charge is delocalised over two, three or four atoms forming a kind of ionised core that is relatively strongly bound (of the order of 0.1–0.3 eV) compared with the van der Waals interactions. This ionic core continues to be a stabilising entity in larger clusters, while the rest of the atoms remain essentially neutral and organise themselves around it. Their positions are determined partly by the polarisation forces induced by the core charge, and partly by their mutual van der Waals interactions, which determine the number of atoms in the surrounding rings. In neon, the ionic core is just Ne_2^+. In xenon, there are two competing types of core, i.e., Xe_3^+ and Xe_4^+. The latter is linearly symmetric (symmetry group D_h) with a charge distribution close to (0.1, 0.4, 0.4, 0.1). This partial localisation of the charge determines in particular the magic numbers related to the first solvation layer, observed in ionised clusters. Argon and xenon clusters thus have particularly stable structures for $n = 13$ (see Fig. 6.10), corresponding to a trimeric core and two rings of 5 atoms around each bond (forming an icosahedron, distorted along the axis), and for $n = 19$ (see Fig. 6.11) with a linear tetrameric core and three rings of 5 atoms (forming a distorted double icosahedron). The magic numbers are thus the same as in the ground state, but for rather different reasons. Indeed, the stability of the ground state is determined by the icosahedral geometric constructions. In the ion, the same magic numbers result at the same time from the electronic stability of the linear ionic core (in which the interatomic separations are very small) and the more favourable geometric arrangement around the core. This is an example of charge inhomogeneity in a system with uniform composition.

Some electronic excitations can also be described by resonant models insofar as the initial and excited orbitals belong to the same entity. One then speaks of an exciton model, which can be considered as an extension of the charge transfer model, but which now describes excitation transfer. This type of excitation is the molecular cluster analogue of Frenkel excitons in the solid. The excited wave functions are expressed in the form

$$\Phi_m^* = \sum_{A\mu\lambda} c_{A\mu\lambda}^m \Phi_A^{\mu\lambda} ,$$

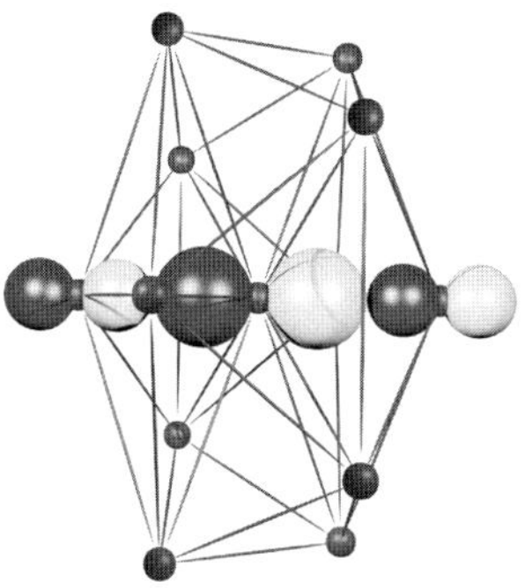

Fig. 6.10. Lowest energy isomer of Ar_{13}^+ and qualitative representation of the singly occupied HOMO

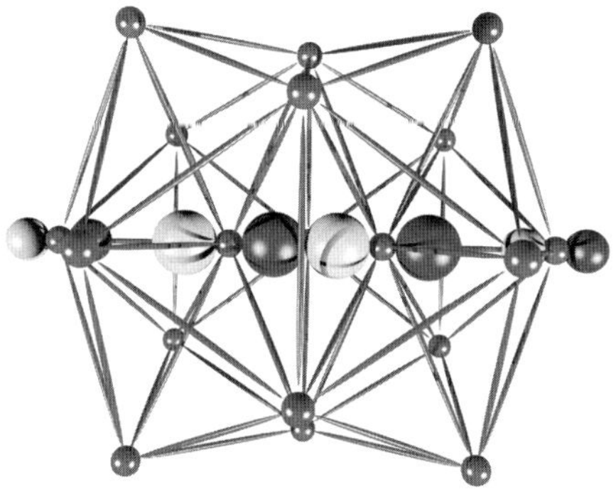

Fig. 6.11. Lowest energy isomer of Xe_{19}^+ and qualitative representation of the singly occupied HOMO

where $\Phi_A^{\mu\lambda}$ represents an excited state localised on A in which one electron has been promoted from the occupied orbital $\phi_{A\mu}$ to the virtual orbital $\phi_{A\lambda}$ in the same molecule. In this basis, the diagonal terms of the Hamiltonian are now the energies of the excited molecule A^* in interaction with the other molecules in their ground states. Couplings or non-diagonal terms are associated with excitation transfer and expressed in terms of two-electron integrals

$$H_{A\mu\lambda,B\nu\gamma} = \int \frac{\phi_{A\mu}(\boldsymbol{r})\phi_{A\lambda}(\boldsymbol{r})\phi_{B\nu}(\boldsymbol{r}')\phi_{B\gamma}(\boldsymbol{r}')}{|\boldsymbol{r}-\boldsymbol{r}'|}\mathrm{d}^3\boldsymbol{r}\mathrm{d}^3\boldsymbol{r}' - \int \frac{\phi_{A\mu}(\boldsymbol{r})\phi_{B\nu}(\boldsymbol{r})\phi_{A\lambda}(\boldsymbol{r}')\phi_{B\gamma}(\boldsymbol{r}')}{|\boldsymbol{r}-\boldsymbol{r}'|}\mathrm{d}^3\boldsymbol{r}\mathrm{d}^3\boldsymbol{r}' .$$

In the dipole approximation, the first integral can be expressed in terms of the dipoles $\boldsymbol{D}_A^{\mu\lambda} = \langle\phi_{A\mu}|\boldsymbol{r}|\phi_{A\lambda}\rangle$ and $\boldsymbol{D}_B^{\nu\gamma} = \langle\phi_{B\nu}|\boldsymbol{r}|\phi_{B\gamma}\rangle$ associated with the electron distributions of intramolecular transitions $\phi_{A\mu}(\boldsymbol{r})\phi_{A\lambda}(\boldsymbol{r})$ and $\phi_{B\nu}(\boldsymbol{r})\phi_{B\gamma}(\boldsymbol{r})$:

$$H_{A\mu\lambda,B\nu\gamma} \approx \frac{\boldsymbol{D}_A^{\mu\lambda}\cdot\boldsymbol{D}_B^{\nu\gamma}}{R_{AB}^3} - 3\frac{(\boldsymbol{D}_A^{\mu\lambda}\cdot\boldsymbol{R}_{AB})(\boldsymbol{D}_B^{\nu\gamma}\cdot\boldsymbol{R}_{AB})}{R_{AB}^5} .$$

It only contributes to singlet states and generates an interaction going as $1/R^3$. The second term corresponds to exchange. It depends on the overlap that comes into play at shorter range. It is the only residual coupling in the case of triplet states for which the contribution of the first integral is zero.

These exciton models assume a concerted localisation of hole and particle. They have been applied to rare gas clusters [101] and more recently to various molecular clusters, in particular those made up of aromatic molecules, e.g., naphthalene and anthracene, and even other complex molecular systems such as double-stranded oligonucleotide fragments of DNA [194]. As for the charge, the coefficients in the expansion can be used to analyse the degree of delocalisation of the excitation over the various sites. The model is inapplicable either when the excited states are charge transfer states (A^+B^-), or when the excitation concerns highly excited (or Rydberg) states which can no longer be represented on local components. Hybrid models using pseudopotential techniques must then be used with an explicit quantum description of the excited electron [195, 196].

6.4.3 Ionic and Ionocovalent Clusters

Ionic crystals are made up of periodic assemblages of ions with opposite charges. The alkali halides (MX, M=metal, X=halogen), which crystallise into simple cubic (e.g., CsCl) or face-centered cubic (e.g., NaCl, NaF, Kbr, LiF) lattices, provide typical examples. One may consider that, in these structures, the valence electron of each alkali metal atom is transferred to a halogen, with all atoms adopting electronic configurations close to those of the electronically similar rare gases, e.g., neon for Na^+ and argon for Cl^- in NaCl. Some of these crystals are characterised by ionic charges $q_{M,X}$ close to ± 1, short distances between ions of opposite charge, and high cohesive energies. For a non-interacting set of n atoms of M and X (i.e., infinitely far apart), the energy cost corresponding to the formation of M^+ and X^- ions is

$$\Delta E(\infty) = \sum_{\mathrm{M,X}} (I_{\mathrm{M}} - A_{\mathrm{X}})$$

where I_{M} is the ionisation potential of the metal atom and A_{X} the electron affinity of the halogen. This energy penalty is balanced by Coulomb interactions between ions, or the Madelung field which guarantees the high stability of the crystal owing to the proximity of oppositely charged ions. At short range, one must also add the repulsive interactions due to interpenetration of the electron clouds of the anions and cations:

$$E = \Delta E(\infty) + V_{\mathrm{rep}} + V_{\mathrm{Mad}} .$$

Many compounds such as metallic halides, hydrides, and of course oxides fall within this category. Clearly, stabilisation of the ionic structure depends on how favourable the difference is between the ionisation potentials and the electron affinities of the constituents, these expressing the ease with which an electron can leave one atom and attach itself to another. It is also determined by the ionic charges which may only be partial, and by Coulomb forces, themselves dependent on the geometry. Similar situations are encountered in stoichiometric clusters and nanoparticles of positive and negative ions, with a complication induced by the disappearance of periodicity. Identical atoms that occupy inequivalent positions feel different electrostatic potentials, depending whether they are at the surface, in the bulk, or included in defects.

A potential model that is widely used to represent the potential surface of the ground state of ionic clusters in the stability region is the so-called rigid ion model [84–89]. The name simply means that a fixed charged is assigned to each ion. This model thus includes a simple Coulomb term depending on the charges of the ions:

$$E = \Delta E(\infty) + \sum_{a<b} \left[\lambda \exp(-\beta_{ab} R_{ab}) + \frac{q_a q_b}{R_{ab}} \right] .$$

The repulsive term has an exponential form here. A more sophisticated model treats the ions as polarisable and adds the polarisation energy induced on each ion by the others, a contribution that can be screened at short range to avoid divergences. These Coulomb-type models can describe perfect or weakly polarised ionic systems, but they are not suitable for ionocovalent systems. These are characterised by partial charges and differ too much from the ideal ionic situation, a (reverse) charge transfer giving significant weight to the covalent situation. Indeed this leads to a certain level of hybridisation in the atomic orbitals of the anions and cations. Charge transfers can be quantified by the electronegativity ξ which is equal to minus the chemical potential μ and measures the capacity of the system to vary the number of electrons within it:

$$\xi = -\frac{\partial E}{\partial N} = -\mu .$$

Discretising the definition, one obtains the ionisation potential and electron affinity

$$I(N) = E(N-1) - E(N) , \qquad A(N) = E(N) - E(N+1) ,$$

respectively.

In the limit of purely ionic systems, the electrons are localised on the atomic orbitals $\chi_{a\mu}$ of the negative ions. The HOMO are thus anionic orbitals, whereas the LUMO are cationic orbitals $\chi_{c\nu}$ that have lost their electrons. In a one-electron model, this means that the hopping integrals between the orbitals of nearest neighbour anions and cations are small compared with the differences between diagonal one-electron levels ϵ_c and ϵ_a of cations and anions, respectively (we take $\epsilon_{c\mu} = \epsilon_c$ and $\epsilon_{a\lambda} = \epsilon_a$, for all μ and λ):

$$|t_{c\mu,a\nu}| \ll |\epsilon_c - \epsilon_a| .$$

Like insulating ionic crystals, these clusters have a large band gap.

In ionocovalent systems, it is important to quantify charge transfer. Since the charge of each ion is a cumulative indicator, the details of the transfer on each bond cannot be specified by analysing the charge alone. A systematic model for analysing transfers has been developed by Finocchi et al. [118–124] in the framework of a DFT ab initio calculation of the electronic structure of small metal oxide clusters. The model for electron transfers Δ_{ac} per bond is based here on the so-called bond electron transfer model (BETM) due to Harisson [26]. Taking a fully ionic situation as reference here, the electron transfer that reestablishes a covalent contribution on a bond can be analysed by perturbative treatment of the hopping integral. At anionic sites, the first order perturbed orbitals are

$$\chi_{a\mu}^{(1)} = \chi_{a\mu} + \sum_{c\nu} \frac{t_{a\nu,c\mu}}{\epsilon_a - \epsilon_c} \chi_{c\nu} ,$$

corresponding to charges

$$q_a = q_a^{(0)} + \sum_c \left[\sum_{\mu\nu} \left(\frac{t_{a\mu,c\nu}}{\epsilon_a - \epsilon_c} \right)^2 \right] .$$

In this approximation, charge transfers form an additive system:

$$q_a = q_a^{(0)} + \sum_c \Delta_{ac} , \qquad q_c = q_c^{(0)} - \sum_a \Delta_{ac} ,$$

where $q_a^{(0)}$ and $q_c^{(0)}$ are the respective charges of anions and cations in the purely ionic limit. The sums are limited to nearest neighbours. Bond electron transfers Δ_{ac} cancel in the purely ionic limit, and depend on the squares of the ratios $|t_{a\mu,c\nu}|/(\epsilon_c - \epsilon_a)$. The effective transfers are obtained by inverting the above equations when the atomic charges are known. One thus obtains information about the ionocovalent nature of each bond.

In the charge-dependent self-consistent tight-binding approximation introduced above, the diagonal elements of the Hamiltonian are the effective levels taking into account the perturbation by the Madelung field:

$$\epsilon_{a\mu} = \epsilon_{a\mu}^{(0)} - U_{aa} q_a - \sum_{b \neq a} q_b U_{ab} = \epsilon_{a\mu}^{(0)} - U_{aa} q_a - V_a .$$

The first correction term U_{aa} comes from the electron–electron repulsion at the site, while the second represents the Madelung potential V_a induced by other ions. The latter depends sensitively on the environment.

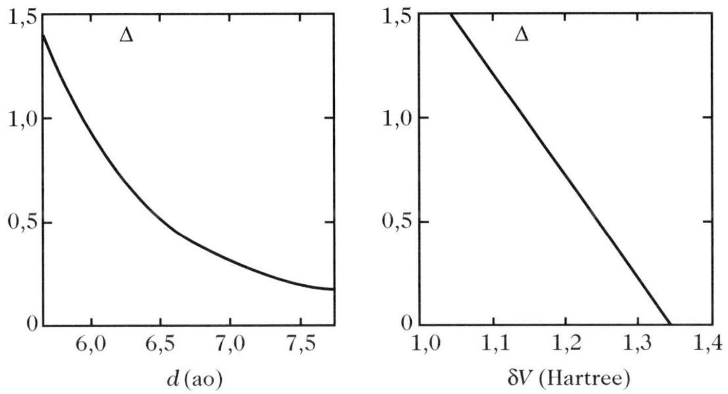

Fig. 6.12. Electron transfer Δ in titanium oxide clusters as a function of the distance Ti–O (*left*) and the electrostatic potential difference δV between Ti atoms and neighbouring O (*right*). Qualitative interpolation of data from Albaret et al. [122–124]

Hence, compared with the energy levels of isolated ions, the levels of the anions of a crystal, carried by the surrounding positive charges to an electrostatic potential $V_a > 0$, are lowered (even though the on-site electron–electron repulsion term acts in the opposite way), whereas the levels of the cations are pushed up to higher energy. It follows that the electronic levels of an isolated atom, a bulk atom, or an atom at the surface of a crystal or cluster can all be very different. The same goes for the atomic electronegativities of the different sites,

$$\xi_a = \xi_a^{(0)} + U_{aa} q_a + V_a \ .$$

This explains why, insofar as the electrostatic potential created by the Madelung field dominates over the interatomic term U_{aa}, the electrons tend to leave anionic sites with lower coordination number, which are less stabilised. Likewise, reducing properties will tend to affect cationic sites subject to weaker electrostatic potentials. Figure 6.12 clearly shows this correlation between coordination and bond electron transfers Δ for titanium oxide clusters Ti_nO_{2n+p} studied by Albaret et al. [122–124]. Another interesting feature is the correlation between transfers and bond lengths. As the coordination number decreases, the bond shortens, and the hopping integral (proportional to the interatomic overlap) and the quotient $|t_{a\mu,c\nu}/(\epsilon_a - \epsilon_c)|$ increase, and so therefore does electron transfer from the anion to the cation. Figure 6.12 shows this anti-correlation between bond lengths and charge transfers. This analysis provides an understanding of the ionisation potentials, electron affinities, and charge transfers. The transfer mechanisms influence the reactivity and acid–base properties. They play an analogous role for clusters, surfaces and nanostructures.

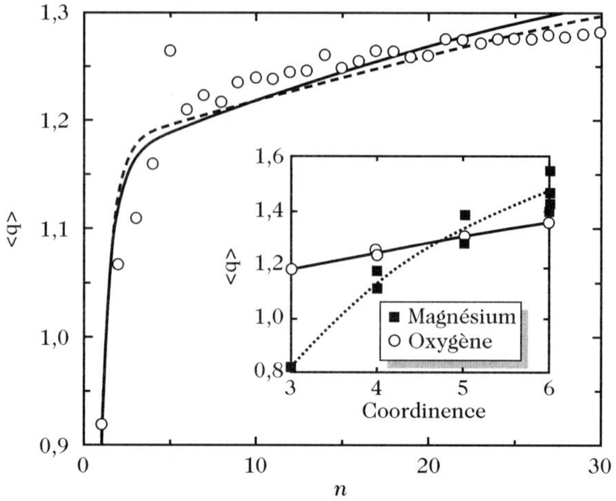

Fig. 6.13. Size dependence of the average charge per atom $\langle q \rangle$ in $(MgO)_n$ clusters. Fluctuating charge model, also including polarisation effects. *Insert*: Dependence of the charge on the coordination number of atoms in $(MgO)_{108}$. Taken from Calvo [183]

For large ionocovalent nanoparticles, more flexible models than the rigid or polarisable ion models can be used. Hence, the fluctuating charge model [179–182] involves minimising the energy

$$E(\boldsymbol{R}) = V_{\mathrm{rep}}(\boldsymbol{R}) + \sum_{k=a,c} \left(\xi_k q_k + \frac{1}{2} U_{kk} q_k^2 \right) + \sum_{k<l} J_{kl}(R_{kl}) q_k q_l + \lambda \left(q - \sum_k q_k \right)$$

with respect to the charge variables q_k, determining the latter at each site (anion a or cation c). The first term V_{rep} is a short-range repulsive function, ξ_k is the electronegativity of ion k ($-\epsilon_a$ or $+\epsilon_c$), U_{kk} is an interatomic electron–electron penalty term, and the functions J_{kl} are two-centre Coulomb interactions screened at short range. The Lagrange multiplier λ fixes the total charge. This model is used to study ionocovalent systems without assumptions about charge transfer. For example, one can then look for an ionocovalent–ionic transition during the growth of an oxide nanoparticle, or study fluctuations in the charge transfer depending on the position of the atoms, in the bulk or at the surface. Figure 6.13 shows the the dependence of ionic charges calculated with the fluctuating charge model in $(MgO)_n$ clusters as a function of the size n. It also shows the fluctuation of the charges as a function of the coordination number of the magnesium or oxygen atoms in the $(MgO)_{108}$ cluster. The metal–oxygen charge transfer is of the order of 0.8 in the diatomic molecule MgO, but close to 2 in the crystal. The ionic character of the bonds is clearly seen to increase continuously with size in the case of clusters.

6.4.4 Covalent Systems

The prototypes of covalent and semiconductor clusters are carbon clusters and silicon clusters. These two families of clusters nevertheless display rather different properties, in particular due to hybridisation, which is almost always sp^3 in silicon clusters. The world of carbon clusters has been the subject of much investigation since the discovery of fullerenes and especially the most remarkable of these, the C_{60} cluster, followed by carbon nanotubes. Reviews can be found in [139, 140, 145].

In all these nanoparticles, the electronic structure of the carbon atoms is very different from that in diamond carbon (sp^3). It is very highly correlated with the geometric structure. In the free state, the smallest fullerene clusters occur in the form of linear or ring-shaped chains, characterised mainly by sp hybridisation. Carbon clusters then adopt 2D arrangements, sometimes distorted (close to spherical caps). Finally, 3D shapes arise, as illustrated by the C_{60} cage with very high molecular symmetry (the icosahedral I_h symmetry group), or C_{70} with a longer structure (the D_{5h} symmetry group). This structural behaviour is highly specific and differs greatly from what is observed in silicon clusters, for example, which do not naturally form hollow cages.

A qualitative approach to the electronic structure of carbon cycles and rings can be obtained using a tight-binding model limited to nearest neighbours and based on the σ–π separation. For linear chains along the z axis, the σ and π orbitals are distributed according to the eigenvalues λ (0 or 1) of l_z. For plane cyclic chains, the separation is defined by the symmetry or antisymmetry of the orbitals with respect to the plane of the cycle. The bonding orbitals of the σ system are usually deeper than those of the π system, while the antibonding orbitals are higher. It is thus the bonding π orbitals and antibonding π^* orbitals that encompass the Fermi level. Let us consider a tight-binding model based on the atomic orbitals χ_r of type $2p$ at each site r. For a cycle, only one function is retained per site, perpendicular to the plane of the cycle. This (highly qualitative) model is then defined topologically by the hopping integrals t between nearest neighbour atoms. The eigenfunctions (the molecular orbitals) of regular rings of n atoms are easily determined using the cyclic symmetry:

$$\phi_k = \sqrt{\frac{1}{n}} \sum_{j=0}^{n-1} \exp\left(2\pi \mathrm{i} \frac{kj}{n}\right) \chi_j \,,$$

leading to pairwise degenerate k and $-k$ levels

$$\epsilon_k = 2t \cos\left(\frac{2\pi \mathrm{i} k}{n}\right) \,,$$

with the exception of the lowest bonding level ($k = 0$), which is non-degenerate, and also the highest antibonding level ($k = n/2$) for even cycles.

In particular, it follows that the HOMO level ($k = \pm n/4$) can contain four electrons.

Linear chains are handled with two p orbitals ($2p_x$ and $2p_y$) per site, perpendicular to the axis z of the chain. The two π systems are clearly isomorphic and the solution is deduced from the previous case by assuming that all atoms feel the same potential (identical hopping integrals between nearest neighbours) with the exception of sites l and n for which outward hopping is not allowed. These boundary conditions imply that the wave function must vanish at 0 and $n+1$, a condition satisfied by

$$\phi_k = \sqrt{\frac{1}{n+1}} \sum_{j=1}^{n} \sin \frac{\pi k j}{n+1} \chi_j \ .$$

For linear chains, each π level is therefore spatially doubly degenerate (orbitals π_x and π_y), with energy

$$\epsilon_k = 2t \cos \frac{k\pi}{n+1} \ .$$

The electronic structure of small linear clusters can then be deduced. A neutral cluster of size n has n valence electrons. It can form $2(n-1)$ σ bonds (hybridisation sp^1) and two free pairs at the ends, making a total of $2n+2$ electrons. The π system thus contains $2n-2$ electrons, bestowing greater stability on odd neutral cycles whose π orbitals are filled, with $2n-2 = 4m$, i.e., $N = 2m+1$.

Rings have no free pair and the number of σ orbitals along C–C bonds is n. Since the other σ orbitals of the plane perpendicular to the bonds and the orbitals of the π system are each doubly degenerate, except for the two lowest, the maximal stability condition is $2n = 4m$, i.e., $n = 2m$, now favouring even cycles.

It is easy to find the HOMO–LUMO gap using this model. For example, considering odd linear chains [$k_{\mathrm{HOMO}} = (n-1)/2$, $k_{\mathrm{LUMO}} = (n+1)/2$], the gap is given by

$$\Delta\epsilon = -2t \left[\cos \frac{(n-1)\pi}{2(n+1)} - \cos \frac{(n+1)\pi}{2(n+1)} \right] = -2t \sin \frac{\pi}{n+1} \ ,$$

and thus decreases as $1/n$ for long chains.

This description remains rather simplistic. In particular, it must be modulated by the alternating character of the bonding distances in chains or cycles, corresponding to the formation of simple and double C–C bonds. Hybridisation also has a non-negligible effect on the positions of the single-electron levels.

Apart from ab initio calculations [146–148], the electronic structure of fullerenes has also been treated using the tight-binding approach. Due to the high level of symmetry of C_{60} (point group I_h) and the delocalised nature of

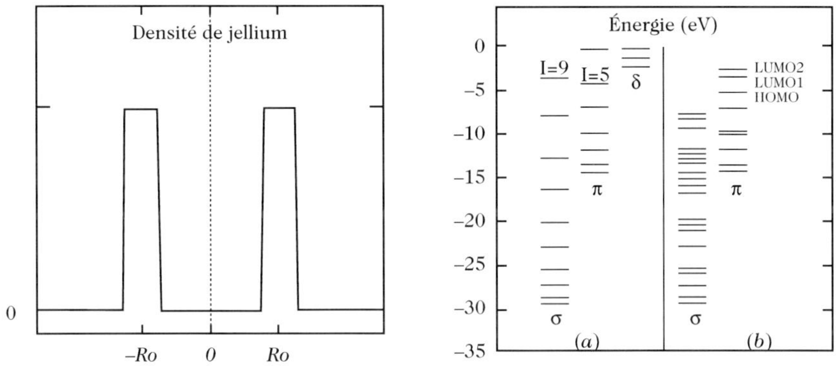

Fig. 6.14. *Left*: Positive charge density in the model of the nuclear skeleton of C_{60} by a hollow spherical jellium distribution. *Right*: Electronic level structure according to Yannouleas and Landman [178], with (**a**) and without (**b**) correction for the crystalline field

the π electrons, it is also instructive to model this system using the hollow jellium approach [149–151, 175, 178], which takes into account the cage structure with 240 valence electrons (four $2s$ and $2p$ electrons per atom). In the calculation by Yannouleas et al. [176, 178], the total charge $Z = 240$ of the jellium is distributed with uniform density in a spherical shell of thickness $2d$ around a sphere of radius R [see Fig. 6.14 (left)]. The effects of the crystalline field representing the pointlike nature of the nuclei and the icosahedral symmetry can be incorporated into the model by adding the following potential to the mean field one-electron operator:

$$U(\boldsymbol{r}) = -v \sum_a \frac{1}{|\boldsymbol{r} - \boldsymbol{R}_a|} ,$$

expanded in spherical harmonics. The crystalline field significantly modifies the electronic structure in the vicinity of the Fermi level.

Figure 6.14 (right) shows the one-electron levels of C_{60}. In the hollow jellium model without ionic structure, and due to the spherical symmetry, the levels (**a**) are characterised by the principal quantum number k associated with the number of nodes of the radial function, i.e., in the direction perpendicular to the fullerene surface, together with the angular momentum l and its projection m_l. Levels are classified according to the values of k, into either σ ($k = 0$), π ($k = 1$) or δ ($k = 2$), and then in each class by the values of l, beginning with $l = 0$. Ignoring the curvature, this σ, π classification is compatible with the one used for plane carbon-bearing molecules in which the σ system is symmetrical with respect to the molecular plane, whereas the π system is antisymmetrical with respect to this plane. When the degeneracy due to the spherical symmetry is partially removed by the crystalline potential (Fig. 6.14b), the HOMO remains ten-fold degenerate (symmetry group h_u), whereas the lowest virtual orbitals resulting from the π spherical levels ($l = 5$ and 6) have symmetry t_{1u} and t_{1g}, and are six-fold degenerate. The σ level ($l = 9$) splits into five sublevels. Two of these (g_u eight-fold degenerate and h_u ten-fold degenerate) are considerably lowered and become closed shell occupied levels, while the others (not shown in the figure) are significantly raised in energy, thereby creating a large gap in the σ symmetry. Starting from the hollow jellium as a zero order model, it is easy to see the role played by the icosahedral symmetry due to the discrete structure of the C_{60} atoms in the orbital energies and degeneracies near the Fermi level.

6.5 Valence Changes

Valence changes, i.e., transitions in the valence electronic function and electronic properties, can occur. On the one hand, the energy bands of the solid build up and broaden with the size of the nanoparticle. Different bands arising from different atomic levels are then likely to meet and interact above certain critical sizes (see Fig. 6.2). The nature of the occupied electronic levels can then be expected to change. This may have consequences for the insulating or metallic character of the nanoparticle, for example, or for its magnetic properties. On the other hand, at a given size, controlled fabrication and manipulation of nanoparticles can be used to vary the composition of mixed systems. In this case it is therefore possible to observe changes in electronic structure as a function of the stoichiometry.

6.5.1 Transitions with Size

Divalent metal clusters (Be_n, Mg_n or Hg_n) are composed of atoms with a closed shell s^2 electronic configuration. In small clusters of divalent metals, the s electrons remain localised on the atoms which are thus bound by van der Waals interactions, and these systems are therefore insulating. The s band is full and the Fermi level is defined by the top of the band. However, in the solid state, these elements have metallic properties and there is therefore an insulator–metal transition. This transition was studied by Pastor et al. in mercury clusters (see Fig. 6.15) [130]. The propensity of an s electron to delocalise to different sites involves promotion into the p band and clearly depends on the energy of the top of the s band relative to states at the bottom of the p band which allow electron transfers. The cost of an electron transfer $s_a \longrightarrow p_b$ on a single bond ab in the free electron approximation is $\Delta\epsilon_{sp} = \epsilon_p - \epsilon_s$. Now the band widths are proportional to $\sqrt{z}$, where z is the coordination number. On the basis of criteria arising from the tight-binding model, one can thus expect a metallic transition when the gap disappears, i.e., when $W_s + W_p = 2\Delta\epsilon_{sp} = 11.6\,\text{eV}$. Assuming uniform densities of states and hopping integrals limited to nearest neighbours, the condition becomes $12(t_{ss}^2 + t_{p\sigma p\sigma}^2 + 2t_{p\pi p\pi}^2)z \approx 4\Delta\epsilon_{sp}^2$, where $t_{ss} = 0.58\,\text{eV}$, $t_{p\sigma p\sigma} = 0.98\,\text{eV}$,

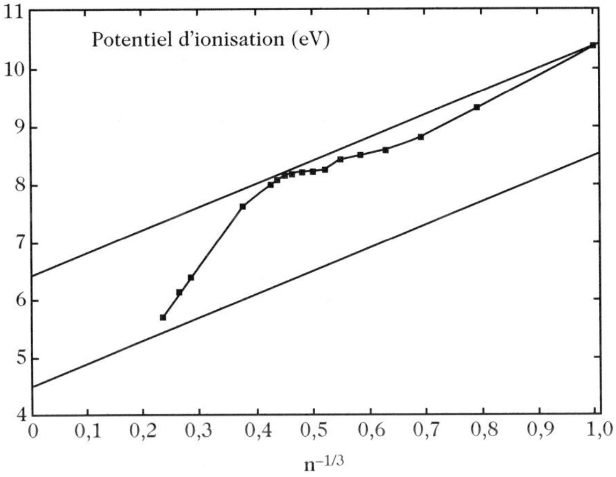

Fig. 6.15. Ionisation potential of mercury clusters as a function of $n^{-1/3}$. From Pastor et al. [130]

$t_{p\pi p\pi} = -0.27$ eV in the solid. This simple calculation predicts a transition for an average coordination of $z \approx 7.79$. The average coordination in an icosahedron of 13 atoms, with 12 atoms at the surface, is slightly greater than 5 (in fact, 5.53). For an icosahedron of 55 atoms, it is 7.96, and for a cuboctahedron, it is 7.85. This allows us to estimate the size range in which the transition is likely to occur. This simple calculation assumes a uniform density of states and does not take into account either the renormalisation of the atomic levels with size, or the change in the hopping integral due to relaxation of distances in the cluster, or indeed electron correlations. In reality, in a Hartree–Fock calculation, the energy difference ΔE_{sp} is not exactly equal to $\Delta\epsilon_{sp}$, but can be expressed in terms of the ionisation potential I, the electron affinity A, and the Coulomb exchange integrals. At a bond ab,

$$\Delta E_{sp} \approx \Delta\epsilon_{sp} + 2U_{sp} - K_{sp} - U_{ss} - 1/R_{ab} = I - A - 1/R_{ab} ,$$

where U_{sp} and U_{ss} are single-centre Coulomb integrals and K_{sp} is a single-centre exchange integral. Going from the atom to the infinite solid, the transition occurs earlier when the electrostatic and exchange terms are explicitly taken into account.

6.5.2 Transitions with Stoichiometry

For constant size n, alkali halide clusters M_nX_{n-p} (M = metal, X = halogen) can make the transition from an insulating salt ($p = 0$) to a pure metal ($p = n$). Likewise for metal oxides with different stoichiometries. Let us consider the alkali hydrides or halides. For $p = 1$, the cluster has the physics of a hybrid system. To begin with, all the metallic valence electrons but one are captured

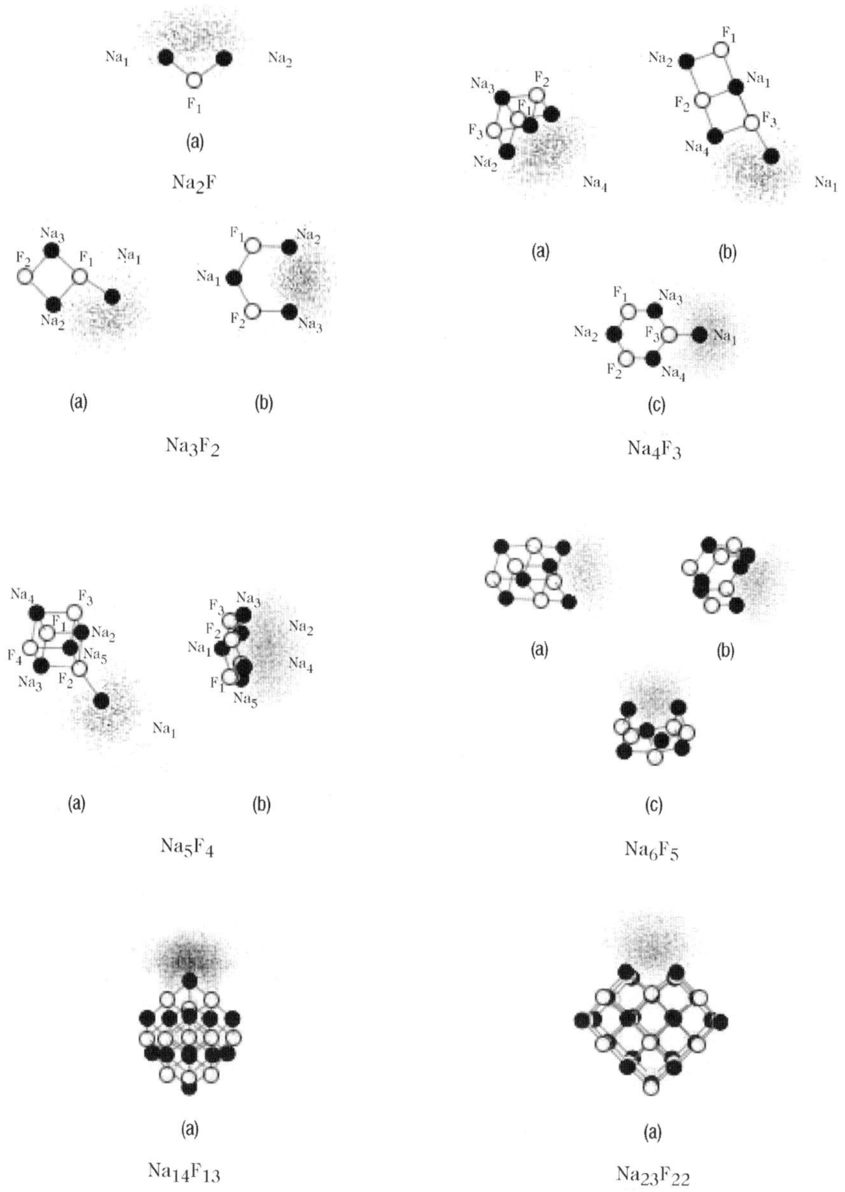

Fig. 6.16. Geometric structure and location of the extra electron in sodium fluoride clusters Na_nF_{n-1}. From Durand et al. [96, 97]

by the fluorine atoms, whereupon they generate ionic bonds between n alkali ions and $n-1$ fluorine ions, generally well described by standard Coulomb potentials. Then the extra electron cannot attach itself to a fluorine atoms. It

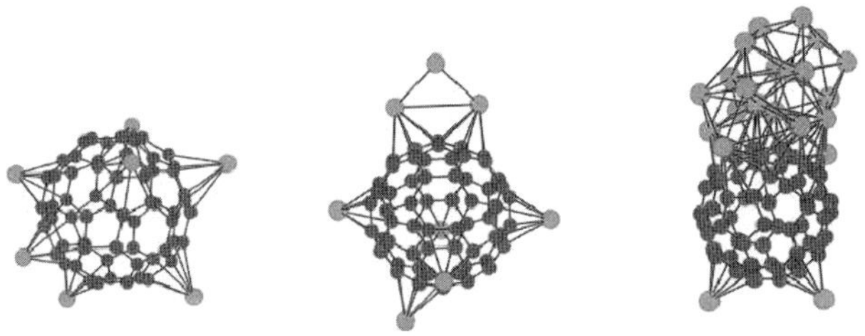

Fig. 6.17. Growth of a sodium cluster Na_n on C_{60}. Equilibrium structures occur for $n = 7$, 8 and 24. From Roques et al. [184, 185]

interacts with all the ions and must be described by quantum mechanics. The interdependence between geometric structure and electronic structure thus gives rise to a range of possible locations for this electron, depending on the size and the types of defect present on the usually cubic, sometimes hexagonal structures. Situations in which the electron replaces the missing fluorine are thus analogous to what happens with colour centres in the solid: in this case an electron fills the space left by a missing fluorine and is thus stabilised by the dominant influence of the strongly positive electrostatic potential.

Within the same family, some sizes may exhibit specific features. Hence, $Na_{14}F_{13}$ has a cubic structure that is slightly distorted along a diagonal of the cube due to an indirect Jahn–Teller effect (via coupling with excited states), where the weakly bound electron remains at the surface and is polarised on a vertex (see Fig. 6.16). In other configurations, the electron is localised on an isolated sodium atom, or replaces a missing edge. This landscape reflects the wide range of possibilities for electron localisation on free or supported nanostructures with defects.

When the number of extra metallic p atoms (and electrons) increases, a metallic part forms that is segregated from the insulating part [83, 90–92, 94, 95]. Finally, when the clusters have a large metal surplus, i.e., $n \gg n - p$, then can behave like a metallic system in which $n - p$ electrons are attached to the halogens. We then return to the properties of simple metals and the shell model, but with electronic magic numbers shifted by $n - p$ with respect to those of a pure metal cluster of size n, reflecting shell closures associated with the p delocalised metallic electrons [83, 159, 160]. The metallic nature of the latter reveals itself through collective excitations.

Finally, other mixed systems, some more complex, provide a good illustration of the interaction between electronic and structural properties. For example, whereas the endofullerenes can trap metallic atoms in the carbon cage, it is also interesting to see how metallic atoms can cover a fullerene. The contrast here is between the segregation of a metallic clump (droplet) or the distribution of metallic atoms around the cage (wetting). This question is closely related to the question of charge transfer between metallic atoms and the fullerene. In the case of $C_{60}Na_n$ studied using a fluctuating charge model,

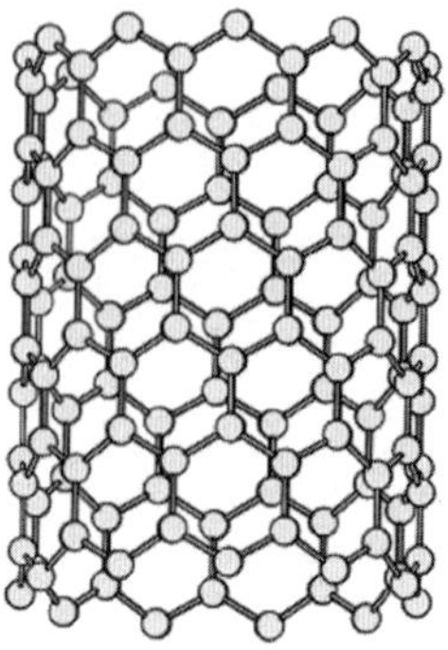

Fig. 6.18. Atomic structure of a carbon nanotube

Roques et al. [184, 185] showed that a transition occurs towards segregation for $n = 8$ (see Fig. 6.17). At smaller sizes, each alkali atom almost completely transfers its electron to the fullerene. The metallic ions distribute themselves uniformly over the surface under the effect of electrostatic repulsion. When the number of metallic atoms is increased, charge transfer saturates. At the same time, the greater number of metallic atoms serves to screen the charges, favouring the formation of metallic binding and a segregated droplet.

6.6 Nanotubes

Single-walled carbon nanotubes are obtained by rolling up a graphite sheet about an axis (see Fig. 6.18). The roll-up axis, which makes the unit cells overlap at an angle of 2π, plays a key role in the electronic structure. Indeed, whereas clusters always have a gap owing to their finite size, carbon nanotubes may or may not possess such a gap, depending on the choice of roll-up axis. The band structure of these objects can be simply understood in terms of the band structure of the graphite sheet.

The unit cell in this 2D lattice comprises two carbon atoms a and b separated by a distance d_{cc}, and has lattice parameter $a = \sqrt{3}d_{\mathrm{cc}}$. If $\boldsymbol{x}$ and $\boldsymbol{y}$ are unit vectors along the axes, the basis vectors of the direct lattice can be chosen as

$$\boldsymbol{u}_1 = a\boldsymbol{x} , \qquad \boldsymbol{u}_2 = \frac{a}{2}\boldsymbol{x} + \frac{\sqrt{3}a}{2}\boldsymbol{y} .$$

Those of the reciprocal lattice are then

$$\boldsymbol{v}_1 = \frac{2\pi}{a}\left(\boldsymbol{x} - \frac{1}{\sqrt{3}}\boldsymbol{y}\right) \qquad \boldsymbol{v}_2 = \frac{2\pi}{a}\frac{2}{\sqrt{3}}\boldsymbol{y}$$

The Brillouin zone is shown in Fig. 6.19.

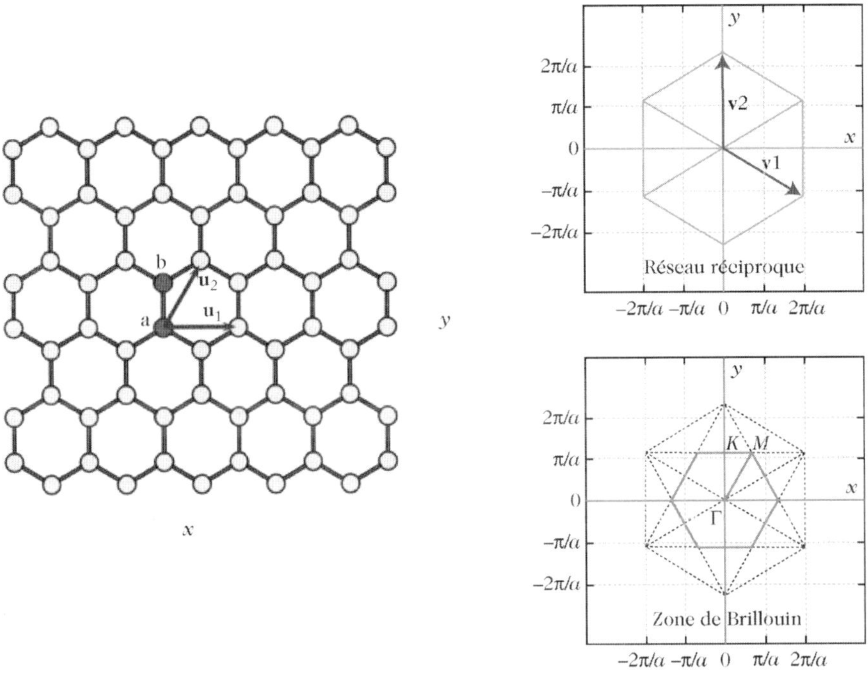

Fig. 6.19. Atomic structure, reciprocal lattice and first Brillouin zone of a graphite sheet. a and b specify the two carbon atoms in the 2D unit cell. $\boldsymbol{u}_1$ and $\boldsymbol{u}_2$ are the basis vectors of the direct lattice, while $\boldsymbol{v}_1$ and $\boldsymbol{v}_2$ are those of the reciprocal lattice. Γ, K and M define the irreducible zone

The electronic structure of graphite can be described by a tight-binding Hamiltonian h. We shall only consider the π electrons with a $\chi(2p_z)$ orbital per site and hopping integrals t restricted to nearest neighbours. If the plane contains n atoms, the π system contains n electrons (half-filled band with two electrons per orbital). The Bloch functions $\phi^{\boldsymbol{k}}_{a,b}(\boldsymbol{r})$ with 2D periodicity are given in terms of the complete set of atomic orbitals $\chi^{\boldsymbol{R}}_{a,b} = \chi_{a,b}(\boldsymbol{r} - \boldsymbol{R})$ generated by translations $\boldsymbol{R}$ of the functions χ_a and χ_b defined in the unit cell (with l_1 and l_2 integer-valued):

$$\phi^{\boldsymbol{k}}_{a,b}(\boldsymbol{r}) = \sqrt{\frac{1}{n}} \sum_{\boldsymbol{R}} \exp(\mathrm{i}\boldsymbol{k} \cdot \boldsymbol{R}) \chi_{a,b}(\boldsymbol{r} - \boldsymbol{R}) \;, \qquad \boldsymbol{R} = l_1 \boldsymbol{u}_1 + l_2 \boldsymbol{u}_2 \;.$$

In this nearest-neighbour tight-binding model, the only nonzero couplings are those between atoms a and b. Each function χ_a of an atom a in the lattice is therefore only coupled by the tight-binding Hamiltonian to the function χ_b of its three nearest neighbours. The terms $h^{\boldsymbol{k}\boldsymbol{k}'}_{ab}$ and $h^{\boldsymbol{k}\boldsymbol{k}'}_{ab}$ are zero, and in the end only the complex conjugate off-diagonal elements $h^{\boldsymbol{k}\boldsymbol{k}'}_{ab}$ and $h^{\boldsymbol{k}'\boldsymbol{k}}_{ab}$ remain:

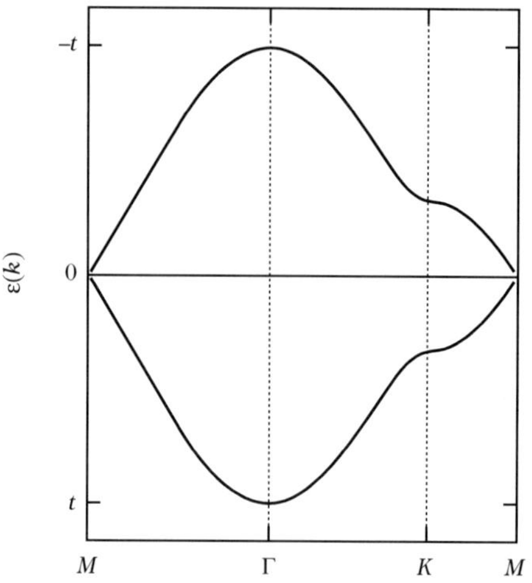

Fig. 6.20. Band structure of π electrons in a graphite sheet according to the tight-binding model

$$
\begin{aligned}
h_{ab}^{\boldsymbol{k}\boldsymbol{k}'} &= \sum_{\boldsymbol{R},\boldsymbol{R}'} \exp(\mathrm{i}\boldsymbol{k}\cdot\boldsymbol{R} - \mathrm{i}\boldsymbol{k}'\cdot\boldsymbol{R}') \left\langle \chi_a^{\boldsymbol{R}} | h | \chi_b^{\boldsymbol{R}'} \right\rangle \\
&= \delta_{\boldsymbol{k},\boldsymbol{k}'} t \left\{ 1 + \exp(-\mathrm{i}\boldsymbol{u}_2\cdot\boldsymbol{k}) + \exp\left[\mathrm{i}(\boldsymbol{u}_1 - \boldsymbol{u}_2)\cdot\boldsymbol{k}\right] \right\} .
\end{aligned}
$$

The Hamiltonian is diagonal in 2×2 blocks. Diagonalisation of each block leads to the energies

$$
\epsilon_{\boldsymbol{k}} = \pm t \sqrt{3 + 2\cos(\boldsymbol{k}\cdot\boldsymbol{u}_1) + 2\cos(\boldsymbol{k}\cdot\boldsymbol{u}_2) + 2\cos\left[\boldsymbol{k}\cdot(\boldsymbol{u}_1 - \boldsymbol{u}_2)\right]} .
$$

Figure 6.20 shows the band diagram in this approximation. The diagram is symmetrical with respect to the zero energy (the energy of the atomic p levels) which corresponds here to the Fermi level since the $n/2$ negative energy levels are filled. The functions $\phi_{\boldsymbol{k}}$ and $\phi_{-\boldsymbol{k}}$ are degenerate.

Clearly there is no gap at the Fermi level and the zero energy level corresponds to states in the Brillouin zone equivalent to the point M in the irreducible zone specified by

$$
\boldsymbol{k} = \frac{1}{3}(\boldsymbol{v}_2 + 2\boldsymbol{v}_1) \equiv \frac{1}{3}(\boldsymbol{v}_2 - \boldsymbol{v}_1) .
$$

Note, however, that these points are isolated on the Fermi surface, in contrast to what happens in metallic systems. The Bloch theorem yields the wave function $\phi_{\boldsymbol{k}}$, $\boldsymbol{k} = k_1\boldsymbol{v}_1 + k_2\boldsymbol{v}_2$, and in particular its phase for any translation of the direct lattice defined by (l_1, l_2) in the graphite plane, given its value in the initial cell:

$$
\phi_{\boldsymbol{k}}(\boldsymbol{r} + \boldsymbol{R}) = \exp(\mathrm{i}\boldsymbol{k}\cdot\boldsymbol{R})\phi_{\boldsymbol{k}}(\boldsymbol{r}) = \exp\left[2\pi\mathrm{i}(k_1 l_1 + k_2 l_2)\right]\phi_{\boldsymbol{k}}(\boldsymbol{r}) .
$$

Turning now to nanotubes, the cyclic periodicity condition due to the fact that the graphite sheet is rolled up superposes the cell (l_1, l_2) on the cell $(0, 0)$. This condition is equivalent to the phase condition $k_1 l_1 + k_2 l_2 = m$, with m a whole number. It follows that nanotubes will maintain a state $\phi_{\boldsymbol{k}}$ of zero energy only for helicities respecting the periodicity condition at the points M of the Brillouin zone. These points are defined by $(k_1, k_2) = (1/3, 2/3)$, i.e.,

$$l_1 + 2l_2 = 3m \,, \qquad \text{i.e.,} \quad l_1 - l_2 = 3p \quad (p \text{ a whole number}) \,.$$

Nanotubes satisfying this roll-up condition thus have a conductance channel at the Fermi level, whereas the others must be considered as insulators at low temperature.

6.7 Prospects

We have discussed several ideas, models, and calculational methods that can be used to study electronic structure in clusters and nanoparticles. With the exception of nanotubes, the examples treated represent the gas phase. Even in the gas phase, this discussion is far from exhaustive and other aspects deserve attention. It is worth mentioning size-dependent valence transitions in mixed valence compounds such as transition metal clusters. In these compounds, several electronic bands, generated by localised (magnetic) d electrons and delocalised s and p electrons, compete with one another. One should also mention the Kondo effect in heterogeneous metallic clusters containing a magnetic impurity (an atom of a d or f metal) coupling with delocalised electrons. These two cases exemplify coupling between highly localised and delocalised electrons [129,216]. A careful study must involve an adequate (explicit) investigation of electronic correlation. An configuration-interaction ab initio calculation remains difficult for this type of cluster when the size becomes to large ($\gtrsim 10$). Model Hamiltonians including the correlation explicitly can be used. An example is the Hubbard Hamiltonian for applications considering magnetic effects in transition metal clusters [129], or the Anderson Hamiltonian to handle the Kondo effect in doped metallic clusters [208].

Many technological applications concern clusters and nanoparticles deposited on a substrate. Some ideas introduced for free clusters may remain valid in this context. For example, electronic shells are not only observed with spherical confinement. When small alkali clusters are deposited on insulating surfaces, they can assume highly flattened shapes under the effects of interaction with the surface. Their electronic structure can be modelled by a 2D jellium in which the shells are characterised by the number of nodes k of the radial function and the eigenvalue of l_z [198]. These 2D clusters prefigure the more extended nanostructures known as quantum wells in which quantum confinement can also be tackled using a 2D jellium model [202]. Obviously, many supporting surfaces are not inert and can fundamentally change the electronic properties of adjacent nanoparticles, capturing electrons from the adsorbed clusters or donating electrons to them. In this case, one must take into account coupling between discrete levels of the clusters and the band structure of the supporting surface (see for example [209]), as well as coupling between electronic structure and geometrical relaxation of the adsorbate and the surface [210]. In the same way, it is still difficult to perform purely ab initio calculations to ascertain the electronic structure of particles in solution or passivated by ligands. Hybrid methods combining quantum mechanics and molecular mechanics (QM–MM methods) are promising for tackling the quantum aspects of reactivity in a complex environment.

As far as the theory and ab initio calculation of electronic structure are concerned, progress is currently being made in several directions. Technical efforts to reduce computation times, and in particular techniques known as linear scaling which seek to make the computation time proportional to the number of atoms, are being developed in each of the CI, DFT, and tight-binding contexts. The density functional still needs to face a certain number of problems, even for calculation of the ground state. One could mention treatment of open shells and spin multiplicity, and improvement of functionals, especially for the treatment of dispersion forces.

A particularly important issue in cluster theory is the integration of electronic structure calculations into molecular dynamics simulations, given the strong coupling between geometric relaxation and electronic properties. It is thus essential to be able to simulate properties at finite temperatures. Density functional calculations of electronic structure can be integrated into standard molecular dynamics codes. One then speaks of ab initio molecular dynamics (AIMD). One commonly used version is Car–Parinello dynamics [48–50], which cleverly couples electronic relaxation and standard nuclear dynamics. In addition, the hybrid QM–MM methods, combining DFT and classic force fields, further extend the range of systems, e.g., to heterogeneous nanoparticles, supramolecular assemblages and proteins, even in the liquid phase. We should also mention the efficiency of molecular dynamics simulations using tight-binding methods (TBMD), and simplified versions of AIMD involving approximations in the treatement of electronic structure [81].

Theoretical study of electronic excitations is also important. A great deal of effort is being made in the context of time-dependent density functional theory (TDDFT) [80, 107, 126, 127], based on solution of the time-dependent Schrödinger equation and response theory. In this formalism, the time dependence of Kohn–Sham electronic orbitals can be described by

$$\mathrm{i}\frac{\partial\phi_j(\boldsymbol{r},t)}{\partial t} = \left\{-\frac{1}{2}\Delta + \int\frac{\rho(\boldsymbol{r}',t)}{|\boldsymbol{r}-\boldsymbol{r}'|}\mathrm{d}^3\boldsymbol{r}' + \frac{\delta E_{\mathrm{xc}}[\rho(r,t)]}{\delta\rho(\boldsymbol{r})} + V_{\mathrm{ext}}(\boldsymbol{r},t)\right\}\phi_j(\boldsymbol{r},t)\ .$$

The energies of the stationary electronic states is obtained by Fourier transforming the dynamical electron observables, in particular, the electric susceptibility. Although it is often the only available approach for large systems, and sometimes quantitatively adequate, TDDFT remains somewhat unsatisfactory. This is due to the functionals currently in use and the corrections required to recover a Coulomb-type effective potential when the electrons are highly excited, but also to the inherent difficulties in treating situations where there is significant electron correlation. TDDFT is still unable to represent excited states of uniform quality over the whole space of geometric configurations, in particular when the dynamical correlation, i.e., the correlation between electron configurations exhibiting fluctuations in the electron distribution at different sites (correlation between charge transfer configurations, for example), becomes significant.

Another point concerns simultaneous propagation of electron dynamics and nuclear dynamics, especially in the study of non-adiabatic processes, photoinduced or resulting from collisional excitation (relaxation, reactivity, fragmentation) and involving excited electronic states. The development of experiments using time-resolved spectroscopy on femtosecond time scales (pump–probe experiments) [215, 217] and the prospect of coupling them with other detection methods such as multi-coincidence fragment detection or electron spectroscopy imaging techniques has opened the way to a fuller characterisation of these processes. Interpretation must be supported by theory. One specific area is that of rapid and/or high-energy excitation processes, e.g., ultrashort and/or ultrastrong laser pulses, collisions, or bombardment by high-energy particles, which can produce nonlinear electron dynamics and ionisation processes in clusters and nanomaterials [152, 153], or multiscale relaxation in the time domain [218]. One should also mention electron transport in nanowires and coupling with phonons and distortions.

In the context of ab initio calculations, coupling between nuclear dynamics (usually a problem of classical molecular dynamics) and non-adiabatic electron dynamics can be handled using mean field (Ehrenfest) methods. These describe the nuclear motion in a force field deriving from an instantaneous mean electron potential energy [81, 152, 213]. Electrons are then treated using DFT–KS or semi-classical techniques [154]. At low energies, this type of combination between electron and nuclear dynamics is likely to cause problems due to the use of a single mean trajectory for nuclei. An alternative solution is the so-called surface hopping method in which the nuclear dynamics is propagated on adiabatic surfaces and quantum jumps mediated by non-adiabatic couplings are carried out [204, 211, 212, 214]. When there are many degrees of freedom, the latter methods require on-the-fly determination, i.e., during propagation, of the electronically excited adiabatic potential surfaces and non-adiabatic couplings. This in-flight dynamics is becoming commonplace with certain Hamiltonian models. Although feasible for very small clusters [111], ab initio non-adiabatic dynamics by surface hopping remains difficult to implement both theoretically and practically in most cases.

Acknowledgements

The author would like to thank those colleagues with whom he has recently collaborated, in particular F. Calvo, M.C. Heitz, G. Durand, and L. Montagnon in the Modélisation, Agrégat Dynamique group at the Laboratoire de Chimie et Physique Quantiques in Toulouse, and C. Meier and P. Labastie at the Laboratoire Collisions Agrégats Réactivité in Toulouse. Their encouragement and generous availability for both organised and spontaneous discussions were greatly appreciated.

References

1. S. Sugano, Y. Nishina, S. Ohnishi (Eds.): *Microclusters*, Springer, Berlin (1987)
2. P. Jena, S.N. Khanna, B.K. Rao (Eds.): *Physics and Chemistry of Finite Systems: From Clusters to Crystals*, Kluwer Academic Publishers, Vols. I and II, Dordrecht (1992)
3. G. Benedek, T.P. Martin, G. Pacchioni (Eds.): *Elemental and Molecular Clusters*, Springer, Berlin (1998)
4. V. Kumar, T.P. Martin, E. Tosatti (Eds.): *Clusters and Fullerenes*, World scientific, Singapore (1993)
5. H. Haberland (Ed.): *Clusters of Atoms and Molecules*, Vols. I and II, Springer Series in Chemical Physics, Berlin (1994)
6. T.P. Martin (Ed.): *Large Clusters of Atoms and Molecules*, NATO ASI Series, Kluwer Academic Publishers, Dordrecht (1995)
7. Y. Kawazoe, T. Kondow, K. Ohno (Eds.): *Clusters and Nanomaterials: Theory and Experiment*, Springer Series in Cluster Physics, Berlin (2001)
8. C. Guet, P. Hobza, F. Spiegelman (Eds.): *Theory of Atomic Clusters and Nanoparticles*, Ecole des Houches, France (2001)
9. A. Szabo, N. Ostlundt: *Modern Quantum Chemistry. Introduction to Advanced Electronic structure Theory*, Dover Publications, Mireola, New York (1996)
10. T. Helgaker, P. Jorgensen, J. Olsen: *Molecular Electronic Structure Theory*, John Wiley and Sons, New York (2000)
11. R.J. Bartlett: Coupled cluster theory: An overview of recent developments. In: *Modern Electronic Structure Theory*, ed. by D. Yarkony, World Scientific (1995) p. 1047
12. P.R. Taylor: Coupled cluster methods in quantum chemistry, *Lecture Notes in Quantum Chemistry II*, *Lecture Notes in Chemistry*, Vol. 64, ed. by B.O. Roos, Springer Verlag (1994) p. 125
13. K. Andersson, B.O. Roos: Multiconfigurational second-order perturbation theory, *Modern Electronic Structure Theory*, ed. by D. Yarkony, World Scientific, Singapore (1995) p. 55
14. T.J. Lee, G.E. Scuseria: Achieving chemical accuracy with coupled cluster theory, *Quantum Mechanical Electronic Structure Calculations with Chemical Accuracy*, ed. by R.S. Langhoff, Kluwer, Dordrecht (1995) p. 47
15. P. Hohenberg, W. Kohn: Phys. Rev. **136**, 864 (1964)
16. M.H. Levy: Proc. Nat. Acad. Sci. USA **76**, 6062 (1979); E.H. Lieb: Phys. Rev. Lett. **140**, 44 (1997)

17. W. Kohn, L.J. Sham: Phys. Rev. **140**, A1133 (1965)
18. R.M. Dreizler, E.K.U. Gross: *Density Functional Theory*, Springer Verlag, Berlin (1990)
19. R.G. Parr, W. Yang: *Density Functional Theory of Atoms and Molecules*, Oxford University Press, Oxford (1989)
20. C. Lee, W. Yang, R.G. Parr: Phys. Rev. B **37**, 785 (1988)
21. B. Miehlich, A. Savin, H. Stoll, H. Preuss: Chem. Phys. Lett. **157**, 200 (1989)
22. A.D. Becke: J. Chem. Phys. **98**, 5648 (1993)
23. J.P. Perdew: Phys. Rev. B **33**, 8822 (1986)
24. A.D. Becke: Phys. Rev. A **38**, 3098 (1986)
25. H. Chermette: J. Comp. Chem. **20**, 129 (1999)
26. W.A. Harisson: *Electronic Structure and the Properties of Solids*, W.H. Freeman, San Francisco (1980)
27. J. Friedel: In: *The Physics of Metals*, ed. by J.M. Ziman, Cambridge University Press, Cambridge (1969)
28. B. Legrand, M. Guilopé: In: *Atomistic Simulations of Materials*, ed. by V. Vitek, D.J. Srolovitz, Plenum Publishing (1989) p. 361
29. C. Mottet: Doctoral thesis, Université d'Aix Marseille II (1997)
30. T.P. Martin, B. Wasserman: J. Chem. Phys. **99**, 5108 (1989)
31. W.D. Myers, W. Swiatewski: Ann. Phys. New York **55**, 395 (1969)
32. M. Brack, C. Guet, H.B. Hakkansön: Phys. Rep. **123**, 275 (1985)
33. M. Seidl: Phys. Rev. B **50**, 5744 (1994)
34. M. Seidl, M. Brack: Ann. Phys. New York **245**, 275 (1996)
35. M. Brack, S. Creagh, P. Meier, S. Reiman, M. Seidl: In [6]
36. V.V. Pogosov: Sol. State. Commun. **75**, 469 (1990)
37. C. Fiolhais, J.P. Perdew: Phys. Rev. B **45**, 6207 (1992)
38. A. Mananes, M. Membrado, A.F. Pacheco, J. Sunado, L.C. Balbas: Int. J. Quantum Chemistry **52,** 767 (1994)
39. W.D. Knight, K.L Clemenger, W.A. de Heer, W.A. Saunders: Phys. Rev. Lett. **52**, 2141 (1985)
40. M. Brack: Rev. Mod. Phys. **65**, 677 (1993)
41. W. de Heer: Rev. Mod. Phys. **65**, 611 (1993)
42. K.L. Clemenger: Phys. Rev. B **32**, 1359 (1985)
43. W. Eckardt: Phys. Rev. B **29**, 1558 (1984)
44. D.E. Beck: Solid State Com. **49**, 381 (1984)
45. H.A. Jahn, E. Teller: Proc. Roy. Soc. London A **161**, 220 (1937)
46. R.D. Woods, D.S. Saxon: Phys. Rev. B **95**, 577 (1984)
47. S.G. Nilsson: Mat. Fys. Medd K. Dan Vidensk. Selsk **29**, No. 16 (1955)
48. R. Car, M. Parinello: Phys. Rev. Lett. **55**, 2471 (1985)
49. R. Car, M. Parinello: Phys. Rev. Lett. **60**, 208 (1988)
50. R. Car, M. Parinello: Phys. Rev. Lett. **60**, 271 (1988)
51. O. Gardet, F. Rogemond, H. Chermette: J. Chem. Phys. **105**, 9933 (1996)
52. J. Blocki, W.J. Swiateski: Nuclear deformation energies to a liquid drop model with a sharp surface. US Department of Commerce (1982)
53. A. Bohr, B.R. Mottelson: *Nuclear Structure*, W.A. Benjamin (1969)
54. J.M. Eisenberg, W. Greiner: *Nuclear Models*, North Holland (1970)
55. P. Ring, P. Schuck: *The Nuclear Many Body Problem*, Springer Verlag, New York (1980)
56. F. Cyrot-Lackman, M.C. Desjonquères: Surf. Sci. **80**, 159 (1979)

57. D. Tomanek, S. Mukherjee, K.H. Benneman: Phys. Rev. B **28**, 665 (1983)
58. A.F. Voter, P. Chen: In: *Characterization of Defects in Materials*, ed. by R.W. Siegal, J.R. Weertman, R. Sinclair, Pittsburg (1987)
59. M. Schmidt, Ph. Cahuzac, C. Bréchignac, H.P. Cheng: J. Chem. Phys. **118**, 1956 (2003)
60. M.S. Daw, M.I. Baskes: Phys. Rev. B **29**, 6443 (1984)
61. Z.B. Güvenc, J. Jellinek: Z. Phys. D **26**, 304 (1993)
62. Y. Li, E. Blaisten-Barojas, D.E. Papaconstantopoulos: Phys. Rev. B **57**, 15519 (1998); Chem. Phys. Lett. **268**, 331 (1997)
63. C. Rey, L.J. Gallego, J. Garcia-Rodega, J.A. Alonso, M.P. Iniguez: Phys. Rev. B **48**, 8253 (1993)
64. I.L. Garzon: Phys. Rev. B **54**, 10362 (1996)
65. E. Curotto, A. Matro, D.L. Freeman, J.D. Doll: J. Chem. Phys. **108**, 729 (1998)
66. F. Calvo, F. Spiegelman: J. Chem. Phys. **120**, 9684 (2004)
67. D.R. Jennison, P.A. Schultz, M.P. Sears: J. Chem. Phys. **106**, 1856 (1997)
68. P. Nava, M. Sierka, R. Ahlrichs: Phys. Chem. Chem. Phys. **5**, 3372 (2003)
69. Y. Xie, J.A. Blackman: Phys. Rev. B **64**, 195115 (2001)
70. J.L. Martins, J. Buttet, R. Car: Phys. Rev. B **31**, 1804 (1985)
71. D. Pavolini, F. Spiegelman: J. Chem. Phys. **89**, 4954 (1988)
72. V. Bonačić-Koutecký, P. Fantucci, J. Koutecký: Phys. Rev. B **37**, 4369 (1988)
73. V. Bonačić-Koutecký, I. Boustani, M. Guest, P. Fantucci, J. Koutecký: J. Chem. Phys. **89**, 4861 (1988)
74. U. Röthlisberger, W. Andreoni: J. Chem. Phys. **94**, 8129 (1991)
75. R. Poteau, F. Spiegelman: J. Chem. Phys. **98**, 6540 (1993)
76. F. Spiegelman, R. Poteau: *Comments in Atomic and Molecular Physics: Nuclear Aspects of Simple Metal Clusters*, Gordon and Breach Science Publishers (1995) pp. 1–14
77. W. Eckardt, Z. Penzar: Phys. Rev. B **43**, 1322 (1991); Z. Penzar, W. Eckardt: Z. Phys. D **17**, 69 (1990)
78. B. Montag, P.G. Rheinardt: Z. Phys. D **33**, 265 (1995)
79. S. Kümmel, M. Brack, P.G. Reinhardt: Phys. Rev. B **62**, 7602 (2000)
80. M.E. Casida: Time-dependent density functional response theory for molecules. In: *Recent Advances in Density Functional Methods*, Part I, ed. by D.P. Chong, World Scientific, Singapore (1995) p. 155
81. T. Frauenheim, G. Seifert, M. Elstner, T. Niehaus, C. Köhler, M. Amkreutz, M. Strenberg, Z. Hajnal, A. Di Carlo, S. Suhai: J. Phys. Cond. Matt. **14**, 3015 (2002)
82. F. Furche, R. Ahlrichs: J. Chem. Phys. **117**, 7433 (2002)
83. V. Bonačić-Koutecký, J. Pittner, J. Koutecký: Chem. Phys. **210**, 313 (1996)
84. D.O. Welch, O.W. Lazareth, G.J. Dienes: J. Chem. Phys. **64**, 835 (1976)
85. J. Diefenbach, T.P. Martin: J. Chem. Phys. **83**, 4585 (1985)
86. E.S. Rittner: J. Chem. Phys. **19**, 1030 (1951)
87. D.O. Welch, O.W. Lazareth, G.J. Dienes, R.D. Hatcher: J. Chem. Phys. **83**, 4585 (1978)
88. S. Franck, N. Malinowski, F. Tast, M. Heinebrodt, I.M.L. Billas, T.P. Martin: J. Chem. Phys. **106**, 6217 (1997)
89. N.G. Phillips, C.W.S. Conover, L.A. Bloomfield: J. Chem. Phys. **94**, 4980 (1991)

90. P. Weis, C. Ochsenfeld, R. Ahlrichs, M.M. Kappes: J. Chem. Phys. **97**, 2553 (1992)
91. C. Ochsenfeld, R. Ahlrichs: J. Chem. Phys. **101**, 5977 (1994)
92. G. Rajagopal, R.N. Barnett, U. Landman: Phys. Rev. Lett. **67**, 727 (1991)
93. K. Hakkinen, R.N. Barnett, U. Landman: Phys. Rev. Lett. **232**, 79 (1995)
94. U. Landman, D. Scharf, J. Jortner: Phys. Rev. Lett. **54**, 1860 (1985)
95. G. Rajagopal, R.N. Barnett, A. Nitzan, U. Landman, E.C. Honea, P. Labastie, M.L. Homer, R.L. Whetten: Phys. Rev. Lett. **64**, 2933 (1990)
96. G. Durand, D. Maynau, J. Giraud-Girard, F. Spiegelman, F. Calvo: J. Chem. Phys. **110**, 7871 (1999)
97. G. Durand, F. Spiegelman, M. Sence, Ph. Poncharal, J.M. Lhermite, P. Labastie: Chem. Phys. **110**, 7871 (1999)
98. R.S. Mulliken: J. Chem. Phys. **23**, 1833 (1955)
99. F. Rabilloud, F. Spiegelman, J.M. Lhermite, P. Labastie: J. Chem. Phys. **11**, 289 (2001)
100. M. Grigorov, F. Spiegelman: Surf. Rev. Lett. **3**, 211 (1996)
101. I. Last, T.F. George: J. Chem. Phys. **93**, 8925 (1990)
102. P.J. Kuntz, J. Valldorf: Z. Phys. D **8**, 195 (1988)
103. T. Ikegami, T. Kondow, S. Iwata: J. Chem. Phys. **98**, 3083 (1990)
104. M. Amarouche, G. Durand, J.P. Malrieu: J. Chem. Phys. **88**, 1010 (1988)
105. F.X. Gadea, M. Amarouche: Chem. Phys. **140**, 385 (1990)
106. D.G. Pettifor, I.I. Oleinik: Phys. Rev. Lett. **84**, 4124 (2000); I.I. Oleineik, D.G. Pettifor: Phys. Rev. B **59**, 8487 (1999); A.P. Horsfield, A.M. Bratkovsky, M. Feam, D.G. Pettifor, M. Aoki: Phys. Rev. B **53**, 12694 (1996)
107. L. Reining, V. Olevano, A. Rubio, G. Onida: Phys. Rev. Lett. **88**, 066404 (2002); F. Sottile, V. Olevano, L. Reining: Phys. Rev. Lett. **91**, 056402 (2003); F. Bruneval, F. Sottile, V. Olevano, R. del Sole, L. Reining: Phys. Rev. Lett. **94**, 186402 (2005)
108. V. Bonačić-Koutecký, P. Fantucci, J. Koutecký: Chem. Rev. **91**, 1035 (1991)
109. M. Gerloch, R.C. Slade: *Ligand Field Parameters*, University Press, London (1973)
110. V. Bonačić-Koutecký, J. Pittner, C. Fuchs, P. Fantucci, M.F. Guest, J. Koutecký: J. Chem. Phys. **104**, 1427 (1996)
111. V. Bonačić-Koutecký, R. Mitrić: Chem. Rev. **105**, 11 (2005)
112. V. Bonačić-Koutecký, J. Pittner: Chem. Phys. **232**, 79 (1997)
113. V. Bonačić-Koutecký, J. Pittner, J. Koutecký: Chem. Phys. **210**, 313 (1996)
114. R.F.W Bader: *Atoms in Molecules: A Quantum Theory*, Oxford University Press, Oxford (1990)
115. R.F.W. Bader: Chem. Rev. **91**, 893 (1991)
116. J.E. Carpenter, F. Weinhold: J. Mol. Struct. (Theochem) **169**, 141 (1988)
117. B.H. Besler, K.M. Merz, P.A. Kollman: J. Comp. Chem. **11**, 431 (1990)
118. F. Finocchi, C. Noguera: Phys. Rev. B **53**, 4989 (1996)
119. F. Finocchi, C. Noguera: Far. Discuss. **106**, 233 (1997)
120. F. Finocchi, C. Noguera: Phys. Rev. B **57**, 14646 (1998)
121. F. Finocchi, C. Noguera: Euro. Phys. J. D **9**, 327 (1999)
122. T. Albaret, F. Finocchi, C. Noguera: Appl. Surf. Sci. **144/145**, 672 (1999)
123. T. Albaret, F. Finocchi, C. Noguera: Far. Discuss. **114**, 285 (1999)
124. T. Albaret, F. Finocchi, C. Noguera: J. Chem. Phys. **113**, 2238 (2000)
125. Y.M. Niquet, X. Gonze: Phys. Rev. B **70**, 254115 (2004)

126. E.K.U. Gross, J.F. Dobson, M. Petersilka: Density-functional theory of time-dependent phenomena, *Topics in Current Chemistry*, Springer Verlag **181**, 81 (1996)
127. W. Runge, E.K.U. Gross: Phys. Rev. Lett. **52**, 997 (1984)
128. T. Kreibig, E.K.U. Gross: Phys. Rev. Lett. **86**, 2984 (2001)
129. G.M. Pastor, R. Hirsch, B. Müschlegel: Phys. Rev. Lett. **72**, 3879 (1994); G.M. Pastor, R. Hirsch, B. Müschlegel: Phys. Rev. B **53**, 10382 (1996); G.M. Pastor: Magnetism of clusters, in [8]; J. Dorantes-Davila, K. Benneman: Phys. Rev. B **70**, 064420 (2004)
130. G. Pastor, K.H. Benneman: Transition from van der Waals to metallic bonding in clusters, in [5]; M.E. Garcia, G.M. Pastor, K. Benneman: Phys. Rev. B **48**, 8388 (1993)
131. M. Madget, C. Guet, W.R. Johnson: Phys. Rev. A **51**, 1327 (1995)
132. D. Scharf, J. Jortner, U. Landman: J. Chem. Phys. **88**, 4173 (1988)
133. M. Bixon, J. Jortner: J. Chem. Phys. **91**, 1631 (1989)
134. I. Rips, J. Jortner: J. Chem. Phys. **97**, 536 (1992)
135. J. Jortner, N. Ben Horin: J. Chem. Phys. **98**, 9436 (1993)
136. H.W. Kroto, J.R. Heath, S.C. O'Brien, R.F. Curl, R.E. Smalley: Nature **318**, 162 (1985)
137. K. Raghavachari, L.A. Curtiss: Accurate theoretical studies of small elemental clusters. In: *Quantum Mechanical Electronic Structure with Chemical Accuracy*, ed. by S.R. Langhoff, Kluwer, Dordrecht (1995)
138. J. Hutter, H.P. Lüthi, H.P. Diederich: J. Am. Chem. Soc. **116**, 750 (1994)
139. M.S. Dresselhaus, G. Dresselhaus, P.C. Ecklund: *Science of Fullerenes and Nanotubes*, Academic Press, San Diego (1996)
140. A. Van Orden, R. Saykally: Chem. Rev. **98**, 2313 (1998)
141. R.O. Jones: J. Chem. Phys. **110**, 5189 (1999)
142. R.O. Jones: Phys. Rev. Lett. **79**, 443 (1997)
143. M.S. Giuffreda, M.S. Deleuze, J.-P. François: J. Phys. Chem. A **103**, 5137 (1999)
144. K. Ragavachari, J.S. Binkley: J. Chem. Phys. **87**, 2191 (1987)
145. W. Weltner Jr., R.J. Van Zee: Chem. Rev. **89**, 1713 (1989)
146. N. Troullier, J.L. Martins: Phys. Rev. B **46**, 1754 (1992)
147. J.L. Martins, N. Troullier, J.H. Weaver: Chem. Phys. Lett. **180**, 457(1989)
148. J. Kohanoff, W. Andreoni, M. Parrinello: Chem. Phys. Lett. **198**, 472 (1992)
149. K. Yabana, G.F. Bertsch: Physica Scripta **48**, 633 (1993)
150. N. Van Giai, E. Lipparini: Z. Phys. D **27**, 193 (1993)
151. M.J. Puska, R.M. Niemienen: Phys. Rev. A **47**, 1181 (1993)
152. E. Suraud, P.G. Reinhardt: Phys. Rev. B **85**, 2296 (2000)
153. F. Calvayrac, P.G. Reinhardt, E. Suraud: Phys. Rev. B **52**, 17056 (1995); F. Calvayrac, P.G. Reinhardt, E. Suraud: Ann. Phys. (San Diego) **255**, 125 (1997); P.G. Reinhardt, E. Suraud: AIP Conf. Proceedings **597**, 235 (2001)
154. A. Domps, P.G. Reinhardt, E. Suraud: Phys. Rev. Lett. A **80**, 5520 (1998); A. Domps, P.G. Reinhardt, E. Suraud: Ann. Phys. (San Diego) **280**, 211 (2000)
155. O. Knospe, J. Jellinek, U. Sallman, R. Schmidt: Phys. Rev. A **61**, 022715 (2000)
156. F. Calvo, S. Tran, S.A. Blundell, C. Guet, F. Spiegelman: Phys. Rev. B **62**, 10394 (2000)

157. C. Bréchignac, Ph. Cahuzac, J.Ph. Roux: J. Chem. Phys. **87**, 229 (1987); C. Bréchignac, Ph. Cahuzac, F. Carlier, J. Leygnier: Phys. Rev. Lett. **63**, 1368 (1989)
158. C. Bréchignac, Ph. Cahuzac, N. Kebaili, J. Leygnier, A. Sarfati: Phys. Rev. Lett. **68**, 3916 (1992)
159. B. Vezin, Ph. Dugourd, D. Rayane, P. Labastie, J. Chevaleyre, M. Broyer: Chem. Phys Lett. **206**, 52 (1993); C. Bréchignac, Ph. Cahuzac, J. Leygnier, A. Sarfati: Phys. Rev. Lett. **70**, 2036 (1993)
160. R. Antoine, Ph. Dugourd, D. Rayane, E. Benichou, M. Broyer: J. Chem. Phys. **107**, 2264 (1997)
161. H. Häkkinen, M. Manninen: J. Chem. Phys. **105**, 10565 (1996); J. Akola, H. Häkkinen, M. Manninen: Phys. Rev. B **58**, 3601 (1998); J. Akola, M. Manninen, H. Häkkinen, U. Landman, X. Li, L.-S. Wang: Phys. Rev. B **62**, 13216 (2000)
162. J. Wörmer M. Joppien, G. Zimmerer, T. Möller: Phys. Rev. Lett. **67**, 2053 (1995)
163. F. Björnholm, F. Fössing, T. Möller: Phys. Rev. Lett. **74**, 3017 (1995)
164. M. Manninen: Effect of the surface and finite temperature on the electronic structure of metal clusters, in [6]
165. L.H. Thomas: Proc. Cambridge Philos. Soc. **23**, 542 (1926); E. Fermi: Z. Phys. 48, 73 (1928)
166. C.H. Hodges: Can. J. Phys. **51**, 1428 (1973)
167. C. Teichteil, L. Maron, V. Vallet: Relativistic pseudopotential calculations for electronic excited states. In: *Relativistic Electronic Structure*, Part II ed. by P. Schwerdtfeger, Elsevier (2004)
168. V.M. Strutinsky: Nucl. Phys. A **122** (1968)
169. N.W. Ashcroft, N.D. Mermin: *Solid State Physics*, Saunders College Publishing (1976)
170. C.F. Von Weiszäcker: Z. Phys. **96**, 431 (1935)
171. C.H. Hodges: Can. J. Phys. **51**, 1428 (1973)
172. C. Kittel: *Introduction to Solid State Physics*, Wiley, New York (1996)
173. M.L. Homer, E.C. Honea, J.L. Persson, R.L. Whetten: results published in [178]
174. C. Yannouleas, U. Landman: Phys. Rev. B **48**, 3876 (1993)
175. C. Yannouleas, U. Landman: Chem. Phys. Lett. **217**, 175 (1994)
176. C. Yannouleas, U. Landman: Phys. Rev. B **51**, 1902 (1995)
177. C. Yannouleas, U. Landman: Phys. Rev. Lett. **78**, 1424 (1997)
178. C. Yannouleas, U. Landman: Shell correction method for clusters, in [6]
179. A.K. Rappé, W.A. Goddard III: J. Phys. Chem. **95**, 3358 (1991)
180. S. Sawada, S. Sugano: Z. Phys. D **20**, 259 (1991)
181. M.C.C. Ribeiro: J. Chem. Phys. **117**, 266 (2002)
182. S.W. Rick, S.J. Stuart, J. Berne: J. Chem. Phys. **101**, 6141 (1994)
183. F. Calvo: Phys. Rev. B **67**, 161403 (2003)
184. J. Roques, F. Calvo, C. Mijoule, F. Spiegelman: Phys. Rev. Lett. **90**, 075505 (2003)
185. J. Roques, F. Calvo, C. Mijoule, F. Spiegelman: Phys. Rev. B **68**, 205412 (2004)
186. M. Fuchs, Y.-M. Niquet, X. Gonze, K. Burke: J. Chem. Phys. **122**, 094116 (2005)

187. P. Claverie: *Intermolecular Interactions: From Diatomics to Biopolymers*, Wiley, New York (1978)
188. A.J. Stone: *Theory of Intermolecular Forces*, Clarendon Press (1997)
189. I.N. Levine: *Quantum Chemistry*, 5th edn., Prentice Hall, New Jersey (2000)
190. A.I. Kitaigorodski: Tetrahedron **14**, 975 (1961)
191. A. Bondi: J. Phys. Chem. **68**, 441 (1964)
192. D.E. Stogryn: Mol. Phys. **22**, 81 (1971)
193. B. Bouvier, V. Brenner, Ph. Millié, J.-M. Soudan: J. Phys. Chem. **106**, 10326 (2002)
194. B. Bouvier, T. Gustavsson, D. Markovitsi, Ph. Millié: Chem. Phys. **275**, 75 (2002)
195. P. Duplaa, F. Spiegelman: J. Chem. Phys. **105**, 1492 (1996)
196. G. Durand, P. Duplaa, F. Spiegelman: Z. Phys. D **40**, 177 (1997)
197. J. Lermé, M. Pellarin, J.-L. Vialle, B. Baguenard, C. Bordas, M. Broyer: Z. Phys. D **26**, 137 (1993)
198. C. Kohl, B. Montag, P.G. Reinhardt: Z. Phys. D **38**, 81 (1996); H. Häkkinen, M. Manninen: J. Chem. Phys. **105**, 10565 (1996)
199. B. Baguenard, M. Pellarin, C. Bordas, J. Lermé, J.-L. Vialle, M. Broyer: Chem. Phys. Lett. **203**, 13 (1993)
200. M. Pellarin, B. Baguenard, C. Bordas, M. Broyer, J. Lermé, J.-L. Vialle: Phys. Rev. B **48**, 1645 (1993)
201. J. Lermé, C. Bordas, M. Pellarin, B. Baguenard, J.L. Vialle, M. Broyer: Phys. Rev. B **48**, 12110 (1993); J. Lermé, M. Pellarin, E. Cottancin, B. Baguenard, J.-L. Vialle, M. Broyer: Phys. Rev. B **52**, 16807 (1995); ibid.: Surf. Rev. Lett. **3**, 217 (1996)
202. S. Reiman, M. Manninen: Rev. Mod. Phys. **74**, 1283 (2002)
203. M.J. Field, P.A. Bash, M. Karplus: J. Comp. Chem. **11**, 700 (1990); G. Monard, M. Loos, V. Thery, K. Baka, J.L. Rivail: J. Comp. Chem. **15**, 269 (1990); J. Gao, P. Amara, C. Alhambra, M.J. Field: J. Phys. Chem. A **102**, 4174 (1998); N. Reuter, A. Dejagne, B. Maifret, M. Karplus: J. Phys. Chem. A **104**, 1720 (2000)
204. J.C. Tully: J. Chem. Phys. **91**, 1061 (1990); J.C. Tully: Int. J. Quantum Chemistry **25**, 299 (1991)
205. C. Voisin, D. Christofilos, P.A. Loukakos, N. Del Fatti, F. Vallée, J. Lermé, M. Gaudry, E. Cottancin, M. Pellarin, M. Broyer: Phys. Rev. B **69**, 195416 (2004); A. Arbouet, C. Voisin, D. Christofilos, P. Langot, N. Del Fatti, F. Vallée, J. Lermé, G. Celep, E. Cottancin, M. Gaudry, M. Pellarin, M. Broyer, M. Maillard, M.P. Pileni, M. Treguer: Phys. Rev. Lett. **90**, 77401 (2003)
206. G. Mie: Ann. der Physik (1908)
207. R. Gan: Ann. der Physik (1915)
208. J.L. Ricardo-Chávez, G.M. Pastor: Eur. Phys. J. D **16**, 169 (2001)
209. M.C. Desjonquère, D. Spanjaard: *Concepts in Surface Physics*, Springer Series in Surface Sciences, Vol. 30, Springer, Berlin (1993)
210. C. Noguera: *Physics and Chemistry at Oxide Surfaces*, Cambridge University Press, Cambridge (1996); C. Noguera: In: *The Chemical Physics of Solid Surfaces*, Vol. 9, ed. by D.P. Woodruf, Elsevier, Amsterdam (2001)
211. D. Bonhommeau, A. Viel, N. Halberstadt: J. Chem. Phys. **123**, 54316 (2005)
212. M. Sizun, A. Aguilon, V. Sidis: J. Chem. Phys. **123**, 74331 (2005)
213. F. Calvo, J. Galindez, F.X. Gadéa: Phys. Chem. Chem. Phys. **5**, 3221 (2003)

214. P. Jungwirth, B. Gerber: J. Chem. Phys. **104**, 5803 (1996)
215. A.H. Zewail: *Femtochemistry, Ultrafast Dynamics of the Chemical Bond*, Vols. I and II, World Scientific, Singapore (1994)
216. G.M. Pastor, J. Dorantes-Davila, K.H. Bennemann: Phys. Rev. B **70**, 064420 (2004); R.A. Guirado-Lopes, J. Dorantes-Davila, G.M. Pastor: Phys. Rev. Lett. **90**, 226402 (2003)
217. M. Chergui: *Ultra Fast Chemical and Physical Processes in Molecular Systems*, World Scientific, Singapore (1996)
218. H.O. Jeschke, M.E. Garcia, K.H. Bennemann: Phys. Rev. **96**, 015003 (2001)

7

Optical Properties of Metallic Nanoparticles

F. Vallée

The bright and changing colours obtained by dispersing metallic compounds in a glass matrix have been known empirically for centuries. Indeed, glasses have been coloured in the bulk by inclusion of metallic powders since ancient times to make jewellery and ornaments (see Chap. 25). Then in the Middle Ages, they were used for stained glass windows and later on for coloured glass artefacts, e.g., ruby red glass objects. However, the role played by nanoparticles in this colouring effect, i.e., the effects of nanoparticles on optical properties, were only first studied scientifically in the nineteenth century, by Michael Faraday [1].

These novel optical properties arise because there is a resonance in the absorption spectrum of the nanoparticles, the surface plasmon resonance. Its characteristics, such as spectral width and position, and sensitivity to light polarisation, depend not only on the intrinsic properties of the nano-objects (composition, structure, size, shape), but also on their environment. In the case of noble metals, it occurs in the visible region of the spectrum, in the blue somewhere near 400 nm and in the green around 520 nm for small silver and gold spheres, respectively, and it produces yellow and red colouring, respectively, of the material in which they are included.

Apart from such decorative applications, this possibility of modifying the optical properties of nanomaterials and maybe even controlling them in such a way as to achieve specific functions, by adjusting their characteristics or those of their surroundings, has generated a good deal of interest recently. The impact of these novel materials in the field of photonics has already made itself felt, and it is only in its early stages, extending to a wide range of different areas, e.g., linear and nonlinear optics, polarisers, solar cells, chemical and biosensors, nano-optical devices, optical switches, local electromagnetic field enhancement, and molecular labelling in biology. In addition, their optical response provides a way of studying the characteristics of nano-objects and hence constitutes a powerful tool for analysing their fundamental properties.

This phenomenon was described theoretically by Mie at the beginning of the twentieth century in the case of spherical inclusions of arbitrary size dispersed in a dielectric medium [2]. For small enough sizes compared with the optical wavelength ($D \leq \lambda/10$), it can be most simply described in the so-called quasi-static approximation [3,4]. Indeed, its physical origin can be related to a local field effect, modelled in other contexts by Lorenz and Lorentz, and by Clausius and Mossotti in the nineteenth century for solids and molecular media, respectively [5,6].

In contrast to semiconducting nanoparticles or quantum dots, the optical properties arising with size reduction in metallic media are due to a classical enhancement effect on the electric field, often called dielectric confinement. The discretisation of electronic states (quantum confinement) which underlies the optical properties of quantum dots only leads to small corrections to the observed responses down to nanometric sizes. Actually, this difference results from the very high electron density in metals ($5 \times 10^{22}\,\mathrm{cm}^{-3}$) compared with typical values in semiconductors ($10^{18}\,\mathrm{cm}^{-3}$). In metallic media of nanometric size, electronic properties stem from high energy quantised states, close to the Fermi level. The separations between these energy levels are very small compared with their widths and the thermal energy (at room temperature). For the nanoparticle sizes considered here, greater than one nanometer, these states can therefore be modelled to a first approximation by a quasi-continuum of states similar to what is observed in the bulk metal [7].

The aim in this chapter is not to provide an exhaustive description of the way size effects influence the optical response of metallic nanoparticles, but rather to bring out the important effects for those sizes most frequently studied, ranging from the nanometer to a few tens of nanometers. We begin by describing the optical response of materials containing metallic inclusions, introducing dielectric confinement in the framework of the quasi-static approximation, valid for very small sizes compared with the optical wavelength. The results of this approach will then be compared with the more general results of the Mie theory. The optical response is directly related to the dielectric constant of the metal. Modifications due to quantum confinement, which affect the response at the smallest sizes, will be introduced as corrections. Finally, we briefly discuss a few consequences of size reduction for the nonlinear optical response.

7.1 Optical Response for Free Clusters and Composite Materials

A plane monochromatic electromagnetic wave of angular frequency ω and wave vector $k = 2\pi/\lambda$ incident on a medium in the z direction, viz.,

$$\boldsymbol{E} = \boldsymbol{E}_0 \exp[\mathrm{i}(kz - \omega t)] + \mathrm{c.c.}\,, \tag{7.1}$$

induces a polarisation which characterises its optical response. (Here c.c. denotes the complex conjugate of the previous term.) The latter can be fully

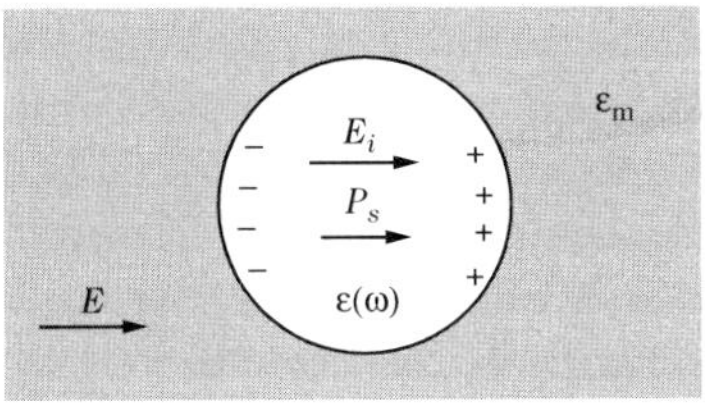

Fig. 7.1. Schematic view of the interaction between a spherical particle of diameter D and an electromagnetic wave in the quasi-static approximation

described by introducing the complex dielectric constant $\varepsilon(\omega) = \varepsilon_1(\omega) + \mathrm{i}\varepsilon_2(\omega)$ of the material, which reduces to a scalar function for an isotropic medium [5]. The intensity transmitted by a medium of thickness L is then given in terms of the complex refractive index $\widetilde{n} = n + \mathrm{i}\kappa = \varepsilon^{1/2}$ of the medium:

$$I_{\mathrm{t}} = I_0 \exp(-\alpha L) , \quad \text{with} \quad \alpha(\omega) = 2\frac{\omega}{c}\kappa(\omega) . \tag{7.2}$$

We consider an ensemble of non-magnetic nanoparticles with dielectric constant ε in the vacuum (free clusters) or dispersed in a transparent dielectric matrix with real dielectric constant ε_{m} (composite material), interacting with an electromagnetic wave (see Fig. 7.1). We shall assume that the level of dilution, characterised by a volume fraction $p = N_{\mathrm{np}}V_{\mathrm{np}}/V$, where N_{np} is the number of nanoparticles, with volume V_{np}, dispersed in the volume V of the material, is sufficient to ensure that the particles can be treated as independent as far as electromagnetic interactions are concerned. We shall consider nanoparticles with sizes greater than two nanometers, i.e., comprising at least a hundred atoms (a silver sphere of diameter 2 nm contains around 250 atoms). To find the optical response of the medium, one must first describe that of a nanoparticle in its dielectric environment, the medium subsequently being described globally by introducing an effective dielectric constant $\widetilde{\varepsilon}$ (Maxwell–Garnett effective medium model) [6, 8].

7.2 Optical Response in the Quasi-Static Approximation: Nanospheres

The scale on which the electromagnetic field varies is characterised by the reciprocal length of the wave vector $\boldsymbol{k}$. For particles with very small diameters, i.e., such that $|k|D \ll 2\pi$ (or equivalently, $|n|D \ll \lambda$, where λ is the wavelength in the vacuum), the field can be treated as uniform within each sphere. The same goes for time, in the sense that the time of propagation in the sphere, i.e., nD/c, is then also small compared with the oscillation period $2\pi/\omega$ of the field. For very small sizes compared with the wavelength, the task reduces to a simple electrostatic problem regarding the polarisation of a sphere placed in a uniform field (see Fig. 7.1). As the electric fields derive from a scalar potential, this problem is easy to solve by writing down continuity conditions at the sphere–matrix interface and boundary conditions far from the sphere (uniform field).

In the composite medium, the interaction of the electromagnetic wave with a sphere can be treated in a slightly more general way, taking into account the influence, assumed slight, of the other particles by introducing the idea of a local field $\boldsymbol{E}_\mathrm{l}$ due to all the other polarisable entities. It is related to the incident field $\boldsymbol{E}$ by the Lorentz–Lorenz expression

$$\boldsymbol{E}_\mathrm{l} = \boldsymbol{E} + \frac{\boldsymbol{P}_\mathrm{sph}}{3\varepsilon_\mathrm{m}\varepsilon_0} , \tag{7.3}$$

where $\boldsymbol{P}_\mathrm{sph} = p\boldsymbol{p}_\mathrm{s}/V_\mathrm{np}$ is the polarisation due to the spheres and $\boldsymbol{p}_\mathrm{s}$ is the dipole moment of a sphere. From the continuity conditions, it can be shown that the field $\boldsymbol{E}_\mathrm{i}$ inside the sphere is given by the sum of the local field and a depolarisation field due to accumulation of charges at the sphere–matrix interface (see Fig. 7.1):

$$\boldsymbol{E}_\mathrm{i} = \frac{3\varepsilon_\mathrm{m}}{\varepsilon(\omega) + 2\varepsilon_\mathrm{m}}\boldsymbol{E}_\mathrm{l} = f(\omega)\boldsymbol{E}_\mathrm{l} . \tag{7.4}$$

Outside the sphere, the field created by these charges is equivalent to that of a dipole placed at the centre of the sphere and with dipole moment

$$\boldsymbol{p}_\mathrm{s} = 3V_\mathrm{np}\varepsilon_0\varepsilon_\mathrm{m}\frac{\varepsilon - \varepsilon_\mathrm{m}}{\varepsilon + 2\varepsilon_\mathrm{m}}\boldsymbol{E}_\mathrm{l} , \tag{7.5}$$

where $\varepsilon = \varepsilon(\omega)$. The total polarisation $\boldsymbol{P}$ of the composite medium is the sum of the polarisation due to the metallic particles and the polarisation due to the dielectric matrix:

$$\boldsymbol{P} = \boldsymbol{P}_\mathrm{m} + \boldsymbol{P}_\mathrm{sph} = \varepsilon_0(\varepsilon_\mathrm{m} - 1)\boldsymbol{E} + \frac{p}{V_\mathrm{np}}\boldsymbol{p}_\mathrm{s} . \tag{7.6}$$

The relation $\widetilde{\varepsilon}\varepsilon_0\boldsymbol{E} = \varepsilon_0\boldsymbol{E} + \boldsymbol{P}$ is used to define the effective dielectric constant $\widetilde{\varepsilon}$ of the medium (the Maxwell–Garnett relation):

$$\frac{\widetilde{\varepsilon} - \varepsilon_\mathrm{m}}{\widetilde{\varepsilon} + 2\varepsilon_\mathrm{m}} = p\frac{\varepsilon - \varepsilon_\mathrm{m}}{\varepsilon + 2\varepsilon_\mathrm{m}} . \tag{7.7}$$

For small volume fractions, $p \ll 1$, (7.7) becomes

$$\widetilde{\varepsilon} = \varepsilon_\mathrm{m} + 3p\varepsilon_\mathrm{m}\frac{\varepsilon - \varepsilon_\mathrm{m}}{\varepsilon + 2\varepsilon_\mathrm{m}} . \tag{7.8}$$

This approximation is equivalent to treating the spheres as independent and identifying $\boldsymbol{E}_\mathrm{l}$ and $\boldsymbol{E}$. To lowest order in p, the real and imaginary parts of the effective dielectric constant are determined by the matrix and the nanospheres, respectively (non-absorbent matrix, ε_m real):

$$\widetilde{\varepsilon}_1 = \varepsilon_{\mathrm{m}} , \qquad \widetilde{\varepsilon}_2 = 9p\varepsilon_{\mathrm{m}}^2 \frac{\varepsilon_2}{(\varepsilon_1 + 2\varepsilon_2)^2 + \varepsilon_2^2} . \tag{7.9}$$

The global absorption coefficient of the medium is then given by

$$\alpha(\omega) = \frac{9p\omega\varepsilon_{\mathrm{m}}^{3/2}}{c} \frac{\varepsilon_2}{(\varepsilon_1 + 2\varepsilon_{\mathrm{m}})^2 + \varepsilon_2^2} = \frac{\omega}{\varepsilon_{\mathrm{m}}^{1/2} c} p|f(\omega)|^2 \varepsilon_2(\omega) . \tag{7.10}$$

It is multiplied by a factor $|f(\omega)|^2$. [Without any enhancement of the internal field (7.4), the absorption would simply be proportional to ε_2 and to the fraction of metal in the matrix, $\alpha_0 = (\omega/\varepsilon_{\mathrm{m}}^{1/2} c) p\varepsilon_2(\omega)$, an expression that can be obtained using (7.2) and $p\varepsilon_2 \ll \varepsilon_{\mathrm{m}}$.] This factor may be resonant, i.e., $|f(\omega)| > 1$, if ε_1 is negative. This condition is satisfied in metals when the contribution to the dielectric constant ε from the conduction electrons is dominant (see below). The composite medium then exhibits an enhancement of absorption near the frequency Ω_{R} for which the denominator is minimum: this is the surface plasmon resonance (see Fig. 7.2). If the imaginary part ε_2 is small or shows little dispersion near Ω_{R}, the resonance condition becomes simply

$$\varepsilon_1(\Omega_{\mathrm{R}}) + 2\varepsilon_{\mathrm{m}} = 0 . \tag{7.11}$$

The surface plasmon resonance is a purely dielectric effect. It results from the enhancement of the amplitude of the internal field $\boldsymbol{E}_{\mathrm{i}}$ by confinement in a nanoparticle, with respect to the incident field $\boldsymbol{E}$ [9], the intrinsic properties of the metal being barely modified by size reduction (see below). This is similar to a local field effect and is often called the dielectric confinement effect [7].

From a classical standpoint, the field induces an oscillation of the electron cloud. The electron density of each cluster oscillates at the optical frequency relative to the lattice ions, creating an oscillating charge at its surface (see Fig. 7.1). This charge generates a restoring force and an electric dipole field, which will react back on the electron motion and the amplitude of the electric field, this being considerably strengthened at resonance (see Fig. 7.2). This oscillation is analogous to the collective oscillation of an electron gas in a bulk system, the plasmon mode at frequency ω_{p}, modified by the presence of interfaces, whence the name of surface plasmon resonance.

The response of the composite medium has been globalised through its effective dielectric constant. The absorption coefficient α defined by (7.2) expresses the extinction of the incident wave by both absorption and scattering by the particles. The scattered field can be calculated from the induced dipole $\boldsymbol{p}_{\mathrm{s}}$. Indeed the scattering cross-section due to one particle is [3]

$$S_{\mathrm{scatt}} = \frac{24\pi^3 V_{\mathrm{np}}^2 \varepsilon_{\mathrm{m}}^2}{\lambda^4} \left| \frac{\varepsilon - \varepsilon_{\mathrm{m}}}{\varepsilon + 2\varepsilon_{\mathrm{m}}} \right|^2 . \tag{7.12}$$

It is proportional to the square of the volume V_{np} of the particle and goes as $1/\lambda^4$ (neglecting dispersion in the dielectric constants), i.e., it corresponds

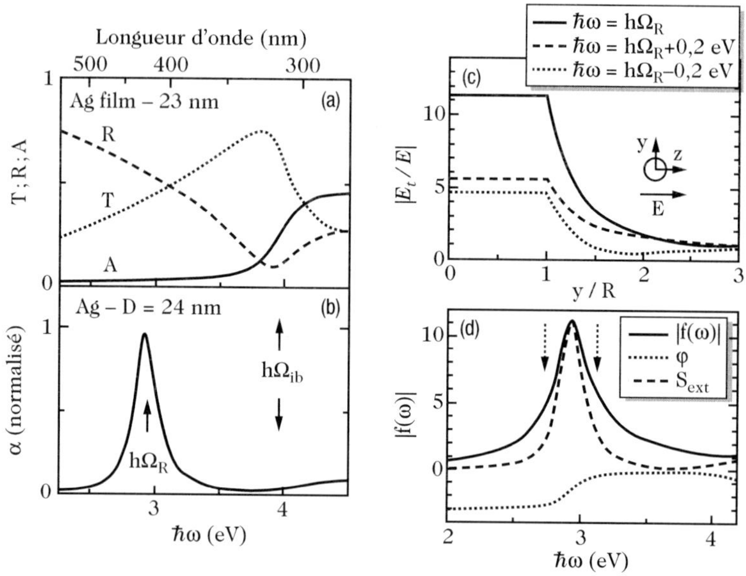

Fig. 7.2. (**a**) Reflection R, transmission T and absorption A of a silver film of thickness 23 nm on a fused silica substrate. (**b**), (**c**) and (**d**) Optical response of silver nanospheres of diameter $D = 24$ nm in a glass matrix (BaO–P_2O_5). (**b**) Absorption spectrum measured near the interband transition threshold $\hbar\Omega_{ib}$ and the surface plasmon resonance $\hbar\Omega_R$. (**c**) Amplitude E_t of the electric field of the total electromagnetic wave, incident plus scattered, normalised to the incident field E, calculated using the quasi-static model at resonance $\omega = \Omega_R$ and at 0.2 eV away from it [*arrows* in (**d**)], in and near a nanosphere as a function of the distance from its centre in the y direction orthogonal to $\boldsymbol{E}$. (**d**) Dispersion of the extinction cross-section S_{ext} (7.13), normalised, and the magnitude and phase φ of the enhancement factor $f(\omega) = E_i/E$ (7.4)

to scattering of Rayleigh type. The extinction cross-section is calculated in a similar way:

$$S_{ext} = \frac{18\pi V_{np}\varepsilon_m^{3/2}}{\lambda}\frac{\varepsilon_2}{|\varepsilon + 2\varepsilon_m|^2} = \frac{2\pi V_{np}}{\lambda\varepsilon_m^{1/2}}|f|\varepsilon_2 , \tag{7.13}$$

and is proportional to the volume of the particle. In the same environment, the ratio of the cross-sections thus varies as the volume of the nanoparticles, i.e.,

$$S_{scatt}/S_{ext} \propto (D/\lambda)^3 . \tag{7.14}$$

In the quasi-static approximation, the scattering term is necessarily small, i.e., $(D/\lambda)^3 \ll 1$. The optical extinction for small nanoparticles is thus due to their absorption. The absorption coefficient for an ensemble of particles is given by

$\alpha = NS_{\text{ext}}$ and it is identical to (7.10). The scattering cross-section has the same denominator $\varepsilon + 2\varepsilon_{\text{m}}$ and is thus also enhanced when the condition (7.11) is satisfied.

As the size increases, scattering becomes more important and in the end it dominates extinction at large particle sizes. The quasi-static approximation, typically valid for $D \leq \lambda/10$, is then no longer applicable and the interaction between the electromagnetic wave and the nanoparticle must be treated in the more general framework of the Mie model.

7.3 Dielectric Constant of a Metal: Nanometric Size Effect

The surface plasmon resonance does not directly involve quantum confinement of the electronic states but is nevertheless influenced by it. Indeed, the dielectric constant ε occurring here is that of the metal in the nanoparticle, and it is modified with respect to that of the bulk metal. In a metal, two electronic mechanisms underlie the optical properties and must be taken into account with regard to light–matter interactions.

The first is the intraband mechanism. This is due to the quasi-free electrons of density n_{e} in the conduction band and it is associated with optical transitions without change of band (see Fig.7.3). Its contribution to the dielectric constant is described by the Drude model [10, 11]:

$$\varepsilon_{\text{Drude}} = 1 - \frac{\omega_{\text{p}}^2}{\omega(\omega + \mathrm{i}\gamma_0)} , \tag{7.15}$$

where $\omega_{\text{p}}^2 = n_{\text{e}} e^2 / \varepsilon_0 m_{\text{e}}$ is the plasma frequency of the metal. The absorption, given by the imaginary part of the dielectric constant, is proportional to the average optical collision rate of the electrons, $\gamma_0(\omega)$. This term corresponds to three-body processes in which a collision occurs with simultaneous absorption of a photon of frequency ω. This reflects the fact that direct absorption of a photon by an electron cannot happen, because the wave vector of the photon is very small compared with that of the electrons and the size of the Brillouin zone. This would correspond to an almost vertical transition with constant electron wave vector (see Fig. 7.3). So absorption must be assisted by a third particle, i.e., another electron or a phonon, in order to guarantee conservation of energy and momentum. (Phonons are normal vibration modes of the periodic lattice of atoms making up a crystal [11].) Electron–phonon interactions dominate at room temperature.

The Drude Model

This model was introduced by P. Drude at the beginning of the twentieth century to describe electrical conduction in metals [10, 11]. It is based on a kinetic approach in which conduction electrons are treated as a free electron gas subject to an external electric field $\boldsymbol{E}$. The motion of an electron of effective mass m_{e} is given classically by Newton's equation

$$m\frac{\mathrm{d}\boldsymbol{v}}{\mathrm{d}t} - \mathrm{i}\gamma\boldsymbol{v} = -\frac{e\boldsymbol{E}}{m} .$$

The motion driven by the electric field is damped by collisions with other particles, i.e., electrons (electron–electron interactions) or ions in the crystal lattice (electron–lattice interactions), or with impurities and defects. This effect, which leads to the resistivity of metals, is introduced phenomenologically by a fluid friction force with damping rate $\gamma = \gamma_{\mathrm{S}}$.

This approach has been generalised to the highest frequencies in the case of a field oscillating at frequency ω in the infrared or visible region of the spectrum. The forced solution at frequency ω yields the induced polarisation $\boldsymbol{P} = -n_{\mathrm{e}}e\boldsymbol{r}$ with $\boldsymbol{v} = \mathrm{d}\boldsymbol{r}/\mathrm{d}t$, and, in association with the constitutive relation $\varepsilon\varepsilon_0\boldsymbol{E} = \varepsilon_0\boldsymbol{E} + \boldsymbol{P}$, defines the dielectric constant of the medium due to free electrons [see (7.15)].

The damping rate depends on the frequency. The static term γ_{S}, i.e., at zero frequency $\omega \approx 0$, which governs the properties of the metal in non-alternating conditions, e.g., conductivity, must then be replaced by the optical rate γ_0, with $\gamma_0 \ll \omega$ [12]. Although it is based on a classical approach that looks at first sight to be poorly suited to the description of electrons in a solid, the Drude model has been successful in describing many phenomena and providing simple physical interpretations for them. In optics, the quantum microscopic description of light absorption does justify it and can be used to calculate the relaxation rate γ_0 explicitly, showing that it exhibits little dispersion in the visible [11]. In a perfect crystal, both the electron–phonon and electron–electron interactions contribute, the latter only slightly at room temperature, i.e., about 10% of the total optical collision rate.

The second mechanism is the interband mechanism, related to optical transitions between two electronic bands, from filled bands to states in the conduction band or from the conduction band to empty bands of higher energy. In the alkali metals, these transitions occur at high frequencies and produce only a small correction to the real part of ε in the optical region. In the noble metals, this contribution is much greater and is essentially due to transitions between the d bands and the s–p conduction band (see Fig. 7.3) [11]. When a photon is absorbed, an electron changes band and can only be excited into an unoccupied state of the conduction band (Pauli exclusion principle), i.e., above the Fermi energy E_{F}. There is therefore a threshold frequency Ω_{ib} for such transitions, $\hbar\Omega_{\mathrm{ib}}$ varying from about 2 eV in copper to 4 eV in silver (see Table 7.1). This absorption causes the characteristic colour of these metals in their bulk solid form.

The presence of bound electrons leads one to write the dielectric constant of the metal as a sum of two terms:

$$\varepsilon = \varepsilon_{\mathrm{Drude}} + \delta\varepsilon_1^{\mathrm{ib}} . \tag{7.16}$$

The interband term dominates the optical response in the visible and ultraviolet, whereas the intraband term is predominant in the red and infrared. It

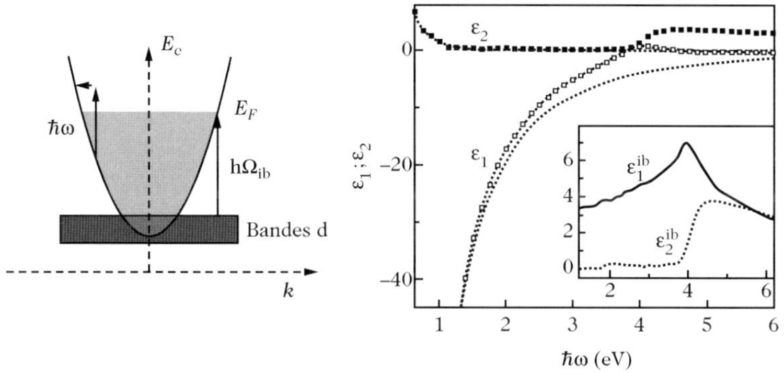

Fig. 7.3. *Left*: Simplified model of the electronic band structure of the noble metals. Energy E_e as a function of the wave vector k (parabolic conduction band $E_e = \hbar^2 k^2 / 2m_e$ filled up to the Fermi energy E_F; filled d bands with low dispersion). *Arrows* indicate (phonon-assisted) intraband transitions at $\hbar\omega$ and the threshold for interband transitions at $\hbar\Omega_{ib}$. *Right*: Real and imaginary parts ε_1 and ε_2, respectively, of the dielectric constant measured in bulk solid silver [13]. *Dotted curves* show intraband contributions [Drude model, see (7.15)] and the *insert* shows interband contributions ($\varepsilon^{ib} = 1 + \delta\varepsilon^{ib}$)

leads to a negative real part of the dielectric constant in this spectral region (see Fig. 7.3 for Ag) [11, 12].

For clusters with sizes greater than 2 nm, the absorption mechanisms are very similar and involve the same processes. Although quantum confinement dominates for semiconductors (confinement of excitons for which the Bohr radius becomes comparable with the particle size and confinement of electronic states with low quantum number, close to the centre of the Brillouin zone in the bulk), it is much less important for metal particles. The properties of the latter are related to the response of electrons in the vicinity of the Fermi level. They involve states of high quantum number for which a continuum of states can be used in conjunction with a similar approach to the one for the bulk medium [13]. Size effects appear as corrections and, in the optical properties, they are revealed mainly through an increase in the optical collision rate of electrons in the Drude model.

Indeed, the presence of surfaces typically influences these properties when the particle size is of the same order or less than the mean free path of the electrons (typical values are 20–50 nm in metals, see Table 7.1). From a classical standpoint, this corresponds to the fact that electron–surface collisions will no longer be negligible compared with the other interaction processes and must therefore be taken into account in calculating the optical collision rate of the electrons. In the context of the Drude model, this reflects the fact that intraband absorption is increased because it can also be assisted by electron collisions with the surface. A simple model thus consists in introducing in a phenomenological way a term proportional to the frequency of collision with the surfaces, i.e., proportional to the reciprocal of the time $t_t = D/v_F$ taken by the electron to cross a particle, where v_F is the Fermi speed of the electrons ($t_t \approx 0.7\,\mathrm{fs/nm}$ in silver). For a sphere of diameter D, the total collision rate is then [9]

$$\gamma = \gamma_0' + g\frac{v_F}{D} , \qquad (7.17)$$

where g is a proportionality factor close to unity and γ_0' is the intrinsic electron collision rate (electron–electron and electron–phonon) similar to the one in the solid and possibly modified by confinement in the nanoparticle [14,15]. As for the bulk metal, the intraband contribution is then given by a Drude-type expression, with a modified electron collision rate. From a classical standpoint, the enhancement of intraband absorption is a consequence of the increased importance of electron–surface collisions which provide a way of conserving momentum during absorption of a photon.

This phenomenological model is justified by quantum theory [7,9,16,17]. In the simplest model, the electrons are situated in the periodic potential of the bulk crystal, bounded by an infinite spherical potential well corresponding to the outer surface of the sphere. The exact expression for the confined wave functions is found by solving Schrödinger's equation and the wave functions are then used to calculate the matrix elements of the dipole transition between confined electron states. From the standpoint of quantum mechanics, the increase in intraband absorption due to confinement results from the appearance of allowed optical transitions between confined states, k no longer being a good quantum number.

Redefining an electron density of states similar to that of the bulk material, it can then be shown that, in the quantum model, the intraband contribution to the dielectric constant can also be rewritten as an expression of Drude type (7.15), with an effective collision rate identical to (7.17). The value of the proportionality factor g, which can be calculated here, depends on the model used. Values between 1 and 3 have been obtained [17,18]. In the framework of the model described above and assuming an isotropic parabolic conduction band in the bulk metal (see Fig. 7.3), it is given by

$$g(\omega) = \frac{2}{\hbar\omega E_F^2}\int_0^{\infty} E_e^{3/2}\sqrt{E_e + \hbar\omega} f_e(E_e)\Big[1 - f_e(E_e + \hbar\omega)\Big]\mathrm{d}E_e , \qquad (7.18)$$

where f_e is the electron distribution function (Fermi–Dirac at temperature T). For $\hbar\omega = 3\,\mathrm{eV}$ (close to $\hbar\Omega_R$ for Ag) and $T = 295\,\mathrm{K}$, we obtain $g \approx 1.4$ (g varies slightly with frequency). For the noble metals, the collision rate due to the surface is of the order of $\hbar g v_F/D \approx 1.3/D\,\mathrm{eV\,nm^{-1}}$.

The interband absorption of gold and silver nanoparticles has been studied as a function of their size [18]. The interband dielectric constant of the metal is slightly modified down to sizes of the order of 2 nm. For the size range discussed here, the main consequence of quantum confinement on the dielectric constant is thus a modification of the optical collision rate:

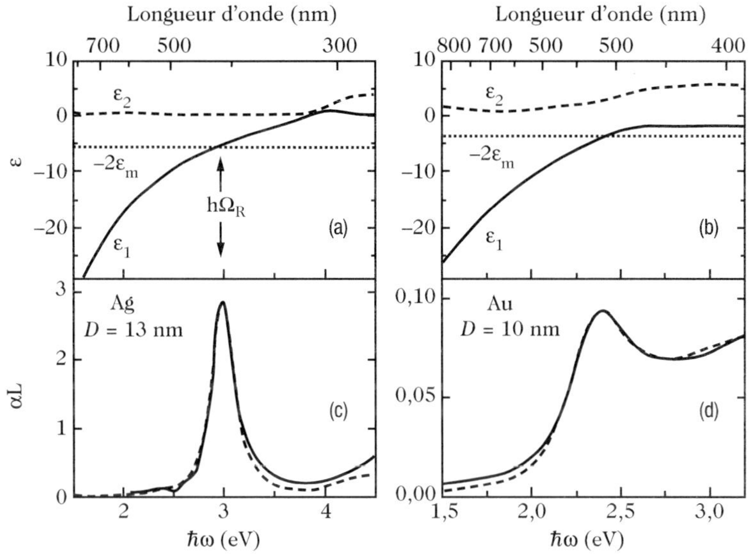

Fig. 7.4. (**a**) Dispersion of the real and imaginary parts of the dielectric constant for silver and for the glass Ba0–P_2O_5, $\varepsilon_m \approx 2.75$. (**b**) Dispersion of the dielectric constant for gold and for water $\varepsilon_m \approx 1.77$. (**c**) and (**d**) Measured (*continuous curves*) and calculated (*dotted curves*) absorption spectra for silver nanoparticles ($D = 13$ nm) dispersed in the glass Ba0–P_2O_5 and gold nanoparticles ($D = 10$ nm) dispersed in water. The approximate resonance condition in the quasi-static approximation is indicated: $\varepsilon_1(\Omega_R) = -2\varepsilon_m$

$$\varepsilon(\omega) = \varepsilon^{ib}(\omega) - \frac{\omega_p^2}{\omega\big[\omega + i(\gamma_0' + g v_F/D)\big]} , \tag{7.19}$$

where $\varepsilon^{ib} = 1 + \delta\varepsilon^{ib}$.

7.4 Surface Plasmon Resonance in the Quasi-Static Approximation: Nanospheres

The dielectric constant of the confined metal written out above can be used to describe the surface plasmon resonance more precisely. In the framework of the quasi-static approximation, i.e., for diameters satisfying $D \leq \lambda/10$, and if the approximate resonance condition $\varepsilon_1(\Omega_R) + 2\varepsilon_m = 0$ is justified, its frequency is given by

$$\Omega_R = \frac{\omega_p}{\sqrt{\varepsilon_1^{ib}(\Omega_R) + 2\varepsilon_m}} . \tag{7.20}$$

This is approximately the case for the noble and alkali metals (see Fig. 7.4).

Table 7.1. Effective electron mass m_e, normalised to the free electron mass m_0, electron density n_e, Fermi velocity v_F, mean free path $l_f = v_F/\gamma_s(0)$ of electrons in the solid (conductivity measurements), bulk plasmon energy $\hbar\omega_p$, interband transition energy $\hbar\Omega_{ib}$, and surface plasmon resonance energy in vacuum $\hbar\Omega_R$ (quasi-static approximation for $\varepsilon_m = 1$) for various metals

	m_e/m_0	n_e [$10^{28}\,\mathrm{m}^{-3}$]	v_F [$10^6\,\mathrm{m/s}$]	l_f [nm]	ω_p [eV]	Ω_{ib} [eV]	$\hbar\Omega_R$ [eV]
Na	1	2.65	1.07	34	5.95	–	3.23
Cu	1.47	8.47	1.05	28	9.04	2.1	2
Au	1.01	5.90	1.40	42	9.01	2.3	2.45
Ag	1.03	5.86	1.39	55	8.98	3.9	3.5

The surface plasmon resonance does not only involve conduction electrons, but also bound electrons in deeper bands (the conduction electrons are so-called in nanoparticles by analogy with the bulk medium). They reveal themselves most clearly by a shift in the resonance frequency relative to that of the free electron gas, denoted by ε_1^{ib} ($\Omega_R = \omega_p/\sqrt{1+2\varepsilon_m}$ for free electrons, $\varepsilon_1^{ib} = 1$, which reduces to $\Omega_R = \omega_p/\sqrt{3}$ in vacuum). For the alkali metals, this shift is very small, e.g., reducing $\hbar\Omega_R$ from $\hbar\omega_p/\sqrt{3} = 3.44\,\mathrm{eV}$ to $3.23\,\mathrm{eV}$ for sodium (assuming $\varepsilon_m = 1$), in agreement with experimental studies. For the noble metals, it is much bigger, e.g., for silver, $\hbar\omega_p/\sqrt{3} = 5.18\,\mathrm{eV}$ and $\hbar\Omega_R = 3.5\,\mathrm{eV}$ (for $\varepsilon_m = 1$). Although gold, silver, and copper have very similar plasma frequencies, their surface plasmon resonances occur at very different frequencies (see Table 7.1). As for the bulk plasmon [19], this shift reflects the response of the bound electrons to the electromagnetic wave: the resonance is associated with polarisations induced by quasi-free and bound electrons.

For a given metal, the resonance frequency of the nanoparticles is modified by their surroundings, the important parameter being its dielectric constant ε_m [see (7.20)]. It is shifted toward the red for matrices with increasing ε_m (see Fig. 7.5). The frequency Ω_R is always well predicted by the simplified condition (7.11) for alkali metals and silver, but not for other metals. This is in particular the case for gold and copper, for which the surface plasmon resonance considerably overlaps the interband transitions for small ε_m values, leading to a significant discrepancy between the position of the maximum of the extinction cross-section (7.13) and Ω_R predicted using (7.11) (see Fig. 7.6).

The physical origin of absorption in the vicinity of Ω_R remains unchanged and is similar to that in the bulk metal at the same frequency. For silver, the resonance is well separated from the interband absorptions (see Fig. 7.4) and appears as an enhancement by dielectric confinement of the intraband absorption, very small at this frequency in the bulk material (see Fig. 7.2). This is also the case in the alkali metals. For gold or copper nanoparticles, the interband transitions are at lower frequencies and are partially superposed on

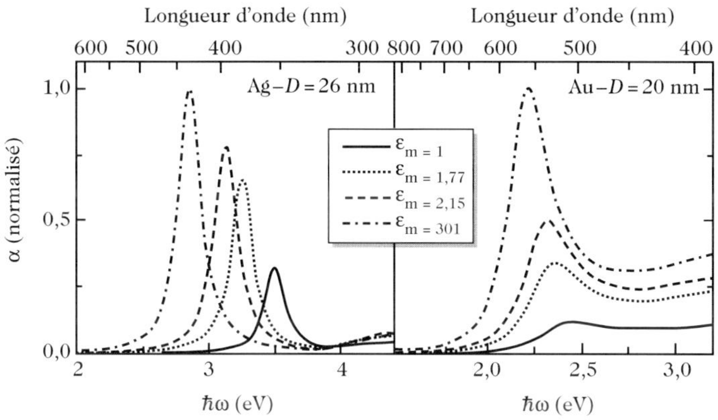

Fig. 7.5. Absorption spectra calculated using the quasi-static approximation for silver and gold nanospheres with diameters $D - 26$ and 20 nm, respectively, in vacuum (cluster beam, $\varepsilon_{\mathrm{m}} = 1$) and in different matrices: water ($\varepsilon_{\mathrm{m}} = 1.77$), silica ($\varepsilon_{\mathrm{m}} = 2.15$), and alumina ($\varepsilon_{\mathrm{m}} = 3.1$). The metal fraction p is constant. Broadening due to interactions with the surface has been included with $g = 3$ for Ag and $g = 1.6$ for Au [see (7.16)]. The increased absorption above 3.9 eV for silver and above 2.5 eV for gold is due to interband transitions

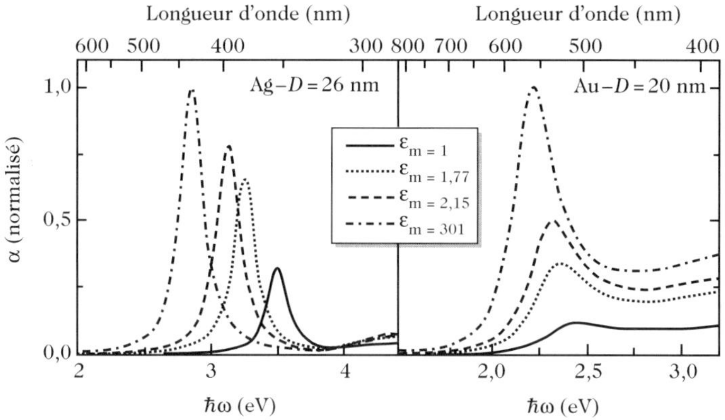

Fig. 7.6. Computed surface plasmon resonance wavelength λ_{R} of a single silver and gold nanosphere of diameter 15 nm as a function of the dielectric constant of the surrounding matrix ε_{m}. The *open dots* are the λ_{R} values calculated using the approximate resonance condition (7.11) and the *dashed line* represents the λ_{R} values maximising the extinction cross-section S_{ext} (7.13)

the surface plasmon resonance (see Fig. 7.4). Its interpretation is thus more complex and absorption involves both intra- and interband transitions, still enhanced by dielectric confinement.

If the dispersion $\varepsilon_1^{\mathrm{ib}}$ is small in the vicinity of the resonance, the expression (7.10) for the absorption coefficient α can be expanded about the frequency

Ω_R using the dielectric constant of the confined metal (7.19) with

$$\gamma_R = \gamma + \omega^3 \varepsilon_2^{ib}(\omega)/\omega_p^2 \ll \Omega_R \,,$$

which yields [9, 20]

$$\alpha(\omega) = \frac{9p\varepsilon_m^{3/2}}{c\omega_p^2} \frac{\omega^2 \Omega_R^4 \gamma_R}{(\omega^2 - \Omega_R^2)^2 + (\Omega_R^2 \gamma_R/\omega)^2} \,. \tag{7.21}$$

If γ_R shows little dispersion near Ω_R, the surface plasmon resonance then has the form of a quasi-Lorentzian line and its width Γ_R can be defined by

$$\Gamma_R \approx \gamma_R(\Omega_R) = \gamma_0'(\Omega_R) + g(\Omega_R)\frac{v_F}{D} + \frac{\Omega_R^3}{\omega_p^2}\varepsilon_2^{ib}(\Omega_R) \,. \tag{7.22}$$

The first two terms correspond to the electron scattering rate in a nanoparticle (intraband term). In a classical picture, this contribution to the width Γ_R is interpreted as a damping of the collective motion of the electron cloud with intraband excitation of an electron: the coupled photon–plasmon excitation is damped with excitation of a single electron whose energy increases by $\sim \hbar\Omega_R$ (Landau-type damping induced by collisions). The last term corresponds to interband absorption at frequency Ω_R. It can be interpreted in a similar way as damping due to interband excitation of an electron.

In the alkali metals and silver, the surface plasmon resonance occurs in a region where the interband absorption is small, i.e., $\varepsilon_2^{ib}(\Omega_R) \approx 0$. It then appears in the absorption spectrum as a relatively narrow line of width Γ_R (Fig. 7.4c). A similar line shape is predicted for the spectrum of the scattered light, using (7.12). As Γ_R is determined by the electron scattering rate (intraband term), it varies linearly with $1/D$, thus leading to an increase in the line width as the size decreases [see (7.22)]. This dependence due to electron–surface interactions has been confirmed experimentally in the noble metals by carrying out measurements of the absorption spectrum as a function of size for a given matrix [9, 16, 18].

However, the above theoretical model predicts a width that should be independent of the matrix, whereas the experimental results show that it varies significantly with the particle environment. This is reflected in a variation of the factor g determined by measurements [see (7.22)], e.g., $g = 2$ and $g = 0.52$ for silver nanoparticles in a solid glassy matrix or a rare gas (Ar or Ne) [18]. This dependence has been attributed to surface modification effects, such as physisorption, trapping, or chemical reactions at the interface [18]. Like all surface effects, their contribution increases as the particle size decreases: to first order, it varies as the surface-to-volume ratio, i.e., as $1/D$. It is therefore often included in the factor g of (7.22), which is used as a parameter to fit the size dependence of the widths Γ_R obtained experimentally. These effects, introduced globally under the heading of chemical damping at the interface, are included in an ad hoc manner and have not yet been explained by detailed models.

For gold or copper, the surface plasmon resonance is around the interband transition threshold. Although the above expansion of the absorption coefficient about Ω_R is no longer strictly valid, ε_1^{ib} then varying more significantly with the frequency near Ω_{ib} (Fig. 7.3b), it does help to understand the shape of the absorption spectrum. The increased interband contribution to γ_R [see (7.22)] leads to a spectral broadening of the resonance which no longer has a quasi-Lorentzian shape, with ε_2^{ib} and γ_R thus exhibiting high dispersion (Fig. 7.3b). The resonance is then less marked and adds to the increase in interband absorption at short wavelengths (see Fig. 7.4d). Although it is difficult to define its width adequately, it broadens as before with size reduction owing to the $1/D$ contribution of the intraband term [see (7.22)].

The resonance appears more and more clearly as its superposition on the interband transitions decreases. This effect is illustrated for gold in Fig. 7.5: when matrices with increasing dielectric constant are used, the resonance becomes better defined because it is redshifted, moving away from the interband transitions. As the other metals have many interband transitions in the visible and near-infrared regions of the spectrum, their resonances are therefore often rather spread out and difficult to observe. The alkali metals and especially, owing to their ease of use in experiments, the noble metals provide model systems in which changes in optical properties due to size reduction are clearly visible. This is why most optical studies and applications have focused on these metals.

7.5 Surface Plasmon Resonance: Quantum Effects for Small Sizes ($D < 5$ nm)

One important conclusion of the quasi-static model is that the frequency of the surface plasmon resonance is independent of size for a given metal provided that its dielectric constant does not vary. This is confirmed experimentally for intermediate sizes (typically between 5 and 30 nm). The upper bound comes from multipole contributions, neglected in this model (see the discussion of the Mie model below), while the lower bound is due to quantum confinement effects, i.e., to the fact that the 'small solid' approximation used up to now is no longer valid. For smaller sizes, the electron wave functions are significantly modified, whereby the electron distribution and polarisability near the surface are also changed.

As the spherical potential associated with the matrix is finite, the electron wave functions extend beyond the sphere defined by the crystal lattice, over a distance of the order of one angstrom. In a simple model, the electron diameter D_e, which defines the region explored by quasi-free electrons, is greater than the diameter D of the nanoparticle, defined by its lattice. We write $D_e = D + d_{so}$. This so-called spill-out effect leads to a reduction in the average electron density n_e^{nano} in the particle, which depends on its size [21]:

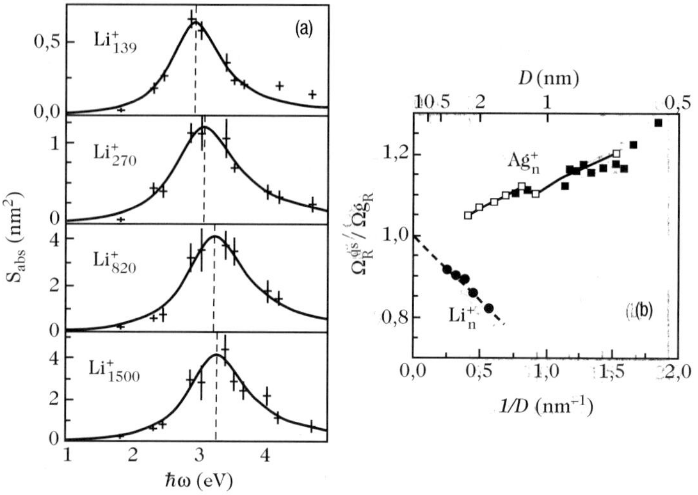

Fig. 7.7. (**a**) Dispersion of the absorption cross-section S_{abs} of free ionised lithium clusters comprising 139, 270, 820 and 1 500 atoms (sizes 0.9, 1.1, 1.6 and 2 nm) [22]. In the quasi-static approximation, the resonance occurs at $\hbar\Omega_{\text{R}}^{\text{qs}} = 3.55\,\text{eV}$. (**b**) Energy of the surface plasmon resonance, normalised to its quasi-static value, for ionised silver clusters (*black squares*: measured values, *white squares*: calculated values with spill-out and confinement of d electrons [23, 24]) and ionised lithium clusters (*black dots*: measured values, *curve*: calculated values with spill-out [22]) as a function of cluster size

$$n_{\text{e}}^{\text{nano}} = n_{\text{e}} \frac{D^3}{D_{\text{e}}^3} \approx n_{\text{e}} \left(1 - \frac{3d_{\text{so}}}{D}\right) , \tag{7.23}$$

where n_{e} is the electron density in the bulk metal. Since the spill-out amplitude d_{so} is barely size dependent, this effect is all the more noticeable when the size decreases, i.e., as surface effects begin to dominate bulk effects.

The plasma frequency ω_{p}, proportional to the square root of the electron density, is thus also reduced, leading to a redshift of the surface plasmon resonance. This shift, linear in $1/D$, has been observed in the alkali metals where the response is dominated by conduction electrons [22]. The extrapolated value for intermediate sizes ($d_{\text{so}}/D \ll 1$) agrees with the quasi-static approximation: $\hbar\Omega_{\text{R}}^{\text{qs}} = \hbar\omega_{\text{p}}/\sqrt{3}$ (see Fig. 7.7 for lithium [22]).

In the noble metals, modifications to the wave functions of bound electrons (d electrons) are also relevant. In contrast to the conduction electrons, they are localised within the inner region of the particle and their polarisability is reduced close to the surface, across a spherical shell of thickness around one angstrom [23, 24]. Their contribution to $\varepsilon_1^{\text{ib}}$ is therefore reduced, and this induces a blueshift in the surface plasmon resonance (7.20). These two quantum effects due to free and bound electrons induce opposite shifts which partly balance one another [23]. A slight shift toward the blue has indeed been observed for small gold and silver nanoparticles, in agreement with the predictions of this model (see Fig. 7.7). A further consequence of these two effects is to reduce screening of the Coulomb interaction potential between electrons near the surface and hence to increase the probability of their interacting. The consequent acceleration of energy exchanges between electrons has been demonstrated using time-resolved optical spectroscopy for gold and silver nanoparticles, the effect appearing very clearly for sizes below 10 nm [14].

7.6 General Case for Nanospheres: The Mie Model

For sizes greater than a few tens of nanometers, the quasi-static approximation is no longer valid and the interaction of an electromagnetic wave with a nanoparticle must explicitly take into account the spatial variations of the field over the size of the object, i.e., it must allow for retardation effects. The problem, although similar, since it requires one to solve Maxwell's equations with the boundary conditions imposed at the object–environment interface, is actually much more involved here because the electromagnetic field is no longer constant in the object. This problem was solved by Mie in 1908 for the interaction of a plane wave with a sphere, in order to understand the colouring effects due to scattering and absorption by colloidal solutions of gold [2]. The model is based on a multipole expansion of the electromagnetic field. Here we shall only give the final result regarding the scattering and extinction cross-sections at frequency ω for a nanosphere of diameter D [3, 16]:

$$\begin{aligned} S_{\text{scatt}} &= \frac{2\pi c^2}{\varepsilon_{\text{m}}\omega^2} \sum_{l=1}^{\infty} (2l+1) \left(|a_l|^2 + |b_l|^2\right) , \\ S_{\text{ext}} &= \frac{2\pi c^2}{\varepsilon_{\text{m}}\omega^2} \sum_{l=1}^{\infty} (2l+1) \text{Re}\left(a_l + b_l\right) , \end{aligned} \tag{7.24}$$

where $S_{\text{abs}} = S_{\text{ext}} - S_{\text{scatt}}$ and the coefficients a_l and b_l are defined by

$$a_l = \frac{m\psi_l(mx)\psi_l'(x) - \psi_l(x)\psi_l'(mx)}{m\psi_l(mx)\xi_l'(x) - \xi_l(x)\psi_l'(mx)} , \quad b_l = \frac{\psi_l(mx)\psi_l'(x) - m\psi_l(x)\psi_l'(mx)}{\psi_l(mx)\xi_l'(x) - m\xi_l(x)\psi_l'(mx)} , \tag{7.25}$$

with ψ_l and ξ_l the Riccati–Bessel functions of order l and primes denoting derivatives. The parameters x and m are the normalised size and refractive index of the nanosphere, viz.,

$$x = \pi\varepsilon_{\text{m}}^{1/2} D/\lambda , \qquad m = (\varepsilon/\varepsilon_{\text{m}})^{1/2} . \tag{7.26}$$

The index l is the order of the multipole expansion, dipole for $l = 1$, quadrupole for $l = 2$, and so on. The various cross-sections can only be calculated numerically.

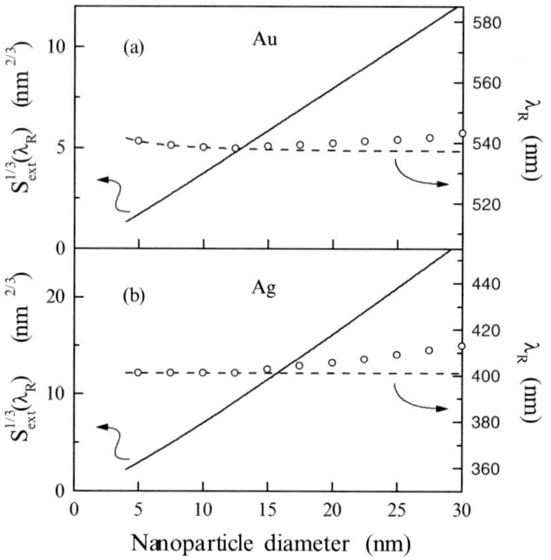

Fig. 7.8. Size dependencies of the surface plasmon resonance wavelength λ_R (*dashed line*), and cube root of the extinction cross-section S_{ext} (*continuous line*) for a gold (**a**) and silver (**b**) nanosphere in glass ($\varepsilon_m = 2.25$) computed in the quasi-static or in the dipole approximation (7.13). *Open dots* are the λ_R values calculated using the multipole expansion of the Mie theory up to fifth order [$l = 5$ in (7.24)]

At intermediate sizes, approximate expressions can be found by expanding the Bessel functions in powers of x. The quasi-static approximation is equivalent to the lowest order expansion in x, i.e., to the dipole approximation of Mie's theory, whereby the expressions (7.12) and (7.13) are recovered. The scattered field is equivalent to that emitted by a dipole at the center of the nanosphere, with an expression identical to $\boldsymbol{p}_s$ [see (7.5)]. For an ensemble of nanospheres embedded in a dielectric matrix, the effective dielectric constant $\tilde{\varepsilon}$ (7.8) of the composite medium can thus be computed as before, using the model of localized dipole induced in the material, yielding a polarization density $\boldsymbol{P}$ (7.6).

The limit of applicability of the dipole approximation in the small size range is illustrated in Fig. 7.8, where the frequency of the surface plasmon resonance computed in this approximation (or in the quasi-static approximation) Ω_R^{qs} is compared with the Ω_R value obtained from Mie's theory by summing the multipole contributions up to $l = 5$. The range of validity depends on the material and is typically a few tens of nanometers.

The results obtained with the two models at intermediate sizes are compared in Fig. 7.9 for silver. They show a significant redshift of the plasmon resonance for sizes typically above 30 nm (Figs. 7.8 and 7.9a). In parallel, the contribution of the scattering cross-section increases (Fig. 7.9b).

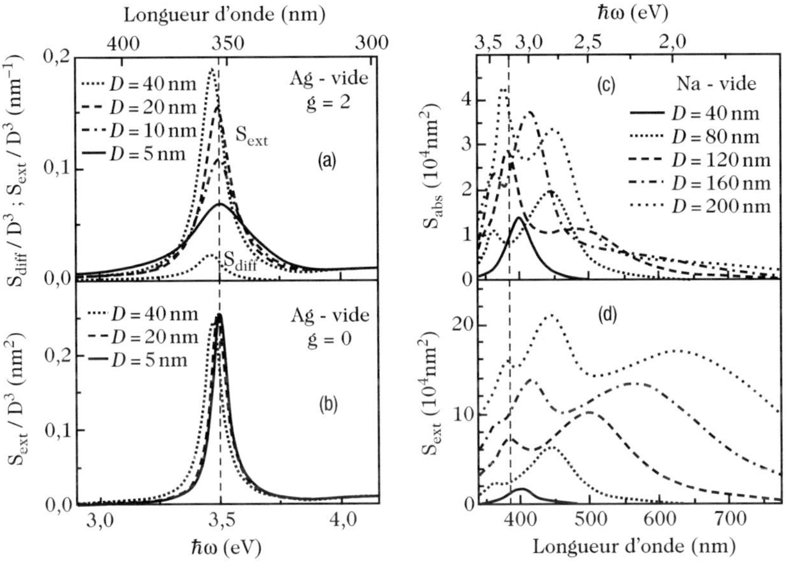

Fig. 7.9. (**a**) and (**b**) Spectra of the extinction cross-section S_{ext} and scattering cross-section S_{scatt} for silver nanospheres of diameters $D = 5$, 10, 20 and 40 nm in vacuum, calculated using the Mie model and a dielectric constant that has been corrected ($g = 2$) or not ($g = 0$) for surface effects [see (7.17)]. The broadening and reduction in maximal amplitude of S_{ext} at small sizes is a consequence of this correction. For $g = 0$, $S_{\mathrm{ext}} \propto D^3$ in the quasi-static approximation. (**c**) and (**d**) S_{abs} and S_{ext} for sodium nanospheres of diameter $D = 40$, 80, 120, 160 and 200 nm [17]. *Vertical lines* indicate $\hbar\Omega_{\mathrm{R}}^{\mathrm{qs}}$, calculated using the Mie theory in the dipole approximation, or equivalently, in the quasi-static approximation

At large sizes, the absorption cross-section reveals a large redshift of the dipole resonance with the appearance of multipole contributions at higher energy. The latter dominate the absorption spectrum at sizes above or of the order of 100 nm (see Fig. 7.9c). The extinction cross-section has a different spectrum (see Fig. 7.9d). While dominated by absorption at small sizes, it is dominated by scattering at large ones ($D \geq 100$ nm). This change of regime agrees with the variation $S_{\mathrm{scatt}}/S_{\mathrm{abs}} \propto V_{\mathrm{np}}$ in the dipole approximation (7.14).

The width of the dipole resonance increases significantly at sizes greater than a few tens of nanometers, in proportion to their volume (Fig. 7.9). This effect, a consequence of the increased scattering, is referred to as radiative damping of the surface plasmon resonance [9]. The coupled electromagnetic wave–electron motion excitation relaxes preferentially by photon emission, whereas relaxation by energy transfer to electrons in the particle dominates at small sizes (non-radiative damping via absorption by the nanoparticle). Variations in the frequency and damping of the surface plasmon resonance described in the last three sections are summarised in Fig. 7.10.

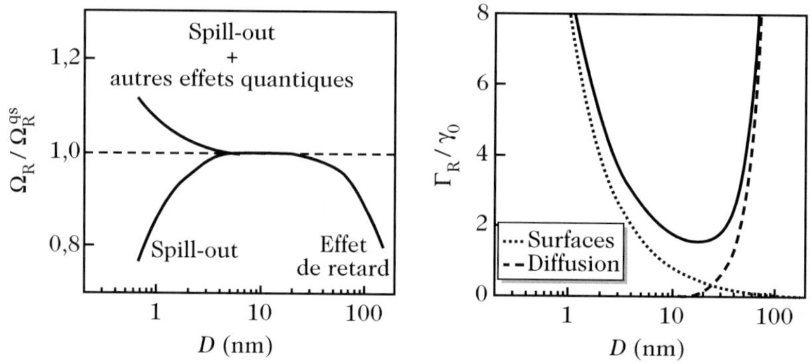

Fig. 7.10. Schematic size dependence of the frequency and width of the surface plasmon resonance (*continuous curves*): they are described by the quasi-static approximation with and without quantum effects at small and intermediate sizes, and including the retardation effect at large sizes (Mie model). The origins of the main effects are indicated. From [17]

7.7 Non-Spherical or Inhomogeneous Nanoparticles in the Quasi-Static Model

7.7.1 Shape Effects: Ellipsoids

The optical response is significantly modified by the shape of the objects. It can be calculated analytically provided that they are symmetrical enough [3, 17]. This is the case for an ellipsoid with dielectric constant ε, defined by its semi-axes a, b and c, where $a \geq b \geq c$. This includes several types of shape, notably the disk ($a = b > c$) and needle ($a > b = c$). In the quasi-static approximation ($a \ll \lambda$), the problem consists once again in using the boundary conditions and neglecting propagation effects. For an incident electromagnetic wave polarised along the principal axis $i = x$, y or z, associated with the semi-axis a, b or c, respectively (see Fig. 7.11), the scattering and extinction coefficients are given by [3, 17]:

$$S_{\mathrm{scatt}}^i = \frac{8\pi^3 V_{\mathrm{np}}^2 \varepsilon_{\mathrm{m}}^2}{3\lambda^4 L_i^2} \left| \frac{\varepsilon - \varepsilon_{\mathrm{m}}}{\varepsilon + \dfrac{1-L_i}{L_i}\varepsilon_{\mathrm{m}}} \right|^2 , \quad S_{\mathrm{ext}}^i = \frac{2\pi V_{\mathrm{np}} \varepsilon_{\mathrm{m}}^{3/2}}{\lambda L_i^2} \frac{\varepsilon_2}{\left| \varepsilon + \dfrac{1-L_i}{L_i}\varepsilon_{\mathrm{m}} \right|^2} , \tag{7.27}$$

where ε_{m} is the dielectric constant of the surrounding medium and V_{np} the volume of the ellipsoid. These expressions are similar to those found for a sphere, with the presence of an additional geometrical coefficient, viz., L_i :

$$L_i = \frac{abc}{2} \int_0^\infty \frac{du}{(u + l_i^2)\sqrt{(u+a^2)(u+b^2)(u+c^2)}} , \tag{7.28}$$

where $l_x = a$, $l_y = b$ and $l_z = c$ (u is an integration variable). Clearly, $L_x + L_y + L_z = 1$, whence $L_i = 1/3$ for a sphere where the three coefficients are equal. The optical response exhibits three surface plasmon resonances (degenerate for the sphere) associated with each of the three axes. It will thus depend on the polarisation of the incident wave, defined with respect to the orientation of the ellipsoid. Along the direction of polarisation i, similarly to (7.11) for a sphere, the resonance frequency Ω_{R}^i is determined to a first approximation by

$$L_i \varepsilon_1(\Omega_{\mathrm{R}}^i) + (1 - L_i)\varepsilon_{\mathrm{m}} = 0 . \tag{7.29}$$

For an ensemble of ellipsoids, polarisation effects persist if they are oriented, e.g., ellipsoids produced by stretching a glass doped with nanospheres. The medium then exhibits polarisation-selective absorption and can be used as a polariser. For N_{np} randomly oriented ellipsoids, the cross-sections of the material are the orientational averages of those for the individual objects:

$$S_{\mathrm{ext,scatt}}^{\mathrm{mat}} = \frac{N_{\mathrm{np}}}{3} \sum_i S_{\mathrm{ext,scatt}}^i . \tag{7.30}$$

For a spheroid, i.e., an ellipsoid in which two semi-axes are equal, L_i can be expressed analytically in terms of the eccentricity e ($e^2 = 1 - c^2/a^2$). For a prolate spheroid ($a \geq b = c$):

$$L_x = \frac{1-e^2}{e^2} \left[-1 + \frac{1}{2e} \ln \left(\frac{1+e}{1-e} \right) \right] , \tag{7.31}$$

and $L_y = L_z = (1 - L_x)/2$ (see Fig. 7.11). For an oblate spheroid ($a = b \geq c$):

$$L_x = \frac{\sqrt{1-e^2}}{2e^2} \left[\frac{1}{e} \left(\frac{\pi}{2} - \arctan \sqrt{\frac{1-e^2}{e^2}} \right) - \sqrt{1-e^2} \right] , \tag{7.32}$$

and $L_y = L_x$, $L_z = 1 - 2L_x$ (see Fig. 7.11). The extinction coefficient then exhibits two resonances whose frequencies depend only on the shape, i.e., on e, and are independent of the size of the object (provided that the quasi-static approximation is justified). Within certain limits, the optical characteristics can thus be adjusted by modifying the shape of the ellipsoids (see Fig. 7.12).

The optical response of a nano-object of arbitrary size and shape, or of interacting nano-objects, cannot generally be calculated analytically. A numerical approach must then be used, e.g., dividing the object up into small polarisable entities and summing their responses [17].

7.7.2 Structure Effects: Core–Shell System

The optical response of a nano-object is also modified when it is structured on the nanoscale, e.g., by putting together several different materials. The

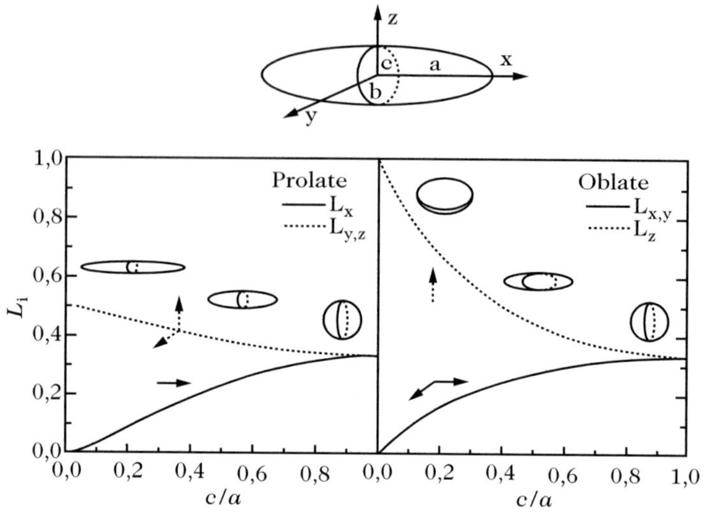

Fig. 7.11. Shape factors for prolate ellipsoids ($a \geq b = c$, $L_x \leq L_y = L_z$) and oblate ellipsoids ($a = b \geq c$, $L_x = L_y \leq L_z$) as a function of the ratio of the minor to major axes c/a. Polarisation directions x, y or z are shown. Different nanoparticle shapes are illustrated

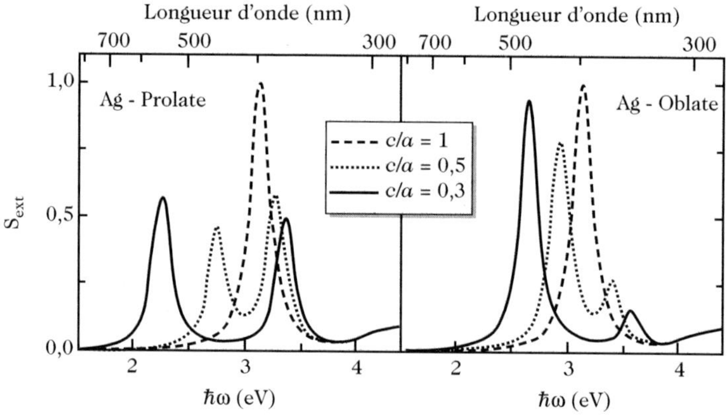

Fig. 7.12. Spectra of the extinction cross-section S_{ext} for a randomly orientated ensemble of prolate or oblate silver nano-ellipsoids with $c/a = 0.5$ and 0.3 and nanospheres ($c/a = 1$) embedded in silica. Resonances shifted toward the red or the blue are associated with electron motions along the largest dimensions (x prolate, x, y oblate) or the smallest (y, z prolate, z oblate) of the ellipsoids, respectively

simplest example is provided by core–shell spheres comprising a nanosphere of diameter D_{i} and dielectric constant ε_{i} surrounded by a shell of nanometric thickness d and dielectric constant ε_{e}. Using the quasi-static approximation (D_{i}, $d \ll \lambda$), the extinction and scattering cross-sections take the following forms [3, 4]:

$$S_{\mathrm{ext}} = \frac{6\pi V_{\mathrm{np}}\varepsilon_{\mathrm{m}}^{1/2}}{\lambda}\mathrm{Im}\left[\frac{(\varepsilon_{\mathrm{e}}-\varepsilon_{\mathrm{m}})(\varepsilon_{\mathrm{i}}+2\varepsilon_{\mathrm{e}})+p_{\mathrm{i}}(\varepsilon_{\mathrm{i}}-\varepsilon_{\mathrm{e}})(\varepsilon_{\mathrm{m}}+2\varepsilon_{\mathrm{e}})}{(\varepsilon_{\mathrm{e}}+2\varepsilon_{\mathrm{m}})(\varepsilon_{\mathrm{i}}+2\varepsilon_{\mathrm{e}})+2p_{\mathrm{i}}(\varepsilon_{\mathrm{i}}-\varepsilon_{\mathrm{e}})(\varepsilon_{\mathrm{e}}-\varepsilon_{\mathrm{m}})}\right],$$

$$S_{\mathrm{scatt}} = \frac{24\pi^3 V_{\mathrm{np}}^2\varepsilon_{\mathrm{m}}^2}{\lambda^4}\left|\frac{(\varepsilon_{\mathrm{e}}-\varepsilon_{\mathrm{m}})(\varepsilon_{\mathrm{i}}+2\varepsilon_{\mathrm{e}})+p_{\mathrm{i}}(\varepsilon_{\mathrm{i}}-\varepsilon_{\mathrm{e}})(\varepsilon_{\mathrm{m}}+2\varepsilon_{\mathrm{e}})}{(\varepsilon_{\mathrm{e}}+2\varepsilon_{\mathrm{m}})(\varepsilon_{\mathrm{i}}+2\varepsilon_{\mathrm{e}})+2p_{\mathrm{i}}(\varepsilon_{\mathrm{i}}-\varepsilon_{\mathrm{e}})(\varepsilon_{\mathrm{e}}-\varepsilon_{\mathrm{m}})}\right|^2, \tag{7.33}$$

where ε_{m} is the dielectric constant of the surrounding matrix, p_{i} is the volume fraction of the inner material defined by $p_{\mathrm{i}} = D_{\mathrm{i}}^3/(D_{\mathrm{i}}+d)^3$, and V_{np} is the total volume of the nanosphere. The resonance has a complex structure depending sensitively on the different materials. It is often spread out with several maxima. In the simple case of a dielectric core with a metallic shell, two resonances may appear, associated with the two dielectric–metal interfaces [17]. These expressions can be extended to ellipsoids and multilayer structures [3]. Naturally, the main difficulty here consists in actually synthesising such objects in a controlled way.

7.8 Optical Response of a Single Metal Nanoparticle

In most optical experiments, a large number of nanoparticles, typically 10^4 to 10^6, are simultaneously observed. Only a global optical response is then obtained, summing over all the individual contributions. Due to the unavoidable size, shape and structural fluctuations of the synthesized particles in a real sample, only averaged information is thus obtained, and this is sometimes difficult to compare with theoretical models. This is a particularly limiting point when one wishes to address nano-objects synthesized with a low yield and coexisting with others of different geometry and structure. The optical study of a single metal cluster is then necessary to avoid statistical fluctuation effects or to address or identify a relevant nano-object.

Different approaches have been developed recently for optically detecting and characterising single semiconductor or metal nano-objects. In the case of luminescent nanoparticles, detection of a single light emitter is now a routine laboratory experiment [25]. These techniques are, however, limited to highly luminescent objects such as semiconductor quantum dots or molecules, and cannot be extended to almost non-luminescent nanoparticles such as the metallic ones.

The optical methods for observing a single weakly luminescent nano-object require detection of its absorption or scattering. In this context, the high spatial resolution of near-field optical techniques is very interesting, although the problem of particle–tip coupling makes it difficult to interpret the experimental data [26]. This has fostered the development of far-field spectroscopy techniques whose lower intrinsic spatial resolution is overcome by using dilute

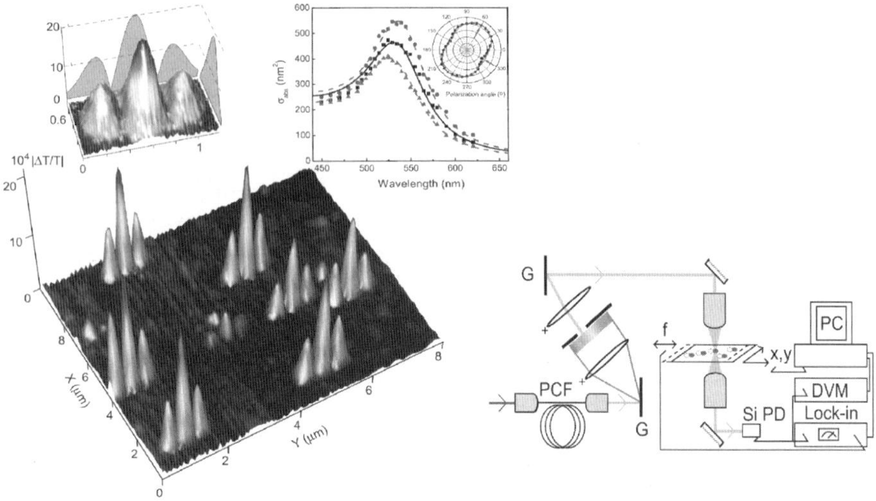

Fig. 7.13. *Right*: Schematic of the SMS setup with a supercontinuum source created in a photonic crystal fibre (PCF). *Left*: Spatial modulation image of single gold nanoparticles deposited on a glass substrate. Each main peak with two satellites corresponds to one particle located at the peak maximum. *Left inset*: Zoom on one particle. *Right inset*: Extinction spectra of a single gold nanoparticle measured for two perpendicular polarizations of the light (*circles* and *triangles*). *Squares* are measured with unpolarized light. The nanoparticle size is about 16 nm [29]

systems so that only one particle is in the probed zone (typically a density of the order of one particle per μm^2 for 2D samples). Scattering-based – dark-field – methods are now well established and have led to the development of very sensitive nano-scatterer sensors. They are limited to 'large' nanoparticles, typically of a few tens of nm for standard approaches and about 5 nm using heterodyne detection [27]. Its main strength is the sensitive spectral investigation of single particles with a high degree of precision and speed, which allows in situ detection of small changes in the plasmon resonance wavelength. This dark-field single particle spectroscopy method is therefore very useful for investigating chemical changes on single nanoparticle objects in real time.

As scattering significantly decreases with the particle size (as the square of their volume), absorption (scaling directly with the volume) dominates the optical response of very small particles [see (7.14)]. It is thus the most sensitive optical property to monitor for very small particles. A photothermal technique has recently been demonstrated, based on detection of the local laser heating of a single metal nanoparticle [28]. However, since absorption is detected indirectly, a large amount of energy must be deposited, limiting the application of this method in spectroscopic studies. It has recently been shown that the weak light absorption of a single nano-absorber can be directly monitored using a spatial modulation technique, allowing detection of single gold nanospheres down to few nm [29]. This approach relies on spatially modulating the position of an isolated particle in the focal spot of a laser beam (Fig. 7.13). Apart from its simplicity, it has the key advantages of requiring very low light power and of yielding the absolute value of the absorption cross-section (down to a few nm^2). This spatial modulation spectroscopy (SMS) technique is particularly well adapted to the investigation of small absorbing nanoparticles (down to a few nanometers). Furthermore, as the optical signature of the nano-object is directly and quantitatively measured, information can be extracted about its geometry (size and shape) and orientation on the surface (Fig. 7.13) [29].

7.9 Electromagnetic Field Enhancement: Applications

The surface plasmon resonance is associated with an enhancement of the electromagnetic field inside the nanoparticles, as compared with the applied field, this by a factor $f(\omega) = 3\varepsilon_m/(\varepsilon + 2\varepsilon_m)$ [see (7.4) and (7.10)]. This effect also modifies responses depending locally on the field in or near the nanoparticle, leading to Raman or luminescence enhancement, increased nonlinear response, etc.

7.9.1 Nonlinear Optical Response

When the electromagnetic field incident on a material is strong enough, the induced polarisation, which reflects the response of the medium, is no longer simply linear but involves terms depending on powers of the field [30]:

$$\boldsymbol{P} = \chi^{(1)}\boldsymbol{E} + \chi^{(2)}\boldsymbol{E}\boldsymbol{E} + \chi^{(3)}\boldsymbol{E}\boldsymbol{E}\boldsymbol{E} + \cdots , \tag{7.34}$$

where $\chi^{(n)}$ is the nonlinear susceptibility of order n, a tensor of order $n+1$. In a material where the optical nonlinearity arises from metallic nanoparticles, the relevant electric field is the one to which they are subjected, i.e., the internal field $\boldsymbol{E}_i$ (see Fig. 7.1). Its enhancement near the surface plasmon resonance (see Fig. 7.2) amplifies nonlinear effects such as second harmonic generation or the optical Kerr effect.

The latter is a third order effect, widely studied in nanomaterials, especially for its applications in the field of telecommunications. It corresponds to a modification of the dielectric constant of the medium, and hence of its refractive index, by an incident wave of intensity I [30]:

$$n = n_L + n_2 I , \quad \text{with} \quad n_2 = 3\tilde{\chi}^{(3)}/4n_L^2\varepsilon_0 c . \tag{7.35}$$

Using a local field approach, the effective susceptibility $\tilde{\chi}^{(3)}$, referred to the amplitude of the incident field, of a medium composed of metallic nanoparticles in a transparent matrix with negligible nonlinear susceptibility is [7,31]

$$\tilde{\chi}^{(3)}(\omega, -\omega, \omega) = pf(\omega)^2|f(\omega)|^2\chi^{(3)}(\omega, -\omega, \omega) , \tag{7.36}$$

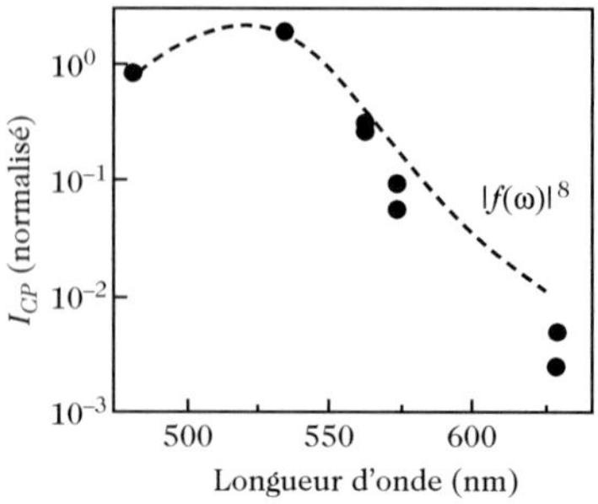

Fig. 7.14. Dispersion of the intensity $I_{\mathrm{pc}}(\omega)$ produced near the surface plasmon resonance in a nonlinear optical study of phase conjugation in a gold colloid solution. I_{pc} is proportional to $\left|\widetilde{\chi}^{(3)}(\omega,-\omega,\omega)\right|^2$ and hence to $|f(\omega)|^8$ [31]

where $\chi^{(3)}$ is the nonlinear response of the metal. Like the dielectric constant of the metal, it is barely altered by confinement down to sizes of nanometric order and its physical origin remains the same as in the bulk solid [7, 31]. Similarly to the linear optical response, the main consequence of size reduction is thus a large increase in the nonlinear response close to the surface plasmon resonance, and this by a factor of $f(\omega)^2|f(\omega)|^2$ (see Fig. 7.14).

More generally, for the third order process producing the frequency $\omega = \omega_1+\omega_2+\omega_3$ via the susceptibility $\chi^{(3)}(\omega_1,\omega_2,\omega_3)$, the field enhancement factor is $f(\omega)f(\omega_1)f(\omega_2)f(\omega_3)$. Results are similar for the second order effects. For example, the creation of the sum frequency 2ω via the second order susceptibility $\chi^{(2)}(\omega,\omega)$ is enhanced by a factor of $f(2\omega)f(\omega)^2$. Since the field enhancement effect extends over a distance of a few nanometers in the vicinity of a metallic nanoparticle (see Fig. 7.2), interfacing with another material makes it possible to envisage induced enhancement of its nonlinear response.

7.9.2 Time-Resolved Spectroscopy

The nonlinear optical response of metals is directly related to their electronic properties so studying this response can provide information about those properties. In particular, this approach has been used in the time domain to analyse electron interactions and vibrational modes of metallic nanoparticles [20]. The idea in these experimental studies is to disturb the equilibrium of the conduction electrons by applying a femtosecond optical pulse and then to monitor their return to equilibrium (redistribution of energy between the electrons, towards the lattice and towards the matrix). To do this, one measures the time evolution of the changes induced in the optical properties of the medium by means of a second, slightly delayed femtosecond pulse. It has thus been possible to carry out direct studies of the electron–electron and electron–vibrational mode (phonon) interactions and to determine their size dependence [14, 15].

Applications of time-resolved optical techniques to the study of the dynamics of metallic nano-objects are still in their early stages. There are good prospects for investigating the size dependence of vibration, spin and electron kinetics, as well as the relation between these properties and the structure and environment of the particles, crucial information for understanding and optimising nanomaterials. In the context of fundamental research, it may be feasible to continuously monitor the evolution of energy exchange and interaction mechanisms from a solid to a cluster containing just a few atoms, which are treated with two quite different formalisms, in the contexts of solid state physics and molecular physics, respectively.

As for linear optical measurements, most time-resolved studies were performed on a large number of nanoparticles. They have only recently been extended to a single nanoparticle, combining a specific optical detection scheme with a femtosecond spectroscopy setup to address its nonlinear response [32–36]. Because time-resolved luminescence studies are limited to resolutions of a few picoseconds (using a streak camera) or require a large luminescence yield, femtosecond investigation of a single weakly luminescent object necessitates measurement of its transient nonlinear absorption or scattering. In the case of metal nanoparticles, near-field optical microscopy has been extended to the femtosecond study of large gold nanorods (180×30 nm) [32]. The difficulties involved in these experiments have made far-field techniques much more interesting. Transient single nanoparticle scattering and absorption have thus been detected in large nanospheres (diameter greater than 50 nm) with picosecond resolution [33]. Scattering-based methods have recently been extended to the femtosecond domain in 80-nm gold nanospheres [34] and in nanorods [35]. Combining the far-field space modulation spectroscopy (SMS) technique (see Sect. 7.8), with a high sensitivity pump–probe setup, the femtosecond response of a single silver nanosphere has been measured down to a size of 20 nm [36]. This is currently the smallest single nano-object investigated with a femtosecond technique.

7.9.3 Local Enhancement of Raman Scattering: SERS

The enhancement by several orders of magnitude of Raman scattering by molecules deposited on a nanopatterned metal substrate or in the vicinity of metal nanoparticles has been observed on many occasions and is known by the name of surface enhanced Raman scattering (SERS) [17, 37]. It has been attributed in part to the strengthening of the local field near nanostructures. This process is similar to the one described above with enhancement of the incident wave at the laser frequency ω_L and also for waves emitted at the Stokes frequencies $\omega_S = \omega_L - \omega_V$, or anti-Stokes frequencies $\omega_A = \omega_L + \omega_V$, where ω_V is the frequency of the relevant vibrational mode.

The enhancement factors for the electric fields of the Stokes and anti-Stokes waves are then approximately given by

$$A_{S,A} = f(\omega_L) f(\omega_{S,A}) , \tag{7.37}$$

and hence by $|A_{S,A}|^2$ for the corresponding intensities. However, the estimated amplification effects are less than those observed experimentally. This additional enhancement has been attributed to a change in the molecular response due to direct molecule–metal coupling and to the collective effect of nanoparticles (giant field enhancement between two interacting particles).

It should be noted that enhancement of the emitted wave has also been observed in the case of luminescence of the metal itself for optical excitation above the interband transition threshold of a noble metal. It was first described for nanopatterned metal films, then for nano-objects [38]. However, luminescence in metals, related to interband radiative relaxation of electrons, remains an extremely small effect compared to the much greater non-radiative relaxation (electron–phonon energy exchange).

Enhancement of the luminescence of a nearby dipole, such as a molecule or an ion, has also been reported. However, this process depends strongly on the metal particle–dipole coupling, and in particular on their separation, strong luminescence quenching also being possible if there is efficient energy transfer to the metal (i.e., for short distances) [39], energy then being damped non-radiatively to the metal lattice.

7.10 Conclusion

Reducing the size of a metallic object to the nanoscale leads to major changes in its optical response as compared with the bulk metal. This response depends on the size, shape, and structure of the nanoparticles, and on their environment. Dielectric confinement, i.e., electromagnetic field enhancement within the particle, lies at the origin of the main size effects and manifests itself in both linear and nonlinear optical properties.

For spherical particles of arbitrary size, it can be treated quite generally in the framework of the Mie model, which is identical to the quasi-static approximation at small sizes, typically $D \le \lambda/10$. In this regime, dielectric confinement is reflected by a large enhancement of optical absorption and scattering near a certain wavelength. This is the surface plasmon resonance. Its spectral position is determined by the nature of the metal and its surrounding matrix through their dielectric constants. At larger sizes, this dipole resonance shifts and broadens with the appearance of multipole resonances (see Fig. 7.10).

Quantum electron confinement manifests itself through a change in the dielectric constant of the confined metal at small sizes, typically $D \le 20$ nm. It leads to a broadening of the surface plasmon resonance as the size diminishes, and then to a spectral shift at smaller sizes, viz., $D \le 5$ nm (see Fig. 7.10). Changes in the optical properties due to dielectric and quantum confinement at the level of single nanoparticles are also observed in ensembles of nanoparticles, provided that their density remains below a few percent (in volume

Table 7.2. Parameters discussed in this chapter and their impact on optical properties

Parameter	Definition	Impact	Equations
λ, ω	Optical wavelength and frequency		(7.1)
α	Absorption coefficient	Optical absorption	(7.2), (7.10), (7.21)
E_F, v_F	Fermi energy and velocity of a metal	Electronic properties of a metal	
n_e, m_e	Electron density and mass	Electronic properties of a metal	
D, V_{np}	Diameter of a nanosphere and volume of a nanoparticle	Size effects	
p, N_{np}	Volume fraction of metal and density of nanoparticles in a nanomaterial	Composition of a composite material	(7.6), (7.30), (7.36)
$\varepsilon = \varepsilon_1 + i\varepsilon_2$	Dielectric constant of the metal	Optical response of a metal	(7.2), (7.16), (7.19)
ε_{Drude}, ε^{ib}	Contribution of free and bound electrons to ε, the effective dielectric constant of a nanomaterial	Optical response of a metal	(7.15), (7.16), (7.19), (7.20)
$\tilde{\varepsilon}$			(7.7), (7.9)
γ_0, γ_0'	Intrinsic optical rate of electronic collision for the bulk metal and a nanoparticle	Intraband absorption ε_{Drude}	(7.15), (7.17), (7.19), (7.22)
g	Electron–surface interaction factor	Intraband absorption of nanoparticles	(7.17), (7.18)
ε_m	Dielectric constant of the matrix	Optical response of a composite material	(7.3)
ω_p	Plasma frequency of a metal	Intraband absorption ε_{Drude}	(7.15)
Ω_{ib}	Threshold frequency of interband transitions	Interband optical response (noble metal)	
Ω_R, Γ_R	Frequency and width of the surface plasmon resonance	Optical response of a composite material	(7.11), (7.20), (7.22), (7.29)
f	Electric field enhancement factor	Enhancement of the optical response, dielectric confinement	(7.4), (7.10), (7.36), (7.37)
S_{ext}, S_{scatt}, S_{abs}	Extinction, scattering and absorption cross-sections of a nanoparticle	Optical response of a nanoparticle	(7.12), (7.13), (7.14), (7.24), (7.27), (7.33)
L_i	Shape factor for an ellipsoid along the direction of polarisation i	Shape effect, optical response of an ellipsoid	(7.27), (7.28), (7.31), (7.32)
$\tilde{\chi}^{(n)}$	Effective nonlinear susceptibility of order n	Nonlinear optical response of a composite material	(7.34), (7.36)

fraction p). At higher densities, typically when the interparticle distance is of the same order as their size, dipole coupling can significantly modify the optical response [17]. At low densities, the nanoparticles can be considered independent. It is under these conditions that optical studies have been carried out, simultaneously observing a very large number of particles, typically some 10^6–10^5. It should be noted that only a global optical response, summing over all the individual contributions, is then obtained. The recent development of far-field optical techniques for studying the scattering and absorption of single metal nanoparticles is of central interest here. It opens up many possibilities for the quantitative investigation of single nanoparticles and for their applications in linear and nonlinear optical devices.

All these effects due to size, shape, structure and environment offer many possibilities for manipulating and controlling the optical properties of metallic nanomaterials. In this context, the synthesis of new types of nano-objects, especially hybrid or organised materials, and the understanding of the physical origins of their novel properties constitute important lines of research in the nanosciences today.

References

1. M. Faraday: Philos. Trans. R. Soc. (London) **147**, 145 (1857)
2. G. Mie: Am. Phys. (Leipzig) **25**, 377 (1908)
3. C.F. Bohren, D.R. Huffman: *Absorption and Scattering of Light by Small Particles*, John Wiley, New York (1988)
4. H.C. Van der Hulst: *Light Scattering by Small Particles*, Dover Publ. Inc., New York (1981)
5. J.D. Jackson: *Classical Electrodynamics*, John Wiley, New York (1962)
6. C.J. Böttcher: *Theory of Electric Polarization*, Elsevier, Amsterdam (1973)
7. C. Flytzanis, F. Hache, M.C. Klein, D. Ricard, P. Roussignol: In: *Progress in Optics*, Vol. XXIX, ed. by E. Wold, North Holland, Amsterdam (1991) p. 321
8. J.C. Maxwell-Garnett: Philos. Trans. R. Soc. (London) **203**, 385 (1904); ibid. **205**, 237 (1906)
9. A. Kawabata, R. Kubo: J. Phys. Soc. Jap. **21**, 1765 (1966); R. Kubo, A. Kawabata, S. Kobayashi: Ann. Rev. Mater. Sci. **14**, 49 (1984)
10. P. Drude: Ann. der Phys. **1**, 566 and **3**, 369 (1900)
11. N.W. Ashcroft, N.D. Mermin: *Solid State Physics*, Holt-Saunders, Tokyo (1981)
12. E.D. Palik: *Handbook of Optical Constants of Solids*, Academic Press, New York (1985) Vol. I; ibid. (1991) Vol. II; P.B. Johnson, R.W. Christy: Phys. Rev. B **6**, 4370 (1972); H. Ehrenreich, H.R. Philipp: Phys. Rev. **128**, 1622 (1962)
13. M. Kaveh, N. Wiser: Adv. in Phys. **33**, 257 (1984)
14. C. Voisin, D. Christofilos, N. Del Fatti, F. Vallée, B. Prével, E. Cottancin, J. Lermé, M. Pellarin, M. Broyer: Phys. Rev. Lett. **85**, 2200 (2000)
15. A. Arbouet, C. Voisin, D. Christofilos, P. Langot, N. Del Fatti, F. Vallée, J. Lermé, G. Celep, E. Cottancin, M. Gaudry, M. Pellarin, M. Broyer, M. Maillard, M.P. Pileni, M. Treguer: Phys. Rev. Lett. **90**, 177401 (2003)
16. J.A.A. Perenboom, P. Wyder, F. Maier: Phys. Reports **78**, 173 (1981); W.P. Halperin: Rev. Mod. Phys. **58**, 533 (1986)
17. U. Kreibig, M. Vollmer: *Optical Properties of Metal Clusters*, Springer Verlag, Berlin (1995); U. Kreibig, L. Genzel: Surf. Sci. **156**, 678 (1985)
18. H. Hovel, S. Fritz, A. Hilger, U. Kreibig, M. Vollmer: Phys. Rev. B **48**, 18178 (1993)

19. J.J. Quinn: Appl. Phys. Lett. **2**, 167 (1963)
20. C. Voisin, N. Del Fatti, D. Christofilos, F. Vallée: J. Phys. Chem. B **105**, 2264 (2001); N. Del Fatti, F. Vallée: C. R. Acad. Sci. **3**, 365 (2002); S. Link, M.A. El-Sayed: J. Phys. Chem. B **103**, 8410 (1999)
21. W. Ekardt: Phys. Rev. B **29**, 1558 (1984)
22. C. Bréchignac, Ph. Cahuzac, J. Leygnier, A. Sarfati: Phys. Rev. Lett. **70**, 2036 (1993); ibid. **68**, 3916 (1992)
23. J. Lermé, B. Palpant, B. Prével, M. Pellarin, M. Treilleux, J.L. Vialle, A. Perez, M. Broyer: Phys. Rev. Lett. **80**, 5105 (1998)
24. A. Liebsch: Phys. Rev. B **48**, 11317 (1993)
25. W.E. Moerner, M. Orrit: Science **283**, 1670 (1999)
26. T. Klar, M. Perner, S. Grosse, G. Von Plessen, W. Spirkl, J. Feldmann: Phys. Rev. Lett. **80**, 4249 (1998); A.A. Mikhailovsky, M.A. Petruska, M.I. Stockman, V.I. Klimov: Opt. Lett. **28**, 1686 (2003); A. Liu, A. Rahmani, G.W. Bryant, L.J. Richter, S.J. Stranick: J. Opt. Soc. Am. A **18**, 704 (2001)
27. C. Sönnichsen, T. Franzl, T. Wilk, G. von Plessen, J. Feldmann: New J. Phys. 4 (2002); K. Lindfors, T. Kalkbrenner, P. Stoller, V. Sandoghdar: Phys. Rev. Lett. **93**, 37401 (2004)
28. D. Boyer, P. Tamarat, A. Maali, B. Lounis, M. Orrit: Science **297**, 1160 (2002)
29. A. Arbouet, D. Christofilos, N. Del Fatti, F. Vallée, J.R. Huntzinger, L. Arnaud, P. Billaud, M. Broyer: Phys. Rev. Lett. **93**, 127401 (2004); O.L. Muskens, N. Del Fatti, F. Vallée, J.R. Huntzinger, P. Billaud, M. Broyer: Appl. Phys. Lett. **88**, 063109 (2006)
30. Y.R. Shen: *Principles of Nonlinear Optics*, John Wiley, New York (1984)
31. D. Ricard: In: *Nonlinear Optical Materials: Principles and Applications*, ed. by V. Degiorgio and C. Flytzanis, IOS Press, Amsterdam (1995) p. 298
32. K. Imura, T. Nagahara, H. Okamoto: J. Phys. Chem. B **108**, 16344 (2004)
33. M.A. van Dijk, M. Lippitz, M. Orrit: Phys. Rev. Lett. (2006)
34. T. Itoh, T. Asahi, H. Masuhara: Appl. Phys. Lett. **79**, 1667 (2001)
35. M. Pelton, M. Liu, S. Park, N.Scherer, P. Guyot-Sionnest: Phys. Rev B (2005)
36. O.L. Muskens, N. Del Fatti, F. Vallee: Nano Letters **6**, 552 (2006)
37. R.K. Chang, T.E. Furtak: *Surface Enhanced Raman Scattering*, Plenum, New York (1982); U. Kreibig: In: *Dynamics on Surfaces*, ed. by B. Pullman, J. Jortner, A. Nitzan, and B. Gerber, Reidel, Dordrecht (1984)
38. G.T. Boyd, Z.H. Yu, Y.R. Shen: Phys. Rev. B **33**, 7923 (1986); O. Varnavski, M. Mohamed, M. El-Sayed, T. Goodson III: J. Phys. Chem B **107**, 3101 (2003)
39. P. Anger, P. Bharadwaj, L. Novotny: Phys. Rev. Lett. **96**, 113002 (2006); S. Kühn, U. Hakanson, L. Rogobete, V. Sandoghdar: Phys. Rev. Lett. **97**, 017402 (2006)

8

Mechanical and Nanomechanical Properties

C. Tromas, M. Verdier, M. Fivel, P. Aubert, S. Labdi, Z.-Q. Feng, M. Zei, and P. Joli

8.1 Macroscopic Mechanical Properties

8.1.1 Introduction

The growing interest in nanomaterials over the past decade or so can be put down to their unique structure, characterised by grains with nanometric dimensions and by a rather high density of crystal defects, which will undoubtedly lead to quite exceptional properties. In particular, extrapolating the constitutive laws of large-grained materials down to the nanoscale leads one to expect interesting mechanical behaviour for nanomaterials. Materials can be produced with high levels of hardness, ductility, and sometimes superplasticity (see Chap. 9) at relatively low temperatures. These characteristics lead to remarkable mechanical performance and machining possibilities, by virtue of which such nanomaterials have immediate scope for technological innovation.

Section 8.1 describes the main mechanical properties associated with nanostructured materials and also the mechanisms so far put forward to relate grain size to observed properties. The discussion here will be restricted to the measurement of structural properties such as elasticity, hardness, and ductility in 'cold' conditions, leaving the description of superplasticity in nanomaterials and properties of moulding in higher temperature regimes to Chap. 9. We shall discuss nanostructured bulk materials, i.e., three-dimensional bodies in which the crystallites have nanometric dimensions, and thin films, which are macroscopic in two dimensions but have submicron thickness and can be made up of grains with sizes much smaller than their thickness, or of successive layers of nanometric thickness (multilayer structures).

8.1.2 Elastic Properties

The elastic limit of a material (the stress beyond which a remanent plastic strain is observed) is intimately related to its elastic constants (see below).

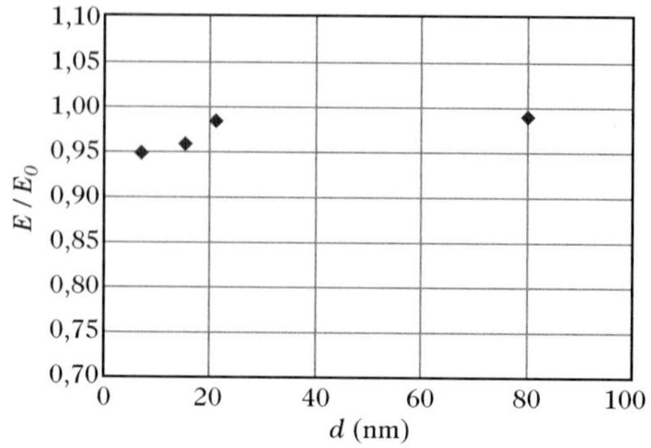

Fig. 8.1. Grain size dependence of the Young's modulus relative to that of large-grained polycrystals for Fe nanocrystals. From [1]

Indeed, a plastic strain only occurs in order to reduce the energy associated with the elastic strain of the material. The elastic constants reflect the nature and density of atomic bonds. In nanomaterials, the high density of structural defects and grain boundaries have an effect on the elastic constants. The first measurements of Young's modulus revealed a significant difference between nanostructured materials and the corresponding large-grained materials. In some cases the modulus was much higher, as for superlattices, and in others much lower, as for materials produced by sintering. However, it was subsequently demonstrated that these early results were due to experimental artifacts and often to the presence of defects, such as fractures and high porosity, introduced during fabrication and not properly taken into account when interpreting the observations. More recent results obtained on dense materials have finally shown that the Young's modulus gradually falls off only for grain sizes below 10 nm, i.e., when the fraction of atoms associated with grain boundary and triple junctions becomes very high (see Fig. 8.1).

However, the change in the modulus measured in this way never exceeds 10–20% of the value for the material with conventional grain size.

Mechanical Behaviour Under Tension

The elastic–plastic behaviour of a material is most commonly investigated by carrying out a uniaxial tensile test on a test bar. The tensile stress σ is defined as the ratio of the tensional force and the cross-sectional area of the test bar, i.e., $\sigma = F/S$. σ is expressed in megapascal ($1\,\mathrm{MPa} = 1\,\mathrm{N/mm^2}$). The strain or deformation of the useful zone of the test bar is defined as $\varepsilon = \ln(l/l_0)$, where l and l_0 are the instantaneous length and the initial length of the bar, respectively. The curve is divided into two parts: an 'elastic' part in which any strain remains completely reversible, and a plastic part in which the test bar retains a permanent residual lengthening even when the load is removed.

Several quantities can be directly measured in a tensile test:

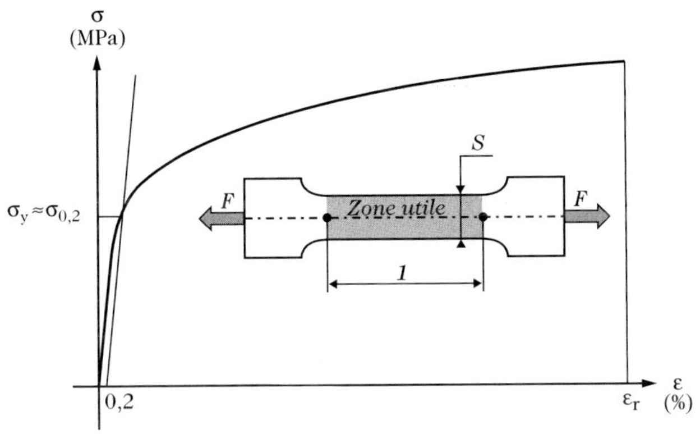

Fig. 8.2. Mechanical behaviour under tension

- The elastic limit σ_y, also called the yield point, beyond which plastic strain occurs. In practice, this transition point is difficult to determine and one measures instead the conventional elastic limit at $\sigma_{0.2}$ for which the plastic strain rate is 0.2%.
- Young's modulus E, the slope of the elastic strain region approximated by a straight line.
- The constriction coefficient Z, defined as the ratio $Z = (S_0 - S_u)/S_0$, where S_0 is the initial cross-sectional area of the useful zone of the test bar and S_u is its final cross-sectional area after rupture. A brittle material will have a value of Z close to zero, while a ductile material, with high strain capacity, will have a value of Z close to unity.
- The maximal strain ε_r sustainable by the test bar before rupture.

8.1.3 Hardness

The (Brinell) hardness H of a material is defined as the ratio between the load F applied to a hard ball and the residual indent A left in the material after withdrawing the load, i.e., $H = F/A$. The hardness is one of the most commonly measured mechanical characteristics of a nanomaterial. It is determined by a nanoindentation test, which is relatively easy to carry out (see below). This characteristic is often very useful for predicting the technological capabilities of the material, e.g., for thin protective films. This quantity is intrinsically related to the ability of the material to deform plastically. To a first approximation it can be related to the yield point σ_y of a material by the empirical relation $H = 3\sigma_y$.

Nanoindentation

The nanoindentation technique is a mechanical test derived from the standard hardness test. The idea is to sink a hard tip of known geometry into the material under

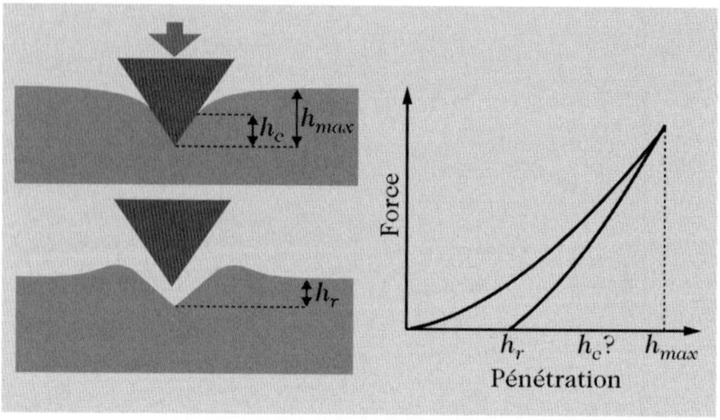

Fig. 8.3. Measuring hardness

investigation and monitor the depth of penetration as a function of the applied load. By carrying out a loading–unloading cycle, a characteristic force–penetration curve is obtained. By analysing the unloading curve and modelling the material as an elastic and isotropic continuous medium, one can deduce the relevant parameter here, which is the true contact area between the indenter and the material under maximal load. One can then deduce not only the hardness, but also the Young's modulus of the material. The applied forces range from a few tenths to several hundred millinewtons, and the depths of penetration from a few nanometers to several microns.

In the case of large-grained materials, plastic strain corresponds to nucleation and/or motion of dislocation lines (see below) in the material. These lines then multiply via the Frank–Read mechanism. The stress needed to activate such a source depends on the distance separating the two points at which the dislocation is anchored. It is thus easy to see that reducing the grain size will also reduce the distances between such points and hence raise the elastic limit. The empirical Hall–Petch relation expresses this effect:

$$\sigma_{\mathrm{y}} = \sigma_0 + \frac{k}{\sqrt{d}} ,$$

where σ_0 is a lattice friction stress, k a constant, and d the grain size. The proportionality between hardness and elastic limit can be used to define a relation of the same type between hardness and grain size. For large-grained materials, the Hall–Petch relation is explained by dislocations piling up along grain boundaries: plastic strain is triggered when the stress at the front of the dislocation pile-up is enough to activate a Frank–Read source in the neighbouring grain.

Extrapolating this law to grains of around ten nanometers, very high values of the hardness are predicted, the only limit being the theoretical limiting stress of a perfect crystal, generally taken as $G/10$, where G is the transverse shear modulus. However, although the hardness does indeed increase at small grain sizes, the observed effects are generally much smaller than would be predicted in this way, and at very small grain sizes, or for very small periods

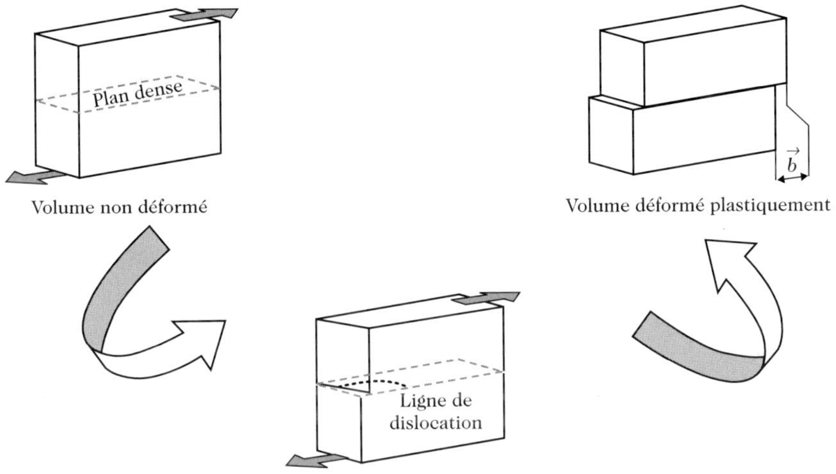

Fig. 8.4. Production of a dislocation

in the case of multilayer films, the opposite trend to the Hall–Petch relation is sometimes observed, i.e., a reduction in hardness with reduced grain size (or period). This 'negative' Hall–Petch effect has nevertheless been explained for nanocrystalline materials that have undergone thermal treatment, by invoking densification or phase transformation phenomena.

Whatever the case may be, for materials with grain sizes below 10 nm, the absence of mobile dislocations means that any models based on the behaviour of dislocations applying to large-grained materials are no longer relevant. The limits of the Hall–Petch effect in 3D nanomaterials have been demonstrated by several computer simulations. For grain sizes below a few nanometers, plasticity is no longer due to the motion of dislocations, but to a very large number of small amplitude slipping motions at grain boundaries.

Dislocations

When a material is subjected to stress, it begins by deforming in a reversible manner (elastic strain), then irreversibly, with permanent consequences (plastic strain). In the case of plastic strain in a crystalline material, the macroscopic change in shape as seen from the outside must occur without altering the periodic arrangement of the crystal lattice on the atomic scale. This is only possible via a series of shears with amplitude proportional to the size of the unit cell in the atomic lattice, in such a way that the crystal structure is reinstated after each shear. However, calculation shows that such shear transformations cannot occur simultaneously throughout the material, because this would require stresses more than a thousand times stronger than those observed. If we imagine a shear with amplitude of the order of the lattice parameter which has propagated over only a part of the crystal, the line defining the boundary of the sheared region is a linear crystal defect called a dislocation.

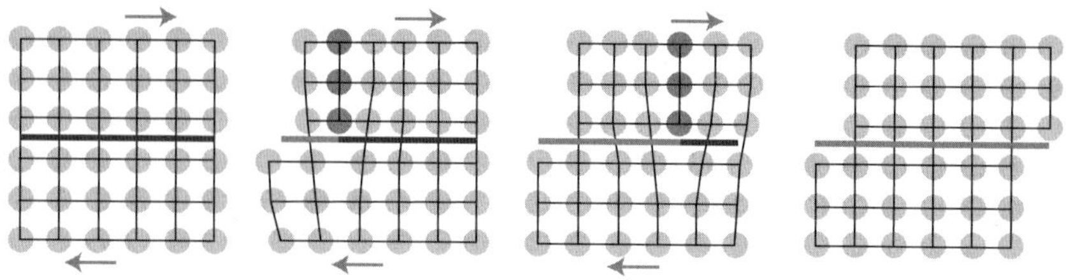

Fig. 8.5. Motion of an edge dislocation. The dislocation line runs perpendicular to the plane of the figure and marks the boundary between the region already sheared (*light grey*) and the unsheared region (*dark grey*)

Figure 8.5 then shows how the displacement of a dislocation will propagate the shear through the crystal, whilst leaving the crystallographic structure intact behind it. This is called an edge or screw dislocation depending on whether the direction ***b*** of the propagated shear is perpendicular or parallel to the dislocation line.

The shear plane in which the dislocation moves is called the slip plane of the dislocation. It corresponds to a close-packed plane of the crystal lattice, i.e., one containing the greatest number of atoms. The amplitude and direction of the shear propagated by the dislocation are characterised by the so-called Burgers vector ***b*** of the dislocation. Dislocations play a major role in the mechanical behaviour of conventional materials through their density, mobility and interactions.

The same kind of behaviour is found in 2D nanomaterials. Hence, in structures composed of layers of different materials, large increases in hardness can be observed, well above those predicted by the law of mixing. These systems have several points in common, such as an increase in hardness when the period of the bilayer is reduced, at least down to 4 nm. This phenomenon can be explained by a behavioural model similar to the Hall–Petch law. These increases in hardness reflect the resistance to motion of dislocations in each of the layers making up the multilayer structure. However, discrepancies are sometimes observed with respect to the Hall–Petch law for layer thicknesses below 20 nm, as illustrated in Fig. 8.6.

Such discrepancies from the Hall–Petch law have not yet been fully explained, but changes in the chemical nature of the interface may be relevant for quasi-monatomic layers. In any case, observations using atomic force microscopy clearly show a change in the plastic strain mode: slip bands visible for thicknesses above 20 nm are no longer observed at smaller thicknesses (see Fig. 8.7).

8.1.4 Ductility

Ductility is the ability of a material to change shape without fracture. This property is extremely important as regards technological applications of materials, whether it be for their performance in use or the ease with which they can be worked. The effect of grain size on ductility is well understood for

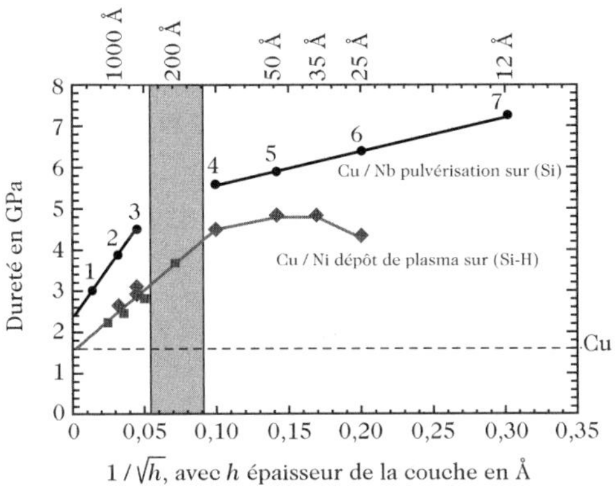

Fig. 8.6. Example of deviation from the Hall–Petch law in the case of alternating Cu/Ni multilayers obtained by plasma deposition and Cu/Nb multilayers deposited on an Ni substrate. Taken from [2]

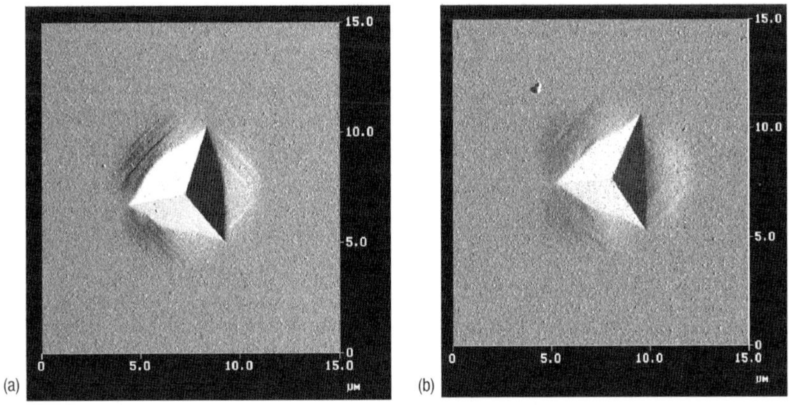

Fig. 8.7. AFM observations of the residual indent from a 150-nm indentation of a Cu/Ni multilayer. (**a**) Alternating layers of thickness 50 nm. (**b**) Layers of thickness 2.5 nm

materials with micrometric grains. For example, the brittle–ductile transition temperature in a steel can be lowered by 40°C by reducing the grain size by a factor of five. Extrapolation to the nanoscale promised extremely good ductility properties for nanomaterials, but it is a delicate matter to characterise a nanocrystalline material in this respect. This explains the many contradictory results encountered in the literature. Indeed, ductility reflects the possibility of plastic strain without fracture. However, fracture behaviour depends sensitively on the way the sample was prepared and on its final surface state, as well as the type of mechanical test used. In fact, fracture mechanisms depend sensitively on the tensile stresses in the material, and compression tests often prove inadequate for establishing the ductility.

For metals, which generally exhibit ductile behaviour when their microstructure comprises large grains, results obtained by tensile tests usually reveal brittle behaviour for grain sizes below 30 nm. This phenomenon can be explained by the major difficulty for dislocations to nucleate and slip inside such small grains.

Ceramic and intermetallic materials are conventionally brittle. The first studies of ceramics with nanocrystalline structure (CaF_2 and TiO_2) suggested that with nanometric grain sizes these materials might exhibit ductile behaviour at low temperatures. The hypothesis then made involved a creep mechanism by diffusion at the grain boundaries that would increase in importance at small grain sizes. Indeed, conventional models of creep predict a relation of the type

$$\dot{\varepsilon} = \alpha \frac{\sigma^n}{d^p}$$

between the plastic strain rate $\dot{\varepsilon}$, the creep stress σ, and the grain size d. For low temperature creep at grain boundaries, called Coble creep, it is generally found that $n = 1$ and $2 < p < 3$. According to this law, dividing the grain size by a factor of two would increase the creep rate by a factor of eight. However, the first results obtained on nanocrystalline CaF_2 and TiO_2 could not be reproduced. It seems likely that the ductile behaviour observed in these early studies were illusory and probably influenced by the high porosity of the samples used.

In the end, nanocrystalline materials did not live up to expectations with regard to improved ductility. However, interesting properties were reported for polyphase materials with nanometric structuring. An increase in the elastic limit is observed in these materials in conjunction with a good level of ductility. These effects were observed for alloys with partially crystallised microstructures, i.e., for nanocrystallites confined within an amorphous matrix, or more recently for micrometric copper grains embedded in a nanocrystalline copper matrix.

8.1.5 Numerical Modelling

It is often a difficult matter to determine mechanical properties of nanomaterials experimentally, and various techniques for producing the samples can lead to a problem of reproducibility in the results. This explains the contradictory results often encountered in the literature. Modern numerical tools provide new possibilities for exploring the mechanical properties of nanomaterials. However, in order to model the mechanical behaviour of nanomaterials correctly, experimentally observed size effects must be reproduced.

A great many numerical models are poorly suited to this task because devoid of any reference length. For example, the finite element methods so widely used in solid mechanics can only be used if the behavioural relations they employ involve a characteristic length able to account for size effects.

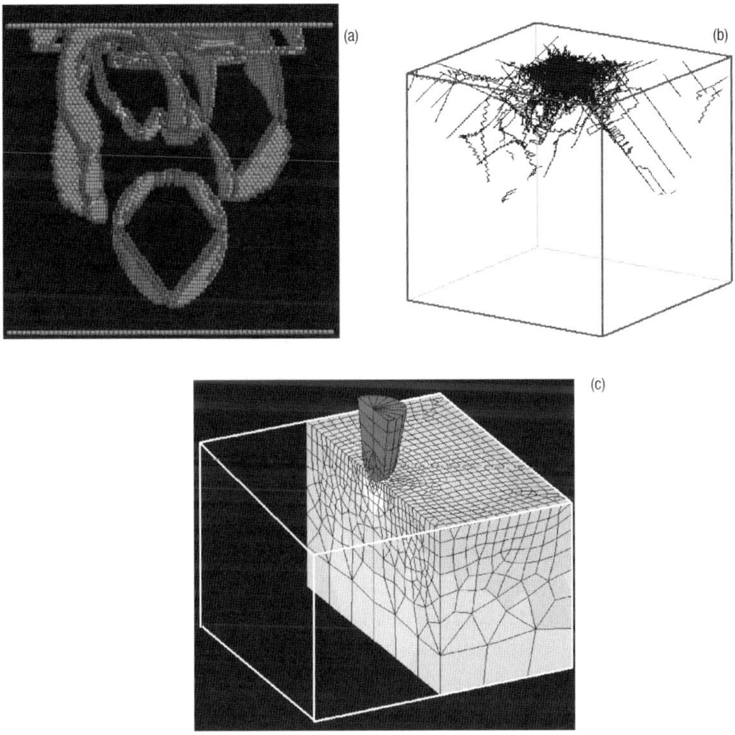

Fig. 8.8. Simulations of a nanoindentation test with a spherical tip. (**a**) Molecular dynamics. (**b**) Dislocation dynamics. (**c**) Finite elements. See also the colour plate

To achieve this, one can for example include strain gradients, and hence the second gradient of the displacement field, when expressing the constitutive laws. This naturally brings an internal length scale into the equations. These are called second gradient methods [3]. Another solution is to take into account the kinematics in a more detailed way, involving the strain modes of the material, e.g., motion of dislocations, lattice rotation, interaction with a magnetic moment, and so on, and also including an intrinsic length scale. This is exemplified by Cosserat media [4, 5] and micropolar media [6], among others. In each simulation based on such a method, the internal length scale must be related to a characteristic length scale of the material, such as grain size, dimensions of microstructural elements, and so on.

On smaller length scales, of the order of the micron, recent models have been developed to simulate the collective behaviour of interacting populations of dislocations. These dislocation dynamics models are particularly well suited to handle plastic strain in nanomaterials when it is due to motion of dislocations. They lead naturally to scaling laws of Hall–Petch type.

At still smaller length scales, the methods of molecular dynamics can be used to calculate the positions of atoms in a crystal as a function of time. The appearance and multiplication of dislocations can then be simulated, together with any other mechanism leading to plastic strain. These methods can then be used to study alternative mechanisms of plastic strain that may come into play when grains have nanometric dimensions.

Figure 8.8 shows simulations of a nanoindentation test carried out using the three main families of simulation methods.

8.2 Nanomechanical Properties

8.2.1 Experimentation

Nanomechanical Characterisation by Nanoindentation: Hardness

Metallurgists have long been concerned with the problem of measuring hardness. Over time, many methods have been developed to measure this property. They can be classified into two main categories of tests [7]: dynamical hardness tests and static indentation tests. In a dynamical hardness test, a fixed load is dropped from a fixed height onto the surface. The hardness is expressed in terms of impact energy and size of indent. The quasi-static indentation method, which remains the most widely used, consists in pressing a very hard object with well defined geometry against the surface of the material under test. Depending on the load used, there are three techniques here: macroindentation (loads greater than 10 N), microindentation (loads in the range 0.1–10 N), and nanoindentation where loads are of the order of the mN.

The first two techniques cannot be used to measure the mechanical properties of very thin films and materials with surface treatments. Indeed, they simply crush and pierce the film and end up testing only the substrate. Today, technical progress in instruments capable of measuring nanometric displacements and others capable of applying and controlling loads of mN order have made nanoindentation into a genuine mechanical microprobe, widely used in the mechanical characterisation of thin films (see Fig. 8.9).

Basic Idea

In conventional indentation techniques, the residual indent is measured when the indenter has been withdrawn. The results of such tests are not easy to interpret. Indeed, there is no way of taking into account any relaxation in the material. Consequently, it is impossible to distinguish the elastic part, and this increases the uncertainty in the hardness measurement of ceramic materials characterised by a high degree of elasticity. In nanoindentation, continuous measurement of the load and displacement has replaced this simplistic measurement of the residual indent.

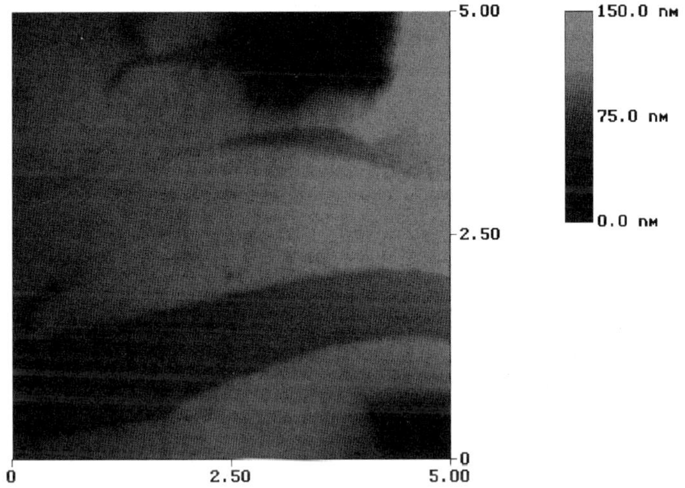

Fig. 8.9. Indentation in a deposit of alumina and titanium oxide obtained by plasma projection. See also the colour plate

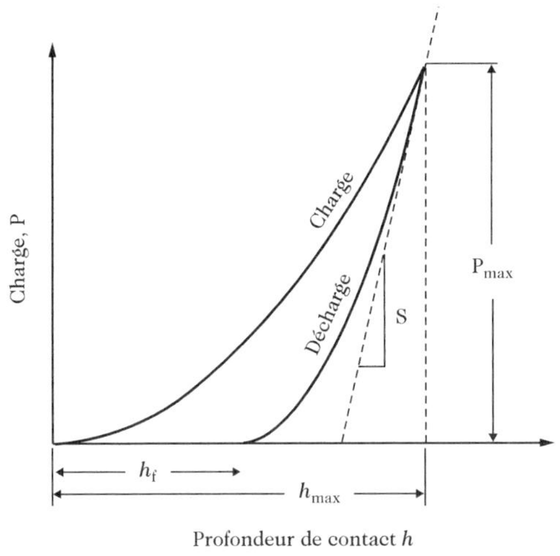

Fig. 8.10. Loading/unloading curve showing the relevant physical quantities

A typical loading/unloading curve is shown in Fig. 8.10. The indenter, initially in contact with the surface, is driven into the material until a predetermined maximal load or depth is reached, using a constant loading rate (loading cycle). The load is then reduced to zero at the same rate (unloading cycle).

The loading curve is characterised by an elastic–plastic regime and can be used to deduce the value of the hardness defined as the ratio of the maximal load to the area of contact between the indenter and the material at this load.

The unloading curve generally corresponds to a purely elastic regime. It can be used to deduce the Young's modulus.

Figure 8.10 shows the depth h_d defined by the intersection of the tangent to the unloading curve taken at the top of the curve with the horizontal axis. This tangent corresponds to the unloading curve that would have been obtained if the indenter had had plane geometry and the same area as the contact area at the maximum of the curve between a real indenter and the material.

Choice of Indenter

The indenter must have a high elastic modulus and no plastic strain at the contact pressures occurring during indentation, as well as a smooth surface and well defined geometry so that the indent itself will be well defined. Diamond is generally chosen for its satisfactory properties with regard to these requirements. In practice, it is very difficult to work indenters into a perfect geometry. The indenter thus exhibits irregularities that must be taken into account when calculating the contact pressure.

A Berkovich geometry is usually used for nanoindenters. This has pyramidal shape with triangular base. The Vickers indenter (pyramidal but with square base) is sometimes used when applied loads approach values used for microindentation. The advantage of the Berkovich indenter comes from the fact that it is more finely tapered than the Vickers indenter. Indeed, the radius of curvature is less than 50 nm for most commercially available Berkovich indenters at the present time. The two types of indenter do share one geometrical similarity, namely they have the same area function $A = f(h)$. For an ideal geometry, the area function is $A = 24.5h^2$.

Analysing the Loading/Unloading Curve

Figure 8.11 shows the response of an elastic–plastic material during indentation by a pyramidal indenter. In this figure, the depth of contact h_c is defined as the depth of penetration when the indenter is in contact with the sample. h is the depth measured during indentation. During indentation, h satisfies the relation

$$h = h_s + h_c \,, \tag{8.1}$$

where h_s is the displacement of the surface on the perimeter of the contact due to elastic depression. The maximal load and depth are denoted by P_{max} and h_{max}, respectively. After unloading and elastic recovery, the final depth of the residual indent is denoted by h_f.

The whole difficulty in interpreting the curve lies in determining the contact area between indenter and material for a given depth of penetration. Now the indenter geometry (area function) can be well established (using a calibration procedure to be described below) and it then suffices to determine

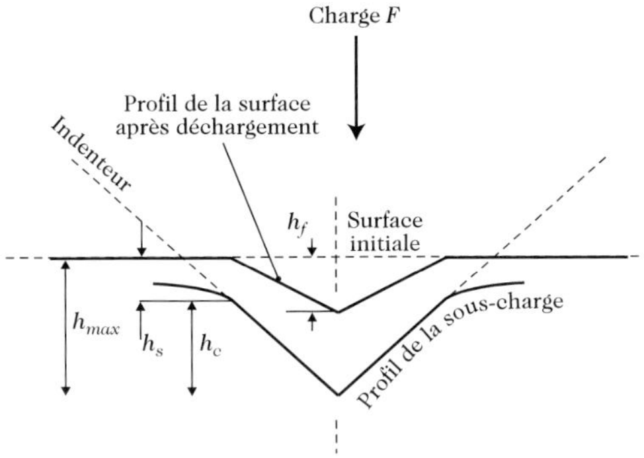

Fig. 8.11. Schematic representation of the indentation process illustrating the depression of the sample and reduction of penetration depth after unloading. From Oliver and Pharr [8]

the depth of contact in order to know the contact area. The problem thus reduces to determining the contact depth h_c.

The analysis described here and used by most commercial nanoindenters is the one developed by Oliver et Pharr [8] for elastoplastic materials.

Measuring Contact Rigidity $S_{\max}$

The unloading curve is analysed using the theory of elastic contact. This problem was treated by Hertz [16] and Boussinesq [17], but the theory most widely used today is the one developed by Sneddon, who established a relation between the load and the displacement:

$$P = Ch^{\mathrm{m}} , \tag{8.2}$$

where C and m are constants ($m = 1$ for a cylindrical geometry, $m = 2$ for conical geometry, and $m = 1.5$ for spherical geometry).

Oliver and Pharr [8] noted that the unloading curve is better described by a power law similar to (8.13) in the term $h - h_{\mathrm{f}}$, viz.,

$$P = \alpha(h - h_{\mathrm{f}})^m , \tag{8.3}$$

where the constants α, m and h_{f} are determined by a simple least-squares fitting method. The initial slope of the unloading curve can then be simply found by calculating the derivative of the curve analytically at maximal depth.

Calculating Young's Modulus

Tabor [9] was the first to use continuous monitoring of the load as a function of the displacement as a way of measuring hardness. One of the most important results was that the effect of the non-rigidity of the indenter can be taken into account by defining a reduced Young's modulus E_{r} by

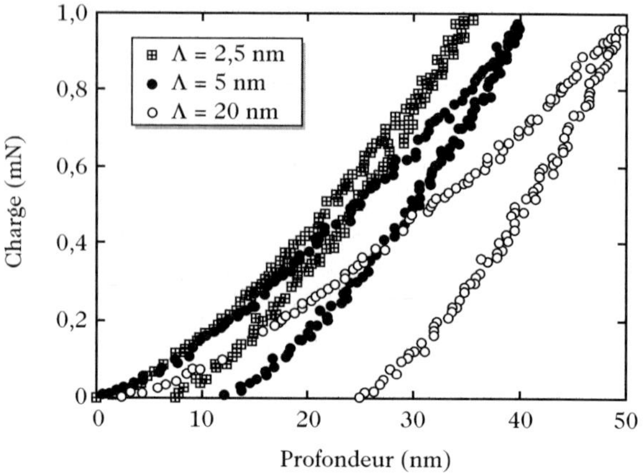

Fig. 8.12. Loading/unloading curve for Ti/TiN multilayers with variable periods at constant load 1 mN. Thickness 300 nm

$$\frac{1}{E_r} = \frac{1 - \upsilon^2}{E} + \frac{1 - \upsilon_i^2}{E_i} , \tag{8.4}$$

where E and υ are the Young's modulus and the Poisson constant of the sample, respectively, and E_i and υ_i are the same parameters for the indenter.

According to Bulychev [10] and Shorshorov [11], motivated by Sneddon's method [12], this reduced Young's modulus can be related to the rigidity S measured at the maximum of the unloading curve by the relation

$$S = \frac{dP}{dh} = \beta \frac{2}{\sqrt{\Pi}} E_r \sqrt{A(h_c)} , \tag{8.5}$$

where β is a constant depending on the indenter geometry:

- $\beta = 1.034$ for a Berkovich indenter,
- $\beta = 1.012$ for a Vickers indenter,
- $\beta = 1.000$ for a spherical indenter.

A is the contact area (for a Vickers indenter) or the projected area of contact (for a Berkovich indenter) of the the elastic contact. The Young's modulus can be directly obtained by calculating the rigidity S:

$$E_r = \frac{\sqrt{\Pi}}{2\beta} \frac{S}{\sqrt{A(h_c)}} , \tag{8.6}$$

where S is the contact rigidity calculated in the first part of the unloading curve ($S = dP/dh$). Equation (8.6) is valid for any indenter with axial symmetry.

In nanoindentation several methods have been developed to determine the contact area. Pethica et al. [13] proposed a simple method based on the area function of the indenter. Examining (8.6), the reduced Young's modulus can be calculated if the contact rigidity S and the contact area at the maximum of the curve can be determined independently. This means that the contact depth h_c must be known.

Determining the Contact Depth

The first approximation for h_c was proposed by Doerner and Nix [14], simply setting the contact depth equal to the plastic depth h_p. This approximation assumes that the unloading curve can be identified with its tangent at maximal load. However, it proved inadequate for acquiring a good hardness measurement. It is preferable to calculate the contact depth at the maximal depth and load for which the material fits perfectly against the shape of the indenter. In this case the contact depth h_c is expressed as a function of h_{max} and h_s:

$$h_c = h_{max} - h_s . \tag{8.7}$$

Since h_{max} is easily measured, the problem here amounts to determining the elastic displacement h_s of the surface. For a conical indenter, Sneddon [12] showed that h_s can be written as a function of the final depth h_f :

$$h_s = \frac{\pi - 2}{\pi}(h_{max} - h_f) . \tag{8.8}$$

This relation is valid only in the elastic part of the displacement. Sneddon [12] also showed that the quantity $h - h_f$ can be written in the form

$$h_{max} - h_f = 2\frac{P_{max}}{S} . \tag{8.9}$$

Oliver and Pharr [8] then showed from (8.8) and (8.9) that h_s could be expressed as

$$h_s = \varepsilon\frac{P_{max}}{S} , \tag{8.10}$$

where $\varepsilon = 0.72$ for a conical indenter.

This result can be generalised to other types of indenter in which only the value of ε changes. Empirically, $\varepsilon = 0.75$ for an indenter in the form of a paraboloid of revolution, and $\varepsilon = 1$ for an indenter with plane geometry.

For greater accuracy, a recent calculation presented by Pharr, Bolshakov et al. [15] has been used to refine the calculation of ε. This new expression can be used to find ε from experimental data via knowledge of the coefficient m. For this calculation, two hypotheses are required: it is assumed that the material is elastic and that the indenter is rigid.

From Sneddon's method [12], one can calculate the pressure P exerted on the sample:

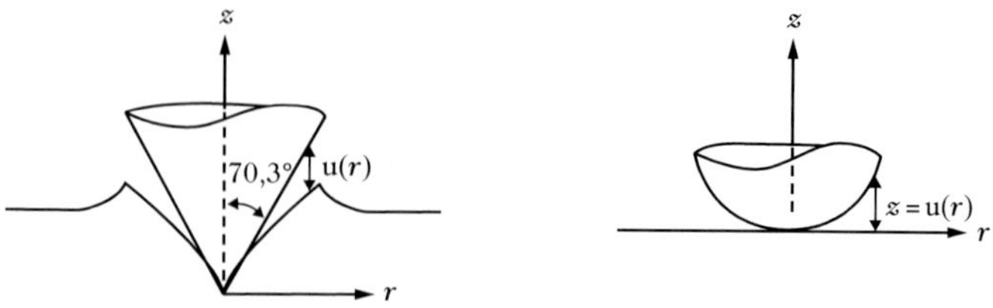

Fig. 8.13. Schematic representation of a non-ideal indenter

$$P = 2E_{\mathrm{r}}\sqrt{\pi}Br^{n+1}\frac{n}{(n+1)}\frac{\Gamma\left(\frac{n}{2}+1\right)}{\Gamma\left(\frac{n}{2}+\frac{1}{2}\right)}, \tag{8.11}$$

where $m = 1 + 1/n$, r is the radius of contact, B is a constant, E_{r} is Young's modulus, and Γ is the gamma function. The radius of curvature r is then given by

$$r = \left[\frac{\Gamma\left(\frac{n}{2}+\frac{1}{2}\right)}{B\sqrt{\pi}\Gamma\left(\frac{n}{2}+1\right)}\right]^{1/n}(h - h_{\mathrm{f}})^{1/n}. \tag{8.12}$$

P can now be expressed in terms of the elastic response $h' = h - h_{\mathrm{f}}$:

$$P = \frac{2E_{\mathrm{eff}}}{(\sqrt{\pi}B)^{1/n}}\left(\frac{n}{n+1}\right)\left[\frac{\Gamma(n/2+1/2)}{\Gamma(n/2+1)}\right]^{1/n}h'^{1+1/n}. \tag{8.13}$$

This expression corresponds to the result obtained by Oliver and Pharr [8]:

$$\alpha = \frac{2E_{\mathrm{eff}}}{(\sqrt{\pi}B)^{1/n}}\left(\frac{n}{n+1}\right)\left[\frac{\Gamma(n/2+1/2)}{\Gamma(n/2+1)}\right]^{1/n}. \tag{8.14}$$

To introduce the plasticity into their model, Oliver, Pharr and Bolshakov considered the deviation of the indenter from the ideal shape, as illustrated in Fig. 8.13. They approximated the shape of the indenter using a paraboloid correction described as illustrated in the figure:

$$Z = Br^n.$$

In this precise case, one may consider that $h_{\mathrm{s}} = h_{\max} - h_{\mathrm{c}} = h_{\max} - Br^n$.

Making the approximation that h_{s} tends to h_{f} and substituting what corresponds to the purely elastic part in the expression (8.13) for P, the following expression for ε is obtained from (8.10):

$$\varepsilon = m\left[1 - \frac{2\Gamma\left(\dfrac{m}{2(m-1)}\right)}{\sqrt{\pi}\Gamma\left(\dfrac{1}{2(m-1)}\right)}\right] . \tag{8.15}$$

As a last remark, note that h_s does not correspond exactly to the elastic part (which is equal to P_{max}/S), but to a fraction ε of the latter.

Calculating the Hardness

The hardness H is defined by the average pressure sustained by the material under the indenter. It is expressed as the ratio of the applied load P to the contact area. In practice,

$$H = P_{max} A(h_c) , \tag{8.16}$$

where P_{max} is the maximal load and $A(h_c)$ is the contact area at maximal load and depth.

Calibrating the Nanoindenter

Determining the Compliance of the Indenter. The compliance is defined as the reciprocal of the rigidity:

$$C = \frac{1}{S} = \frac{dh}{dP} . \tag{8.17}$$

The instrument influences the total compliance and interpretation of loading/unloading curves. Indeed, the measured displacement is the sum of the true displacement and a component due to the compliance of the instrument.

Modelling the instrument and sample by two springs in series shows that the total or measured compliance can be written

$$C = C_{instr} + C_{sample} , \tag{8.18}$$

where C_{sample} is the compliance of the sample calculated from (8.6):

$$C_{sample} = \frac{\sqrt{\pi}}{2\beta E_r} \frac{1}{\sqrt{A(h_c)}} . \tag{8.19}$$

The total compliance is then given by

$$C = C_{instr} + \frac{\sqrt{\pi}}{2\beta E_r} \frac{1}{\sqrt{A(h_c)}} = C_{instr} + \frac{\sqrt{\pi}}{2\beta E_r} \frac{\sqrt{H}}{\sqrt{P_{max}}} , \tag{8.20}$$

where $S_{instr} = 1/C_{instr}$ is the rigidity of the instrument, $S = 1/C_{sample}$ is the contact rigidity, S_{spring} is the rigidity of the column spring, and C_{damp} is the damping coefficient.

If the Young's modulus is constant, C_{instr} can be obtained from a plot of $C = f(1/\sqrt{P_{max}})$. The procedure then is to carry out several tests on a calibrating material (see Fig. 8.14).

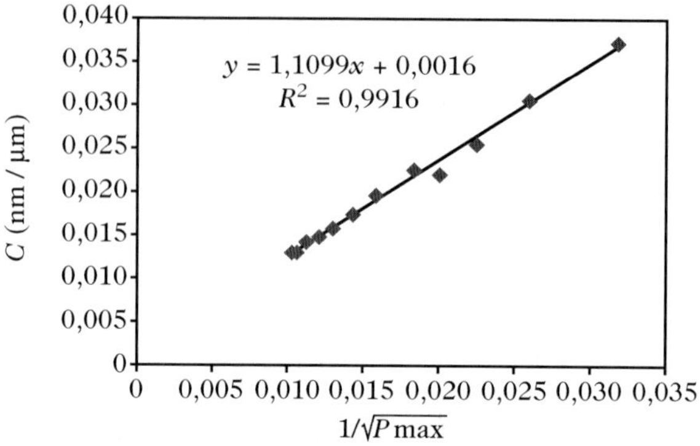

Fig. 8.14. Determining the compliance. Here $C_{\text{instr}} = 1.6\,\text{nm/mN}$

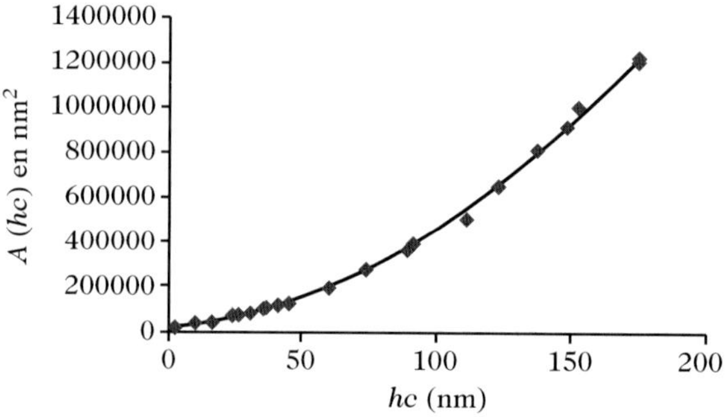

Fig. 8.15. Area function for a Berkovich indenter

Determining the Area Function $A = f(h)$. For an ideal indenter, the area A can be related to the plastic depth h_p by $A(h_c) = 24.5h_c^2$. The contact area is generally written in the form of a polynomial expansion:

$$A(h_c) = 24.5h_c^2 + C_1h_c^1 + C_2h_c^{1/2} + C_3h_c^{1/4} + \cdots + C_8h_c^{1/128}\,, \qquad (8.21)$$

where the coefficients C_1, C_2, ..., C_8 are constants to be determined by a procedure described below. The procedure consists in making a large number of indents at different penetration depths on a calibration material (see Fig. 8.15). The contact area is calculated from (8.18) in each case.

Causes of Error and Precautions: The Pile-up Phenomenon

The so-called pile-up phenomenon [18] occurs in soft materials such as aluminium and leads to matter rising up around the indent, as shown in Fig. 8.16. This produces an error in the depth measurement, and in particular in the measurement of h_c. This kind of effect can lead to an underestimate of the contact area by a factor of as much as 40% and hence to an overestimate of the hardness and the Young's modulus.

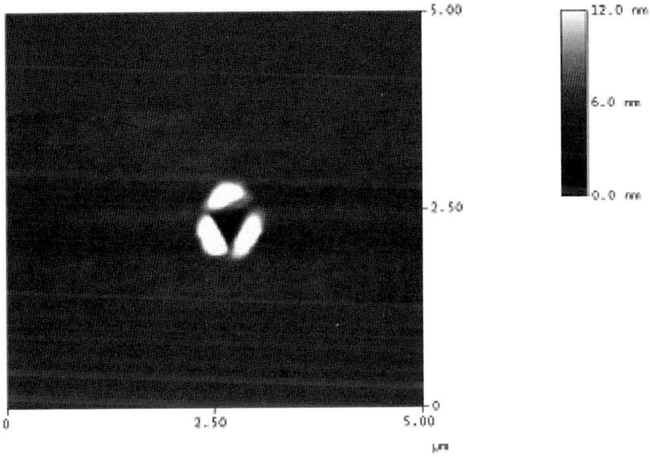

Fig. 8.16. Pile-up phenomenon as viewed by atomic force microscopy

A simple way of checking whether pile-up has occurred is to calculate the ratio h_f/h_{max}, easily deduced from the loading/unloading curve. This ratio lies between 0 and 1. The minimal value corresponds to the perfectly elastic case and the maximal value to the case where no elastic recovery is observed. Using finite element analysis, Bolshakov et al. [19] showed that pile-up can be significant.

When there is a high level of pile-up, h_c can be calculated from a formula due to Loubet et al. [20, 21] which takes into account the ridges pushed up around the edge of the indent. These authors consider (by definition) that the plastic part satisfies

$$h_r = h - P/S . \tag{8.22}$$

Experimentally, they find that the plastic part h_r is proportional to the rigidity S. Applying the same approach as Sneddon [12], one then obtains

$$h_c \propto S \quad (\text{because } h_c \propto \sqrt{A}) . \tag{8.23}$$

The main assumption in this calculation is that Young's modulus E should be constant throughout the thickness affected by the indentation.

Assuming that the indenter tip is not perfect, the rigidity can be rewritten as a function of h_i (see Fig. 8.17), which gives

$$S = B(h_r + h_i) . \tag{8.24}$$

Finally, these authors define

$$h_c = \alpha(h_r + h_i) , \tag{8.25}$$

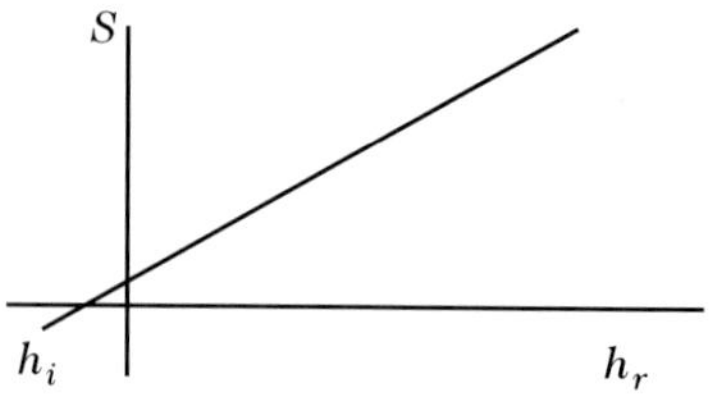

Fig. 8.17. Rigidity as a function of the plastic strain h_r

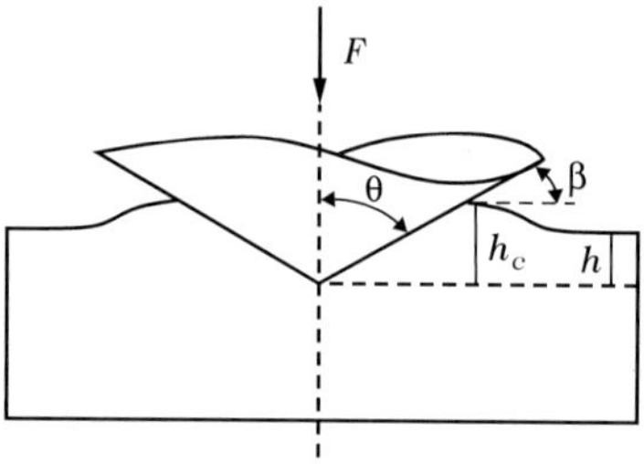

Fig. 8.18. Definition of β

where h_i corresponds to the tip imperfection and α is a constant with value $\alpha = 1.2$ for a Berkovich tip.

From this value of h_c, one can deduce the contact area

$$A = \beta\left[\alpha^2(h_r + h_i)^2\right] , \tag{8.26}$$

where β is called a shape factor. The Sneddon formula then gives the rigidity S and Young's modulus E_r as [12]

$$S^2 = \frac{4E_r^2 A}{\pi} , \tag{8.27}$$

$$E_r = \frac{B\sqrt{\pi}}{2\sqrt{\beta}\alpha} . \tag{8.28}$$

It should be noted that β comes from Sneddon's calculation and α and B from Loubet's. Bucaille et al. [21] used simulation to obtain the value $\alpha = 1.2$. In this case, using (8.25), h_c can be recalculated to give

$$h_c = \alpha(h_r + h_i) = \alpha(h - P/S) , \quad \text{where} \quad \alpha = 1.2 ,$$

for a Berkovich tip, which corresponds to the situation described in Fig. 8.18.

It is important to note that, in this model, h_c is proportional to the plastic part and that, in contrast to Oliver and Pharr's model [8], the whole elastic part is taken into account to obtain the plastic part. Moreover, the ridge effect is such that h_c is greater than h_{max}.

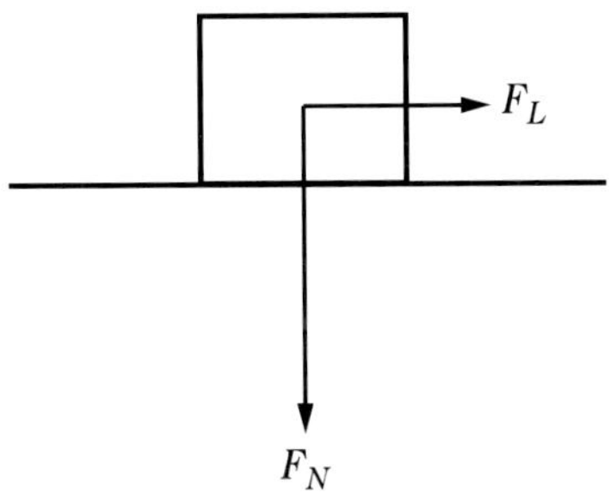

Fig. 8.19. Macroscopic forces relevant to contact between two solids

To conclude, in the presence of an elastoplastic material, one must use the model (8.14) due to Oliver and Pharr. In the presence of a soft material with a sizeable ridge around the edge of the indent, the expression for h_c given by the model due to Loubet and Bucaille [21] comes closer to the mark.

Mechanical Characterisation in Tribology. Nanoscratch Tests and Friction

Definition of Friction

The friction between two solid surfaces is defined by a coefficient. There corresponds a value to each regime, static or dynamic. We thus define:

- for the static regime, $\mu_s = F_{LS}/F_N$,
- for the dynamic regime, $\mu_d = F_{LD}/F_N$,

where F_{LS} and F_{LD} are the lateral forces required to initiate slipping and maintain the motion, respectively, and F_N is the applied normal force (see Fig. 8.19). For dry contact, $\mu_s < \mu_d$. In the present discussion, we shall be mainly concerned with the dynamic regime.

To characterise friction, the tools used to measure quantities such as forces or displacements depend mainly on the structure of the systems under consideration. Hnece, for a homogeneous solid medium, a conventional triboscope is used, such as the pin-on-disk tribometer. For a nanostructured sample, one seeks information on the scale of the structure, i.e., the nanoscale. One can therefore define three length ranges appealing to very different apparatus (see Fig. 8.20).

Standard Model Due to Bowden and Tabor

Physical modelling of the phenomena underlying friction was first undertaken around 1950 by Bowden and Tabor [22]. This model takes into account the roughness of the surfaces in contact, which gives rise to shearing of adhesive microjunctions and ploughing by surface asperities. This is expressed by two contributions to the lateral force:

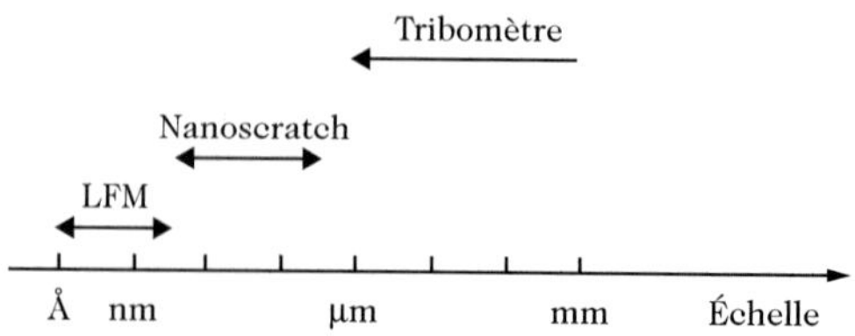

Fig. 8.20. Three length scales for tribological measurements

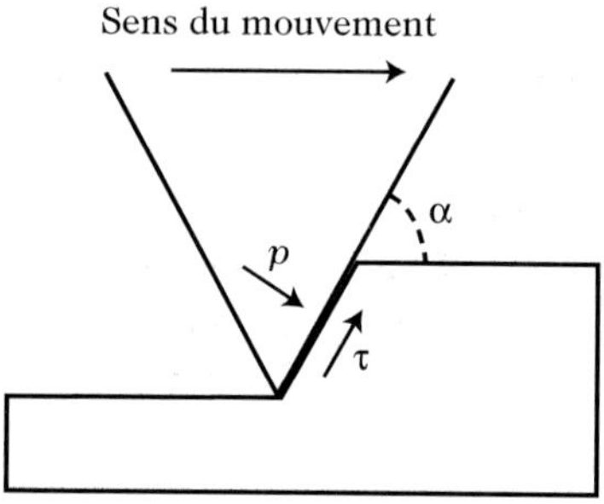

Fig. 8.21. Indenter in scratch mode

$$F_{\mathrm{L}} = F_{\mathrm{LS}} + F_{\mathrm{LP}} = \tau A_{\mathrm{TC}} + \mu_{\mathrm{L}} F_{\mathrm{N}} ,$$

where F_{LS} and F_{LP} are the lateral forces related to shear and ploughing, respectively, τ is the shear stress, and A_{TC} is the true area of contact. A_{TC} decreases as the roughness increases. The apparent contact area A_{AC} is always greater than A_{TC}.

When asperities have rather flattened profiles, the ploughing phenomenon can be neglected in comparison to shear. In this case, we have

$$\mu = \frac{F_{\mathrm{L}}}{F_{\mathrm{N}}} \approx \frac{\tau A_{\mathrm{TC}}}{H A_{\mathrm{TC}}} = \frac{\tau}{H} ,$$

where H and τ are then the plastic hardness and the shear stress of the least hard material.

Scratch Test

The plastic strain mode is conditioned by the shape of the indenter (see Fig. 8.21). Hence, depending on its apex angle, the predominant mechanism will be either shear or ploughing. Indeed, the friction coefficient can be expressed in terms of the shear and compression stresses, τ and p, respectively:

$$\mu = \frac{F_{\mathrm{L}}}{F_{\mathrm{N}}} = \frac{p \sin\alpha + p\cos\alpha}{p\cos\alpha - p\sin\alpha} .$$

So we have a context of pure shear when the angle α tends to zero, and ploughing when τ tends to zero.

Fig. 8.22. Lateral force map obtained in LFM mode on a graphite surface. The scanned area is 2 nm × 2 nm. From [24], ©American Physical Society 1987

Lateral Force Microscope (LFM)

This mode of operation of the atomic force microscope (AFM) is used to probe outer surface tribological properties. It was developed at the end of the 1980s [23, 24]. In topographic imaging mode, a system of photodiodes is used to measure the change, in the vertical direction, of the path of the laser beam reflected by the cantilever arm holding the AFM tip. In LFM, in contact mode, a normal force is applied and the tip scans the surface, whilst measuring the change in path of the same laser beam but this time in the horizontal direction. This change of path is cause by torsion in the cantilever arm. Given the mechanical characteristics of the cantilever (stiffness constant, etc.), information can be deduced concerning the lateral force. Finally, as the tip scans the surface, a map of the lateral force can be produced on the atomic scale. This is illustrated for a graphite surface in Fig. 8.22.

Figure 8.23 shows the change in the lateral force as a function of the lateral position x on the same sample. The oscillation phenomenon known as stick–slip motion is clearly visible. As the name suggests, it corresponds to motion in a series of stops and starts, caused directly by the crystal structure of the material. The tip–surface interaction remains elastic and no wear is observed. The phenomenon of hysteresis observed between the outward and return journey of the tip is due to energy dissipation [21]. Indeed, the measured response signal contains a reversible elastic part and a dissipative part. The image obtained depends closely on the crystal structure of the surface under investigation. The measured lateral force can be described by a surface potential of the form

$$V_{\mathrm{S}} = V_0 \cos(k_x x) \mathrm{e}^{-k_z z} .$$

This gives $F_{\mathrm{L}} = k_x V_0 \sin(k_x x) \mathrm{e}^{-k_z z}$.

Although this experiment provides local information, it remains semi-quantitative. Indeed, it is a delicate matter to describe the mechanical be-

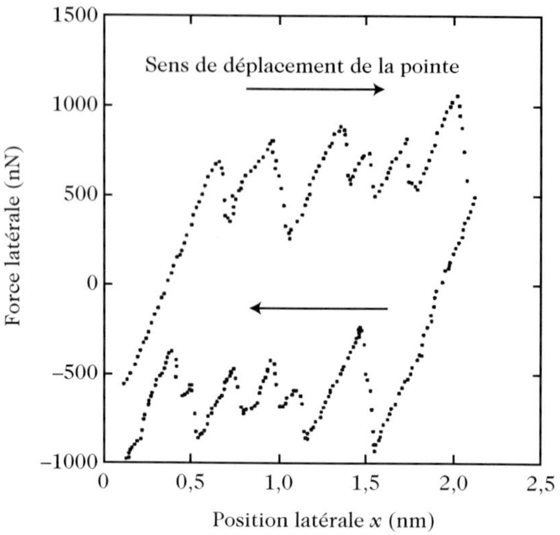

Fig. 8.23. Change in lateral force as a function of the lateral position of the tip [20]

haviour of the cantilever with any accuracy, and all the more so when it is under torsion. This is why other methods have been developed. These are systems in which the tip is no longer carried by a cantilever, such as the nanoscratch experiment, even though the spatial resolution is no longer as good as with LFM.

Nanoscratch Test

This type of measurement uses the same type of tip as nanoindentation. This may be an axially symmetric (e.g., conical, spherical) or pyramidal (Berkovich) tip. A first fundamental difference with respect to the LFM method is the characteristic size of the 'active' part of the tip. Whereas the tips used in LFM, made from silicon nitride, have a radius of curvature at the apex of nanometric order, which corresponds to a few atomic sites, the tips used in nanoscratch tests, generally made from diamond, have characteristic size around a hundred nanometers. A direct consequence is the loss of atomic lateral resolution. On the other hand, normal forces applied in the nanoscratch test can reach 1 mN. With such forces, these tests are able to investigate both plastic strain and wear. For a metallic film, the penetration depth can be as much as 100 nm.

The system developed by Hysitron is based on a double capacitive sensor, as shown in Fig. 8.24. Each sensor comprises three plates, one of which is mobile and connected to the tip. The first sensor is used to exert a normal load by applying a tension between the moving plate on which the tip is mounted and a fixed plate. The second capacitive sensor imposes the lateral displacement of the tip and measures the lateral resistance force. This provides quantitative information concerning normal and lateral forces and displacements.

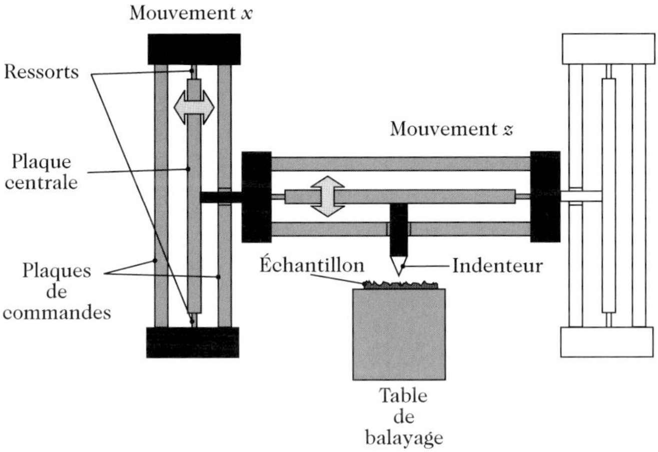

Fig. 8.24. Mode of operation of the double capacitive sensor

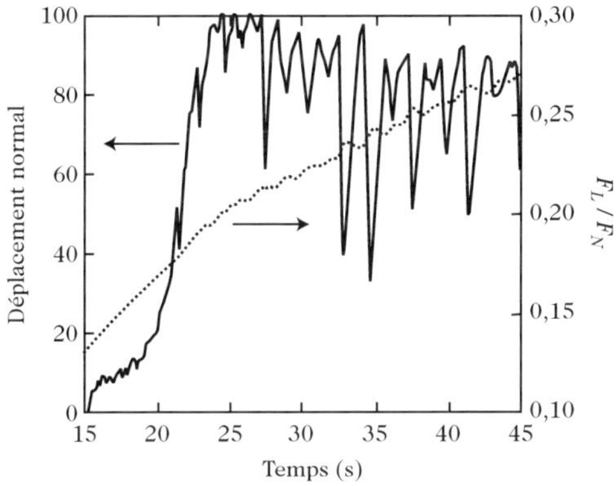

Fig. 8.25. Time dependence of the normal displacement (*left-hand scale*) and the ratio F_L/F_N (*right-hand scale*) during a nanoscratch test on a Ti/TiN multilayer

In the case of nanomaterials and in particular multilayers with nanometric periods, this system can provide local tribological information. Figure 8.25 shows the time variation of the ratio F_L/F_N and the normal displacement of the tip when the latter is subject to a lateral displacement at constant speed of the order of 0.5 μm/s and a normal load varying from 0 to 1 500 μN. The case illustrated is a multilayer made by PVD and comprising ten Ti/TiN layers. Each layer of Ti/TiN has thickness 5 nm + 5 nm = 10 nm. The observed oscillations can thus be directly related to the nanometric strata of the multilayer. It would have been impossible to observe this phenomenon using conventional tribology.

Using this method, one can also study plastic deformation mechanisms during the test. The importance of the way the normal load is applied can

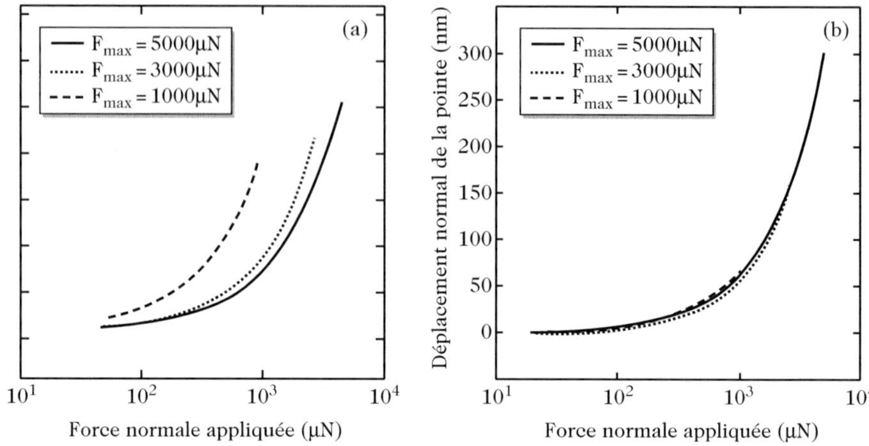

Fig. 8.26. (**a**) Variation of the normal displacement of the tip as a function of the normal force applied during the nanoscratch test: dx/dt constant and dF_N/dt variable. (**b**) Same quantities measured under the same deformation conditions but with dx/dt variable and dF_N/dt constant

thus be demonstrated. Figure 8.26 shows the normal displacement curves as a function of the applied normal load during nanoscratch tests over a time span that is held constant but with different maximal normal loads of 1 000, 3 000, 4 000 and 5 000 μN. The sample here is a titanium nitride film of thickness 300 nm deposited by PVD on a silicon (100) substrate. The logarithmic scale of the normal force (horizontal axis) suggests the possibility of superposing all the curves by simple translation. On a linear scale, this amounts to applying a multiplicative factor. The calculation shows that this multiplicative factor amounts in turn to imposing a variation of the normal force that is constant in time, from one test to the next. This demonstrates the importance of the strain and wear regime during measurements.

8.2.2 Computer Modelling

Introduction

The development of nanomaterials used in nanotechnology as assembly structures demands an understanding of their mechanical properties on the nanoscale. Evaluating physical properties like the elastic modulus, the elastic limit under tension or compression, or the buckling stress constitutes a real challenge for nanomechanics research owing to their experimental inaccessibility. Moreover the experimental validity of such results can only be upheld if the measurements are fully repeatable, and this involves large numbers of samples and a consequent high cost in many cases. To help experimenters and theoreticians alike to improve understanding of these nanomaterials by means of computer simulations, it is essential to develop well-suited models, i.e., models with reasonable numerical cost.

When analysing mechanical behaviour, there are two main approaches to modelling: either using direct methods based on molecular dynamics, or using 'continuum' methods developed to study the mechanics of solids or structures. These two approaches make different use of results obtained in molecular mechanics, which is an empirical method resulting from fitting simple mathematical functions to experimental results. Molecules are treated as assemblies of atoms subject to repulsive and attractive force fields resulting from interaction potentials between these atoms. This theoretical framework deviates in several ways from the one provided by quantum mechanics, but it is much less costly in calculation time and molecules comprising several thousand atoms can be modelled in this way.

Molecular dynamics provides a nanoscale description of the motions and spatial evolution of molecules. Each atom is treated as a point mass vibrating about its equilibrium position under the effects of thermal agitation and force fields defined by molecular mechanics. In practice, the idea is to solve Newton's equations numerically using the Verlet algorithm [26]. Displacements are calculated explicitly, requiring very short time steps, of picosecond order, given the high frequencies contained in the model. Simulation periods are thus very short and the numerical cost soon becomes prohibitively high when large assemblies of molecules containing millions or even billions of atoms are considered to reproduce behaviour on microscopic or macroscopic scales. Recently, with the development of parallel computing, it has been possible to simulate behaviour on larger scales [27–29].

To characterise the mechanical behaviour, e.g., by identifying the elastic modulus or mechanical resistance, a quasistatic approach is preferable, in which vibrational dynamics is ignored. To reduce computation time still further, the interaction forces between atoms and their nearest neighbours can be modelled by beam-type assembly forces, so that the system becomes equivalent to a truss structure as conventionally studied in mechanics [30]. However, when studying large samples, the numerical cost is still rather large and a continuum approach is then required.

The continuum methods provide a macroscopic description of matter. Above a certain characteristic volume, the material has homogeneous behaviour defined by the laws relating strains to internal stresses on the macroscopic scale. These constitutive laws governing mechanical behaviour derive from a nonlinear hyperelastic potential which can be estimated from the interaction potentials between atoms. The solution is then obtained by the finite element method which discretises the continuum model into a set of elements of finite volume with dimensions greater than the characteristic volume. A finite element may correspond to several hundred or even thousand atoms depending on its dimensions, and this considerably cuts down the number of degrees of freedom of the system under investigation in comparison with a molecular dynamics model. In the case of crystals with sheet-like structure such as graphene, it is possible to use finite surface elements. With these methods, one can no longer monitor the motions of individual atoms; it is rather the global behaviour of the whole structure that is under investigation. These studies may be either quasistatic or dynamic, and in the latter case only low frequency phenomena are accessible. Simulation times can be greater than in molecular dynamics, thereby reducing the numerical cost.

In the next section, we review the basic ideas of the finite element method, illustrating with examples of applications to the micro- and nanoscale. In these examples, one begins with a priori knowledge of the behaviour of the material on the macroscopic scale, without taking molecular mechanics into account. What is revealed here is the feasibility of the finite element method for investigating small scale multiphysical phenomena. In the second section, we discuss carbon nanotubes as a particularly well-developed field of industrial application in nanomechanics and in which the finite element method has been put to use by integrating the laws of molecular mechanics. In the third section, we give a brief discussion of work combining both molecular mechanics and finite elements in the context of multiscale approaches.

Application of the Finite Element Method

Engineering sciences such as solid and fluid mechanics or heat transfer provide ways of describing how physical systems behave by means of partial differential equations. The finite element method is one of the most widely used methods today for effective solution of such equations. It is a very general method that can be applied to most problems encountered in practice: stationary or non-stationary, linear or nonlinear, defined for arbitrary geometries. Furthermore, it can be easily adapted to heterogeneous media and multiphysical phenomena. Since Clough [31] first introduced the finite element method in 1960, many problems of mechanics and civil engineering have been solved using this approach. A great deal of work has been devoted to developing the theory. One should mention in particular the work of Marçal and King [32], who introduce the formulation of elastoplastic finite elements for small strains, Hibbitt et al. [33], McMeeking and Rice [34], who introduced the Lagrangian formulation for elastoplastic finite elements in the presence of large strains, and Zienkiewiez and Owen [35] who introduced the formulation of elasto-viscoplastic finite elements.

The constitutive equations most often used to model large strains in solids are the differential equations relating stress to strain rates. These laws must satisfy the principle of incrementally objective integration. Nagtegaal [36] and Hughes and Winget [37] proposed integration schemes for the constitutive equations in the presence of large deformations which respect incremental objectivity during finite time steps.

Among the various techniques used to integrate the constitutive equations, one in particular has steadily grown to become almost unanimously accepted: this is the radial return algorithm, developed by Wilkins in 1964 [38] for the theory of small strains and then resurrected and generalised by Krieg and Krieg in 1977 [39]. Moreover, in this field, the idea of consistent linearisation introduced by Nagtegaal [36] and extended by Simo and Taylor [40] has been used to generate genuinely efficient tangential stiffness matrices for both small and large strain problems.

The finite element method is based on one simple idea: to discretise a complex geometrical shape by subdividing it into a large number of elementary subregions with simple geometrical shape (finite elements), interconnected at points called nodes. We consider the mechanical behaviour of each element separately, then piece these elements together in such a way that forces balance and displacements are compatible at each node.

In each element, simple approximations are used for the unknown variables to transform the partial differential equations into algebraic equations. The nodes and elements do not necessarily have any specific physical meaning but are based on considerations of accuracy in the approximation.

In a general context, the equation of motion of a deformable body can be expressed in the following matrix form:

$$M\ddot{\boldsymbol{u}} + C\dot{\boldsymbol{u}} + \boldsymbol{F}_{\text{int}} - \boldsymbol{F}(t) = 0 ,$$

where $\boldsymbol{F}_{\text{int}}$ is the internal force vector and $\boldsymbol{F}$ the external load vector, possibly a function of time t. M is the mass matrix, C the damping matrix, $\dot{\boldsymbol{u}}$ the velocity vector, and $\ddot{\boldsymbol{u}}$ the acceleration vector. In the particular case of linear static analysis in solid and structural mechanics, we have

$$K\boldsymbol{u} = \boldsymbol{F} ,$$

where K is the rigidity matrix of the system, $\boldsymbol{u}$ the vector of unknown variables, i.e., displacements of the nodes, and $\boldsymbol{F}$ the vector of known loads applied at the nodes.

The logical stages of the calculation by finite elements can be summarised as follows:

- Define nodes and elements (determine the mesh).
- For each element, establish the elementary matrices relating the nodal degrees of freedom to the forces applied at the nodes.
- Assemble elementary vectors and matrices into a global system in such a way as to satisfy equilibrium conditions at nodes.
- Modify the global system to account for boundary conditions.
- Solve the modified global system to obtain the displacements of the nodes.
- Calculate the gradients (strains and stresses) within the elements and the reactions at the nodes upon which boundary conditions are imposed.

From a practical point of view, calculation codes generally comprise three functional processors:

Preprocessor

- Choose the type of elements.
- Enter the geometrical properties.
- Enter the physical parameters.
- Create the geometric model.
- Create the mesh by defining nodes and elements.
- Apply loads.
- Impose boundary conditions.

Solver

- Choose the type of analysis (static, dynamic, etc.).
- Construct elementary vectors and matrices.
- Assemble elementary vectors and matrices to obtain a global system.
- Take into account boundary conditions.
- Solve the global system of equations.
- Calculate additional variations (stresses, reaction forces, etc.).

Post Processor

- Present the results in an intelligible and synthesised way in numerical or graphical form.
- Carry out complementary functions such as combination, animation, interpolation, or interpretation.

The finite element method has recently been applied to problems at small scales. We shall now present a few examples of these applications.

Plasma Projection

Plasma projection techniques have been successfully used in industry for many years to produce coatings that resist corrosion, oxidation, wear, and so on. The formation of a coating is a multiphysical problem, involving impacts of molten particles on the substrate, heat transfer between particles and substrate, and flattening out and solidification of the particles. The first example is a simulation of this multiphysical problem by the finite element method [41]. In order to provide an adequate treatment of the large strains occurring during impact, we have developed a remeshing technique. Figure 8.27 shows the temperature field of the liquid particle and the substrate at different times. The two curves show the time variation of the temperature at the centre of and outside the particle. Note that the rates of flattening and heat transfer are very high and that the particle is very small, having a diameter of only 50 μm.

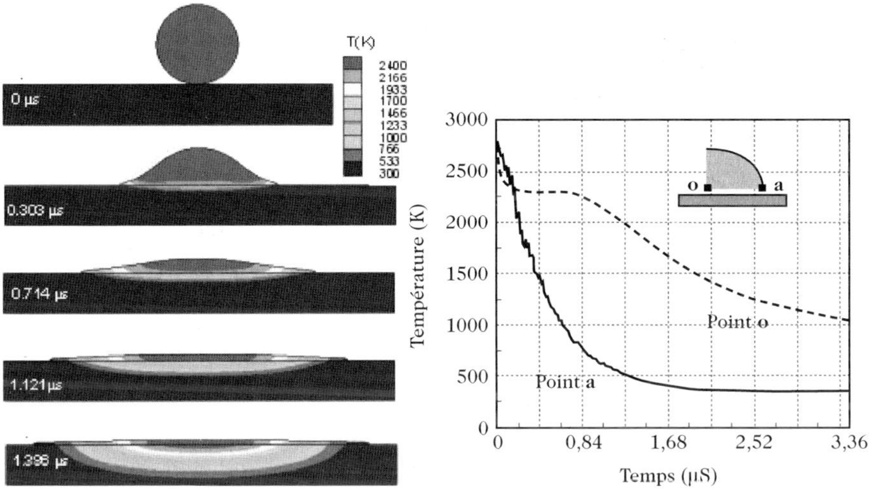

Fig. 8.27. Temperature field at different times during flattening. See also the colour plate

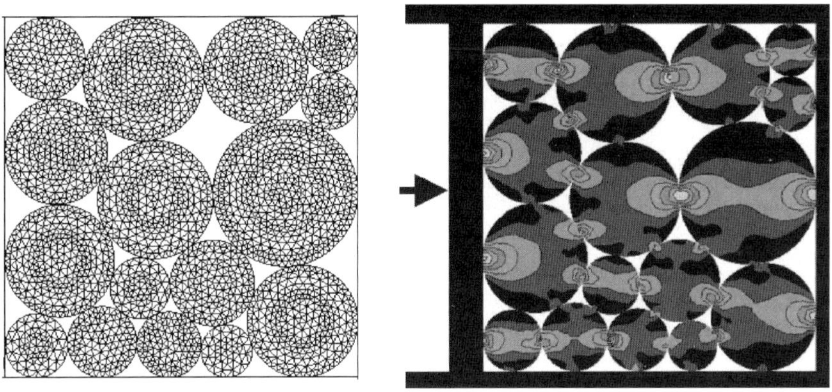

Fig. 8.28. Multiple contacts between particles. Mesh and constant value contours of the von Mises stresses

Representative Volume Element

The representative volume element (RVE) plays a key role in determining the effective properties of heterogeneous materials. Numerical computation is an essential tool for establishing the characteristic behaviour within this volume under external loading. The second example thus deals, for the first time, with this problem of multiple contacts between deformable particles, where the behaviour of each particle can be either elastic or hyperelastic. The finite element method is used to discretise the particles and obtain the stress field within each one, as shown in Fig. 8.28. In order to treat the problem of

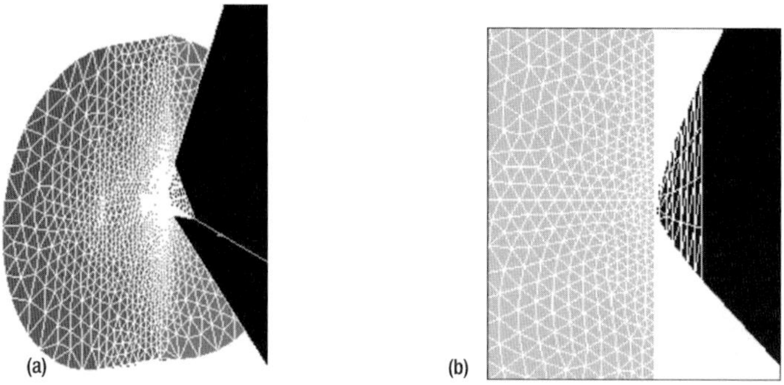

Fig. 8.29. Nanoindentation. Mesh

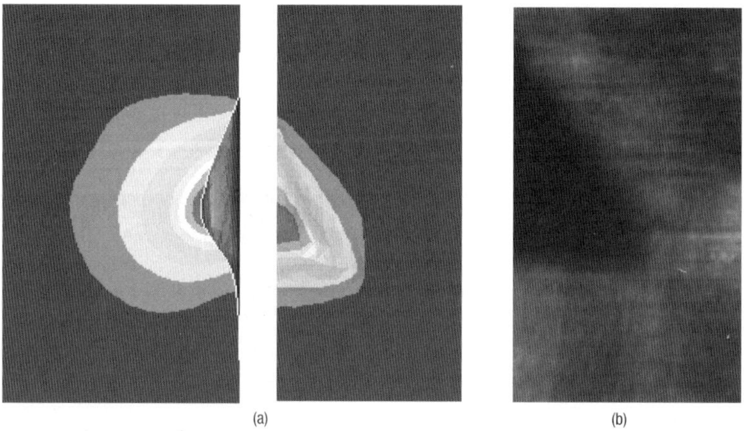

Fig. 8.30. Nanoindentation. (**a**) Numerical simulation. (**b**) Experimental result. See also the colour plate

frictional contact between the particles in an adequate way, a new approach was developed by Feng [42].

Nanoindentation

The third example is a model for nanoindentation. A tetrahedral (Berkovich) indenter comes into contact with the surface of a thin film (thickness 300 nm) on a substrate (different types of film being tested on different types of substrate). The thin film is assumed to have elastoplastic behaviour. The indenter is assumed to be rigid and to have a blunted tip with radius of curvature about 50 nm. Figure 8.29 shows the 3D meshed model and a more detailed 2D model. Note that the symmetry of the model is used to reduce computation time. Figure 8.30a shows the indent left on the surface and the stress distribution around the indentation region. The triangular shape of the indent faithfully reproduces what is observed experimentally on these scales (Fig. 8.30b).

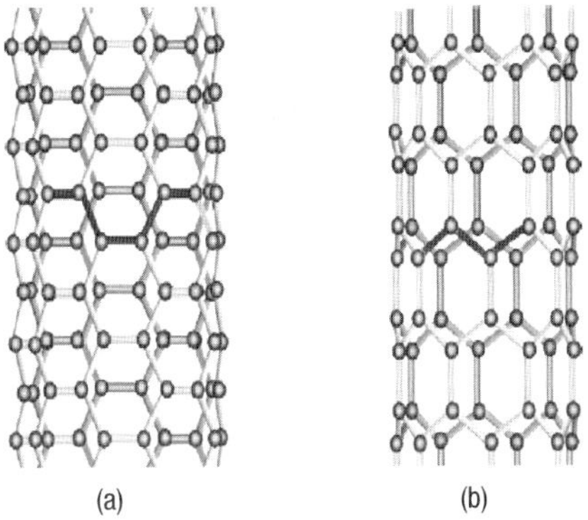

Fig. 8.31. Carbon nanotubes. (**a**) Armchair configuration. (**b**) Zigzag configuration

Simulation of Carbon Nanotubes

Carbon nanotubes are one of the main industrial products in the field of nanotechnology. They appear spontaneously when graphite is evaporated under the effect of an electrical discharge between two electrodes placed in a rare gas such as helium. These objects have molecular dimensions and are formed from one or more tiny carbon sheets arranged in concentric cylinders of diameter 1–10 nm (depending on the number of sheets) and length several microns, closed at the ends by spherical caps. The crystal structure of these carbon sheets, called graphenes, is hexagonal, while the crystal structure of the caps is pentagonal. Sumio Iijima, a Japanese scientist, was the first to observe them under the electron microscope in 1991 [43]. The way the graphene rolls up defines a parameter called the helicity which characterises the different types of nanotube. The helicity is specified by the chiral angle θ between the axis of the cylinder and the direction of one side of the hexagon. The configurations known as zigzag and armchair nanotubes are characterised by the values $\theta = 0$ and $30°$, respectively (see Fig. 8.31). Given the hexagonal symmetry of the crystal lattice, θ lies between 0 and $30°$.

Carbon nanotubes have remarkable physical properties with a wide range of potential applications in industry. For example, their elastic modulus is of the order of 1–5 TPa, well above the values for conventional carbon fibres, whilst remaining more flexible too [44]. The electrical properties of carbon nanotubes vary enormously with the diameter and helicity parameter, ranging from a semiconducting state to a metallic state. The helicity parameter also influences the mechanical properties [45].

Several theoretical models have been developed in the literature using the molecular dynamics approach [46, 47]. However, computation times can be prohibitive with these models, e.g., when simulating the behaviour of a bundle of multiwalled tubes involving a very large number of atoms. Another discrete approach has been put forward, treating the carbon nanotube as an

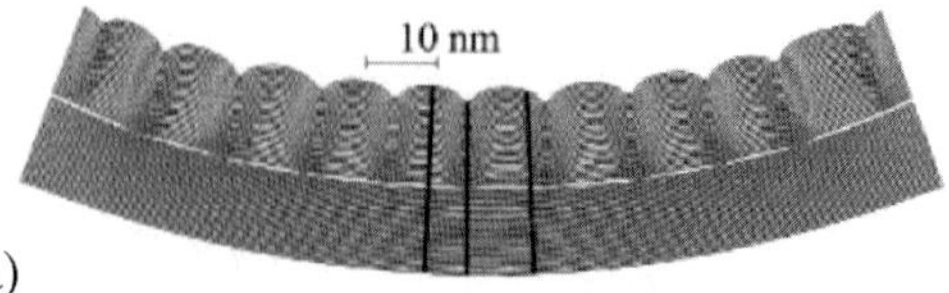

(a)

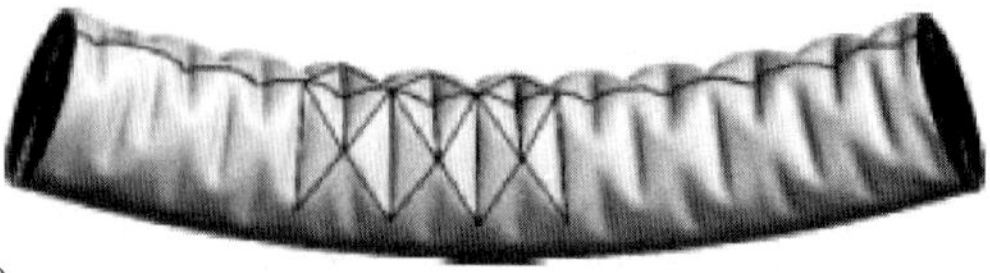

(b)

Fig. 8.32. Undulations of a 34-walled carbon nanotube. (**a**) Experimental observation in the plane of the tube axis [49]. (**b**) 3D simulation of the same [48]

assembled beam structure known as a truss in mechanics [30], in which the beam elements represent covalent binding forces between atoms. This type of model provides a static study and reduces calculations compared with the strictly molecular approach by restricting the atomic interaction of each atom to its nearest neighbours.

A continuum approach to modelling carbon nanotubes by the finite element method has been achieved by Arroyo and Belytschko [48]. It is based on a membrane model of strains in the carbon nanotube. Strains are calculated using the exponential Cauchy–Born rule to account for the effects of curvature of the tube. On the macroscopic level, stresses derive from a potential energy resulting from the sum over each finite element of the interaction energies between atoms bound according to the Tersoff–Brenner model of molecular mechanics. The interaction energy between atoms that are not bound together is also taken into account to model van der Waals forces according to the classical Lennard-Jones potential model, also from the field of molecular mechanics. These van der Waals forces operate between the carbon sheets in the case of multiwalled carbon nanotubes and between nanotubes when they form bundles.

Arroyo and Belytschko [48] present the results of a simulation of a carbon nanotube of length 124 nm, comprising 34 concentric carbon sheets, which corresponds to around 6 million atoms. The finite element model contains just 100 000 nodes, considerably reducing the size of the problem to be solved in comparison with a molecular dynamics simulation (by a factor of 60). The simulation reproduces the undulations observed experimentally in images obtained by transmission electron microscope (TEM) and published in [49] when the carbon nanotube is subject to flexion (see Fig. 8.32).

Figure 8.33 shows simulation results for a carbon nanotube under torsion, using molecular dynamics and finite elements. Good agreement is obtained between the two approaches.

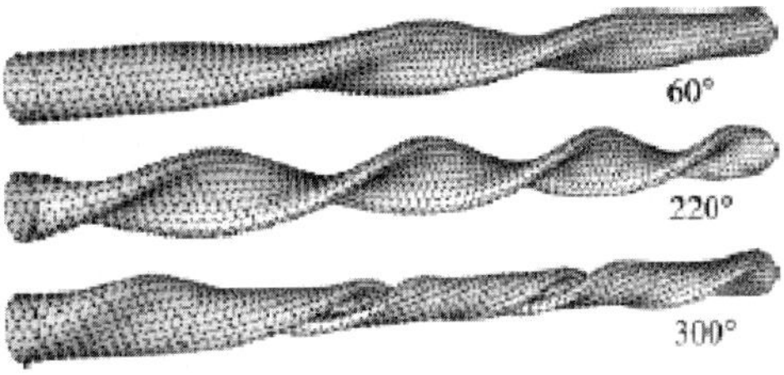

Fig. 8.33. Simulation of a carbon nanotube under torsion: molecular dynamics (*black dots*) and finite element (*grey surface*). Taken from [48] with kind permission of John Wiley & Sons

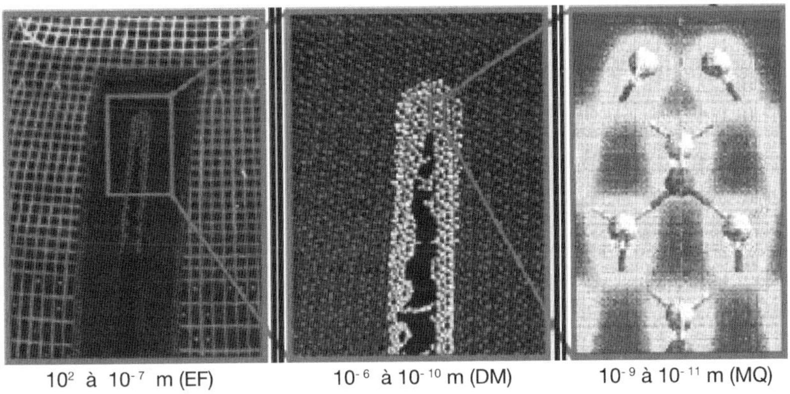

Fig. 8.34. Multiscale simulation. From [51], with the kind permission of Elsevier. See also the colour plate

Multiscale Simulations

The so-called multiscale simulation techniques were developed to obtain local quantum and molecular descriptions while retaining the continuum description on the global level, as illustrated in Fig. 8.34.

Several studies [50,51] have proposed ways of combining quantum mechanics (QM), molecular dynamics (MD), and finite elements (FE). In regions where the atoms obey the laws of continuum mechanics, the finite element method is used due to the advantages in terms of calculation time. However, in critical areas such as the extremity of a fracture, molecular dynamics and even quantum mechanics [e.g., ab initio, or density functional theory (DFT)] are required to obtain a more detailed study of the fracture process. The transition from the global to local levels involves a change of scale. More recently, Xiao and Belytschko [52] have proposed a way of improving the numerical compatibility between regions modelled by molecular dynamics and those modelled using the finite element method. At the interface between these two types of region, significant reflection effects are observed in the stress and

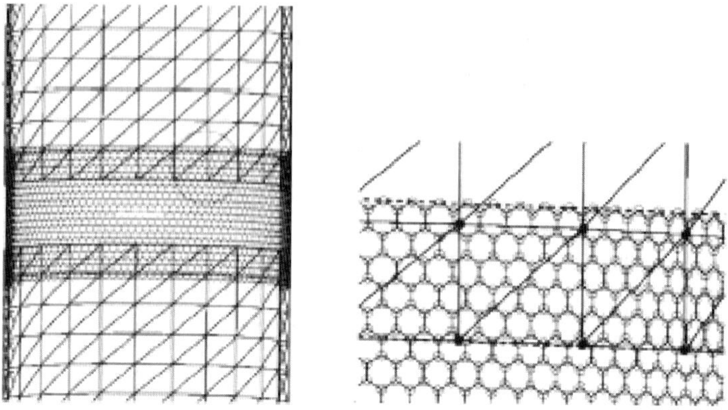

Fig. 8.35. Combined finite element and molecular dynamics (FE/MD). From [52], with the kind permission of Elsevier

strain fields, leading to unphysical distortion of the solution. The method suggested consists in introducing a rather broad transition region by superposing the finite element mesh of the continuum region on the atomistic structure of the molecular dynamics region, as can be seen in Fig. 8.35. In this region, the high frequencies generated by molecular dynamics are absorbed by the low frequencies of the continuum region.

Conclusion

Simulation is an important, even essential tool for aiding scientists to obtain a better understanding of the physical phenomena operating within nanomaterials. The examples discussed above, i.e., plasma projection, nanoindentation, and carbon nanotubes, show how one can get access to multiphysical phenomena at very short length scales (micron or nanometer) that are so difficult to observe experimentally. We have seen that, at these scales, a continuum approach by the finite element method can give excellent results, provided that certain precautions are taken. A good level of a priori knowledge concerning the constitutive laws and also a carefully chosen mesh are essential here. Understanding of the constitutive laws can be acquired a posteriori using molecular mechanics and/or quantum mechanics. The advantage with the finite element method as compared with molecular dynamics is that one can considerably reduce the number of degrees of freedom of the system under investigation, thereby reducing the numerical cost. The disadvantage is that one obtains a global description of the behaviour, thus losing any phenomenological description on the interatomic level, e.g., in problems of fracturing where propagation mechanisms must be determined). This is why the multiscale approach seems to be a promising solution for simulating the behaviour of nanomaterials, combining as it does the finite element and molecular dynamics methods.

References

Macroscopic Mechanical Properties

1. T.D. Shen, C.C Koch, T.Y. Tsui, G.M. Pharr: On the elastic moduli of nanocrystalline Fe, Cu, Ni and Cu–Ni alloys prepared by mechanical milling/alloying, J. Mater. Res. **10**, No. 6, 2892 (1995)
2. M. Verdier, M. Fivel, B. Gilles: Some investigations on the effect of length scale on mechanical properties, Adv. Eng. Mat. **3**, No. 8, 597–601 (2001)
3. R.D. Mindlin: Second gradient of strain and surface tension in linear elasticity, Int. J. Sol. Struct. **1**, 417–418 (1965)
4. E. Cosserat, F. Cosserat: *Théorie des corps déformables*, A. Herman et Fils, Paris (1909)
5. J. Besson, G. Cailletaud, J.-L. Chaboche, S. Forest: *Mécanique non-linéaire des matériaux*, Hermes Science Publication, Paris (2001)
6. A.C. Eringer: *Polar and Non-Local Field Theories*, Continuum Physics, ed. by A.C. Eringer, Vol. IV, Academic Press (1976)

Nanomechanical Properties: Experimentation

7. B. Bhushan: *Micro/Nanotribology*, CRC Press, Florida, USA (1995)
8. W.C. Oliver, G.M. Pharr: J. Mater. Res. **7**, 1564 (1992)
9. D. Tabor: *The Hardness of Solids*, Oxford, Clarendon Press (1951)
10. S.I. Bulychev, V.P. Alekhin, M.Kh. Shorshorov, A.P. Ternovskii, G.D. Shnyrev: Zavod. Lab. 1137 (1975)
11. M.Kh. Shorshorov, S.I. Bulychev, V.P. Alekhin: Sov. Phys. Dokl. **26**, 769 (1982)
12. I.N. Sneddon: Int. J. Eng. Sci. **3**, 47 (1965)
13. J.B. Pethica, R. Hutchings, W.C. Oliver: Philos. Mag. A **48**, 593–606 (1983)
14. M.F. Doerner, W.D. Nix: J. Mater. Res. **1**, 845–851 (1986)
15. G.M. Pharr, A. Bolshakov: J. Mater. Res. **17**, No. 10, 2660–2669 (2002)
16. H. Hehtz, J. Reine: Angewandte Mathematik, **92**, 156 (1882)
17. J. Boussinesq: *Applications des potentiels à l'étude de l'équilibre et du mouvement des solides élastiques*, Gauthier-Villars, Paris (1885)
18. B. Bushan, S. Sundararajan: Acta Mater. **46**, 3793 (1998)
19. A. Bolshakov, W.C. Oliver, G.M. Pharr: Pittsburgh, Materials Research Society, 141–146 (1997)
20. G. Hochstetter, A. Jimenez, J.L. Loubet: J. Macromol. Sci. Phys. B **38** (5 and 6), 681–692 (2004)
21. J.L. Bucaille, S. Stauss, E. Felder, J. Michler: Acta Materialia **51**, 1663–1678 (2003)
22. F.P. Bowden, D. Tabor: *The Friction and Lubrication of Solids*, Clarendon Press, Oxford (1950–1954)
23. J.-M. Georges: *Frottement, usure et lubrification*, Sciences et techniques de l'ingénieur, CNRS Editions, Eyrolles (2000)
24. C.M. Mate, G.M. McClelland, R. Erlandsson, S. Chiang: Phys. Rev. Lett. **59**, 1942–1945 (1987)
25. O. Pietrement: Doctoral thesis, University of Reims (2000)

Nanomechanical Properties: Computer Modelling

26. L. Verlet: Computer 'experiments' on classical fluids. I. Thermodynamical properties of Lennard-Jones molecules, Phys. Rev. **159**, 98–103 (1967)
27. P. Vashishta, R.K. Kalia, A. Nakano: Multimillion atom molecular dynamics simulations of nanoparticles on parallel computers, J. Nanoparticle Research **5**, 119–135 (2003)
28. M.J. Buehler, A. Hartmaier, H. Gao, M. Duchaineau, F.F. Abraham: Atomic plasticity: Description and analysis of a one-billion atom simulation of ductile materials failure, Comput. Methods Appl. Mech. Engrg. **193**, 5257–5282 (2004)
29. H. Auradou, M. Zei, E. Bouchaud: Numerical study of the temperature and porosity effects on the fracture propagation in a 2D network of elastic bonds, Eur. Phys. J. B **44**, 365–372 (2005)
30. C. Li, T.W. Chou: A structural mechanics approach for the analysis of carbon nanotubes, Int. J. Solids Struc. **40**, 2487–2499 (2003)
31. R.W. Clough: The finite element in-plane stress analysis, Proc. ASCE Conf. on Electronic Computation (1960)
32. P.V. Marçal, I.P. King: Elastic plastic analysis of two dimensional stress systems by the finite element method, Num. Meth. Ind. Form. Swansea **9** (1967)
33. H.D. Hibbitt, P.V. Marçal, J.R. Rice: A finite element formulation for problems of large strain and large displacement, Int. J. Solids Struc. **6**, 1069–1086 (1970)
34. R.M. McMeeking, J.R. Rice: Finite element formulation for problems of large elastic–plastic deformation, Int. J. Solids Struc. **11**, 601–616 (1975)
35. O.C. Zienkiewicz, D.R.J. Owen: Analysis of viscoplastic effects in pressure vessels by the finite element method, Nuclear Engn. Des. **28** (1974)
36. J.C. Nagtegaal: On the implementation of inelastic constitutive equations with special reference to large deformation problems, Comp. Methods Appl. Mech. Eng. **33**, 469–484 (1982)
37. T.J.L. Hughes, J. Winget: Finite rotation effects in numerical integration of rate constitutive equations arising in large deformation analysis, Int. J. Num. Meth. Eng. **15**, 1862–1867 (1980)
38. M.L. Wilkins: Calculation of elastoplastic flaws. In: *Methods in Computational Physics*, ed. by B. Alder, Academic Press (1964)
39. R.D. Krieg, B.D. Krieg: Accuracies of numerical solution method for the elastic–perfectly plastic model, ASME, J. Pressure Vessels and Piping Div. **99**, 510–515 (1977)
40. J.C. Simo, R.L. Taylor: Consistent tangent operator for rate-independent elasto-plasticity, Comp. Methods Appl. Mech. Eng. **48**, 101–118 (1985)
41. Z.Q. Feng: Sur la modélisation d'un problème multiphysique: la projection à plasma, 7th Colloque National en Calcul des Structures, Giens, France (17–20 May 2005)
42. Z.Q. Feng: 2D or 3D frictional contact algorithms and applications in a large deformation context, Comm. Num. Meth. Eng. **11**, 409–416 (1995)
43. S. Iijima: Helical microtubules of graphitic carbon, Nature **354**, 56–58 (1991)
44. M.M.J. Treacy, T.W. Ebbesen, T.M. Gibson: Exceptionally high Young's modulus observed for individual carbon nanotubes, Nature **381**, 680–687 (1996)
45. V.N. Popov, V.E. Van Doren, M. Balkanski: Elastic properties of single-walled carbon nanotubes, Phys. Rev. B **61**, 3078–3084 (2000)

46. B.I. Yakobson, M.P. Campbell, C.J. Brabec, J. Bernholc: High strain rate fracture and C-chain unraveling in carbon nanotubes, Computational Materials Science, **8**, 341–348 (1997)
47. E. Hernandez, C. Goze, P. Bernier, A. Rubio: Elastic properties of C and BxCyNz composite nanotubes, Phys. Rev. Lett. **80**, 4502–4505 (1998)
48. M. Arroyo, T. Belytschko: Finite element methods for the non-linear mechanics of crystalline sheets and nanotubes, Int. J. Numer. Meth. Engng. **59**, 419–456 (2004)
49. T. Kuzumaki, T. Hayashi, H. Ichinose, K. Miyazawa, K. Ito, Y. Ishida: In-situ observed deformation of carbon nanotubes, Phil. Mag. A **77** (6), 1461–1469 (1998)
50. F.F. Abraham, J.Q. Broughton, N. Bernstein, E. Kaxiras: Spanning the length scales in dynamic simulation, Comput. Phys. **12** (6), 538–546 (1998)
51. S. Ogata, E. Lidorikis, F. Shimojo, A. Nakano, P. Vashishta, R.K. Kalia: Hybrid finite-element/molecular-dynamics/electronic-density-functional approach to materials simulations on parallel computers, Comp. Phys. Comm. **138**, 143–154 (2001)
52. S.P. Xiao, T. Belytschko: A bridging domain method for coupling continua with molecular dynamics, Comput. Methods Appl. Mech. Engrg. **193**, 1645–1669 (2004)

9

Superplasticity

T. Rouxel

9.1 Introduction

Permanent deformation of a material through flow, e.g., creep, viscosity, viscoplasticity, gets easier as the grain size in the material gets smaller. In the most spectacular cases, relative extensions greater than 100% (nominal strain > 1) can be obtained at relatively low temperatures compared with the temperatures usually required to observe creep in materials: this is the effect known as superplasticity. Typically, superplasticity only occurs in fine-grained dense materials (grains < 0.01 mm for metals, < 1 μm for ceramics), barely affected by strain localisation effects, or striction, at temperatures $> 0.5T_{\text{melting}}$, when such a temperature has any meaning (materials sometimes decomposing before melting). Even in ancient times, smiths made good use of this remarkable property to forge tough, hard steel blades. The steel used by the Persians at the time of the crusades, and by Saladin's armies, or Damascus steel, is one of the greatest achievements of metallurgy and the forge, where the choice of alloy at the outset (in this case a steel with a high carbon content, known as wootz, from India) and the masterly control of a judicious forging cycle (the thickness of the initial ingot was first reduced by a factor of about 10 by hammering) produced a material with ideal fine microstructure for making sharp cutting blades that could also resist mechanical shocks. Figure 9.1 illustrates the phenomenon of superplastic behaviour for a steel containing 1.6% carbon (ultrahigh carbon steel), with a fine microstructure, close to Damascus steel, which seems to have been produced first in India in the fourth century BC.

Since the early work by Pearson [1] who obtained 1 950% relative extension for a Sn–Bi alloy, the phenomenon of superplasticity has been the subject of many studies and is today widely used in making items with complex architecture, while metal alloys with ever higher performance have been developed over the last three decades, e.g., aluminium and titanium alloys and intermetallic materials (see Fig. 9.2).

It has only recently been discovered that superplasticity can occur in ceramics (see Fig. 9.3): 1980 for magnesia (MgO) [2], 1985 for zirconia (Y-

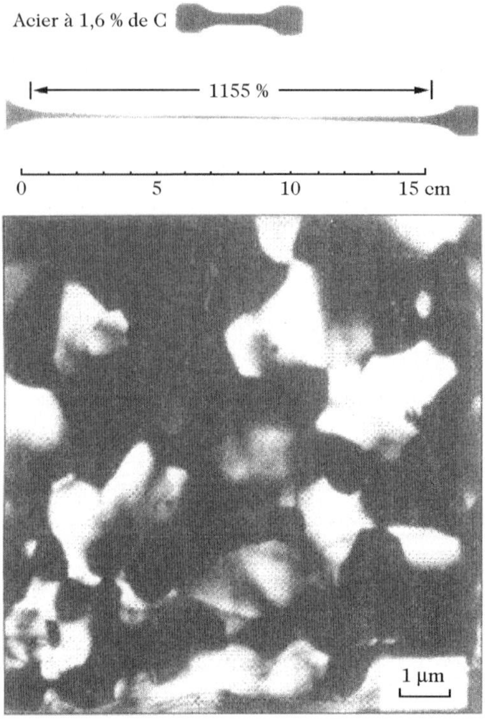

Fig. 9.1. Example of superplastic behaviour of an ultrahigh carbon steel. In this case, superplasticity is caused by the presence of fine spheroidal precipitates of cementite (Fe_3C) dispersed in a fine-grained ferritic matrix. *Top*: Test bar under tension before and after tests at 973 K with an imposed strain rate of $1.7 \times 10^{-4}\,s^{-1}$. *Bottom*: Associated microstructure. Black particles are iron carbide and the light background is made from pure iron (α iron) or ferrite (solid α solution). Taken from [12]

TZP) [3], compounds with mainly ionic bonding, and 1990 for silicon nitride (Si_3N_4) [4], with highly covalent bonding. Feasibility has been demonstrated not only for forming by hot deformation, the most immediate application of superplasticity, but also for solid state diffusion welding, and is not limited to structural materials but also concerns functional ceramics for electronics, e.g., $PbTiO_3$, ZrO_2, or for medicine, e.g., hydroxyapatite. The basic condition in all cases is that the grain size should be less than the micron, and typically of the order of a hundred nanometers.

9.2 Mechanism

The semi-empirical theory most often used to describe the stationary flow regime (constant strain rate) was introduced by Muckherjee et al. [5], in which

Fig. 9.2. Example of an aluminium alloy formed using superplasticity in the automobile industry. © Morgan Cars

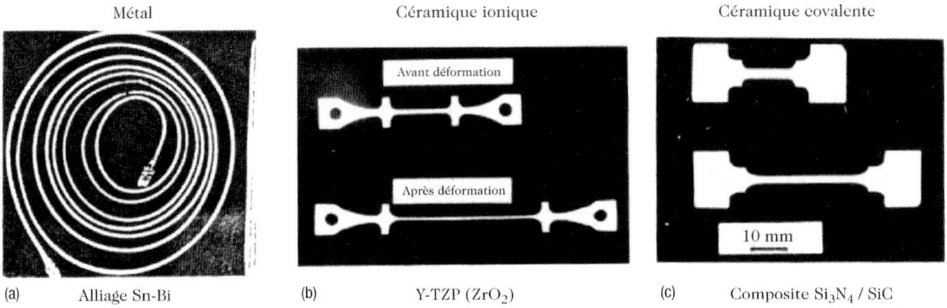

Fig. 9.3. Superplasticity in a metal alloy (**a**), in an oxide ceramic Y-TZP (ionic bonding) (**b**), and in a non-oxide ceramic, a nanocomposite Si_3N_4/SiC (covalent bonding) (**c**). From [13]

the dependence of the strain rate on the stress σ and grain size d is given as σ^n/d^p. The exponents n and p can assume widely different values from one material to another, and with the exception of the pairs $(n, p) = (1, 2)$ and (1,3) corresponding respectively to the early Herring–Nabarro model [6, 7] (diffusion across grains) and the Coble model [8] (diffusion along grain boundaries), the other values of n and p determined experimentally – and there are many – have led to a rather excessive proliferation of models over the past four decades.

In the superplastic deformation regime, the dominating effect is sliding at grain boundaries, in which grains move over distances close to or greater than the grain size, without major change in their geometry. Superplasticity

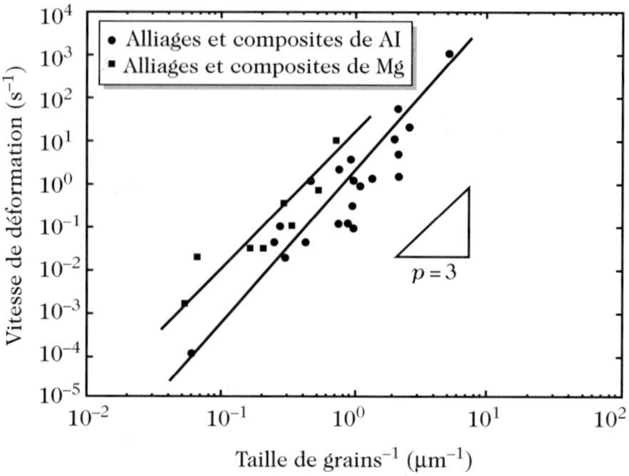

Fig. 9.4. Dependence of the strain rate (in the stationary creep regime) on grain size for aluminium and magnesium alloys and composites

is characterised by a non-Newtonian flow ($n \neq 1$), generally rheofluidifying ($n > 1$), sometimes accompanied by a dynamical enlargement of the grains favoured by stresses. Hence the strain rate generally varies as σ^2/d^2, where σ is the flow stress and d the average grain size, but sometimes as σ^2/d^3 (see Fig. 9.4), while a dependence of type σ/d^2 or σ/d^3 is more commonly observed in the diffusional creep regime.

Taking as a starting point the kinds of argument used to interpret diffusional creep, several physical models have been developed to account for superplastic behaviour. Keeping to the main features of the theory and simplifying somewhat, the argument goes as follows: the deformation is explained by the diffusion of atoms (from faces under compression toward faces under tension) and the diffusion of cavities in the opposite direction. The latter effect is activated by the cavity concentration gradient which results from application of a stress. A surface subject to tensile stress is assumed to favour cavity formation (cavity concentration per unit volume C^+), whereupon cavities subsequently diffuse in a natural way toward compressed surfaces, or surfaces under lower tension (concentration C^-) to reduce the imbalance (concentration C at equilibrium). (This argument can also be formulated in terms of chemical potentials, although it is not easy to give meaning to this quantity in a solid.) One then has

$$C = C_0 \exp\left(- \Delta G/kT \right) , \tag{9.1}$$

$$C^+ = C \exp\left(+ \sigma \Omega/kT \right) , \tag{9.2}$$

$$C^- = C \exp\left(- \sigma \Omega/kT \right) . \tag{9.3}$$

This amounts to considering that the stress, depending on its sign, lowers or raises the free enthalpy of cavity formation (ΔG) by the value $\pm\sigma\Omega$, where Ω is the volume of a cavity and T the temperature. The induced flow of cavities is then

$$J = -D\nabla C , \tag{9.4}$$

where D is the coefficient of diffusion and ∇ the gradient operator.

Assuming homogeneous diffusion throughout the volume of the grain, and assuming also that the latter has equiaxial geometry, the number of cavities transported per unit time across an area taken equal to d^2 to simplify can be written

$$\phi = -Jd^2 . \tag{9.5}$$

Then by (9.1)–(9.4),

$$\phi = -2D\mathrm{d}C \sinh\left(\sigma\Omega/kT\right) . \tag{9.6}$$

Diffusion of one cavity produces on average a displacement of amplitude u of the surface perpendicular to itself, where

$$u = -\Omega/d^2 . \tag{9.7}$$

The quantity u is counted negatively because it is associated here with the flow of cavities (leading to a loss of matter) and it is thus of opposite sign to the stress. The flow of atoms varies in the opposite direction and corresponds obviously to an increase of matter on the faces undergoing tension. Assuming that this cavity travels a distance d, this gives an elementary strain $\varepsilon = u/d = -\Omega/d^3$.

Finally, the strain rate $\mathrm{d}\varepsilon/\mathrm{d}t = \phi\varepsilon$ can be written

$$\frac{\mathrm{d}\varepsilon}{\mathrm{d}t} = 2DC\Omega\frac{\sinh\left(\sigma\Omega/kT\right)}{d^2} . \tag{9.8}$$

Expressing C in terms of the atomic fraction of cavities ν, i.e., $C = \nu/\Omega$, and noting that for small values of its argument $\sinh x \approx x$, which corresponds to the common case where $\sigma\Omega \ll kT$, the last expression reduces to the characteristic relation for Nabarro–Herring creep by bulk diffusion:

$$\frac{\mathrm{d}\varepsilon}{\mathrm{d}t} = 2D\nu\frac{\Omega\sigma}{kTd^2} , \tag{9.9}$$

where the temperature dependence is essentially contained in the diffusion coefficient $D = D_0 \exp\left(-\Delta G_0/kT\right)$.

When diffusion paths are limited to grain boundaries, only a fraction of the volume of the material need be taken into account, and a factor δ/d must be applied to the last expression, where δ is the boundary thickness, in order

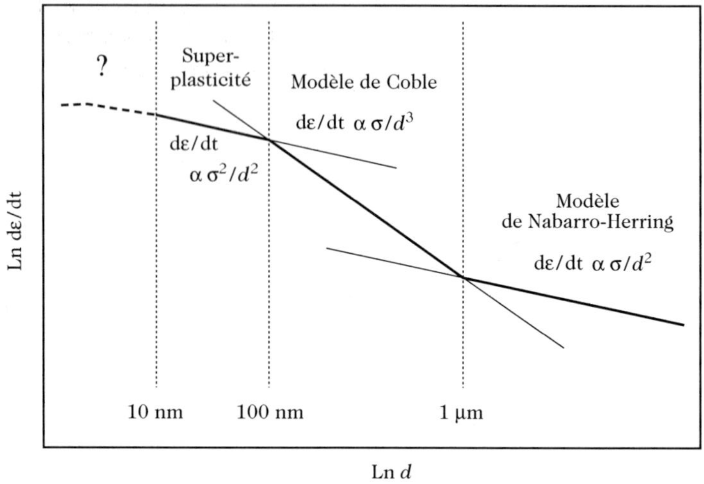

Fig. 9.5. Grain size dependence of the strain rate

to obtain an expression for the rate. One then obtains a similar expression to the one proposed by Coble on the basis of deformation via diffusion in grain boundaries:

$$\frac{\mathrm{d}\varepsilon}{\mathrm{d}t} = 2D_{\mathrm{b}}\nu\frac{\Omega\sigma\delta}{kTd^3} \,. \tag{9.10}$$

In the latter case, it is the diffusion coefficient D_{b} in the grain boundaries that must be taken into account.

It is interesting to note that, when the grain size tends to the interatomic distance a, equation (9.9) leads to

$$\eta \propto \frac{kT}{aD} \,, \tag{9.11}$$

where $\eta \propto \sigma/(\mathrm{d}\varepsilon/\mathrm{d}t)$ is the coefficient of viscosity. This therefore leads to an expression similar to the Stokes–Einstein equation [9], which governs viscous flow, and this suggests that to a first approximation a glass can be viewed as a polycrystalline material in which the grain size is on the scale of the interatomic distance. Figure 9.5 gives a schematic representation of the various deformation regimes.

Comments

- In most cases, deformation by creep results from a combination of various types of creep, i.e., Nabarro–Herring, Coble, etc., and the total strain can be interpreted as the sum of the components associated with diffusion mechanisms, dislocation movements, sliding at grain boundaries, and

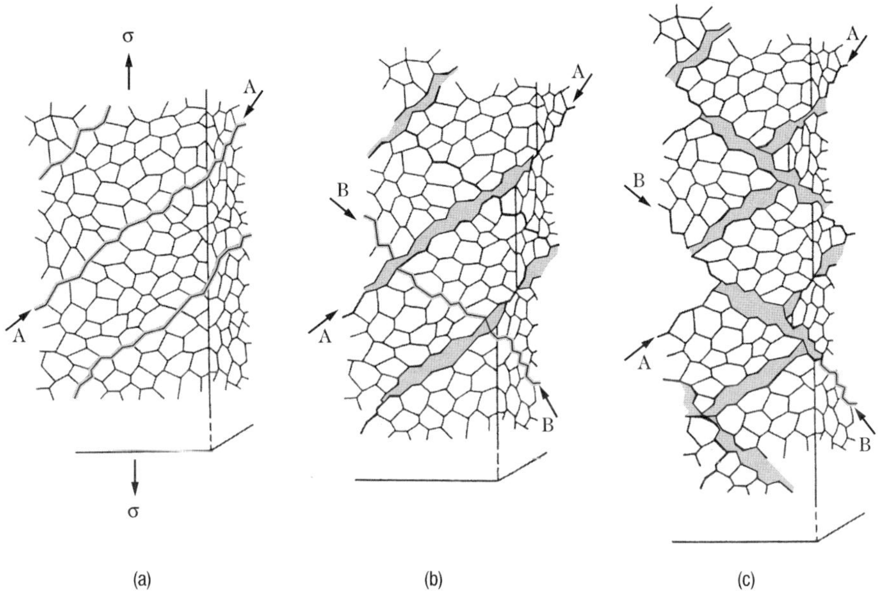

Fig. 9.6. Formation of flow superstructures during superplastic deformation of fine-grained materials [11]. Sliding is localised along boundaries of regions indicated by *arrows* A and B. The superstructures grow more easily visible as the relative extension increases [progression from (**a**) to (**c**), via (**b**)]

damage (cavitation, microfracture). In the case of superplastic flow, it has been shown that intergranular sliding represents the main contribution to the deformation. This sliding depends a great deal on the nature of the grain boundaries (amorphous, crystallised, thin, thick) and the crystalline misorientation angle between adjacent grains. Hence an amorphous film and/or a random distribution of this angle favour sliding. Now the frequency of randomly oriented boundaries is proportional to d^2, so that there is probably an ideal grain size for obtaining large deformations: too small a size favours the appearance of grain boundaries with particular coincidence angles, while too large a size will reduce the total surface area of the grains and thus make sliding more difficult [10].

- Relations (9.8) and (9.9) predict a deformation rate that tends to infinity as the grain size tends to zero. Glass, which can be viewed as an extreme case, can easily be used to obtain large deformations: several kilometers of optical fibre can be drawn out from a silica glass preform of centimeter dimensions. However, when the constitutive laws determined experimentally for materials with grain sizes of the order of a hundred nanometers, are extrapolated down to grain sizes of a few nanometers, the predicted creep rates are significantly overestimated and temperatures that should be suitable for observing a flow (for which $\mathrm{d}\varepsilon/\mathrm{d}t > 10^{-8}\,\mathrm{s}^{-1}$) are impossibly low.

It is found, for example, that a silicon nitride ceramic with nanometric grains would flow at room temperature, whereas even glasses barely flow below their glass transition temperature ($T_g \approx 818\,K$ for window glass).

- Superplasticity generates large relative displacements of the grains. Imagining all these displacements occurring throughout the three dimensions of a polycrystalline material, one realises how important it is for the material to be able to accommodate these relative motions in such a way as to avoid the appearance of cavities and fractures. The mechanism for accommodating the flow is thus just as fundamental as the deformation mechanism associated with it. An optimal range of temperatures and strain rates is often observed for obtaining large extensions.
- Recent observations seem to show that the deformation generates a flow superstructure with relative displacements of groups of grains forming rigid clusters (see Fig. 9.6). If this is the case, is it a good policy to try at any price to produce nanophase materials as a way of reducing the temperature at which they can be formed by plastic deformation?

9.3 Superplastic Nanostructured Materials

The fabrication of dense fine-grained materials is certainly the most delicate stage in the lead-up to a forming process based on superplasticity. In order to develop new alloys with grain sizes of a few tens of nanometers, metallurgists have turned to cold working processes by severe plastic deformation, torsion straining (TS), and forced shear obtained by extrusion through a channel bent through an angle of as much as 90° (equal channel angular extrusion ECAE), as shown in Fig. 9.7, or again to mechanosynthesis. These processes have led to a certain degree of success, especially for light alloys, and are able to obtain submicron grain sizes. However, they cannot produce large solid items. Today, aluminium, zinc and titanium alloys (e.g., Al–Mg, Al–Zn, Ti–Al) are the main metal alloys used in superplastic forming processes.

Research on ceramics has led to ingenious ways of synthesising materials with submicron grains. Among the exotic methods used are hydrothermal synthesis and the sol–gel method, which raise serious difficulties with regard to preparation time, drying, or again the quantities of material that can be synthesised; and sintering of ultrafine powders at high pressure (the powders coming in some cases from organometallic precursors), which raise the problem of first deagglomerating and then dispersing the powders in a liquid medium. Several sintering cycles can be imposed to avoid grain enlargement during the consolidation phase. In all these cases, as the grain size is refined, it gets more and more difficult to obtain dense materials and the processes become more costly. Moreover, these materials, still rather uncommon today, exhibit large residual stresses which result mainly from drying and/or sintering phases, and these make them unstable in use, especially at higher

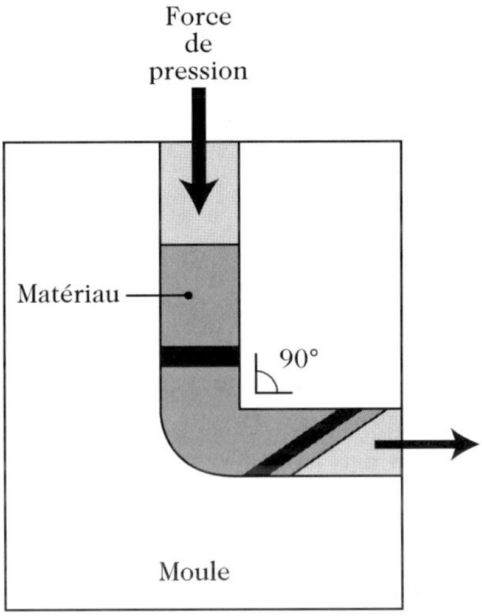

Fig. 9.7. Extrusion device with a right-angle bend in the die (ECAE). Several passes are generally required to produce submicron grain sizes in an alloy initially obtained in the form of raw foundry ingots. The deformation undergone by the material in each pass is close to 100% ($\varepsilon \approx 1$)

temperatures. Some examples of microstructures with grain sizes of the order of a hundred nanometers are illustrated in Fig. 9.8.

In the face of the difficulties encountered in fabricating dense nanostructured materials, two questions arise:

- Are dense materials with fine, nanometric, grains always superplastic?
- Is it possible to consolidate the material after forming in such a way as to give it a satisfactory creep behaviour in use?

The first, essentially fundamental question demands deeper reflection. The second is of considerable practical importance and should be made the subject of a thorough investigation, in parallel with the study of superplasticity, to allow industry to draw proper conclusions concerning the prospects for these processes.

9.4 Industrial Applications

Superplasticity provides a rapid way of obtaining items with complex geometry in a single forming operation and with relatively low flow stresses. It is thus easy to understand the growing interest shown by the world of industry for this property and the superplastic forming (SPF) and diffusion bonding (DB) processes it makes possible, which reduce the required quantities of matter and fabrication costs. However, industrial applications are still restricted mainly to the aeronautic industry [16] and on the whole concern light

Fig. 9.8. Examples of microstuctures with submicron grain sizes, demonstrating the feasibility of synthesising dense nanostructured materials, either metallic or ceramic. (**a**) Armco iron after severe plastic deformation and annealing at 247 K. (**b**) Monoclinic zirconia (pure) produced below 727 K from a powder obtained by hydrothermal synthesis [14]. (**c**) Yttrium oxide (Y_2O_3) sintered with 1% MgO. From [15]. Reproduced with the kind permission of Macmillan Publishers © 2000

aluminium and titanium alloys (see Fig. 9.9). Titanium alloys are the most widely used today because these were the superplastic alloys first developed industrially, with TA6V being the main representative. Superplastic forming can be used to obtain in a single operation an element that would otherwise have required the assembly of several parts by riveting. Assembly times are thereby greatly reduced and, by avoiding extra thickness due to assembly, superplastic forming can also make lighter elements. Although superplasticity has been demonstrated in fine-grained ceramics, and in particular in stabilised tetragonal phase zirconia (Y-TZP), the fabrication of ceramic items using this property remains a laboratory curiosity that has barely entered the research and development stage. There are nevertheless several remarkable examples: the fabrication of missile nozzles and axially symmetric elements by the US group Lockheed (see Fig. 9.9b) [16], and the forming of parabolic sheaths to protect the leading edges of helicopter rotor blades at the National Industrial Research Institute of Nagoya (NIRIN, Japon).

Concepts and Terminology

Strain. Ratio of the change in length to the initial length of a structure, e.g., $\varepsilon = (L - L_0)/L_0$ (relative extension). For large strains, we write

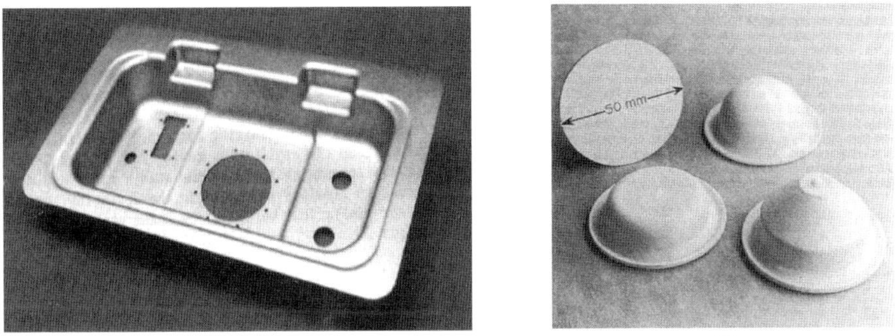

(a) (b)

Fig. 9.9. (**a**) Service panel made from the titanium alloy Ti6Al4V (TA6V) for the Airbus A300/A310 [16]. (**b**) Stabilised zirconia (Y-TZP) elements produced by Lockheed Martin Missiles & Space Co. (USA), obtained by superplastic forming under a gas pressure of 690 kPa at 1 277 K [17]

$$\varepsilon = \int_{L_0}^{L} \frac{\mathrm{d}L}{L} = \ln \frac{L}{L_0} \, .$$

Stress. Ratio of a force to the projected area of the surface on which it is applied, viz., $\sigma = F/S$. Note that, for large strains, the change of area generated by the deformation cannot be neglected. For a uniaxial strain with the volume assumed constant, $\sigma = \sigma_0 \exp \varepsilon$, where $\sigma_0 = \sigma(L = L_0)$. The unit of stress is the pascal (Pa).

Striction. Sudden narrowing of the cross-section of a test bar under tension resulting from strain localisation and leading to rupture.

Creep. Change in the geometry of a structure under constant load. A thermally activated phenomenon.

Cavity. Void or point defect left by the departure or absence of an atom in an atomic lattice.

Coefficient of Viscosity. A coefficient expressing the proportionality between the shear stress and the resulting rate of angular distortion in Newton's law of viscous flow, viz.,

$$\tau = \eta \frac{\mathrm{d}\gamma}{\mathrm{d}t} \, .$$

By abuse of language, η is often called the viscosity.

Glass. Non-crystalline (amorphous) solid in which the glass transition occurs. Material obtained by quenching a liquid, i.e., freezing in the disordered structure.

Ceramic. Non-metallic inorganic solid. Material with ionocovalent interatomic bonding obtained by a high temperature processing stage.

References

1. C.E. Pearson: J. Inst. Metals **54**, 111–116 (1934)
2. J. Crampon, B. Escaig: J. Am. Ceram. Soc. **63**, 11–12 (1980)
3. F. Wakai, S. Sakaguchi, Y. Matsuno: Advanced Ceram. Mat. **1**, 3, 259–263 (1986)
4. F. Wakai, Y. Kodama, S. Sakaguchi, N. Murayama, K. Izaki, K. Niihara: Nature Lett. **344**, 421–423 (1990)
5. A.K. Mukherjee et al.: Trans. ASM **62**, 155–179 (1969)
6. F.R. Nabarro: Report of a conference on the strength of solids, London Phys. Soc. 75–90 (1948)
7. C. Herring: J. Appl. Phys. **21**, 437–445 (1950)
8. R.L. Coble: J. Appl. Phys. **34**, 6, 1679–1682 (1963)
9. See for example: Y. Quéré: *Physique des matériaux*, Edn. Ellipses, Paris (1988) p. 21
10. T. Watanabe: Mater. Sci. Forum., Trans Tech. Pub., Suisse, 304–306, 421–430 (1999)
11. M.G. Zelin et al.: Acta Met. and Mat. **40**, 11, 2943–50 (1994)
12. J. Wadsworth: MRS Bulletin **27**, 12, 980–987 (2002)
13. F. Wakai: *Ceramics Today – Tomorrow's ceramics*, ed. by P. Vincenzini, Elsevier Science (1991) pp. 61–76
14. T. Rouxel et al.: Mat. Sci. Forum, Trans Tech. Pub. Suisse, 243–245, 245–250 (1997)
15. I.W. Chen et al.: Nature **404**, 168–171 (9 March 2000)
16. H.E. Friedrich et al.: *Superplasticity in Advanced Materials*, ed. by S. Hori, M. Tokizane, and N. Furushiro, Japan Society for Research on Superplasticity (1991) pp. 601–610
17. J. Wittenauer: Mat. Sci. Forum, Trans Tech. Pub., Suisse, 243–245, 653–662 (1997)

10

Reactivity of Metal Nanoparticles

J.-C. Bertolini and J.-L. Rousset

This chapter will be concerned with the parameters governing interactions between metallic nanoparticles and a reactive surrounding material, and hence with the catalytic properties of such nanoparticles. Indeed, most industrial metal catalysts contain very costly metals, such as the precious metals at the end of the transition series and the noble metals, and must therefore have as large a surface-to-volume ratio as possible in order to economise the number of atoms required. This is the solution provided by nanoparticles.

In Sect. 10.1 we shall see how chemical reactivity can be significantly modified by using materials in the form of nanoparticles rather than bulk materials as a way of reducing the number of atoms used. Such modifications are associated with the intrinsic effects of using very small particles, such as non-standard structure and specific electronic properties, and also with the presence of a large number of low-coordination sites such as corners and edges. These kinds of effects are usually grouped together under the heading of size effects in catalysis, but they are sometimes also referred to as structural sensitivity [1, 2].

To implement such nanoparticles, one requires a powdered support with high specific area. The choice and preparation of this support must be tailored to the type of reactor and the desired reaction. The support can induce electronic and structural effects on the supported nanoparticles which will modify their reactivity. It may itself be directly involved in the catalytic process. Migration from the support to the particle may also favour the formation of a compound with novel properties via the strong metal support interaction (SMSI). We shall refer to these as support effects, discussed in Sect. 10.2

The performance of a catalyst in terms of activity, selectivity and stability are also modified by alloying effects when a second metallic partner is introduced [3], and this is discussed in Sect. 10.3. Apart from intrinsic changes in reactivity due to modifications induced by chemical binding between the two partners, other parameters may prove relevant in a catalytic process. Indeed, the active site often comprises not just one but a group of atoms and the activity will be altered by the effects of dilution, even if the partner is inactive [4].

This parameter may influence the selectivity when several reactions occurring in parallel do not require the same number and/or the same arrangement of surface atoms at the active site. Complementary chemical reactivity of the different elements present on the particle surface with regard to the various reactants can also, by synergy, greatly amplify the reactivity of small alloy particles.

These effects of size, support and alloying can be exploited to design and/or synthesise tailor-made catalysts with better performance, i.e., more active and/or more selective, but there can be drawbacks, e.g., catalyst grains may be partly deactivated in certain reactions. In each case, the effects will be illustrated by examples of relevant catalytic reactions.

10.1 Size Effects

The transition from the cluster containing several atoms to the nanoparticle comprising several tens of atoms to several hundred atoms will be discussed first. We shall then consider a few specific properties of nanoparticles which govern their chemical reactivity in chemisorption and catalysis through their intrinsic electronic and geometrical properties.

10.1.1 Structural Properties

From Molecular Clusters to Particles

For very small sizes, of nanometric order, a cluster is not necessarily anything like a simple chunk cut off from the corresponding crystal. In its quest for stability, the nanoscale cluster may even assume non-crystalline structures, including the icosahedron (with a C_5 symmetry axis) and the truncated decahedron amongst others (see Fig. 10.1).

As the particle size increases, structural changes thus occur so as to leave the system in the most favourable energy state. It should be noted that the size corresponding to such modifications depends to a large extent on the metal. For example, in the case of copper or silver, the icosahedral form is the most stable up to several hundred atoms, whereas the fcc structure (i.e., the structure of the bulk crystal) of the truncated cubo-octahedron is already the most stable for a few tens of atoms in the case of Pd, Pt or Au [5].

Dispersion and Equilibrium Shape of Face-Centered Cubic Nanoparticles

As discussed above, the equilibrium shape expected for a metal particle with fcc structure is generally a cube truncated to form a cubo-octahedron, as shown in Fig. 10.2 for a nanoparticle containing 586 atoms. Recall that the fcc

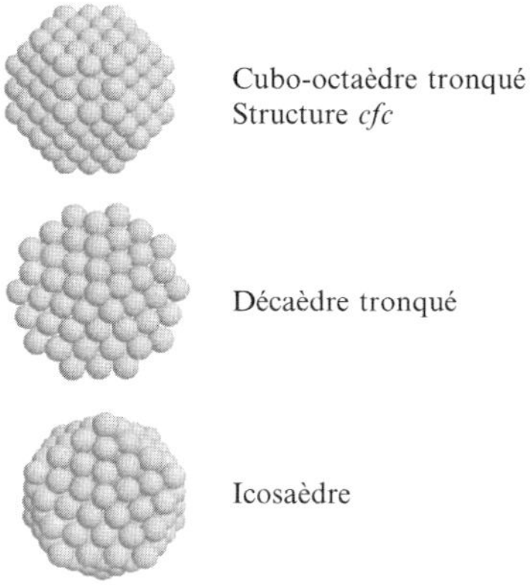

Fig. 10.1. Equilibrium shapes of small fcc metal particles [5]

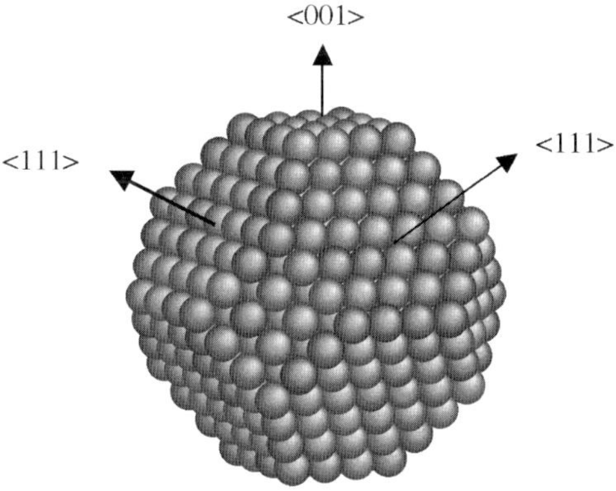

Fig. 10.2. Truncated cube or cubo-octahedron, the equilibrium shape of an fcc crystal comprising 586 atoms

structure is the one found in the noble metals (Cu, Ag, Au) and the *d* metals at the end of the transition series (Ni, Pd, Pt), widely used in catalysis.

As the particle size changes, the dispersion, i.e., the ratio of the number of surface atoms to the total number of atoms, will change, and so will the nature and number of sites with specific positions on faces, edges and corners. These sites have different coordination numbers, i.e., different numbers of nearest neighbours. The coordination, with value 12 for an atom in the bulk, can take a range of values at the surface: 9 for close-packed (111) facets, 8 for (100) faces, 7 for atoms on edges and 6 for atoms at corners in fcc metals. The variation in the number of atoms with different coordinations as a function of the total number of atoms is given in Table 10.1 for a cubo-octahedron of an fcc crystal.

Table 10.1. Numbers of atoms in different positions in a cubo-octahedral particle and considering successive filled shells, according to Van Hardeveld and Hartog [6]. Z is the coordination number of the atoms under consideration

Total number of atoms	Number of surface atoms	Number of corner atoms $Z = 6$	Number of edge atoms $Z = 7$	Number of atoms of type (100) $Z = 8$	Number of atoms of type (111) $Z = 9$
38	32	24	0	0	8
201	122	24	36	6	56
586	272	24	72	24	152
1 289	482	24	108	54	296
2 406	752	24	144	96	488
4 033	1 082	24	180	150	728
6 266	1 472	24	216	216	1 016
9 201	1 922	24	252	294	1 352
12 234	2 432	24	288	384	1 736
27 534	3 632	24	360	600	2 648
46 929	5 882	24	468	1 014	4 376

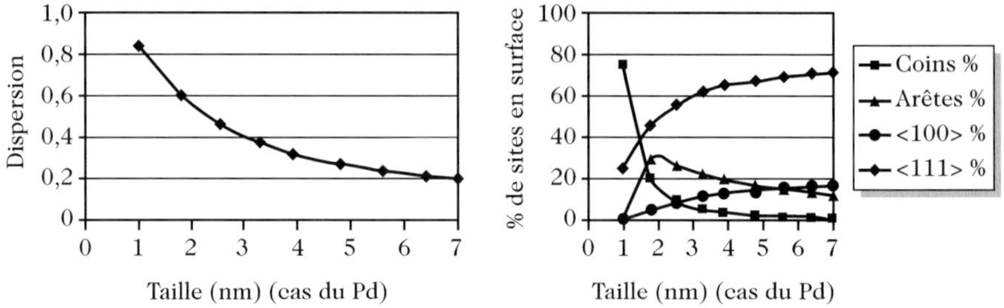

Fig. 10.3. Size dependence of the dispersion and relative frequency of various surface sites for a Pd particle (fcc metal with lattice parameter 0.389 nm)

Figure 10.3 shows the size dependence of the dispersion and surface percentage of atoms with different coordinations for cubo-octahedral nanoparticles of the fcc metal palladium with lattice parameter 0.389 nm (see also Fig. 10.2).

Moreover, the environment of the different surface atoms with the same coordination can vary depending on their position. This is the case for the atoms with coordination 7 that can be seen in Fig. 10.2, considering on the one hand the edges joining two close-packed facets of type (111) where there is a hexagonal arrangement of surface atoms, and on the other hand, the edges joining a (111) facet with a truncation of the cubo-octahedron of type (100) where there is a square arrangement of atoms.

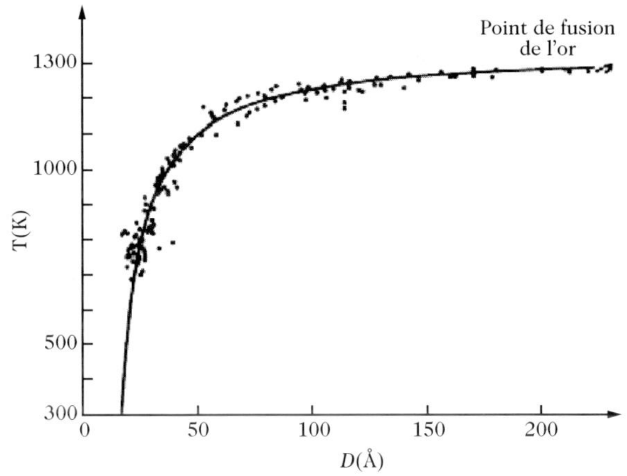

Fig. 10.4. Size dependence of the melting temperature of gold

At the surface, interatomic distances are generally observed to contract [7]. This is a consequence of the fact that bonds are broken at the surface. The contraction between surface atoms and atoms in the immediately underlying plane depends on the coordination number Z of the surface atoms, increasing as the coordination number of the atoms decreases. This has also been shown for monocrystals, e.g., in nickel, the contraction is close to 10% for the (110) face where $Z = 7$ [8], whereas it is only 1–2% for the close-packed (111) faces with $Z = 9$.

The adsorption of reactants will saturate any dangling bonds, which then means that interatomic distances will return to values closer to those measured in the bulk 3D crystal [7].

Effect of Size on Melting Temperature

The contribution of the surface to the energy of the system also tends to facilitate melting as the system dimensions decrease. The reduction in melting temperature has been observed experimentally in several cases [9] and has been predicted in many theoretical studies. The physical reason for this reduction can be explained qualitatively by the Lindemann criterion, which stipulates that a bulk material will have melted when fluctuations (due to the temperature) in interatomic distances reach a certain value, viz., about 10% of the lattice parameter. If this criterion is assumed to remain valid for nanosystems, it is clear that the many surface atoms, being less restricted in their thermal motions, will fluctuate more easily spatially, thereby lowering the melting temperature. For example, Buffat and Borel [9] have shown by electron diffraction that a 2-nm gold particle melts at 573 K, i.e., at a temperature 700 degrees lower than solid gold. The change in the melting temperature of gold with the particle size is illustrated graphically in Fig. 10.4.

This property is only very rarely considered with regard to the reactivity of small metal particles. However, it may lead to quite unexpected consequences in reactions occurring at high temperatures, carried out below and/or above the melting temperature. Hence, the different morphologies of carbon nanotubes formed by catalytic decomposition of polyethylene at 700°C and 800°C is explained by the fact that the nickel catalyst particles go from the solid to the liquid state between these two temperatures [10].

Another important consequence in catalysis is the greater mobility of smaller supported particles in the various heat treatments (calcination, reduction, reaction) of catalysts and hence a greater tendency to sintering, an undesirable effect in catalysis.

10.1.2 Electronic Properties

There are two different ways to assess the properties of clusters with dimensions between the simple atom and the macroscopic crystal. Small clusters are treated theoretically as large molecules and the techniques of quantum chemistry are brought to bear to describe their magnetic, optical and electronic properties. Larger particles are often treated as part of the bulk material, specific electronic properties being related to the large proportion of surface atoms and their positions at the surface of the nanoparticle.

Very small metal clusters exhibit characteristics that change radically and rapidly with size. As regards electronic properties, the first task is often to identify the size, or size range, at which the metal–insulator transition occurs. To answer this question, the energy difference between the highest occupied molecular orbital (HOMO) and the lowest unoccupied molecular orbital (LUMO) is traditionally considered to be analogous to the band gap separating valence and conduction bands in the bulk material. The reduction in the HOMO–LUMO energy difference down to a value close to zero is thus commonly taken to indicate the transition to the metallic state. Figure 10.5 compares values of the HOMO–LUMO energy difference deduced by calculation with experimental values for magnesium clusters [11]. A fast overall decrease is indeed observed in the HOMO–LUMO gap, even though this decrease is far from monotonic and depends to a large extent on the exact number of atoms contained within the structure.

For larger clusters, the variation in electronic characteristics becomes monotonic, and more importantly, more gradual. The overall properties now tend slowly toward those of the bulk solid, and understanding the electronic properties of active sites reduces to understanding the electronic structure of surface atoms. It is in fact the number of surface atoms and their lack of saturation, which vary one way or the other depending on the nanoparticle diameter, that will determine local electronic properties at the surface and the consequent chemical reactivity.

In the framework of the tight-binding model [12, 13], well suited to the transition metals, the width of the valence band falls off with the coordination

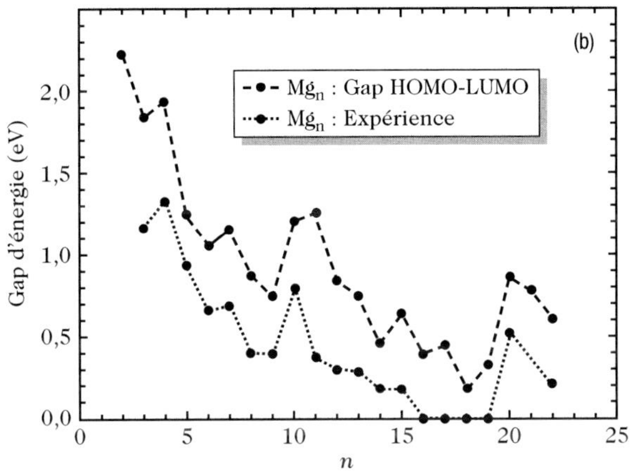

Fig. 10.5. Calculated and experimental values of the energy difference between the HOMO and LUMO as a function of the number of atoms contained in Mg clusters [11]

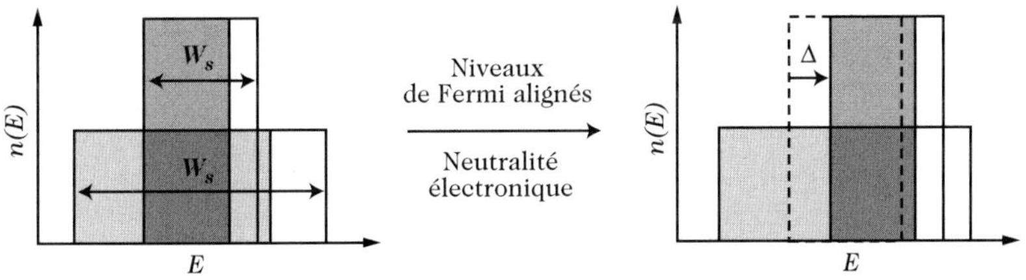

Fig. 10.6. Schematic representation of the narrowing of the surface band width (*left*) and shift of the surface valence band imposed by charge neutrality (*right*). The illustration is made for a relatively full d band, i.e., a metal at the end of the transition series

number Z in proportion to $\sqrt{Z}$. The valence band is thus narrower at the surface. This narrowing of the band width at the surface, coupled with the requirement of local electronic neutrality, leads to a shift in the surface valence band to align the Fermi level fixed by the density of states of the bulk atoms (see Fig. 10.6). The direction of this shift depends on the filling of the d band and it vanishes for a half-filled band.

It should also be noted that the fine structure of the valence band undergoes further modifications depending on the coordination number.

Local electronic densities are therefore modified by the different coordinations of surface atoms, leading to differences in reactivity. Indeed, as has been shown by several groups [14–16], the position of the centre of gravity of the band is an extremely important parameter as regards the interaction between the surface and the molecular orbitals of adsorbates, partly through the donation of electrons from filled orbitals of the molecules to the surface, and partly through back-donations of electrons from the solid to the antibonding orbitals of the molecule.

For example, on a platinum surface, the adsorption energy of molecules like CO, NO, oxygen, or hydrogen increases as the valence band shifts upward, i.e., as the coordination number decreases [16, 17].

More generally, the elementary processes involved in catalysis, i.e., adsorption, dissociation, recombination, and desorption (see the Appendix at the end of the chapter) will all depend on the coordination numbers of surface atoms.

10.1.3 Reactivity in Chemisorption and Catalysis of Monometallic Nanoparticles

We have seen how surface sites can vary with particle size, and thereby constitute specific sites with regard to chemical reactivity. It is when the particle dimensions are reduced to the nanoscale that the relative variation of the different sites becomes most marked. It is also in this size range that the physical and electronic properties of the particles vary the most. Specific chemisorption and catalytic properties can thus be expected in the 'nano' range of particle sizes. We shall see below how the coordination of surface atoms and their geometric configuration, together with the associated modifications in electronic properties, can predetermine these chemisorption and catalytic properties.

Reactivity of Free Clusters

Beams of transition metal clusters with very small dimensions have been produced in some laboratories, and their reactivity with small molecules measured. This reactivity depends on the number of atoms contained in the cluster. For example, the dependence of the rate of addition of hydrogen H_2 on free iron clusters varies considerably with the number of atoms contained in the particle (Fig. 10.7). The activity, zero below four or five atoms, increases quickly, stabilises, and then drops rapidly between Fe_{15} and Fe_{18} before assuming a high, stable value above Fe_{23} [18, 19]. The reactivity of iron clusters with water H_20 and ammonia NH_3 follow practically the same variations [18, 19].

Strictly speaking, one cannot use the term 'magic numbers', but one can say that there are significant structural changes in clusters around Fe_{14} and Fe_{15}, arising in such a way as to minimise the free energy of the cluster. It is also tempting to suggest a correlation between the reactivity and the ionisation potential in these small clusters (see Fig. 10.7). In this context, the reactivity of gold clusters with O_2 has been investigated both experimentally and theoretically [20–24]. The various authors cited have shown that the adsorption of O_2 on very small Au_n clusters is different from (less than) adsorption on Au_n^-, indicating that the charge increases the reactivity of the cluster with respect to oxygen. More precisely, the activation of O_2 is a consequence of charge transfer from the gold into the antibonding π^* orbital of the oxygen molecule. The binding energy of the oxygen with Au_n or Au_n^- thus depends

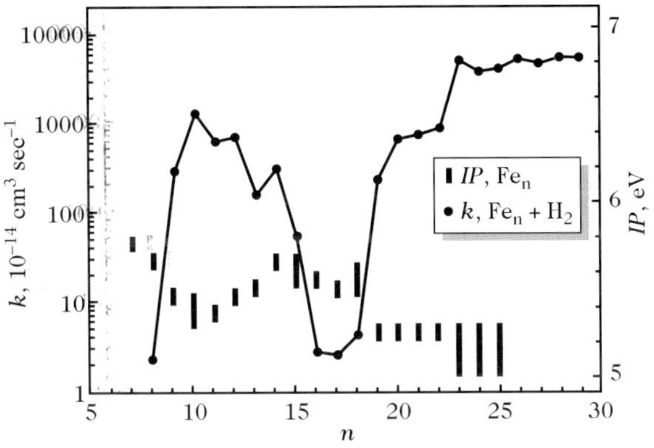

Fig. 10.7. Dependence of the hydrogen addition rate (k) and ionisation energy (IP) on the number of atoms in iron clusters [18]

on the parity of the number of electrons in the cluster. The binding is strong when the number of electrons is odd. A possible explanation [24] is that the cluster with an odd number of electrons has a smaller ionisation potential, thereby increasing the availability of the HOMO electrons. Other measurements carried out on Au_n^- ($n = 2, \ldots, 22$), have shown similar behaviour [22], also explained by a correlation between the weak electron affinity of the cluster and its high reactivity with oxygen. Naturally, addition of an electron to a small cluster with discrete and well separated electron energy levels can greatly affect the energy level of the HOMO. On the other hand, it should be noted that, for clusters comprising a hundred or more atoms, the gap between electron levels close to the Fermi level can be considered as a continuum and an additional electron will not significantly modify their Fermi levels. In any case, this shows, if such was necessary, that the reactivity and the electronic (and atomic) structure are related.

It is quite clear that free clusters cannot be used as such in catalysis, i.e., they must be supported if they are to be exploited as catalysts.

Reactivity of Monometallic Catalysts

Some work has been done on the reactivity of mass-selected supported metal particles with very small sizes. It has been shown that the chemisorption properties of supported clusters, in the size range from a few atoms to several tens of atoms, varies with the number of constitutive atoms [25, 26]. Their reactivity with regard to bond breaking (dissociation of CO [25, 26]) or bond formation (trimerisation of acetylene [27]) also varies significantly with the number of atoms making up the cluster. It seems as though the reactivity of very small clusters (at most a few tens of atoms), even supported, is related to their particular and distinct electronic properties for clusters containing a well-defined number of atoms.

However, the vast majority of industrial catalysts prepared in a more conventional way have sizes between one and a few nanometers, i.e., they comprise several hundred or several thousand atoms. We shall choose our examples from amongst these, considering metals supported on supports that are generally considered to be 'neutral', i.e., having no decisive effect on the properties of nanometric catalysts. It is then the electronic and structural properties of the surface atoms that control the reactivity.

Chemisorption Properties

Chemisorption without dissociation is an elementary step which provides a measure of the reactivity of a solid surface. It is also a first decisive step in the catalytic process: by destabilising the molecule to be transformed, the catalytic solid succeeds in lowering the energy barrier to be overcome in order to modify it and thereby accelerates the process at a given temperature (see the Appendix at the end of the chapter).

Furthermore, as we saw earlier, the whole chemisorption phenomenon (the nature of the adsorption site and the number of atoms it contains, the adsorption force, and so on) must depend on the structure of the surface sites and hence on the size of the nanoparticles. Apart from the reactivity itself, the energetic or spectroscopic characterisation of molecules adsorbed onto surfaces with given crystallographic orientation and onto particles with perfectly controlled morphology provides a way of assessing surface sites on conventional catalysts, exploiting the idea of a probe molecule.

Consider for example the case of CO adsorbed onto Pd, for which many experimental and theoretical studies have been carried out on monocrystals with various orientations and on well characterised particles deposited on different supports [28–32]. The adsorption energy of the CO falls rapidly when the Pd particle size increases from 1.5 to 5 nm, the size for which the values become comparable with those obtained for bulk Pd (see Fig. 10.8) [28, 29]. This is true not only for supported Pd particles made by the usual means for preparing catalysts on different supports, such as SiO_2, Al_2O_3, SiO_2–Al_2O_3, or TiO_2 (see Fig. 10.8 left), but also for model catalysts prepared by atomic beam on a plane monocrystalline MgO support (see Fig. 10.8 right).

IRRAS data (infrared reflection absorption spectroscopy) of CO adsorbed at room temperature on model catalysts made from small Pd particles deposited by atomic beam on a plane support [nanometric film of alumina obtained by oxidising an NiAl (110) monocrystal, see Fig. 10.9 left], and IRTF (infrared transmission) data on a powder catalyst (see Fig. 10.9 right) are very similar.

From the data obtained by vibrational spectrometry of the CO adsorbed on Pd monocrystals, and on the basis of detailed knowledge of the morphology of the Pd particles deposited on model supports, Freund et al. [31, 32] have suggested attributing each line to a different adsorption site: low frequency vibrations correspond to bridged and/or triply bound CO molecules adsorbed

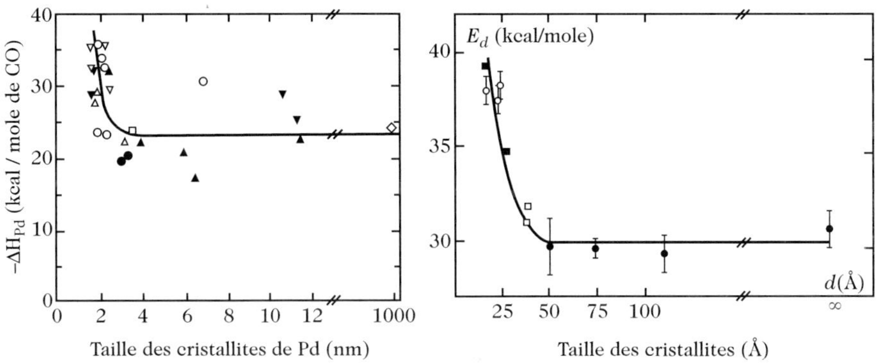

Fig. 10.8. Energy of adsorption of CO on supported Pd particles prepared chemically on various oxide supports (*left*) [28] and by atomic beam on MgO (100) (*right*) [29]

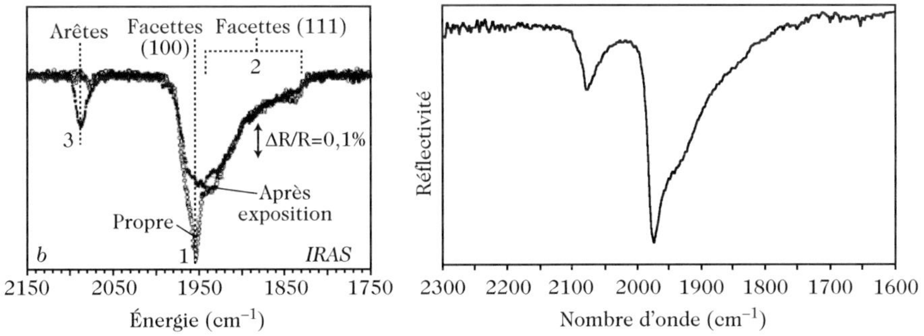

Fig. 10.9. *Left*: IRRAS spectrum of CO adsorbed at 300 K on clean Pd/Al_2O_3/NiAl (110) (*white circles*) and after exposure to methanol (*black circles*) [31, 32]. *Right*: IRTF spectrum of CO adsorbed on a Pd catalyst (0.5 wt%)/Si_3N_4 (average diameter 4–5 nm)

on (100) and (111) facets, while vibrations measured at high energies (2 060–2 080 cm^{-1}) would be associated with CO species adsorbed linearly on edges. It can be observed that the spectrum of chemisorbed CO on the supported Pd/Si_3N_4 catalyst, made from 4–5 nm particles, is similar to the spectrum measured on the model particles; the morphology of the Pd particles is thus the same. It is in fact the cubo-octahedral shape with (100) and (111) facets joined by edges and corners which dominates.

The species adsorbed on edges and/or corners are more stable and desorb at higher temperatures [30, 32]. This observation is in good agreement with the increased adsorption energy measured on very small particles (see Fig. 10.8), which include a large percentage of atoms with a small number of nearest neighbours (see Fig. 10.3), an increase in energy that is expected from theoretical calculations [16, 17].

The dependence of chemisorption properties on the coordination of surface atoms is manifest here. We shall now examine the consequences for catalytic processes, illustrating with several examples.

Catalytic Properties

We shall discuss here some examples of the particular reactivity of small metal particles in catalysis. Size reduction can prove useful for catalytic processes, by inducing better activity and/or better selectivity in the target product and/or better stability in the catalyst. On the other hand, it may also induce undesirable effects.

The higher reactivity of atoms with lower coordination (situated on edges or corners in the case of nanoparticles or on steps in the case of monocrystals) with regard to the breaking of chemical bonds, i.e., dissociation, has been demonstrated on many occasions [33], and is now treated as acquired knowledge. One may thus expect higher activity of smaller particles in various catalytic reactions involving bond breaking, such as:

- Synthesis of ammonia ($N_2 + 3H_2 \longrightarrow 2NH_3$) on Fe or Ru, a reaction in which the very strong N–N bond of N_2 must be broken.
- Production of H_2 and synthesis gas (mixture of CO and H_2) by decomposition of CH_4 in the presence of oxygen on Pt or Rh.
- Hydrogenolysis reactions (breaking of C–C bonds in hydrocarbons) on Ni.
- Generation of carbon nanotubes by catalytic decomposition of hydrocarbons or CO on Fe, Co or Ni.

The advantages of having such highly active sites are nevertheless offset by a common effect, namely the blocking of active sites (deactivation) by strongly adsorbed residues. Consider for example the decomposition of methanol (CH_3OH) on Pd. This is easy on small particles, but the deposited carbon remains adsorbed on the surface and thus blocks the highly reactive sites on edges [32]. This has been clearly demonstrated by IRRAS of adsorbed CO as probe molecule, where it is found that the CO no longer adsorbs onto edges when the catalyst has been exposed to methanol (see Fig. 10.9 left). Here is a sign of the conflict between the fact that a particular site can be highly reactive with respect to the target reaction, in this case decomposition, and the fact that it can be blocked by very strong preferential bonding with residues in the same decomposition, leading to deactivation of the catalyst.

To end, let us illustrate the size effect for palladium in hydrogenation reactions of unsaturated hydrocarbons, i.e., hydrogen addition reactions [34]. In Fig. 10.10, it can be observed that the activity of surface atoms with regard to hydrogenation can be independent of the particle size (as happens with olefins like butene-1), but it can also be strongly dependent on the dispersion (as happens with the alkynes and conjugated dienes such as butadiene-1,3). Once again, variations are particularly clear between 1.5 and 4 nm.

The selective hydrogenation of alkynes and dienes in the presence of olefins is in fact an important reaction in industry, used to eliminate impurities, e.g.,

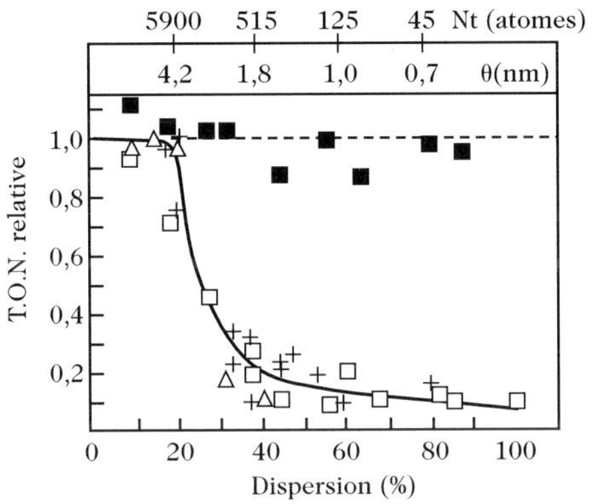

Fig. 10.10. Relative variation of the turn-over number (TON), i.e., the number of molecules transformed per surface atom and per unit time in given reaction conditions, as a function of the dispersion for hydrogenation of butene-1 (*black squares*), butyne-1 (*white squares*), butadiene-1,3 (*plus signs*), and isoprene (*white triangles*). θ is the particle size and N_t is the total number of metal atoms in a particle of size θ [34]

alkynes or dienes, present in small proportions in olefins produced by the petrochemical industry. Any hydrogenation of the target product, in this case the olefins, precursors for many polymers, should be avoided since saturated hydrocarbons resulting from total hydrogenation have almost no market value. Looking at the two graphs in Fig. 10.10, it is clear that the very small Pd particles will be less selective and hence less effective for this catalytic purification of the olefins. However, they may well prove more useful in other reactions.

10.2 Support Effects

The presence of a support can influence particle morphology. While cubo-octahedral particles are generally formed (see Fig. 10.11a), particles interacting with a support can assume different equilibrium shapes from those expected for the free nanoparticle of the same size (see Fig. 10.11b).

The interaction with the support depends on the metal–support pair. A priori, two effects are therefore expected, depending directly on this interaction:

- Electron transfer to or from the support, so that the particles have either an excess or a deficit of electrons and consequently have less tendency to accept or donate electrons, respectively.
- An epitaxial stress that can modify the particle structure (see Fig. 10.11), the lattice parameter, and/or the particle morphology.

These effects will be more marked for smaller particles and will of course modify their properties with regard to chemical reactivity. The support may

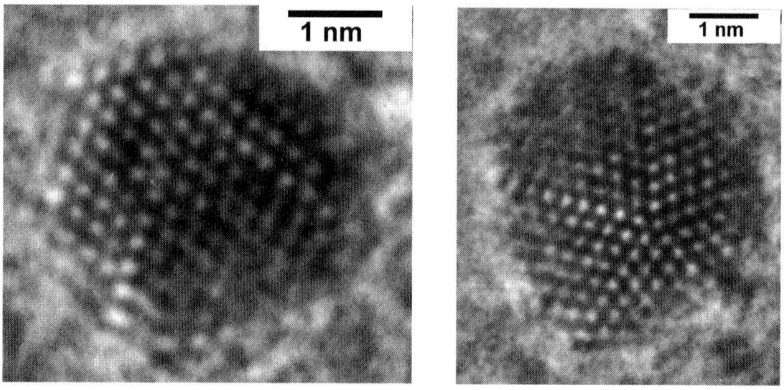

(a) (b)

Fig. 10.11. High-resolution TEM images of Pd particles deposited by atomic beam on amorphous carbon, illustrating the different shapes: (**a**) cubo-octahedral (predominant) and (**b**) icosahedral (with C_5 axis perpendicular to the support plane). Reprinted from [35] with the kind permission of Elsevier

also play a role in the catalytic process if it is itself involved in the reaction. This is bi-site catalysis.

These different influences can be used to explain the unexpected properties of small gold particles when catalysing the oxidation of CO at low temperatures, as observed by Haruta and summarised in the review article by Bond and Thomson [36]. It was at first thought that the observations could be explained by a specific property of the very small gold particles in contrast to the bulk metal. However, as we shall see below, for particles of the same size, the activity depends on the support.

In order to make direct observations of the contribution of the support to the system reactivity, clusters of the same size and the same morphology must be deposited on different supports. This can be done by laser vaporisation, for example (see below). Gold nanoparticles of average diameter 3 nm deposited on TiO_2, ZrO_2 and Al_2O_3 powders exhibit very different catalytic activities with regard to oxidation of carbon monoxide (see Fig. 10.12) [37].

It is sometimes suggested that electronic effects and morphological and/or structural changes in the gold particles induced by interaction with the support are what lead to differences of activity in this reaction. The modified gold is then assumed to be able to dissociate oxygen, a basic step required in the reaction and known to be impossible at the relevant temperatures when using bulk gold. However, the support itself can be invoked as the source of oxygen, the role of the gold being to 'activate' the CO (bi-site catalysis). Indeed, the results can be explained by the fact that TiO_2 (and to a lesser extent ZrO_2) are easily reduced and therefore capable of stocking or supplying oxygen during the catalytic process.

Note that this bi-site catalysis indirectly induces a size effect quite independently of the particular electronic properties of the clusters under consid-

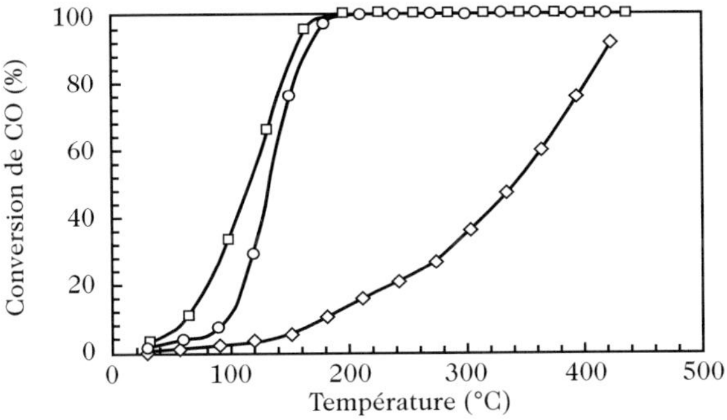

Fig. 10.12. Temperature dependence of the conversion percentage for oxidation of CO on gold catalysts: Au/Al_2O_3 (*diamonds*), Au/ZrO_2 (*circles*), and Au/TiO_2 (*squares*)

eration (see the last section). In this case the relevant parameter which can explain the variations in catalytic activity is simply the perimeter of the cluster where active sites for oxygen adsorption would be located, the CO being adsorbed on the gold [38].

In parallel with such interactions specific to the metal–support pair, a true chemical reaction may occur between support and particle under certain reaction conditions, leading to coating and/or compound formation (often restricted to the particle surface) in the place of the metal particle. This is the effect known as strong metal support interaction (SMSI) in catalysis. Such compounds generally form via surface migration of the support modified by the reactive medium on the metal nanoparticle. It can be reversible.

The best known classic example in catalysis concerns Pt supported on TiO_2. Annealing above 450°C, it is observed that the Pt is encapsulated by a layer of titanium oxide. The smallest particles are encapsulated most quickly. The SMSI phenomenon reduces the chemisorption capacity of the catalyst [39]. The formation of surface compounds has now been observed in many other cases. An example is the formation of transition metal silicides observed during processes with Ni or Pd catalysts supported on SiO_2 at high temperatures in a reducing medium. The fully or partially reduced silica becomes mobile and migrates across the particle.

10.3 Alloying Effects

In alloys, several further parameters must be taken into account: the surface composition, the relative positions of the two components, and the associated changes in electronic properties.

10.3.1 Effect of Surface Segregation

The surface composition is controlled by the effects of segregation of one of the partners on the surface, this happening so as to minimise the energy at thermodynamic equilibrium. It is usually the element with the lowest surface tension which segregates at the surface. This phenomenon is enhanced when the segregating element has the largest atomic radius. However, the phenomenon of surface segregation is moderated for A–B alloys with exothermic formation energy, i.e., those with

$$\Delta H_{\mathrm{A-B}} > \frac{1}{2}(\Delta H_{\mathrm{A-A}} + \Delta H_{\mathrm{B-B}}) ,$$

but enhanced for endothermic alloys with a demixing tendency [40].

Moreover, surface segregation effects are amplified on more 'open' faces, i.e., those composed of surface atoms with smaller coordination number. For example, in fcc systems, surface segregation effects are greater on less dense faces than on very close-packed (111) faces: (110) > (100) > (111) [40, 41]. On nanoparticles, one consequence of this effect is that an excess concentration of atoms that segregate towards the surface is expected at edges and corners. This is clearly demonstrated by Monte-Carlo-type statistical simulations applied to bimetallic nanoparticles in combination with a broken-bond thermodynamic model for extended surfaces (see Fig. 10.13) [42].

In the case of nanoparticles, the law of mass action must of course be taken into account, i.e., one must consider the fact that the atoms that segregate at the surface will be depopulated in the core of the particle. This phenomenon will become more important as the particle gets smaller, i.e., when the dispersion (ratio of surface atoms to total number of atoms) is high.

Segregation phenomena sometimes suffice to explain the reactivity of bimetallic catalysts. For example, two important results have been demonstrated from studies of the hydrogenation of toluene and tetralin [43] in the presence of sulfur on PdPt catalysts with different compositions (Pd_{100}, Pt_{100}, $Pd_{65}Pt_{35}$, $Pd_{17}Pt_{83}$) combined with low energy ion back-scattering measurements (a technique for identifying surface composition) and Monte Carlo simulations (see Fig. 10.13). Firstly, the activity of $Pd_{65}Pt_{35}$ is comparable with that of pure palladium. Given that the alloy $Pd_{65}Pt_{35}$ only has Pd atoms in the first layer, as shown by low energy ion back-scattering, but with Pt nearest neighbours located in the underlying layer, any electronic effect of the platinum on the palladium can be excluded. Concerning the catalyst $Pd_{17}Pt_{83}$, which has 50% Pd at the surface, its reactivity is intermediate between that of Pd and that of Pt. The results can be perfectly well explained by the additivity of the catalytic properties of the corresponding pure metals, weighted by the respective surface concentrations of the two constituent elements.

The example of reforming methane on Au–Ni catalysts illustrates the consequences of the specific localisation of certain elements on edges and corners. This hydrogen production reaction, in which water and methane are brought

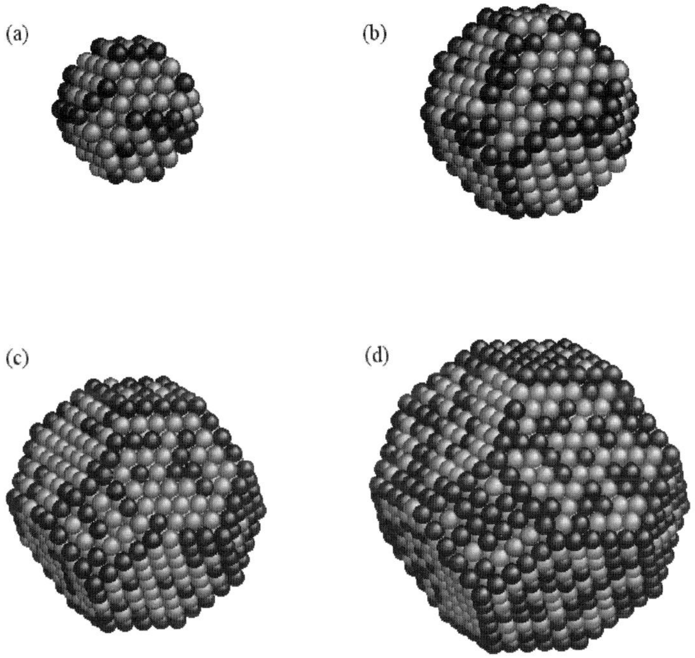

Fig. 10.13. Monte Carlo simulations of bimetallic clusters $Pd_{17}Pt_{83}$ containing (**a**) 201, (**b**) 586, (**c**) 1289 and (**d**) 2406 atoms. *Dark grey* and *light grey spheres* represent Pd and Pt atoms, respectively [42]

into contact at high temperature to give CO_2 and H_2, is catalysed by Ni. However, a carbon-bearing deposit is often observed to form, deactivating the catalyst. The carbon-bearing deposits nucleate at corners and edges of the nickel particles [44]. Adding gold to the nickel makes the catalyst much more resistant to carbon deposits [44]. In fact, the gold segregates at the surface mainly on the weakly coordinated sites, i.e., edges and corners and 'blocks' the sites responsible for deactivation of the catalyst [45,46].

In other cases, in fact most cases, the catalytic modifications observed by associating a second metal can also be related to effects that are readily divided into geometric and electronic effects. However, electronic and geometric influences can only rarely be completely separated as independent parameters. For example, increasing the size of a metal particle causes a broadening of the valence band, but also modifies the structure of the planes and the surface topology.

10.3.2 Geometric Effects

Geometric effects are related on the one hand to dilution of an active metal by a less active metal with respect to some given reaction (dilution effect) and on the other hand to structural changes, such as the appearance of an ordered arrangement of the partners at the surface (surface structures) and/or surface reconstructions.

The dilution effect can change the distribution of sites on the surface compared with the distribution for pure metals (ensemble effect), but it can also modify the interactions between molecules adsorbed on the surface. The first work showing the existence of dilution effects dates back a long way [47, 48]. It was shown that certain reactions require several adjacent atoms in order to proceed. This geometric model has recently been revived. Consider for example the work by Martin et al. [49] who showed that the dilution of Ni atoms by Cu atoms leads to a considerable reduction of activity, as compared with the results for pure Ni, in the hydrogenolysis of saturated hydrocarbons and the hydrogenation of benzene. These authors also show that three adjacent Ni atoms are needed for the latter reaction. Likewise, Sachtler et al. [50] observed that the substitution of Au atoms for Pt atoms at the surface increases the isomerisation of n-hexane and decreases aromatisation and hydrogenolysis reactions of n-hexane.

One consequence of this ensemble effect is to eliminate all catalytic reactions that require a large number of contiguous atoms of the same kind as active site, while other reactions are considerably less affected. The ensemble effect can thus be decisive for controlling the selectivity of a catalytic reaction.

10.3.3 Electronic Effects

This heading refers to electronic modifications induced either by interaction between the two metals, e.g., formation of a chemical bond between the two partners, charge transfer, polarisation, etc., or by the stress exerted on the catalytically relevant metal by its partner. Once again, these influences will modify the valence orbitals of the surface sites of the catalyst, and hence also the interactions between the molecular orbitals of the reactants and the reaction products.

Most fundamental studies regarding electronic effects are carried out on alloy monocrystals or on metal monolayers deposited on a metallic substrate. In this case, the electronic properties of surface atoms, the main actors in catalytic reactions, can be probed and dilution effects neglected.

For example, Rodriguez et Goodman [51] have carried out parallel investigations of electronic and chemisorption (with respect to CO) properties of bimetallic systems obtained by depositing Pd on another transition metal, viz., Ta(110), W(110), Ru(0001), Mo(110), Pt(111), Re(0001), and Rh(100). All these deposits display pseudomorphic growth, in which deposited atoms adopt the lattice parameter of the substrate. The study shows that the increased binding energy of the $3d_{5/2}$ core levels of the deposited atoms, measured by XPS, is accompanied by a decrease in the CO desorption temperature, i.e., a decrease in the CO adsorption energy. On the other hand, an increase in the CO desorption temperature is observed for a fall in binding energy of the core levels of the deposited atoms. Note that a modification of the binding energy in the core levels of an atom is in fact the signature of a modification in its valence levels [52]. The electronic structures of metal atoms deposited on another metal can thus be altered, whereupon they will display different chemisorption properties. However, it would seem difficult in this case to separate electronic effects due to stress and ligand effects between the partners.

Some experimental [53, 54] and theoretical [55–57] work nevertheless suggests that, even without the ligand effect, stress alone has an influence on the adsorption of CO. Variations due to tensional stress would appear to be associated with a narrowing of the band width when the distance between atoms decreases, followed by a change in the centre of gravity of the band, and hence a change in interactions with the reactants (see Sect. 10.1.2). The opposite happens for a compression stress.

All the above considerations take into account the relative variations in the energy position of the centre of gravity of the valence band with respect to the Fermi level, in order to predict or explain the differences in chemical reactivity as compared with the pure metals. However, in absolute terms, the Fermi level of the alloy differs from that of the pure metals and this energy parameter must also be taken into account, in addition to the relative variations of the electronic band structure. This is a difficult problem and often neglected.

As we have just seen, it is no easy matter to distinguish the various alloying effects (dilution effects and electronic modifications induced by ligand effects and stresses) occurring on model monocrystalline surfaces. The direct application to real supported catalysts is greatly complicated by the multiplicity of sites, not to mention in many cases a heterogeneity of size and composition and an interaction with the support that cannot be neglected.

10.4 Preparation and Implementation in the Laboratory and in Industry

There are many ways of preparing supported metallic catalysts, with a view to producing small particles. One can distinguish model catalysts and industrial catalysts.

Model catalysts are generally obtained by atom beams on plane supports, particle sizes and morphologies being governed by the general laws of nucleation and growth, i.e., the nature and structure of the substrate, the presence of defects, temperature, etc. [30, 32, 58]. There have been several conclusive trials with direct deposition of clusters previously formed in the effusion chamber of a plasma, obtained for example by laser vaporisation from a mono- or bimetallic rod. The main interest of this technique is in preparing alloys, insofar as the deposited nanoparticles, with average dimensions around 3 nm, will have exactly the composition of the evaporated rod of alloy (see Fig. 10.14) [45].

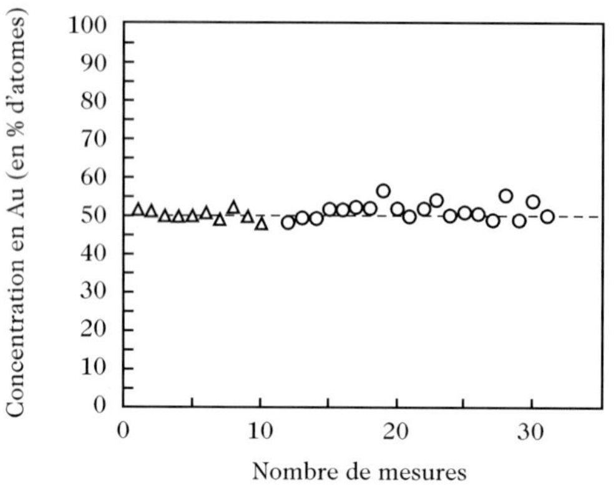

Fig. 10.14. Concentration of gold in a collection of Au–Ni clusters produced by laser vaporisation from an $Au_{50}Ni_{50}$ rod (in % of atoms). *Triangles* and *circles* correspond to analysis of broad regions or individual particles, respectively [45]

Industrial metallic catalysts are generally composed of nanoparticles deposited on a powdered support with high specific area. While these are often used as such in (packed bed) laboratory reactors, they are usually prepared in some specific way for use in industrial catalysis. They can be produced in the form of pellets, or microbeads for use in fluidised bed reactors, or they can be smeared on the walls of a tubular reactor (e.g., for use as a catalytic burner in energy production or as a catalytic converter for car exhaust).

There are many ways of preparing catalysts. We shall not give an exhaustive description of the techniques used, but simply list the main methods. We shall also mention several modern approaches that could supplant current techniques in the future. Chemical channels remain the most common for preparing metallic catalysts. They are generally implemented in solution. Examples are:

- Impregnation, which involves thermal decomposition of precursors (salts, organic and organometallic compounds) deposited by evaporation of the solvent on a support. The latter do not interact strongly with the support. Particle formation is governed by the temperature and mode of decomposition of the precursors, and the diffusion of complexes and/or surface atoms and their meeting, i.e., nucleation and growth.
- Exchange, a method in which the support is 'functionalised', i.e., specific anchoring sites are created on the surface, at which the precursor can fix itself relatively firmly before being decomposed to leave only the metal on the surface.
- Sol–gel methods.

Techniques for preparation in the gas phase, such as CVD, follow the same principles but are much less widely used.

Although techniques for preparing catalysts have barely evolved over the past few years, knowhow has improved enormously, notably through in situ observation of the various stages using different spectroscopic techniques. For example, it is known how to fabricate very small (2–5 nm) supported gold nanoparticles, something which could not be done only a decade ago. As we have already seen, gold nanoparticles deposited on certain supports, such as TiO_2, Fe_2O_3, and others, have proved to be the best catalysts for oxidising carbon monoxide at low temperatures, i.e., at around room temperature. Note here that this reaction is extremely useful for removing traces of CO in confined spaces or in energy production vectors, such as in hydrogen used for fuel cells where CO is an inhibitor.

Despite this, there are some new trends in the definition of catalysts. Consider first the application of combinatorial methods. The best catalysts are in fact produced by combining the wide range of parameters we have been discussing up to now – size, support, alloying and/or additive effects – which means that their optimisation is complex and time-consuming. Combinatorial catalysis which associates fast preparation of various combinations, rapid performance tests and statistical analysis of results should make it possible to determine effective catalytic phases on a much shorter time scale. This technique nevertheless remains empirical and should not be allowed to replace the rational search for new catalytic phases on the basis of ideas put forward by fundamental research.

Lithographic techniques originating in the microelectronics industry are also envisaged now for preparing particle lattices with catalytic properties [59]. For example, on lattices of Ni, Fe, or Co particles, carbon nanotubes can be grown by catalytic decomposition of CO or hydrocarbons. These nanotubes are sources of electronic emission under strong electric fields, with potential applications in the fabrication of flat screens and in multiprobe nanolithography [60].

Appendix: Schematic View of the Catalytic Process

The role of a heterogeneous catalyst is to lower the activation energy barrier that must be overcome to carry out a chemical reaction. It will thus accelerate the process, i.e., make the reaction more likely at lower temperatures. After the reaction, the catalyst regenerates without modification, otherwise there is deactivation.

There are many types of catalytic reaction and various reaction mechanisms. Consider for example the Eley–Rideal mechanisms, in which an adsorbed atom reacts directly with a gas molecule or very weakly adsorbed molecule; or the Langmuir–Hinschelwood mechanism, in which the various reactants are chemisorbed. This is the most common mechanism, where the catalytic process involves chemisorption of the reactants, the meeting of adsorbed species, the reaction itself, and the desorption of the products. If the reaction is to proceed in the best possible conditions, the reactants must be

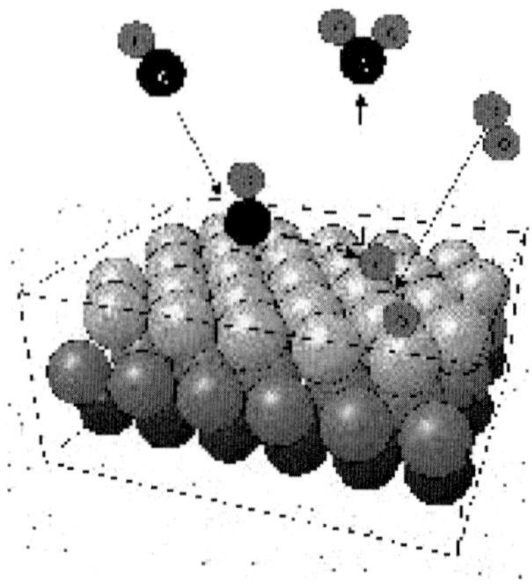

Fig. 10.15. Oxidation of carbon monoxide

adsorbed strongly enough to be 'activated', but not too strongly, because they would then be too stable to react. Desorption is of course easier if the product is only weakly coupled to the surface.

As an example, consider the oxidation of carbon monoxide (see Fig. 10.15):

$$CO + \frac{1}{2}O_2 \longrightarrow CO_2 .$$

This is a simple but important reaction for eliminating CO from various environments. It is extremely exothermic and catalysed by metals at the end of the transition series, such as Rh, Pd, and Pt, but also by gold in the form of nanoparticles supported on certain oxides.

Oxidation of carbon monoxide requires:

- associative chemisorption of carbon monoxide,
- dissociative (bond breaking) adsorption of the molecule O_2,
- diffusion of the adsorbates CO_{ads} and O_{ads} towards each other, until they meet,
- insertion of an oxygen atom in the CO, which is the target reaction,
- desorption of the carbon dioxide CO_2.

References

1. M. Che, C.O. Bennett: Adv. Catal. **36**, 55 (1989)
2. C.O. Bennett, M. Che: J. Catal. **120**, 293 (1989)
3. V. Ponec, G.C. Bond: Studies in surface science and catalysis, in: *Catalysis by Metals and Alloys*, ed. by B. Delmon and J.T. Yates, Elsevier 1995
4. G.A. Martin: Catal. Rev. Sci. Eng. **30**, 519 (1988)
5. F. Baletto, R. Ferrando, A. Fortunelli, F. Montalenti, C. Mottet: J. Chem. Phys. **116**, 3856 (2002)

6. R. Van Hardeveld, F. Hartog: Surf. Sci. **15**, 189 (1969)
7. G. Bergeret, P. Gallezot: Z. Phys. D – Atoms, Molecules and Clusters **12**, 591 (1989)
8. Y. Gauthier, R. Baudoing, Y. Joly, C. Gaubert, J. Rundgren: J. Phys. C **17**, 4547 (1984)
9. P. Buffat, J.P. Borel: Phys. Rev. A **13**, 2287 (1976)
10. F. Kukoviltsky, S.G. L'vov, N.A. Sainov, V.A. Shustov: Appl. Surf. Sci. **215**, 201 (2003)
11. J. Jellinek, P.H. Acioli: J. Phys. Chem. A **106**, 10919 (2002)
12. M.C. Desjonquères, F. Cyrot Lackmann: J. Phys. (France) **36**, L45 (1975)
13. G. Tréglia: In: *Catalysis by Metals*, ed. by A.J. Renouprez and H. Jobic, Springer 1996
14. J.A. Rodriguez, D.W. Goodman: Science **257**, 897 (1992)
15. B. Hammer, Y. Morikawa, J.K. Norskov: Phys. Rev. Lett. **76**, 2141 (1996)
16. B. Hammer, O.H. Nielsen, J.K. Norskov: Catal. Lett. **46**, 31 (1997)
17. B. Hammer, J.K. Norskov: Adv. in Catal. **45**, 71 (2000)
18. E.K. Parks, B.H. Weiller, P.S. Bechthold, W.F. Hoffman, G.C. Nieman, L.G. Pobo, S.J. Riley: J. Chem. Phys. **88**, 1622 (1988)
19. S.J. Riley: Z. Phys. D **12**, 537 (1989)
20. B.E. Salisbury, W.T. Wallace, R.L. Whetten: Chem. Phys. **262**, 131 (2000)
21. A. Sanchez, S. Abbet, U. Heiz, W.D. Schneider, H. Hakkinen, R.N. Barnett, U.J. Landman: J. Phys. Chem. A **103**, 9573 (1999)
22. W.T. Wallace, R.L. Whetten: J. Am. Chem. Soc. **124**, 7499 (2002)
23. B. Yoon, H., Häkkinen, U. Landman: J. Phys. Chem. A **107**, 4066 (2003)
24. G. Mills, M.S Gordon, H. Metiu: J. Chem. Phys. **118**, 4198 (2003)
25. U. Heiz: Appl. Phys. A: Mater. Sci. and Proc. A **67**, 621 (1998)
26. U. Heiz, F. Vanolli, A. Sanchez, W.D. Schneider: J. Am. Chem. Soc. **120**, 9668 (1998)
27. U. Heiz, S. Abbet, H. Hakkinen, U. Landman: Materials Research Society Symposium Proceedings **648** (Growth, Evolution and Properties of Surfaces, Thin Films and Self-Organized Structures) P9.1/1–P9.1/10 (2001)
28. P. Chou, M.A. Vannice: J. Catal. **106**, 17 (1987)
29. C.R. Henry, C. Chapon, C. Goyhenex, R. Monot: Surf. Sci. **272**, 283 (1992)
30. C.R. Henry: Surf. Sci. Rep. **31**, 231 (1998)
31. J. Freund, M. Bäumer, J. Libuda, T. Risse, G. Rupprechter, S. Shaikhutdinov: J. Catal. **216**, 223 (2003)
32. J. Freund, J. Libuda, M. Bäumer, T. Risse, A. Carlsson: Chemical Record **3**, 181 (2003)
33. R.C. Egeberg, S. Dahl, A. Logadotir, J.H. Larsen, J.H. Norskov, I. Chorkendorff: Surf. Sci. **491**, 183 (2001)
34. J. Boitiaux, M. Cosyns, S. Vasudevan: Appl. Catal. **6**, 41 (1983)
35. B. Tardy, C. Noupa, C. Leclercq, J.C. Bertolini, A. Hoareau, M. Treilleux, J.P. Faure, G. Nihoul: J. Catal. **129**, 1 (1991)
36. G.C. Bond, D.T. Thomson: Catal. Rev. Sci. Eng. **41**, 319 (1999)
37. S. Arrii, F. Morfin, A.J. Renouprez, J.L. Rousset: J. Am. Chem. Soc. **126**, 1199 (2004)
38. M. Haruta, M. Daté: Appl. Catal. A **222**, 427 (2001)
39. F. Pesty, H.P. Steinbrück, T.E. Madey: Surf. Sci. **359**, 83 (1995)
40. J.L. Rousset, J.C. Bertolini, P. Miegge: Phys. Rev. B **53**, 4947 (1996)

41. S. Helfensteyn, J. Luyten, L. Feyaerts, C. Creemers: Appl. Surf. Sci. **212–213**, 844 (2003)
42. J.L. Rousset, A.J. Renouprez, A.M. Cadrot: Phys. Rev. B **58**, 2150 (1998)
43. J.L. Rousset, L. Stievano, F.J. Cadete Santos Aires, C. Geantet, A.J. Renouprez, M. Pellarin: J. Catal. **197**, 335 (2001); ibid. J. Catal. **202**, 163 (2001)
44. H.S. Bengaard, J.K. Nørskov, J. Sehested, B.S. Clausen, L.P. Nielsen, A.M. Molenbroek, J.R. Rostrup-Nielsen: J. Catal. **209**, 365 (2002)
45. J.L. Rousset, F.J. Cadete Santos Aires, B.R. Sekhar, P. Mélinon, B. Prevel, M. Pellarin: J. Phys. Chem. B **104**, 5430 (2000)
46. A.M. Molenbroek, J.K. Norskov, B.S. Clausen: J. Phys. Chem. B **105**, 5450 (2001)
47. N.I. Kobozev: Acta Physicochim. URSS **9**, 1 (1938)
48. O.M. Boronin, V.S. Poltorak: Int. Chem. Eng. **7**, 452 (1967)
49. G.A. Martin, J.A. Dalmon: J. Catal. **75**, 233 (1982)
50. J.W.A. Sachtler, G.A. Somorjai: J. Catal. **81**, 77 (1983)
51. J.A. Rodriguez, D.W. Goodman: J. Phys. Chem. **95**, 4196 (1991)
52. D. Spanjaard, C. Guillot, M.C. Desjonquères, G. Tréglia, J. Lecante: Surf. Sci. Rep. **5**, 1 (1985)
53. E. Kampshoff, E. Hahn, K. Kern: Phys. Rev. Lett. **73**, 704 (1994)
54. J.C. Bertolini: Applied Catalysis **191**, 15 (2000)
55. M. Mavrikakis, B. Hammer, J.K. Norskov: Phys. Rev. Lett. **81**, 2819 (1998)
56. A. Ruban, B. Hammer, P. Stolze, H.L. Skriver, J.K. Norskov: J. Mol. Catal. A **115**, 421 (1997)
57. V. Pallassana, M. Neurock: J. Catal. **191**, 301 (2000)
58. A.K. Santra, D.W. Goodman: J. Phys.: Condensed Matter **15**, 31 (2003)
59. A.S. Eppler, G. Rupprechter, G.A. Somorjaï: J. Phys. Chem. B **101**, 9973 (1997)
60. V.I. Merkulov, M.A. Guillorn, D.H. Lowndes, M.L. Simpson: Appl. Phys. Lett. **79**, 1178 (2001)

11

Inverse Systems – Nanoporous Solids

J. Patarin, O. Spalla, and F. Di Renzo

11.1 Introduction

Many natural materials are characterised by an inorganic framework, generally negatively charged, containing cavities, cages, or tunnels in which inorganic (charge-balancing) cations and/or water molecules are occluded. Among these materials, the zeolites form one large family of crystalline porous materials (from the Greek *zein* meaning 'to boil' and *lithos* meaning 'stone'). Pore sizes in these aluminosilicates are generally of nanometric order. Because of their specific properties, the synthesis of zeolites, and more generally, zeolitic materials (zeolites and related solids) has considerably increased over the last few years. Indeed, applications are many and varied. They are relevant not only to the chemical industry (or more precisely, the petrochemical industry), but also to our everyday lives (phosphate-free washing powders, double-glazing insulation, and many others). By virtue of their porous structure and the mobility of the cations and water molecules occluded within their porous structures, these materials can be used as highly selective cation exchangers and adsorbents.

A major step forward in the synthesis of porous materials was obtained by introducing self-assembling organic entities (surfactant micelles) into reaction gels. This was the step from crystalline microporous materials to ordered mesoporous materials.

After discussing the nomenclature of porous materials in Sect. 11.2, the main families of porous solids are described in Sects. 11.3–11.5, paying particular attention to solids with ordered pore systems such as zeolites and ordered mesoporous solids.

11.2 Nomenclature: The Main Families of Porous Materials

Depending on the pore size, there are three main categories of porous solid. These are illustrated in Fig. 11.1 together with a number of examples. The

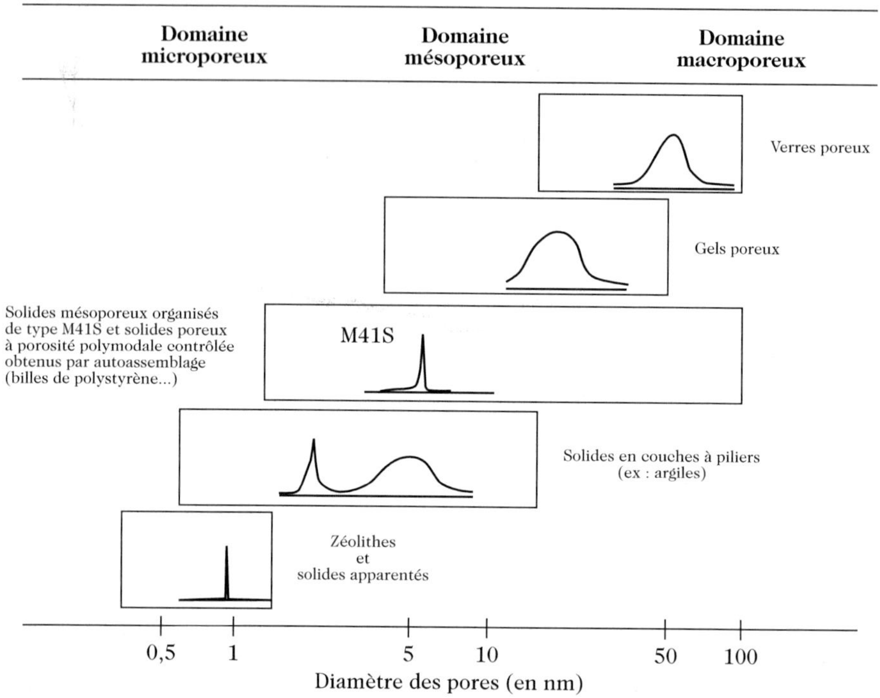

Fig. 11.1. Definition of micro-, meso- and macroporous materials with representative examples and pore size distribution. Adapted from [3]

mesoporous solids, with pore sizes between 2 and 50 nm according to the IUPAC classification [1], form the intermediate category (whence the name 'mesoporous') between the microporous solids (pore diameter < 2 nm) and the macroporous solids (pore diameter > 50 nm). All these solids, in which pore sizes vary between the nanometer and a few tens of nanometers, can be grouped together in the family of nanoporous solids.

In the present chapter, we shall be concerned primarily with the crystalline nanoporous solids with their 3D inorganic framework, and more particularly with the zeolites and ordered nanoporous solids of the M41S family [2]. Indeed, these two families of porous solids are the only ones to have very narrow pore size distributions. We shall then describe other solids with less ordered nanopore systems that are commonly used in industry.

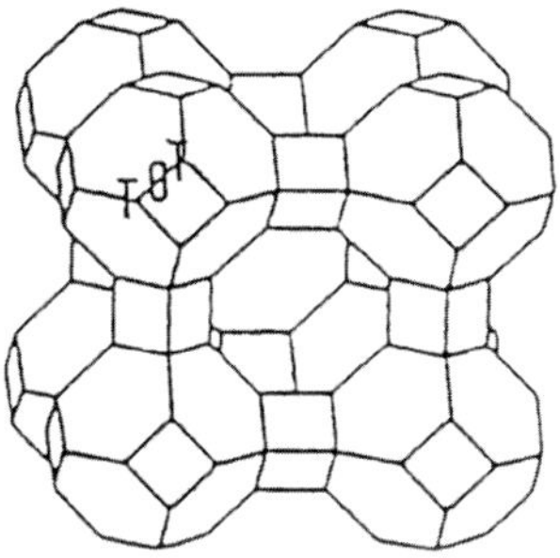

Fig. 11.2. Structure of the LTA (Linde type A) zeolite. In this representation, the T elements (Si, Al) of the framework are located at the intersection of the edges, while oxygen atoms are located in the middle of the edges. This structure constitutes a cage (LTA cage) in which apertures measuring 0.41 nm are marked out by eight T atoms. Cations (balancing the negative charge of the inorganic framework) and water molecules occluded within the porous structure are not shown

11.3 Zeolites and Related Microporous Solids. Definition and Structure

The word 'zeolite' refers to natural crystalline aluminosilicates.[1] Their inorganic framework (Si–O–Al) can be described as an assembly of tetrahedra TO_4, where T stands for Si or Al, in which the vertices are occupied by oxygen atoms and each oxygen atom is shared by two tetrahedra. These assemblies determine channels and cavities of molecular dimensions, with precisely defined sizes varying from 0.3 to 0.8 nm and connected with the surrounding medium. Water molecules and cations (alkali metals and alkaline earth metals), present within the porous structure, balance the negative charge of the aluminosilicate inorganic framework.[2] As an example, the structure of the LTA (Linde type A) zeolite is shown in Fig. 11.2.

Quite generally, the chemical formula of these aluminosilicates can be expressed in oxide form by

$$M_{2/n}O\,,\ Al_2O_3\,,\ x SiO_2\,,\ y H_2O\,,$$

[1]Zeolites are crystalline products. The atoms making up the inorganic structure are perfectly ordered and form rows and planes defining a pattern built up from the unit cell which is repeated in all three space directions and satisfies specific rules of symmetry. Crystalline solids have the particularity of producing a well-defined diffraction pattern when illuminated by an X-ray beam, comprising a series of peaks and troughs that provide a sort of identity card for the given material.

[2]The aluminium atoms do not exactly balance the negative charge carried by the oxygen atoms around them. The inorganic structure (framework or skeleton) of the zeolites is thus negatively charged and electrical neutrality of the structure as a whole is achieved by the presence of alkali cations, e.g., sodium, potassium, or alkaline earth cations, e.g., calcium, magnesium, occluded within the structure.

where M is one or more cations of valence n, and x is a number greater than or equal to 2.

The introduction of organic species as templates in reaction media has opened the way to synthesising a great many new zeolitic structures, and even a whole new family of materials, isostructural[3] or otherwise with respect to the zeolites. This is the case in particular of the microporous metallophosphates, e.g., aluminophosphates with inorganic framework Al–O–P (T = Al, P), gallophosphates with framework Ga–O–P (T = Ga, P), ferrophosphates with framework Fe–O–P (T = Fe, P), and so on.

Today, the word 'zeolite' is no longer restricted to the aluminosilicates, but refers to any silica-based crystalline microporous solid in which some of the silicon is replaced by other T elements such as the trivalent elements T = Al, Fe, B, Ga, etc., or tetravalent elements T = Ti, Ge, etc. Depending on the molar ratio Si/T, the following terms are used:

- zeolites $Si/T < 500$,
- zeosils $Si/T > 500$.

Zeosils and clathrasils[4] form the two classes of silicic materials in the porosil family.

Phosphate-type crystalline microporous solids are commonly called related microporous solids. A relatively exhaustive review of these microporous phosphate materials can be found in [4].

Quite generally, the microporous solids are thus characterised by:

- a 3D framework resulting from the stringing together of tetrahedral units TO_4 with shared oxygen vertices (T = Si, Al, P, Ge, Ga, etc.);
- channels and/or cavities with molecular-sized cross-sections, connecting with the surrounding medium.

Real structures may deviate from this ideal definition by the presence of units with non-tetrahedral coordination, e.g., TX_4, TX_5, or TX_6 polyhedra, where X = O, F, or non-bridging neighbouring units, e.g., T–OH end groups.

A code comprising three capital letters (structure code), defined by the structure commission of the International Zeolite Association (IZA), is attributed to each structure obtained in this way [5]. As an example, the zeolite faujasite and its synthetic equivalents X and Y correspond to the FAU structure type. At the present time, counting all compositions of the inorganic framework, there are 167 different structure types [6]. Table 11.1 gives several examples of crystalline microporous solids with their structure codes, the dimensionality of their channel network and their pore sizes. The latter is usually determined by the number of T elements (T = Si, Al, P, Ga, etc.) defining the aperture.

---

[3] Having the same structure.

[4] Clathrasils are silicic materials containing cavities or cages but in which apertures are too small to allow access to small molecules (diameters less than 0.3 nm).

Table 11.1. Examples of crystalline microporous solids, giving the dimensionality of their channel network and pore sizes

Name of material	Structure code	Element T in inorganic framework	Dimensionality of porous system	Aperture of largest pores [nm] (number of T atoms per aperture)
Faujasite	FAU	Si, Al	3D	0.74 (12)
Zeolite β	BEA	Si, Al or Si	3D	0.67 (12)
Zeolite A	LTA	Si, Al or Al, P or Ga, P	3D	0.41 (8)
UTD-1F	DON	Si or Si, Al	1D	0.82 (14)
Cloverite	-CLO	Ga, P	3D	1.32 (20)
Chabazite	CHA	Si, Al or Al, P or Ga, P	3D	0.38 (8)
Cancrinite	CAN	Si, Al	1D	0.59 (12)
ZSM-5 (silicalite-1)	MFI	Si, Al Si	3D	0.56 (10)

For a given zeolitic structure, the composition of the inorganic framework can vary. Hence, in Table 11.1, the MFI (Mobil FIve) structure can be obtained in aluminosilicate form (Si–O–Al framework, zeolite ZSM-5) or in silicic form (Si–O–Si, silicalite-1). This situation leads to isostructural materials which nevertheless display different characteristics. The aluminosilicate framework (ZSM-5) has an affinity for water molecules and is said to be hydrophilic, whereas silicalite-1 can, depending on conditions of synthesis, be totally hydrophobic, i.e., water hating. In Chap. 30, we shall see that this property may be put to use in the preparation of molecular springs.

11.4 Ordered Mesoporous Solids

A new class of mesoporous silicates and aluminosilicates called M41S has been synthesised by introducing into reaction gels, not individual organic species, but entities able to self-assemble, viz., surfactant micelles[5] [2]. One of the main phases of this family is MCM-41, where MCM stands for Mobil Composition of Matter. This is characterised by an almost regular hexagonal arrangement of cylindrical mesopores, with an extremely narrow pore size distribution. In contrast to the crystalline microporous solids (zeolites and related solids), the

[5]Surfactant molecules, i.e., molecules of a substance capable of altering the surface tension, can group together to form micelles, which are clusters of molecules minimising exposure to the solvent of the part of the molecule that is least soluble.

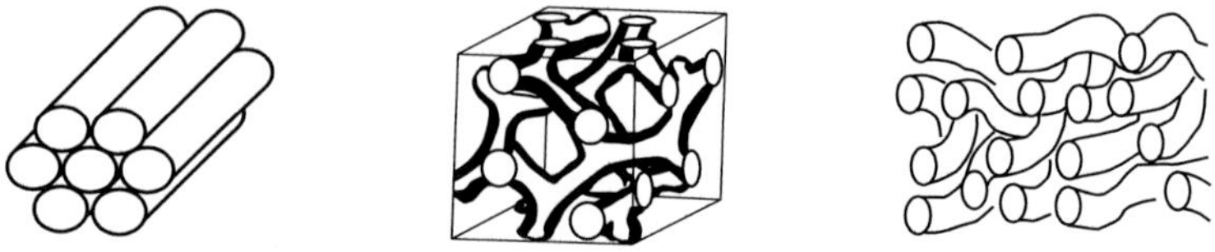

Fig. 11.3. Different types of porosity observed in ordered mesoporous materials. *Left*: Hexagonal arrangement of pores (material of type MCM-41). *Centre*: Cubic arrangement of pores (material of type MCM-48). *Right*: Wormlike arrangement of pores (material of type MSU)

Table 11.2. A selection of ordered mesoporous solids and their most common names

Name of material	Meaning of name	Symmetry of channel system (see footnote 1 and Fig. 11.3)
MCM-41	Mobil Composition of Matter no. 41	Hexagonal
MCM-48	Mobil Composition of Matter no. 48	Cubic
SBA-1	Santa BArbara no. 1	Cubic
SBA-3	Santa BArbara no. 3	Hexagonal
SBA-15	Santa BArbara no. 15	Hexagonal
MSU-x	Michigan State University	None (wormlike pore system)
HMS	Hexagonal Mesoporous Silica	None (wormlike pore system)

inorganic walls of this new class of materials are amorphous.[6] The M41S family also includes among its members a phase with cubic symmetry called MCM-48, and lamellar phases such as MCM-50 [7]. Since these discoveries, many other amphiphilic molecules[7] have been introduced into synthesising media. The pore system of this type of material (see Fig. 11.3) can be spatially ordered (e.g., hexagonal arrangement of cylindrical tubes) or disordered (vermicular or wormlike arrangement). Many acronyms are used to identify these materials. Table 11.2 gives the names of some of the more commonly encountered examples. It should be noted that, although the first ordered mesoporous solids were mainly silica-based, methods of synthesis were soon extended to include other oxides, such as Al_2O_3, TiO_2, ZrO_2, etc., or mesoporous sulfides. Ordered mesoporous silicas are extremely useful as host lattices for elaborating nanoporous carbons by a negative replica process.

[6]By analogy with a crystal compound (see footnote 1), an amorphous compound does not in general diffract X rays and the diffraction pattern of this type of compound does not usually exhibit diffraction peaks. However, the diffraction patterns of ordered mesoporous silicates do exhibit such peaks, even though they are composed of amorphous inorganic walls. These peaks reveal the perfect ordering of the cylindrical channels.

[7]An amphiphile is a molecule or compound with a hydrophobic part and a hydrophilic part.

11.5 Disordered Nanoporous Solids

Among those materials with industrial potential, there are two main classes of porous solids with no long-range periodicity:

- porous materials formed by aggregation and partial sintering of particles;
- materials whose pores result from selectively dissolving part of some initial solid.

In the first class of solids, the pores are just the cavities remaining between the aggregated particles. The smaller the particles making up the material, the greater will be its specific surface area; the better the particles are compacted together, the greater will be the coordination number, i.e., the number of nearest neighbours of each particle, and the smaller will be the pore volume of the solid. Silica gels and most porous ceramics belong to this class, which contains mainly mesoporous and macroporous materials. Only a few materials in this class, obtained by sol–gel processes, can be considered amongst the nanoporous solids, and the reader is referred to more specialised works [8].

More relevant to the present discussion are those porous solids obtained by selectively dissolving part of some initially non-porous solid. The example that is currently most promising for industrial applications is provided by glasses with tailored pore systems, amongst which the best known is the porous glass Vycor. This glass is prepared by exploiting a solubility gap in the phase diagram of the Na_2O–B_2O_3–SiO_2 glasses. A glass with suitable composition can be prepared at high temperature (around 1 400°C) and cooled to a temperature between 450°C and 700°C, a temperature at which it separates into two liquid (glassy) phases. If the glass is cooled sufficiently, the separation of the two phases will not occur by an activated nucleation and growth process, but rather through spinodal decomposition, a phenomenon in which the phase separation occurs spontaneously throughout the solid. The result is a solid comprising two entangled networks, one made from 96% silica, and the other composed essentially of sodium borate. The latter is highly soluble in acids, leaving a wormlike pore structure that is clearly visible in the scanning electron microscope image of Fig. 11.4.

Glasses with tailored pore sizes, for which the glass Vycor was the first example, developed in 1938, are not ordered, although they exhibit a well-defined characteristic dimension, namely the spinodal wavelength, which decreases as subcooling increases. The pore size can be controlled by adjusting the temperature and duration of phase demixing, as well as the severity of the extraction by acid, or possibly base solutions. The interested reader is referred to the specialised literature [9–11].

There are important applications of porous glasses in filtration, especially for the food and pharmaceutical industries, owing to the ease with which they can be shaped into macroscopic objects, and also the high level of chemical inertness of the silicic glasses. These materials have a high ion exchange capacity with solutions by virtue of the very high specific surface areas that can be

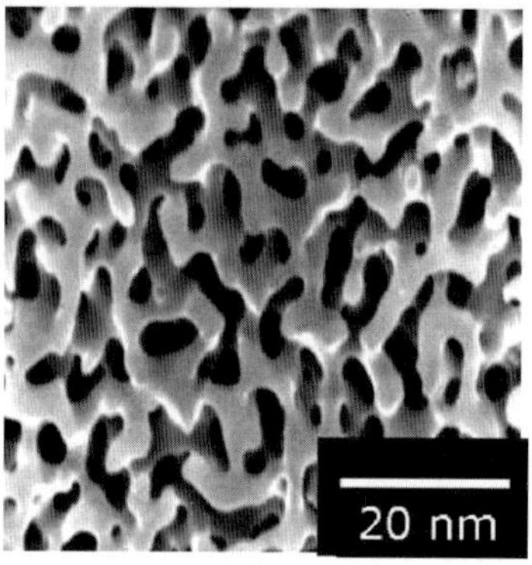

Fig. 11.4. SEM image of the porous glass Vycor obtained by spinodal decomposition of a borosilicate glass and subsequent acid extraction of the sodium borate. After B. Lefèvre

Fig. 11.5. Scanning electron microscope image of the surface of a glass treated with water at 90°C

achieved (several hundred m^2/g). Analogous materials may also result from natural degradation processes. Indeed, prolonged contact of various synthetic materials such as glass, cement or steel, with water leads to the formation of a nanoporous surface layer. An example of such a surface film is shown in Fig. 11.5, for a glass monolith after several months in contact with pure water. The layers formed in this way are of particular interest, because they are due to preferential dissolution of certain elements in the initial solid matrix. However, they thicken as time goes by and can eventually form protective layers that stop water from penetrating the material.

Glasses have been the subject of many studies because they are relevant to a wide range of industrial activities. One example is provided by the search

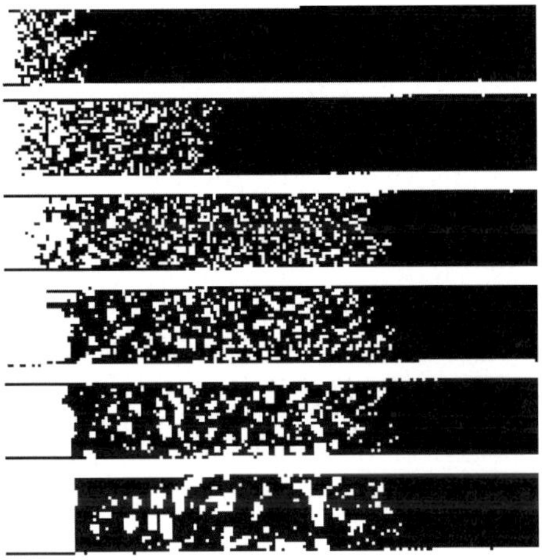

Fig. 11.6. Monte Carlo simulation of the dissolution of a two-component glass: A (soluble) and B (partially soluble). *White regions* correspond to water and *black regions* to the solid. The elements can dissolve and recondense at a different point on the solid matrix. The nanoporous layer is thus observed to age by exchange of species B with the solution. Extract from Devreux [12]

for glass fibres (for insulation) that can dissolve very quickly if accidentally inhaled into the lungs. On the other hand, for glasses designed to confine nuclear or chemically toxic waste, where the elements to be fixed are distributed throughout the bulk of the monolith (which may have a mass of several hundred kilos), the idea is obviously to achieve the greatest possible durability for the glass monolith. In this context it has been shown that the formation of a modified surface film on the glass can reduce the rate of degradation of a storage glass for nuclear waste by several orders of magnitude. Moreover, this film conserves very poorly soluble elements in its skeleton, e.g., radioactive elements in the case of nuclear waste glasses. It is thus particularly important to study the morphology of this type of nanoporous layer.

Today these studies are carried out by combining X-ray scattering experiments with Monte Carlo simulations [12]. When a glass containing several elements is considered, some of these elements (A) being fully soluble and others (B) only partially soluble, the redeposition of some dissolved B elements on the skeleton of the deteriorated layer induces a regular reconformation of the morphology of this layer which slows down, or even completely halts the progress of the dissolution front in some cases. A series of Monte Carlo simulations by F. Devreux [12] is shown in Fig. 11.6. Of particular interest in this example is the arrest of the dissolution front and a coalescence of growing pores. This behaviour has been corroborated experimentally by small-angle X-ray scattering observations.

References

1. K.S.W. Sing, D.H. Everett, R.A.W. Haul, L. Moscou, J. Pierotti, J. Rouquerol, T. Siemieniewska: Pure Appl. Chem. **57**, 603 (1985)
2. C.T. Kresge, M.E. Leonowicz, W.J. Roth, J.C. Vartuli, J.S. Beck: Nature **359**, 710 (1992)
3. P. Behrens: Adv. Mater. **5**, 127 (1993)
4. J. Patarin, J.L. Paillaud, H. Kessler: Synthesis of $AlPO_4$s and other crystalline materials. In: *Handbook of Porous Solids*, Sect. 4.2.3, ed. by F. Schüth, K.S.W. Sing, and J. Weitkamp, Wiley-VCH, Germany, Vol. 2, 815 (2002)
5. `http://www.iza-structure.org/databases/`
6. C. Baerlocher, W.M. Meier, D.H. Olson: *Atlas of Zeolite Framework Types*, 5th edn., Structure Commission of the International Zeolite Association, Elsevier, Amsterdam (2001) Zeolites, p. 302
7. J.C. Vartuli, C.T. Kresge, W.J. Roth, S.B. McCullen, J.S. Beck, K.D. Schmitt, M.E. Leonowicz, J.D. Lutner, E.W. Sheppard: Prepr. Am. Chem. Soc. Div. Pet. Chem. **40**, 21 (1995)
8. C.J. Brinker, G.W. Scherer: *Sol–Gel Science: The Physics and Chemistry of Sol–Gel Processing*, Academic Press, Boston (1990)
9. F. Janowski, D. Enke: Porous glasses. In: *Handbook of Porous Solids*, Sect. 4.6, ed. by F. Schüth, K.S.W. Sing, and J. Weitkamp, Wiley-VCH, Germany, Vol. 3, 1432 (2002)
10. J. Zarzycki: *Les verres et l'état vitreux*, Masson, Paris (1982) p. 392
11. S.R. Elliott: *Physics of Amorphous Materials*, 2nd edn., Longman, UK (1990) p. 481
12. F. Devreux, P. Barboux, M. Filoche, B. Sapoval: J. Mat. Sci. **36** (6), 1331 (2001)

12

Inverse Systems – Confined Fluids: Phase Diagram and Metastability

E. Charlaix and R. Denoyel

When a fluid is confined by solid surfaces, its equilibrium properties are modified by several mechanisms. One point is that surface effects become quantitatively significant compared with bulk effects, and for this reason, solid–fluid interactions begin to compete with fluid–fluid interactions in the determination of the phase diagram of the fluid. If the fluid–solid attraction is sufficient, confinement can stabilise a liquid phase that would have evaporated in the absence of the solid. The liquid–vapour transition is modified by confinement. The same goes for the melting–solidification transition. For high levels of confinement, such that the mean size of pores becomes of the same order of magnitude as density inhomogeneities in the fluid, the liquid can no longer be distinguished from its vapour: the critical point is modified by confinement, the critical temperature decreasing with pore size. Finally, for the ultimate confinement, in which pore sizes are comparable with the size of molecules in the fluid, the very notion of fluid is no longer applicable. In the following, we shall review these various degrees of confinement and the resulting modifications of physical properties.

12.1 Displacement of First Order Transitions: Evaporation and Condensation

12.1.1 Adsorption Isotherms

When porous nanomaterials are put in contact with a vapour phase at a temperature below the critical temperature, they strongly adsorb the molecules of that vapour phase and it is then easy to determine an adsorption–desorption isotherm representing the quantity of molecules adsorbed as a function of the vapour pressure. Several examples are given in Fig. 12.1 for materials in the mesoporous category defined by IUPAC (see Chap. 11). On these adsorption isotherms can be seen a first region at low pressures where adsorption is reversible and corresponds to the gradual formation of an adsorbed film.

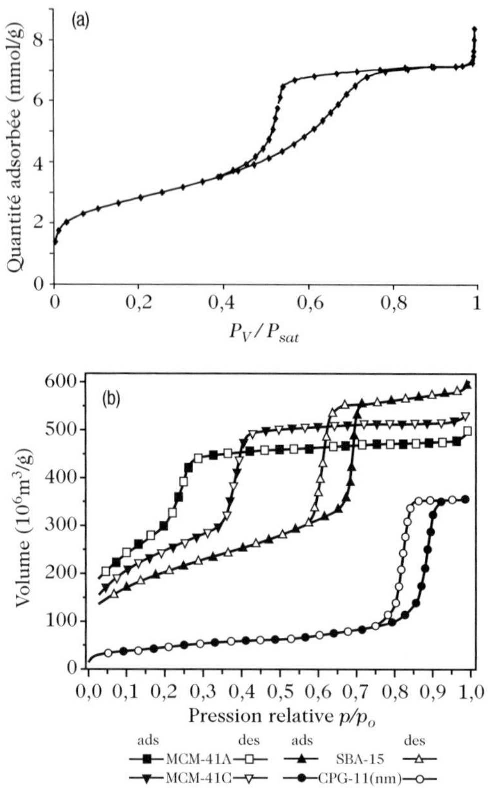

Fig. 12.1. Nitrogen adsorption isotherms at 77 K. (**a**) Porous glass Vycor. After Torralvo et al. [4]. This isotherm belongs to the category H2 defined by IUPAC. (**b**) Ordered mesoporous materials and CPG-type mesoporous glass. After Thommes et al. [5]

Then, on the adsorption branch, there is a clear point of inflection which corresponds to large-scale condensation of liquid in the pores. This is called capillary condensation, discussed further in the Sect. 12.1.2. It is important to note that this condensation occurs at a lower vapour pressure than the saturated vapour pressure, showing that the liquid–vapour equilibrium is shifted inside the pores. At the end of the capillary condensation stage, there is a saturation plateau, indicating that the pores are filled. This plateau is all the more level as the outer surface area is small, i.e., the surface area of the grains is negligible compared with the surface area inside the pores, and as the number of macropores (with sizes greater than 50 nm) becomes negligible. At desorption, hysteresis occurs, showing that the pores empty out at a pressure below the pressure at which they filled up. The form and amplitude of the hysteresis effect depends on the temperature, as well as the size and arrangement of the pore system. Hysteresis is discussed further in Sect. 12.1.6.

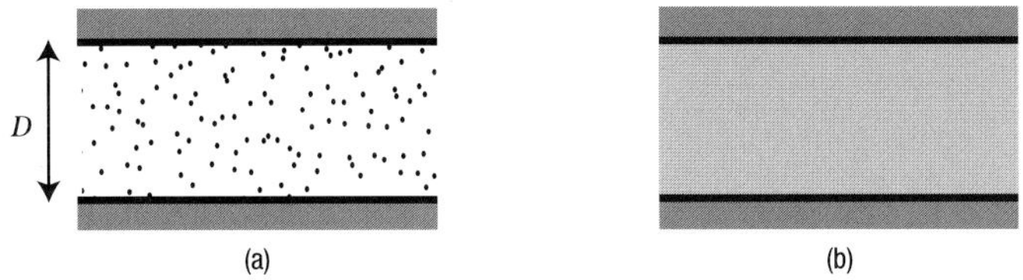

Fig. 12.2. Planar pore in vapour-filled (**a**) and liquid-filled (**b**) configurations

12.1.2 Capillary Condensation

We consider here the simple case of a planar pore bounded by two parallel solid surfaces separated by a distance D, as shown in Fig. 12.2. The space between the surfaces is in contact with a reservoir of vapour at pressure P_V and temperature T. The fluid–solid interactions are characterised by the surface energies γ_{SV} of the solid–vapour interface and γ_{SL} of the solid–liquid interface (see below). Clearly, if $\gamma_{SL} < \gamma_{SV}$, the solid favours the presence of liquid in the confined space. One should therefore ask for what degree of confinement the solid can successfully stabilise a liquid phase, when it is the vapour phase that is thermodynamically stable in the given conditions of temperature and pressure. To answer this question, one must compare the grand canonical potential of the configuration 'liquid-filled interstice' (which we shall call the condensed state) with that of the configuration 'vapour-filled interstice' (the non-condensed state). Indeed, the grand canonical potential is the thermodynamic potential that is minimum at equilibrium for the imposed conditions: open system of fixed area $2A$, volume $V = AD$, and temperature T, in contact with a reservoir of fluid at the chemical potential $\mu = \mu_V(P_V, T)$.

To begin with, we assume that the fluid–solid interactions are such that partial wetting occurs, as given by the Young–Dupré equation (see below for details):

$$0 < \gamma_{SV} - \gamma_{SL} = \gamma_{LV} \cos\theta < \gamma_{LV} \;, \tag{12.1}$$

where the angle θ is the contact angle of the liquid on the solid surface. The grand canonical potentials of the condensed and non-condensed states shown schematically in Fig. 12.2 are thus, per unit area A of the system,

$$\Omega_{\text{non-condensed}} = 2A\gamma_{SV} - DAP_V(\mu) \;, \quad \Omega_{\text{condensed}} = 2A\gamma_{SL} - DAP_L(\mu) \;. \tag{12.2}$$

For vapour pressures not too far from coexistence [characterised by $P_L(\mu_{sat}) = P_V(\mu_{sat}) = P_{sat}$], the difference $P_L(\mu) - P_V(\mu)$ can be approximated by the thermodynamic relation $\partial P/\partial\mu = \rho$, where ρ is the concentration (number of moles per unit volume). Hence, the liquid phase is thermodynamically stable when the gap D is less than the critical gap D_c given by

$$D < D_c(\mu) = \frac{2(\gamma_{SV} - \gamma_{SL})}{P_V(\mu) - P_L(\mu)} \simeq \frac{2\gamma_{LV}\cos\theta}{\Delta\rho\Delta\mu} \;, \tag{12.3}$$

where $\Delta\mu = \mu_{\mathrm{sat}} - \mu$ is the undersaturation and $\Delta\rho = \rho_{\mathrm{L}} - \rho_{\mathrm{V}}$ is the concentration difference.

Conversely, for a given pore size D, the liquid–vapour transition in the confined space does not occur at the saturated vapour pressure, but at the chemical potential $\mu = \mu_{\mathrm{sat}} - 2\gamma_{\mathrm{LV}} \cos\theta / D\Delta\rho$. This is indeed what is observed on the adsorption isotherms.

To estimate the order of magnitude of the pore size at which this effect occurs, consider the case of water at room temperature: $\gamma = 72\,\mathrm{mJ/m^2}$, $\Delta\rho \simeq \rho_{\mathrm{L}} = 5.5 \times 10^3\,\mathrm{mol\,m^{-3}}$, and assume a contact angle $\theta = 30°$. At a standard humidity of $P_{\mathrm{V}}/P_{\mathrm{sat}} = 50\%$, one has $D_{\mathrm{c}} = 13\,\mathrm{nm}$. The length scale at which the first order transition is significantly shifted is indeed the nanoscale. This reflects the fact that, if surface effects are to significantly modify the thermodynamic properties of the fluid, the fluid molecules interacting with the solid surface must correspond to a non-negligible fraction of all the molecules present.

Surface Tension

The surface tension of a fluid interface, in this case a liquid–vapour interface, is defined in terms of the work required to increase the area A_{LV} of the interface, while holding the other parameters constant, viz.,

$$\gamma_{\mathrm{LV}} = \left(\frac{\partial F}{\partial A_{\mathrm{LV}}}\right)_{N_{\mathrm{L}},V_{\mathrm{L}},N_{\mathrm{V}},V_{\mathrm{V}},T} ,$$

where F is the free energy of the two-phase system, and N_{L}, V_{L}, N_{V}, V_{V} are the number of molecules present in and the volume of each phase, respectively. The fact that work is required to increase the area of the liquid–vapour interface leads to a discontinuity in the pressure on either side of a curved interface. This is Laplace's law of capillarity:

$$P_{\mathrm{int}} - P_{\mathrm{ext}} = \frac{\gamma_{\mathrm{LV}}}{R} ,$$

where R is the average radius of curvature of the interface. The pressure is always higher on the concave side. For a solid surface, one can likewise define a difference of surface tension in terms of the work required to wet a fraction A_{SL} of the solid surface:

$$\gamma_{\mathrm{SL}} - \gamma_{\mathrm{SV}} = \left(\frac{\partial F}{\partial A_{\mathrm{SL}}}\right)_{N_{\mathrm{L}},V_{\mathrm{L}},N_{\mathrm{V}},V_{\mathrm{V}},N_{\mathrm{S}},V_{\mathrm{S}},T} .$$

It can be shown in thermodynamics that the surface tension is also a grand canonical excess potential per unit interfacial area. Since the grand canonical potential $\Omega = -PV$ of a homogeneous phase is minus the product of the pressure and the volume, the total grand canonical potential of a multiphase system is

$$\Omega = -P_{\mathrm{V}}V_{\mathrm{V}} - P_{\mathrm{L}}V_{\mathrm{L}} - P_{\mathrm{S}}V_{\mathrm{S}} + \gamma_{\mathrm{LV}}A_{\mathrm{LV}} + \gamma_{\mathrm{SL}}A_{\mathrm{SL}} + \gamma_{\mathrm{SV}}A_{\mathrm{SV}} .$$

Thermodynamic arguments show that the grand canonical potential is the potential energy of an open system, i.e., that its variation is equal to the work done on the system during a transformation and that it is minimal at thermodynamic equilibrium.

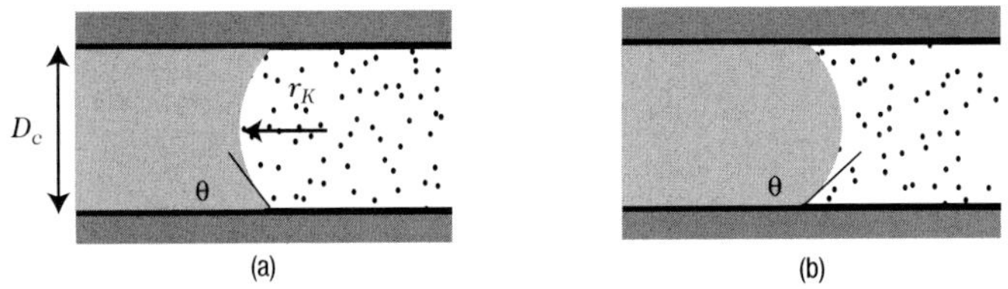

Fig. 12.3. Liquid–vapour coexistence in the planar pore. (**a**) Partially wetting liquid. (**b**) Non-wetting liquid

Young–Dupré Law of Partial Wetting

The diagram of Fig. 12.3 shows a translation dx of the meniscus. At equilibrium, the system is invariant under translation and the grand canonical potential does not change under this transformation:

$$d\Omega = 0 = -P_L dV_L - P_V dV_V + (\gamma_{SL} - \gamma_{SV}) dA_{SL} ,$$

or

$$P_V - P_L = \frac{\gamma_{SV} - \gamma_{SL}}{D} .$$

But according to Laplace's law of capillarity,

$$P_V - P_L = \frac{\gamma_{LV}}{R} ,$$

where the radius of curvature R of the interface is related to the contact angle θ of the liquid on the solid by the geometrical relation $R = D/2\cos\theta$. We deduce the Young–Dupré law of wetting:

$$\gamma_{LV} \cos\theta = \gamma_{SV} - \gamma_{SL} .$$

12.1.3 Capillary Pressure and the Kelvin Radius

When $D = D_c$, at the liquid–vapour equilibrium, a confined liquid–vapour interface has radius of curvature $r_K = D_c/2\cos\theta = \gamma_{LV}/\Delta\rho\Delta\mu$. This curvature of the interface at the liquid–vapour interface is thus related to the undersaturation potential by the so-called Laplace–Kelvin relation

$$\frac{\gamma_{LV}}{r_K} = P_V(\mu) - P_L(\mu) \simeq \Delta\rho\Delta\mu . \qquad (12.4)$$

The pressure difference $P_V(\mu) - P_L(\mu)$ between the liquid and vapour phases is thus nothing other than the capillary pressure. The displacement of the liquid–vapour equilibrium in the confined space is therefore related to the fact that the two coexisting phases are not at the same pressure, because the interface separating them must curve in order to satisfy the wetting conditions imposed by the solid wall. This effect does not occur when the interface is flat, i.e., in the absence of the solid surface, or if the latter does not favour either of the two fluid phases ($\gamma_{SV} = \gamma_{SL}$, i.e., $\theta = \pi/2$).

The radius of curvature r_K defined by the Laplace–Kelvin relation is called the Kelvin radius. It is worth noting that the notions of capillary pressure and Kelvin radius can be used to generalise the result of (12.3) to arbitrary pore geometries. In the general case, the chemical potential of the phase change is obtained by applying the Laplace–Kelvin relation (12.4) with an interfacial radius of curvature calculated from the size and geometry of the pore, together with the wetting conditions on the solid wall. In the case of a cylindrical pore, for example, the critical radius of capillary condensation is given by

$$r_c = \frac{2\gamma_{LV}\cos\theta}{\Delta\rho\Delta\mu} . \tag{12.5}$$

Note that the distance between the surfaces at which condensation occurs for a given undersaturation is twice that for the planar pore, because a meniscus in a cylindrical pore is curved in both space directions.

12.1.4 Non-Wetting Fluid

Most applications of the shift in the liquid–vapour transition considered so far concern the condensation of fluid in nanoporous media when the liquid wets the solid. However, there are interesting applications with non-wetting liquids. In this case, the Young–Dupré relation between the surface tensions is

$$\gamma_{SV} - \gamma_{SL} = \gamma_{LV}\cos\theta < 0 ,$$

and the contact angle of the liquid–vapour interface on the wall is $\theta > 90°$ (see Fig. 12.3). This can be arranged, for example, by functionalising a nanoporous silica with alkyl chains in such a way that the surface becomes non-wetting with regard to water, or else by using mercury.

In this configuration, (12.3) gives a negative chemical potential $\Delta\mu$, which corresponds to supersaturation. The vapour does not therefore condense spontaneously in the porous medium. Indeed, an overpressure must be applied to make the liquid enter into the pores. The Laplace–Kelvin equation gives the required value of this overpressure: for a pore of radius 2 nm and a contact angle $\theta = 120°$, the penetration pressure for water is about 200 bar. This property also underlies the porosimetry with respect to mercury, commonly used to study the pore size distribution in porous materials. An interesting example involving nanoporous materials is discussed in Chap. 26.

12.1.5 Perfectly Wetting Fluid

When the condition $\gamma_{SV} < \gamma_{LV} + \gamma_{SL}$ is not satisfied, the affinity of the solid for the fluid is such that a single solid surface can stabilise a liquid phase in equilibrium with an undersaturated vapour in the form of a wetting liquid

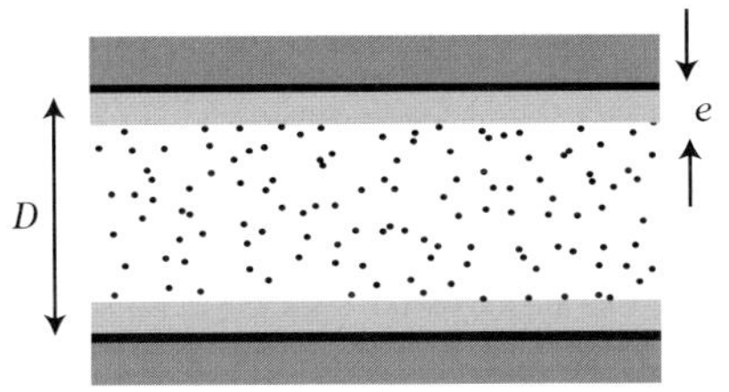

Fig. 12.4. Planar pore in contact with a vapour for which the liquid wets the surfaces

film, without any need for confinement. Indeed, the energy γ_{SV} of the solid interface in the presence of the vapour is greater than that, viz., $\gamma_{LV} + \gamma_{SL}$, of the solid–liquid and liquid–vapour interfaces formed by the film. The existence of these liquid films wetting solid surfaces must be taken into account when determining the liquid–vapour equilibrium in a confined space.

In the theory of wetting, wetting films are described by the idea of interface potential. When the thickness of these films is not too great compared with the range of molecular interactions, the excess potential per unit area does not reduce to the sum $\gamma_{SL} + \gamma_{LV}$ of the surface tensions. A further excess must be taken into account, corresponding to the fact that the molecular interactions which generate the surface tension do not operate over a thickness of liquid phase that can be considered infinite here. The excess grand canonical potential of a solid surface of area A covered with a liquid film is then

$$\frac{\Omega_{SV}}{A_S} = \gamma_{LV} + \gamma_{SL} + W_{SLV}(e) , \tag{12.6}$$

where the interface potential vanishes for a macroscopic film and takes the value $W_{SLV}(0) = \gamma_{SV} - \gamma_{LV} - \gamma_{SL}$ for a film of zero thickness. In the case of van der Waals forces, for example, the interface potential results from dipolar interactions going as $1/r^6$ between the relevant molecules. It varies as $1/e^2$ and can be written in the form

$$W_{SLV}(e) = -\frac{A_{SLV}}{12\pi e^2} . \tag{12.7}$$

The constant A_{SLV} is called the Hamaker constant. It has dimensions of energy and typically varies between 10^{-21} and 10^{-18} J. It has negative sign when the liquid wets the solid, i.e., in the case where the interface potential is positive.

The notion of interface potential enables one to describe the equilibrium thickness e of the liquid film in equilibrium with the undersaturated vapour. Indeed, in the non-condensed configuration, the grand canonical potential when wetting films are taken into account is (see Fig. 12.4)

$$\frac{\Omega_{\text{non-condensed}}}{A} = 2\left[\gamma_{LV} + \gamma_{SL} + W_{SLV}(e)\right] - 2eP_L(\mu) - (D-2e)P_V(\mu) . \tag{12.8}$$

The equilibrium thickness corresponds to the minimum of this grand canonical potential and satisfies

$$-\frac{\mathrm{d}W_{\mathrm{SLV}}}{\mathrm{d}e} = P_{\mathrm{V}}(\mu) - P_{\mathrm{L}}(\mu) = \Delta\rho\Delta\mu \,. \tag{12.9}$$

As for capillary condensation, the pressure in the liquid phase is not the same as in the vapour phase. This pressure difference $\Pi_{\mathrm{d}}(e) = P_{\mathrm{L}}(\mu) - P_{\mathrm{V}}(\mu)$ between the liquid film and the surrounding medium is called the disjoining pressure. It is thus equal to minus the derivative of the interface potential:

$$\Pi_{\mathrm{d}}(e) = -\frac{\mathrm{d}W_{\mathrm{SLV}}}{\mathrm{d}e} \,.$$

The thickness of the adsorbed film thus increases with the vapour pressure, until it reaches a macroscopic value (on a single solid surface) when the vapour becomes saturated.

In a planar pore, the condensed state is stable when its grand canonical potential is less than that for the non-condensed state. According to (12.2) and (12.8), the liquid phase is thus stable when the width of the pore satisfies

$$D - 2e < D'c = \frac{2\gamma_{\mathrm{LV}}}{\Delta\rho\Delta\mu} + \frac{2W_{\mathrm{SLV}}(e)}{\Delta\rho\Delta\mu} \,. \tag{12.10}$$

The difference with the partially wetting case does not reduce to decreasing the available interstice to twice the film thickness. Hence for van der Waals forces, (12.7) and (12.9) determine the thickness $e = (-A_{\mathrm{SLV}}/6\pi\Delta\rho\Delta\mu)^{1/3}$ of the wetting film. The liquid phase is then stable when

$$D - 3e < \frac{2\gamma_{\mathrm{LV}}}{\Delta\rho\Delta\mu} = 2r_{\mathrm{K}} \,,$$

where $e = (-A_{\mathrm{SLV}}/6\pi\Delta\rho\Delta\mu)^{1/3}$. These effects for adsorbed films become quantitatively important when determining pore radii from the advance of capillary condensation in adsorption isotherms.

12.1.6 Hysteresis, Metastability and Nucleation

Adsorption isotherms show that the phenomenon of capillary condensation is accompanied by a hysteresis loop: the same quantity condensed in the medium can be obtained for two different vapour pressures. For a given pressure, the amount adsorbed is lower on the adsorption branch than on the desorption branch.

The same effect is observed when studying condensation in a single planar pore, using equipment to measure surface forces. With this technique, the separation of two surfaces can be adjusted to angstrom accuracy in an atmosphere of fixed vapour pressure, measuring a quantity proportional to

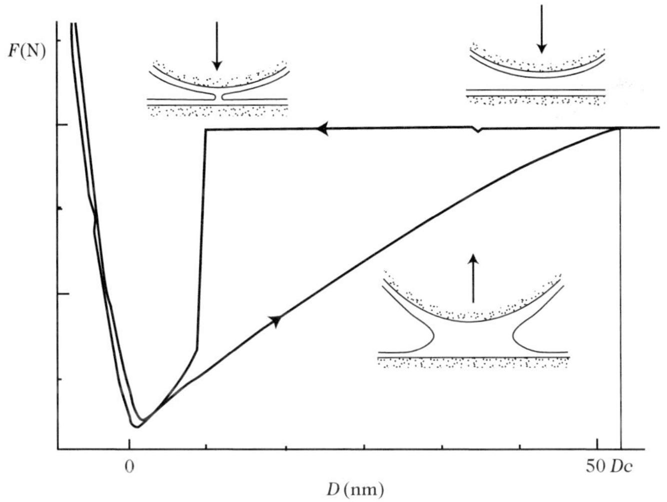

Fig. 12.5. Interaction force between a sphere and a plane as a function of their separation D in the presence of a vapour at fixed pressure P_V. Since the sphere radius is macroscopic, it can be shown that the interaction force is proportional to the grand canonical potential of the planar pore formed by two parallel surfaces with separation D. When the surfaces come together, the capillary condensation is manifested via a discontinuity in the force, which becomes attractive. When the surfaces are moved apart after the liquid phase has condensed, the latter remains stable up to a separation D_c

the potential energy of two parallel plane surfaces as a function of their separation. When the surfaces are brought closer together, capillary condensation is manifested by a sudden drop in the energy, as would be expected since we have seen that, when $D < D_c$, the grand canonical potential of the liquid-filled pore is lower than that of the vapour. However, when the surfaces are moved apart, the liquid phase remains present between the surfaces for positions where it did not during the approach stage (see Fig. 12.5). As with the adsorption isotherms, such experiments show that the capillary condensation phenomenon is accompanied by hysteresis.

Moreover, since the measured quantity is proportional (up to a constant) to the grand canonical potential of a planar pore of separation D, these measurements show that it is indeed the liquid phase that is is thermodynamically stable throughout the hysteresis range.

Capillary condensation is possible because the wetting phase lowers the surface energy when it fills the interstice. However, before it can arise at all, this wetting phase must nevertheless fill the whole interstice, in such a way that there is no further fluid interface increasing the energy of the system. Now in undersaturated conditions, i.e., when the non-wetting phase is stable in the bulk, the wetting phase can only arise a priori in small volumes, thus bounded

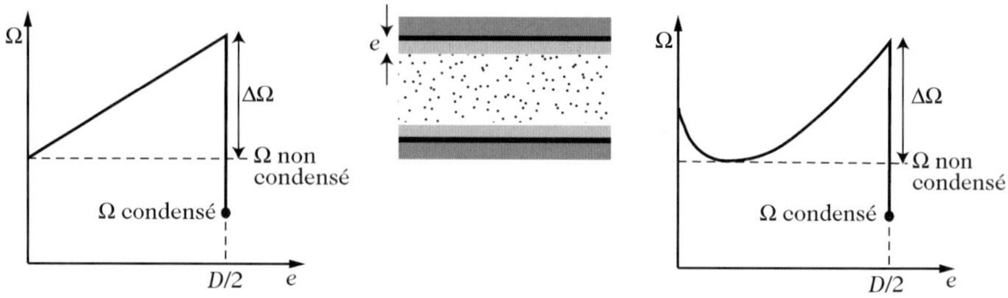

Fig. 12.6. Grand canonical potential of a planar pore partially filled with liquid as a function of the thickness e of the liquid layer coating the walls. The separation D of the two surfaces is such that the condensed state is stable. The *left-hand diagram* shows the partially wetting case, and the *right-hand diagram* the totally wetting case. In each case, the system must overcome an energy barrier $\Delta\Omega$ for the liquid phase to condense

by a liquid–vapour interface. There is therefore a problem of nucleation for this phase.

To illustrate this problem, let us return to the case where the liquid wets the wall, discussed in Sect. 12.1.5. If the vapour pressure is gradually increased from zero, a point is reached where $D_c(P_V) = D$. From this point on, it is the liquid that becomes the stable phase in the interstice. However, in order to fill this interstice, a volume $A(D-2e)$ of liquid must suddenly condense. If we imagine a condensation mechanism via growing thickness of the films coating the walls, the grand canonical potential of the partially filled system, given by (12.8), increases with the film thickness almost up to the point where these films touch one another. Growth is therefore not favoured, and the films tend rather to evaporate. At the point where the liquid–vapour interface disappears, the grand canonical potential drops suddenly and the liquid phase is stabilised. Following this path, an energy barrier $\Delta\Omega$ has to be crossed, of the order of

$$AD\big[P_V(\mu) - P_L(\mu)\big] \simeq AD\Delta\rho\Delta\mu \ ,$$

in order to reach the stable configuration (see Fig. 12.6).

This mechanism is of course too approximate to estimate a realistic order of magnitude for the value of the nucleation barrier. A more reasonable estimate can be made by considering the growth of a ring-shaped bulge in a cylindrical pore (see Fig. 12.7). It turns out that the critical germ is a bulge with radius of curvature equal to the Kelvin radius r_K, which is a quite general result. The nucleation energy barrier is then given by

$$\Delta\Omega = \Delta\rho\Delta\mu V_{\text{germ}} + \gamma_{SL}A_{SL} + \gamma_{SV}A_{LV} \ ,$$

which involves the volume of the critical germ, its free surface area and its area of contact with the solid.

The nucleation phenomenon only occurs a priori in one direction, i.e., the one in which the wetting phase appears. Indeed, when P_V is reduced from

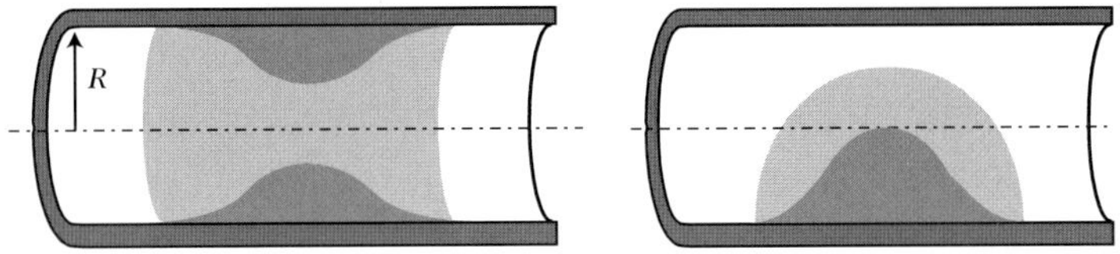

Fig. 12.7. Nucleation of a ring-shaped bulge (*left*) or a droplet (*right*) in a cylindrical pore of radius R. In both cases, the free surface of the critcal germ has a radius of curvature equal to the Kelvin radius. The energy barrier to be overcome in order for the liquid phase to condense has the form $\Delta\Omega \simeq \Delta\rho\Delta\mu K_1 R^3 + \gamma_{\rm LV} K_2 R^2$, where the constants K_1 and K_2 depend only on the contact angle between the liquid and the solid

the saturation value (desorption branch of the isotherm), the configuration is that of a liquid-filled pore surrounded by vapour. When the vapour reaches the equilibrium pressure in the pore, the liquid can in principle evaporate directly into the atmosphere from the ends of the pore. For pores with simple geometries, we may thus consider that it is the desorption branch of the isotherm that represents the liquid–vapour equilibrium branch in the confined space, and that is used to determine pore sizes. For more complicated geometries, such as bottle-shaped pores, energy barriers for formation of the vapour phase are sometimes encountered.

It should be noted that the metastability of the vapour and the nucleation barrier which hinder the appearance of the liquid phase are not of course related to their vapour or liquid nature, but just to the extent to which they are wetting. It is always the least wetting phase that is metastable, and the nucleation problem arises for the appearance of the wetting phase. In the case where it is the liquid phase that is less wetting, it is this phase that is metastable, once it has been forced to penetrate the medium. Its withdrawal from the medium is then dependent on nucleation of vapour bubbles inside the pores.

12.2 Melting–Solidification

When a solid is in a divided state, its free energy is modified by the presence of an extended interface. This is reflected by a change in its melting–solidification temperature, as can be observed either for small clusters (see Chap. 3), or in the case where the solid is confined within a porous matrix. As for the liquid–vapour transition, let us seek the equilibrium condition between the liquid-filled and solid-filled pore. When the pore, assumed planar, is filled with liquid, the free energy of the system can be written in the form

$$F = 2\gamma_{\rm WL} A + \mu_{\rm L}\rho_{\rm L} DA\ .$$

When it is filled with the solid phase, we have

$$F = 2\gamma_{\mathrm{WS}} A + \mu_{\mathrm{S}} \rho_{\mathrm{S}} D A \,.$$

In these equations, γ_{WS} and γ_{WL} are the wall–solid and wall–liquid interface tensions, μ_{S} and μ_{L} are the chemical potentials of the homogeneous solid and liquid at the relevant temperature, and ρ_{L} and ρ_{S} are the densities of the liquid and the solid, respectively. Assuming the densities of the liquid and solid to be similar, we obtain at equilibrium

$$2(\gamma_{\mathrm{WS}} - \gamma_{\mathrm{WL}}) = (\mu_{\mathrm{S}} - \mu_{\mathrm{L}}) \rho_{\mathrm{L}} D \,.$$

The difference between the chemical potentials can be obtained from the average value of the enthalpy of melting of the solid between the melting temperature T^{o} and the temperature T at which melting is observed in the pore [13]:

$$\mu_{\mathrm{S}} - \mu_{\mathrm{L}} = \Delta H_{\mathrm{melt}} \frac{T - T^{\mathrm{o}}}{T^{\mathrm{o}}} \,, \qquad (12.11)$$

whence

$$\frac{\Delta T}{T^{\mathrm{o}}} = \frac{2(\gamma_{SM} - \gamma_{\mathrm{WL}})}{\rho_{\mathrm{L}} D \Delta H_{\mathrm{melt}}} \,. \qquad (12.12)$$

This is the Gibbs–Thompson equation, which tells us how the temperature of the phase transition will vary as a function of the solid–liquid and solid–wall interface tensions. In some cases, the melting–solidification temperature will increase (the only cases cited correspond to organic molecules confined within microporous carbons). One may then consider that the pore wall is a nucleation point for the solid phase.

However, in most cases, the melting–solidification temperature is observed to decrease. This reflects a situation in which the liquid–wall interaction is favoured over the solid–wall interaction, and we encounter the phenomenon of surface melting. Indeed, most experimental investigations show that, when a confined liquid solidifies after a fall in temperature, there remains between the solid and the wall a layer that has not changed phase and which is thus in a state close to liquid. The main experiment for studying this type of phenomenon is called scanning calorimetry. An example is given in Fig. 12.8. This type of experiment reveals two important effects:

- The temperature of the solid–liquid transition can be modified by several tens of degrees by the confinement effect.
- There is a significant hysteresis effect between the melting and solidification temperatures.

Furthermore, experiments like NMR show that there is a mobile liquid at a temperature below the melting–solidification temperature. This leads to the representation in Fig. 12.9 for a pure compound confined within a porous matrix at low temperature. We consider here only a cylindrical pore of radius r_{o}. The compound is in the solid state at the centre of the pore and there is

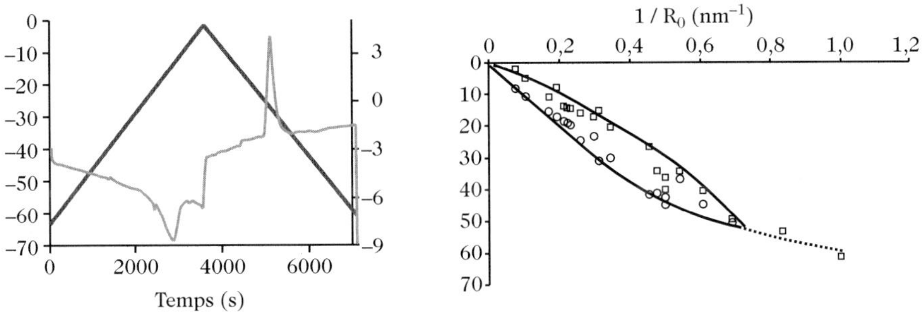

Fig. 12.8. Scanning calorimetry experiment. *Left*: temperature ramp (*black*) applied to a water-saturated nanoporous solid of type MCM-41. Heat exchanged with the surroundings (*light grey*) shows a first peak corresponding to endothermic melting and a second peak corresponding to exothermic solidification. *Right*: melting and solidification temperatures plotted as a function of the reciprocal of the pore radius of the material. After Denoyel et al. [3]

a liquid film between the solid core and the pore wall. When the pore is filled with liquid, the free energy of the system can be written in the form

$$F = \gamma_{\mathrm{WL}} A + \frac{1}{2}\mu_{\mathrm{L}}\rho_{\mathrm{L}} r A' \ .$$

When it is filled with solid with a liquid boundary layer,

$$F = \gamma_{\mathrm{WL}} A + \gamma_{\mathrm{SL}} A' + \frac{1}{2}\mu_{\mathrm{S}}\rho_{\mathrm{S}} r A' \ .$$

Here r is defined as in Fig. 12.9, A is the pore area, and A' is the area of the surface at radius r. Assuming once again that the densities of the liquid and solid are close, we obtain at equilibrium

$$\gamma_{\mathrm{SL}} = -\frac{1}{2}(\mu_{\mathrm{S}} - \mu_{\mathrm{L}})\rho_{\mathrm{L}} r \ .$$

Substituting for the chemical potentials as prescribed by (12.11), we obtain

$$\frac{1}{r} = -\frac{\overline{\Delta H}_{\mathrm{melt}}\Delta T}{2\gamma_{\mathrm{SL}} v_{\mathrm{L}} T^{\circ}} \ , \qquad (12.13)$$

where v_{L} is the molar volume of the liquid.

This type of equation has been used by Brun et al. to implement a method known as thermoporometry. This technique uses a differential scanning calorimeter (DSC) to determine melting and solidification thermograms. For a point in the solidification region of the thermogram, the heat rate is proportional to the quantity of matter changing phase, and hence also the pore volume in which melting takes place. The corresponding temperature is directly related to the pore size by the relation (12.13). The difficulty in using

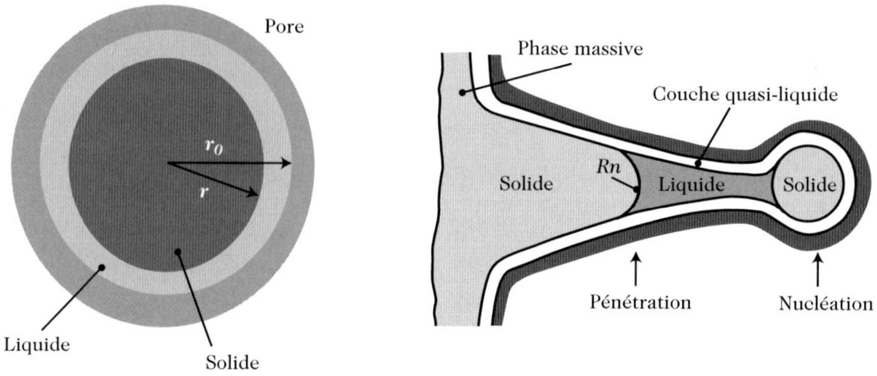

Fig. 12.9. *Left*: cross-section of a cylindrical pore. In general, the liquid phase wets the wall and the solid phase therefore appears at the pore centre. *Right*: longitudinal cross-section of a pore with complex geometry. After Quinson et al. [7]

these equations lies in determining the various parameters they involve. On the one hand, there are the interface terms which cannot be measured directly in the case of a solid–liquid interface, and on the other hand, there are the values of the melting–solidification enthalpies, which can only be found by extrapolating into the supercooled region.

Brun et al. thus examined known samples to calibrate the different parameters, using capillary condensation as reference method. Their work focused on mesoporous media for the main part. Water and organic molecules are generally used, e.g., benzene, decane, especially in the case of hydrophobic solids.

As mentioned earlier, melting–solidification is a phenomenon exhibiting a high degree of hysteresis. The technique of thermoporometry is therefore only feasible if applied to a hysteresis branch close to equilibrium. Brun et al. suggested using the thermal path shown in Fig. 12.8. An excess amount of liquid is used, and the system cooled to obtain a solid throughout. The temperature is then ramped up until a melting peak is obtained, but stopping before reaching the melting temperature of the homogeneous phase so as to keep the solid phase outside the pores. The system is then cooled once more to analyse the solidification process. In this method, it is assumed that the porous medium is filled near equilibrium by advance of a solid front. Since the liquid phase is the wetting phase, there is no delay to solidification due to the nucleation problem. However, it is clear that, in the case of a complex porous medium at equilibrium (see Fig. 12.9), the advance of the solid front is unable to fill cavities connected to the surroundings by small openings. As in the case of capillary condensation, it is then difficult to relate the pore size distributions obtained in this way to a real distribution. Regardless of the complexity of the medium, there is a hysteresis effect intrinsic to the investigated phenomenon, which is also produced in a simple cylindrical pore, as we have seen for capillary condensation.

Let us return to the model proposed in Fig. 12.9. For an infinite cylinder, melting occurs by raising the temperature when the liquid film reaches a thickness such that the system becomes unstable. On the other hand, solidification can only occur via nucleation of the solid somewhere in the pore. The curvatures of the solid–liquid interface are thus very different in the two situations, and they cannot therefore occur at the same temperature. A factor of two in the curvatures leads, to a first approximation, to a factor of two in the difference between the melting and solidification temperatures for a confined geometry, as compared with the melting temperature in the homogeneous phase. This hysteresis is clearly demonstrated experimentally by results obtained using nanoporous samples in which the cylindrical pores are as perfect as possible (silicas of types MCM-41 and SBA-15). In Fig. 12.8, the difference between the melting and solidification temperatures in the porous material are plotted as a function of the reciprocal of the pore radius for the case of water. Indeed, for large pores, i.e., low values of $1/r$, there is a factor of about two between the values corresponding to solidification and those corresponding to melting. However, as the pore size decreases, this difference decreases, and for the smallest pore sizes (diameters of about 2 nm), the phenomenon becomes reversible. In the next section, we shall discuss the disappearance of hysteresis related to the first order transition, when confinement becomes severe.

12.3 Modification of the Critical Temperature

The hysteresis observed in the phenomenon of capillary condensation is characteristic of a first order transition: nucleation effects give rise to an energy barrier due to the fact that the surface tension of the two phases present is nonzero. In a confined geometry, one should determine the range of validity of this description. Two key results should be taken into account (see Fig. 12.10):

- When the temperature is raised at constant pore size, hysteresis disappears.
- When the pore size is decreased at constant temperature, hysteresis also disappears.

It would thus appear that the first order nature of the transition can disappear under the influence of these two parameters. Reducing the system size or increasing the temperature amounts to making the system size and the correlation length of the fluid converge to similar values. The correlation length is the distance over which fluctuations are correlated in the fluid. This correlation length is directly related to the thickness of the liquid–vapour interface. In the bulk of the fluid, the correlation length diverges when the temperature approaches its critical value. It is infinite for an infinite system at the critical

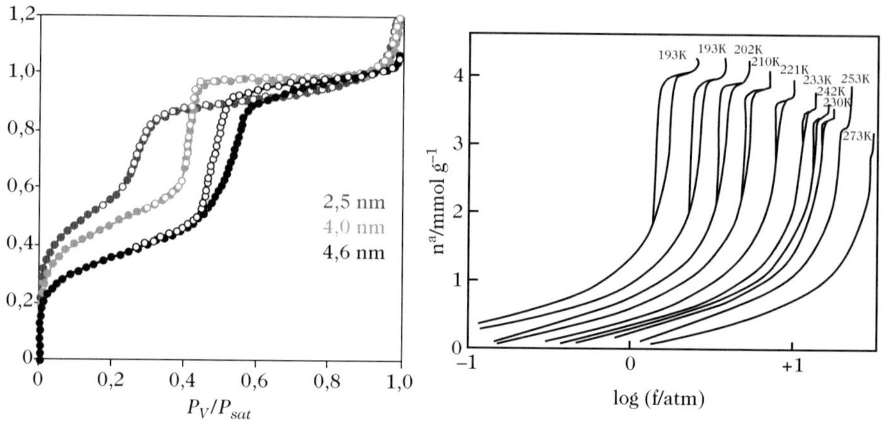

Fig. 12.10. Disappearance of hysteresis for evaporation–condensation. *Left*: adsorption isotherm for nitrogen (normalised at the saturation plateau) at 77 K on mesoporous solids of type MCM-41 with different pore widths. After Llewellyn et al. [8]. *Right*: adsorption isotherm for xenon on a Vycor glass. After Burgess et al. [9]

temperature. But as in any critical phenomenon, size effects must be taken into account. In a confined system, the critical point will in fact be reached when the correlation length of the fluid becomes of the same order as the space available between the walls. Indeed, when the size of fluctuations becomes greater than that of the sample, one can no longer say whether the fluid is a liquid or a vapour. One can therefore, in principle, define a critical pore temperature, denoted by T_{cp}, lower than the critical temperature T_{c}.

For a temperature greater than the critical pore temperature, capillary condensation can no longer occur, and the adsorption isotherms are no longer expected to exhibit hysteresis. To illustrate this, one can plot a hysteresis diagram in confined geometry from the two branches of the adsorption isotherm (see Fig. 12.11). This diagram looks rather like the phase diagram of the bulk fluid, with a lower critical temperature.

In order to make an exact determination of the temperature T_{ch} at which hysteresis disappears, adsorption isotherms must be obtained very accurately. One point that has not yet been dealt with is the experimental determination of the critical temperature. This is defined a priori as the temperature at which the discontinuity in the capillary condensation disappears. It is thus determined from the way the slope of the adsorption isotherm varies with temperature at the point of inflection in the advance of capillary condensation. Figure 12.11 shows these two temperatures for different fluids as a function of the ratio of the molecule size to the pore size. In the mesoporous media with model geometries referred to as MCM-41, these two temperatures are different: condensation–evaporation hysteresis disappears before the critical point, while the liquid and vapour phases are still distinct. We shall return to

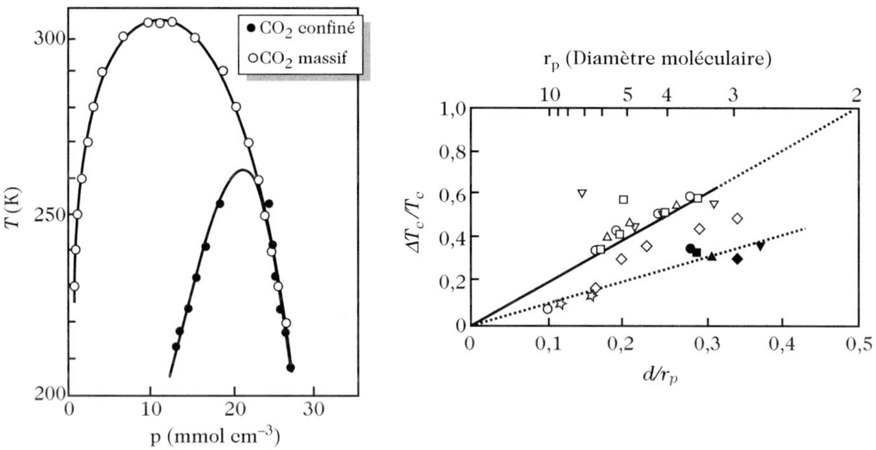

Fig. 12.11. *Left*: comparison between the phase diagram of CO_2 and its hysteresis diagram in Vycor. The hysteresis diagram represents the average density of the fluid adsorbed in the two branches of the isotherm. *Right*: temperature $(T_{ch} - T_c)/T_c$ at which hysteresis disappears and critical pore temperature $(T_{cp} - T_c)/T_c$ as a function of the reciprocal of the pore radius scaled by the size of the adsorbed fluid molecule. The different fluids are Ar (○), N_2 (△), O_2 (□), C_2H_4 (▽), and Xe (∗). *Black symbols* represent T_{cp} and *white symbols* T_{ch}. The four points for the small values of d/r lying significantly lower than the straight line fitted to the other values of T_{ch} obtained in MCM-41 materials, correspond to interconnected mesoporous materials, Vycor and silica gels. From Morishige et al. [10]

this point in the next section. However, for mesoporous materials with more complex pore geometries such as Vycor, this difference is no longer visible. This illustrates the fact that, for more complex pore geometries, there are hysteresis mechanisms related to the shape and connectivity of the pores.

12.4 Ultraconfinement: Microporous Materials

As the pore size is reduced, any description of observed phenomena in terms of parameters that can be measured on the macroscopic scale, such as the surface tension, or use of a simple representation of the type 'adsorbed film plus confined liquid', becomes less and less quantitative. This arises because, when the pore size is made very small, the range of the interactions produced by the wall becomes comparable with the system size, while the binding energies between confined molecules decrease. Adsorption becomes reversible and interactions between the walls and adsorbed molecules dominate. One must then consider a more microscopic description of the interactions. A typical example of an interaction potential that takes into account the van der Waals interactions and the repulsion of electronic orbitals at contact is the Lennard-Jones potential:

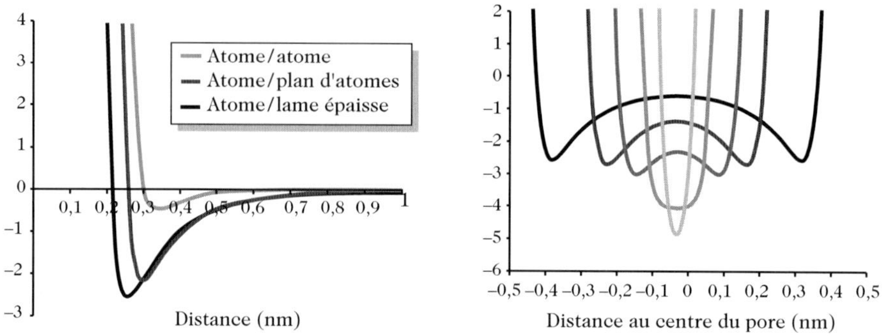

Fig. 12.12. Potential due to Lennard-Jones fluid–wall interaction. *Left*: potential energy between two atoms, an atom and a plane of atoms, and an atom and a thick slab. *Right*: potential energy of an atom in a pore

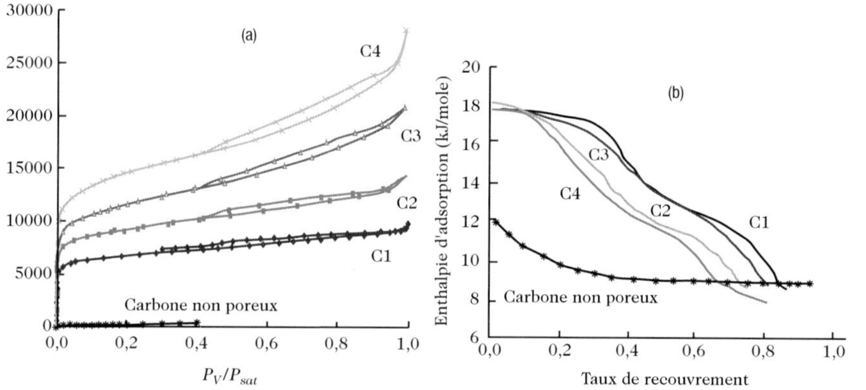

Fig. 12.13. Nitrogen adsorption isotherms at 77 K and corresponding adsorption enthalpies on a series of active carbons. After Denoyel et al. [11]

$$u(r) = 4\pi\varepsilon\left(\frac{\sigma^{12}}{r^{12}} - \frac{\sigma^6}{r^6}\right) ,$$

where ε is the depth of the potential well, r is the distance between two atoms, and σ is their size. Figure 12.12 (left) shows the interaction energy between two atoms, between an atom and a slab of atoms, and between an atom and a semi-infinite wall, calculated assuming that contributions from each pair of atoms to the total interaction are additive. A significant deepening of the potential well is observed. One can then consider a planar pore between two walls as in Fig. 12.12 (right).

When there is a clear separation between the two walls, two minima are observed in the potential close to the walls, with an almost-zero adsorption potential at the centre. It is thus easy to see that, when a pore of this size is filled with liquid at a pressure well below the saturation pressure, molecules at the pore centre are largely 'held' by interactions with their neighbours. As

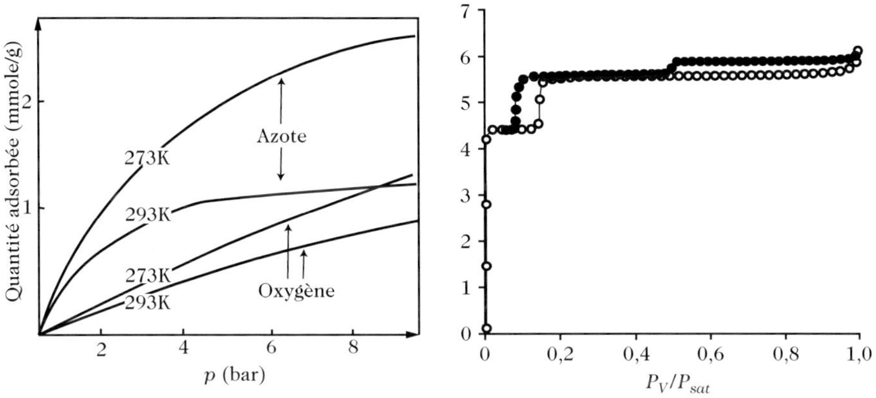

Fig. 12.14. *Left*: Nitrogen and oxygen adsorption isotherms on a zeolite 5A at temperatures close to room temperature. After Kirkby et al. [12]. *Right*: Nitrogen adsorption isotherm at 77 K on silicalite

the pore width gradually diminishes, the potential minima grow deeper and the adsorption potential at the pore centre becomes more negative. For very small sizes, the two minima merge to form one that is about twice as deep as the one close to a surface not located in a pore. The microporous structure thus introduces a highly negative energy component into the free energy of adsorption, causing adsorption to occur for a much more negative chemical potential than it would for a non-porous surface (hence at much lower pressures). Experimentally measured adsorption energies are also much higher. Figure 12.13 shows the adsorption isotherms and corresponding enthalpies for adsorption of argon on a non-porous carbon-bearing surface and on a microporous carbon-bearing surface. It can be seen that adsorption occurs at very low pressure on the microporous sample, and that the adsorption enthalpies at low coverage are almost twice as big.

In these microporous systems, adsorption potentials are so high that the physisorption of gases like nitrogen and oxygen becomes possible at room temperature. For example, Fig. 12.14 shows nitrogen and oxygen adsorption isotherms on a zeolite. The different affinities of these two molecules for this sample make it possible to use adsorption–desorption cycles to separate these two gases (pressure swing adsorption). We mentioned that adsorption in microporous systems is generally reversible. However, there are some exceptions which arise due to the fact that the surface field can create different types of organisation in the adsorbed phase, which depend on the adsorbed molecule and the arrangement of the 3D pore network. This is the case for adsorption of nitrogen in silicalite (see Fig. 12.14), which exhibits hysteresis in the low pressure range.

References

Further Reading

1. J.K. Israelachvili: *Intermolecular and Surface Forces*, Academic Press, (1992)
2. J.L. Barrat, J.P. Hansen: *Basic Concepts for Simple and Complex Liquids*, Cambridge University Press (2003)
3. F. Rouquerol, J. Rouquerol, K. Sing: *Adsorption by Powders and Porous Solids*, Academic Press San Diego (1999)

Figures from the Literature

4. M.J. Torralvo, Y. Grillet, P.L. Llewellyn, F. Rouquerol: J. Colloid Interface Sci. **206**, 527 (1998)
5. M. Thommes, R. Köhn, M. Fröba: Appl. Surf. Sci. **196**, issues 1–4, 239–249 (2002)
6. R. Denoyel, R.J.M. Pellenq: Langmuir **18**, 2710–2716 (2002)
7. J.F. Quinson, M. Astier, M. Brun: Applied Catalysis **30**, 123–130 (1987)
8. P.L. Llewellyn, Y. Grillet, F. Schuth, H. Reichert, K.K. Unger: Microporous Materials **3**, issue 3, 345–349 (1994)
9. C.G.V. Burgess, D.H. Everett, M. Nutall: Pure Appl. Chem. **61**, 1845 (1989)
10. K. Morishige, M. Shikimi: J. Chem. Phys. **108**, 7821 (1998)
11. R. Denoyel, J. Colinas, Y. Grillet, J. Rouquerol: Langmuir **9**, 515 (1993)
12. N.F. Kirkby: Membranes in gas separation and enrichment, Special Publication 62, Royal Society London (1986) p. 221
13. R. Defay, I. Prigogine: In: *Surface Tension and Adsorption*, Longmans, London (1966)

13

Supramolecular Chemistry: Applications and Prospects

N. Solladié and J.-F. Nierengarten

13.1 From Molecular to Supramolecular Chemistry

Supramolecular chemistry is an interdisciplinary field of research, reaching across from chemistry to the physics and biology of chemical species more complex than the molecules themselves. These molecular structures are held together and organised by non-covalent intermolecular interaction forces. Supramolecular chemistry is rooted in the chemistry of organic synthesis for the construction of molecules, and in the field of coordination chemistry for the assembly of multimolecular entities. But it also extends to physical chemistry in the theoretical and experimental investigation of the interactions that come into play, and biochemistry via the biological processes of recognition and binding of a substrate which originally inspired its development, and finally the science of materials where one investigates the novel physicochemical properties of these new chemical species.

Supramolecular chemistry is based on three concepts: binding, recognition, and coordination [1]. Paul Ehrlich first introduced the idea of a receptor (or host) molecule when he stated that a molecule is only active if it binds (*corpora non agunt nisi fixata*) [2]. Then in 1894, Emile Fischer put forward the notion of a lock and key, based on the idea of geometric or steric complementarity, thereby laying the foundations for the study of molecular recognition [3]. Finally, insofar as any attachment of molecules requires interactions between the partners, Alfred Werner introduced the notion of coordination, thus establishing supramolecular chemistry as a generalisation of coordination chemistry [4].

13.2 Molecular Recognition

Molecular recognition is a process of coordination and selection of a substrate by a given molecular host, and coordination alone does not constitute an example [1]. Molecular recognition involves both the storage of information on

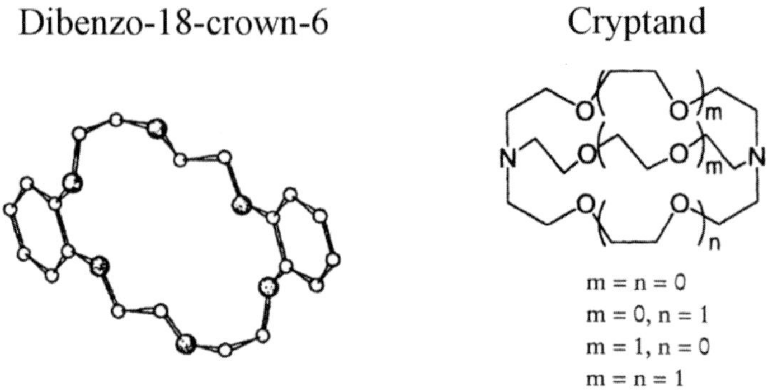

Fig. 13.1. X-ray structure of the crown ether dibenzo-18-crown-6 and schematic representation of a family of macrobicyclic compounds known as cryptands. From J.M. Lehn et al. [5, 6]

the molecular level and the reading of that information on the supramolecular level. Information can be stored in the very architecture of the host or in its coordination sites, while it is read via the formation and dissociation of oligomolecular structures called supermolecules. Information is a key idea in supramolecular chemistry, which can thus be considered as the science of chemical information. Molecular recognition certainly attains its most sophisticated expression in biology, where it leads to a highly selective coordination of substrates resulting in a variety of reactions, and transport, regulation and other mechanisms.

A molecular receptor (or host) is defined as a covalent organic structure capable of coordinating ionic or molecular substrates in a selective way through intermolecular interactions, and leading to the assembly of two or more molecules into a supermolecule. The chemistry of molecular hosts is therefore an extension of coordination chemistry, usually limited to transition metals, to all types of substrates, i.e., cationic, anionic, and neutral organic, inorganic or biological species.

There are several types of molecular recognition, namely the recognition of spherical cations, the recognition of ammonium cations by macrocyclic polyethers or tetrahedrally by cryptands, and the recognition of neutral molecules. Of these the most widely known is undoubtedly the recognition of spherical substrates. Macrocyclic polyethers like the crown ethers form complexes with various metallic cations depending on the size of their cavity. As an example, the complexation of the cation Rb^+ by the crown ether dibenzo-18-crown-6 is shown in Fig. 13.1 [5]. An important step forward was made with regard to size selectivity of the substrate when J.-M. Lehn synthesised the macrobicyclic compounds known as cryptands (see Fig. 13.1) [6], work which earned him the Nobel Prize for Chemistry in 1987. The cryptands provide a 3D cavity of spherical shape, capable of accurately selecting a guest molecule in terms

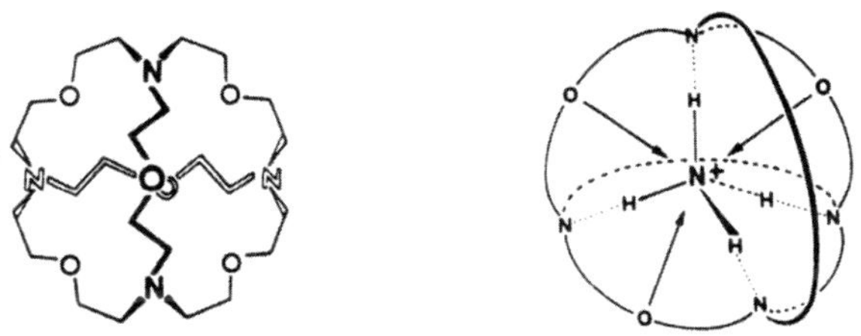

Fig. 13.2. Macrotricycle with a tetrahedral cavity capable of forming a particularly stable complex with an ammonium substrate

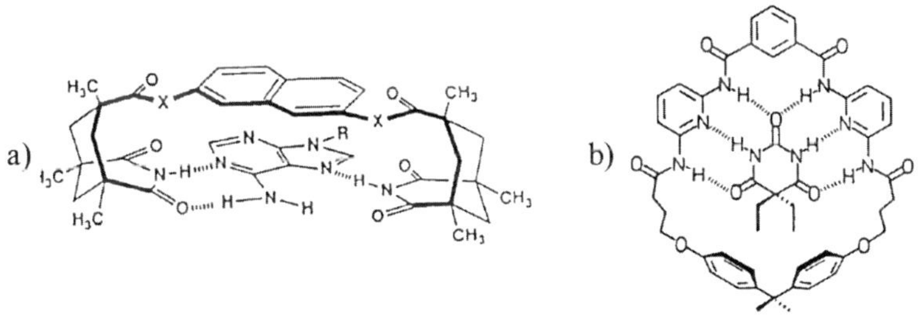

Fig. 13.3. (**a**) Adenine molecule complexed by an appropriate receptor using a network of hydrogen bonds. (**b**) Barbituric acid molecule coordinated by a cyclic receptor using a network of hydrogen bonds

of its size. When the arms of the macrobicycle grow longer, the size of the cavity gradually increases and the most strongly complexed cations are Li^+, Na^+, and K^+, depending on the extent to which the arms have grown.

In contrast, tetrahedral molecular recognition occurs via macrotricyclic cryptands. The selective coordination of tetrahedral substrates requires the construction of a molecular host possessing a tetrahedral recognition site. This can be achieved by placing four coordination sites at the four corners of a tetrahedron and joining them by six bridges. As an example, the macrotricycle shown in Fig. 13.2 contains four nitrogen atoms corresponding to the four corners of the tetrahedron and six oxygen atoms incorporated into the six bridges connecting the four corners. This cryptand complexes the tetrahedral cation NH_4^+ exceptionally well (compared with the cation K^+), demonstrating the excellent complementarity with regard to both structure (shape and size) and interactions between the substrate NH_4^+ and its host.

Finally, it is worth saying a few words about the recognition of neutral molecules. This type of recognition makes use of electrostatic interactions of donor–acceptor or hydrogen bond type, although the latter is the most common. The recognition of a given substrate by this process results by establishing a network of hydrogen bonds between complementary units, in an analogous way to the hydrogen bonds stabilising the formation of a DNA double helix. Functional groups likely to establish such bonds are judiciously

Bis-tren cryptand

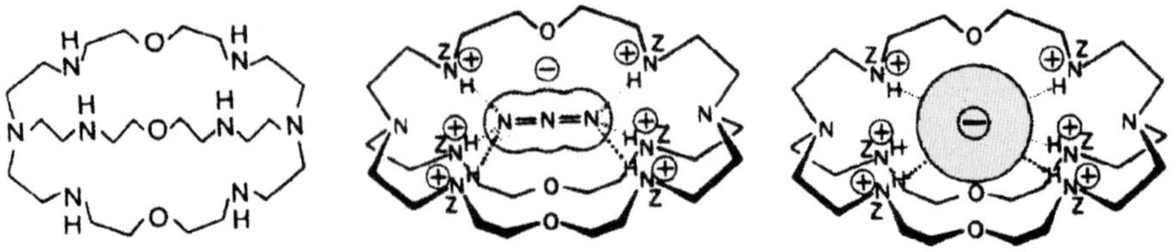

Fig. 13.4. Coordination of the anion N_3^- and a halide-type spherical anion by the hexaprotonated bis-tren cryptand [9]

distributed within the receptor, thereby defining a suitably sized cavity for recognition of the given substrate. Figure 13.3 shows two examples of molecular recognition via hydrogen bonds. In Fig. 13.3a, an adenine molecule is complexed by an appropriate receptor [7, 8], and in Fig. 13.3b, a barbituric acid molecule is recognised by a cyclic receptor [8].

13.3 Anionic Coordination Chemistry and Recognition of Anionic Substrates

Although anionic entities play an important role in chemistry, as they do in biology, their complexation has not been widely investigated in comparison to the complexation of cations or cationic molecules [1]. As an example, the molecular recognition of anions can be achieved by protonated macropolycyclic polyamine receptors. Figure 13.4 shows the complexation of two anions of different shapes by the same polycyclic polyamine: one anion is linear, viz., N_3^-, and the other is spherical of halide type [9]. The ellipsoidal shape of the hexaprotonated bis-tren cryptand allows it to recognise linear anions like N_3^- with excellent affinity by virtue of an optimal accordance in terms of size, shape and complementarity between receptor and substrate sites. However, the lack of complementarity between this polycyclic polyamine and halide-type spherical anions leads to poor recognition of this type of anion.

13.4 Multiple Recognition

Once the recognition units have been identified for each functional group, one can consider the possibility of combining several recognition sites within the same host, which is then called a polytopic coreceptor molecule. These coreceptors thus contain several coordinating sites that can either cooperate to form complexes with several substrates at the same time, or recognise a single polyfunctional species (called a polyhapto) via several coordinations. A distinction is made between cosystems in which the coordination of different substrates is commutative, i.e., interchangeable, and cascade systems in

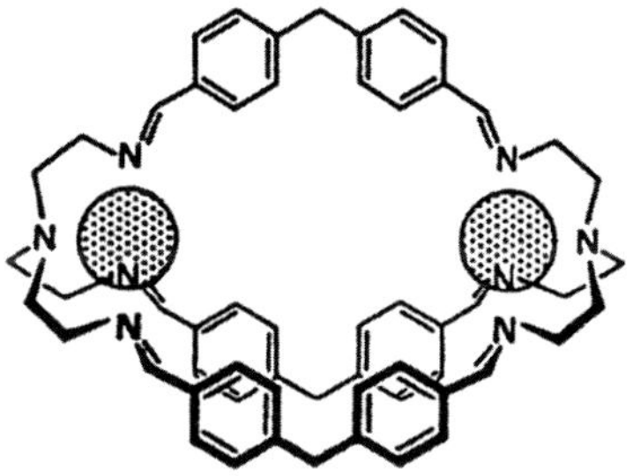

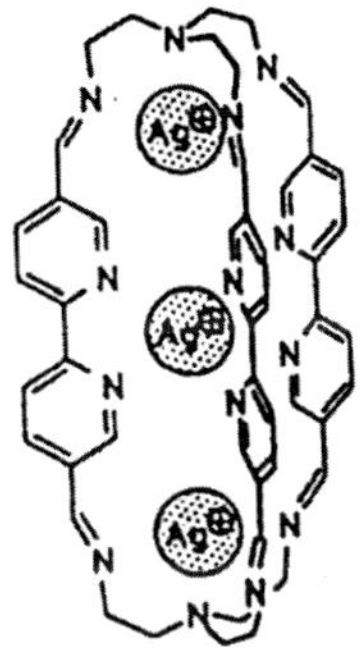

Fig. 13.5. Homotopic dinuclear cryptate and heterotopic trinuclear cryptate [10]

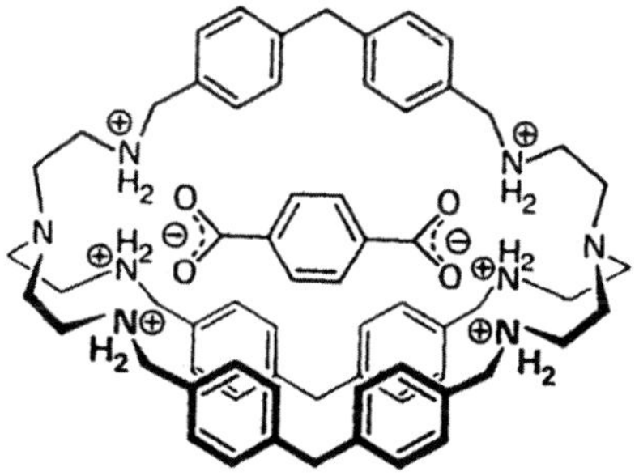

Fig. 13.6. Homotopic coreceptor selectively coordinating terephthalate substrates in terms of their size and carboxylate groups [11]

which the different complexations occur according to a well-defined sequence. Polytopic receptors are said to be homotopic (or autotopic) if they possess identical coordinating units, and heterotopic (or allotopic) if these units are different.

These coreceptors can form complexes with different metallic cations, organic substrates, or both, depending on the nature of their coordinating units. Hence, the macropolycyclic coreceptors carrying at least two metallic cation complexing units can form polynuclear cryptates in which the separation and arrangement of coordinated cations within the host cavity can be controlled by modifying the nature of the ligand. Figure 13.5 shows a dinuclear and a trinuclear cryptate.

These are hexamine-type coreceptors that can coordinate two metallic cations such as Cu(I) (homotopic coreceptor) or three metallic cations such as Ag(I) (heterotopic coreceptor), depending on the nature of the aromatic subunits incorporated in the macrocycles [10]. Indeed, three bipyridine moieties are needed to coordinate a third Ag(I) cation in the middle of the cavity. A ditopic coreceptor can be used for specific recognition of a substrate of appropriate size. Indeed, a homotopic coreceptor carrying two coordinating

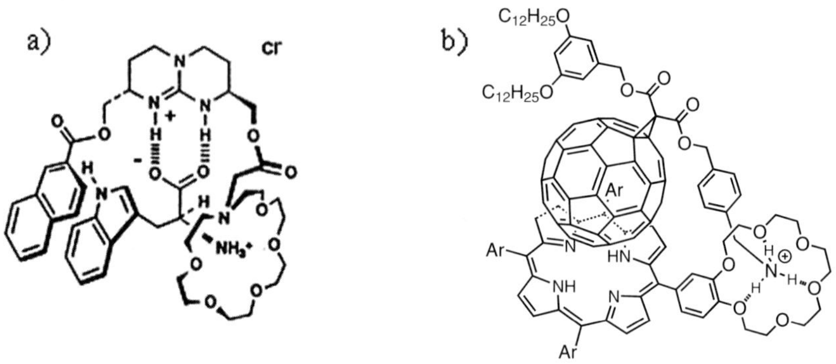

Fig. 13.7. (**a**) Enantioselective coordination of aromatic amino acids by a tritopic and acyclic heterotopic coreceptor [12]. (**b**) Recognition of a fullerene derivative substituted by an ammonium, by a ditopic coreceptor [13]

units located at its two ends will preferentially complex a substrate carrying two complementary groups located at the right distance from one another. Hence the hexaprotonated macrobicyclic polyamine shown in Fig. 13.6 specifically recognises terephthalate guest compounds [11]. Combining coordinating units of different kinds leads to heterotopic coreceptors capable of complexing substrates by simultaneous interaction with anions, cations and neutral molecules, using electrostatic interactions, van der Waals forces, donor–acceptor interactions, or hydrophobic effects.

Enantio- or diastereoselective molecular recognition is also possible using chiral hosts. A particularly interesting example concerns the enantioselective coordination of aromatic amino acids by the tritopic and acyclic heterotopic coreceptor shown in Fig. 13.7a. Enantioselective recognition occurs between the guanidinium function, the macrocycle and the naphthalene moiety of the coreceptor and the carboxylate, the ammonium and the aromatic group of the amino acid substrate, respectively. Another example is the recognition of a fullerene derivative substituted by an ammonium, achieved by a ditopic coreceptor carrying a crown ether and a porphyrinic aromatic macrocycle, as shown in Fig. 13.7b. Two types of cascade interactions come into play here: firstly, the coordination of the substrate ammonium by the crown ether of the coreceptor, and then the occurrence of π-interactions between the substrate fullerene and the porphyrin of its partner. These π-interactions considerably increase the stability of the complex formed between the ditopic coreceptor and its fullerene-based substrate.

Finally, it is worth mentioning the metalloreceptors. These are heterotopic coreceptors capable of coordinating both metallic cations and organic molecules. As an example, Fig. 13.8a shows a coreceptor with two porphyrin macrocycles able to complex metallic cations and two macrocycles recognis-

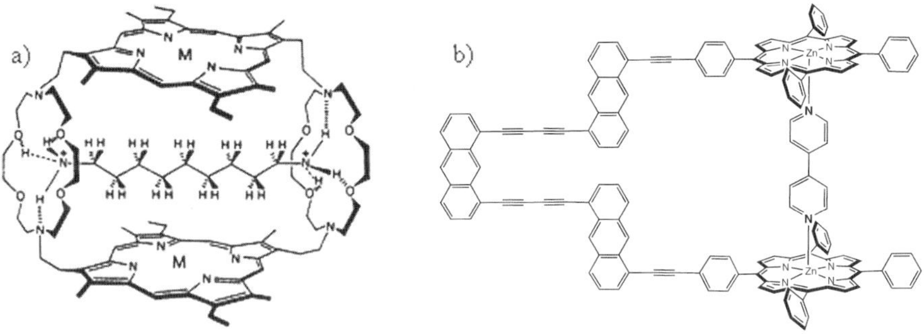

Fig. 13.8. (**a**) Heterotopic coreceptor that can coordinate both metallic cations and an organic molecule [14]. (**b**) Heterotopic coreceptor able to coordinate both metallic cations and dinitrogenated organic molecules with sizes in the range 3–12 Å [15]

ing ammonium groups. These coreceptors form supermolecules with mixed substrates by simultaneous coordination of metallic cations and an organic molecule functionalised by two ammonium cations. Another example is provided by a bis-porphyrin receptor carrying two porphyrins joined together by a spacer that can hold the porphyrins face to face but at a variable distance of 5–15 Å by rotation about acetylene junctions. This receptor with its adjustable cavity can thus host a whole range of substrates with sizes, i.e., separations between the two coordinating nitrogens, between 3 Å and 12 Å (see Fig. 13.8b).

13.5 Applications

There are applications of supramolecular chemistry in many different fields, including catalysis, transport processes, the synthesis of supramolecular devices, and structures self-assembled by means of programmed supramolecular systems [1].

Supramolecular Reactivity and Catalysis

Since enzyme reactions involve the coordination and reaction of a given substrate, they have all the characteristics of supramolecular chemistry. Supramolecular reactivity and catalysis involve two main steps: selective coordination of a given substrate and transformation of the latter within the thus formed supramolecular complex. In contrast to simple molecular reactivity, we have here a molecular recognition stage operating through coordination. For its part, supramolecular catalysis involves a third step in which the substrate is released. Selection of an appropriate substrate is not the only function of the coordination stage. The latter also serves to impose a conformation on the substrate that is close to the transition state of a given reaction. Consequently, a good supramolecular catalyst must have a greater tendency to form complexes with molecules in their transition state than in the conformation adopted in the absence of any coordination.

Transport Processes and Synthesis of Transporters

The organic chemistry of transmembrane transport processes has only recently attracted interest, even though the physicochemical and biological characteristics of these phenomena have been well known for many years now. The synthesis of molecular hosts able to selectively coordinate a given substrate, be it organic or inorganic, has made it possible to develop compounds which, made soluble in the membranes, behave as transporters, inducing a selective transport by making the membranes permeable to the complexed species. Like molecular recognition and catalysis, these transport processes represent one of the key functions of supramolecular chemistry. The chemistry of transport systems has three main goals: to synthesise suitable transporters, to invent new transport processes, and to study the applications of such processes in chemistry and biology.

Molecular and Supramolecular Devices

A molecular device is a chemical species with an organised structure endowed with a given function. It is made up of specific components arranged in an appropriate way and can be a supramolecular structure. The function carried out by the device consists in integrating together the various elementary operations achieved by each component. A device is said to be photonic, electronic, or ionic depending on whether the relevant components are photoactive, electroactive, or ionic, respectively, i.e., depending on whether photons, electrons, or ions are involved. There are two main types of component: active components carrying out a given operation (accepting, donating, transferring, etc.) with photons, electrons, or ions; and structural components contributing to the constitution of the supramolecular structure and the positioning of the components, in particular those involved in molecular recognition processes. Auxiliary components can be added to modulate the properties of the other two types of component. The formation of supramolecular entities from photoactive components may perturb the properties of the ground state and excited states of individual species, giving rise to novel properties; this is the domain of supramolecular photochemistry. As in catalysis, there are three key steps: coordination of a substrate by the host, implementation of a photochemical process, and finally either the restoration of the initial state or the realisation of a reaction.

Self-Assembly and Programmed Supramolecular Systems

Self-assembly is without doubt the most recently developed application in the field of molecular recognition, and forms a major part of current research into the control of information on the molecular level. Molecular recognition is used here as a way to control the construction of supramolecular species from their components, in such a way that they self-assemble. The formation of double helices is the most notable example. Self-assembly has generally been studied in biology and physics, but supramolecular chemistry provides a new opportunity for the chemical sciences to explore this area in a new light, making use of its ability to create sophisticated molecular architectures. Hence, supramolecular self-assembly consists in the spontaneous association of several components leading to the formation of discrete oligomolecules or polymolecular aggregates such as films, membranes, and so on.

Self-assembly via directed molecular recognition can be considered as an information-handling process. The information required for a given process to occur must therefore be stored at the molecular level. Systems with these characteristics are called programmed systems. Molecular programming involves incorporating at the molecular level information regarding the formation of a given supramolecular species. Three levels of information should be distinguished: selective coordination of a substrate at the molecular recognition level, an appropriate orientation for directing growth by self-assembly, and control of the end of the process by incorporating a suitable stop signal to indicate when self-assembly has been completed.

13.6 Prospects

From Structure to Information

Today, chemists are able to synthesise species with ever more sophisticated architecture and behaviour, and the language of information extends to the study of structure and constitution. In this respect, supramolecular chemistry opens the way to a better understanding of information science. Although it has up to the present time been developed in line with its primary definition, i.e., as the chemistry of species built by non-covalent interactions, molecular recognition and self-assembly now make it possible for supramolecular chemistry to tackle the concepts of information and programmed systems. Supramolecular chemistry is gradually becoming a branch of information science, concerned with data storage on the molecular level, and the possibility of reading such data, transferring it, and transcribing it on the supramolecular level.

A Step Towards Complexity

Going beyond programmed systems, a further step can be taken towards complexity through the synthesis of chemical systems able to 'learn', i.e., systems that are not just programmed to carry out some predetermined function, but which can be trained to accomplish a well-defined task and are capable of modifying themselves and adapting in response to some outside stimulus. This opens the way to evolution, i.e., the observation of gradual changes in the internal structure under the influence of environmental factors.

Chemistry and Biology, Creativity and Art

This sets the challenge, showing that a door has been opened between the animate and inanimate worlds, between the living and the non-living, via the control of supramolecular structures, functions and organisations. The highest level of complexity is expressed in living beings, and in particular in the brain, the nervous system, consciousness and thought. As has been shown on several occasions throughout this chapter, supramolecular chemistry has many points in common with biology. While many studies have been concerned with the synthesis of biomimetic species, chemists have also often applied biological processes in chemistry, as was the case for the use of enzymes as reactants, the synthesis of catalytic antibodies, or the control of genetic expression. The French chemist Marcellin Berthelot said: "chemistry creates its own object". In this sense, the essence of chemistry is not only to discover, but also to invent, and above all, to create.

References

1. J.-M. Lehn: Chem. Eur. J. **6**, 2097–2102 (2000)
2. P. Ehrlich: *Studies on Immunity*, Wiley, New York (1906)
3. E. Fischer: Ber. Deutsch. Chem. Ges. **27**, 2985 (1894)
4. A. Werner: Zeitschr. Anorg. Chem. **3**, 267 (1893)
5. B. Dietrich, J.-M. Lehn, J.-P. Sauvage: Tetrahedron Lett., 2885–2889 (1969); B. Dietrich, J.-M. Lehn, J.-P. Sauvage, J. Blanzat: Tetrahedron **29**, 1629 (1973); B. Dietrich, J.-M. Lehn, J.-P. Sauvage: Tetrahedron **29**, 1647 (1973)
6. J.-M. Lehn: Struct. Bonding **16**, 1 (1973)
7. J. Rebek: Topics Curr. Chem. **149**, 189 (1988); Acc. Chem. Res. **23**, 399 (1990); J. Rebek: Angew. Chem. Int. Ed. Engl. **29**, 245 (1990)
8. A.D. Hamilton: Bioorg. Chem. Frontiers **2**, 115 (1991)
9. J.-M. Lehn, E. Sonveaux, A.K. Willard: J. Am. Chem. Soc. **100**, 4914 (1978); B. Dietrich, J. Guilhem, J.-M. Lehn, C. Pascard, E. Sonveaux: Helv. Chim. Acta **67**, 91 (1984); for other examples of macrobicyclic receptors, see also: M.W. Hosseini, J.-M. Lehn: Helv. Chim. Acta **71**, 749 (1988); B. Dietrich, M.W. Hosseini, J.-M. Lehn, R.B. Sessions: Helv. Chim. Acta **68**, 289 (1985)
10. J. Jazwinski, J.-M. Lehn, D. Lilienbaum, R. Ziessel, J. Guilhem, C. Pascard: J. Chem. Soc. Chem. Commun. 1691 (1987); J. de Mendoza, E. Mesa, J.-C. Rodriguez-Ubis, P. Vazquez, F. Vögtle, P.-M. Windscheif, K. Rissanen, J.-M. Lehn, D. Lilienbaum, R. Ziessel: Angew. Chem. Int. Ed. Engl. **30**, 1331 (1991)
11. J.-M. Lehn, R. Meric, J.-P. Vigneron, I. Bkouche-Waksman, C. Pascard: J. Chem. Soc. Chem. Commun. 62 (1991)

12. A. Galan, D. Andreu, A.M. Echavarren, P. Prados, J. de Mendoza: J. Am. Chem. Soc. **114**, 1511 (1992)
13. N. Solladié, M.E. Walther, M. Gross, T.M. Figueira Duarte, C. Bourgogne, J.-F. Nierengarten: Chem. Commun. 2412–2413 (2003)
14. A.D. Hamilton, J.-M. Lehn, J.L. Sessler: J. Chem. Soc. Chem. Commun. 311 (1984); A.D. Hamilton, J.-M. Lehn, J.L. Sessler: J. Am. Chem. Soc. **108**, 5158 (1986)
15. R. Rein, M. Gross, N. Solladié: Chem. Commun. 1992–1993 (2004); L. Flamigni, A.M. Talarico, B. Ventura, R. Rein, N. Solladié: Chemistry, A Eur. J. **12**, 712 (2005)

14

Nanocomposites: The End of Compromise

H. Van Damme

14.1 Composites and Nanocomposites

Increase the Young's modulus of a glassy resin by a factor of ten without making it heavier, for a new ski design, for example? Triple the rupture strength of an elastomer? Improve the thermal behaviour of an object made from a thermoplastic polymer by 100 degrees, to make a car dashboard, for example, or a part for the engine? Double the fire resistance time for the sheath around an electricity cable? Reduce the oxygen permeability of a film by a factor of ten, to make long conservation food packaging? All these things have been made possible by incorporating a few percent of inorganic nanoparticles in a polymer matrix. Figures 14.1 and 14.2 illustrate two such nanocomposites: the first was obtained by incorporating lamellar clay particles, and the second by incorporating fibrous nanoparticles, in fact, carbon nanotubes.

Polymers now represent, in volume, something like a third of all the materials we use, in the form of objects made from 'plastic' or 'rubber' (elastomers), glues, resins, composites, paints, gels, and so on. In contrast to common opinion, the polymer materials used today in solid (or soft solid) form are rarely homogeneous. In actual fact, they are almost invariably composite materials, in the sense that they almost always comprise a dominant, continuous phase called the matrix, modified by incorporation of inorganic 'fillers'. The aim may simply be to reduce the cost of the material, but usually, it is rather to enhance specific properties. In common parlance, the word 'composite' is often reserved for materials made from two or more phases with characteristic sizes approaching the macroscale, i.e., the scale of the objects themselves, e.g., plywood. In the world of polymer materials, the standard example is provided by the epoxy resins, strengthened by long glass or carbon fibres. In reality, the idea of composite materials is much more general than this. Indeed, it is implemented whenever one attempts to modify some property by mixing, whatever the size and shape of the phases that happen to be mixed together.

The designer of composite materials aims basically to achieve a compromise. Addition of high-modulus but fragile carbon fibres to a low-modulus

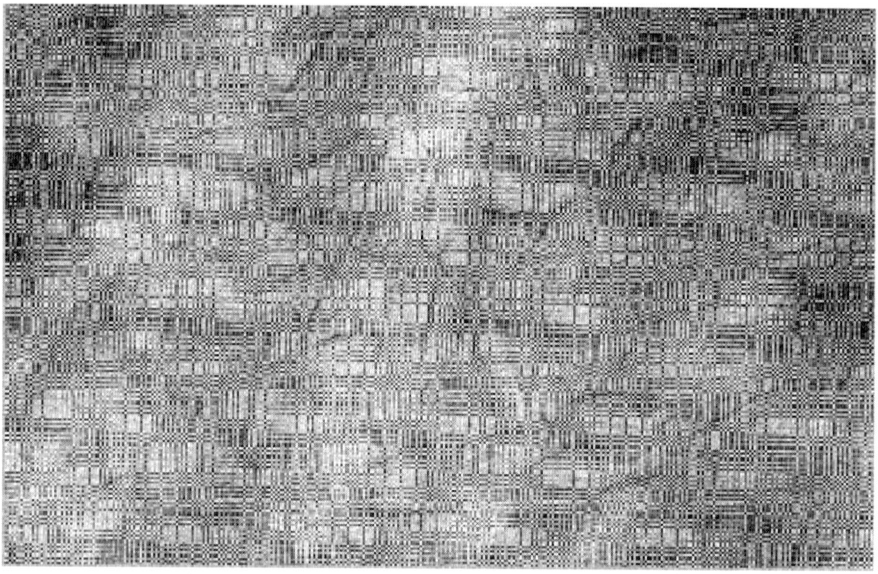

Fig. 14.1. Transmission electron microscope (TEM) image of a nanocomposite obtained by dispersing a smectite clay in a polyamide matrix. From Wang et al. [4]. The clay platelets (a montmorillonite), with thickness about 1 nm, are remarkably well dispersed, but are locally aligned by excluded volume effects due to their high anisotropy. The thickness of polymer separating them is only a few tens of nanometers. Scale bar 100 nm

polymer matrix leads to a light, rigid composite which may indeed possess a certain resilience, but at a cost in terms of manufacture. Loading a thermoplastic polymer with very rigid particles (or dispersoids) will increase the modulus, but it will also reduce the shock resistance. Dispersing rubber inclusions in a hard, transparent but fragile matrix will confer a better shock resistance to it, but the price to pay is often a loss of transparency. The great difference when using very small particles, or at least, particles with one nanoscale dimension, is that one may thereby overcome the need for this kind of compromise.

A first rather trivial reason is that, on this size scale, certain unwanted phenomena begin to tail off. An example is the scattering of light; or the concentration of stresses around inclusions, which is less pronounced in the case of very small inclusions, therefore limiting any weakening of the material. The strengthening of the electric field reacts in the same way as the size is reduced, thus improving the resistance to dielectric breakdown.

The second, more subtle reason is that, insofar as the properties of the polymer matrix are modified over a non-negligible distance by the proximity of a solid surface (over a few nanometers, say), the properties of the matrix are soon modified throughout. The enormous interface area between nanofiller

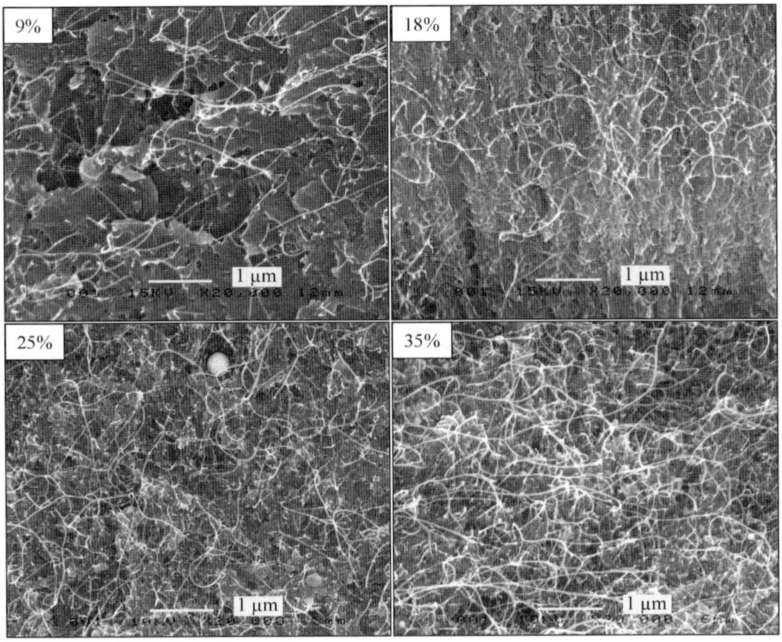

Fig. 14.2. TEM image of a nanocomposite obtained by dispersing single-wall carbon nanotubes (SWNT) in an epoxy matrix, for different mass fractions of nanotubes. The nanotubes are entangled due to their high flexibility. From L. Vaccarini [5]

and matrix means that, if it is given a certain thickness (the thickness of the interface or interphase region), the matrix properties can be modified throughout a significant fraction of its total volume. The logic is then no longer that of a simple mixture. One no longer disperses very rigid and dense inclusions in a softer and less dense matrix in order to produce a 'relatively' rigid and 'relatively' light mixture; on the contrary, one uses the inclusion–matrix interface to modify the properties of the matrix and thereby make it more rigid, without significantly altering its density. In other words, a matrix with bulk properties has been replaced by one with largely interfacial properties.

The notion of interface region is not a new one. It has been known since Gibbs that interfaces are not strictly 2D spaces, and that modified composition and properties extend over a certain distance. A direct determination of these variations is now possible, and it is observed that, in some cases, the interface region can be several tens of nanometers thick. The glass transition temperature is one of those properties influenced over considerable distances by the solid surface, as we shall see later. In such conditions, the combination of a large interface area and a thick interface region can result in a large volume throughout which properties differ from those of the pure polymer. The curves in Fig. 13.3 show the evolution of the specific interface area, i.e., the

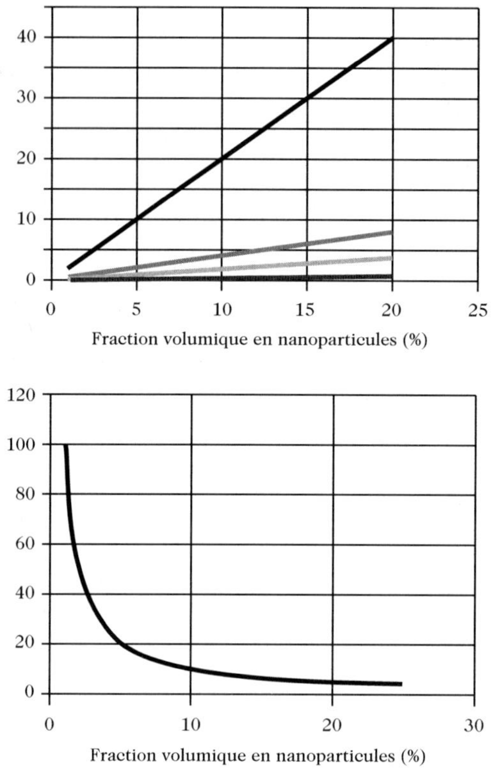

Fig. 14.3. *Top*: evolution of the interfacial area per unit volume of matrix (in $10^7\ m^3/m^2$) with the particle thickness (1, 5, 10 and 50 nm, from top to bottom), for a composite composed of platelets well-dispersed, i.e., not stuck together, throughout the matrix. *Bottom*: Evolution of the average distance (in nm) between platelets of thickness 1 nm (clay sheets, for example), as a function of the volume fraction of the platelets

interface area per unit volume of composite, and the average distance between particles, as a function of the particle size and their volume fraction in the case of platelets. It is easy to see how a few percent of particles with high specific area can modify properties throughout the volume of the matrix.

There is a third explanation, rather close to the last one, for the efficiency of nanofillers as compared with traditional fillers or strengtheners: this is the confinement effect, particularly strong in the case of particles in the form of platelets. When these have nanoscale thickness, a few percent volume fraction is enough to achieve average distances between particles of the same order of magnitude as the radius of gyration of the macromolecules. The polymer matrix is as though sliced up into ever finer slices within which the polymer molecules are confined, in interaction with two solid surfaces. The number of conformations that the molecules can adopt, and hence their entropy, is all the more reduced as this confinement increases. Confinement also imposes complex diffusion paths, much longer than the straight line distance. In fact, the medium behaves as an extremely tortuous porous medium.

There are potentially a great many fillers that would lend themselves to the synthesis of nanocomposites. To begin with, there is everything that can be synthesised in the form of approximately spheroidal nanoparticles, either crystalline or amorphous: metals, oxides, chalcogenides, and so on.[1] Such particles, with sizes as small as the nanometer, can in some cases be synthesised in situ in the polymer, or more often, in the monomer prior to polymerisation. They are often reserved for special applications, e.g., polymers with specific electrical, magnetic or optical characteristics.

Regarding thermomechanical properties or barriers (permeability) mentioned at the beginning of this chapter, the main part of the research effort has concerned lamellar silicas and silicates.

Besides these, there is another special case which has little regard for cost (at least, up to the present time): this concerns the carbon nanotubes. Despite some rather disappointing results, the properties of carbon nanotubes – and in particular, the so-called single-walled nanotubes (SWNT) – are a priori so interesting that their use in nanocomposites demands further exploration.

In this chapter, we shall focus on lamellar silicate nanocomposites and, to a lesser degree, carbon nanotubes. We shall show, for several properties such as strengthening, impermeability, dimensional stability, and fire resistance, how the notions of interface region and confinement can help us to understand what might be called the nanocomposite effect. At the same time, the nanotubes will illustrate some of the difficulties that remain to be overcome.

14.2 Introduction to Polymers

Before tackling the main themes here, it will be useful to recall some of the basic ideas and physical properties required to understand polymers. For brevity and efficiency, we adopt a somewhat textbook approach in this section. Recall to begin with that polymer materials are composed of macromolecular chains, either linear or branching (ramified), themselves composed of a succession of basic units called monomers, each one bound to the next. In the solid state, the chains can be organised in a locally ordered way (one then speaks of semi-crystalline polymers such as polyethylene, polypropylene, or nylon), or totally amorphous (e.g., PMMA or plexiglas). In this latter state, the material is a random tangle of chains, rather well parallelled by a plate of spaghetti. The number of different conformations that a macromolecular chain can adopt is

[1] We shall not consider here the case, discussed in Chap. 30, of carbon blacks and silicas, used traditionally to strengthen elastomers, e.g., in tyres. These particles, with their rather special characteristics as fractal clusters, are situated at the upper end of the nanometric scale, which certainly does nothing to stop them being remarkably effective!

in principle enormous. In a crystalline solid phase, only very local movements are possible. In an amorphous solid phase, whole segments of the chains can move in a cooperative manner, these segments being longer as the temperature increases. In the molten state, whole molecules are able to move in a diffusion mode known as reptation. Each molecule follows a sort of tube formed by the many other molecules that lie along or across it. The same thing happens in concentrated solutions.

14.2.1 Ideal Chains

Taken individually, each polymer chain in an amorphous polymer fills up space in a rather inefficient way. At first sight, one might think that its conformation would depend on the details of the chemical structure of the molecule. In fact, and fortunately, this is not the case, provided that one considers the structure of the chain at a resolution that is low enough to wipe out local details. Then all random chains exhibit universal properties. But at what resolution should one observe a chain in order to fulfill these conditions? Answering this question amounts to choosing the size of the chain 'pixels' that will be used to describe the various conformations. Expressed like this, the answer is straightforward: it suffices to choose as chain pixels segments of length Δl that are long enough to ensure that the orientations of the unit vectors $\boldsymbol{e}(l)$ and $\boldsymbol{e}(l+\Delta l)$ located at the two ends of a segment are uncorrelated. Between the beginning and end of the segment, the chain loses any memory of its initial orientation. The required length scale is called the persistence length and denoted by l_{p} (see Fig. 14.4). Naturally, this length will depend on the rigidity of the chemical groups comprising the chain.

Equipped with this characteristic length, it is now possible to construct a representation of the chain in which its true contour is replaced by a succession of segments each longer than l_{p}. By doing this, resolution has been lost, but a complicated object has been replaced by a simple mathematical object, viz., a succession of randomly orientated segments of equal length. In other words, the polymer has been represented by a random walk, i.e., a succession $(\boldsymbol{r}_1, \boldsymbol{r}_2, \boldsymbol{r}_3, \ldots, \boldsymbol{r}_N)$ of vectors whose orientations are completely uncorrelated, as shown in Fig. 14.4. The simplest model for describing a polymer, i.e., the ideal chain, is thus a random walk. And the simplest model for a polymer melt, an amorphous solid polymer or concentrated polymer solution is an ensemble of entangled random walks.

Consider the random walk model. The vector joining the ends of the path can be written

$$\boldsymbol{R}_N = \sum_{i=1}^{N} \boldsymbol{r}_i \ ,$$

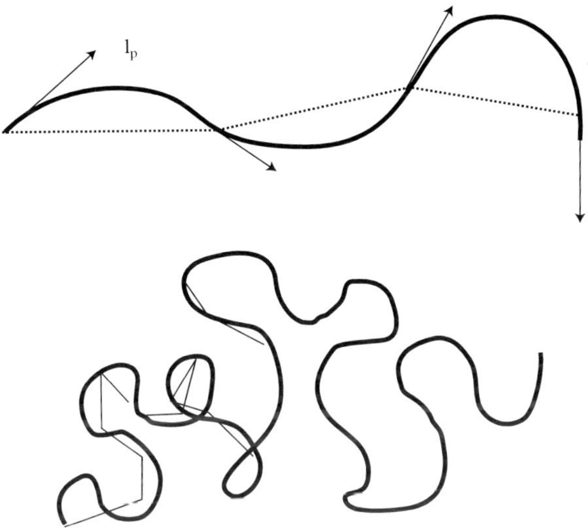

Fig. 14.4. At length scales greater than the persistence length l_{p}, a polymer chain loses all memory of its local orientation. At these scales, the chain can be represented by a random walk with step at least equal to l_{p}

where N is the number of segments of the walk or the polymer chain.[2] The average of $\boldsymbol{R}_N$ is not a useful quantity: it is zero, since all orientations are equiprobable. One therefore defines the mean squared length R_0^2, the average of the square of the end-to-end distance, by

$$R_0^2 \equiv \langle \boldsymbol{R}_N^2 \rangle = \left\langle \left(\sum_1^N \boldsymbol{r}_i \right)^2 \right\rangle = \left\langle \sum_1^N \boldsymbol{r}_i \cdot \boldsymbol{r}_j \right\rangle .$$

All the steps of the random walk representing the chain have the same length l and are strictly uncorrelated. Under these conditions, all the scalar products $\boldsymbol{r}_i \cdot \boldsymbol{r}_j$ with $i \neq j$ are zero and we have

$$R_0^2 = N \left\langle |\boldsymbol{r}_i|^2 \right\rangle = Nl^2 = Ll ,$$

where $L = Nl$ is the total length of the path. Finally, we thus have

$$R_0 \equiv \langle \boldsymbol{R}_N^2 \rangle^{1/2} = N^{1/2} l .$$

The radius of an ideal statistical coil thus varies as the square root of its mass, since the mass is proportional to N. Inverting the last relation, we arrive at

[2]Note that, for a given macromolecule, N depends on the length of the segments, which is arbitrary, provided that it remains greater than the persistence length. However, the following argument is independent of this choice.

the strange conclusion that the mass of polymer contained in a coil of mean squared radius R_0 varies as the square of the radius, i.e.,

$$N \propto R_0^2 \, , \quad \text{i.e., mass} \propto R_0^2 \, .$$

It is as though the coil, viewed with resolution l, were an object of 'effective' dimensionality $d = 2$, since the mass of a 1D object goes as its size, that of a 2D object goes as the square of its size, and that of a 3D object goes as the cube of its size. An ideal chain taken individually thus occupies space very inefficiently. Moreover, its density decreases as its size increases, i.e., the object contains more and more empty space as it grows larger. Indeed, the envelope volume occupied by the chain varies as the cube of its radius, whereas its mass varies as N, which itself varies as the square of the radius. Hence,

$$\rho_{\text{app}} = \frac{\text{mass}}{\text{volume}_{\text{app}}} \propto \frac{N}{R_3^N} \propto \frac{R_N^2}{R_N^3} = R_N^{-1} \, .$$

The density of a chain thus decreases as its length increases, and an infinite chain would therefore be infinitely diffuse. In a concentrated solution, a polymer melt or an amorphous solid polymer, these surprising properties are hidden by entanglement. The number of monomers of a given chain in a sphere of radius R always varies as R^2, but the total number of monomers belonging to different chains varies as R^3. The medium itself is 'normal'.

14.2.2 The Glass Transition

At high temperatures (let us say, temperatures above the melting temperature T_{m} of the crystalline phase, if it exists), an amorphous thermoplastic polymer, i.e., one which softens when heated, is nothing but a very viscous liquid in which the chains move around by reptation. The viscosity depends on the temperature through a standard relationship known as an Arrhenius law:

$$\eta = \eta_0 \exp \frac{\Delta G_{\text{a}}}{k_{\text{B}} T} \, ,$$

where ΔG_{a} is the free activation enthalpy. The viscosity increases gradually as the temperature decreases. At a temperature close to 0.7 or $0.8 T_{\text{m}}$, the viscosity begins to rise much more quickly and the polymer more clearly displays its viscoelastic properties. In general, this increase in viscosity is well described by an empirical law called the Vogel–Fulcher–Tammann or VFT law:

$$\eta = \eta_0 \exp \left(\frac{\text{const.}}{T - T_0} \right) \, ,$$

where T_0 is a temperature at which the viscosity would diverge. At the same time, elastic properties are displayed at ever lower frequencies as the temperature falls. The temperature T_0 cannot in fact be reached. Large-amplitude

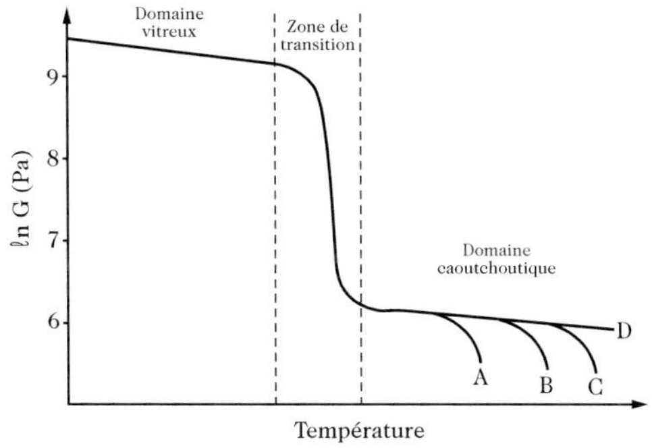

Fig. 14.5. Temperature dependence of the modulus of an amorphous polymer (schematic). At temperatures above the glass transition, the length of the rubber plateau where rubber elasticity is displayed increases with the cross-linking (or reticulation) from A to D

motions are totally blocked before it can be attained and the material goes into a glassy state. Mechanically, the latter then behaves as a hard and brittle elastic solid. The transition temperature is called the glass transition temperature, denoted by T_g.

The temperature dependence of the shear modulus as shown in Fig. 14.5 directly reflect this evolution. At temperatures below T_g, the modulus is typically of the order of a few GPa. During the transition, it drops by several orders of magnitude and stabilises temporarily in the MPa region. This is the rubber plateau, the soft solid region, which extends over a temperature range that increases as the chains are made longer (and friction increases during reptation). Finally, the chains flow. An important exception occurs when the chains are cross-linked: flow is then impossible and the polymer must break up before it can flow. This is the case for elastomers.

The glass state is basically a metastable frozen state. A simple model which shows why the system blocks is the free volume model. The argument goes as follows. In an amorphous polymer in any molten or supercooled (or any liquid) state, there is an unoccupied volume called the free volume, denoted by v_f, which can be redistributed without changing the total energy of the system. Locally, such a redistribution can generate spaces that are large enough for chain elements to carry out movements of similar amplitude to their size, thus conferring a fluid character on the system. The critical point is that the free volume depends sensitively on the temperature and vanishes at a finite temperature. There is thus a temperature where even the most extreme redistribution of the total available free volume is not sufficient to generate the space required for movement.

The fact that there really does exist unoccupied volume between the ideal chains entangled together in an amorphous polymer is easily demonstrated by dissolving in a solvent. The final volume is less than the sum of the initial volumes of the two constituents. However, the notion of free volume is a little more subtle than this. The problem was formulated by M. Cohen and D. Turnbull [1]. Each structural unit of the polymer is situated in a cage whose walls are formed by neighbouring units (of the same or different chains). The function describing the interaction potential of a unit in its cage is of Lennard-Jones type, with a point of inflection at a certain distance which separates a short-range zone (small cage) from a long-range (large cage) zone. Imagine two neighbouring cells attempting to exchange unoccupied volume. This can be done by a motion of the separating wall such that the volume of one of the cages increases at the expense of the volume of the other. If the two cages are initially in the compact configuration, energy would have to be supplied to redistribute the unoccupied volume, because the energy recovered by contracting one of the cages is less than the energy cost in expanding the other. On the other hand, if the two cages are in the broad configuration, such a rearrangement can be carried out without net energy expenditure, and even with a gain in energy. Cohen and Turnbull referred to the free volume as that part of the unoccupied volume that can be redistributed without energy expenditure. Since the average unoccupied volume of the cages increases with temperature in proportion with the thermal expansion, there is therefore a temperature T_0 such that, for $T \leq T_0$, the unoccupied volume is no longer free volume, but simply expanded solid volume, while for $T \geq T_0$, a fraction of the unoccupied volume is genuine free volume. One thus has

$$v_{\mathrm{f}} = v_0(\alpha_{\mathrm{liq}} - \alpha_{\mathrm{glass}})(T - T_0) \,,$$

where v_0 is the minimal volume of a cage and α_{liq} and α_{glass} are the coefficients of thermal expansion of the liquid and the glassy state, respectively.

At temperatures above T_0, the distribution of the cage volume broadens significantly, since there is no energy cost, and the configuration entropy grows in parallel. One can calculate the most probable free volume distribution (the one that maximises the number of possible configurations) using standard methods of statistical physics:

$$p(v) = \frac{\gamma}{v_{\mathrm{f}}} \exp\left(-\frac{\gamma v}{v_{\mathrm{f}}}\right) \,,$$

where $p(v)$ is the probability of the existence of a cage of free volume in the range from v to $v + \mathrm{d}v$, and v_{f} is the average free volume per cage, while γ is a parameter close to unity which accounts for a certain overlap between free volumes. If it is assumed that a minimum free volume v^* is required for a rearrangement by diffusion or reptation to occur, we obtain by integration the following dependence of the diffusion coefficient and fluidity (reciprocal viscosity):

$$D \text{ or } \phi = \frac{1}{\eta} \propto \exp\left(-\frac{\gamma v^*}{v_{\mathrm{f}}}\right) \,.$$

Given the expression for the variation of the average free volume, this equation agrees with the empirical VFT expression.

14.2.3 Entropic Elasticity

The very large number of conformations available to a polymer chain has a rather unexpected consequence: it exhibits a form of elasticity which, unlike the elasticity observed in most other materials, is absolutely unrelated to the attractive interactions between its units. This elasticity occurs in elastomers (rubber). An elastomer is an amorphous polymer material at a temperature for which the chains have high mobility. It is thus in a state close to the molten state, but in contrast to that state, mobility is not total because some covalent bonds have been created between the chains. An elastomer is thus a sort of macromolecular net, each element of which is a portion of the ideal chain.

In a crystal, elasticity is caused by the fact that a macroscopic deformation, either compressional or tensional, shifts the atoms or molecules from their equilibrium position at the minimum of the intermolecular potential, and this generates a restoring force proportional in the first approximation to the deformation. It is thus the intermolecular potential, i.e., the intermolecular potential energy, which lies at the origin of elasticity in crystals. But in elastomers, the elasticity arises because a deformation reduces the number of different conformations accessible to the macromolecules, and thereby reduces their entropy, which is unfavourable. When the stress is released, the macromolecules return to their most probable state, i.e., the state of higher entropy, reassuming their initial dimensions and their statistical coil conformation. The whole process occurs at constant energy.

To bring out this difference between enthalpic and entropic elasticity, consider the differential $\mathrm{d}U$ of the internal energy U of a piece of matter of volume V, composed of a tangle of ideal chains and subjected to a force F producing an elongation $\mathrm{d}x$. According to the second law of thermodynamics, we have

$$\mathrm{d}U = \mathrm{d}q + \mathrm{d}w = T\mathrm{d}S - P\mathrm{d}V + F\mathrm{d}x \; .$$

In this expression $\mathrm{d}q$ is the heat supplied to the system and $\mathrm{d}w$ is the work done on it. Under reversible conditions, $\mathrm{d}q = T\mathrm{d}S$, where T is the temperature and S the entropy, while $\mathrm{d}w$ can be decomposed into a bulk component $-P\mathrm{d}V$, with P the pressure, and an elongation component $F\mathrm{d}x$.

At constant temperature and volume, we have

$$\left(\frac{\partial U}{\partial x}\right)_{T,V} = T\left(\frac{\partial S}{\partial x}\right)_{T,V} + F \; .$$

Note that the assumption of constant volume is a reasonable one, implying that the sample narrows as it is extended. Now the extension of this set of mobile chains occurs without change in the internal energy. The extension of the sample in no way shifts the chains from their energetic equilibrium position in the intermolecular potential. We thus have

$$\left(\frac{\partial U}{\partial x}\right)_{T,V} = 0 \qquad \text{and} \qquad F = -T\left(\frac{\partial S}{\partial x}\right)_{T,V} .$$

It thus remains to evaluate the partial derivative of the entropy with respect to the elongation. According to the Boltzmann formula, the entropy is related to the number Ω of different configurations of the chains by

$$S = k_{\mathrm{B}} \ln \Omega ,$$

where k_{B} is the Boltzmann constant. The number of configurations is a function of the stretching of the chain. The more it is stretched, the lower will be the number of different conformations leading to the same distance between the endpoints. This number can be assumed proportional to the probability that the chain endpoints are separated by a distance R. This probability is a Gaussian function of the distance (the distance between the endpoints is a sum of random variables and any sum of random variables leads to a Gaussian function):

$$P(R) = A \exp\left(-\frac{3R^2}{2\langle R_N^2\rangle}\right) = A \exp\left(-\frac{3R^2}{2Nl^2}\right) ,$$

where N is the number of segments in the chain and l is their length, as before. The prefactor A is equal to

$$\left(\frac{3}{2\pi\langle R_N^2\rangle}\right)^{3/2} = \left(\frac{3}{2\pi N l^2}\right)^{3/2} .$$

The entropy can thus be written

$$S = \ln A - k_{\mathrm{B}}\left(\frac{3R^2}{2Nl^2}\right) .$$

To obtain the relation between the force and the elongation, it remains to find the relationship between the macroscopic extension and the stretching of the chains. If the chains were not cross-linked, the fact of exerting a tensional force on the sample would simply serve to disentangle the chains by reptation of the chains in their tubes, unless by chance one happened to take hold of the two ends of the same chain. By virtue of the cross-links, the stress on the macroscopic sample is transmitted to all the chains. The derivative with respect to x can thus be replaced up to a factor by a derivative with respect to R, and we then have

$$F = -T\left(\frac{\partial S}{\partial x}\right)_{T,V} \propto -T\left(\frac{\partial S}{\partial R}\right)_{T,V} = 3k_{\mathrm{B}}T\frac{R}{Nl^2} .$$

This expression shows that the force is proportional to the length R. It is therefore an analogous relation to Hooke's law. It also shows that the force is proportional to the temperature. Indeed, it is well known that, when an elastic band is heated, it grows shorter.

14.3 Nanofillers

We shall restrict the discussion here to two types of non-standard nanofiller (i.e., not silicas or carbon blacks) which are currently stimulating the most research interest: clays, especially those in the smectite family, and carbon nanotubes.

14.3.1 Clays

From a mineralogical point of view, clays form an extremely broad family. They are all phyllosilicates in which the main structural framework is a continuous 2D layer of tetrahedra made from oxygen atoms coordinating silicon and sometimes aluminium or iron atoms. This tetrahedral layer is bound to a layer with octahedral coordination in which oxygens or hydroxyls (OH) coordinate Al^{3+}, Mg^{2+}, Fe^{3+}, Fe^{2+}, or Li^{+} cations. The structure formed by binding an octahedral layer to a tetrahedral layer is called a 1:1 sheet, and it has thickness about 0.7 nm. By binding a tetrahedral layer on either side of an octahedral layer, one obtains a 2:1 sheet, which has thickness around 0.9 nm (see Fig. 14.6).

1:1 and 2:1 sheets are not necessarily electrically neutral. Ionic substitutions in the crystal lattice can induce a deficit of positive charge, reflected by a net negative charge. Electrical neutrality is then maintained by species between the sheets, which may be anhydrous cations such as K^{+}, as in the micas, hydrated cations as in the smectites, or a continuous layer of hydroxides as in the chlorites. The type of sheet, its electrical charge and the nature of the interlayer species can thus be used to classify the different kinds of clay. Each family possesses specific structural, morphological, and colloidal characteristics. Here are a few examples.

Kaolinite

Kaolinite is a typical 1:1 clay, with zero lattice charge. There are no interlayer species and the stacks of sheets are bound together by hydrogen bonds and van der Waals forces. The crystallites resulting from these stacks are rigid platelets with typical lateral extensions of a few hundred nanometers, and a thickness of a few tens of nanometers. Water cannot enter the interlayer space in kaolinites. Despite the absence of a permanent lattice charge, the acid–base equilibria of the –OH groups present on the lateral surfaces of the platelets

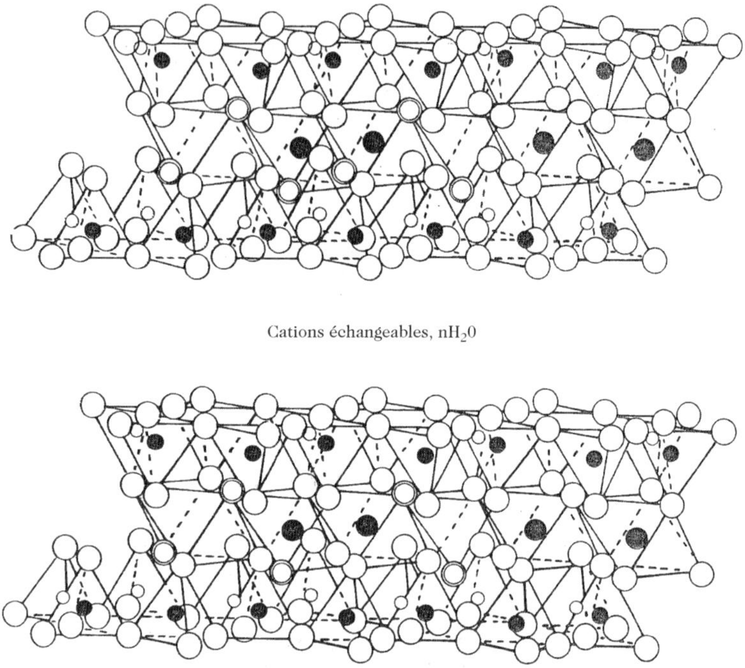

Fig. 14.6. Structure of dioctahedral 2:1 phyllosilicate sheets, e.g., montmorillonite. The sheets carry a negative charge due to substitutions of cations of given valence by cations of lower valence. The interlayer space is occupied by water molecules and, in natural clays, by Na^+, Ca^{2+} or Mg^{2+} cations

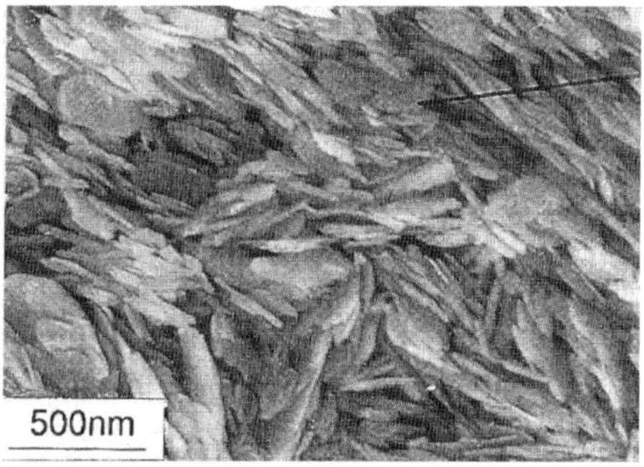

Fig. 14.7. Scanning electron microscope (SEM) image of a kaolinite. From D. Tessier, INRA

confer a small electric charge on kaolinite particles which depends on the pH: positive in an acid medium, when these groups have the form $-OH_2^+$, and negative in an alkali medium, when they have the form $-O^-$. Kaolinite is not a genuine nanofiller. Platelet thicknesses are typically in the range 50–100 nm

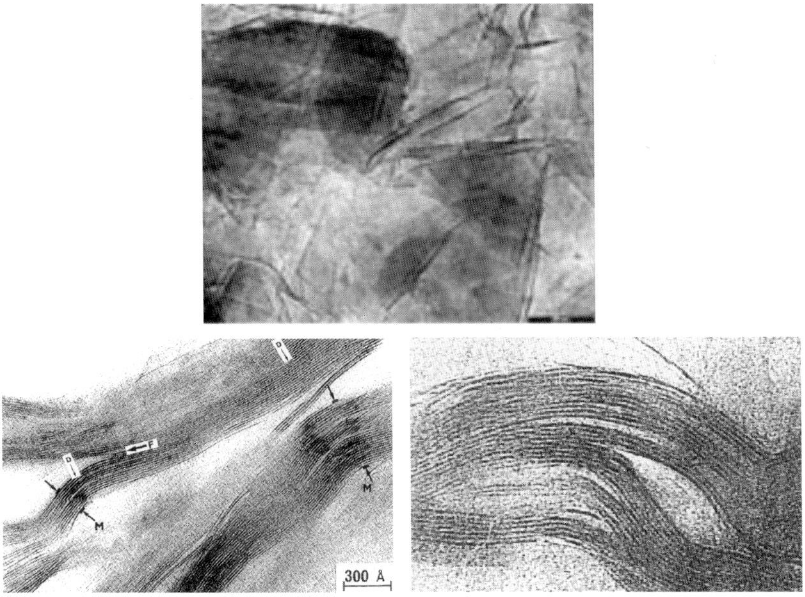

Fig. 14.8. TEM images of montmorillonite. *Upper*: top view showing large lateral extension (∼ 1 μm) of the sheets. From R. Mulhaupt. *Lower*: ultrathin cross-sections, after exchanging water for organic solvents and including within a resin. The sheets are seen to form bundles (*arrows*), which split apart, then join up with others. The art of manufacturing nanocomposites is to disperse all sheets individually. Lower left image from H. Gaboriau (BRGM), C. Clinard (CNRS), F. Bergaya (CNRS), and C.H. Pons (University of Orléans). Lower right image from D. Tessier (INRA)

and the specific surface area is of the order of 10–20 m^2/g (see Fig. 14.7). However, it is still a very finely divided filler, with a high shape factor.

Smectites

Smectites have practically the opposite properties to kaolinites. There is a significant electrical charge on their 2:1 sheet, of the order of 0.7 negative charge per unit cell, or about 1 meq/g. It is balanced by hydrated interlayer cations which, in natural smectites, are usually calcium, magnesium, or sodium ions. The large hydration energy of these ions allows water to enter the interlayer space, causing intracrystalline swelling, and this makes them easy to exchange with other metallic or organic cations. This swelling may proceed to very wide separation distances due to repulsive forces of entropic origin when the compensating cations form a double diffuse layer around the particles (osmotic swelling). Each 2:1 sheet has an extremely large lateral extension in comparison to this thickness. It can reach as much as 1 μm, which corresponds to an aspect ratio of 10^3. Taken alone, the sheets are extremely flexible, although this flexibility decreases when they are stuck together by hydrated

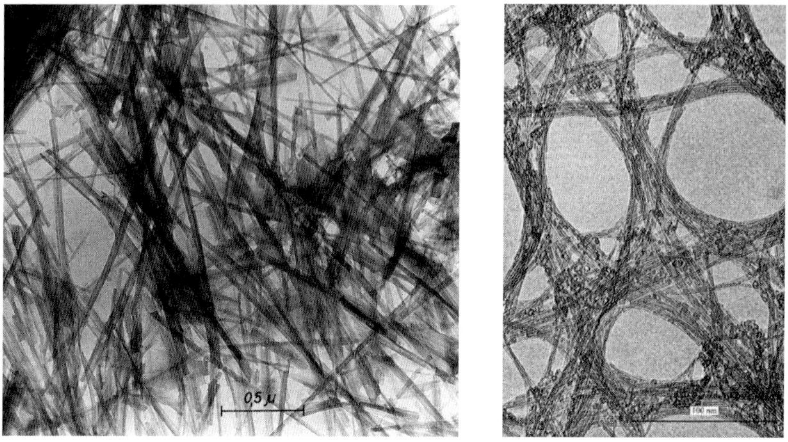

Fig. 14.9. TEM image of two fibrous clays. *Left*: sepiolite. From E. Ruiz, C.S.I.C., Madrid. *Right*: imogolite. From M. Nakai and N. Yoshinaga, International Conference on Clays, 1978. It can be seen that the imogolite fibres are in fact nanotubes

interlayer cations. The most common representatives of the smectite family are the montmorillonites (see Fig. 14.8), which have the trade name of Bentonite for a certain minimal level of purity, and the hectorites, for which a commercially produced synthetic form is Laponite.

Fibrous Clays

The fibrous clays form a third important group. Sepiolite and palygorskite (or attapulgite) are the best known members of the family. Their crystal structure is made from 2:1 sheets, but with lateral extensions of only a few tetrahedra and octahedra. This means that the sheets are more like ribbons with rectangular cross-section which associate with one another to form nanoporous fibres. A typical sepiolite is made from fibres of length 2–3 µm and diameter 50 nm (see Fig. 14.9). The 1:1 sheets can also form nano-objects with fibrous morphologies by rolling themselves up. They thus form silicate nanotubes of diameter 2 nm and micrometric lengths. This is the case in imogolite, common in volcanic soils (see Fig. 14.9).

Natural or synthetic clays are hydrophilic. This means that a good dispersion in polymers is not favoured, since the latter are generally non-polar. Clays must therefore by subjected to a surface treatment making them compatible with the matrix. By far the most common method is cation exchange with an amphiphilic organic cation. The most widely used are alkylammonium ions. The clay surface grows less polar as the alkyl chain is made longer and the charge density (and hence the number of organic cations) is raised. At high charge densities, the chains adopt a paraffin-type organisation, as shown in Fig. 14.10.

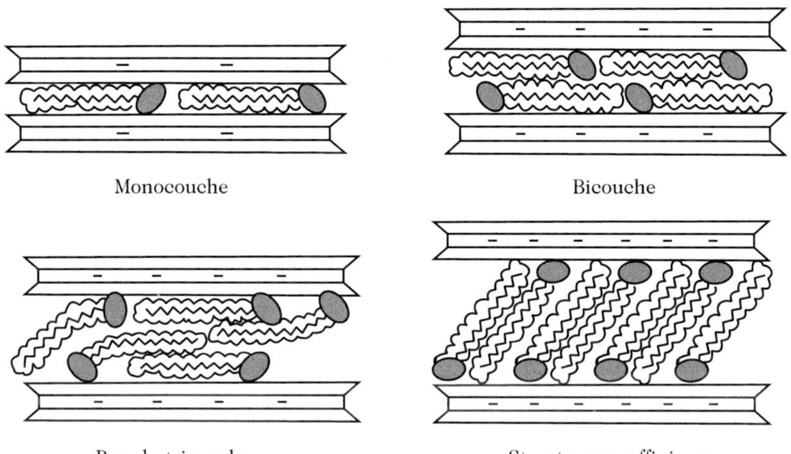

Fig. 14.10. Schematic representation of the arrangement of alkylammonium ions in the interlayer space of smectites at increasing charge densities. At high charge densities, the chains adopt a paraffin-type organisation

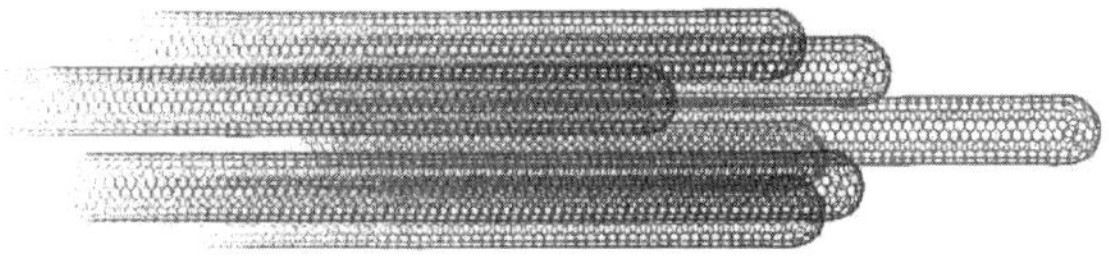

Fig. 14.11. Schematic representation of a bundle of single-walled nanotubes. From L. Vaccarini [5]

14.3.2 Carbon Nanotubes

With their unique association of strength, rigidity, and flexibility at densities as low as 1.8 g/cm^3, carbon nanotubes may prove to be the ideal strengthening fibre. Carbon nanotubes (CNT) are hollow cylinders with walls formed from carbon atoms. They have diameters of 0.7–10 nm, and lengths of micrometric order, although much longer ones can be grown on suitable walls. A CNT consists of a graphene sheet, i.e., an elementary sheet of carbon atoms with hexagonal symmetry in which atoms are bound together by sp^2 bonds, rolled up into a cylindrical shape, and in some cases closed at each end by a half-fullerene. A graphene sheet is extremely rigid under tension, but it is also very flexible owing to the ability of the lattice of sp^2 hybridised atoms to rehybridise when the sheet is curved, the degree of sp^2/sp^3 rehybridisation being proportional to the curvature. A CNT can be considered as a 1D structure, since the ratio of length to diameter can reach values as high as 10^4 or 10^5. Considering the graphene structure, CNTs differ by their symmetry, diameter, and in the relevant cases, the way they are closed at the ends. The symmetry is determined by the helicity or chirality, measuring the orientation of the carbon hexagons of the CNT surface with respect to the tube axis. The helicity is very important for the electronic properties, but not for mechanical properties. In fact, CNTs tend to gather together and exist in two different forms: multi-walled nanotubes (MWNT), consisting of several coaxial nanotubes fitted one inside the next, and single-walled nanotubes (SWNT), generally organised in bundles (or ropes) of aligned nanotubes (see Fig. 14.11).

On the mechanical level, a nanotube should therefore combine rigidity and flexibility, thus forming a supple and strong nanofibre. Electron microscope observations do indeed provide a glimpse of this flexibility (see Fig. 14.12). Many simulations of nanotubes have been carried out to determine their mechanical properties under axial, torsional and bending stresses. Measurements have also been made, either on single isolated nanotubes or on nanotubes within composites. It is observed that, under axial loads, SWNTs do indeed have an extremely high Young's modulus E, of terapascal order. This increases with the tube diameter to reach 1.2 TPa for tubes with diameters greater than 1 nm, agreeing with simulations. The Poisson coefficient stabilises at around 0.3. Apart from this exceptionally high rigidity, carbon nanotubes are very strong. The compression strength of SWNTs is estimated to be in the range 100–150 GPa, well above that of conventional carbon fibres (of the order of 1 GPa). Under tension, the strength may be as high as 50 GPa or more, with relative extensions of more than 10% at rupture. When subject to bending or torsion stresses, nanotubes are remarkably elastic, capable of withstanding large strains without showing signs of fragility, plasticity, or atomic rearrangement. Manipulations under atomic force microscopy (AFM) have shown that MWNTs can be bent repeatedly to large strains, of the order of 3°/nm, without breaking, as predicted by simulations. Beyond a critical stress, an elbow forms in the nanotube, without rupture. Under torsion, all strains are reversible up to torsion angles of 20–30°/nm, demonstrating the considerably elasticity of nanotubes in this strain mode.

14.4 Strengthening and Permeability Control: Models

Before examining some examples of experimental results demonstrating the specific features of nanocomposites, we need to describe the expected behaviour in a classical context. We shall consider two simple models, one for the Young's modulus, and the other for permeability. These will then be used to evaluate the effectiveness of nanofillers in actual materials.

14.4.1 Strengthening: Increasing the Modulus

Calculating the properties of a disordered heterogeneous or composite medium is an old problem. The first general solutions worthy of the name go back to the

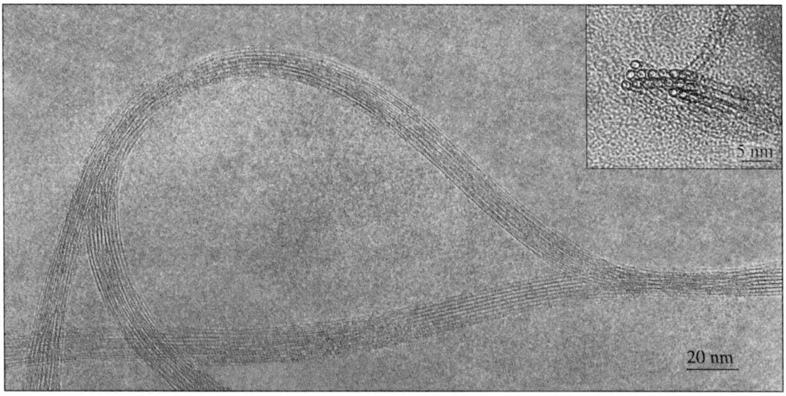

Fig. 14.12. TEM images of bundles of single-walled carbon nanotubes. Filaments are easily deformed under flexion and torsion. From L. Vaccarini [5]

nineteenth centure. The general strategy is to define a homogeneous 'effective' medium, endowed with properties, also qualified as effective, that are identical to those of the true medium at the same scale. A first strategy consists in directly calculating these effective properties on the basis of a structural model. Another approach, which can be very simple, does not attempt to calculate an effective quantity directly, but rather seeks to obtain upper and lower bounds for it.

We shall select this latter approach, in its simplest form, which involves finding the so-called series and parallel bounds. It consists in replacing the true disordered and statistically isotropic composite with n phases by a stack of n layers in which one calculates properties parallel and perpendicularly to the stacking axis. It can be shown (Wiener, 1912) that the value of a property of the true composite in an arbitrary direction necessarily lies between the two values obtained for the stacking model, as illustrated in Fig. 14.13. As regards the modulus, it can be expressed in the following way:

$$\left(\sum_i \phi_i \frac{1}{E_i}\right)^{-1} \leq E_{\text{eff}} \leq \sum_i \phi_i E_i \ ,$$

in which the ϕ_i are the volume fractions of the different phases.

We shall illustrate this general principle in the simple case of a two-phase medium. The property we are concerned with here is the elastic modulus. Each phase has a modulus M defined by a linear stress–strain relation:

$$\sigma_1 = M_1 \varepsilon_1 \quad \text{and} \quad \sigma_2 = M_2 \varepsilon_2 \ .$$

Considering a rigid plate, we apply a uniform compression stress to the model medium in the two extreme configurations, as shown in Fig. 14.14. In the parallel configuration, the two phases suffer the same displacement: $\langle \varepsilon \rangle =$

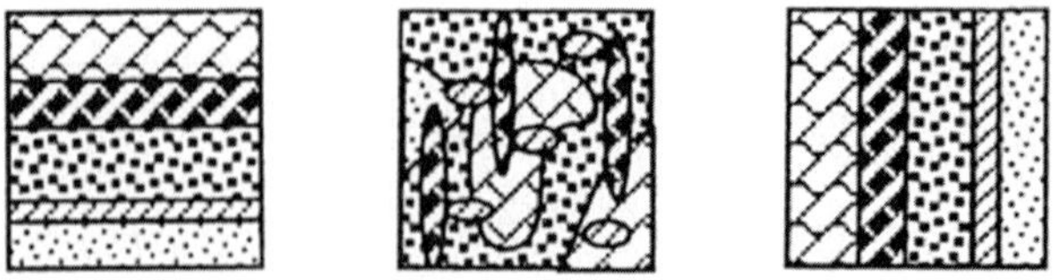

Fig. 14.13. Schematic representation of the series bound (*left*) and parallel bound (*right*) methods for a material comprising several phases. The medium is represented by a succession of layers each of which has a volume fraction corresponding to that of the phase it represents. The properties of the actual material necessarily lie between those of its two extreme representations

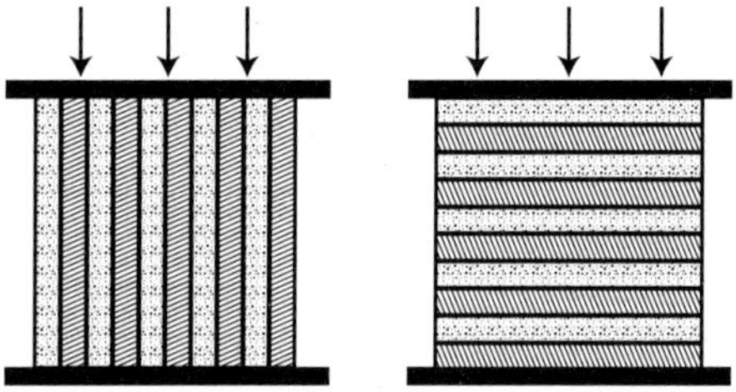

Fig. 14.14. Schematic representation of conditions for calculating the so-called series and parallel bounds for a modulus. In the parallel configuration (**a**), the strain is the same for all phases and the total modulus is just the sum, weighted by the volume fractions, of the moduli of the different elements. In the series configuration (**b**), it is the stress that is the same for all phases and the total compliance is the weighted sum of the compliances of the various elements

$\varepsilon_1 = \varepsilon_2$. In the series configuration, it is the stress that is equal in the two phases: $\langle\sigma\rangle = \sigma_1 = \sigma_2$.

In both cases, an effective modulus is defined by

$$\langle\sigma\rangle = M_{\text{eff}}\langle\varepsilon\rangle \,,$$

where the averages are expressed by

$$\langle\sigma\rangle = \phi_1\sigma_1 + \phi_2\sigma_2 \quad \text{and} \quad \langle\varepsilon\rangle = \phi_1\varepsilon_1 + \phi_2\varepsilon_2 \,,$$

with $\phi_1 + \phi_2 = 1$. Introducing these expressions into the definition of M_{eff} and taking into account the conditions expressing equality of strains or stresses described above, this leads to:

- For the parallel configuration,

$$M_{\text{parallel}} = \phi_1 M_1 + (1 - \phi_1) M_2 \,.$$

- For the series configuration,

$$\frac{1}{M_{\text{series}}} = \phi_1 \frac{1}{M_1} + (1 - \phi_1) \frac{1}{M_2} \, .$$

The first expression represents the upper bound, and the second the lower bound:

$$M_{\text{series}} \leq M_{\text{eff}} \leq M_{\text{parallel}} \, .$$

The advantage of this calculation of parallel and series bounds is that it is simple and rigorous. The main disadvantage is that it predicts a range of possible values that broadens rapidly when the two phases of the composite become very different. However, with a little intuition, one can predict which of the two bounds is likely to be closer to reality. For example, a composite made from an insulating matrix and highly dispersed conducting inclusions (that do not touch one another) will remain insulating until ϕ_2 is very close to unity, since even a very thin film of insulator around each conducting grain will suffice to maintain the insulating property of the whole. It is thus the lower (series) bound of the conductivity that will be closer to reality. On the other hand, a composite made from a conducting matrix with well-dispersed insulating inclusions will display a conductivity that is essentially proportional to the amount of matrix material. It is thus the upper (parallel) bound that will mimic the properties of the composite to a good approximation. This same argument can be applied to the Young's modulus (or the viscosity) of a mixture of a rigid (or high-viscosity) phase and a soft (or low-viscosity) phase. If the rigid (or high-viscosity) phase is the matrix, it is the upper bound that will be more accurate. If it is the soft (or low-viscosity) phase that is the matrix, as in polymer matrix nanocomposites, it will be the lower bound.

14.4.2 Impermeability: Reducing the Diffusivity

The method described in the last section can also be used to bound the value of the diffusion coefficient D of a solute in the composite. In reality, in the case we are concerned with here, where a polymer contains nanoparticles that are totally impermeable to the solute, the lower bound $D = 0$ is of no use whatever. The upper (parallel) bound, for its part, corresponds to a situation in which the diffusion cross-section has simply been reduced by a factor of $1 - \phi$, where ϕ is the volume fraction of the filler. The diffusion coefficient of the composite is reduced in the same proportions. In reality, the reduction is always greater because the particles constitute obstacles that the solute must find their way around, and this lengthens their diffusion paths, especially when the particles have anisometric shapes. To account for this, a tortuosity coefficient τ is introduced, with the property that

$$D = D_{\text{m}} \frac{1 - \phi}{\tau} \, ,$$

where D_{m} is the diffusion coefficient of the solute in the pure polymer matrix. The parameter τ thus summarises all the geometrical complexity of the

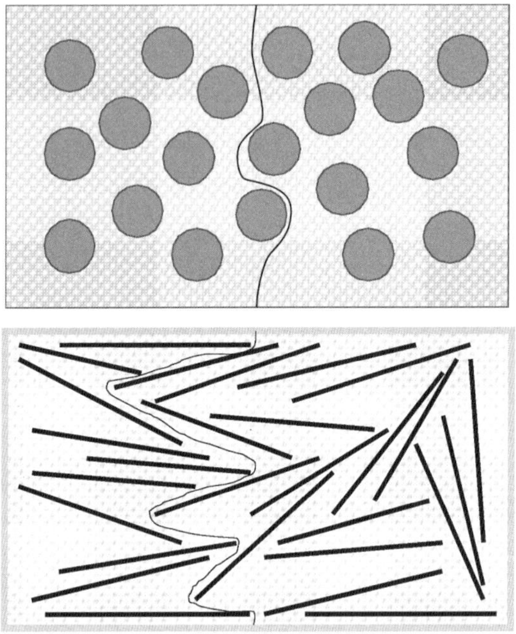

Fig. 14.15. Increase in the tortuosity coefficient when spherical particles are replaced by the same volume fraction of platelets

medium. It depends to a large degree on the anisometry of the particles, as is easily understood by comparing the two diagrams in Fig. 14.15. For the same total volume fraction, very thin disks represent a quite different obstacle to spheres. For randomly oriented disks of thickness e and diameter L, calculation gives

$$\tau = 1 + \frac{L}{2e}\phi \,.$$

When the volume fraction becomes large, excluded volume effects arise (as in liquid crystals); the particles begin to align themselves and the tortuosity increases still further.

If we are concerned with the impermeability of a packaging or container, the above argument is not always sufficient. It contains only half of the information required to characterise the material. Consider for example a TV dinner packaged in nitrogen at atmospheric pressure. The aim here is to limit the penetration of oxygen which would deteriorate the product by oxidation, and also to restrict the amount of water vapour that can get out, to avoid the product drying up. Fick's equations, which involve the diffusion coefficient D discussed above, provide a good description of the flow and time dependence of the oxygen and water concentrations within the polymer film:

$$J_i = -D_i \frac{\partial c_i}{\partial x} , \qquad \frac{\partial c_i}{\partial t} = \frac{\partial}{\partial x}\left(D_i \frac{\partial c_i}{\partial x}\right) ,$$

where D_i is either D_{ox} or D_{water}. (It is assumed that the presence of dissolved oxygen does not affect the diffusion coefficient of the water and vice versa, which is not strictly true.)

The concentration gradient appearing in these equations is the concentration gradient within the film. In reality, it is generally difficult to measure these concentrations. In fact, one usually only has access to the partial pressure of the relevant compound outside the film, on either side of it. The equations then have to be solved in terms of the partial pressures. In particular, for the second of Fick's equations, if the diffusion is treated as resulting from a partial pressure gradient, this amounts to writing

$$J_i = -D_i \left(\frac{\partial c_i}{\partial p_i}\right) \frac{\partial p_i}{\partial x} , \qquad \left(\frac{\partial c_i}{\partial p_i}\right) \frac{\partial p_i}{\partial t} = \frac{\partial}{\partial x}\left[D_i \left(\frac{\partial c_i}{\partial p_i}\right) \frac{\partial p_i}{\partial x}\right] .$$

The transfer coefficient D_i', called the diffusivity, obtained by solving these equations with respect to p_i, is in fact the product of the diffusion coefficient D_i and a differential solubility that proves to be nothing other than the slope of the sorption isotherm:

$$D_i' = \left[D_i \left(\frac{\partial c_i}{\partial p_i}\right)\right] .$$

It should be noted that, in analogy with the Darcy equation which relates fluxes to pressure gradients, D_i' is often called the permeability. If the differential solubility of the diffusing species is unchanged by the presence of particles, the upper bound for diffusivities (or permeabilities) can be written in an analogous form to the one we have seen above for the diffusion coefficients: $D' = D_{\mathrm{m}}'(1-\phi)/\tau$.

14.5 Strengthening and Permeability of Nanocomposites: Facts and Explanations

We are now sufficiently well equipped to examine the extent to which nanocomposites live up to our expectations. In the two cases detailed below, we shall compare experimental results with predictions (bounds) from the mixing models.

14.5.1 Strengthening: Successes and Failures

A Success Story

Figure 14.16 shows the dependence of the Young's modulus of nylon–montmorillonite nanocomposites on the volume fraction ϕ of clay. The clay is extremely well dispersed in this matrix (see Fig. 14.1). The Young's modulus

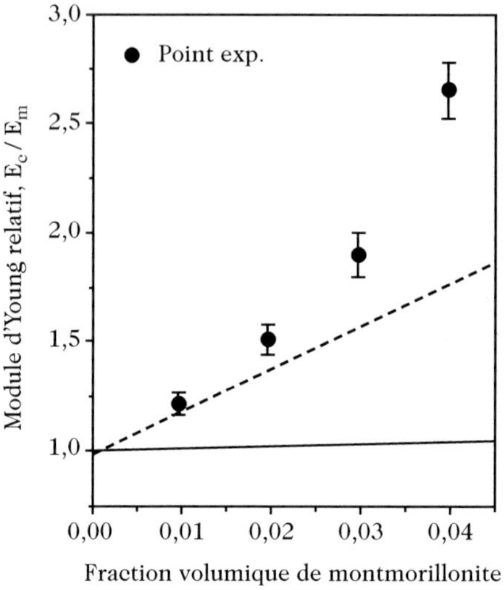

Fig. 14.16. Dependence of the relative Young's modulus of nylon–montmorillonite nanocomposites on the volume fraction of nanoparticles. Also indicated are the predictions corresponding to the lower and upper bounds deduced from a mixing law. From [6]

goes from $E_m \approx 0.5\,\text{GPa}$ in the pure polymer to $E_{nc} \approx 1.4\,\text{GPa}$ at $\phi = 0.04$. As can be seen, the measured values of the Young's modulus are incomparably greater, by a factor of the order of 10^2, than the lower (series) bound deduced from a mixing law. More surprising still is the fact that they are also much greater than the upper bound, even for very low clay contents, e.g., twice the maximum predicted by a mixing law at 4% clay content. The conclusion is inescapable: the nanocomposite is not a simple, even optimised mixture. It really is a new material, in which the polymer is no longer in the state it was before the clay was added, at least in the vicinity of the nanoparticles.

To account for these results, the authors constructed a three-phase model: the matrix as before, the particles, and an interface region. In the latter, they assumed that the modulus of the polymer increases gradually as one approaches the interface, to reach a given value on contact with the particles. The parameters for this model are thus the thickness t of the interface region, the contact value E_c of the modulus, and the form of the gradient. Assuming the latter to be linear, a good agreement was obtained for $t = 7\,\text{nm}$ and $E_c = 12 \times E_m$, where E_m is the modulus of the matrix (while the modulus of the clay particles is some 40 times higher).

A Less Successful Case

The example described above is a particularly successful case of strengthening, or more precisely, rigidification. Here we shall outline a case in which strength-

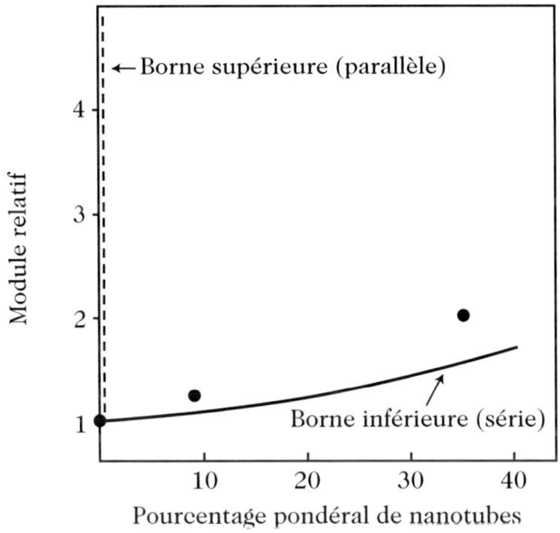

Fig. 14.17. Dependence of the relative Young's modulus of epoxy resin–SWNT nanocomposites on the volume fraction of nanoparticles. Also indicated are the predictions corresponding to the lower and upper bounds deduced from a mixing law. From L. Vaccarini [5]

ening remains rather modest despite the remarkable properties of the nano-objects used. It concerns epoxy resin–SWNT nanocomposites (see Fig. 14.2). Figure 14.17 shows the increase in modulus obtained by adding single-walled carbon nanotubes, up to weight fractions of 35%. Taking into account the voids between nanotubes within nanotube bundles (see Figs. 14.11 and 14.12) and in the core of each nanotube, these weight fractions correspond roughly to the volume fractions effectively occupied by the nanotubes in the composite. However, the modulus goes from $E_m \approx 2.4\,\mathrm{GPa}$ to only $E_{nc} \approx 5\,\mathrm{GPa}$, so that the strengthening factor E_m/E_{nc} is only about 2. The strengthening corresponding to the upper bound of the mixing law would be several hundred times greater! In reality, the observed strengthening barely reaches the value corresponding to the lower (parallel) bound, whereas the orientation of the tubes in the matrix, according to electron microscope images, seems to be thoroughly randomly distributed in space. The conclusion is the same as in the last case: the polymer in this nanocomposite is no longer quite the same as in the bulk polymer, at least at the interfaces. But this time, instead of being more rigid, it is actually softer. Another reason for the low level of strengthening is related to the poor penetration of the polymer into the interstices separating the nanotubes within bundles.

Explaining these Opposite Types of Behaviour

What is the physical basis for the 3-phase model, in which one interface phase displays a gradient of the relevant properties? Why should the rigidity of a polymer change in the vicinity of an interface? At the present time, there is no

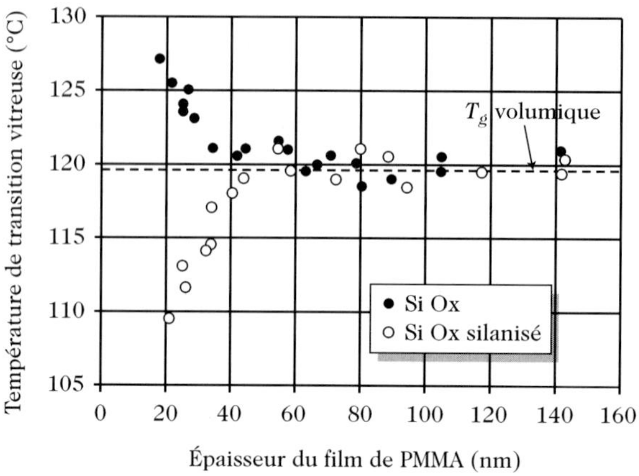

Fig. 14.18. Dependence of the glass transition temperature of thin PMMA films on their thickness. On a substrate interacting weakly with the polymer, such as a silanised Si wafer, T_g decreases as the film thickness is reduced. On a strongly interacting substrate such as an Si wafer, the opposite is observe. From [7]

direct experimental answer to this question (although one should not be long in coming). However, there are many clues showing that the mobility and/or macromolecular density vary close to an interface. One of these clues is the variation of the glass transition temperature in very thin polymer films (see, for example, [2]). In unsupported films, T_g is generally observed to decrease in films of less than a few tens of nanometers, suggesting increased mobility and a larger free volume than in the bulk polymer. In thin films mounted on a solid substrate, everything depends on the interaction between polymer and substrate (see Fig. 14.18). If the interaction is weak, T_g generally decreases. However, if the interaction is strong, T_g can increase. In terms of wetting, the first case is that of a non-wetting film, and the second, a wetting film (see Chap. 12 for more on wetting). Roughly speaking, the wetting case corresponds to the situation in which the solid is more polarisable than the polymer and the non-wetting case to the opposite situation. A polymer molecule then prefers to surround itself with its own kind rather than with atoms from the solid.

In order to understand how these variations of T_g correspond to mechanical properties, recall that the glass transition reflects the change in free volume, and hence the density of the material (see Sect. 14.2.3). The glass transition temperature is just the temperature at which the density of the molten material (characterised by a thermal expansion coefficient α_l) becomes too great to allow significant molecular rearrangements. It thus comes to the same thing to speak of the glass transition temperature or the glass transition density ρ_g. The question is then: why would a weakly interacting interface reduce the local density of a polymer, and why would a strongly interacting interface increase it? There is still no general agreement about this, but there is one explanation which has the advantage of being simple and at the same time correctly accounting for the main phenomena. This is the one put forward by Long and Lequeux [3]. The argument depends on whether the interface is weakly or strongly interacting. In both cases, it is assumed that the molecular weight is not too high, so that chain radii are not greater than the film thickness.

- *Non-Wetting Film (Weakly Interacting Interface).* One piece of experimental data that needs to be taken into account is that, in the extreme case of a weakly interacting interface represented by an unsupported thin film, no measurement has ever revealed a reduction in mean density. The subtlety of Long and Lequeux's reasoning is to consider that what counts in reality are density fluctuations. It is known from the theorem of equipartition of energy that any elastic body with incompressibility modulus K and volume V undergoes volume fluctuations with relative amplitude $\Delta V/V$ given by

$$\left(\frac{\Delta V}{V}\right)^2 = \left(\frac{\Delta \rho}{\rho}\right)^2 = \frac{T}{KV} .$$

In the case we are considering (weak interaction with the surroundings), each half of the film, of thickness $D = h/2$, fluctuates independently and undergoes thickness fluctuations with amplitude given by

$$\frac{\Delta D}{D} = \left(\frac{T}{KD^3}\right)^{1/2} .$$

Let us examine the effect of these fluctuations on the glass transition temperature. At temperatures greater than T_g, the film density is less than ρ_g. This density increases if the temperature goes down and, at temperatures below T_g, becomes equal to ρ_{bulk}, higher than ρ_g. However, this density fluctuates between two bounds, $\rho^+ = \rho_{bulk} + \delta\rho$ and $\rho^- = \rho_{bulk} - \delta\rho$. For the lower bound, the temperature dependence is

$$\rho^- = \rho_{bulk} - \left(\frac{8T}{Kh^3}\right)^{1/2} .$$

While these fluctuations take the film temporarily into a density state below ρ_g, it cannot be in the glassy state. For the film to be glassy, the lower value of the film density must remain greater than the density of the bulk polymer. A new glass transition temperature $T'_g \leq T_g$ can thus be defined, such that

$$\rho_{bulk}(T'_g) - \delta\rho = \rho_g .$$

The variation of the glass transition temperature relative to that of the bulk polymer becomes

$$\frac{\Delta T_{\mathrm{g}}}{T_{\mathrm{g}}} \propto \frac{-1}{\alpha_{\mathrm{T}}} \frac{\delta\rho}{\rho_{\mathrm{bulk}}} = -\frac{1}{\alpha_{\mathrm{T}}} \left(\frac{8T}{Kh^3}\right)^{1/2} ,$$

where the parameters (thermal expansion coefficient α_{T} and modulus K) are those of the bulk polymer. The last relation predicts $\Delta T_{\mathrm{g}}/T_{\mathrm{g}} \propto h^{-3/2}$, in reasonable agreement with experiment.

- *Wetting Film (Strongly Interacting Interface).* The case of a polymer film in strong interaction with a rigid substrate (in fact, a strongly adsorbed film) follows a different logic. Indeed, the argument given above cannot explain an increase in T_{g}. At best, it can explain a smaller decrease than for a free film, since only one side of the adsorbed film (the one that is not 'bound' to the substrate) can be the scene of density fluctuations. In order to explain an increase, one must appeal to an analogy between a strongly adsorbed layer and a grafted, i.e., covalently bound, layer. In such layers, the density of the layer is modified by an elongation energy. Indeed, the entropic elasticity (see Sect. 14.2.3) tends to compress the layer, thus increasing its density compared with the bulk polymer. This effect increases with the graft density Σ, i.e., the number of strongly interacting monomers per unit area. It can be shown that this leads to an extra contribution δK to the elastic modulus of the layer, given by

$$\delta K = 3T\Sigma^2 a ,$$

where a is the monomer size. On contact with the surface, the graft density in the limiting case of very strong adsorption is equal to a^{-2} (so that all monomers are in contact with the surface). To calculate the contribution over the thickness of the layer, one must take into account the reduction of Σ as one moves away from the substrate, viz., $\Sigma(z) \propto a/z$. The average increase in the modulus calculated over the whole thickness h of the adsorbed layer is then

$$\delta K = \frac{3k_{\mathrm{B}}T}{a^2 h} .$$

This leads to a change in the free energy per monomer and a change $\delta\rho$ in the average density as compared with the value ρ_{bulk} for the bulk polymer at the same temperature:

$$\frac{\delta\rho}{\rho_{\mathrm{bulk}}} = \frac{\delta K}{K} = \frac{3k_{\mathrm{B}}T}{a^2 hK} .$$

The (positive) relative variation of the glass transition temperature then follows immediately:

$$\frac{\Delta T_{\mathrm{g}}}{T_{\mathrm{g}}} = \frac{1}{\alpha_{\mathrm{T}}} \frac{\delta K}{K} = \frac{1}{\alpha_{\mathrm{T}}} \frac{3k_{\mathrm{B}}T}{a^2 hK} .$$

Feeding in typical numerical values, this formula predicts increases in T_{g} of the order of $0.1T_{\mathrm{g}}$, i.e., several tens of kelvins for a strongly bound

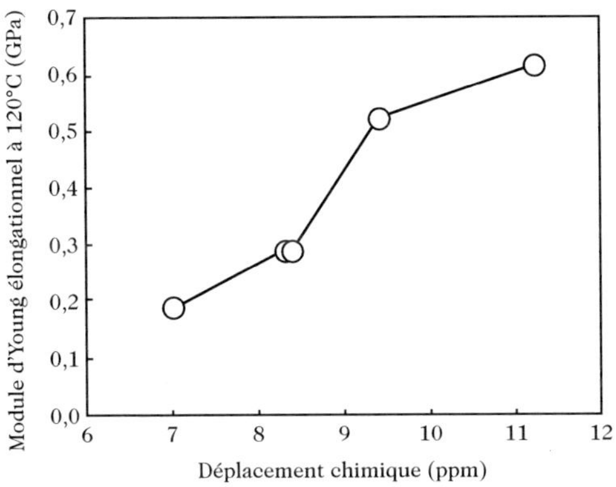

Fig. 14.19. Relation between the elongational Young's modulus of nylon–montmorillonite nanocomposites and the strength of the polymer–clay interaction, measured by monitoring the chemical shift of the NMR signal of the ^{15}N nucleus. From [8]

film with a thickness of about 10 nm when the bulk T_g is of the order of 400 K (e.g., PMMA), in agreement with experiment. The increase in the incompressibility modulus nevertheless remains relatively small, of the order of 10^7 Pa.

Let us return to the two families of nanocomposites discussed as examples above: nylon–montmorillonite and epoxy resin–nanotubes. The first example is clearly a case of strong wetting, thereby producing a rigidification of the interface region. The link between the strengthening and the level of polymer–surface interaction has even been directly demonstrated by NMR studies of the ^{15}N nucleus of the polymer (see Fig. 14.19). The modulus grows with the chemical shift. In the second family of nanocomposites, wetting is poor, as would be expected given the low surface energy of graphite. This results in a softening of the polymer at the interface.[3] The moral here is clear: nanomaterials are not a miraculous solution for all strengthening problems; it all depends on the wetting.

However, not everything is completely clear. A shift in the glass transition temperature only has a significant effect on the mechanical properties of a composite if this shift occurs near the operating temperature, so that a certain volume of polymer transits from the glassy state to the rubbery state, or vice versa. An elastomeric matrix in which the interface region goes into the glassy state is an ideal case. But what happens if the matrix is already in the glassy state? An increase in T_g should only have a rather modest effect,

[3]It should not be concluded that carbon nanotubes are of no use as strengtheners. When their surface is modified, or in the form of MWNTs, and used with other matrices than those considered here, or for other thermomechanical properties than a simple increase of the modulus, they still harbour a considerable potential.

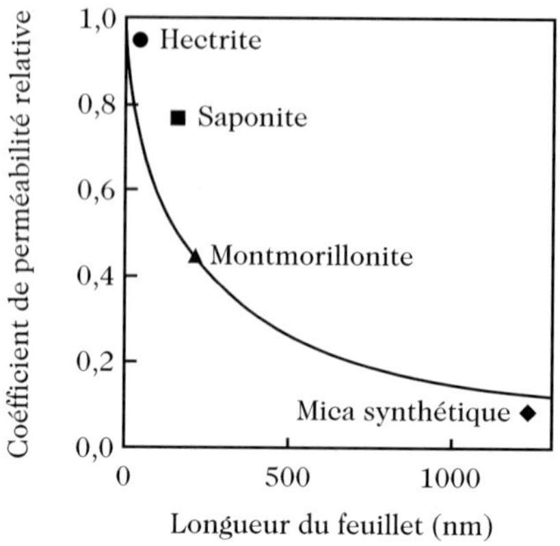

Fig. 14.20. Relative diffusivity of water vapour in clay–polyimide nanocomposites for a clay weight fraction of 2% (i.e., about 0.8% volume fraction) as a function of the lateral extent of clay sheets. From [9]

because the increase in the modulus as the temperature falls, below T_g, is itself rather weak (see Fig. 14.5). However, large increases in the modulus have been obtained for this type of nanocomposite; a spectacular specific case is the one in Fig. 14.16. Further research must therefore be done to understand the properties of polymers at interfaces and in confined media. One hypothesis that gains weight, at least in the case of semicrystalline nanocomposites, is that clay sheets set up 'bridges' between crystalline regions.

14.5.2 Impermeability

As we saw in Sect. 14.3.2, dispersing particles throughout a matrix can provide a very efficient way of making the matrix impermeable. The effect can in principle be significant even at very low volume fractions, provided that platelets with high aspect ratio are used. This is illustrated in Fig. 14.20, which compares the relative efficiency, measured by the ratio D'/D'_m, of phyllosilicates with increasing aspect ratio (the thickness is held constant at about 1 nm, while the lateral dimensions of the sheets are increased). Increasing the lateral dimensions of the sheets leads to a spectacular reduction in diffusivity. This reduction follows the expected behaviour for a random distribution of platelets, i.e., $D'/D'_m = (1 - \phi)/\tau$, with $\tau = 1 + (L/2e)\phi$ (see Sect. 14.3.2), which means that in this case, the solubility of the diffusing species in the matrix has not been affected by the clay. For 1 μm sheets and a volume fraction ϕ less than 1%, the diffusivity in the nanocomposite is only one tenth of what it was in the pure matrix, whence an order of magnitude is gained in impermeability. For a given particle type, the relative diffusivity decreases very quickly as ϕ increases (see Fig. 14.21). In the case illustrated here, the decrease also follows the relations given above.

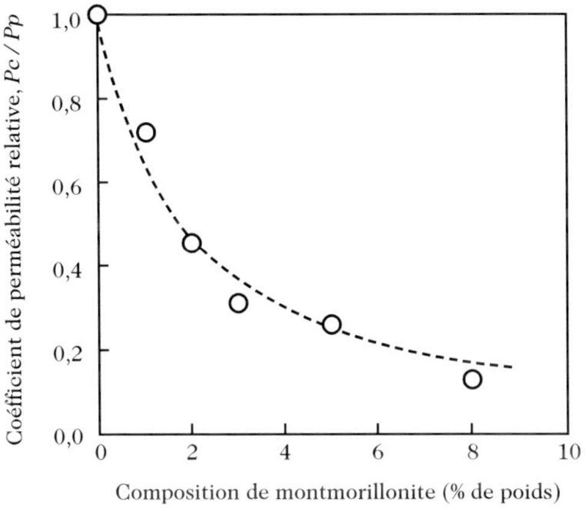

Fig. 14.21. Dependence of the relative diffusivity of water vapour in montmorillonite–polyimide nanocomposites on the weight fraction of clay. From [9]

In the example above, the observed effect can be adequately explained by the mixing law and the theoretical tortuosity coefficient. Hence the particles do not induce any significant modification in the solubility. But this is not always the case. In nylon, for example, the incorporation of clay would appear also to reduce the solubility of water. This is important because water uptake is a limiting factor in the mechanical performance of nylon.

14.5.3 Dimensional Stability

The deformation of a polymer object under the effects of temperature or a liquid is generally an undesirable effect. In general, one aims to make objects or parts that are relatively insensitive to their environment. Regarding thermal effects, one thinks of certain rather harmless problems such as the deformation of a dashboard exposed to the full heat of the sun, but also of the more serious difficulties posed by the ultra-precise assembly of printed circuits or microprocessors by superposing layers. Concerning chemical effects, one thinks of all the rubber shock-absorbing parts which react so poorly to contact with solvents. The incorporation of fillers with low expansion coefficient or low sensitivity to solvents is a standard way of limiting such effects. In both cases (high temperatures or presence of solvents), upper and lower bounds can be calculated as before. The upper bound is now the series average. Indeed, in this configuration, each slice of the composite is able to expand or contract freely. This corresponds to a highly expandable matrix (under either temperature increases or inclusion of a solvent) and inclusions that have little tendency to expand. In the parallel configuration, it is the least expandable component that dominates here by preventing its neighbours from expanding freely (at the expense of internal stresses).

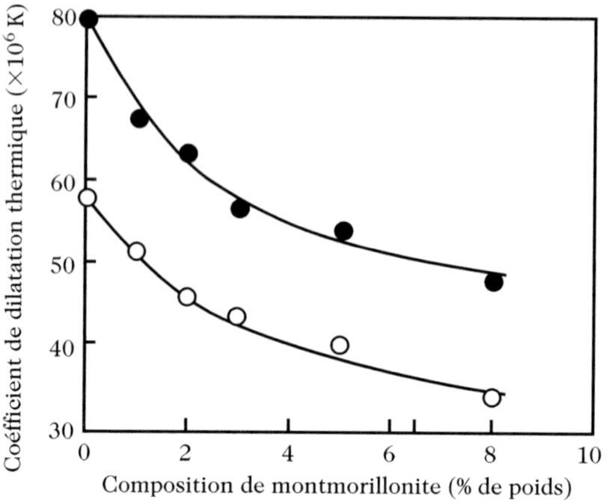

Fig. 14.22. Influence of the clay volume fraction on the thermal expansion coefficient of a polyimide–montmorillonite nanocomposite. From [10]

For certain nanocomposites, the thermal behaviour is 'normal', with expansion coefficients exhibiting a modest reduction and, at low volume fractions of the filler, a linear variation in the latter, agreeing with the prediction of the series model. For others, the effect is much greater and is explained by taking into account the interfacial gradient of the relevant properties. This is the case, for example, for the thermal expansion coefficent of polyimide–montmorillonite nanocomposites (see Fig. 14.22). The decrease is not only very large, but also highly nonlinear. It is tempting to think that, by virtue of the interface region which 'thickens' the clay sheets, regions with low expansion coefficient (clay particles and interface regions) percolate through the matrix. A simple calculation shows that it would suffice to have $\phi \approx 0.01$ in order to form a percolating low-expansion network. In this configuration, it is the 'hard' phase that determines the behaviour and it is the parallel bound that should be considered.

For objects made from elastomers, it is rather the swelling in the presence of solvents that is critical. One still has the two types of behaviour found when considering moduli: weakly interacting interfaces (poor elastomer–filler wetting), where swelling in the presence of the filler is greater than in the pure elastomer; and strongly interacting interfaces (good elastomer–filler wetting), where swelling is considerably reduced. An example of the first kind of behaviour is provided by the PDMS–kaolinite composites (kaolinite platelets are electrically neutral). An example of the second kind of behaviour is provided by the PDMS–montmorillonite nanocomposites (charged sheets). PDMS–carbon black nanocomposites obey the classical (series) dependence.

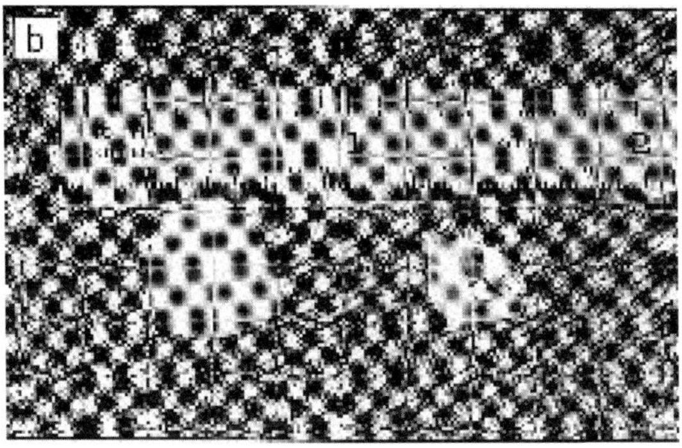

Fig. 14.23. Comparison between a nylon–montmorillonite nanocomposite component and a pure nylon component under exposure to an oxygen plasma. From [11]

14.5.4 Fire Resistance

Of the different properties considered here, fire resistance is certainly the most complex. Combustion or pyrolysis involve thermoactivated chemical degradation, gas transfer, and heat transfer. The incorporation of nanofillers can modify these phenomena and globally improve fire resistance. We have already seen in Sect. 14.3.2 that the diffusivity of volatiles (produced by degradation of the polymer) towards the surface, or of oxygen towards the interior is slowed down by the presence of nanoparticles. The latter have another beneficial effect: when the surface layers decompose, the sheets join together into bundles (as in Fig. 14.8), and this strengthens and renders impermeable the tars resulting from pyrolysis. These tars no longer form droplets. Instead, they form a crust which prevents erosion, further delaying the destruction of the material. Figure 14.23 compares a nanocomposite with a pure polymer.

14.6 Conclusion

The enormous interface area and interparticle confinement generated by the dispersion of nanoparticles in a polymer matrix can considerably alter the properties of part, or even the whole of the polymer via interface effects. When these interactions are manifested, e.g., in the Young's modulus, by swelling in a solvent, or thermal expansion, the nanocomposite is no longer a simple mixture, even a very intimate mixture, whose properties could be described as a compromise between those of its constituents; rather, it is a new material. The physical origin of interface effects and confinement, which depend on wetting conditions, remains to be elucidated, although it is now clear that molecular mobility plays a key role.

References

Specific References

1. M.H. Cohen, D. Turnbull: Molecular transport in liquids and glasses, J. Chem. Phys. **31**, 1164–1169 (1959); Free volume model of the amorphous phase: Glass transition, J. Chem. Phys. **34**, 120–125 (1961)
2. J.A. Forrest, K. Dalnoki-Veress: Adv. Coll. Interf. Sci. **94**, 167–196 (2001)
3. D. Long, F. Lequeux: Heterogeneous dynamics at the glass transition in van der Waals liquids, in the bulk and in thin films, Eur. Phys. J. E **4**, 371–387 (2001)
4. Wang et al.
5. L. Vaccarini: Doctoral thesis, University of Montpellier II (2000)
6. X.L. Ji, J.K. Jing, W. Jiang, B.Z. Jiang: Polym. Eng. Sci. **42**, 983–993 (2002)
7. D.S. Fryer, P.F. Nealey, J.J. de Pablo: Macromol. **33**, 6439–6447 (2000)
8. A. Usuki, A. Koiwai, Y. Kojima, M. Kawasumi, A. Okada, T. Kurauchi, O. Kamigaito: J. Appl. Polym. Sci. **55**, 119 (1995)
9. K. Yano, A. Usuki, A. Okada: J. Polym. Sci. Part A: Polym. Chem. **35**, 2289 (1997)
10. K. Yano, A. Usuki, A. Okada, T. Kuraychi, O. Kamigaito: J. Polym. Sci. Part A: Polym. Chem. **31**, 2493 (1993)
11. R. Vaia, E. Giannelis: Mater. Res. Soc. Bull. 394–401 (2001)

General References

12. P.M. Ajayan, L.S. Schadler, P.V. Braun: *Nanocomposite Science and Technology*, Wiley-VCH, Weinheim (2003)
13. J. Perez: *Matériaux non-cristallins et science du désordre*, Presses Polytechniques et Universitaires Romandes, Lausanne (2001)

Part III

Synthesis of Nanomaterials and Nanoparticles

15

Specific Features of Nanoscale Growth

J. Livage and D. Roux

15.1 Introduction

It is no easy matter to make nanoscopic objects. In fact, starting from a macroscopic sample, as one does when making objects with centimeter or millimeter dimensions, one soon comes up against technical problems. As we shall see, there are techniques for overcoming them, but they require a certain art to implement them successfully.

Apart from those techniques used to divide up macroscopic materials into small particles, i.e., foaming for gases, emulsification for liquids, and grinding for solids, there are more 'gentle' techniques in which the growth of one phase in another is carefully controlled during a phase separation.

We begin with a highly simplified discussion of the physics of phase transitions, considering both static and dynamic features, then explain how, by controlling the kinetics of such transitions, it is possible to arrest them and thereby obtain nanoscale objects. To achieve this end, one can make use of physical processes such as nucleation and growth and/or spinodal decomposition, or one can exploit a chemical reaction leading to a compound that is insoluble in the medium in which it is synthesised. Having summarised a few simple ideas about phase transitions, we explain the underlying principles of some methods used to arrest growth, thereby obtaining small particles.

15.2 Thermodynamics of Phase Transitions

To set the scene, imagine two typical types of phase transition:

- the transition that occurs in a paramagnetic material, with no permanent magnetisation, when the temperature is reduced, to yield a ferromagnetic material, with permanent magnetisation;
- the phase transition from a homogeneous binary mixture of two liquids to a phase separation between these two liquids when the temperature is reduced.

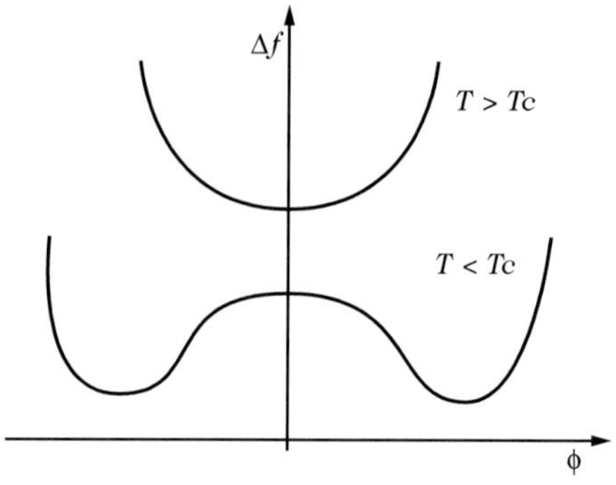

Fig. 15.1. Free energy density Δf as a function of the order parameter φ (schematic). Above the critical temperature, the function remains permanently concave, while below this temperature, a convex region appears. The latter corresponds to unstable solutions. The transition from $T > T_c$ to $T < T_c$ corresponds to a phase transition

The simplest and most general model of phase transitions is the one due to Landau. This is based on the observation that two different phases can be characterised by an order parameter φ which takes different values in each phase. For example, the transition between a paramagnetic material and a ferromagnetic material will be characterised by the value of the magnetisation, i.e., zero in the first phase and different from zero in the second.

In the second example, the difference between the single-phase and two-phase cases will be characterised by a concentration, or more precisely, a difference of concentration. In a certain type of approximation called the mean field approximation, and assuming that the value of this order parameter (or its difference between the two phases) can be small enough, the free energy density of the phases can then be expressed as a truncated expansion in this order parameter (or its difference):

$$\Delta f(\varphi) = \Delta f_0 + \mu\Delta\varphi + a\Delta\varphi^2 + b\Delta\varphi^4 + O(\Delta\varphi^5) + \cdots .$$

The third order term can be eliminated by a suitable choice of origin for the order parameter and a renormalisation of the coefficients. a and b correspond to successive derivatives of the free energy with respect to the order parameter. They correspond to well defined thermodynamic quantities. μ is the value of the external magnetic field (if there is one) in the case of the paramagnetic–ferromagnetic transition, and the value of the chemical potential for a phase separation transition.

In this model, the phase transition is driven by the sign of the compressibility a. The sign of b is considered to be positive to ensure thermodynamic stability of the phases. Supposing generically that $a = a_0(T - T_c)$, where T_c is a critical temperature for which the sign of a changes, we obtain in a very simple manner the curves corresponding to the variation of the free energy

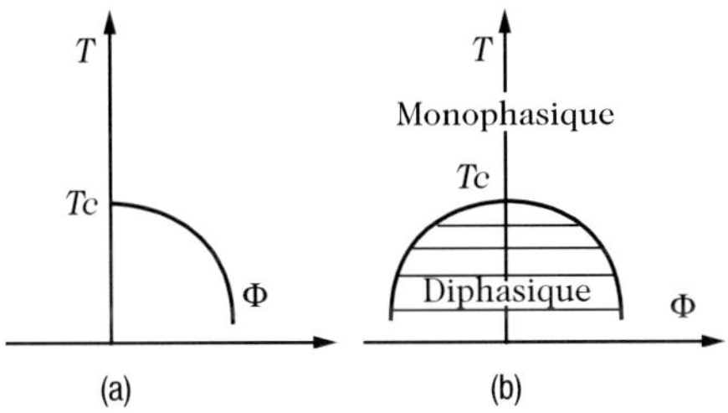

Fig. 15.2. Phase diagram as a function of temperature for non-conserved (**a**) and conserved (**b**) order parameters. In the first case, the value of the order parameter, e.g., the average magnetisation, goes from zero to a nonzero value when T drops below T_c. In the second case, a single-phase state goes to a two-phase state in which two phases with different concentrations coexist

density shown in Fig. 15.1. For simplicity, it will be assumed that the chemical potential μ is zero (as for the magnetisation when there is no external field).

When using this theory, the phase diagrams are obtained in a slightly different way depending on the nature of the order parameter. Indeed, if the order parameter is not conserved, i.e., there is no restriction on the values it can assume, as happens for the paramagnetic–ferromagnetic transition, then this theory will describe the transition from a phase in which $\varphi = 0$ ($T > T_c$) to a phase in which φ differs from 0 ($T < T_c$), corresponding to the respective minima of the free energy. If the order parameter is conserved, i.e., a constraint is imposed via the conservation of matter, as happens in the case of a density or a concentration, this theory will describe the transition from a single-phase state to a two-phase state in which two phases with different compositions will coexist.

The two cases are illustrated schematically in Figs. 15.2a and b. These show the phase diagrams for a paramagnetic–ferromagnetic transition and for the separation of a binary mixture, respectively. Since the aim here is to understand how to exploit a phase separation to control the size of objects it produces, we shall only be concerned in the following with the case of a conserved order parameter and transitions corresponding to a phase separation in which a single-phase system moves towards a two-phase system. The problem is the same whether this separation involves two liquid phases, or a solid phase in a liquid phase, or a solid phase in a gas phase, or indeed two solid phases, one within the other. φ thus represents either a concentration or a density.

15.3 Dynamics of Phase Transitions

When a phase separation is induced, e.g., by a change in temperature, it will not occur instantaneously. Depending on the position of the end point in the phase diagram, there are two mechanisms controlling the transition from one

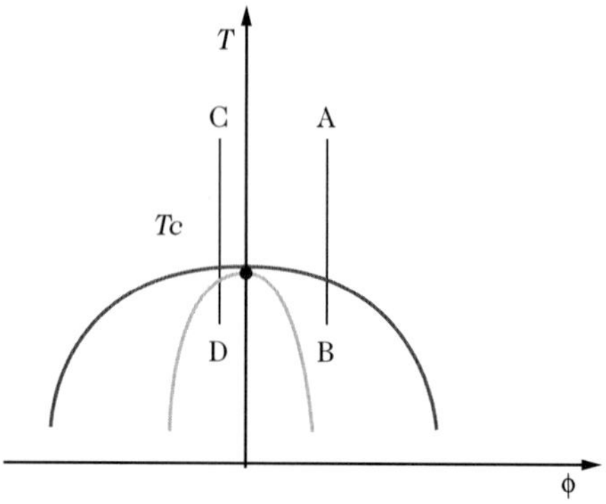

Fig. 15.3. A change in a thermodynamic variable causing transition from a homogeneous phase to a two-phase state, represented in a (T, ϕ) version of the phase diagram: spinodal decomposition (CD) or nucleation–growth (AB). *Black curve*: boundary for coexistence of the two phases. *Grey curve*: spinodal curve

phase to the other: the phenomenon known as nucleation and growth and the phenomenon of spinodal decomposition. In both cases, growth kinetics results from competition between the thermodynamic effects causing the phase transition and the diffusion of species which slows down changes in local concentration.

The different scenarios in spinodal decomposition and nucleation–growth result from the presence or otherwise of an energy barrier along the path of the phase transition. To understand this, we must define the spinodal curve obtained as the set of points where the second derivative of the free energy density is zero.

Figure 15.3 shows this curve in the previous phase diagram. When the end point of the temperature change corresponds to a point located inside the spinodal curve (path CD), the phase separation occurs without energy barrier via the destabilisation of the collective diffusion mode. A bicontinuous structure then appears in the sample.

When the end point is located between the spinodal curve and the coexistence curve (path AB), a more local phenomenon occurs, in which a new phase is nucleated by overcoming an energy barrier. The concentration (or the density) of the new phase is very close to its equilibrium concentration and the system recovers its equilibrium state by growth of nuclei. Figure 15.4 shows schematically the spatial variation of the concentration as a function of time in each case.

15.3.1 Thermodynamics of Spinodal Decomposition

The kinetics of spinodal decomposition can be described by writing down a non-equilibrium free energy describing the thermodynamics of the phase

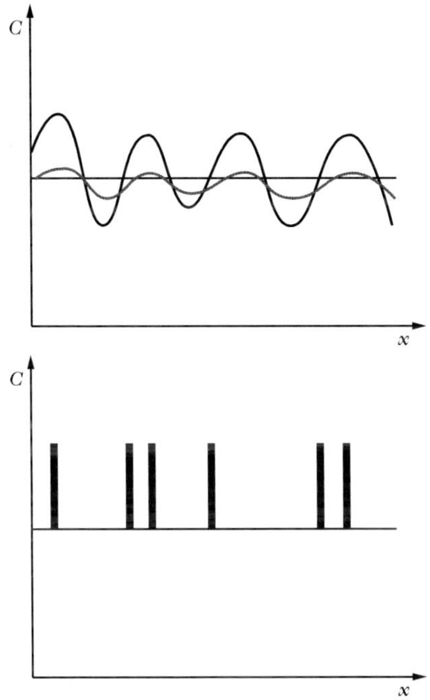

Fig. 15.4. Time dependence of the spatial variation of the concentration for spinodal decomposition (*top*) and nucleation–growth (*bottom*). In the first case, the evolution of the concentration difference is continuous at a given wavelength, while in the second, regions appear directly with the concentration of the new phase and grow in time

transition and diffusion of species. Just after the transition, the free energy density is given by

$$f(\varphi) = a\varphi^2 + K\left(\frac{\mathrm{d}\varphi}{\mathrm{d}x}\right)^2 ,$$

where a is negative (we are inside the spinodal curve and the second derivative of the free energy is negative), and K represents the equivalent of an interface energy resisting local concentration variations ($K > 0$). In terms of the Fourier transform [$\varphi_q = \varphi_0 \cos(qx)$], we obtain for each mode q,

$$f_q = (a + Kq^2)\varphi_q^2 .$$

The modes are unstable for $q < q_\mathrm{c}$, where $q_\mathrm{c}^2 = -a/K$. The kinetics of this transition is driven by the diffusion of species. Fick's law gives

$$\frac{\partial \phi}{\partial t} + \nabla J = 0 ,$$

where J is the concentration current of diffusing species, viz.,

$$J = M\nabla\mu = M\nabla\frac{\partial f}{\partial t} ,$$

so that

$$\frac{\partial\phi}{\partial t} + M\nabla^2\frac{\partial f}{\partial\phi} = 0 ,$$

or

$$\frac{\partial\phi}{\partial t} = M\nabla^2(a - K\nabla^2)\phi .$$

Using the Fourier transform once again, and assuming that the relaxation modes of the concentration are exponential, we obtain

$$\varphi(q,t) = \varphi(q,0)\exp(-\omega_q t) ,$$

where ω_q corresponds to the characteristic growth frequency of the mode with wave vector q. We obtain

$$\omega_q = MKq^2(q^2 - q_c^2) .$$

We see that the most rapidly growing mode is the one corresponding to the maximum of this function, i.e., $q = q_c/2$.

15.3.2 Thermodynamics of Nucleation–Growth

The energy of a nucleus of the new phase of size R appearing in the homogeneous mixture can be described as a sum of two terms:

- a bulk term expressing the energy gain involved in the appearance of a new phase, viz., $-\Delta\mu R^3$, where the sign indicates that energy is gained in forming the new phase;
- a surface term describing the relative part of the energy due to exposure of molecules of the new phase to contact with those in the original phase, viz., γR^2, corresponding to a surface tension.

Hence, we have

$$\Delta E(R) = -\Delta\mu R^3 + \gamma R^2 .$$

The curve $\Delta E(R)$ is shown schematically in Fig. 15.5.

It can be seen that there is a barrier E_c of the order of $\gamma^3/\Delta\mu^2$ for a critical size R_c of the order of $\gamma/\Delta\mu$. If the nucleus obtained during a fluctuation is smaller than R_c, the surface energy dominates and it dissolves in the initial phase; but if it is larger than R_c, the bulk term dominates and it will grow larger as time goes by.

After a temperature change, nuclei of the new phase will be seen to appear and grow in time. This growth occurs via diffusion of species and aggregation for solids (coalescence for liquids).

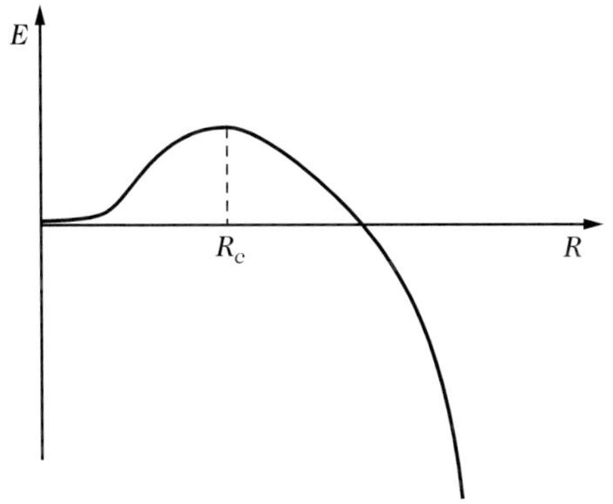

Fig. 15.5. Energy of a nucleus of a new phase appearing after a phase transition, as a function of the size R of the nucleus. When the size of the new phase corresponds to an energy value that is smaller than the maximum during a fluctuation, the latter will decrease as time goes by and the phase will become locally homogeneous once more. On the other hand, when the fluctuation corresponds to a size greater than R_c, the new phase will grow in time to produce a bigger and bigger particle

15.4 Size Control

In both cases, a characteristic small size appears when the system is first destabilised: the magnitude of the wave vector of the most rapidly growing mode in the case of spinodal decomposition and the critical radius R_c in the case of nucleation. In both situations, these sizes are very small, in fact, only a few nanometers. The idea thus arises of arresting the growth sufficiently early to be able to conserve this characteristic size, thereby producing nanoscale objects. A simple way of approaching this problem is to consider the thermodynamic stability of an ensemble of n colloidal objects of characteristic size R in a liquid phase. Since the size is assumed to be very small, these objects will be Brownian.

In this case the free energy will have two terms:

- a mixing entropy term, viz., $nk_BT \ln \varphi$, that favours dispersion of the objects for a sufficiently dilute suspension;
- a surface energy term $n\gamma R^2$, where φ is the volume fraction of objects, i.e., $\varphi = nv/V$, with V the total volume of the phase and v the volume of an object, i.e., $v = R^3$.

The free energy density is thus

$$f = \frac{F}{V} = \frac{nk_BT}{V} \ln \varphi + \frac{n\gamma R^2}{V} ,$$

for a given volume fraction φ of the phase (fixed by the position in the phase diagram). As a function of the size R, this yields

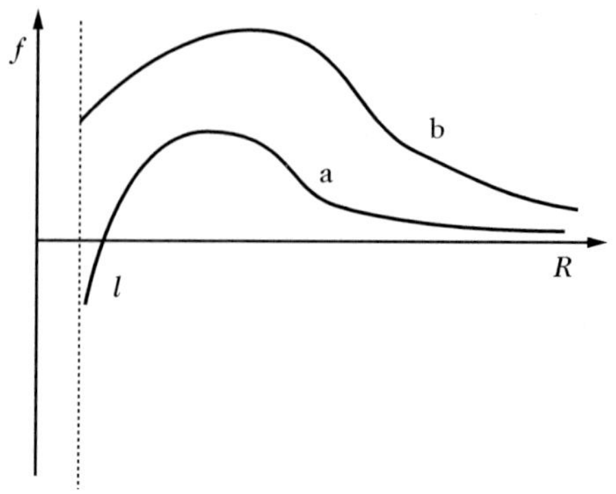

Fig. 15.6. Size dependence of the free energy density of an ensemble of particles of size R occupying a volume fraction φ. (**a**) $\varphi < \varphi_\mathrm{c}$. (**b**) $\varphi > \varphi_\mathrm{c}$. In the first case, the equilibrium size corresponds to the cutoff size, i.e., the size of the molecule, while in the second, the equilibrium size is infinite and corresponds to a phase transition with a macroscopic phase that separates

$$f(\varphi) = \frac{k_\mathrm{B}T\varphi \ln \varphi}{R^3} + \frac{\gamma}{R\varphi} .$$

This function looks something like the curves in Fig. 15.6. Note that $\ln \varphi$ is negative since by definition $\varphi < 1$. There is a cutoff size $R = e$. Indeed, R cannot be smaller than the size e of one molecule. There are two cases. When

$$\varphi < \varphi_\mathrm{c} = \exp\left(-\frac{\gamma e^2}{k_\mathrm{B}T}\right) ,$$

the minimum of this function corresponds to the cutoff size. The mixture is composed of isolated molecules of size R soluble in the phase. When $\varphi > \varphi_\mathrm{c}$, the absolute minimum of this function corresponds to an infinite size R. This allows one to recover the solubility limit of the molecule,

$$\varphi_c = \exp\left(-\frac{\gamma e^2}{k_\mathrm{B}T}\right) ,$$

but also shows that, as soon as the system corresponds to a two-phase state ($\varphi > \varphi_\mathrm{c}$), the most stable thermodynamic state corresponds to a single object of infinite size.

Returning to our original idea of arresting a phase transition for a given small object size, it is clear that the only way is to place an energy barrier along the phase separation path, i.e., to find a kinetic way of preventing the new phase from growing into an object of infinite size. Figure 15.7 shows such an evolution schematically.

There are two generic methods for introducing such an energy barrier. When working with an aqueous medium, one can arrange for the surface to build up an electrostatic charge during the growth process, either naturally

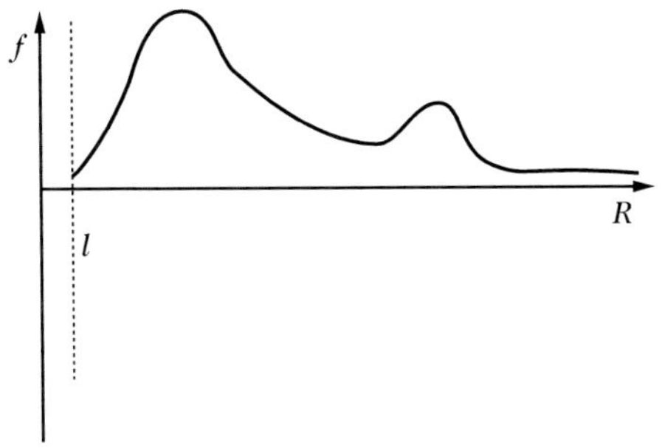

Fig. 15.7. Free energy density of a volume fraction φ of particles as a function of their size R. In contrast to the last case, the introduction of an energy barrier along the growth path defines a local minimum corresponding to a finite size of particle. This general principle is applied when nanoparticles are prepared in suspension in a solvent

by chemical ionisation, or by adsorption of ions present in the medium. The presence of such charges causes a repulsive interaction to develop when two particles move too close together, thereby preventing coalescence or aggregation. Another, more specific approach is to cause the adsorption of molecules such as polymers, proteins, or surfactant organic molecules on the surface, thereby forming a steric barrier. In the rest of this book, we shall meet some of the practical applications of these ideas.

15.5 Triggering the Phase Transition

In order to fabricate very small particles, we have seen that it is possible to trigger a phase transition which leads to the growth of one phase in the other, starting from a characteristic size in which only a few atoms or molecules are involved. Then, to control this size, the particles must be stabilised by developing a repulsive interaction between the interfaces resulting from the appearance of this new phase.

Whatever the nature of the phase transition (liquid–liquid, liquid–solid, liquid–gas, solid–gas), the conceptually simplest way to do this is to change the temperature. Hence, to synthesise solid nanoparticles, one can imagine going from a temperature above the solubility limit of an inorganic compound to a temperature below this limit, so that precipitation occurs. The nucleation–growth process then begins after this temperature change.

However, other techniques can be used based on a change of concentration, by mixing one solution with another, e.g., causing a change of pH or mixing two ionic solutions to produce an insoluble salt. In these cases, nucleation is controlled by the method of mixing constituents. Nuclei then grow by diffusion of species. The additive used to arrest growth of the nano-objects is usually included in one or other of the two solutions.

Another technique consists in triggering a chemical reaction leading to an insoluble species. This is how nanolatex is fabricated, for example. In this case, the starting point is an emulsion of monomers in water. The polymerisation reaction is started by the presence of a catalyst dissolved in the water. The insoluble polymers are thus synthesised outside the emulsion droplets. The size is controlled by the presence of a surfactant which, by adsorbing on the surface of the polymer particles, provides the repulsive interaction required to stabilise the objects.

15.6 Application to Solid Nanoparticles

In the case of solid inorganic particles, the general mechanisms are as described above. The precipitation of solid particles is achieved by starting with soluble precursors when the latter reach the critical supersaturation threshold. Two phenomena then lead to the growth of the solid particles:

- nucleation–growth processes which control the size of the primary particles;
- aggregation of these particles which leads to the precipitation of grains of varying sizes.

15.6.1 Controlling Nucleation

The formation of critical germs occurs as soon as the concentrations of the reactants reach a value corresponding to the critical supersaturation threshold. To arrange for this to happen, one reactive is generally added to the solution containing the second partner. For example, zirconia is prepared by precipitation, by adding a base to a solution of a zirconium salt, and $BaSO_4$, by adding sulfate ions to a solution of a barium salt. A simple mixture usually leads to large concentration gradients and hence to a large disparity in the sizes of the particles produced. To avoid this, the precursors can be generated in situ by simple modification of the physicochemical conditions: thermohydrolysis, decomposition of complexes to gradually release the anions or cations that must react, and so on. Two examples are:

- precipitation of zirconia by thermohydrolysis of an acidic solution of $ZrOCl_2$,
- decomposition of urea, which gradually increases the pH without having to add a base.

A large number of nuclei can thus be formed simultaneously in this way. The next problem is to control their growth.

15.6.2 Controlling Growth

In order to achieve efficient control of particle growth, one must find a way of decoupling the nucleation and growth processes. Primary particles grow by addition of soluble species coming from the solution. These species attach themselves at the surface and, if necessary, diffuse along the surface to the position corresponding to a minimum energy.

This process can be slowed down by complexing the surface sites to increase their reactivity with regard to the growth processes. This method is commonly used to control the formation of titanium oxide particles by hydrolysis of alkoxide precursors. The addition of complexing agents such as acetylacetone or acetic acid has the effect of complexing the titanium and thus limiting hydrolysis and condensation reactions. The sizes of the particles obtained then depend on the amount of complexing agent added. Complexation does not only alter the size of the particles. It also plays an important role in their morphology. The role played by anions in the shape of precipitates is well known. In fact, anions can attach themselves to the reacting cationic species and modify their functionality. In this way, they guide the growth processes and alter the morphology of the crystals that precipitate out.

15.6.3 Controlling Aggregation. Stability of Colloidal Dispersions

Nanometric particles are subject to Brownian motion and hence often collide with one another. They then tend to associate via weak bonds (agglomeration) or strong bonds (aggregation) to form larger particles. This is the well-known phenomenon leading to flocculation in colloidal dispersions.

To avoid such processes, which lead to the formation of polydispersed powders, collisions between primary particles must be avoided. Two methods are widely used:

- Steric hindrance, which involves surrounding the primary particles by an inert layer, e.g., surfactant, complexing agent, etc., thereby preventing contact between the surfaces of colliding particles.
- Electrostatic repulsion, which involves working well away from the zero charge point, in such a way that the particle surfaces are highly charged. Electrostatic repulsion then prevents the particles from coming too close to one another.

15.7 Breaking Matter into Pieces

Here we shall discuss the physical principles of emulsification. This can be considered as a way of fabricating liquid nanoparticles within another liquid. The different techniques for fabricating solid structures will be described later in the book.

When we wish to make small liquid particles, the simplest method is to shake one insoluble liquid within another. This is what happens when water is mixed in oil. Shaking applies shear stresses to the droplets, which break into smaller droplets. In remarkable work carried out in the 1930s, Taylor laid down the basic criteria relating droplet size to the applied shear. The size above which a droplet of radius R subjected to a shear $\dot{\gamma}$ breaks into smaller droplets is given by equating the viscous pressure felt by the droplet, related to the flow of fluid around it, with the Laplace pressure resisting deformation:

$$\eta\dot{\gamma} = \frac{\sigma}{R} \, .$$

We thus obtain

$$R = \frac{\sigma}{\eta\dot{\gamma}} \, .$$

The shear rate $\dot{\gamma}$ is a parameter that can typically range over three orders of magnitude, while the viscosity η can range over six orders of magnitude. One must therefore adjust these two parameters in such a way as to produce small sizes. Taking into account the fact that the surface tension σ of an interface between water and oil covered with surfactants is of the order of $0.03\,\mathrm{N\,m}$, it turns out that emulsion droplets with nanometric sizes (a few hundred nanometers) can be prepared for viscosities of the order of $1\,000\,\mathrm{Pa\,s}$ and shear rates of the order of $100\,\mathrm{s}^{-1}$.

References

1. J.-P. Jolivet: *De la solution à l'oxyde*, InterEditions/CNRS Editions, Paris (1994)
2. J.W. Cahn: On spinodal decomposition, Acta Metallurgica **9**, 795–801 (1961)
3. B. Cabane, S. Hénon: *Liquides: solutions, dispersions, émulsions, gels*, Editions Belin, Paris (2003)

16

Gas Phase Synthesis of Nanopowders

Y. Champion

16.1 Introduction

The aim of gas phase synthesis is to achieve the condensation of nanopowders with maximal specific surface area (see below for the definition), or in other words, to obtain nanoparticles that are not connected by strong chemical bonds (intergrain regions called grain boundaries; see, for example, the discussion of mechanical grinding in Chap. 19). When the contact region is partial, it is called a neck or bridge, a term taken from the field of sintering (see Chap. 21).

In contrast to the neck are weak physical bonds holding the particles together by adhesion alone. These physical forces are of van der Waals type or magnetic in the case of magnetic materials.

Apart from situations in which the nanopowder is destined for direct use, e.g., in catalysis (surface properties), or for abrasion (rheological properties), nanometric powders are synthesised with a view to further transforming them to obtain bulk objects of varying nanostructure and density. This is referred to as bottom-up synthesis, as opposed to top-down synthesis. The nanometric matter is fabricated by assembling nanograins. In principle, this provides a way of obtaining a very broad range of nanostructures: mesoporous, porous, or dense bulk materials, nanostructured coatings, either pure or in mixtures.

Using various techniques of synthesis already optimised, a wide variety of compounds can be made with controlled purity: transition metals such as Cu, Ni, Co, Fe, and Al, and refractory metals such as Mo and W, oxides such as Al_2O_3 and ZrO_2, carbides such as SiC, nitrides such as Si_3N_4, and semiconductors such as Si. Several types of nanometric powders are shown in Figs. 16.1 and 16.2.

Specific Surface Area

The specific surface area S_p is the ratio of the total surface area A of a given quantity of powder to its mass m_p :

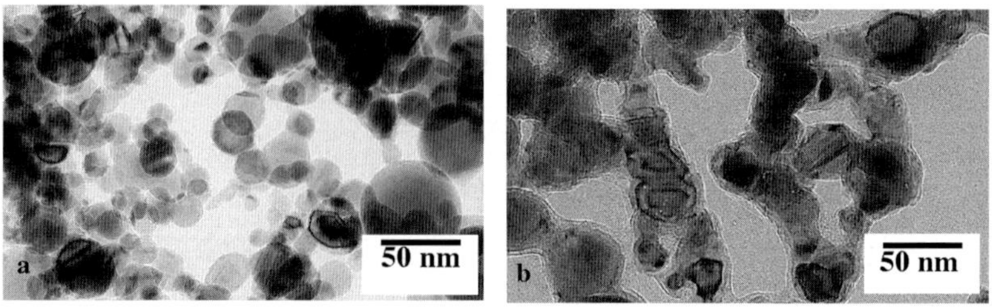

Fig. 16.1. Nanometric copper powders synthesised by evaporation and condensation. (**a**) Dispersed. (**b**) Partially agglomerated. From [3]

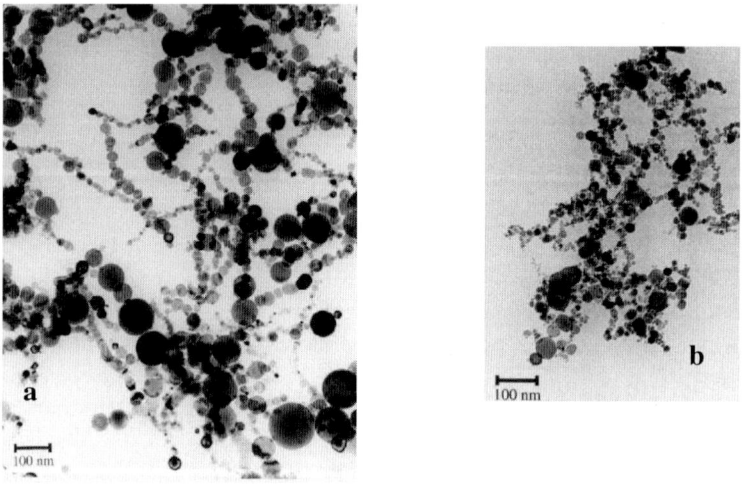

Fig. 16.2. Nanometric powders synthesised by evaporation and condensation. (**a**) Pure ferromagnetic iron. Particles agglomerate under the effect of magnetic forces. (**b**) Small chromium particles. Courtesy of J. Bigot, Centre d'Etudes de Chimie Métallurgique, CNRS, Vitry-sur-Seine

$$S_{\mathrm{p}} = \frac{A}{m_{\mathrm{p}}} \,. \tag{16.1}$$

The specific surface area can be measured experimentally by techniques using the adsorption of an inert gas [1].

If the size x_i of a representative number n of particles of mass m_i is known, one can deduce the theoretical value A_i of the potential free surface area, and the potential specific surface area of the particle ensemble will be

$$S_{\mathrm{p}}^{\mathrm{t}} = \frac{\sum_n A_i}{\sum_n m_i} \,. \tag{16.2}$$

The size x_i is measured by local examination of the powder, without taking into account connections between particles. This can be done by scanning or transmission electron microscopy, depending on the size range of the particles. (TEM requires x_i to be less than 100–150 nm, depending on the material.)

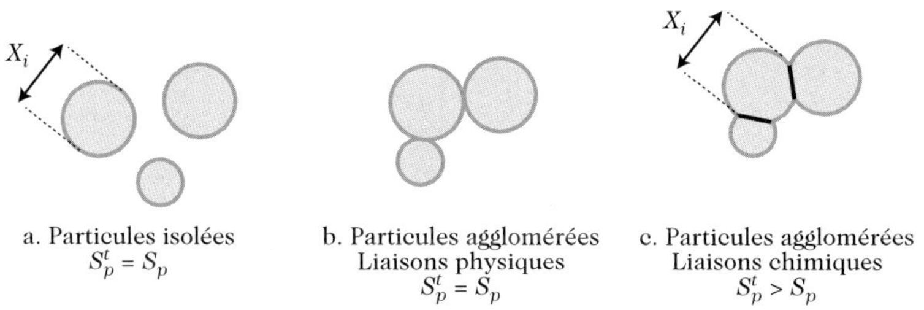

Fig. 16.3. Schematic representation of the level of particle agglomeration. The free surface area, which can be measured experimentally by gas adsorption, is shown in *dark grey*. Agglomeration with strong bonds results from the transformation of free surface into grain boundaries, shown in *black* in (**c**)

The particles are free, i.e., the specific surface area will be maximal, if $S_\mathrm{p}^\mathrm{t} = S_\mathrm{p}$. The particles are connected by chemical bridges if $S_\mathrm{p}^\mathrm{t} > S_\mathrm{p}$. The ratio $S_\mathrm{p}/S_\mathrm{p}^\mathrm{t}$ accounts for the degree of agglomeration by strong bonds between particles. This parameter is of the utmost importance: it influences all subsequent stages in the transformation of the powder batch, as well as the reproducibility, the quality, and the properties of the final material. In general, a high level of agglomeration with chemical bonds ($S_\mathrm{p}/S_\mathrm{p}^\mathrm{t}$ low) is undesirable. A high quality powder will have $S_\mathrm{p}^\mathrm{t} \approx S_\mathrm{p}$.

16.2 The Need for Gas State Processing

Still considering metals to exemplify the discussion, a great many components are fabricated from powders. For example, sintered steels are used for specific parts with properties well suited for use in the automobile industry. Hence, industrial techniques for manufacturing metal and alloy powders already exist. One of the most effective in terms of yield, and especially in terms of quality of the powders it produces, is atomisation.

The idea of atomisation is to form a spray made up of fine droplets of metal or alloy under the action of a high-pressure fluid on a flux of liquid metal or alloy. The fluids used are either inert gases such as nitrogen, helium, or argon, or water. Gases generally give better results, i.e., finer, more spherical powder particles. The effect of the fluid is to fragment the liquid metal in order to produce fine droplets which then solidify to yield the powder. Figure 16.4 shows a gas atomisation tower.

The underlying physical principle is the action of a force exerted by the high-pressure fluid on an element of the liquid metal, breaking it up into several dispersed pieces (fragmentation). This force must overcome surface tensions which tend to hold the bulk metal element in a single lump. The fragmentation process is shown schematically in Fig. 16.5.

The liquid metal exhibits several states depending on its position at the nozzle outlet, finally yielding spherical particles (see Fig. 16.6). The phenom-

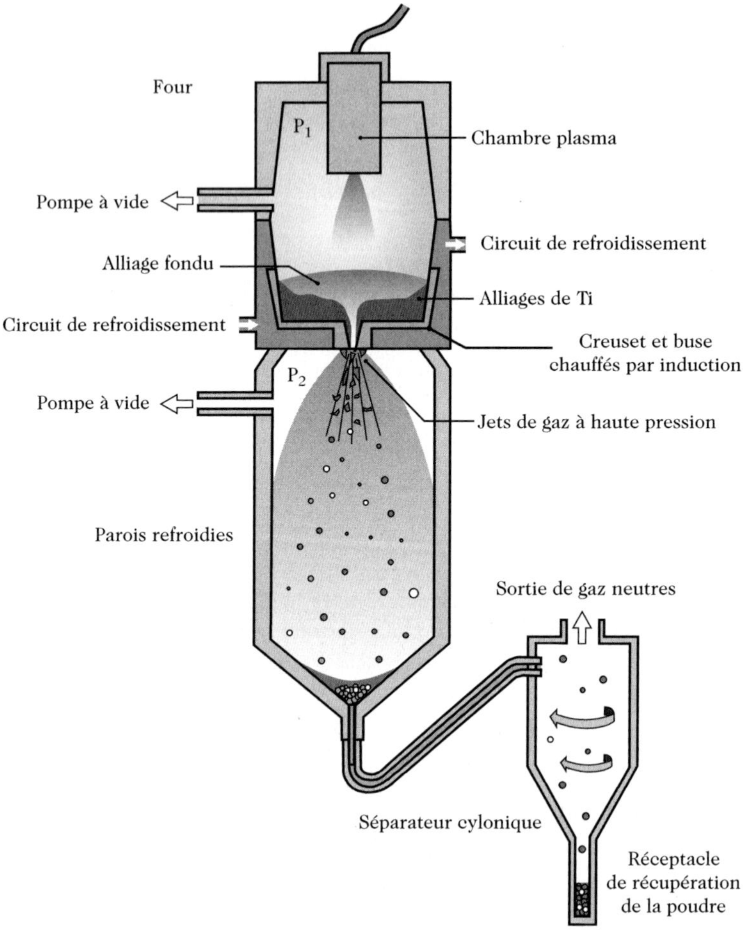

Fig. 16.4. Gas atomisation tower. Copyright IPM Laboratory, University of Virginia, all rights reserved [2]

ena controlling the liquid state are complex and difficult to model, and equipment for fabricating high quality powders is designed empirically. However, specialists predict that the sizes x of powder particles depend on experimental parameters and are related to the physical characteristics of the metal according to

$$x \approx \frac{C}{V_{\mathrm{g}}} \left(\frac{\gamma}{\rho_{\mathrm{m}}}\right)^{0.22} \left(\frac{\eta_{\mathrm{m}}}{\rho_{\mathrm{m}}}\right)^{0.57} , \tag{16.3}$$

where C is a constant determined by the nozzle geometry, V_{g} is the gas speed, and regarding the metal, γ is the surface energy, ρ_{m} is the density, and η_{m} is the viscosity. Equation (16.3) shows above all that the sizes of particles that

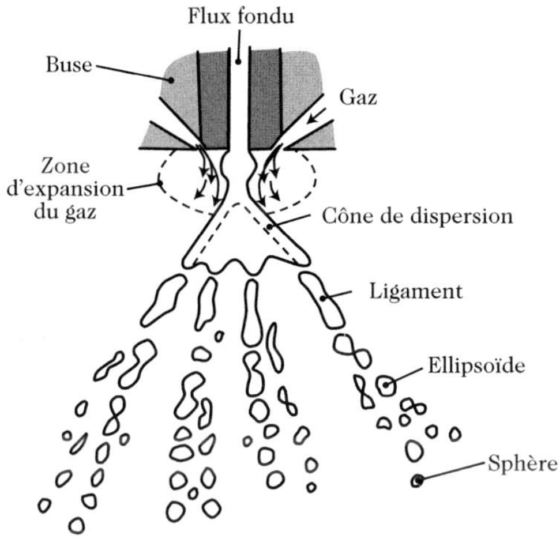

Fig. 16.5. Schematic view of powder formation at the nozzle outlet. From [4]

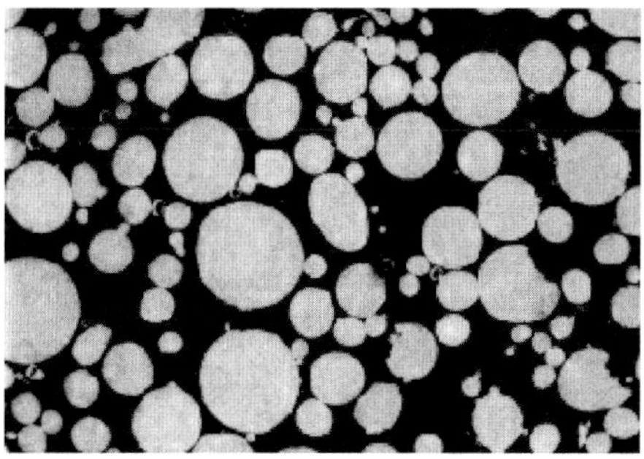

Fig. 16.6. Stainless steel powder prepared by gas atomisation. Magnification ×50. Copyright IPM Laboratory, University of Virginia, all rights reserved [2]

can be made by atomisation depend sensitively on the gas ejection speed. With $\gamma = 1\,\mathrm{J/m^2}$, $\rho_\mathrm{m} = 7\,000\,\mathrm{kg/m^3}$ and $\eta_\mathrm{m} = 0.003\,\mathrm{kg/m\,s}$, and with a feasible ejection speed for industrial application, viz., $V_\mathrm{g} = 100\,\mathrm{m/s}$, a rough estimate is

$$x \approx 3.5 \times 10^{-7} C \,. \tag{16.4}$$

With $C > 1$, the best possible conditions lead to particles of a few tenths of a micrometer. Industrially, powders of Cu, Fe, Sn, and steel are made with particle sizes in the range 20–100 μm, a long way above the nanoscale.

Liquid state fragmentation involves too large a mechanical stress to achieve nanoscale results. An alternative is to reconstruct nanoparticles from atoms of the appropriate matter in the gas state.

16.3 Main Stages of Gas Phase Synthesis

There are three stages in the synthesis of nanoparticles in the gas phase, requiring the implementation of three types of physical and physicochemical phenomena. Generically, these three stages are:

- production of the compound in vapour form,
- condensation in the form of nanoparticles, with a chemical reaction in some cases, e.g., oxidation,
- control and preservation of the dispersed nanocrystalline state.

Following this, the nanoparticles are collected and stored, and depending on the relevant application, subjected to various post-collection processes. These processes will affect subsequent transformation and use of the nanopowder.

It is very important to understand the physical and/or physicochemical principles of the process stages listed above. Indeed, this understanding is essential for:

- assessing the feasibility of synthesising a material in nanoparticle form,
- assessing general trends in a semi-quantitative manner in terms of production yield and accessible nanoparticle sizes.

As we shall see below, the stages in the formation of nanoparticles, the feasibility, the yield and the quality of nanopowders will all depend sensitively on the physicochemical properties of the phases one hopes to produce.

The main stages in the problem we are concerned with here, i.e., the synthesis of nanoparticles, are the control of condensation and the product formed subsequently. Whatever the compound, these stages are treated in the same way and require reference to theories of the formation and behaviour of aerosols (see below).

The way the vapour is formed will depend for its part on the material one hopes to obtain in the nanometric state. In particular, one must decide whether the material can be evaporated by standard or induction heating. If the material is refractory, one must use a plasma or start with a precursor and decompose it by laser pyrolysis. The formation of the vapour raises more technological problems. We shall discuss this aspect, illustrating with nanopowder syntheses, after considering the details of condensation.

Definition of an Aerosol

An aerosol is a dispersion of particles with small enough masses to be held in suspension in a gaseous atmosphere. Aerosols are classified in terms of the way the particles are formed, whether they are solid or liquid, and in terms of their concentration. For more details on metallic aerosols, the reader is referred to the review article by Buckle [5] and the book by Flagan and Seinfeld [6]. These aerosols are often present near industrial sites, as effluents of different metallurgical processes, e.g., foundry, milling, or grinding activities. A major effort was made in the 1950s to understand

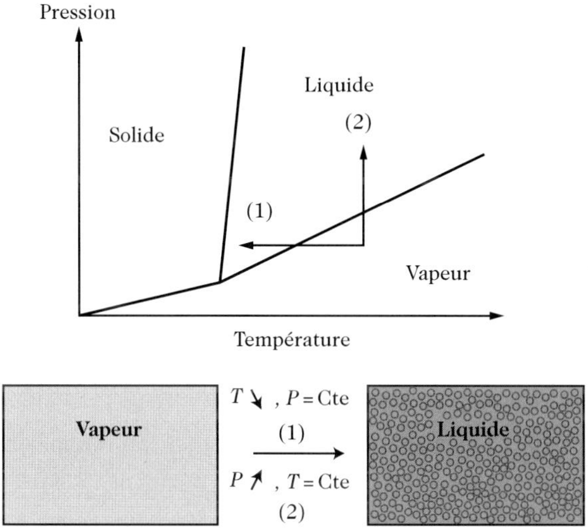

Fig. 16.7. (P, T) phase diagram of a pure substance showing the two routes leading to a supersaturated vapour state: (1) isobaric cooling, (2) isothermal compression

aerosol formation, with a view to controlling them and limiting their impact on the environment. Since the beginning of the 1970s, this interest has been extended to the way they are produced, owing to the exceptional properties of highly divided or nanostructured materials.

16.4 Spontaneous Condensation of Nanoparticles: Homogeneous Nucleation

It is difficult to model the formation of an aerosol by condensation owing to the large number of experimental parameters and the difficulty in controlling them. A simplified model of the phenomenon of homogeneous or standard nucleation is discussed below. The reader is also referred to the very detailed theoretical work of Bradley [7], Buckle [8], and Oxtoby [9].

Condensation occurs when a vapour phase is taken to a supersaturated state (see below), either by rapid isobaric cooling, i.e., at constant pressure P, or by isothermal compression, i.e., at constant temperature T. The phenomenon is shown schematically in Fig. 16.7, which represents the equilibrium (P, T) diagram of a pure substance. In the absence of any foreign germ or seed, the supersaturated metastable state is broken by statistical fluctuations which thus trigger the phenomenon of nucleation and the spontaneous appearance of a new phase in the form of dispersed aggregates.

$$\begin{array}{l} A + A \longleftrightarrow A_2 \\ A_2 + A \longleftrightarrow A_3 \\ \vdots \\ A_i + A \longleftrightarrow A_{i+1} \\ \vdots \\ A_{n-1} + A \longleftrightarrow A_n \\ \hline nA \longleftrightarrow A_n \end{array}$$

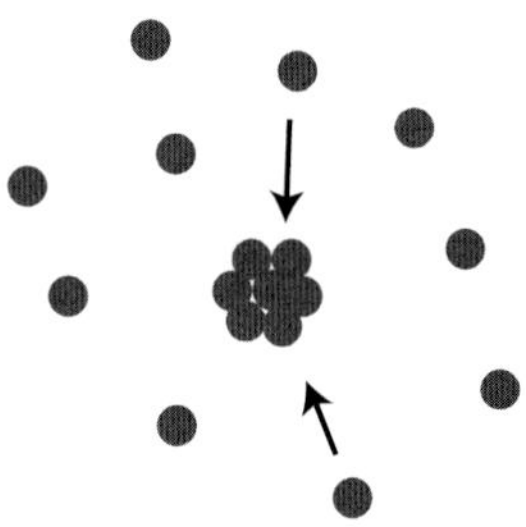

Fig. 16.8. Schematic view of the atomic aggregation process

Saturated Vapour Pressure and Supersaturated State

The saturated vapour pressure P_0 is the pressure of the vapour of the compound in thermodynamic equilibrium conditions. The vapour pressure is related to the absolute temperature T by the Clapeyron relation

$$\ln P_0 = A - \frac{B}{T_0} , \tag{16.5}$$

where A and B are constants depending on the kind of vapour. If a vapour is under a pressure P when the equilibrium condition gives P_0 such that $P > P_0$, the vapour is said to be in a non-equilibrium supersaturated state. This state will be broken to reach equilibrium under the effect of statistical fluctuations, and this will lead to the appearance of the liquid phase. The quantity $S = P/P_0$ is called the degree of supersaturation.

The aim here is to determine the key parameters controlling the size of nanoparticles and the yield, and also those allowing one to control the nanometric state.

Homogeneous nucleation is modelled in a similar way to the simple analysis of polymer formation, by considering a succession of combinations between an atom A and an atomic aggregate A_i formed from i atoms. The sum of the elementary equilibria gives the global equilibrium for formation of an aggregate A_n comprising n atoms (see Fig. 16.8).

The equilibrium constant for this is $K = c_n/c^n$, where c_n is the concentration of aggregates A_n and c the concentration of atoms A. This yields the free enthalpy of formation of the aggregate A_n as

$$\Delta G_n = -k_{\mathrm{B}} T \ln \frac{c_n}{c^n} , \tag{16.6}$$

where k_{B} is the Boltzmann constant. Aggregate sizes are much greater than the size of the atom, i.e., n is very big. It is easy to see that the number of aggregates each containing many atoms is small compared with the number of atoms in the aggregate–vapour mixture at equilibrium. Hence, $n_A \gg n_{A_n}$, where n_A is the number of atoms and n_{A_n} the number of aggregates containing n atoms, and this justifies the approximation

$$c = \frac{n_A}{n_A + n_{A_n}} \approx \frac{n_A}{n_A} = 1 ,$$

so that the concentration of aggregates containing n atoms is

$$c_n = c \exp\left(-\frac{\Delta G_n}{k_B T}\right) . \tag{16.7}$$

Conditions for Germ Formation

The free enthalpy of formation of an aggregate is the sum of the enthalpy of the reversible reaction in which n atoms condense into an aggregate and the surface energy of the aggregate thereby formed. Hence, if P_0 is the saturated vapour pressure of the compound, P the vapour pressure, and dn_1 and dn_2 the change in the number of atoms in the aggregate and in the vapour, respectively, then the change in the free energy of the system in the condensation process is given by

$$dG = \sum \mu_i dn_i = (\mu_0 + k_B T \ln P_0) dn_1 + (\mu_0 + k_B T \ln P) dn_2 , \tag{16.8}$$

where μ_0 is the standard chemical potential.

For a condensation, $dn_1 = -dn_2 = dn$, whence

$$\frac{dG}{dn} = -k_B T \ln \frac{P}{P_0} . \tag{16.9}$$

For the formation of an aggregate of n atoms, there are $n-1$ combination stages, each one involving an energy reduction of dG/dn. Adding the surface energy of the aggregate to this term, we obtain the free enthalpy of formation:

$$\Delta G_n = \gamma S_n - (n-1) k_B T \ln \frac{P}{P_0} . \tag{16.10}$$

The surface energy of the aggregate is the product of its area S_n and the energy per unit area.

The expression for the free enthalpy shows the competition between the negative condensation energy and the positive surface energy (but see the note below in the case of a supersaturated state). This means that, if the aggregate is too small, the surface energy will be greater than the condensation energy. The free enthalpy will be positive and the germ will evaporate.

Note. The free enthalpy varies as $\ln(P/P_0)$. Of course, if this quantity is negative, i.e., if $P < P_0$, there will be no stable germ. In order to form stable germs, P must be sufficiently greater than P_0 to compensate the surface energy that is created. Previously, we defined the degree of supersaturation as $S = P/P_0$. The formation of stable germs is thus more likely for higher degrees of supersaturation (see Fig. 16.9).

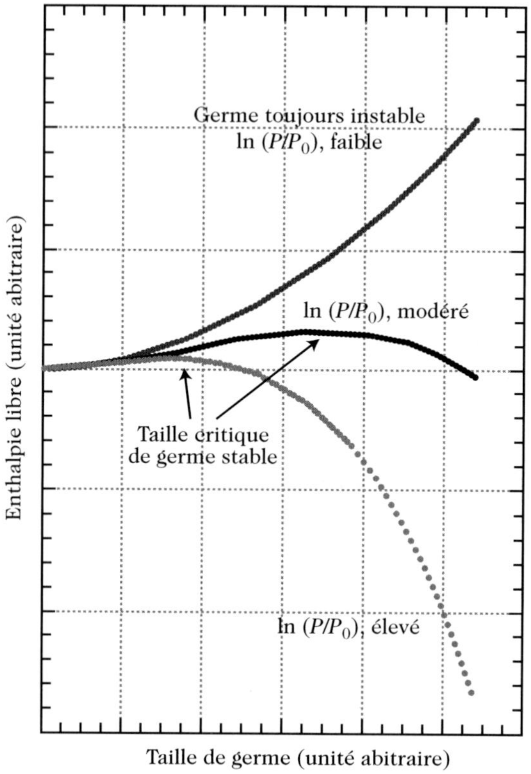

Fig. 16.9. Relation between germ stability and the degree of supersaturation, illustrated by representing the free enthalpy of the germs as a function of their size. If the ratio P/P_0 is too small (*upper curve*), there are no stable germs and condensation does not proceed. When the ratio P/P_0 is high enough, stable germs can form. Their size, given at the maximum of the free enthalpy curve, will decrease as P/P_0 increases (from the *middle curve* to the *lower curve*)

Critical Germ

Examining the expression for the free enthalpy, we deduce that there is a critical number of atoms in the germ for which the condensation energy is greater than the surface energy. For a spherical aggregate,

$$S_n = \pi x^2 , \tag{16.11}$$

$$V_n = \pi \frac{x^3}{6} = \frac{m_n}{\rho} = \frac{nM}{\rho N} , \tag{16.12}$$

where M is the molar mass of the compound, N is Avogadro's number, and ρ is the density. Hence,

$$x = \sqrt[3]{\frac{6nM}{\pi\rho N}} \tag{16.13}$$

$$\Delta G_n = \gamma\pi\left(\frac{6nM}{\pi\rho N}\right)^{2/3} - (n-1)k_\mathrm{B}T\ln\frac{P}{P_0}\,. \tag{16.14}$$

The critical size x^* of the stable germ is given by the minimum of the free enthalpy:

$$\frac{\mathrm{d}\Delta G_n}{\mathrm{d}n} = \frac{2}{3}\gamma\pi\left(\frac{6M}{\pi\rho N}\right)^{2/3} n^{-1/3} - k_\mathrm{B}T\ln\frac{P}{P_0} = 0 \tag{16.15}$$

and

$$\frac{4\gamma M}{\rho N}\frac{1}{x^*} - k_\mathrm{B}T\ln\frac{P}{P_0} = 0\,. \tag{16.16}$$

The critical size of a stable spherical germ is given by

$$x^* = \frac{4\gamma M}{\rho N k_\mathrm{B}T\ln(P/P_0)}\,. \tag{16.17}$$

Once again, we obtain a dependence of the size, here the critical size of the stable germ, on the degree of supersaturation.

Returning to the expression for the concentration of aggregates, we obtain the concentration c_n^* of germs of critical size, where n^* is the number of atoms in the germ of critical size x^* :

$$n^* = \frac{x^{*3}\pi\rho N}{6M}\,. \tag{16.18}$$

Returning to (16.16), viz.,

$$k_\mathrm{B}T\ln\frac{P}{P_0} = \frac{4\gamma M}{\rho N x^*}\,, \tag{16.19}$$

and substituting in (16.14),

$$\Delta G_n^* = \gamma\pi x^{*2} - \frac{x^{*3}\pi\rho N}{6M}\frac{4\gamma M}{\rho N x^*}\,. \tag{16.20}$$

Then substituting into the expression (16.7) for the germ concentration, we end up with

$$c_n^* = c\exp\left(-\frac{\gamma\pi x^*}{3k_\mathrm{B}T}\right)\,. \tag{16.21}$$

This relation shows, as could have been expected intuitively, that for a given initial vapour pressure, the concentration of germs of critical size increases as the critical germ size decreases.

To conclude, when fabricating nanoparticles, one must favour the initial formation of stable critical germs of the smallest possible size. An important result then follows from the above analysis, namely that the size will be smaller if the degree of supersaturation is made higher.

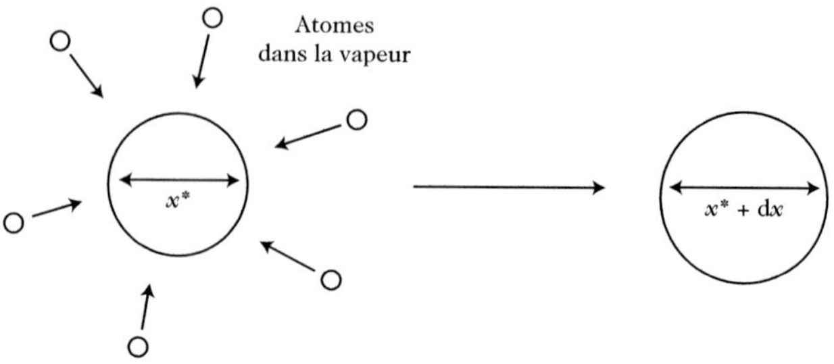

Fig. 16.10. The germ of critical size x^* becomes permanently stable by addition of atoms to increase its size to $x^* + dx$. The time required for this process to occur depends on the flux of gas atoms towards the surface of the critical germ. Determining the nucleation rate amounts to counting the number of these processes occurring per unit volume and per unit time

Nucleation Rate

The above analysis gives the dependence of the main quantities, such as germ size and concentration, on experimental parameters like the degree of supersaturation. However, none of the relations involves the time parameter from which the rate of production of nanoparticles could be estimated.

Combining this analysis with the kinetic theory of gases, a key quantity known as the nucleation rate J can be deduced. This is defined as the number of stable nucleation centers produced per unit volume and per unit time (see Fig. 16.10).

We consider a stationary state in which the vapour pressure and degree of supersaturation are constant. Then the distribution of concentrations of germs with size smaller than the critical size of the stable germ is also constant:

$$c_n = \text{constant}\,, \quad \forall n \text{ such that } n < n^*\,.$$

J is defined as the number of germs of critical size disappearing due to increasing size, per unit volume and unit time. J is thus equal to the product of the number of atoms striking in an effective way per unit time, the total surface area of germs of critical size thereby becoming permanently stable. The total surface area of all germs of critical size in a unit volume is given by the product of the area of a critical germ and the concentration of critical germs, viz., $S_n^* c_n^*$. The amount of atoms β striking a unit area per unit time is given by the Hertz–Knudsen relation, derived from the kinetic theory of gases:

$$\beta = \frac{P}{\sqrt{2\pi m k_{\mathrm{B}} T}}\,. \tag{16.22}$$

Hence,

$$J = \beta S_n^* c_n^*\,, \qquad J = \frac{P}{\sqrt{2\pi m k_{\mathrm{B}} T}} S_n^* c \exp\left(-\frac{\Delta G_n^*}{k_{\mathrm{B}} T}\right)\,. \tag{16.23}$$

Noting in particular that $c = P/k_BT$, and using (16.10), we obtain

$$J = \frac{P}{\sqrt{2\pi m k_B T}} S_n^* \frac{P}{k_B T} \exp\left[-\frac{\gamma S_n^*}{k_B T} + (n-1)\ln\frac{P}{P_0}\right] . \tag{16.24}$$

Then

$$J = \frac{P^2}{\sqrt{2\pi m (k_B T)^3}} S_n^* \exp\left(-\frac{\gamma S_n^*}{k_B T}\right) \times \left(\frac{P}{P_0}\right)^{n-1} , \tag{16.25}$$

and

$$J = \frac{P_0^2}{\sqrt{2\pi m (k_B T)^3}} S_n^* \exp\left(-\frac{\gamma S_n^*}{k_B T}\right) \times \left(\frac{P}{P_0}\right)^{n+1} . \tag{16.26}$$

It is useful to bring the size of the critical germ and the degree of supersaturation into the expression for J, whence

$$J = \frac{P_0^2}{\sqrt{2\pi m (k_B T)^3}} \pi \left[\frac{4\gamma M}{\rho N k_B T \ln(P/P_0)}\right]^2 \exp\left(-\frac{\gamma \pi x^{*2}}{k_B T}\right) \times \left(\frac{P}{P_0}\right)^{n+1} . \tag{16.27}$$

For very small critical germs of the kind required here, the exponential tends rapidly to unity and we thus observe that the nucleation rate also depends sensitively on the degree of supersaturation:

$$J \propto \frac{(P/P_0)^{n+1}}{\left[\ln(P/P_0)\right]^2} .$$

In this expression, when n is large, the degree of supersaturation $S = P/P_0$ becomes the dominating term.

To conclude, in the initial stages of nanoparticle formation by condensation, for a given pressure, the number of particles formed per unit volume and per unit time will increase and the size of particles will decrease when the degree of supersaturation is made larger.

Illustration: Fog Formation

The formation of fog, a well known atmospheric phenomenon feared by aircraft pilots and yachtsmen, is a typical example of aerosol formation by condensation. The amount of water vapour stored in the air decreases with the temperature. If the air contains a large amount of water vapour close to saturation, a small but sudden drop in temperature (as happens at sunrise, for example) will lead to the air being supersaturated in water vapour. The vapour will then condense into fine droplets, the fog reducing visibility to less than 1 000 m. Rather than characterising the humidity of the air by the saturated vapour pressure, meteorologists characterise it by temperature at which fog will form, called the dew point, which can be deduced from the Clapeyron relation (16.5).

16.5 Undesirable Post-Condensation Effects and Control of the Nanometric State

It is now clear what must be done to form large quantities of particles with nanometric sizes per unit volume of vapour. A vapour must be produced with high pressure in a high state of supersaturation. These are of course thermodynamic conditions well outside thermodynamic equilibrium and with extreme kinetics, ensuring extremely fast synthesis on the one hand, but hard to control on the other. We are a long way from the conditions of soft chemistry, where the kinetics of diffusion in a liquid medium, more or less at room temperature, are perfect for controlling the size of slowly formed nanoparticles.

Two harmful effects are liable to occur as soon as stable germs have formed:

- The growth of the nanoparticle, whose gaseous environment is still at high pressure (at least, in a supersaturated state).
- Coalescent coagulation of the nanoparticles, i.e., the melting of particles after collision. The region in which nanoparticles form is at very high temperature, so there is a high probability of collision between particles undergoing Brownian motion (see below), followed by immediate coalescence.

Note also a third possible harmful effect, namely, non-coalescent coagulation in colder regions where particles aggregate by neck formation. This rigid aggregate structure (see, for example, Fig. 16.1b) is detrimental for certain subsequent processes using the nanopowders, e.g., densification after compaction and sintering.

Experimental investigations [10, 11] show that the dominating effect in the size increase of nanopowders is coalescent coagulation. The growth of a particle by gas condensation on its surface occurs in the very first phase of size increase or in the case of very dilute systems. The two effects are discussed separately below. The shape of the nanoparticle size distribution obtained by gas phase condensation and coalescence can be estimated theoretically.

Brownian Motion

Brownian motion is the ceaseless, random motion of small particles suspended in a fluid under the action of collisions with the fluid molecules. This phenomenon was first observed by the botanist R. Brown in 1827 and fully explained in terms of entropy fluctuations by A. Einstein in 1910. Aerosols can be considered as particle–fluid systems in which the particles are subjected to Brownian motion. The random displacement of the particles is a key factor in the evolution of an aerosol.

The Log-Normal Distribution

The statistical theory of coalescence (each event consisting in the combination of two particles) predicts that the logarithm of the volume distribution of the particles

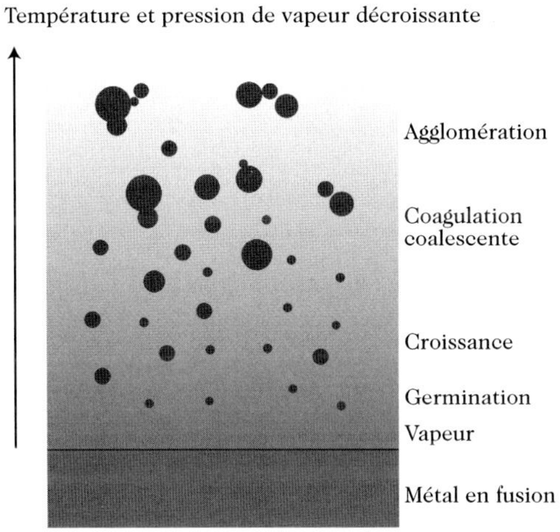

Fig. 16.11. Successive stages in the condensation of a gas when it occurs in a temperature gradient

will be of Gaussian type. It follows that, for spherical particles, the distribution of the diameters x will obey a log-normal distribution given by

$$P(x) = \frac{B}{(2\pi)^{1/2}\sigma(x/\overline{x})} \exp\left\{-\frac{\left[\ln(x/\overline{x}) + \sigma^2/2\right]^2}{2\sigma^2}\right\} , \tag{16.28}$$

where $P(x)$ is the fraction of particles of size x, B is a normalisation factor, $\overline{x}$ is the average size of the particles, and σ is the standard deviation calculated from the variance Var of the following distribution:

$$\overline{x} = \frac{\sum_i (n_i x_i)}{\sum_i n_i} , \qquad \sigma = \left[\ln(\text{Var}/\overline{x}^2) + 1\right]^{1/2} , \tag{16.29}$$

with n_i the number of particles of size x_i.

16.5.1 Why Do These Effects Occur?

Figure 16.11 illustrates why spontaneous condensation by formation of a supersaturated state is difficult to achieve. A phase state characterised in the phase diagram by coordinates $(P_\mathrm{i}, T_\mathrm{i})$ is suddenly placed under different conditions $(P_\mathrm{f}, T_\mathrm{f})$, as far removed as possible. The model assumes that the pressure and temperature are homogeneous at all points of the system, an ideal situation that cannot be achieved experimentally. Recall that the aim here is the continuous and if possible large-scale fabrication of a nanopowder. So no matter how the vapour is obtained, condensation is brought about in a temperature gradient, and hence in a pressure gradient of the relevant vapour.

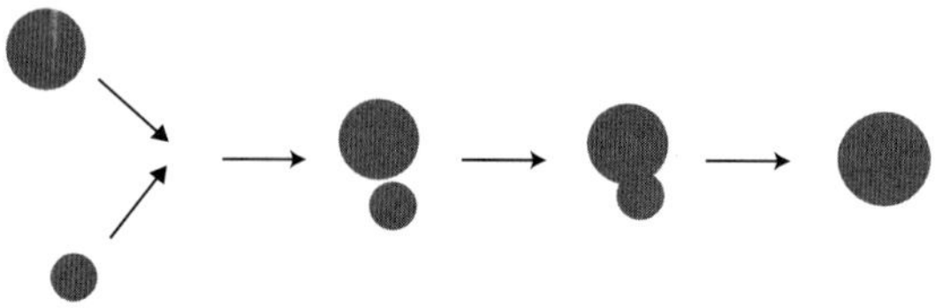

Fig. 16.12. Coalescence of two particles

Figure 16.11 shows a snapshot of the evolution expected during the condensation, growth, coalescent coagulation, and aggregation phenomena. The simple case of a vapour gradient formed by evaporation by standard heating is described below.

16.5.2 Particle Growth by Gas Condensation

Once a stable germ has formed, it is surrounded by a metallic vapour which still has a very high pressure and it will thus continue to grow by condensation of atoms on its surface. This is represented in Fig. 16.10. If we consider a single particle, the flux of atoms condensing on its surface (assuming total adsorption efficiency) is given by the Hertz–Knudsen relation (16.22). Hence the number of atoms striking the surface is directly proportional to the pressure of the surrounding metallic gas. In order to maintain the nanometric state, it is essential to limit the time spent by the particles in regions of high metallic pressure.

Returning to the Hertz–Knudsen formula, it can be shown that the rate of change of the size of an individual particle is constant:

$$\beta = \frac{P}{\sqrt{2\pi m k_{\mathrm{B}} T}} = \frac{\mathrm{d}n}{S\mathrm{d}t} = \frac{\pi\rho N}{6M}\frac{1}{\pi x^2}\frac{\mathrm{d}(x^3)}{\mathrm{d}t} = \frac{\rho N}{3M}\frac{\mathrm{d}x}{\mathrm{d}t}\,, \tag{16.30}$$

where M is the molar mass of the compound, N is Avogadro's number, and ρ is the density. In the period immediately following the formation of the critical germ, $P \approx$ const. and hence $\mathrm{d}x/\mathrm{d}t \approx$ const.

Furthermore, particles suspended in a fluid undergo Brownian motion, and the distribution of the growth time t is normal:

$$f(t) = \frac{1}{\sqrt{2\pi}} \exp\left(-\frac{t^2}{2}\right)\,. \tag{16.31}$$

As time goes by, the population will thus adopt a Gaussian size distribution. The approach described here is very simple and agrees well with some experiments [12]. However, a controversy has nevertheless arisen concerning the growth mode and the dependence of the particle size on the elapsed time (see, for example, [13]).

Let us assume that each atom at the surface of the particle catches a constant number of atoms coming from the gas phase, whatever the size of the particle. Hence, in each time step, the particle grows exponentially, i.e.,

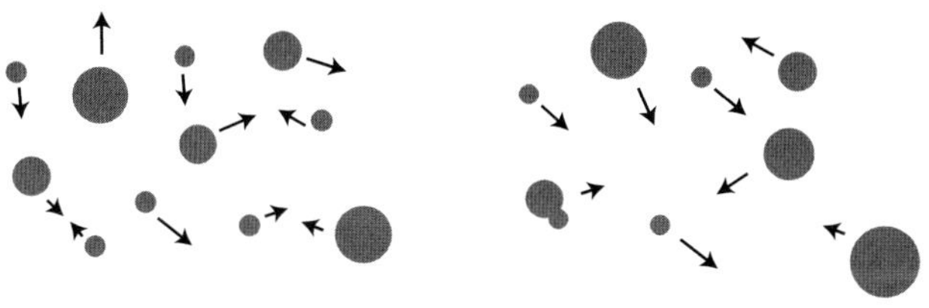

Fig. 16.13. Evolution of a population of particles undergoing Brownian motion and liable to coalesce

$$x = x^* \exp(At) , \tag{16.32}$$

and the growth rate becomes

$$\frac{\mathrm{d}x}{\mathrm{d}t} = Ax , \tag{16.33}$$

for a constant A. Combining (16.32) and (16.33), a log-normal particle size distribution (16.28) is obtained.

16.5.3 Coalescent Coagulation

Coalescence is the combination of several particles after a collision, yielding a single particle under surface tension effects (see Fig. 16.12). This occurs in regions where the partial pressure of the metallic vapour is now almost zero, but where the particle temperature is still fairly high. The particles are still liquid in the very first stages, and the particle temperature is roughly half the melting temperature of the compound. Coalescence then occurs with atomic diffusion in the solid state.

Mean Free Path and Collision Frequency

In order to coalesce, particles must first meet one another (see Fig. 16.13). The collision will occur on average after a certain lapse of time τ, or a certain distance l travelled freely (without collision) by either of the particles. Consider two particles of the same size x, moving at speeds v_1, v_2, hence with the relative speed $v_\mathrm{r} = v_1 - v_2$. When a particle moves at a constant average relative speed $\langle v_\mathrm{r} \rangle$ for a time τ, it meets another particle if it sweeps out a cylindrical region of base area πx^2 and height $\langle v_\mathrm{r} \rangle \tau$. This cylindrical region contains a particle and hence, if c_p is the number of particles per unit volume, we obtain

$$\pi x^2 \langle v_\mathrm{r} \rangle \tau c_\mathrm{p} = 1 . \tag{16.34}$$

The time τ before a collision is the ratio of the mean free path l to the average speed $\langle v\rangle$ of a particle, i.e., $\tau = l/\langle v\rangle$, whence

$$\pi x^2 \frac{\langle v_\mathrm{r}\rangle}{\langle v\rangle} l c_\mathrm{p} = 1 \ . \tag{16.35}$$

Note also that the ratio of the average relative speed to the average speed of a particle is simply obtained from

$$v_\mathrm{r} = v_1 - v_2 \ , \tag{16.36}$$

$$v_\mathrm{r}^2 = v_1^2 + v_2^2 - 2v_1v_2 \ , \tag{16.37}$$

$$\langle v_\mathrm{r}^2\rangle = \langle v_1^2\rangle + \langle v_2^2\rangle = 2\langle v^2\rangle \ , \tag{16.38}$$

and hence the mean free path is

$$l = \frac{1}{\pi x^2 \sqrt{2} c_\mathrm{p}} \ . \tag{16.39}$$

The distance over which a particle can move freely is inversely proportional to the concentration of the particles making up the aerosol. This concentration is once again related to the initial metallic vapour pressure [see (16.7)].

The notion of mean free path makes no mention of kinetic parameters, but it is easy enough to accept that, the higher the metallic vapour pressure, the shorter will be the mean free path, and hence the more likely will be collisions and coalescence, rapidly destroying the nanometric state.

Kinetics of Coalescent Coagulation. Particle Displacement in an Aerosol

Coalescent coagulation is the dominating phenomenon to impair the nanometric state of a particle suspension. It is thus useful to approximate a time parameter in this process, or at least to try to identify the physical quantities affecting it, in order to understand the main trends.

We previously defined a mean free path related to the particle concentration. This parameter can be related to the collision frequency $1/\tau$ by $\langle v\rangle = l/\tau$. The average speed with which particles move in the suspension is the parameter that needs to be examined in order to understand the kinetics of coalescence. This speed is found by integrating the equation of motion for a particle of mass m_p:

$$m_\mathrm{p} \frac{\mathrm{d}v}{\mathrm{d}t} = \sum_i F_i \ , \tag{16.40}$$

where v is the speed of the particle and F_i the force of the mechanism i acting on the particle. In a typical situation, the particles are mainly subjected to frictional forces F_f between the particle and its fluid environment, and the force F_b resulting from impacts when atoms and molecules collide with the particle surface, the origin of the random Brownian motion.

Fluid Flow Around a Particle

The force of friction is found by analysing the flow of a fluid around a rigid, spherical particle. The reader is referred to [14] for a detailed analysis of this physical problem. Integrating the continuity equation and the Navier–Stokes equation, we obtain Stokes' law, in the usual framework of fluid mechanics. This law is not perfectly well suited to the problem of aerosols that we are concerned with here, because the particles (nanoparticles with sizes less than 100 nm) are sensitive to the atomic or molecular nature of the surrounding gas and to discrete events associated with them. The relevant parameter to compare with the particle size, taking into acount the particle–gas interaction, is the mean free path λ_g of the atoms or molecules of the gas, deduced from the kinetic theory of gases:

$$\lambda_g = \frac{\mu}{0.499 P_g \sqrt{8M/\pi RT}} , \tag{16.41}$$

where μ is the viscosity, M the molar mass of the gas, P_g the pressure of the gas, R the perfect gas constant, and T the absolute temperature.

For example, for nitrogen at 1 000°C, $\mu = 4.63 \times 10^{-5}\,\mathrm{kg\,m^{-1}s^{-1}}$, $M = 30\,\mathrm{g\,mol^{-1}}$, $P_g = 1\,\mathrm{atm}$, it follows that $\lambda_g = 80\,\mathrm{nm}$, which is of the same order of magnitude as the nanoparticles one hopes to make. More precisely, concerning the mean free path, we shall make the following assumptions with regard to transport properties for a particle of size x:

- if $x > 250\,\mathrm{nm}$, the regime is continuous,
- if $x < 10\,\mathrm{nm}$, transport properties are governed by the kinetic theory of free molecules,
- if $10 < x < 250\,\mathrm{nm}$, the regime is transient.

These considerations lead to a correction to the Stokes law given the friction force exerted by the fluid on a particle. For the systems we are interested in here, this gives

$$F_f = \frac{3\pi\mu v_r x}{C_c} , \tag{16.42}$$

where v_r is the relative speed of the particle and fluid, and C_c is a correction factor following from the above analysis, taking values as follows:

$$C_c = \begin{cases} 1 + 1.257\dfrac{2\lambda_g}{x} & x \gg \lambda_g , \\[2ex] 1 + 1.657\dfrac{2\lambda_g}{x} & x \ll \lambda_g . \end{cases} \tag{16.43}$$

Returning to the example of particles in nitrogen at 1 000°C and 1 atm, we have $C_c = 27.5$ for $x = 10\,\mathrm{nm}$ and $C_c = 1.02$ for $x = 10\,\mu\mathrm{m}$.

Note. The ratio $2\lambda_g/x = K_n$ is called the Knudsen number.

Force Due to Molecular Bombardment of the Particle Surface

In a first approach, the force causing Brownian motion can be treated in a much simpler manner, since it is good enough to express it in terms of the random accelerations resulting from molecular bombardment of the particle surface: $F_b = m_p\alpha(t)$, where $\alpha(t)$ is the acceleration of the particle, varying randomly over time, and m_p is the mass of the particle.

Returning to (16.40) and (16.42),

$$m_p \frac{dv}{dt} = \frac{-3\pi\mu x}{C_c} v + m_p \alpha(t) , \tag{16.44}$$

and

$$\frac{dv}{dt} = -\frac{1}{\tau} v + \alpha(t) , \tag{16.45}$$

where τ is the relaxation time of a particle. Equation (16.45) is known as the Langevin equation. The displacement, and hence the trajectory, of a particle subjected to the acceleration $\alpha(t)$ is random. In order to study Brownian motion and solve the Langevin equation, we therefore consider a population of particles, so that an average solution can be defined.

Consider a particle with position $r(0)$ at $t = 0$. Its coordinate in the same direction at time t is $r(t)$. For a very large number of particles with random trajectories, the average displacement over the whole population is $\langle r(t) \rangle = 0$. However, for the mean squared displacement of the population, we find $\langle r(t)^2 \rangle \neq 0$.

In 1905, A. Einstein derived an expression for $\langle r(t)^2 \rangle$ from the Langevin equation, later improved by Seinfeld, who took into account the correction factor C_c to Stokes' law. In the continuous regime, a solution is

$$\langle r(t)^2 \rangle = \frac{2k_B T C_c}{3\pi\mu x} t . \tag{16.46}$$

If the motion of the particle population is described as a diffusion process, we define a Brownian diffusion coefficient D from Fick's second law by

$$\frac{\partial N}{\partial t} = D \nabla^2 N . \tag{16.47}$$

For a population of N particles, all released at the origin at time $t = 0$, it follows that [15]

$$\frac{\partial \langle r(t)^2 \rangle}{\partial t} = 2D . \tag{16.48}$$

Integrating and combining with (16.46), we find

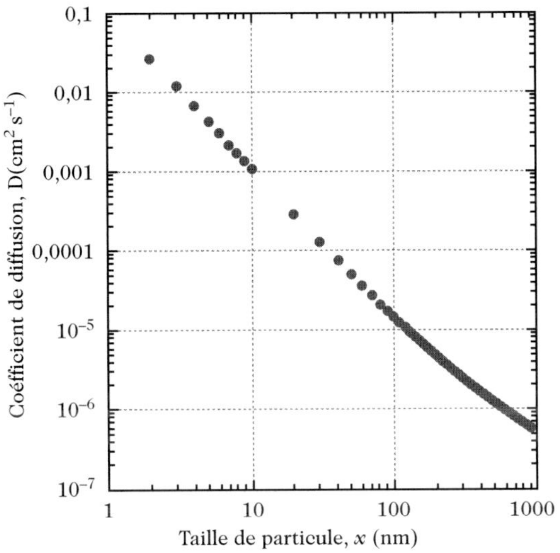

Fig. 16.14. Size dependence of the Brownian diffusion coefficient under the following conditions: nitrogen at 1 000°C, $\mu = 4.63 \times 10^{-5}\,\mathrm{kg\,m^{-1}s^{-1}}$, $M = 30\,\mathrm{g\,mol^{-1}}$, $P_{\mathrm{g}} = 1\,\mathrm{atm}$

$$D = \frac{k_{\mathrm{B}} T C_{\mathrm{c}}}{3\pi\mu x} . \tag{16.49}$$

The diffusion coefficient D accounts for the mobility of the particles in the fluid and the consequent effects on the collision frequency and coalescence of particles.

Returning to (16.43), it follows that

$$\begin{cases} x \gg \lambda , & C_{\mathrm{c}} = 1 + 1.257\dfrac{2\lambda}{x} \approx 1 , \qquad D \approx \dfrac{k_{\mathrm{B}} T}{3pmx} , \\ x \ll \lambda , & C_{\mathrm{c}} = 1 + 1.657\dfrac{2\lambda}{x} \approx 1.657\dfrac{2\lambda}{x} , \qquad D \approx \dfrac{3.314 l k_{\mathrm{B}} T}{3pmx^2} . \end{cases}$$

Note that D goes as x^{-2} for nanometric particles. Figure 16.14 shows the size dependence of the Brownian diffusion coefficient under conditions as indicated in the caption. Note that, for nanometric particles ($x < 100\,\mathrm{nm}$), the mobility becomes very high.

Once again, the nanometric state is very quickly destroyed by coalescent coagulation. The whole art of fabricating nanopowders lies in identifying and implementing techniques to combat the exceedingly rapid growth phenomena. There are three possibilities:

1. to produce extremely widely dispersed aerosols,
2. to continually cool the nanoparticles very quickly as soon as they are formed,

3. to continually disperse the nanoparticles after their formation using a high-speed gas flow.

Case 1 has the disadvantage of very low yields, while cases 2 and 3 involve skill in a wide range of technologies, dealing with high temperatures, cryogenics, and gas flow.

16.6 Vapour Formation and the Production of Nanopowders

In the last few sections, we have discussed the different stages in the formation of a nanopowder from a vapour phase. We have tried to understand the mechanisms whereby the particles form and to identify the main parameters controlling the dispersed nanometric state. The problem now is to implement various strategems in the technological setup to account for these parameters.

This last section is devoted to the way the vapour is produced, which appeals to old and well-tested technology. As a consequence, we have chosen to discuss vapour formation (without claiming to be exhaustive) by describing various techniques for fabricating nanopowders already used in research laboratories and in industry.

We shall see that the method of synthesis to be used is imposed by physico-chemical properties of the nanometric phase one hopes to fabricate. As an example, although copper lends itself well to direct evaporation by heating the bulk metal, there is no way that tungsten could be treated by this same process. This is soon demonstrated by measuring the saturated vapour pressures of the compounds as a function of the temperature, i.e., by determining the parameters in the Clapeyron relation for the two compounds. At 1 700 K, a temperature that can be reached by induction heating or electron bombardment, the saturated vapour pressure of Cu is 0.1 torr, while that of W is 10^{-11} torr!

16.6.1 Physical Processes

Direct Heating

We shall describe here the evaporation–condensation process, known as inert gas condensation, used by H. Gleiter and coworkers (originally at the university of Saarland at Saarbruck, Germany). The underlying principle, to evaporate then condense the phase in the form of small particles, has been the subject of much experimental investigation by many research groups in the US, Japan, Germany, France, and so on, from as early as the 1960s.

Since condensation occurs after vapour formation by cooling due to the reduction of the kinetic energy of the atoms by successive collisions with atoms or molecules of the inert gas, research aimed to investigate the effect of the

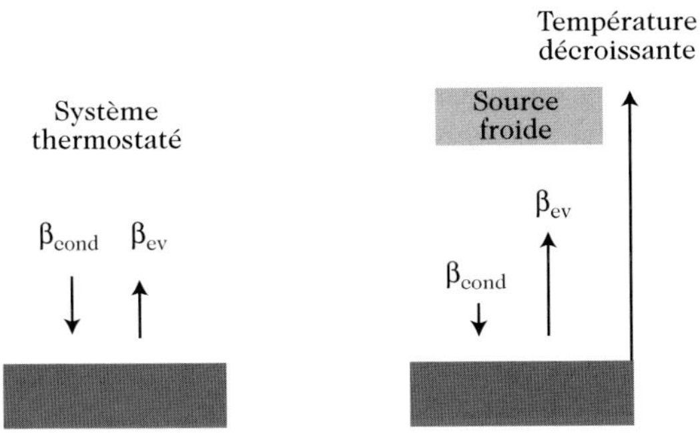

Fig. 16.15. Evaporation. At the equilibrium between the solid or liquid phase and the gas, the flow of atoms evaporated is balanced by the flux of atoms condensed and there is no evaporation. If a cold source is placed nearby, an imbalance is set up and evaporation takes place

type of gas on the size, morphology, and agglomeration rates of the resulting particles [16]. The main point here is that the average particle size is greater when the inert gas is heavier.

Although the basic idea of fabrication was very simplistic, Gleiter's group was able to set up a laboratory experiment that produced nanometric matter, thereby revealing its great potential.

The vapour is produced from a compound either by standard heating by the Joule effect, or by induction heating if the compound is suitable. The idea is illustrated schematically in Fig. 16.15. Above the free surface of a solid or liquid, there is always a certain amount of vapour. At thermodynamic equilibrium at a temperature T, this vapour has a constant pressure, the saturated vapour pressure, given by the Clapeyron relation. On the atomic scale, there are in fact two opposing processes that balance one another: atoms in the liquid or solid leave the surface to join the vapour, while atoms in the vapour condense back onto the surface. The rate of evaporation per unit area is equal to the rate of condensation per unit area. If an imbalance is set up, for example by placing a cold source near the surface or creating a low pressure zone, convection will occur and the evaporation rate will become greater than the condensation rate. The flow of atoms actually evaporated is given by

$$\beta = \beta_{\text{evap}} - \beta_{\text{cond}} \,. \tag{16.50}$$

These phenomena have been studied in great detail, in particular by I. Langmuir, research scientist at General Electric and winner of the Nobel Prize for Chemistry. The flow rates (number of atoms per unit area and per unit time) are given by the kinetic theory of gases, and in particular, by the Hertz–Knudsen relation.

Langmuir established the relation between the amount of matter evaporated per unit area and per unit time and the resulting vapour pressure:

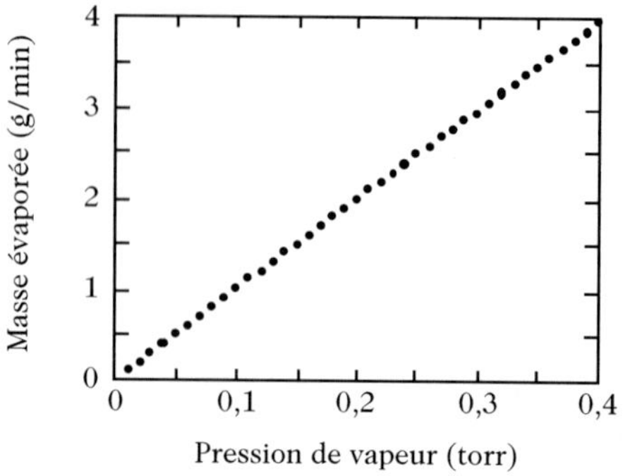

Fig. 16.16. Amount of matter evaporated as a function of the vapour pressure

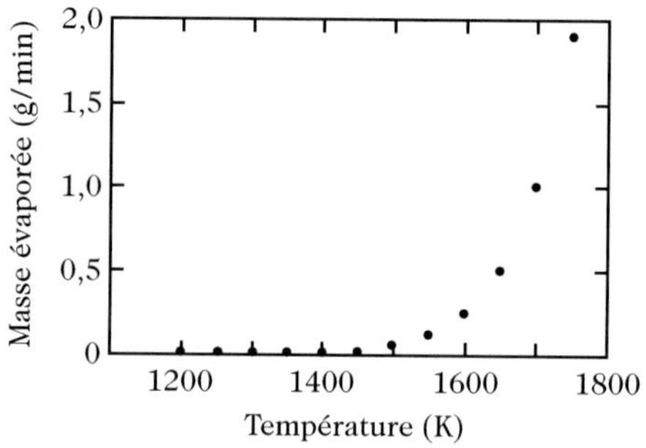

Fig. 16.17. Evaporated mass of copper as a function of the temperature for an evaporation area of $15\,\mathrm{cm}^2$

$$Q = \frac{P\sqrt{m}}{\sqrt{2\pi k_{\mathrm{B}} T}} \, . \tag{16.51}$$

This relation is illustrated in Fig. 16.16 for copper and an evaporation area of $15\,\mathrm{cm}^2$.

Note that a vapour pressure of 0.1 torr must be generated to achieve an acceptable yield of 1 g/min. If we now assume that all the vapour at the surface at equilibrium can be used and continually renewed, we can then replace the pressure in the Langmuir relation by the saturated vapour pressure. Hence, combining the Langmuir and Clapeyron relations, we can estimate the amount of matter as a function of the temperature, as shown in Fig. 16.17.

It should be borne in mind that this is just an estimate of the lower limit of the temperature. For one thing, the system is rarely in a state of total evaporation (as in the case of a very high vacuum), and for another, there is an inert gas which reduces the saturated vapour pressure. However, the saturated vapour pressure of a compound is a good parameter for estimating the feasibility in terms of vapour production.

Table 16.1 shows the values of the saturated vapour pressure of different metals at 1 700 K. This temperature is chosen in relation to copper, taken as

Table 16.1. Saturated vapour pressures of various metals

Element	Cu	Fe	Ni	Co	Al	Ti	W
Saturated vapour pressure at 1 700 K (torr)	0.1	0.01	0.001	0.001	0.1	10^{-4}	$< 10^{-11}$
Measured yield in evaporation–cryocondensation (g/min)	1	1	1	1	0.1	0	–

a reference, for which a yield of 1 g/min is obtained for an evaporation area of 15 cm^2. Fe, Ni, and Co are metals susceptible to evaporation–condensation methods. There is no hope of synthesising W in this way, since it would have to be heated to around 4 000 K!

Such a value of the saturated vapour pressure is of course prohibitive for this vapour production process. However, other physicochemical parameters can be relevant to experimental conditions on a laboratory or industrial scale; parameters that are difficult to control and which inhibit the process, even making it impossible. Table 16.1 gives the vapour production yields measured during so-called evaporation–cryocondensation experiments. This technique was developed by J. Bigot at the Centre d'Etudes de Chimie Métallurgique, a laboratory of the French national research institute (CNRS) [17]. The table shows that, despite the differences in the saturated vapour pressures, Cu, Fe, Ni and Co have comparable production yields. This is related to the fact that, for these metals, it is easy to reach high enough temperatures to obtain the required pressure of 0.1 torr. On the other hand, although high enough temperatures can be reached experimentally, of the order of 2 500 K for Ti, and likewise for Zr and Hf, the evaporation yield remains non-existent! This fact is due to the very high affinity of these metals for oxygen, which is always present, at least as a trace element, in places where matter is transformed. Ti is a genuine oxygen pump (which is why it is used as a 'getter' to clean metallurgical processing gases), capturing oxygen on its surface and considerably reducing the saturated vapour pressure of the metal.

Aluminium is well-suited to illustrate these phenomena, since it has a theoretical saturated vapour pressure that is identical to that of copper, chosen here as standard. The yield of aluminium vapour is around 10% of that obtained for copper by evaporation–cryocondensation. Once again, Al traps oxygen, considerably reducing the saturated vapour pressure of the metal. However, compared with Ti, Al has a very low melting temperature (933.5 K compared with 1 693 K for Ti). A matrix of aluminium oxide forms around the molten liquid. This matrix can crack and partially dissolve when the liquid metal is considerably superheated. This can be done because of the low melting point of aluminium.

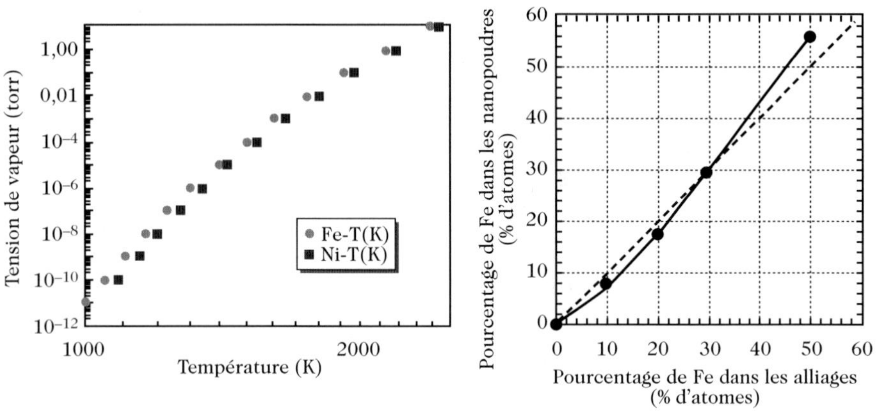

Fig. 16.18. (**a**) Comparison between the saturated vapour pressures of Fe and Ni on a log–log scale. (**b**) Composition of vapours produced by Fe_xNi_{1-x} alloys in evaporation–cryocondensation conditions. From [19]

Nanopowders of Alloys

Up to now, we have only been considering the evaporation of a monatomic compound. What happens when we attempt to evaporate alloys? We shall not discuss the case of processes applied during condensation, whereby it is possible, for example, to form metal oxides during the condensation of the metal in an atmosphere containing a controlled partial pressure of oxygen (see the example of the process developed by Nanophase Technologies Corporation, illustrated in Fig. 16.22). We shall discuss here the evaporation of a phase containing at least two different metal atoms.

Figure 16.18a compares the saturated vapour pressures of pure Fe and Ni, taken separately, on a log–log scale. The difference is very small and almost non-existent at very high temperatures ($T > 3\,000$ K). However, this difference is enough to produce a drift in the chemical composition of powder particles during continuous production. Such a drift cannot be estimated quantitatively from the curves in Fig. 16.18a, because these are the saturated vapour pressures of the pure compounds. They do not take into account the mixing energy of the two phases, controlling the partial pressures of each compound, themselves related to the composition of the alloy. In order to take this effect into account, a diagram giving the correspondence between the alloy composition Fe_xNi_{1-x} and the composition $Fe_{x'}Ni_{1-x'}$ of the vapour they produce has been established in evaporation–cryocondensation conditions [18].

The curve in Fig. 16.18b shows a depletion of Fe particles for atomic compositions containing less than 29.5% Fe and an enhancement for higher Fe contents. In evaporation–cryocondensation conditions, the 29.5% composition behaves like a pure metal. Particles obtained by this technique (see Fig. 16.19)

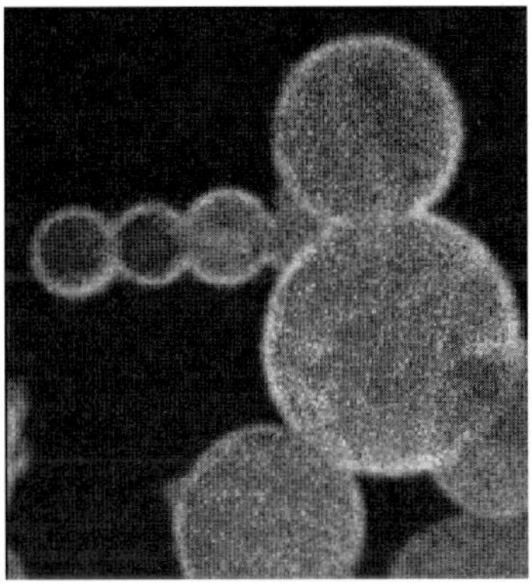

Fig. 16.19. Nanometric particles with atomic composition $Fe_{7.75}Ni_{92.95}$. See also the false colour image in the colour plates. Green (*lighter region* here) represents the oxygen present in surface oxides on the particles. Purple (*darker region* here) corresponds to the Fe/Ni ratio. This representation was made by energy-filtered imaging using a transmission electron microscope. The smallest particles measure 20 nm across. (Nanopowder synthesised at CECM-CNRS, Vitry, France. Courtesy of P. Bayle, CEA-CENG, Grenoble)

are chemically uniform and spherical, the sizes being distributed with a log-normal distribution and average 40 nm.

Controlling the Nanometric State

Thermophoresis

Under direct heating, the vapour forms in an extended region and all the processes described in Fig. 16.11 are likely to occur. In their original experiment, Gleiter's group sorted the tiny particles by a thermophoretic convection process. This exploits an imbalance in the Brownian motion of a particle whereby it moves from a hot region towards a colder one. The thermophoretic force acting on the particle is proportional to its size and the strength of the temperature gradient. The latter is generally rather small, except for very tiny particles with masses that cannot resist this force. In thermophoretic sorting, a very small percentage of the matter (a few milligrams per hour), consisting of particles with sizes of a few nanometers, can be collected (see Fig. 16.20).

Cryogenic Condensation

This technique was first developed by J. Bigot at CECM-CNRS (Vitry, France) and further refined to synthesise different metals and alloys, e.g., Cu [20] and Al [21]. Powders are produced in an evaporation chamber from a levitating superheated metal droplet by high frequency (HF) induction heating. High frequency magnetic induction transfers energy to a material through

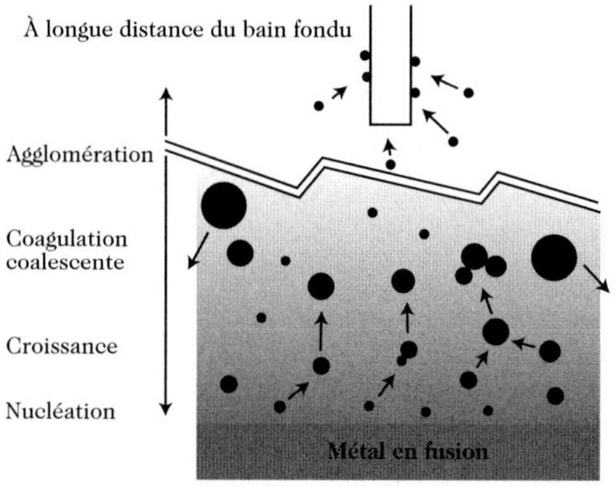

Fig. 16.20. Schematic view of thermophoretic sorting

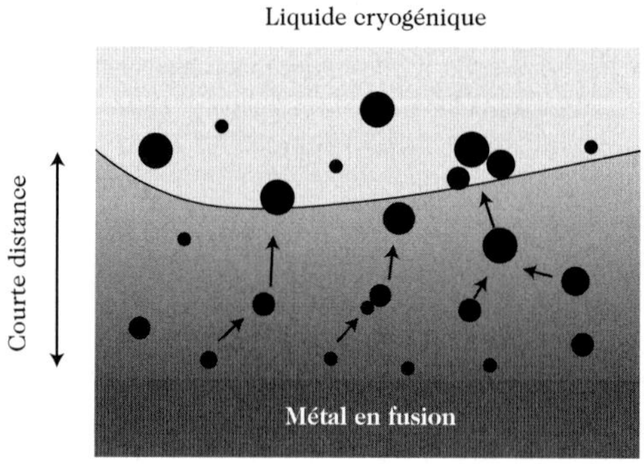

Fig. 16.21. Schematic view of sorting by cryogenic condensation

a medium without requiring heat transfer through this medium. It means that temperature variations can be obtained in the sample with little inertia, while reaching very high temperatures. With no crucible resisting reaction with the metal at the working temperature, high frequency induction also allows use of a levitation device. The form of the inducer depends on the physical properties of the metal, the temperature of the liquid metal, and the mass to be levitated. When the metal is superheated, the chamber is filled with a cryogenic liquid, e.g., N_2 or Ar, at constant flux. The molten metal droplet is thus immersed in the cryogenic fluid. The metal vapour produced is then instantaneously 'quenched', thereby limiting the phenomena of growth and coalescent coagulation (see Fig. 16.21).

Alternative Heating Methods

Table 16.1 and Fig. 16.18 show that the possibility of obtaining any kind of vapour (including refractory metals and alloys) by a physical method depends on being able to reach extremely high temperatures. The relevant substance

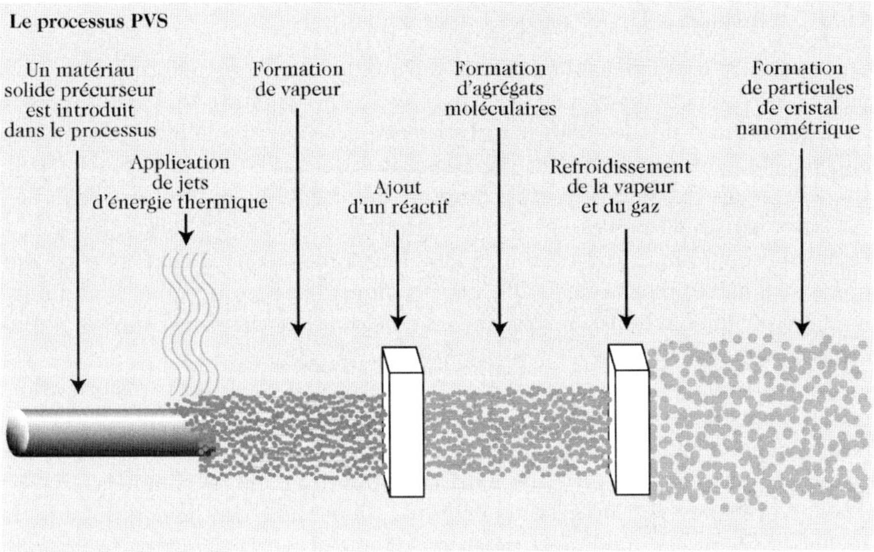

Fig. 16.22. Schematic view of a plasma evaporation process (Nanophase Technologies Corporation)

is directly sublimed to form a uniform spray at very high temperature. Various heating systems are available industrially and can be adapted to produce nanopowders. One method is electron bombardment, in which an electron beam of 200–1 500 kW is used to achieve temperatures up to 3 000°C. More widely used is an argon plasma produced by electric arc at 5–10 kW, sometimes coupled with high frequency induction (50 kW), which can achieve temperatures in the range 3 000–14 000°C!

The synthesis of different binary alloy nanopowders (containing Al, Fe, Si, Cu, Ni, or Ti) is reported in the literature [22]. The authors synthesise these compounds by injecting microscale powders of alloy into a plasma flow produced by electric arc (plasma torch). Other groups report the synthesis of W nanopowders by the same process but coupled with induction heating to 10 000 K [23].

In a nanopowder fabrication and treatment process, Nanophase Technologies Corporation, for example, use the plasma to produce the vapour (see Fig. 16.22). This is subsequently cooled by injecting an inert gas, thereby forming the first nucleation centers. It is then possible to add oxygen, for example, to produce metal oxide nanopowders. In this process, condensation occurs under a forced flow that disperses the germs, thereby limiting Brownian motion and coalescent coagulation.

16.6.2 Chemical Processing: Laser Pyrolysis

Once again, although nanomaterials have only recently come into the spotlight, the technique known as laser pyrolysis is not a new one, having already been introduced in the 1980s [24]. In France, this technique has been studied and developed in particular at the Laboratoire Francis Perrin (CEA-CNRS), the aim being to produce useful compounds for energy applications [25].

Laser pyrolysis makes use of the interaction between a precursor, i.e., a chemical compound, and a laser beam. It is thus essential that at least one of the chemical precursors should have a vibrational mode equal to the resonant frequency of the laser. We shall see below that this constraint is not as restrictive as it sounds.

In contrast to the previous cases, there is now a chemical transformation, rather than initial formation of a vapour via some physical principle. There is thus some hope of gaining access to other nanometric phases, in particular refractory phases that would be hard to obtain by physical channels.

The precursor can be gaseous or liquid, depending on the compounds. In the latter case, the liquid must be atomised in the synthesis region to improve interaction with the laser beam. To this end, a special setup is used to form an aerosol from the compound. The precursors are hydrides, halides, carbonyls, or organic compounds. When decomposed, these supply the chemical elements making up the phase of the nanopowder to be synthesised. For example, the following are found in a certain number of processes: SiH_4 (silane) gives Si, NH_3 (ammonia) gives N, C_2H_4 (ethylene) gives C, $Fe(CO)_5$ (iron carbonyl) gives Fe, and so on.

The light sources used are CO_2 lasers, tunable in the range 9.3–11.6 μm. Lasers are used rather than other optical sources because of the available range of wavelengths, their very small spectral widths, and the intensity of the emitted light.

In the process (see Fig. 16.23), the laser beam is positioned on the path of the precursor flux. The molecules are decomposed by excitation of their vibrational levels. In the decomposition zone, the atoms formed 'heat up' the molecule population by collisions and an incandescent flame is produced, in which nanoparticles form by condensation. Due to confinement and very rapid energy transfer, the reaction zone is permanently thermalised. This means that only one of the chemical compounds needs to have a vibrational mode corresponding to the laser, since the other reactants react under the effect of the resulting heat. Moreover, the time spent by the nanoparticles in the confined region of the reaction is extremely short. Outside this zone, the temperature drops substantially and there is little risk of the nanoparticles undergoing the kind of harmful effects described in Sect. 16.5.

Let us illustrate this with the case of compounds containing silicon. The hydride SiH_4 gives a significant absorption efficiency close to 10.6 μm. Si nanoparticles are thus easily produced according to the reaction

$$SiH_{4(g)} \xrightarrow{h\nu} Si_{(s)} + 2H_{2(g)} \; .$$

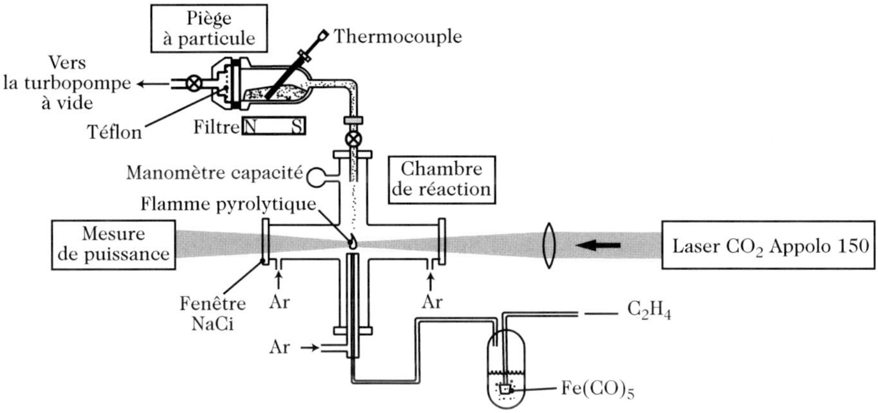

Fig. 16.23. Setup for synthesising a nanopowder by laser pyrolysis [26]. Synthesis of iron and carbon compounds

It is absolutely essential here that the SiH_4 should absorb the radiation. However, in the syntheses

$$
\begin{aligned}
3SiH_{4(g)} + 4NH_{3(g)} &\xrightarrow{h\nu} Si_3N_{4(s)} + 12H_{2(g)} \, , \\
2SiH_{4(g)} + C_2H_{4(g)} &\xrightarrow{h\nu} 2SiC_{(s)} + 6H_{2(g)} \, , \\
SiH_{4(g)} + CH_{4(g)} &\xrightarrow{h\nu} SiC_{(s)} + 4H_{2(g)} \, ,
\end{aligned}
$$

NH_3 and C_2H_4 moderately absorb the radiation, while CH_4 does not absorb it at all. It is interesting to note that H_2 does not absorb the radiation and hence does not consume energy in the process.

Some of the compounds accessible by this method are, without claiming to be exhaustive, metals like Cu, Fe, Ni, Co, Ag, refractory metals like Mo, silicon bases like Si, SiC, Si_3N_4, Si/C/N, Si/C/O, SiO_2, other carbides and borides like FeC, B_4C, TiC, WC, TiB_2, ZrB_2, and oxides like Al_2O_3, TiO_2, ZrO_2.

Yields are of the order of 100 g/hour, in principle, continuously and for unlimited amounts (depending on how the nanopowder is collected). The results are highly repeatable in terms of quality. Depending on the compound, particles have sizes in the range 15–50 nm. Size distributions are log-normal and usually very narrow, with standard deviations of the order of a few nm to 15–20 nm. The nanoparticles are spherical for metals or polyhedral for ionocovalent compounds, and nanopowders are homogenous with a high level of chemical purity, despite the residues produced by pyrolytic decomposition.

16.7 Conclusion

The aim in this section is to summarise the results and the main observations regarding nanopowder production by gas condensation. The goal of this kind of synthesis is to fabricate nanopowders comprising particles with sizes smaller than 100 nm, with a maximal specific surface area. Standard techniques based on fragmentation of the liquid state cannot achieve nanometric dispersions due to the large surface tensions on this scale, so there is no option but to pass by the gaseous state.

The different stages in the synthesis of nanopowders are:

- Formation of the vapour using various techniques depending on the physicochemical properties of the compound to be vaporised. (This question is discussed at the end of the chapter, because the formation of the vapour is not a decisive step with regard to our main theme here, i.e., the formation and control of nanopowders.)
- Condensation of the vapour to obtain a large number of tiny particles, considering two quantities in particular: the critical germ size x^* and the nucleation rate J, both depending on a key parameter for the condensation process, viz., the degree of supersaturation $S = P/P_0$:

$$x^* = \frac{4\gamma M}{\rho N k_{\mathrm{B}} T \ln(P/P_0)} , \qquad J \propto \frac{(P/P_0)^{n+1}}{\left[\ln(P/P_0)\right]^2} .$$

- Post-condensation control of the nanometric state, a critical stage in the formation of nanopowders. In general, and notably in industrial conditions, the nanopowders are produced in a temperature gradient, with a corresponding vapour density gradient. The predominant harmful effect destroying the nanometric state is coalescent coagulation, which occurs all the more rapidly as the particle density is high and localised in the hottest regions.

Finally, let us note that a great many nanopowders are currently commercialised, including transition metals, refractory metals, oxides, nitrides, carbides, and semiconductors, and are now beginning to be used as precursors for the fabrication of a wide range of materials and products.

References

1. Y. Champion et al.: Materials Science Forum **426–432**, 2411–2416 (2003)
2. `http://www.ipm.virginia.edu/research/posters/djs8pweb1/atomizer.htm`
3. J. Charpin, F. Rasneur: Mesure des surfaces spécifiques, Techniques de l'Ingénieur, 11-1997, Vol. PE2, PE 1045
4. R.M. German: *Powder Metallurgy Science*, ed. by Metal Powder Industries Federation, Princeton, New Jersey (1984) p. 77

5. E.R. Buckle, P. Tsakiropoulos, K.C. Pointon: International Metals Reviews **31**, 258 (1986)
6. R.C. Flagan, J.H. Seinfeld: *Fundamentals of Air Pollution Engineering*, Prentice Hall, Englewood Cliffs, New Jersey (1988) Chap. 5 Aerosols, pp. 290–356
7. R.S. Bradley: Quarterly Reviews **5**, 315 (1951)
8. E.R. Buckle: Trans. Faraday Soc. **65**, 1267 (1969)
9. D.W. Oxtoby: J. Phys.: Condensed Matter **4**, 7627 (1992)
10. C.G. Granqvist, R.A. Buhrman: J. Appl. Phys. **47**, No. 5, 2200–2219 (1976)
11. R. Flagan, M. Lunden: Mat. Sci. Eng. **A204**, 113–124 (1995)
12. K. Kimura: Bull. Chem. Soc. Jpn. **60**, 3093–3097 (1987)
13. R.R. Irani, C.F. Callis: *Particle Size: Measurement, Interpretation and Application*, Wiley, New York (1963) Chap. 4
14. R.B. Bird, W.E. Stewart, E.N. Lightfoot: *Transport Phenomena*, Wiley, New York (1960)
15. R.C. Flagan, J.H. Seinfeld: *Fundamentals of Air Pollution Engineering*, Prentice Hall, Englewood Cliffs, New Jersey (1988) Chap. 5 Aerosols, pp. 310–311
16. C. Kaito: Jap. J. of Appl. Phys. **17**, 601 (1978)
17. J. Bigot, A.G. Goursat, G. Vernet, J.F. Rimbert, J. Foulard, T. Sarle: Patent (France) No. Z8307414 (1983)
18. C. Duhamel: Third year dissertation, Ecole Nationale Supérieure de Chimie de Paris, CECM-CNRS, Vitry, France (2002)
19. Y. Champion, J-L Bonnentien, C. Langlois, C. Duhamel, J. Moulin, F. Mazaleyrat, P. Bayle, M.J. Hytch: Materials Science Forum **426–432**, 2411–2416 (2003)
20. Y. Champion, J. Bigot: Scripta Materialia **35**, No. 4, 517–522 (1996)
21. Y. Champion, J. Bigot: Nanostructured Materials **10**, No. 7, 1097–1110 (1998)
22. M. Umemoto, M. Udaka, K. Kawasaki, X.D. Liu: J. Mater. Res. **13**, 1511–1516 (1998)
23. Y. Moriysohi, M. Futaki, S. Komatsu, T. Ishigaki: J. Mater. Sci. Lett. **16**, 347–349 (1997)
24. W.R. Cannon, S.C. Danforth, J.H. Flint, J.S. Haggerty, R.A. Marra: J. American Ceramic Soc. **65** (7), 324–331 (1982)
25. R. Dez et al.: J. of the European Ceramic Soc. **22**, 2969–2979 (2002)
26. X.-X. Bi: J. Mater. Research **8**, No. 7, 1666–1674 (1993)

17

Synthesis of Nanocomposite Powders by Gas–Solid Reaction and by Precipitation

C. Laurent

17.1 Introduction

Nanometric particles of transition metals generally react very strongly with air. One solution for stabilising them is to disperse them in an oxide matrix, either to study them or to benefit from their properties in genuine nanocomposite materials.

Rustum Roy [1] introduced the notion of nanocomposite to refer to a new class of materials defined as follows: any combination of two or more phases of which at least one is nanometric in at least one dimension. This rather general definition includes all associations between metals, ceramics and polymers, encompassing what have become known recently as hybrid materials, either in the form of powders, thin films, or dense materials. Systems comprising discrete metal particles dispersed in a continuous ceramic matrix thus constitute a particular class of nanocomposite.

Niihara [2] proposed a classification of metal–ceramic and ceramic–ceramic nanocomposites into four groups (see Fig. 17.1):

- intragranular nanocomposites, in which the metallic or ceramic nanoparticles are dispersed inside grains of the ceramic matrix,
- intergranular nanocomposites, in which the nanoparticles are distributed at the grain boundaries of the matrix (which corresponds in the case of powders to dispersion at the surface of the grains in the matrix),
- hybrid intra–inter composites,
- nano–nano composites, in which both phases have nanometric dimensions.

However, the number of classes can be reduced to just two. Indeed, there are no pure intragranular nanocomposites, because in any real material, there is always a certain proportion of metal particles at the surface, so that it is actually in the hybrid intra–inter class. The nano–nano composites must also be considered as a special case of the intergranular class.

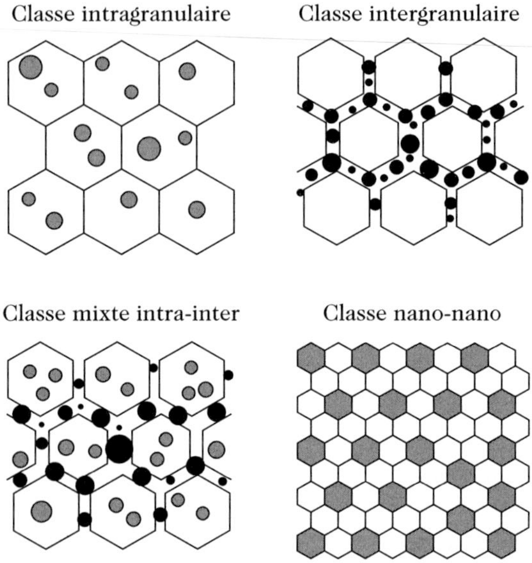

Fig. 17.1. Different classes of nanocomposites according to the Niihara classification [2]. *White*: grains of oxide matrix. *Black*: metallic surface particles. *Grey*: intragranular metallic particles or metallic particles in the nano–nano class

17.2 Synthesis of Nanocomposite Powders by Gas–Solid Reactions

We shall discuss here the different ways of synthesising metal–oxide nanocomposite powders, emphasising ways of controlling the micro/nanostructure.

17.2.1 Synthesis of Intergranular Nanocomposite and Nano–Nano Composite Powders

Naturally, intergranular and nano–nano composite powders can be produced directly by simple mixing, mechanosynthesis, or impregnation of ceramic membranes by metal nanoparticles. However, the most widely used method is probably impregnation of a ceramic powder by a salt of the metal to be dispersed (see Fig. 17.2). In particular, this approach is used to make many supported catalysts (which could be classified among the intergranular nanocomposites).

The main steps are as follows:

- suspension of the ceramic powder and impregnation by a metal salt;
- drying and if necessary annealing in air to decompose the dispersed salt into an oxide,
- reduction, generally in pure hydrogen H_2 or in a mixture with N_2 or Ar, to obtain the dispersoid in metallic form.

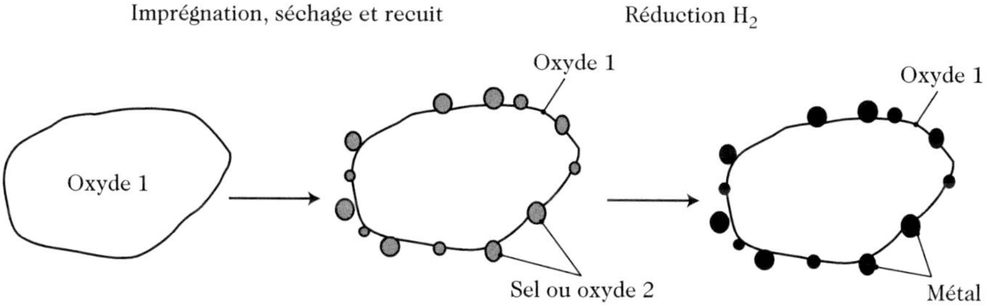

Fig. 17.2. Schematic illustration of the impregnation + reduction method. The diagram shows a cross-section of the material

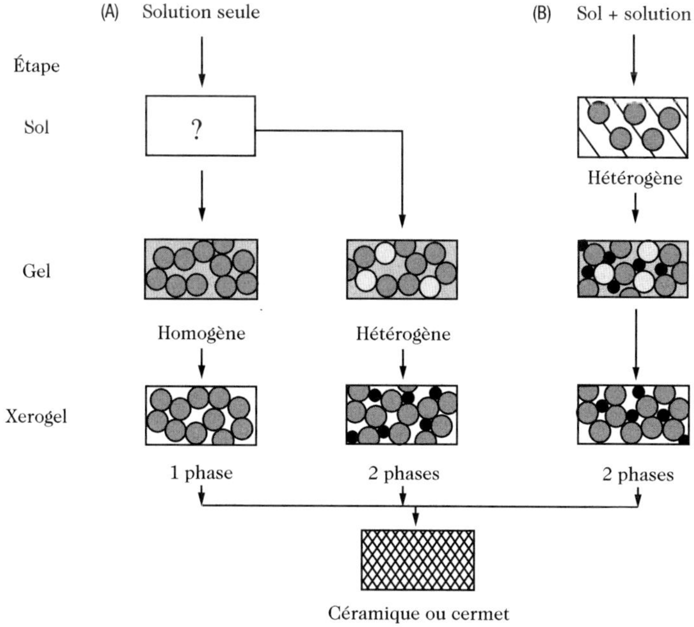

Fig. 17.3. Methods for preparing two-phase xerogels [1]. (A) All constituents are mixed together at the same time. (B) A solution is added to a sol. The *question mark* indicates that the sol can lead to a homogeneous or a heterogeneous gel depending on the chemical composition or the temperature (see text for details)

The size of the metal nanoparticles, which can be controlled in the range 3–30 nm, and the size distribution (usually log-normal) depend mainly on the reduction conditions, i.e., temperature and duration, the initial metal content, and in particular, the concentration of metal in the precursor, i.e., at the surface of the oxide powder. This last parameter is so important that a great deal of the literature is concerned with strategies for synthesising the precursor material. One of the most important strategies conceptually is the one put forward by Roy [1, 3] using a sol–gel process, not to produce highly uniform compounds, but rather to synthesise two-phase xerogels with a maximal level of heterogeneity (see Fig. 17.3).

For channel A in Fig. 17.3, a sol can lead to either a homogeneous or a heterogeneous gel depending on the chemical composition or the temperature (represented by the question mark in the figure). If the constituents are mutually soluble and remain so during growth of the sol particles, a homogeneous gel will form, i.e., all the solid gel particles will have a similar composition. If there are fluctuations in composition during growth of the sol particles, a heterogeneous gel will be obtained.

For channel B in Fig. 17.3, one first prepares a sol that is a precursor of the oxide matrix, then adds a solution containing ionic precursors of the metallic phase. Oxide–oxide nanocomposites can then be prepared by heating such a gel (homogeneous or heterogeneous) at a temperature well below the temperature required to transform a single-phase gel. Moreover, the subsequent reduction stage can be carried out at a rather moderate temperature (200–700°C). The metal–ceramic powders obtained in this way also have an excellent dispersion.

This sol–gel + reduction technique can be used to synthesise almost any combination of noble or transition metal with an oxide matrix. Several examples are given below to illustrate the potential of the method, and others can be found in the review articles [3–7]:

- Roy and Chakravorty [8] have shown that it is possible to control the size of iron particles dispersed in silica by altering the duration of the reduction process at 600°C: the average size is 3.8, 4.8, 5.3, 7.3 and 10.2 nm for durations of 10, 15, 20, 25 and 30 minutes, respectively.
- A higher reduction temperature favours the formation of larger nickel particles in silica: 3, 5, 7 and 10 nm for 600, 700, 800 and 900°C, respectively [9].
- An increase in the nickel content does not cause an increase in the average size of particles dispersed in silica, but rather a broadening of the size distribution [9], in contrast to what is observed for materials prepared by impregnation [10].
- Metal nanoparticles are generally spherical, but platelets of platinum have been observed in Pt–Al_2O_3 powders [11].
- Particles of γ-Fe–Ru and γ-Fe–Ni alloys have been dispersed in α-Al_2O_3 [12].

During the annealing stage in air, a certain proportion of the transition metal ions in the dispersoid phase may diffuse into the lattice of the oxide 'matrix'. These ions are then well diluted in the host lattice and become rather difficult to reduce at the relatively modest temperatures used to reduce surface species. However, if higher temperatures are used, they can be reduced and contribute to the formation of intragranular nanoparticles.

17.2.2 Synthesis of Intragranular and Hybrid Nanocomposite Powders

To obtain metal nanoparticles dispersed intragranularly within the oxide matrix, the corresponding metal ions must first be diluted in the majority phase. Instead of trying to obtain, as before, a well dispersed two-phase precursor system, e.g.,

$$(x)\alpha\text{-Fe}_2\text{O}_3 + (1-x)\alpha\text{-Al}_2\text{O}_3 \ , \quad \text{with } 0 < x < 1 \ ,$$

one must therefore prepare a single-phase system containing the relevant ions, i.e., a solid solution, e.g.,

$$\alpha\text{-Al}_{2-2x}\text{Fe}_{2x}\text{O}_3 \ , \quad \text{with } 0 < x < 0.1 \ .$$

In this particular system, beyond $x = 0.1$, demixing occurs, with the formation of an intimate mixture of two solid solutions [13], one rich in alumina (the majority phase) and the other rich in hematite (the minority phase), with the second being nevertheless relatively well dispersed at the surface of grains of the first:

$$\alpha\text{-Al}_{2-2y}\text{Fe}_{2y}\text{O}_3 + \alpha\text{-Al}_{2-2z}\text{Fe}_{2z}\text{O}_3 \ , \quad \text{with } \begin{cases} 0 < y < 0.1 \ , \\ 0.8 < z < 1 \ . \end{cases}$$

The main condition for subsequently obtaining intragranular metal particles is thus that there should be relatively extensive regions with the appropriate composition for which it is possible to form a solid solution. There are many methods for preparing solid solutions containing a refractory oxide (mainly Al_2O_3, Cr_2O_3, MgO, $MgAl_2O_4$) and a transition metal oxide (mainly Cr, Fe, Co, Ni, or Cu), in which the required elements are mixed on the atomic scale. It may be preferable to choose a technique with moderate working temperature, e.g., methods using coprecipitation and decomposition of precursors, or a relatively fast process, e.g., self-catalysing combustion, so that the specific surface area is not too low, which would constitute an obstacle to reduction.

There are two methods for synthesising intragranular nanocomposite powders:

- selective reduction, in which nanoparticles form by a chemical reaction,
- internal reduction, in which nanoparticles are precipitated out.

We shall concentrate here on the first method, involving a gas–solid reaction, since internal reduction is generally applied to mono- or polycrystals rather than powders.

Synthesis of Intragranular Nanocomposite Powders by Selective Reduction

The transition metal oxide is more easily reduced than the matrix oxide. [The latter metal/oxide pair is lower in the Ellingham diagram, which plots $\Delta G^0 =$

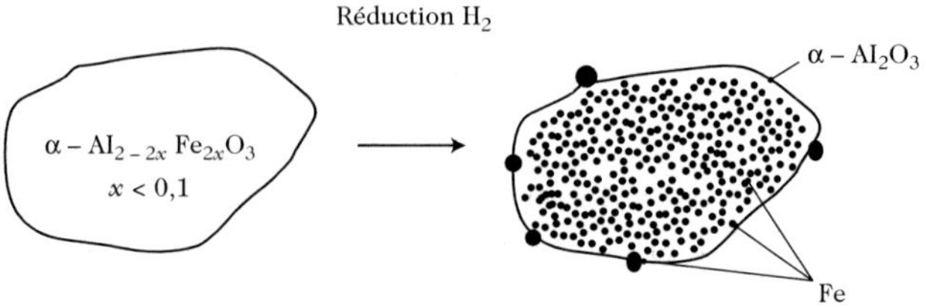

Fig. 17.4. Schematic for the solid solution + reduction method. The diagram represents a cross-section of the material. The most easily reduced ions, in this case the Fe^{3+} ions, are reduced to metallic Fe by H_2 and form metal nanoparticles on the surface of oxide grains (*large black dots*) and intragranular metal particles (*small black dots*). Compare with Fig. 17.2

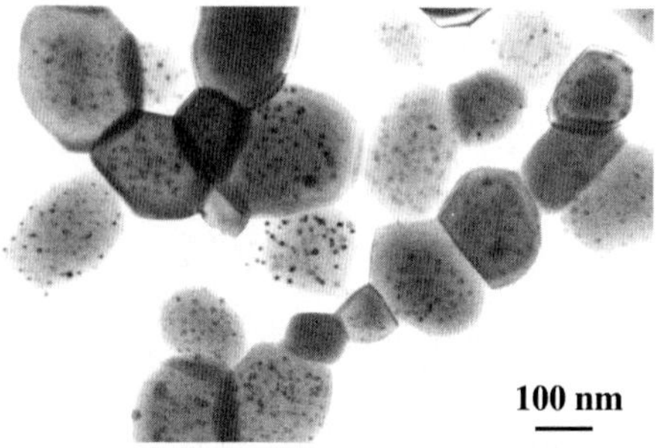

Fig. 17.5. TEM image of an Fe/Cr–Cr_2O_3 powder. From [14], with kind permission of Elsevier. The alloy nanoparticles are observed in the form of *small black dots*, while the oxide grains are *grey*. Compare with Fig. 17.4. Note that this type of image cannot be used to prove that the metal particles are intragranular

$f(T)$ for reactions $2(x/y)M + O_2 \longrightarrow (2/y)M_xO_y$.] Nucleation and growth of metal particles occurs in situ inside the lattice of the matrix. Figure 17.4 illustrates schematically, and Fig. 17.5 shows the image of a Fe/Cr–Cr_2O_3 powder synthesised by reduction of $Cr_{1.8}Fe_{0.2}O_3$ in H_2 [14].

Depending on the system studied, experimental conditions can be adjusted to control the size, the size distribution, and the composition of the resulting nanoparticles, as well as the chemical and crystallographic characteristics of the interface with the oxide matrix.

In the case of M–Al_2O_3 and M–Cr_2O_3 (M = Cr and $Fe_{1-x}Cr_x$) composites, the size distribution of the metal particles depends on the mutual solubility of the oxides making up the initial solid solution [14, 15]. Selective reduction of a single-phase initial material produces metal particles with unimodal size distribution (average size close to 10 nm). For example, the solubility of oxides like Cr_2O_3 and Fe_2O_3 for the whole range of compositions produces very narrow particle size distributions in the composites Fe/Cr–Cr_2O_3, whatever the iron content. In contrast, in the case of a two-phase initial powder, the particle size distribution after reduction is bimodal. This is why, in the system

α-Al_2O_3–Fe_2O_3 discussed above, demixing occurs for iron cation contents greater than 10% ($x > 0.1$), leading to the production of two phases, the first rich in alumina and a source of small metallic iron particles (10 nm), the second rich in hematite and a source of larger iron particles ($\approx$ 30 nm) in the final composite Fe–Al_2O_3. However, the additional presence of Cr_2O_3 extends the range of solubility of Fe_2O_3 in α-Al_2O_3 [16]. Reduction then produces iron–chromium alloy particles.

In Fe–Al_2O_3 and $Fe_{0.8}Cr_{0.2}$–Al_2O_3 nanocomposites, the size of the metal nanoparticles is directly related to the specific surface area of the solid oxide solution before reduction and also to the uniformity of the solid solution [15].

In Fe–MgO composites, for metal cation contents above 10%, the size of metal particles obtained after total reduction at 1 300°C is mainly controlled by the solubility of the oxides making up the initial solid solution [17]. On the other hand, for low iron cation contents (< 10%), the size of metal particles seems to be controlled by the size of the crystallites of the initial solid solution [17, 18]. Üstündag et al. have synthesised Ni–$NiAl_2O_4$–Al_2O_3 [19] and Ni–Al_2O_3 nanocomposites [20] by partial or total reduction of $NiAl_2O_4$. However, for such a high nickel content, a very large proportion of the the particles are not nanometric.

The reduction temperature has a rather complex effect on what happens. In α-$Al_{2-2x}Fe_{2x}O_3$ systems with $0 < x < 0.1$, the use of a temperature below 1 000°C leads to the formation of iron nanoparticles encased in a layer of spinel $FeAl_2O_4$ a few nanometers thick, while no interphase is formed between particles and matrix if reduction is carried out at or above 1 000°C [21–23]. In the first case, the reduction path of iron (III) to iron (0) goes through an iron (II) intermediate stage, whereas the reduction is direct in the second case, but renders the 'reduction temperature' parameter somewhat ineffective for controlling the size of the resulting iron nanoparticles. The epitaxial relationship between the iron nanoparticles and the matrix has been established when the metal is formed directly [21–23]. Since the mismatch between the corresponding crystallographic distances is rather large, there are stresses at the interface. Similar results have been reported for Cr–Al_2O_3 powders densified before reduction in hydrogen [24].

On the other hand, if total reduction is easier to obtain, and if there are no possible intermediate species, e.g., for Co–$MgAl_2O_4$ and Ni–$MgAl_2O_4$ powders (4% mass of metal), it can be shown that the reduction temperature plays a role in the nanoparticle size [25]. As long as the reduction is only partial (about 800–900°C for Co and 800–1 000°C for Ni), the average particle size can be controlled between 8 and 15 nm (see Fig. 17.6). Applying a higher temperature produces a rather clear increase in the average size. The latter thus depends on the degree of reduction, and an appropriate choice of reduction temperature and duration should therefore provide a good level of control.

The different reducibilities of certain cations can also be exploited to control the composition of nanoparticles at a given temperature. When the reduction temperature is increased, the composition distribution narrows and

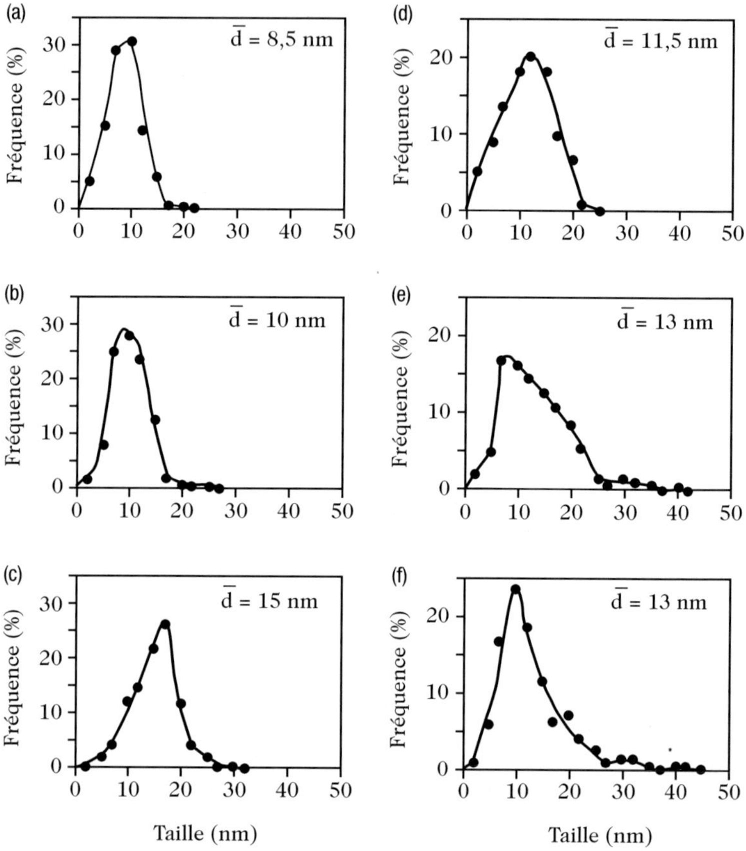

Fig. 17.6. Size distribution of metal particles in 4% mass Co–$MgAl_2O_4$ composite powders synthesised by reduction at (**a**) 800, (**b**) 900, and (**c**) 1 000°C, and 4% mass Ni–$MgAl_2O_4$ composite powders synthesised by reduction at (**d**) 800, (**e**) 900, and (**f**) 1 000°C. The average size $\bar{d}$ is indicated [26]

the average composition of the particles gradually grows richer in the metal whose cations are less easily reduced under such conditions, until a value very close to the target composition is reached, determined by the stoichiometry of the initial oxide [18, 26, 27]. The composition of particles may also be size dependent [28].

It should be noted that it is no easy matter, even impossible, to demonstrate the intragranular character of metal nanoparticles in powders on the basis of TEM images. However, thermogravimetric analysis of oxidation peaks at very high temperatures (800–1 200°C) can give an estimate of the intragranular metal content [14].

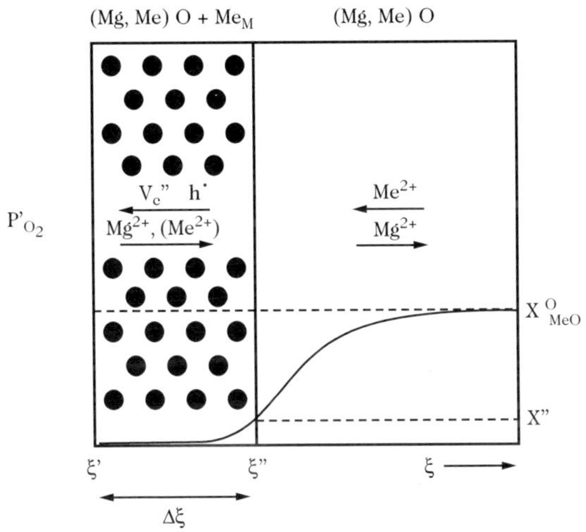

Fig. 17.7. Schematic view of diffusion processes during internal reduction of a solid (Mg, Me)O solution [30]

Synthesis of Intragranular Nanocomposite Powders by Internal Reduction

Chakravorty [29] synthesised metal–glass nanocomposite powders by mixing an organometallic compound with a low-melting-point glass (B_2O_3). After a tempering process, metal particles with diameters of 5–50 nm are dispersed in an amorphous matrix. He notes, however, that this technique can only incorporate very low metal contents (less than 0.5% mass).

Schmalzried and coworkers [30–33] synthesised metal–oxide nanocomposites (metal Cu, Cr, Ni, or Fe, and oxide Al_2O_3 or MgO) by the internal reduction method in ultrahigh vacuum or in CO/CO_2. Schmalzried and Backhaus-Ricoult [32] published a detailed article about this process, applied to mono- or polycrystals rather than powders. Figure 17.7 [30] gives a schematic view of the diffusion processes occurring during the internal reduction of a solid (Mg, Me)O solution (Me = Fe, Co, Ni, Cu, etc.).

If such a solid solution in equilibrium with a partial oxygen pressure P_{O_2} is subjected to another, much lower partial pressure P'_{O_2}, new equilibrium concentrations of cations and cation voids is set up at the surface. This equilibrium is reached within the solid by transport of matter and point defects (diffusion of cations Mg^{2+} and Me^{2+}, cation voids V''_c, and electron holes $h^\bullet$). The most noble metal is then precipitated in the form of discrete metal nanoparticles dispersed in the MgO matrix.

At the interface ξ'', the following reaction occurs:

$$Me^{2+} \longrightarrow Me^0 + \left[V''_c + 2h^\bullet\right] ,$$

and at the surface ξ', we have:

Table 17.1. Synthesis of metal–oxide nanocomposite powders by gas–solid reduction

Microstructure	Origin	Method of synthesis	Possible systems
Intergranular nanocomposite	Polyphase oxide precursor (intimate mixture)	Impregnation or chemical process (soft chemistry, sol–gel, etc.), followed by reduction (H_2)	No restriction
Intragranular nanocomposite	Single-phase oxide precursor (solid solution)	Chemical process (soft chemistry, sol–gel, combustion, etc.), followed by reduction (H_2)	Restriction imposed by existence of precursor

$$\mathrm{MgO} + \left[\mathrm{V}_{\mathrm{c}}'' + 2\mathrm{h}^{\bullet}\right] \longrightarrow \mathrm{Mg}^{2+} + \frac{1}{2}\mathrm{O}_2 \,.$$

In addition to a population of micrometric precipitates distributed in the grain boundaries and junctions in the polycrystals, several populations of intragranular particles are observed, each one characterised by its morphology and a low-energy epitaxial crystallographic relation with the matrix lattice. These results are similar to those reported for reduction of previously sintered powders in hydrogen [24]. Attempts have been made to model nucleation and growth mechanisms for chromium nanoparticles in an alumina matrix [34], making assumptions about the type of majority point defect, at low or high partial oxygen pressures.

17.3 Conclusion

Metal–oxide nanocomposite powders, in which nanometric metal particles are dispersed in an oxide matrix, can be prepared by partial reduction of solid oxide solutions or intimate mixtures of oxides (see Table 17.1). In intergranular nanocomposites, the metal nanoparticles are at the surface of grains of the oxide matrix. In intragranular nanocomposites, the vast majority of the metal nanoparticles are located inside the oxide grains. By appropriate choice of starting material and reduction conditions, the size, size distribution, and composition of the nanoparticles can be fairly well controlled.

References

1. R.A. Roy, R. Roy: Mater. Res. Bull. **19**, 169–177 (1984)
2. K. Niihara: J. Ceram. Soc. Jpn. **99**, 974–982 (1991)
3. R. Roy: Ceramic Microstructures '86. Role of interfaces, ed. by J.A. Pask and A.G. Evans: Mat. Sci. Res. **21**, 25–32 (1986)
4. R. Roy: Mat. Res. Symp. Proc. **286**, 241–250 (1993)

5. D. Chakravorty: Bull. Mater. Sci. **15**, 411–420 (1992)
6. S. Komarneni: J. Mater. Chem. **2**, 1219–1230 (1992)
7. Ch. Laurent, A. Rousset: Key Eng. Mater. **108–110**, 405–422 (1995)
8. S. Roy, D. Chakravorty: J. Mater. Res. **9**, 2314–2318 (1994)
9. C. Estournès, T. Lutz, J. Happisch, T. Quaranta, P. Wissler, J.L. Guille: J. Magn. Magn. Mater. **173**, 83–92 (1997)
10. M. Boudart, A. Delbouille, J.A. Dumesic, S. Khammouna, H. Topsoe: J. Catal. **37**, 486–502 (1975)
11. G.N. Subbanna, C.N.R. Rao: Mater. Res. Bull. **21**, 1465–1471 (1986)
12. M. Verelst, K.R. Kannan, G.N. Subbanna, C.N.R. Rao, M. Brieu, A. Rousset: Mater. Res. Bull. **28**, 293–303 (1993)
13. A. Muan, C.L. Gee: J. Am. Ceram. Soc. **39**, 207–214 (1955)
14. Ch. Laurent, Ch. Blaszczyk, M. Brieu, A. Rousset: Nanostruct. Mater. **6**, 317–320 (1995)
15. X. Devaux, Ch. Laurent, A. Rousset: Nanostruct. Mater. **2**, 339–346 (1993)
16. A. Muan, S. Somiya: J. Am. Ceram. Soc. **42**, 603–613 (1959)
17. V. Carles, M. Brieu, J.-J. Demai, A. Rousset: Fourth Euro Ceramics 1, ed. by C. Galassi, Faenza Editrice Spa, Faenza (1994) pp. 323–330
18. V. Carles, M. Brieu, A. Rousset: Nanostruct. Mater. **8**, 529–544 (1997)
19. E. Üstündag, R. Subramanian, R. Dieckmann, L. Sass: Acta. Metall. Mater. **43**, 383–389 (1995)
20. E. Üstündag, R. Subramanian, R. Dieckmann, L. Sass: Mater. Sci. Eng. A **195**, 39–50 (1995)
21. X. Devaux, Ch. Laurent, M. Brieu, A. Rousset: J. All. Comp. **188**, 179–181 (1992)
22. A. Marchand, B. Barbara, P. Mollard, G. Fillion, X. Devaux, A. Rousset: J. Magn. Magn. Mater. **116**, 64–66 (1992)
23. Ch. Laurent, A. Rousset, M. Verelst, K.R. Kannan, A.R. Raju, C.N.R. Rao: J. Mater. Chem. **3**, 513–518 (1993)
24. C.A. Handwerker, T.J. Foecke, J.S. Wallace, U.R. Kattner, R.D. Jiggets: Mater. Sci. Eng. A **195**, 89–100 (1995)
25. O. Quénard, Ch. Laurent, M. Brieu, A. Rousset: Nanostruct. Mater. **7**, 497–507 (1996)
26. O. Quénard, E. De Grave, Ch. Laurent, A. Rousset: J. Mater. Chem. **7**, 2457–2467 (1997)
27. V. Carles, Ch. Laurent, M. Brieu, A. Rousset: J. Mater. Chem. **9**, 1003–1009 (1999)
28. Ch. Laurent, J.J. Demai, A. Rousset, K.R. Kannan, C.N.R. Rao: J. Mater. Res. **9**, 229–235 (1994)
29. D. Chakravorty: Sadhana **13**, 13–18 (1988)
30. D.L. Ricoult, H. Schmalzried: Phys. Chem. Minerals **14**, 238–244 (1987)
31. D.L. Ricoult, H. Schmalzried: J. Mater. Sci. **22**, 2257–2266 (1987)
32. H. Schmalzried, M. Backhaus-Ricoult: Prog. Solid State Chem. **22**, 1–57 (1993)
33. H. Schmalzried: Ber. Bunsen-Ges. Phys. Chem. **88**, 1183–1191 (1994)
34. M. Backhaus-Ricoult, S. Hagège, A. Peyrot, P. Moreau: J. Am. Ceram. Soc. **77**, 423–430 (1994)

18

Colloidal Methods and Shape Anisotropy

D. Ingert

18.1 Introduction

Over the last decade, the new category of materials formed by nanostructures has stimulated growing interest in the field of fundamental and applied research. However, before such structures can be studied and used, ways must first be found to actually make them. A great many chemical and physical techniques have been developed, including methods using colloidal solutions which have proved themselves extremely effective for controlling the size, but also also the shape of the resulting nanoparticles.

A colloidal system is a multiphase system in which at least one of the phases is composed of particles with sizes less than 1 μm. Many research centres have used reverse micelles (water droplets in oil, stabilised by a surfactant) to exercise very accurate control over the size of spherical nanoparticles. The properties of these droplets mean that they can be considered as nanoreactors. Indeed, the aqueous core of the droplet provides a nanoscale reactive medium with controlled volume in the range 1–500 nm^3. However, several exceptions have been brought to light. From published results regarding shape control, it is very difficult to deduce the relation between the shape of nanocrystals and the shape of the colloidal system used to make them. The key parameters for controlling particle shape are hard to establish. They depend on the shape of the nanoreactor in which the synthesis takes place, but also on the adsorption of ions or molecules onto growing crystal faces and the procedure adopted for synthesis. Size and shape control in nanometric nanomaterials thus form the theme of this chapter.

We begin by describing colloidal solutions, and in particular the manner in which surfactants can self-assemble, as well as the properties of such solutions. We then discuss the synthesis of nanocrystals in colloidal solutions and the function of the colloidal 'template' (the word 'template' is used in analogy with a cake, which can be round or rectangular, depending on the cake dish in which it is cooked) in controlling the size, shape and crystal properties of

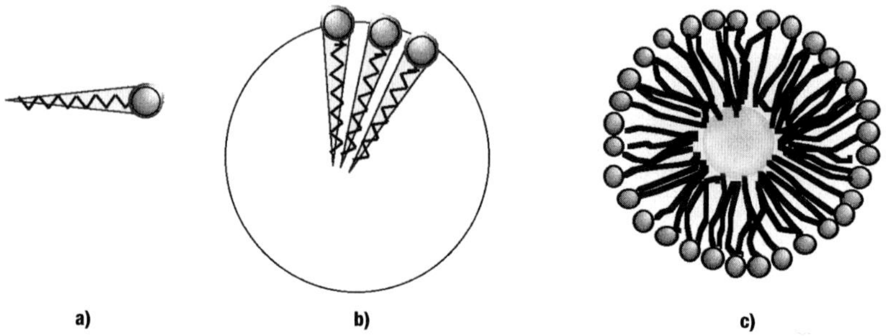

Fig. 18.1. Formation of a micelle. (**a**) Surfactant with bulky polar head and short chain. (**b**) Surfactants join together. (**c**) A micelle forms

inorganic materials, as well as the various parameters affecting the size and shape.

18.2 Surfactants

Surfactants are molecules with a hydrophilic polar head and a hydrophobic hydrocarbon chain. Such molecules tend to spontaneously position themselves at the interface separating two immiscible media, such as water and oil, to form different aggregates [1]. The very composition of these molecules induces a reduction in surface tension between oil and water. Naturally, if the solvent is capable of simultaneously solubilising the polar head and the alkyl chains, no aggregate will form. The shape of the surfactant will play an important role in the kind of assembly that eventually forms. If the surfactant has a very bulky polar head with a short chain, the surfactant then has a cone shape; the chains will associate together to form spherical aggregates called micelles (see Fig. 18.1). When micelles form at low concentrations, they are spherical and their diameter is determined by the length of the hydrocarbon chain and the polar head. The system is dynamic, although the micelle structure is always conserved. Surfactants in micelles can leave the aggregate and are replaced by others that move freely in the aqueous phase. After a few microseconds, all surfactants making up a micelle have been replaced by others.

If the surfactant has a small polar head and a branching (ramified) hydrocarbon chain, it has the shape of a champagne cork and the aggregates it forms are spherical droplets of water in oil. These aggregates are commonly called reverse micelles (see Fig. 18.2). A remarkable property of reverse micelles is that these droplets, subject to Brownian motion, can collapse, exchange their aqueous content and then reform two reverse micelles (see Fig. 18.3). Reverse micelles have another remarkable property which distinguishes them from micelles, namely that their size varies linearly with the amount of water added to the system, from 4 to 18 nm [3] (see Fig. 18.4). The water content of the system is defined as $w = [H_2O]/[SA]$, where [SA] is the surfactant concentration.

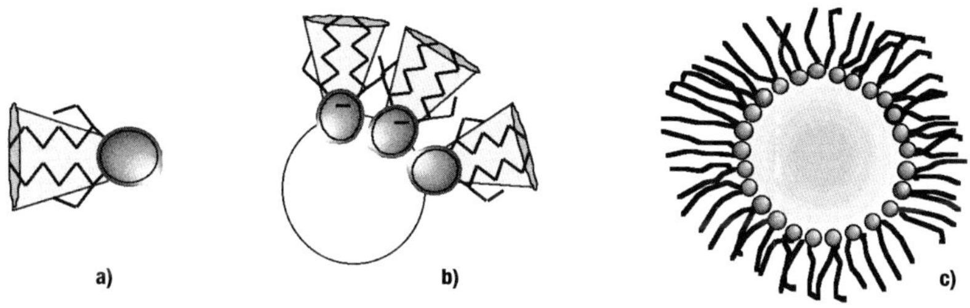

Fig. 18.2. Formation of a reverse micelle. (**a**) Surfactant with small polar head and ramified chain. (**b**) Surfactants join together. (**c**) A reverse micelle forms

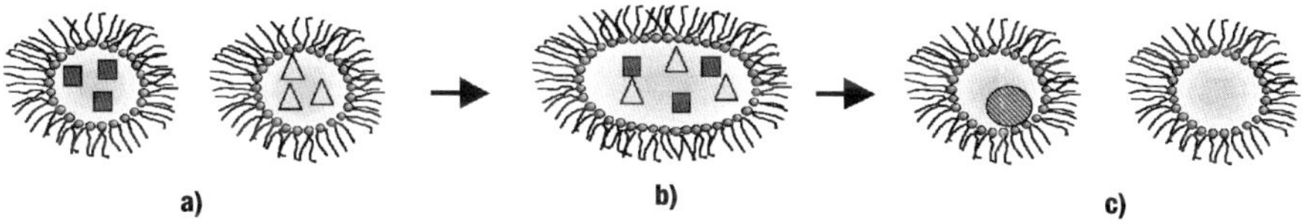

Fig. 18.3. Exchange of aqueous cores. (**a**) Reverse micelles subject to Brownian motion. (**b**) Collapsed reverse micelles and mixing of aqueous contents. (**c**) Formation of two micelles, identical to the initial micelles

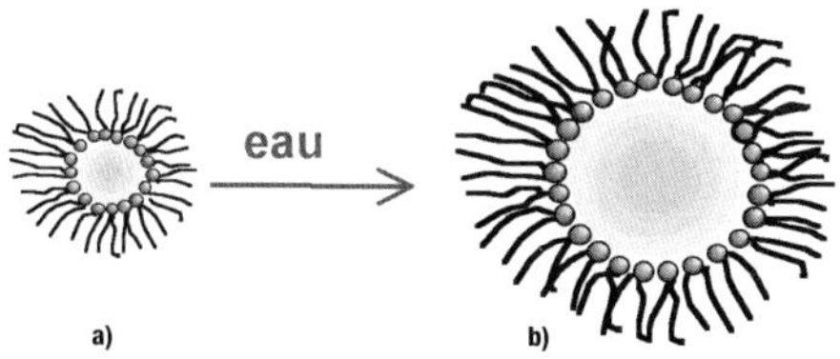

Fig. 18.4. Controlling the size of reverse micelles by water content w. (**a**) $w = 2$. (**b**) $w = 20$

For surfactants in an oil-rich phase with rather large quantities of water, there are changes in the shape and dimensions of aggregates with the formation of interconnected water channels. Space is divided into two regularly alternating volumes [4]. These aggregates are interconnected cylinders with continuous water and oil structures that alternate with one another. Adding more water induces a further phase transition. The system becomes opalescent and birefringent, i.e., light is not extinguished by two crossed polarisers. A lamellar phase forms and surfactant molecules arrange themselves into a planar film. Recently, it has been shown that the water–oil–surfactant ternary system can form thermodynamically stable emulsions. A superaggregate then forms [5], viz., a lamellar phase that is no longer flat but which has an onion shape, containing interconnected cylinders in the internal and external phases (see Fig. 18.5). The formation of such structures can be explained by geometrical factors related to the shape of the surfactant molecule [1,4].

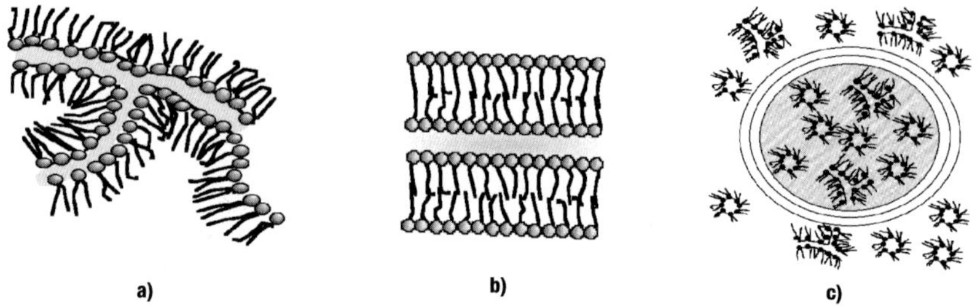

Fig. 18.5. Different structures that can form in the water–oil–surfactant ternary system. (**a**) Interconnected cylinders. (**b**) Lamellar phase. (**c**) Superaggregate

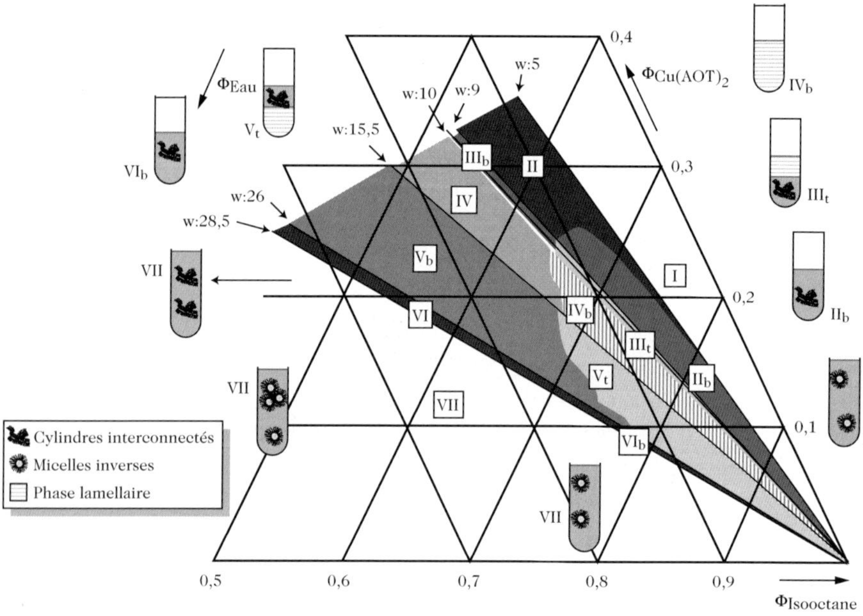

Fig. 18.6. Phase diagram of the water/Cu(AOT)$_2$/isooctane ternary system. By gradually increasing the amount of water in a solution of Cu(AOT)$_2$ in isooctane, seven regions appear in the phase diagram. Each corresponds to a different association of phases. The boundaries of each region appear along a line of constant $w = [H_2O]/[SA]$, i.e., a line of constant water content. These transitions correspond to discontinuities in the level of surfactant hydration in the complex fluid. They are numbered from I to VII, with two- and three-phase systems indicated by b or t, respectively

The most commonly used surfactants are sodium dodecylsulfate (SDS), cetyltrimethylammonium chloride (CTAC) or bromine (CTAB), and bis(2-ethylhexyl) sodium sulfosuccinate Na(AOT) (see Fig. 18.7 for chemical formulas).

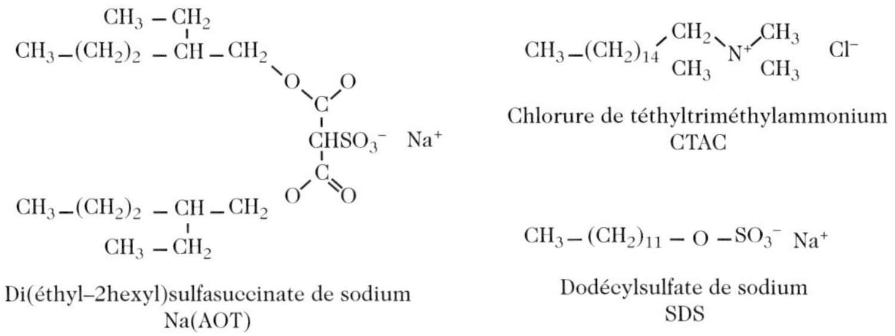

Fig. 18.7. Chemical formulas of the most commonly used surfactants

The dynamics of these structures has not yet been completely understood and it is impossible to know how they evolve as time goes by. The various colloidal self-assemblies described above are valid in the case of dilute solutions. These structures can still evolve by increasing the surfactant concentration and prove to be rather complex.

18.3 Reverse Micelles: Spherical Nanoreactors

About fifteen years ago, it was shown that reverse micelles were good candidates for use as a template [6]. The two properties described above, i.e., size control and exchange of aqueous contents, justify the idea of reverse micelles as variable-size nanoreactors. Indeed, consider two reactants A and B solubilised separately in two micelle solution. When the two solutions are mixed, aqueous core exchange induced by Brownian motion brings the two reactants A and B into contact so that they can react (see Fig. 18.3). This has made it possible to fabricate many nanosized compounds [7], such as semiconductors, metals, oxides, and alloys (although in some cases it has not been possible to obtain nanocrystals, whereas they exist in the bulk state [8]). The size of the template is controlled by adjusting the water content, whereupon the size of the resulting spherical nanocrystals can also be controlled [7] (see Fig. 18.8).

Note that these nanoreactors are able to synthesise metal nanocrystals without oxide (at least, not at detectable levels). Several research groups claimed that nanocrystals obtained by the micelle approach, i.e., at standard temperature and pressure, were amorphous [9]. However, since then it has been clearly demonstrated that nanocrystals synthesised by colloidal channels display very good crystalline characteristics. Indeed, silver nanocrystals display various crystal structures, e.g., decahedron, icosahedron, cubo-octahedron, and yet nevertheless result from the same synthesis [10]. The key to obtaining well-crystallised nanocrystals lies in the use of functionalised surfactants: one of the reactants is the counterion associated with the surfactant molecule, e.g., Ag(AOT) or $Cu(AOT)_2$. If the two reactants are simply solubilised in

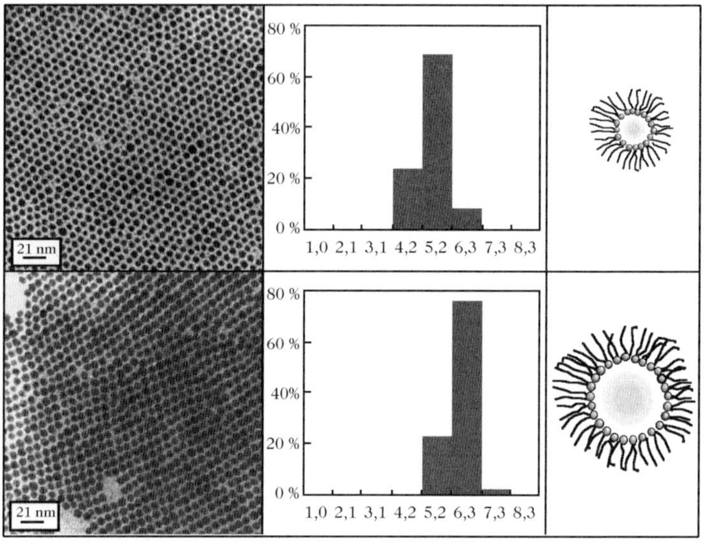

Fig. 18.8. Controlling the size of silver nanocrystals via the size of the nanoreactor. Nanocrystals with average diameter (**a**) 5.2 nm and (**b**) 6.3 nm, together with the corresponding size histograms

the droplets, e.g., the aqueous phase contains dissolved $CuSO_4$, the materials formed are amorphous, and in the case of metals, they are highly oxidised.

It was thus established that reverse micelles are very good nanoreactors for obtaining spherical nanocrystals. However, the size of the resulting nanocrystals is not that of the template, i.e., the size of the reverse micelle. In fact, for II–VI semiconductors such as CdS, CdTe, CdMnS, or ZnS [11], particle sizes vary from 2 to 4 nm, whereas for metals [12], it lies in the range 2–6 nm for silver and 2–10 nm for copper and silver sulfide. Size control of these nano-objects is obtained for the smallest water droplets in oil (between 0.6 and 6 nm), whereas for the largest templates (6–12 nm), no change is observed in the size of the resulting nanoparticles. This can be explained in terms of the structure of the water contained within the micelle [12]. Indeed, for low water contents ($w = 10$), the water molecules in the micelle core are strongly associated with counterions. The water is said to be bound. When the water content is increased ($w > 10$), the water is no longer primarily involved in a hydration process and it is said to be free. However, one exception should be noted, concerning nanocrystals of silver sulfide, whose sizes are not only similar to the micelle size, but also vary linearly with it [13].

Other exceptions are also observed. Mixing two solutions of reverse micelles made from CTAB/butanol/octane, one containing $MnCl_2$ in the aqueous core and the other containing KF, cubic nanocrystals of $KMnF_3$ are produced [14]. In the same way, by bubbling H_2S through a solution of spherical reverse micelles comprising a functionalised surfactant (the reactant Cd^{2+} is associated

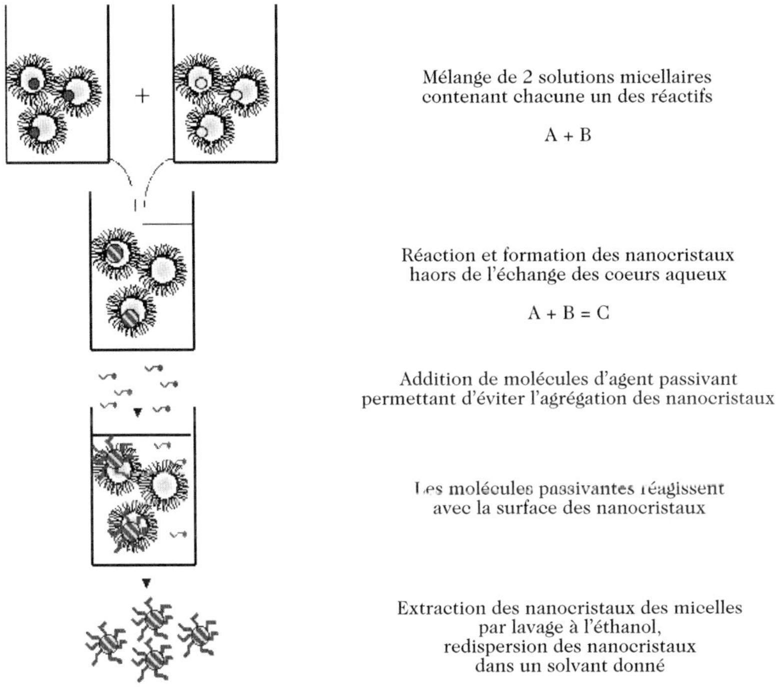

Fig. 18.9. Synthesis and extraction of nanocrystals

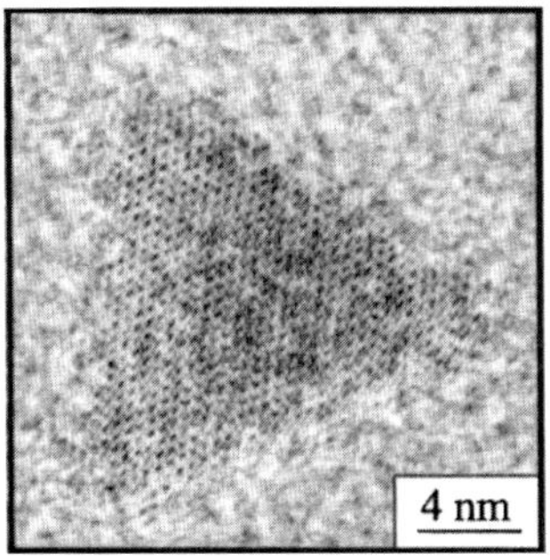

Fig. 18.10. Triangular CdS nanocrystal obtained using a reverse micelle system

with the negatively charged polar head of the surfactant), flat equilateral triangles of CdS are obtained [15], whereas spheres are produced when Na_2S is dissolved in the aqueous core of the micelles (see Fig. 18.10).

If a functionalised surfactant with barium ions as counterions [16] is used in a reverse micelle phase, when the relative fraction of barium and chromate ions increases, cylindrical nanocrystals are produced. Several groups [17–19] have synthesised materials from $BaCrO_4$ or $CaCO_3$ with different shapes using reverse micelle solutions comprising a functionalised surfactant with Ba^{2+} and

Ca^{2+} ions. This may be due to some specific adsorption of impurities and/or residues from the chemical reaction occurring in the reverse micelles. In fact, it is very difficult to obtain extremely pure functionalised surfactants and the presence of salt residues can lead to two situations: the first is a significant change in the phase diagram of the colloidal solution [20], and the second is involvement of such residues as impurities perturbing the crystal growth.

It thus seems reasonable to conclude that reverse micelles can be used as nanoreactors to fabricate nanocrystals. In most cases, a spherical template produces spherical particles. The size of the nanocrystals can be controlled through the water content of the ternary system. However, the production of different species during the chemical reaction and/or the presence of impurities plays a significant role in the crystal growth, inducing the formation of particles with different shapes.

18.4 Factors Affecting Shape Control

Controlling the shape of nanocrystals is a real challenge and much more information needs to be gathered than is presently available if we are to understand the general principles for explaining the formation of different shapes. The main difficulty probably lies in the fact that these shape anisotropic materials are in a thermodynamically unstable state.

18.4.1 Effect of the Colloidal Template on Shape Control

From the results concerning reverse micelles discussed above, and since the same ternary system leads to self-assemblies of surfactants with different shapes, many research groups tried to demonstrate that these systems served as templates and hence imposed their shape upon the nanocrystals. The first results were extremely encouraging. It was shown that the shape of the colloidal template composed of functionalised surfactants was partially responsible for imposing the shape of the nanocrystals [21]. Figure 18.11 shows that a template of interconnected cylinders produces spherical copper nanocrystals, but also some cylindrical ones [22]. In the region of the phase diagram composed of spherulites (onions), containing interconnected cylinders both inside and outside, a wide range of different nanocrystal shapes is observed.

Other experiments lead to the same kind of result [23]. Indeed, addition of phosphatidylcholine to reverse Na(AOT) micelles induces a structural change in the self-assembly, with a change in curvature that leads to wormlike structures [24]. The nanocrystal morphology can be altered by synthesising CdS nanocrystals in such colloidal assemblies, producing rods rather than spheres. This change of shape is accompanied in this case by a change in structure, from the cubic to hexagonal phase.

Notwithstanding, the role of the template is not as clear as all this would suggest. Indeed, adsorption of ions or molecules must also be taken into account.

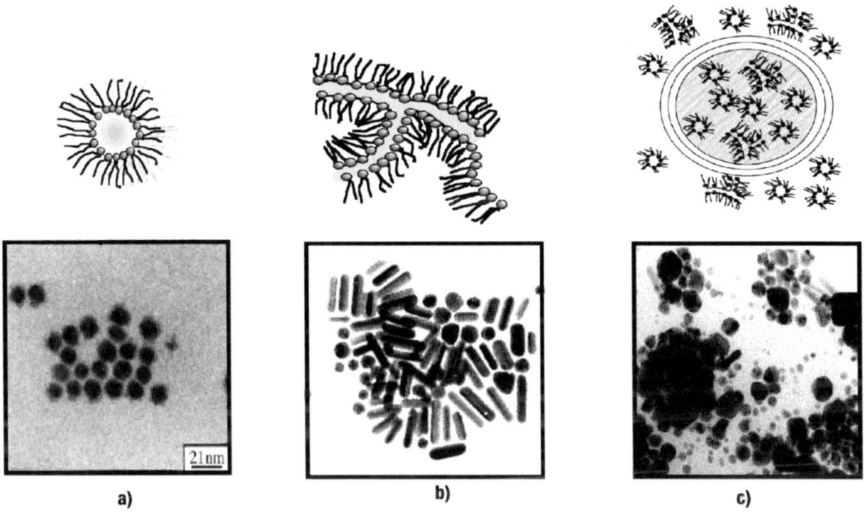

Fig. 18.11. Copper nanocrystals synthesised in different phases of the water/surfactant/oil system. (**a**) Spherical nanocrystals synthesised in reverse micelles. (**b**) Spherical and cylindrical nanocrystals synthesised in the interconnected cylinder phase. (**c**) Mixture of nanocrystals of different shapes synthesised in superaggregates

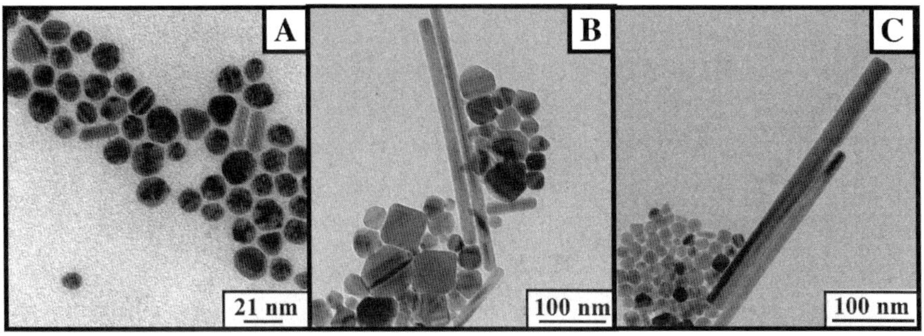

Fig. 18.12. Effect of NaCl concentration on the aspect ratio of copper nanocylinders. (**A**) [NaCl] = 0 M. (**B**) [NaCl] = 5×10^{-4} M. (**C**) [NaCl] = 1.1×10^{-3} M

18.4.2 Effect of Anions on Nanocrystal Growth

Let us consider the ternary system $Cu(AOT)_2$–water–isooctane. In the interconnected cylinder region, small cylindrical nanocrystals are obtained (see Fig. 18.12A). Adding less than 2×10^{-3} M of chlorine ions to the colloidal solution leads to the production of long copper rods, as shown in Figs. 18.12B and C, with aspect ratios (the ratio of the longest to the shortest dimensions of the nanocrystal) varying in relation to the amount of chlorine ions added to the microphase [25, 26]. These long rods have the same crystal structure as the small cylinders.

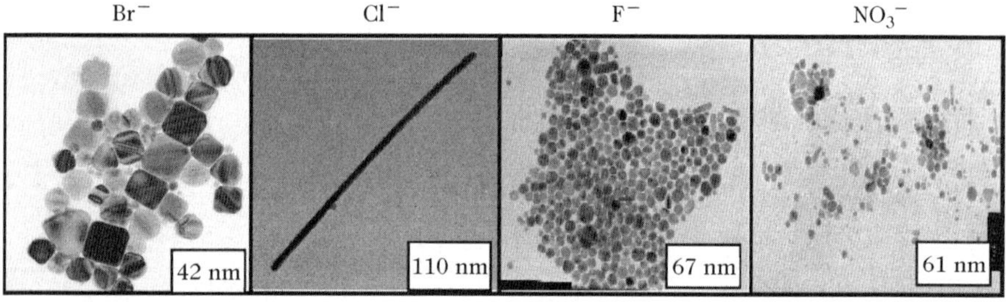

Fig. 18.13. Effect of different anions on the shape of copper nanocrystals

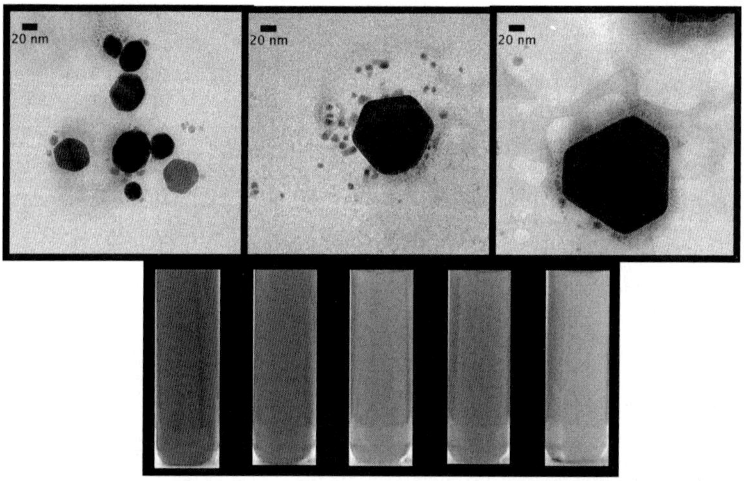

Fig. 18.14. Controlling the size of silver nanodisks by the amount of reducing agent. Solutions of silver nanodisks of different sizes. See also the colour plate

If NaBr is added instead of NaCl, nanocubes are obtained (see Fig. 18.13). However, in contrast to what is observed with NaCl, their size does not vary with the bromine ion concentration. With the addition of NO_3^-, a whole range of nanocrystal shapes and sizes is obtained, while with F^-, small cubes are synthesised [27].

It might be thought that the addition of such salts would induce changes in the structure of the colloidal template. However, this is not so. In fact, structural examination of the colloidal template shows that the salt has a negligible effect. Recall that the concentrations used are very low, viz., [salt] $= 2 \times 10^{-3}$ M [25]. This control of the nanocrystal shape is explained in terms of selective adsorption of ions on the crystal faces during growth. This idea can also explain the formation of cubic $KMnF_3$ nanocrystals in reverse micelles, as discussed earlier [14]. The micelle solution is made from a surfactant (CTAB) with Br^- as counterion. Cubic nanocrystals of CdS [28] are also obtained using self-assembled monolayers (SAM) of surfactant grafted to a surface.

These monolayers are first immersed in a solution containing cadmium ions (bound to the surfactant by electrostatic interactions), then in a solution containing CTAB and Na_2S. Cubic nanocrystals of Cu_n, $KMnF_3$, or CdS are thus obtained in the presence of bromine ions in different environments (colloidal solutions, monolayers).

However, the presence of bromine ions cannot be put forward as a general principle for obtaining cubes. Indeed, bubbling hydrogen through an aqueous solution containing $PtCl_4^{2-}$ ions leads to the formation of cubic nanocrystals of platinum [29]. In time, chlorine and hydrogen ions are formed in solution and the cubic platinum nanocrystals precipitate out. The size of these cubes is controlled by adding acetate ions or polymers [30].

In a similar way, gold nanorods are obtained by UV–visible irradiation of a gold salt (HAuCl) solubilised in the aqueous phase of normal micelles made from CTAC molecules [31]. The authors of this study attribute the role of template to the micelles. However, it is difficult to understand this role. The gold ions do not interact with the micelle solution and are reduced photochemically. On the other hand, the chlorine ions coming from the counterion of the surfactant (CTAC) can play the role of surface impurity in the nanocrystal growth process. In view of these results, it may be concluded that the formation of shape anisotropic nanocrystals is more closely connected to the selective adsorption of ions during growth than to any template effect.

18.4.3 Effect of Molecular Adsorption on Nanocrystalline Growth

Various examples clearly indicate that the selective adsorption of molecules during the nanocrystal growth process is also a key parameter in shape control. Hence, nanodisks are produced in the presence of surfactants that no longer form definite templates [32], while nanospheres are obtained in reverse micelles [10] (see Fig. 18.14). Indeed, to produce disks, a large amount of reducing agent is used, thus destroying the reverse micelle system. The size of the nanodisks depends on the amount of reducing agent, hydrazine, present in the solution [33]. Note that this is the first case where it has been possible to control the size of the synthesised disks. The size change is accompanied by a modification of optical properties: the solutions go from a reddish colour to grey. As the amount of surfactant remains unchanged, the formation of disks is explained by taking into account the adsorption of hydrazine and/or hydrogen, and hydroxyl ions on the faces.

Another experiment providing convincing evidence that molecular adsorption affects the growth process concerns the fabrication of CdTe and ZnTe nanowires using solvothermal processes [34, 35]. The metals are solubilised in hydrated hydrazine. The authors attribute the role of template to the hydrazine in this homogeneous medium. However, selective adsorption of molecules can also induce preferential growth directions leading to shape anisotropies. Salts and polymers must be present in order to produce PbS nanorods [36] from surfactants solubilised in chloroform, whereas only spheres are obtained in their absence. In the latter case, the surfactants do not self-assemble in the chloroform, because not only the alkyl chains, but also the polar heads are miscible in this solvent. Consequently, this system cannot be considered as a template. In an analogous way, shape control of nanocrystals synthesised by inorganic chemistry may involve a mixture of two surfactants [37–39]. This has been demonstrated for various materials with cylindrical shapes, such as CdSe, cobalt, or iron, for which the percentage of surfactant controls their aspect ratio. In fact, the presence of a single surfactant may be sufficient to control the dimensions of silver or gold nanorods [40,41]. In the latter case, the ratio of concentrations of growth seeds and bases is the determining factor controlling aspect ratios of these nanorods. The surfactant plays a minor role, as stabilising agent, but does not serve as a template.

18.5 Conclusion

Quite generally (but with a small number of exceptions), it may be concluded that reverse micelles are good nanoreactors which can be used to control the size of spherical nanocrystals [7, 12]. However, they are not the determining factor when it comes to controlling the shape of the nanocrystals. Crystal growth on the nanoscale seems to obey the same rules as it does for bulk materials [42]. Various elements (ions, molecules, etc.) can play the role of impurities, slowing down growth on some crystal faces. Changes in shape are due to the existence of adsorption layers of varying thicknesses on the crystal. These layers are composed of solvent, impurities, or salts. The specific way they act remains to be determined. Shape changes arise through the different growth rates of different crystallographic faces. The 'template' is not therefore the shape-determining parameter.

From these remarks, one may ask why surfactant templates are so effective for controlling the size of nanospheres, when so many exceptions are observed with regard to shape anisotropy. This probably arises because colloidal 'templates' are dynamical systems. The energy required to produce spherical nanocrystals is lower than the energy needed to produce compounds with shape anisotropy. No general method has yet been found for controlling the shape of nanocrystals produced by soft chemistry. In order to reach a definite conclusion, more data is required, comparing the fabrication of anisotropic nanocrystals of a range of materials. It is likely that other approaches will also be needed to understand such systems. For example, in a biological medium, ferrite particles can also be obtained with one long axis. This mechanism is still poorly understood, but is very different from the examples discussed in this chapter. In order to reproduce the works of nature, very large concentrations of base are required in vitro [43], thereby destroying the biological medium.

References

1. D.F. Evans, D.J. Mitchell, B.W. Ninham: J. Phys. Chem. **90**, 2817–2825 (1986)
2. M.P. Pileni: *Reverse Micelles*, Elsevier, Amsterdam, New York, Oxford, Shannon, Tokyo (1989)
3. M.P. Pileni, T. Zemb, C. Petit: Chem. Phys. Lett. **118**, 414–420 (1985)
4. S. Hyde, S. Andersoon, K. Larsson, Z. Blum, T. Landh, S. Lidin, B.W. Ninham (Eds.): *The Language of Shape*, Elsevier, Amsterdam, New York, Oxford, Shannon, Tokyo (1997)
5. P. André et al.: Adv. Mat. **12**, 119–123 (2000)
6. C. Petit, M.P. Pileni: J. Phys. Chem. **92**, 2282–2286 (1988)
7. M.P. Pileni: J. Phys. Chem. **97**, 9661–9668 (1993)
8. D. Ingert, M.P. Pileni: Adv. Funct. Mater. **11**, 136–139 (2001)
9. H. Weller: Angew. Chem. Int. Ed. Engl. **32**, 41–53 (1993)
10. A. Courty, I. Lisiecki, M.P. Pileni: J. Chem. Phys. **116**, 8074–8078 (2002)
11. M.P. Pileni: Catalysis Today **58**, 151–166 (2000)
12. M.P. Pileni: Langmuir **13**, 3266–3276 (1997)
13. L. Motte, F. Billoudet, M.P. Pileni: J. Mat. Science. **31**, 38 (1996)
14. F. Agnoli, W.L. Zhou, C.J. O'Connor: Adv. Mat. **13**, 1697–1699 (2001)
15. N. Pinna, K. Weiss, J. Urban, M.P. Pileni: Adv. Mater. **13**, 261–264 (2001)
16. J.D. Hopwood, S. Mann: Chem. Mater. **9**, 1819–1828 (1997)
17. M. Li, H. Schnablegger, S. Mann: Nature **402**, 393–396 (1999)
18. G.D. Rees, R. Evans-Gowing, S.J. Hammond, B.H. Robinson: Langmuir **15**, 1993–2002 (1999)
19. L. Qi, J. Ma, H. Cheng, Z. Zhao: J. Phys. Chem. B. **101**, 3460–3463 (1997)
20. A. Filankembo et al.: Colloids and Surfaces A: Physicochemical and Engineering Aspects **174**, 221–232 (2000)
21. M.P. Pileni: Langmuir **17**, 7476–7486 (2001)
22. I. Lisiecki et al.: Phys. Rev. B. **61**, 4968–4974 (2000)
23. B.A. Simmons et al.: Nanoletters **2**, 263–268 (2002)
24. R. Scartazzini, P. Luisi: J. Phys. Chem. **92**, 829–933 (1988)
25. J. Tanori, M.P. Pileni: Change in the shape of copper nanoparticles in ordered phases, Adv. Mat. **7**, 862–864 (1995)
26. A. Filankembo, M.P. Pileni: J. Phys. Chem. B. **104**, 5865–5868 (2000)
27. A. Filankembo, al.: J. Phys. Chem. B **107**, 97 (2003)
28. C.C. Chen, J.J. Lin: Adv. Mat. **13**, 136–139 (2001)
29. A. Henglein, M. Giersig: J. Phys. Chem. **104**, 6767–6772 (2000)
30. T.S. Ahmadi, Z.L. Wang, A. Heinglein, M.A. El Sayed: Chem. Mat. **8**, 1161–1163 (1996)
31. K. Esumi, K. Matsuhira, K. Torigoe: Langmuir **11**, 3285–3287 (1995)
32. M. Maillard, S. Giorgio, M.P. Pileni: Adv. Mat. **14**, 1084–1086 (2002)
33. M. Maillard, S. Giorgio, M.P. Pileni: J. Phys. Chem. B. **107**, in press (2003)
34. Y. Li, Y. Ding, Z. Wang: Adv. Mat. **11**, 847–850 (1999)
35. Y. Li, H. Liao, Y. Ding, Y. Fan, Y. Zhand, Y. Quian: Inorg. Chem. **38**, 1382–1387 (1999)
36. S. Wang, S. Yang: Langmuir **16**, 389–397 (2000)
37. L. Manna, E.C. Scher, A.P. Alivisatos: J. Am. Chem. Soc. **122**, 12700–12706 (2000)
38. V.F. Puntes, K.M. Kristhnan, A.P. Alivisatos: Science **291**, 2115–2118 (2001)

39. S.J. Park, S. Kim, S. Lee, Z.G. Khim, K. Car, T. Hyeon: J. Am. Chem. Soc. **122**, 8581–8582 (2000)
40. N.R. Jana, L. Gearheart, C.J. Murphy: Chem. Comm. 617–618 (2001)
41. N.R. Jana, L. Gearheart, C.J. Murphy: J. Phys. Chem. B. **105**, 4065–4067 (2001)
42. R. Boistelle: *Industrial Crystallization*, ed. by J.W. Mullin, Plenum Press, New York (1976) pp. 203–214
43. M.P. Pileni: Adv. Funct. Mat. **11**, 323–333 (2001)

19

Mechanical Milling

E. Gaffet and G. Le Caër

19.1 Introduction

19.1.1 Mechanosynthesis

Mechanical processing can be used to obtain metastable crystalline phases, e.g., phases that can only be reached at equilibrium using high temperatures and/or pressures, or even amorphous phases from crystalline phases that are stable at room temperature and pressure. Such transformations, also called mechanical alloying, can be produced by ball milling.

In the 1970s, mechanical synthesis was limited to the fabrication of superalloys. However, it is used today to produce nanocrystalline materials, usually in the form of powders with micrometric particles made up of nanometric grains of average size around 10 nm. In the case of crystalline materials, each particle thus contains some 10^6 nanograins per μm^3 and can be considered as a polycrystal of equiaxial grains with a high density of grain boundaries.

This mechanical process provides a way of mixing and combining solids from the scale of a powder particle down to the atomic scale, while remaining in the solid state. Mixtures of powders made from pure or prealloyed elements are generally comilled without the presence of liquids in a high energy mill.

19.1.2 Mechanical Activation

One of the great discoveries of the fourteenth century was that the explosive power of gunpowder was improved when it was very finely ground down. This process, also called incorporation, gives gunpowder the same consistency as talcum powder. Many hours of grinding resulted in the fine particles of saltpeter and sulfur entering the microscopic interstices of the charcoal. Some trees like willow were particularly appreciated because the charcoal was more porous. The original mixture – one measure of charcoal, one measure of sulfur, and six measures of saltpeter – came from China. The process called incorporation was improved around the year 1400 by adding water to the mixture to make a paste which was then left to dry. The explosive power was greatly increased. The water added here partially dissolved the saltpeter, which was then deposited inside the pores of the charcoal, carrying with it insoluble particles of soda ash (Na_2CO_3).

The modern version of mechanical activation is applied to various processes in powder metallurgy, such as reactive sintering and self-sustaining reactions. The milling equipment designed for mechanosynthesis is also used for activation of solids, but milling processes are then shorter in order to conserve the original powder constituents in the same structural state while modifying only their distribution. The phases, separated in the generally micrometric particles of the initial mixture, are intimately mixed on a scale of a few nanometers in the micrometric particles of the milled powders. The complex, tortuous, and convoluted morphologies of the milled phases and the defects produced by milling reduce reaction temperatures and increase reaction rates. In some cases, the final result is nanostructured.

19.2 Ball Mills

The high energy ball mills most commonly used in research laboratories comprise one or more containers in which the powder and balls are placed. Several different mills are shown schematically in Fig. 19.1:

- *Attrition Mill.* Also known as the attritor or stirred ball mill. The balls are set in motion by rotation of the central shaft on which secondary arms are fixed. The cylinder itself is fixed.
- *Horizontal Mill.* The cylinder rotates around its horizontal axis. The combined effects of the centrifugal force induced by this rotation and gravity cause the balls to rise and fall onto the powder particles.
- *1D Vibratory Mill.* Also known as a shaker mill. The vessel is set in vertical oscillatory motion. Under this action, the 1-kg ball rises then falls back onto the powder particles.
- *Planetary Mill.* The containers are fixed on a table which rotates until the centrifugal acceleration reaches 30 to 50 times the acceleration due to gravity. The containers themselves also rotate in modern mills, this rotation being either coupled or uncoupled with respect to the rotation of the table.
- *3D Vibratory Mill.* Also known as a three-axis shaker. These operate according to the same principle as the 1D vibratory mill, but this time in a more complex way due to the 3 vibrational degrees of freedom. The balls collide with the side walls of the container (friction and impacts), but also with its floor and ceiling.

High energy ball mills, in which the balls are in permanent relative motion, differ from the mills traditionally used in industry to grind materials in either

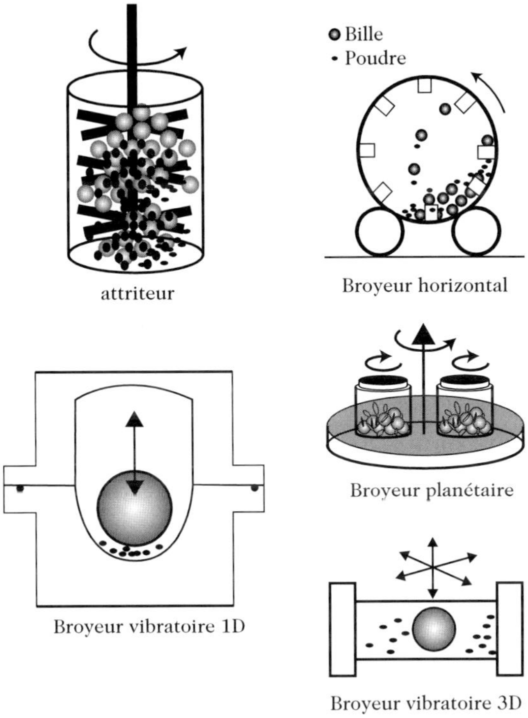

Fig. 19.1. Different types of ball mill: attrition mill, horizontal mill, planetary mill, and the 1D and 3D vibratory mills

dry or wet conditions. Low energy mills can be used to fabricate certain materials by mechanosynthesis, but the milling time then becomes excessively long, from a few hundred to a thousand hours, which significantly reduces their industrial potential, compared with process times of a few hours or a few tens of hours for high energy mills. However, the production capacity of high energy mills needs to be adapted to suit industrial requirements. If we consider the planetary mills developed in Eastern Europe, this process time can be reduced to a few minutes.

The characteristics of the load placed in the container, i.e., the number, size and density of the balls and the amount of powder, depend on the mill used to process it. The containers are vigorously shaken, usually at room temperature, so that the balls reach speeds of a few m/s and impacts between them occur at frequencies of a few hundred Hz.

The material used to make the milling balls is usually steel (type 100C6), but other materials like tungsten carbide or zirconia are sometimes necessary to increase the impact energy, or to reduce the effects of contamination by abrasion. The ratio of the mass of powder to the mass of balls is generally somewhere in the range 1/5 to 1/50 for laboratory milling equipment. The

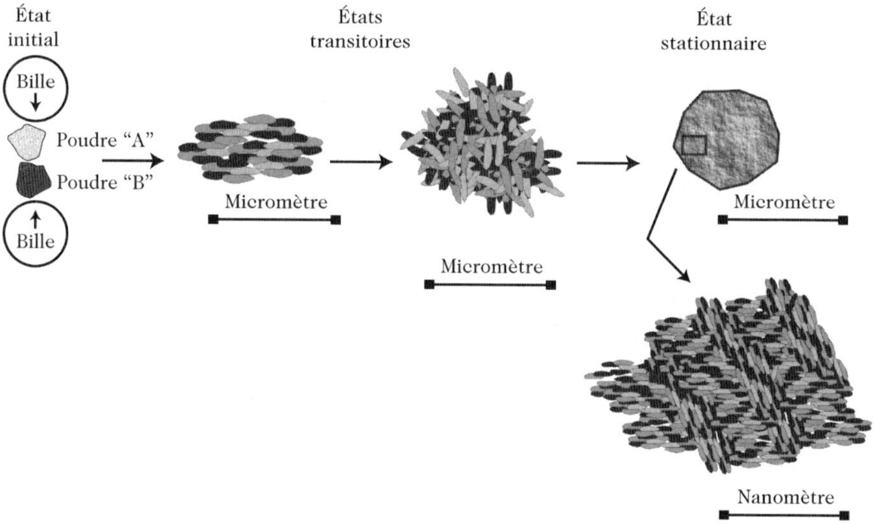

Fig. 19.2. Different stages in the evolution of elementary powders during milling

milling atmosphere is often argon, an inert gas. However, reactive milling can be done in the presence of ammonia.

19.3 Mechanisms

19.3.1 Reducing Cristallite Sizes

As shown in Fig. 19.2, struck in flight between balls and container walls, the powder particles which transiently form small agglomerates are broken up and the various fragments are then fused together again. This sequence of events produces a continual exchange of matter between particles during the process. The elements mixed together in this way combine through diffusion and shearing on impact. To ensure a suitable mix of the elements initially comilled, there must be an equilibrium between fracture and fusing of the particles.

During high energy milling, powders of ductile elements A and B (see Fig. 19.2) are subjected to very large strains at rates of the order of 10^3 to $10^4\,s^{-1}$. Repeated impacts then produce alternating lamellae of A and B, with initial thicknesses of the order of a few microns. This 3D lamellar structure grows finer as time goes by, until it reaches a characteristic length of nanometric order. When metal powders are milled, this refinement involves the creation of dislocation networks in shear zones, followed by their reorganisation into grain boundaries, with low then high degrees of disorientation. The different stages shown in Fig. 19.2 are as follows:

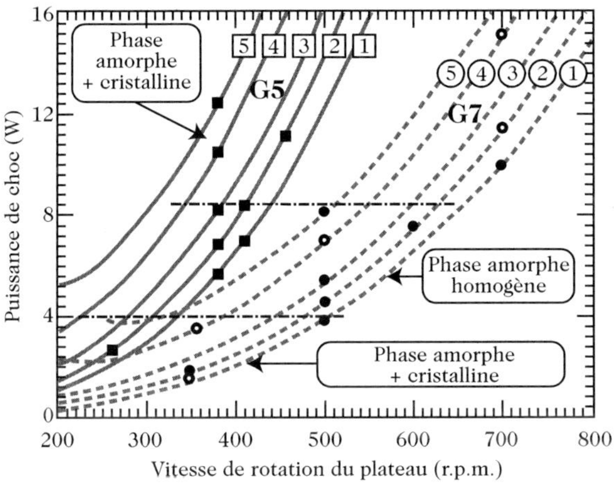

Fig. 19.3. Determining the parameters leading to amorphisation of $Ni_{10}Zr_7$. Curves showing the shock power as a function of the rotation of the turntable (revolutions per minute) and the angular speed of the containers (see Fig. 19.1d) are plotted for two planetary mills, G5 and G7. The G5 mill has a bigger table than the G7. The parameter ranges leading to a homogeneous amorphous phase and/or two-phase constituents (amorphous and crystalline phases) have been determined experimentally [1]

- *Initial State.* The materials placed in the mill can be powders of prealloyed or elementary constituents.
- *Transitory States.* Subject to mechanical impacts, a process of fracture and fusing leads to the formation of alternating lamellae of the initial components.
- *Stationary State.* After a few hours to a few hundred hours of milling, the milled product ceases to evolve further, either in terms of the phases making it up or their spatial distribution (the 'microstructure').

19.3.2 Parameters Relevant to Mechanical Alloying and Activation

Milling conditions such as dynamical parameters characterising impacts, temperature of the milling medium, pressure of the milling atmosphere, and so on, cannot be continuously controlled in a precise way, and so cannot be adjusted in real time to optimise the different stages in mechanical synthesis. However, progress has been made recently in the development of new mills, some carrying instrumentation that makes it possible to determine phase diagrams under dynamical loading. These diagrams are crucial for understanding and modelling phenomena on all scales, from microscopic to macroscopic.

It has been shown in this context that the mechanical power injected is one of the relevant parameters controlling transitions from crystalline to

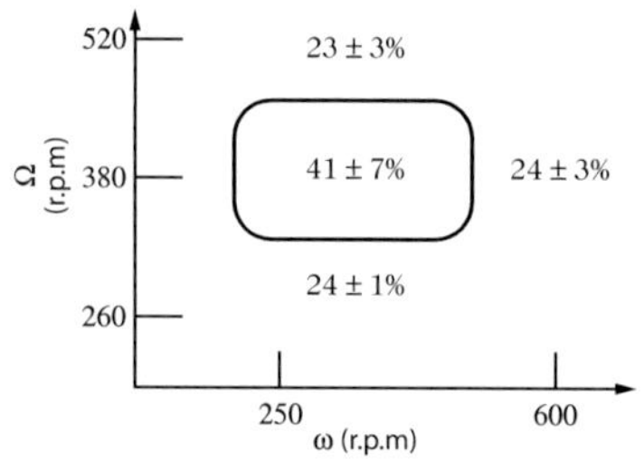

Fig. 19.4. Determining the parameter range leading to amorphisation of Ni_3Al, for a planetary mill. The rate of rotation of the turntable (revolutions per minute) is plotted on the *vertical axis*. Volume fractions of the amorphous phase are indicated directly on the figure. These fractions are determined by X-ray diffraction [2]

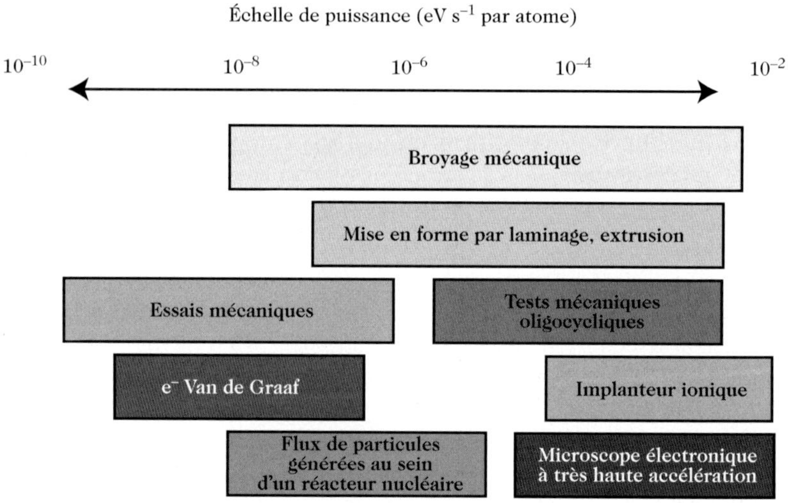

Fig. 19.5. Comparison of injected power for ball milling and other forms of dynamical loading [3, 4]

amorphous phases for Ni–Zr and Ni–Al systems (see Figs. 19.3 and 19.4). It is defined as the power transferred to unit mass of the powder, found by multiplying the kinetic energy of a ball at the moment of impact by the collision frequency.

Some research groups have considered the relevance of momentum, while others have put forward and corroborated analogies between phenomena induced on the atomic scale by collisions and phenomena induced by ion and/or electron irradiation. This approach is illustrated in Fig. 19.5, which shows the scales of mechanical power injected into these high energy ball mills, compared with other means of dynamical loading, including irradiation. The distribution of transferred energy, and not just its average, must be taken into account for more detailed comparison.

Recent work both in mechanical alloying and mechanical activation of powder metallurgical processes [5–7] has shown that it is essential to consider not only the injected mechanical power but also the way this power is injected, in particular, the ratio of the normal and tangential components of impact strengths. Gaffet and Bernard have also shown that the frictional mode should be favoured in order to activate self-propagating exothermic reactions [8–11].

19.3.3 Mechanics of Mechanical Alloying

As stressed throughout this chapter on mechanical alloying and activation, these two processes now constitute a well-established means of synthesis in solid/solid, solid/liquid, solid/gas, and liquid/gas phases. If we are concerned with the kinematics of ball mills, mechanisms of transfer of mechanical energy to the material, and the response of the material to this type of mechanical loading, we must distinguish at least two levels of description of the phenom ena arising during a process of mechanical synthesis [12]: a global level and a local level. The latter describes the interaction between the balls and powder. A third level can be introduced to take into account the behaviour of the material under mechanical loading and its evolution due to energy transfer.

Macroscopic Level

This level concerns the kinematics of the ball mill on a characteristic scale from the millimeter to the decimeter. The aim is to determine the speeds, energies, frequencies and angles of impact for the collisions between balls and/or against the inner container walls.

Mesoscopic Level

This level analyses the elementary collision in the presence of the powder. The latter is considered to be made up of a large number of powder grains, assumed to be statistically identical and to exhibit homogeneous mechanical behaviour. During an elementary interaction, we distinguish the impact itself, and the sliding and rolling of the balls one against the other or against the container walls.

The impact of a ball on a powder particle is not a frequent event, occurring once every 100 to 1 000 s, and it lasts something like 10 ms, although the duration depends on whether it is purely elastic or plastic. In the absence of exothermic chemical reactions between comilled phases, the temperature increase remains rather low or moderate, according to Koch. In fact, the local temperature increase ΔT is a few hundred degrees at the very most, and typically $\Delta T \leq 300°$C for metals and alloys such as Fe–C alloys.

Since collisions are of an elastoplastic nature, the idea of energy restoration can be introduced and the interball collision mechanism can be modelled in two or three dimensions by combining simple rheological models, e.g., the Kelvin–Voigt model, and taking into account normal and tangential viscoelastic stresses like friction. Other more local approaches, taking into consideration specific features of the impact (balls, walls and material to be ground), have been the subject of much semi-empirical work based on the Hertz elastic contact model.

Microscopic Level

This level [11, 12] is the scale of morphological changes, e.g., changes in mechanical properties with grain size, and structural transformations. The powder can now be considered as a composite with two components: the so-called parent phase, corresponding to the initial state, and the daughter phase, corresponding to the transformed state resulting from repeated collisions. The volume fraction of the daughter phase, representing the hard part, increases during the process and can be used to characterise the transformation kinetics [13].

19.4 Materials and Their Applications

19.4.1 Mechanical Alloying

All sorts of materials, from metallic to ionic, as well as organic and organometallic materials, can be synthesised in the form of alloys, compounds, and composites. Various materials with improved mechanical, magnetic, or catalytic properties have clearly established the similarity of a wide range of properties (magnetic, electrical, etc.) of materials, e.g., amorphous alloys, with the same composition, prepared by varied techniques: mechanical alloying, sputtering, chemical vapour deposition, ion implantation, solid state diffusion, and rapid quenching.

In the field of mechanical alloying, it should always be borne in mind that there is a risk of contaminating powders by elements from the atmosphere, the balls, the container walls, or even from milling agents, sometimes added to promote the synthesis. However, work by Goodwin and Ward Close for aeronautic applications has shown that technical solutions can easily be implemented to preserve the potential of untreated powders for mechanosynthesis.

19.4.2 Mechanical Activation

Synthesis by Mechanically Activated Annealing (M2A)

Single phase nanostructured powders can be obtained by first milling the elementary powder mixtures and then carrying out isothermal annealing. This has been shown for $MoSi_2$, which then forms at 800°C, while the conventional industrial process requires heating a mixture of micrometric Mo and Si particles to 1 200°C to synthesise this same silicide [6]. This mechanical activation can also be used to synthesise the compound WSi_2 at 800°C with a yield close to 90% [14].

Application of the M2A process to synthesis of iron disilicide has revealed the importance of milling conditions on the kinetics of silicide formation during low temperature isothermal annealing. For a planetary mill, these conditions can be controlled by varying the rotational speeds of the containers relative to the table that carries them. M2A has been used to stabilise iron disilicide in its high temperature form, but at temperatures 400°C below the lowest temperature at which it can exist in thermodynamic equilibrium [5].

Mechanically Activated Sintering

Mechanical activation can improve the density of a W–(Ni, Fe, Co) composites by almost 10% by solid-phase sintering. The tensile strength of the composite material has been increased by almost 30% by virtue of this improvement in the sintering process [15].

More recent work on the densification of PZT-type oxides has shown that such mechanical activation prior to the sintering stage can lead to densities greater than 90% of the theoretical density at 1 000°C and greater than 95% for sintering temperatures of 1 100°C. Such a density increase leads to a twofold increase in relative permittivity and a reduction of dielectric losses by a factor of 40.

Mechanically Activated Redox Reactions

The temperature of a redox reaction such as $Al+Cr_2O_3 \longrightarrow 2Cr+Al_2O_3$ can be significantly reduced by first carrying out mechanical activation of the mixture of oxide and reducer powders [16]. Carbothermal reduction, e.g., the reaction of TiO_2 with carbon, can also be mechanically activated.

Mechanically Activated Self-Heat-Sustaining Reaction (MASHS)

French research teams have shown that preliminary mechanical activation stages can accelerate the processes leading to self-heat-sustaining (SHS) reactions. In particular, the work presented in [9, 10, 17] has demonstrated that a mechanical activation stage carried out with a decoupled planetary mill (the Vario Mill manufactured by Fritsch), which favours the friction mode in which the tangential shock component is predominant, was much more effective than standard milling in shock mode in which the normal shock component is predominant. This was shown for the synthesis of FeAl, $MoSi_2$, $FeSi_2$, and $NbAl_3$ using the MASHS process.

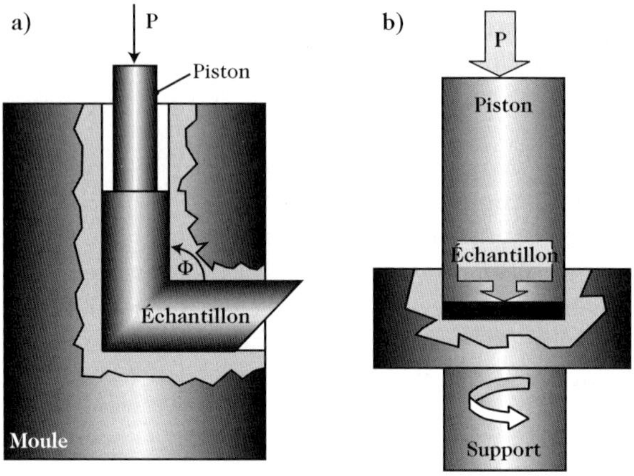

Fig. 19.6. Devices for severe plastic deformation. (**a**) Using an angled tube. The work piece is usually introduced in the form of a bulk material. (**b**) Using high-pressure rotation. The material is introduced in the bulk or powder form. One of the two pistons is then made to rotate. The main difficulty is to prevent the work piece from slipping during rotation

19.5 Shaping and Densifying Nanomaterials

19.5.1 Standard Processes

Various ways of obtaining dense nanostructured materials have been explored. The key parameters are the temperature and pressure: the lower the temperature, the higher must be the applied pressure. This pressure can in extreme cases be applied dynamically, by explosion.

Groza [18] has classified the different processes available: hot isostatic compaction, extrusion forging, hot isostatic compaction followed by quenching, severe plastic deformation (see Fig. 19.6), and dynamical consolidation by explosion. These methods generally require time-consuming adjustment before a truly dense material can be produced, i.e., with a density greater than 99.5% of the theoretical density, with crystallite sizes well below a hundred nanometers.

19.5.2 Mechanically-Activated Field-Activated Pressure-Assisted Synthesis (MAFAPAS)

The idea here is to begin by activating powders of elementary components using a planetary ball mill with decoupled table and containers. The milled powders are then placed in a die and subjected to a uniaxial compression and electrical discharges. This process has been applied to FeAl [19,20], $MoSi_2$ [21], and $NbAl_3$ [22]. Figure 19.7 shows the main features of the equipment used for synthesis of $MoSi_2$.

There are four stages in the MAFAPAS process:

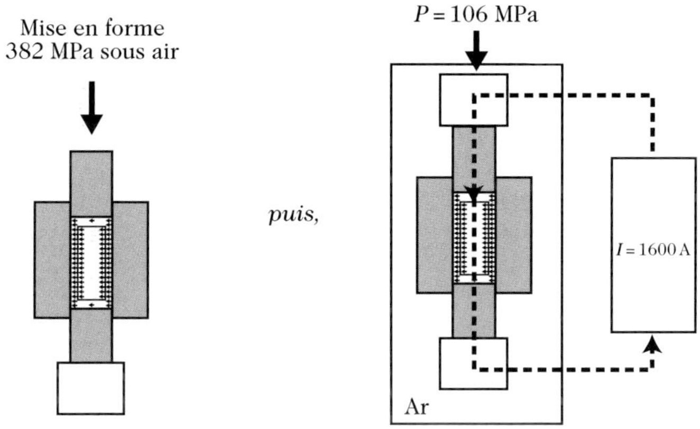

Fig. 19.7. Schematic of the FAPAS setup

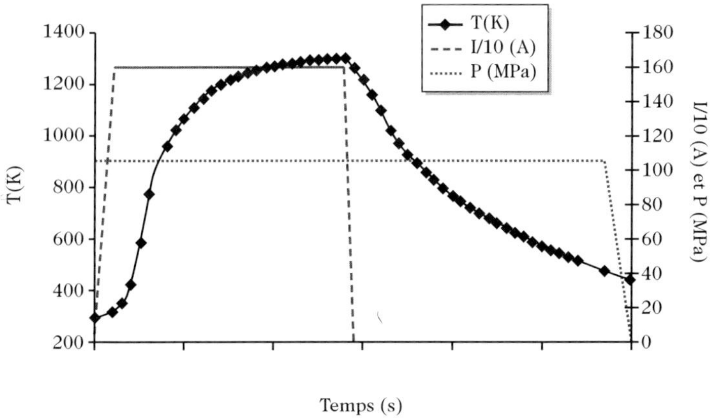

Fig. 19.8. Operating conditions during the MAFAPAS synthesis of $MoSi_2$

1. Mechanical activation of powders.
2. Preliminary compaction in air to obtain samples with density between 40% and 60%. This step is carried out directly in a graphite matrix.
3. The whole work piece (sample + matrix) is placed in a modified uniaxial press in order to:
   - carry out reactions in an inert gas (Ar pumping–filling cycle),
   - pass a large current (> 1 000 A) through the press pistons,
   - apply a load to the graphite matrix.
4. The applied force is defined by the oil pressure in the hydraulic circuit.
5. A uniaxial pressure is maintained and a large current is passed through the sample–matrix system. Zero time $t = 0$ s is defined as the moment when the current generator is switched on. The large current passes through the work piece, heating it by the Joule effect. The reaction begins when the internal temperature goes above the ignition temperature.

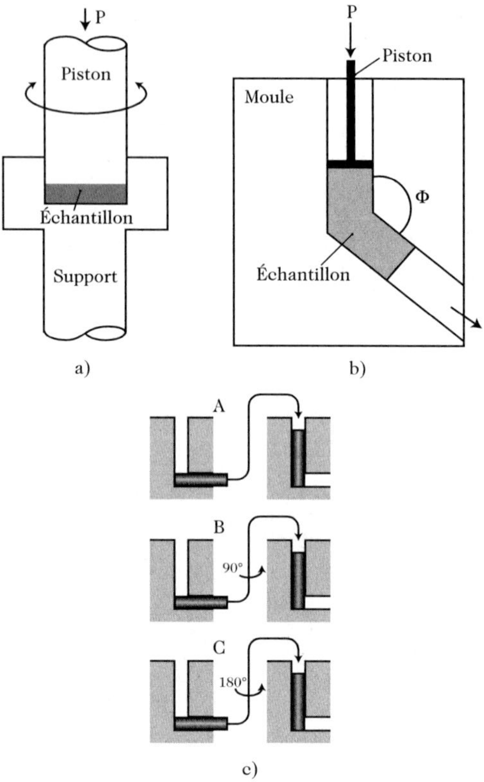

Fig. 19.9. Equipment for high-strain processing (**a**) under torsion, (**b**) by the angled tube method. (**c**) Different channels for the angled tube method. Channel A: direct. Channel B: with 90° rotation. Channel C: with rotation through 180°. From [23]

Figure 19.8 shows the time dependence of the temperature T, the current $I/10$, and the pressure P during a MAFAPAS sequence. The temperature is measured in contact with the graphite die using a type K thermocouple. The pressure is maintained throughout the experimental process. However, a small decrease appears in the applied load, of the order of 5% of the starting value. The maximum rate of heating at the surface of the graphite matrix lies in the range 15–20 K/s for all experiments carried out in this configuration. All operations are performed in a gaseous argon atmosphere (about 1 bar) and lead (on the laboratory scale) to dense bulk samples of FeAl, $MoSi_2$, and $NbAl_3$.

19.6 Severe Plastic Deformation (SPD)

The reader is referred to the recent review articles [23–26] for a detailed discussion of these processes for producing nanomaterials via large plastic strains.

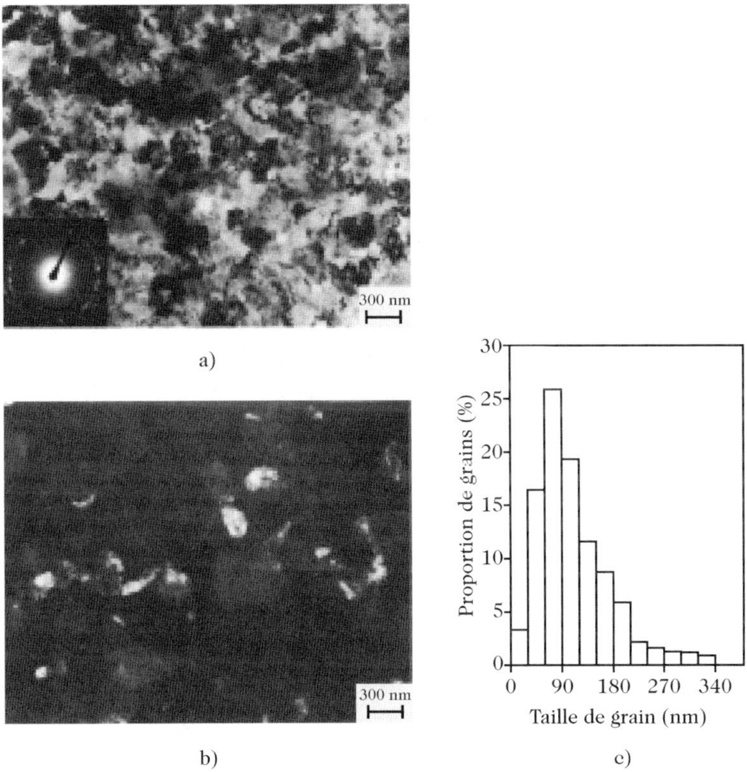

Fig. 19.10. Typical observations of a structure obtained by torsion in the case of copper. (**a**) Bright field observation and associated diffraction by submicron crystallites. (**b**) Dark field. (**c**) Histogram of size distribution. From [23], with the kind permission of Elsevier

However, it should be noted that the materials obtained by these methods contain crystallites with dimensions in the upper bracket of the region that is usually qualified as nanocrystalline, i.e., sizes lie in the range 100–200 nm. The first studies were begun about a decade ago by R.Z. Valiev and coworkers. There are two main processes in this field, viz., high-pressure torsion and an angled tube technique (see Fig. 19.9).

19.6.1 High-Pressure Torsion (HPT)

Figure 19.10 shows the microstructure of a bulk copper material obtained by high-pressure torsion at room temperature. This process can be used to densify powders. For example, starting with Ni powders synthesised by mechanical milling, a material with 95% density is produced, in which crystallites have sizes of the order of 20 nm.

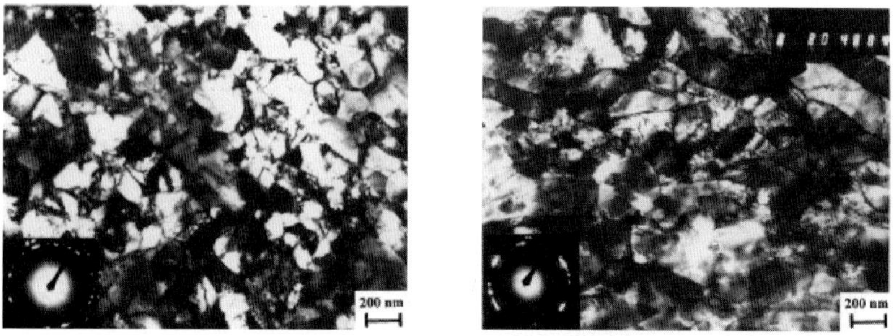

a) b)

Fig. 19.11. Typical TEM image of a Cu sample obtained by ECAP. Submicron crystallites are observed: (**a**) channel B, (**b**) channel C, where the channels are illustrated in Fig. 19.9c

19.6.2 Equal Channel Angular Pressing (ECAP)

According to R.Z. Valiev [23], Segal et al. were the first to work on this technique at the beginning of the 1980s, producing deformations in bulk materials by pure shear. At the beginning of the 1990s, this method was developed to obtain materials with submicron and nanometric crystallites (see Fig. 19.11). Since then, it has been successfully applied to many materials. The reader is referred to the article by R. Srinivasan [27] for a computer simulation of this process.

19.7 Bulk Mechanical Alloying

The method we shall now describe was developed by T. Aizawa et al. [28–30]. According to its inventors, it has the advantages of being free from contamination, easily controllable and fast. Figure 19.12 illustrates the general idea.

19.8 Synthesis of Nanocomposites by Extrusion, Drawing, and Embossing

According to Russell et al. [32], the first work carried out at the beginning of the 1960s on refinement of the hypereutectoid perlitic structure can be attributed to Embury and Fisher [33]. This work was continued at the beginning of the 1970s on a material obtained by spinning and comprising a Cu matrix and Nb filaments with nanometric diameter. Other work achieved the same structure for the same material but the final microstructures turned out to be less fine than by spinning [34, 35].

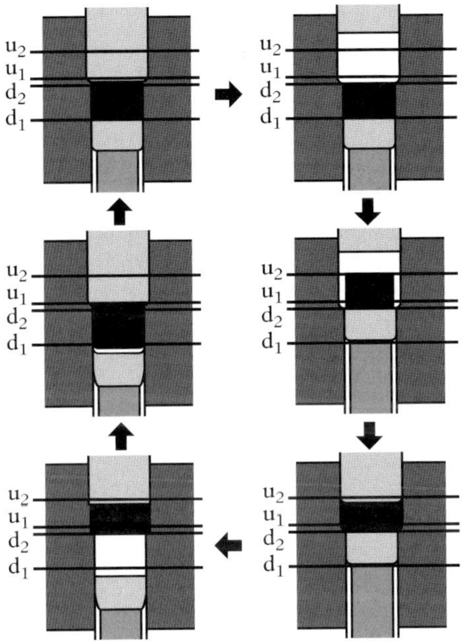

Fig. 19.12. Bulk mechanical alloying. By successive displacements of the upper and lower pistons, the material (*black*) is successively flattened, then stretched in the spaces left free by the displacements of the two pistons. $u_{1,2}$ and $d_{1,2}$ correspond to successive positions of the piston heads causing the material to deform. From [31]

Acknowledgements

One of the authors (EG) would particularly like to thank his students N. Malhouroux, M. Abdellaoui, L. Yousfi, A. Malchère, S. Paris, F. Charlot, C. Gras, V. Gauthier, and H. Souha for their contributions to the research and publications quoted in this chapter. The MAFAPAS process was developed in collaboration between UMR CNRS 5060 (E. Gaffet), UMR 5613 CNRS, Université de Bourgogne (F. Bernard), and University of Davis, USA (Z.A. Munir).

References

1. M. Abdellaoui, E. Gaffet: A mathematical and experimental dynamical phase diagram for ball milled $Ni_{10}Zr_7$, J. Alloys and Compounds **209**, 351–361 (1994)
2. E. Gaffet: Nanocrystalline and amorphous mechanically alloyed Ni–Al–M (M = Fe, Zr) phases, Mater. Sci. For. **225–227**, 429–434 (1996)
3. E. Gaffet, M. Abdellaoui, N. Malhouroux-Gaffet: Nanostructural materials formation induced by mechanical processing, Mater. Trans. JIM **36** (2), 198–209 (1995)

4. G. Martin, E. Gaffet: Mechanical alloying: Far from equilibrium phase transitions, J. Phys. Suppl. Coll. C4/71–C4/77 (1990)
5. N. Malhouroux-Gaffet, E. Gaffet: Solid-state reaction induced by post-milling annealing in the Fe–Si system, J. Alloys and Compounds **198**, 143–154 (1993)
6. E. Gaffet, N. Malhouroux-Gaffet: Nanocrystalline $MoSi_2$ phase formation induced by mechanically activated annealing, J. Alloys and Compounds **205**, 27–34 (1994)
7. E. Gaffet, F. Bernard: Mechanically activated powder metallurgy processing: A versatile way towards nanomaterials synthesis, Annales de Chimie/Science des Matériaux, **27** (6), 47–59 (2002)
8. Ch. Gras, F. Bernard, F. Charlot, E. Gaffet, Z.A. Munir: Simultaneous synthesis and consolidation of nanostructured $MoSi_2$, J. Mater. Research, **13** (3), 542–549 (2002)
9. F. Bernard, E. Gaffet: Mechanical alloying in the SHS research, Int. J. SHS, **10** (2), 109–132 (2001)
10. M. Zeghmati, E. Duverger, E. Gaffet: Mechanically activated self-propagating high temperature synthesis in the Fe–Al system, Proc. CANCAM 95, 15th Can. Cong. Appl. Mech., ed. by B. Tabarrock and S. Dost, **2**, 952 (1995).
11. F. Bernard, F. Charlot, E. Gaffet, J.C. Niepce: Optimization of MASHS parameters to obtain a nanometric FeAl intermetallic, Int. J. Self Propagating High Temp. Synth. **7** (2), 233–247 (1998)
12. D. Choulier, R. Rahouadj, E. Gaffet: Mechanics of mechanosynthesis: Overview and prospectives, Les Annales de Chimie **22**, 351–361 (1997)
13. R. Rahouadj, E. Gaffet: Shock transfer in ball-milling: Nanocomposite mechanical approach, Mater. Sci. For. **225–227**, 249–254 (1996)
14. E. Gaffet, N. Malhouroux-Gaffet, M. Abdellaoui, A. Malchère: Transitions de phases sous sollicitations mécaniques: Elaboration par mécanosynthèse de matériaux à nanostructures (alliages métalliques, semi-conducteurs, céramiques), Rev. de Métal. 757–769 (1994)
15. C. Chausse, F. Nardou, E. Gaffet: Modification induced by milling on liquid phase sintering, Mater. Sci. For. 391–396 (1995)
16. P. Matteazzi, G. Le Caër: Synthesis of nanocrystalline alumina-metal composites by room temperature ball-milling of metal oxides and aluminum, Journal of the American Ceramic Society **75**, 2749–2755 (1992)
17. E. Gaffet, G. Le Caër: Mechanical processing for nanomaterials, *Encyclopedia of Nanoscience and Nanotechnology*, American Scientific Publishers (2003)
18. J.R. Groza: *Powder Consolidation, Non-Equilibrium Processing of Materials*, Pergamon Materials Series (1999) pp. 347–374
19. Z.A. Munir, F. Charlot, E. Gaffet, F. Bernard: One-step synthesis and consolidation of nano-phase materials, US Pat. Ser. 09/374,09, filed 13 August 1999, published under US 6,200,515 on 13 March 2001 and extended by PCT WO 0112366 (22/02/01) Canada/Japan/Europe
20. F. Charlot, E. Gaffet, F. Bernard, Z.A. Munir: One-step synthesis and consolidation of nanophase materials, J. American Ceramics Society, **84** (5), 910–914 (2001)
21. Ch. Gras, F. Bernard, F. Charlot, E. Gaffet, Z.A. Munir: Simultaneous synthesis and consolidation of nanostructured $MoSi_2$, J. Mat. Res. **73** (2), 1–8 (2002)
22. V. Gauthier, F. Bernard, E. Gaffet, Z. Munir, J.-P. Larpin: Synthesis of nanocrystalline $NbAl_3$ by mechanically activated field activated pressure-assisted technique, Intermetallics **9** (7), 571–580 (2001)

23. R.Z. Valiev, R.K. Islamgaliev, I.V. Alexandrov: Bulk nanostructured materials from severe plastic deformation, Prog. Mat. Sci. **45**, 103–189 (2000)
24. Matériaux à grains ultrafins produits par hypercorroyage, Ann. Chim. Sci. Mat. **21** (6–7) (1996)
25. R.Z. Valiev, N.A. Krasilnikov, N.K. Tsenev: Mater. Sci. Eng. A **137**, 35 (1991)
26. M. Furukawa, Y. Ma, Z. Horita, M. Nemoto, R.Z. Valiev, T.G. Langdon: Proc. Int. Conf. on Thermomechanical Processing of Steels and other Materials, ed. by T. Chandra and T. Sakai, 1875 (1997)
27. R. Srinivasan: Computer simulation of the equichannel angular extrusion (ECAE) process, Scripta Mater. **44**, 91–96 (2001)
28. T. Aizawa, C. Zou: Adv. Eng. Mat. **1**–**2**, 29 (2000)
29. J. Kihara, T. Aizoacée, O. Kobayashi: Mater. Sci. For. **235**, 41 (1997)
30. T. Aizawa, J. Kihara, D. Benson: Materials Trans. JIM **36** (2), 138 (1995)
31. T. Aizawa, K. Kondoh: Scripta Mater. **44**, 1751–1755 (2001)
32 A.M. Russell, L.S. Chumbley, Y. Tian: Adv. Eng. Mat. **2** (1–2), 11 (2000)
33. J.D. Embury, R.M. Fisher: Acta Metall. **14**, 147 (1966)
34. J. Bevk, J.P. Harbison, J.L. Bell: J. Appl. Phys. **49** (12), 6031 (1978)
35. C.L. Trybus, W.A. Spitzig: Acta Metall. **37** (7), 1971 (1989)

Other References

36. E. Gaffet, G. Le Caër: Mechanical processing for nanomaterials, *Encyclopedia of Nanoscience and Nanotechnology*, American Scientific Publishers (2003)
37. E. Gaffet, F. Bernard, J.-C. Niepce, F. Charlot, C. Gras, G. Le Caër, J.L. Guichard, P. Delcroix, A. Mocellin, O. Tillement: Recent developments in mechanical alloying and in mechanical activation, J. Mat. Chem. **9**, 305–314 (1999)
38. E. Gaffet, O. Tillement: Mécanochimie et activation mécanique, Les Annales de Chimie **22**, 417–422 (1997)
39. J.R. Groza, A. Zavaliangos: Sintering by external electrical field, Mater. Sci. Eng. A **287**, 171–177 (2000)
40. Annales de Chimie – Sciences de Matériaux, Mécanosynthèse **22**, 6, 341–433 (1997)

20

Supercritical Fluids

A. Taleb

Processes involving supercritical fluids have taken on a considerable importance over the last few years. Indeed, the physical and chemical characteristics of this kind of medium, together with its exceptional transport properties, promise a whole range of potential applications. In the field of nanomaterials, the use of supercritical fluids opens the way to new opportunities for synthesis. We shall restrict the discussion here to what is required for a good understanding of this particular context. Definitions are mentioned briefly, and the interested reader is directed to the literature for more detail.

20.1 Definition

Historically, the discovery of supercritical fluids can be linked to the identification of the critical point by Charles Cagniard de La Tour in 1821 [1]. He was studying the equilibrium of liquid and gas phases in carbon dioxide and was surprised to observe that the meniscus between the two phases disappeared for a certain temperature and pressure. In fact, he had just exceeded the temperature and pressure of the critical point (T_c, P_c). The precise meaning of this point was established by the Irish physicist Thomas Andrews (1813–1885), when he defined the critical point as the end point of the liquid–vapour coexistence curve (see Fig. 20.1). For values of the temperature and pressure beyond the coordinates of this particular point, there is an intermediate state between liquid and gas which is said to be supercritical.

The pressure–temperature phase diagram of a substance is used to indicate the boundaries of the solid (S), liquid (L), and gas (G) regions. The transition from one of these states to another is reflected physicochemically by a large change in the density (see Fig. 20.1). These three states of matter are found together in equilibrium at the triple point (T) which is the intersection of the vaporisation, liquefaction, and sublimation curves. The other end, which bounds the vaporisation curve, defines the critical point (C), beyond which the fluid is both liquid and gas at the same time, i.e., as dense as the liquid

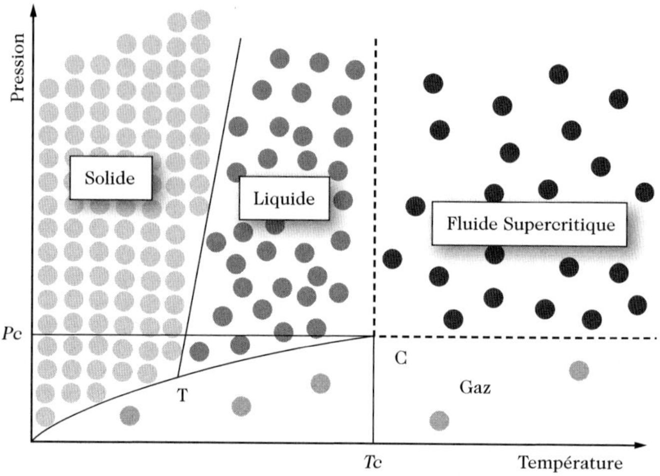

Fig. 20.1. Pressure vs. temperature phase diagram of a pure substance. T = triple point, C = critical point

Table 20.1. Critical coordinates (T_c, P_c) of a few fluids

Compound	T_c [°C]	P_c [bar]	ρ_c [kg/m^3]
CO_2	31.2	73.8	468
NO_2	36.4	72.4	457
NH_3	132.4	112.9	235
H_2O	374.1	22.1	317
C_2H_4	9.5	50.6	220
C_2H_6	32.5	49.1	212
C_3H_8	96.8	42.6	225
C_5H_{12}	196.6	33.7	232
C_6H_{12}	279.9	40.3	270
C_6H_6	289.5	49.2	304
C_7H_8	320.8	40.5	290
CH_4O	240	79.5	275
C_2H_6O	243.1	63.9	280
C_3H_8O	235.6	53.7	274
C_3H_6O	235	47.6	273

while conserving certain properties of the gas. Table 20.1 shows the critical temperatures and pressures defining the critical point for several commonly encountered compounds [2].

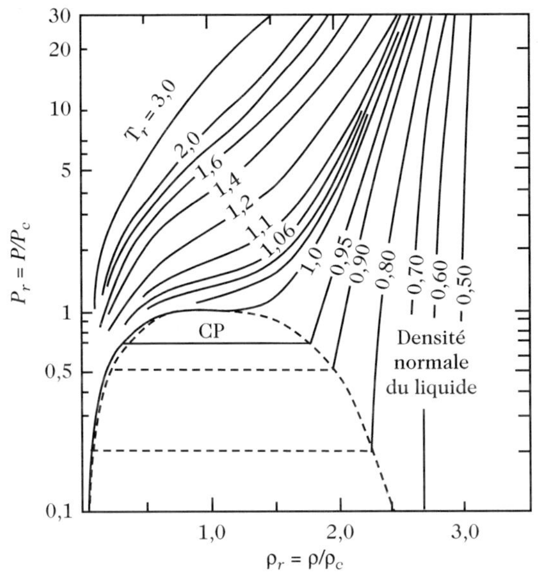

Fig. 20.2. Pressure–density diagram for CO_2 [3]

20.2 Physicochemical Properties

Near the critical point, the properties of fluids change radically. The supercritical fluid becomes much more compressible than the gas and denser than the liquid. The properties of these fluids can be adjusted via the pressure and temperature to favour either the gaseous or the liquid characteristics. Thus, at constant temperature, an increase in pressure leads to an increase in density ρ and hence a densification of the fluid. Approaching the critical point, the difference in density between the liquid and the gas state of a pure substance drops to reach zero at the critical point ρ_c itself (see Fig. 20.2).

The density of a supercritical fluid is closer to that of a liquid than that of a gas. Figure 20.2 shows the dependence of the reduced density $\rho_r = \rho/\rho_c$ on the temperature and the reduced pressure $P_r = P/P_c$ for carbon dioxide [3]. These anomalous properties have important consequences for the behaviour of supercritical fluids, in particular the solubility, viscosity, and diffusion. This diagram clearly shows the existence of a region close to the critical point (CP) where a slight change in the pressure can induce a large change in the density.

20.2.1 Solubility

The ability of a fluid to solubilise a given substance depends on the solute–solvent interactions, i.e., dipole–dipole interactions, hydrogen bonds, and sometimes complexation reactions by charge transfer [4]. For fluids in the supercritical state, the solubility parameter is defined by the dissolving power.

Table 20.2. Dissolving power of several fluids

Fluid	δ (cal/cm^3)$^{1/2}$
CH_3OH	14.4
NH_3	13.2
CO_2	10.7
N_2O	10.6

The latter depends on the physical state of the fluid described by its density, pressure, and temperature, and also its chemical state as determined by its polarity, acid–base properties, and tendency to form hydrogen bonds. The dissolving power δ_f of these fluids at a given temperature and pressure can be expressed as follows [5]:

$$\delta_f = 1.25 P_c^{1/2} \frac{\rho(T,P)}{\rho_{liq}} , \tag{20.1}$$

where P_c is the pressure of the supercritical fluid, ρ_c is the density of the fluid in g/ml, and ρ_{liq} is the density of the fluid in the standard liquid state, also in g/ml. In this expression, the term $1.25P_c^{1/2}$ accounts for the chemical nature of the fluid, while the term ρ/ρ_{liq} accounts for its physical state. Given that P_c and ρ_{liq} are constants for a fluid and that the density depends on the temperature T and pressure P, (20.1) can be written

$$\delta_f = f(T,P) .$$

The dissolving power of a fluid is thus directly proportional to the temperature and pressure. Table 20.2 gives the maximal values of the dissolving power of several fluids. These values correspond to the maximal values that the dissolving power can reach at very high pressures, assuming that $\rho = \rho_{liq}$.

The solubility of solutes in a liquid is predicted by the theory of the dissolving power, which says that a solute will become more soluble when its solubility parameter approaches that of the solvent. For a non-polar solute, Small [6] introduces the solubility parameter δ_s given by

$$\delta_s = \frac{\rho}{M}(\Delta H_r - RT) = \frac{\rho}{M}\sum G ,$$

where ΔH_r is the heat of vaporisation of the solute (cal/mol), G is the molar attraction constant, ρ is the fluid density (g/ml), M is the molar mass (g/mol), R is the perfect gas constant, and T is the temperature.

Comparison of solubilities does not provide a general rule for predicting the behaviour of solutes. In the case of CO_2 as a supercritical solvent, a certain number of rules have been determined by studying the solubility of different families of compounds in CO_2 [7,8]. However, the more massive the molecule, the smaller its solubility. Other rules have been established by taking into account the branching, polarity, and saturation of the solute.

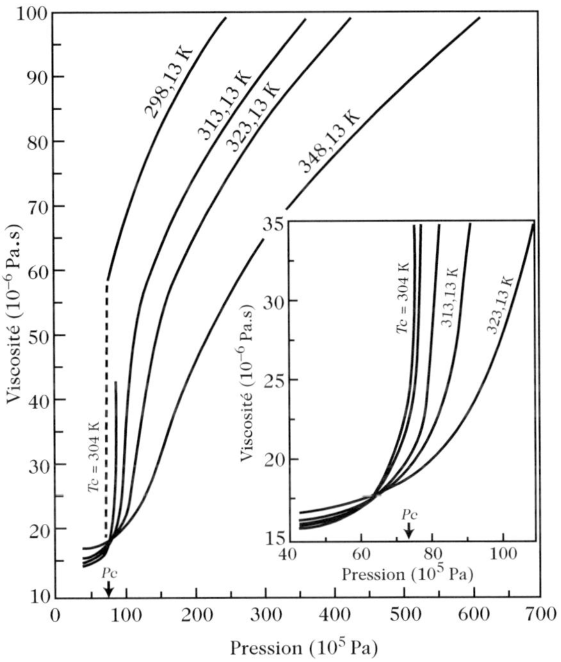

Fig. 20.3. Pressure dependence of the viscosity of CO_2 at different temperatures. *Insert*: Close-up near the critical point [9]

20.2.2 Viscosity

The resistance of fluids to flow is called viscosity. It reflects the extent to which the molecule is able to move around in an environment containing other molecules. For a liquid, the viscosity decreases with increasing temperature, as certain bonds contributing to its cohesion are broken. The viscosity of a supercritical fluid lies in the range 10^{-4} to 10^{-3} Pa s (see Fig. 20.3) [9]. It is higher than the value for gases and 10 to 100 times lower than the value for liquids. Hence, the coefficient of diffusion of a solute is higher in a supercritical fluid than in a liquid. Figure 20.3 shows the pressure dependence of the viscosity of CO_2 at different temperatures. Near the critical point, it can be observed that the viscosity varies rapidly with changes in pressure.

20.2.3 Diffusion

Near their critical point, fluids display transport properties intermediate between those of a gas and those of a liquid (see Fig. 20.4) [10]. Table 20.3 shows the orders of magnitude of the diffusion coefficient for liquids, gases, and supercritical fluids. There are two transport mechanisms for the solute in a supercritical fluid, viz., convection and diffusion. During a physical or

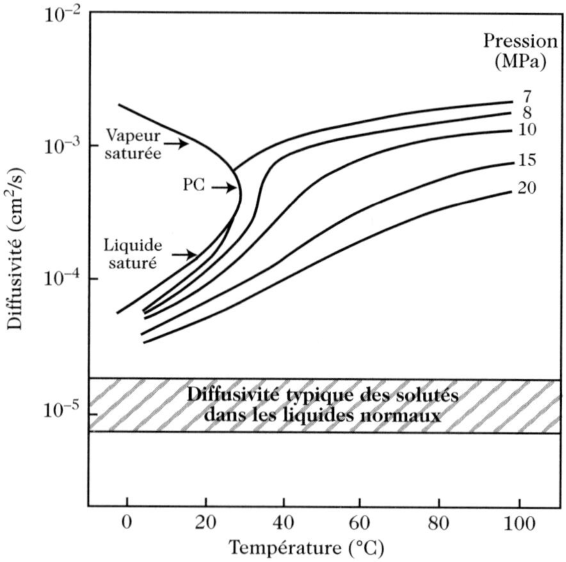

Fig. 20.4. Temperature dependence of the diffusivity at different pressures for a solute in CO_2 [10]

Table 20.3. Orders of magnitude of diffusion coefficients in supercritical fluids

	Gas	Supercritical fluid	Liquid
Diffusion coefficient D [cm^{2}s^{-1}]	1–4 $\times 10^{-1}$	10^{-4}–10^{-3}	0.2–2 $\times 10^{-5}$

chemical transformation, temperature and concentration gradients often arise in the medium, and these modify both the temperatures and the compositions locally. Solutes then diffuse from more concentrated to less concentrated regions. Bulk transport by convection is favoured by thermal vibrations, which are more significant in warm regions, inducing diffusion of molecules toward colder regions.

Figure 20.4 shows values of the diffusivity that can be reached for a solute in supercritical, liquid, or gaseous CO_2. Note that, in the supercritical region, the diffusivity increases when the pressure falls and the temperature rises.

Diffusion and Brownian Motion

Imputing the diffusion to Brownian motion and treating each solute particle as a sphere, the diffusion coefficient D and the mobility of the particle can be found from the Stokes–Einstein relation [11]:

$$D_i = \frac{RT}{6N\pi\eta r_i} ,$$

where D_i is the diffusion coefficient or diffusivity of solute i, R is the perfect gas constant, T is the temperature, and η is the dynamic viscosity of the solution. Brownian motion of the particles in the medium is given by [12]

$$\langle x^2 \rangle = 2Dt ,$$

where $\langle x^2 \rangle$ is the mean free path of the particles and t is the time.

20.2.4 Thermal Conductivity

Supercritical fluids are the scene of a strange heat transport phenomenon by a kind of piston effect [13]. This heat transport mode has recently been identified theoretically as a transport mechanism that could be qualified as the fourth heat transport mode, after conduction, convection, and radiation. Heating one wall of the closed cell containing the fluid, the layer of fluid in contact with the wall heats up and expands, thereby compressing the rest of the fluid, which in turn heats up, very quickly.

20.3 Applications

Supercritical technology is more and more widely and diversely used in industry. This development can be put down to the excellent solvent properties of supercritical fluids and the possibility of adjusting the dissolving power by means of such a simple parameter as the pressure. In addition to this, it is easy to eliminate the supercritical solvent by a straightforward expansion of the system.

Applications of supercritical fluids can be divided into roughly two groups: one concerns purification and extraction, and the other synthesis. Several industrial sectors have now integrated supercritical processes. The technology associated with extraction using supercritical fluids is applied in the food industry to extract antioxidant compounds or aromatic substances from plants or certain fruits [14]. In the brewing industry, it has totally replaced solvents for extracting the bitter component from hops. Most decaffeinated coffee produced in the world is obtained by extracting the caffeine using supercritical CO_2 [15]. This process is also used in the petrochemical industry, cosmetics, pharmaceutics, and leather tanning [16–20]. Its use in organometallic synthesis [21] and organic synthesis [22] has certain environmental advantages.

20.3.1 Purification and Extraction

Extraction is a process that is often required in industry. For a heterogeneous mixture such as a dispersion, extraction involves mass transfer. Supercritical extraction only differs from more conventional means of extraction by the nature of the solvent used. The solubility of conventional liquid solvents depends only on the temperature, whereas that of supercritical solvents depends on both the temperature and the pressure. The extraction process comprises two stages: extraction of the solute soluble in the supercritical solvent followed by separation or purification of the solvent. The latter stage is carried out by simply expanding the solvent below the critical pressure, so that it becomes a gas. The solubility of the solute then drops sharply, thereby inducing its spontaneous separation from the supercritical solvent. Carbon dioxide is the most widely used supercritical solvent due to many advantages: relatively low temperature and pressure, low reactivity with regard to many chemical compounds, low inflammability, and low cost.

20.3.2 Synthesis

The use of supercritical fluids (SCF) to generate materials by synthesis is rather recent. Pure products can be produced in the form of fine, dry powders with narrow size distribution. The procedures used to do this can be divided into two main groups: physical procedures appeal to variations in the physical parameters, viz., temperature, pressure, and volume, while chemical procedures make use of chemical reactions [21, 23].

Physical Procedures

Rapid Expansion of a Supercritical Solution (RESS)

The first step in implementing this procedure is to solubilise the solute (metal compound, polymer, etc.) in the supercritical fluid phase. The next step is to depressurise the SCF solution, leading to a sudden drop in solubility. This favours supersaturation of the solute and formation of a particle precipitate (metal, polymer, molecular cluster, etc.). Figure 20.5 shows the equipment used for this process. The solvent is brought into a high pressure cell using a high pressure pump. Raised to a pressure and temperature above the critical coordinates, the CO_2 goes into the supercritical state in which the solute already introduced into the cell is dissolved. The SCF solution is then transferred to the precipitation chamber where it undergoes depressurisation (expansion), inducing precipitation of the solute. A whole range of particle sizes and morphologies can be obtained by this process, depending on the chosen experimental conditions. Nucleation and growth during the expansion are determining factors in the morphology of the final deposit. Minimising these stages by using a dilute solution favours the formation of a thin film of the solute [24]. With a concentrated solution, nucleation and growth are significantly favoured and particle powders are obtained [25].

One advantage with this procedure is the uniformity of particle sizes in the final product. However, it can only be used with products that are soluble in the supercritical fluid phase.

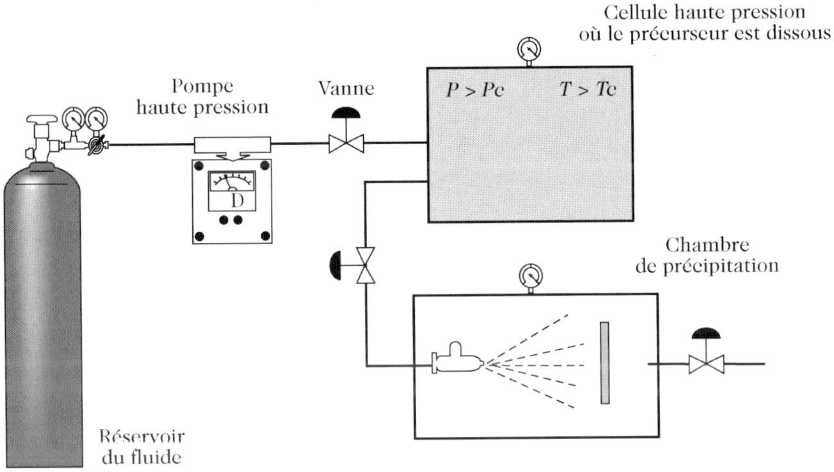

Fig. 20.5. Equipment for the RESS process

Physical Process Without Expansion

Supercritical fluids can be used as solvents for dispersing solutes in matrices, which must also be soluble in the supercritical phase. This is possible in part by virtue of the low viscosity and high diffusivity of these fluids, which means that they have a high degree of penetration into the matrix. At the present time, very few processes are available for obtaining such solute/matrix composites simply and directly. The synthesis of platinum/poly(4-methyl-1-pentene) exemplifies the use of this process in the synthesis of this type of composite [26].

Antisolvent Process with Supercritical Fluids

This process is based on a reduction of the solute solubility by addition of the supercritical fluid. Adding a certain amount of the supercritical fluid will lead to supersaturation, then precipitation of the solute. This process has the advantage of causing precipitation of the solute and elimination of the solvent in a single step, something that cannot be done using conventional processes. A whole range of micrometric and nanometric particles can be obtained by this process. Recently, fluorenes have been obtained using CO_2 as antisolvent [27]. Low yields and the need for an organic solvent have considerably limited the development of this process.

Chemical Procedures

In the RESS physical procedure, expansion of the supercritical solution is carried out in vacuum or in air. In contrast, in the chemical procedure, expansion is carried out in solution, thereby triggering a chemical reaction [28].

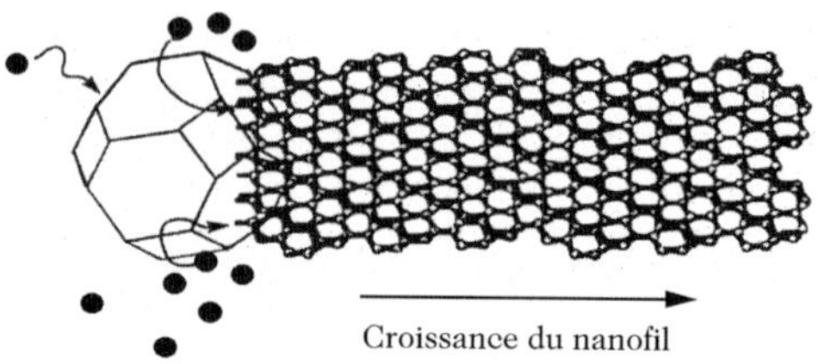

Fig. 20.6. Orientation of the growth of a silicon wire by self-assembly of atoms on a gold particle

The chemical procedure can be achieved without expansion, using the supercritical phase as reaction medium. Particles form by chemical transformation of a precursor in the supercritical fluid medium [29]. This type of synthesis is characterised by high yields and a high level of purity in the final product.

Recently, Holmes et al. have developed an approach to synthesis in a supercritical fluid using this process to synthesise silicon wires (see Fig. 20.6). Gold particles coated with dodecane thiol were dispersed in hexane in supercritical phase (500°C and 270 bar) in the presence of the silicon precursor diphenylsilane [30]. At these temperatures, diphenylsilane decomposes into silicon atoms. The gold and silicon form an alloy in equilibrium with the solid silicon. When the concentration of silicon relative to the concentration of gold goes above 18.6% under certain conditions, the silicon atoms are dissolved in the gold particles. At saturation, the silicon atoms are repelled toward the surface, where they begin to grow in the form of a wire.

Several materials have been synthesised using carbon dioxide as reactive medium. Parag et al. used silver particles from an organometallic silver precursor, silver acetylacetone or Ag(acac), which is reduced by hydrogen H_2 at temperatures between 60 and 100°C [31]. The particles formed are crystallised, as can be seen from the high-resolution transmission electron microscope image in Fig. 20.7a. The resulting particles have a fairly narrow size distribution (see Fig. 20.7b).

Supercritical water features among other solvents often used for synthesis. Crystallised copper particles have been synthesised in this medium by straightforward thermal decomposition of the precursor $Cu(NO_3)_2$ in the presence of dodecane thiol (see Figs. 20.8a and b) [32]. TEM imaging shows that these particles are stable and do not coalesce.

Micelle Systems

Various chemical methods have been developed using supercritical fluids to synthesise nanometric and micrometric particles. Amongst these, one of particular interest involves dispersed systems, and in particular, reverse micelle systems. Although a mixture of water and oil is naturally unstable, it can

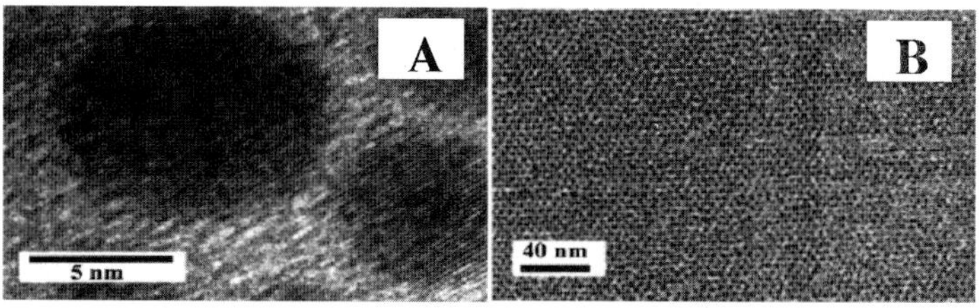

Fig. 20.7. Silver particles in a supercritical fluid. (**a**) High-resolution TEM image showing the crystalline nature of these nanoparticles. (**b**) With their rather narrow size distribution, the nanoparticles arrange themselve into a pattern

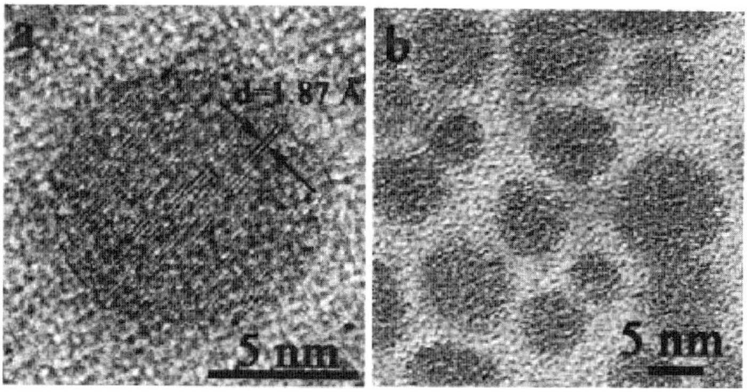

Fig. 20.8. TEM image of copper particles coated with dodecane thiol molecules. (**a**) High resolution showing the crystalline nature of these nanoparticles. (**b**) TEM image showing the size polydispersity of these nanoparticles

be stabilised by adding surfactants. The latter, located at the water–oil interface, have the property of reducing the interface tension. Surfactants are amphiphilic molecules comprising a hydrophobic part that is soluble in oil and a hydrophilic part that is soluble in water. For certain temperature and concentration conditions, and for certain solvents and structures, these molecules can self-assemble to form organised structures called micelles. When the medium is essentially apolar, discrete water regions appear in the oil. The micelle then comprises an aqueous core isolated from the apolar solvent by the hydrophobic chains. The polar heads of the surfactant point toward the inside and one has a reverse micelle.

In the case of supercritical fluids, for reasons of solubility, the hydrocarbon chain is replaced by a fluorocarbon chain [33] in the case of CO_2 or other chains which ensure solubility in the supercritical fluid. The size of the reverse micelles is largely controlled by the amount of water solubilised relative to the amount of surfactant used [34]. More recently, a combination of the properties of reverse micelles and supercritical fluids has opened up new prospects. The supercritical fluid is a medium in which diffusion is 100 times greater than in

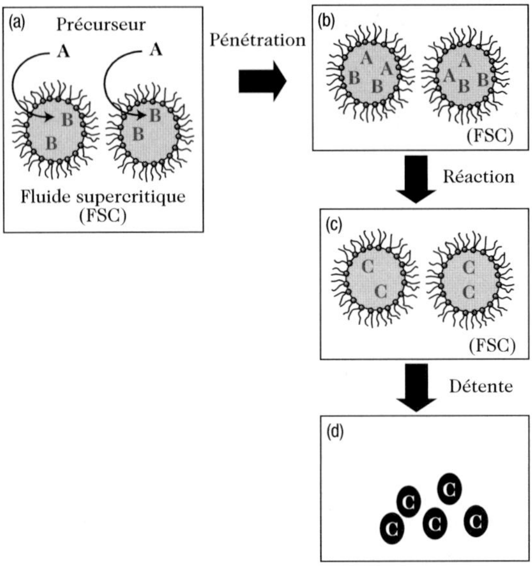

Fig. 20.9. Reaction mechanism involving reverse micelles in a supercritical fluid

a conventional liquid, which makes the kinetics of diffusion-limited reactions more important. These media also offer the possibility of using the pressure as a new parameter for size control [35]. Indeed, an increase in pressure induces an increase in the amount of water contained in the micelles, and hence an increase in the size of the micelles. With these media, very high temperatures can be reached whilst maintaining a degree of stability of the micelle system.

Figure 20.9 shows a typical reverse micelle reaction in a supercritical fluid. In Fig. 20.9a, the precursor reagent A is injected into the micelle solution containing reagent B. The latter dissolves in the supercritical fluid. Hence, in the next stage (Fig. 20.9b), reagent A enters the micelle and the reaction with reagent B takes place to form product C (Fig. 20.9c). The reaction products are separated from the solvent by simple demixing, in which a drop in pressure induces the supercritical state to switch to the liquid or gaseous state.

By combining the properties of reverse micelles and supercritical fluids, a whole range of different particles can be synthesised. Cason et al. [36] obtained copper particles with diameters less than 20 nm in mixed reverse micelles of $Cu(AOT)_2$ and Na(AOT) [bis(2-ethylhexyl) sodium sulfosuccinate] in a mixture of compressed propane and supercritical ethane.

References

1. C. Cagniard de la Tour: Ann. Chem. Phys. **2**, 127 (1822)
2. N.B. Vargaftik: *Table on the Thermophysical Properties of Liquids and Gases*, Hemisphere Publishing corporation, London (1975)

3. E. Kiran, J.M.H. Sengers (Eds.): *Supercritical Fluids. Fundamentals for Application*, NATO, ASI Series, Series E: Applied Sciences, Kluwer, Dordrecht, Vol. 273 (1997)
4. P. Carles: Doctoral thesis, INPT Grenoble, France (1995)
5. J.C. Giddings, N. MacLaren, R.A. Keller: Science **162**, 67 (1968)
6. P.A. Small: J. Appl. Chem. **3**, 71 (1953)
7. D.K. Dandge, J.P. Heller, K.V. Wilson: Ind. Eng. Chem. Prod. Dev. **24**, 162 (1985)
8. A.W. Francis: J. Phys. Chem. **58**, 1099 (1945)
9. K. Stephan, K. Lucas: *Viscosity of Dense Fluids*, Plenum Press, New York (1979)
10. R.C. Reid, J.M. Prausnitz, B.E. Poling: *The Properties of Gases and Liquids*, 3rd edn., MacGraw Hill (1977) p. 391
11. R.H. Perry, D. Green: *Perry's Chemical Engineers' Handbook*, 6th edn., MacGraw Hill (1987)
12. V. Pessey, R. Garriga, F. Weill, B. Chevalier, J. Etourneau, F. Cansell: Ind. Eng. Chem. Res. **39**, 4714 (2000)
13. P. Carles: Doctoral thesis, Institut National Polytechnique de Toulouse, France (1995)
14. D.R. Heldman, D.B.Lund (Eds.): *Handbook of Food Engineering* (1992)
15. P. Mafart, E. Béliard (Eds.): *Génie industriel alimentaire*, Vols. I and II (1992)
16. P.G. Debenedetti: NATO, ASI Series E273, ed. by E. Kiran and J.M.H. Levelt Sengers, Kluwer, Dordrecht (1994) p. 719
17. A. Akgerman, G. Madras: NATO, ASI Series E273, ed. by E. Kiran and J.M.H. Levelt Sengers, Kluwer, Dordrecht (1994) p. 669
18. B. Subramaniam, R.A. Rajewski, K. Snavely: J. Pharm. Sci. **86**, 886 (1997)
19. E.J. Beckman: Nature **271**, 613 (1996)
20. M.A. McHugh, V.J. Krukonis (Eds.): *Supercritical Fluid Extraction: Principles and Practice*, Butterworths, Stoneham, MA (1988)
21. F. Cansell, B. Chevalier, A. Demourgues, J. Etourneau, C. Even, Y. Garrabos, V. Pessy, S. Petit, A. Tressaud, F. Weill: J. Mat. Chem. **9**, 67 (1999)
22. H. Ksibi, P. Subra, Y. Garrabos: Adv. Powder Technol. **6**, 25 (1995)
23. X. Ye, C.M. Wai: J. Chem. Educ. **80**, 2 (2003)
24. J.I. Brand, D.R. Miller: Thin Solid Films **166**, 139 (1988)
25. D.W.Matson, J.L. Fulton, R.C. Petersen, R.D. Smith: Ind. Eng. Chem. Res. **26**, 2298 (1987)
26. J.J. Watkins, T.J. McCarthy: Chem. Mater. **11**, 1991 (1995)
27. P. Chattopadhyay, R.B. Gupta: Ind. Eng. Chem. Res. **39**, 2281 (2000)
28. Y.P. Sun, R. Guduru, F. Lin, T. Whiteside: Ind. Eng. Chem. Res. **39**, 4663 (2000)
29. M. Kryszewski: Polimery **43** (2), 65 (1998)
30. J.D. Holmes, P.K. Johnston, C.R. Doty, A.B. Korgel: Science **287**, 1471 (2000)
31. P.S. Shah, S. Husain, K.P. Johnston, B.A. Korgel: J. Phys. Chem. B **105** (39), 9433 (2001)
32. K.J. Ziegler, R.C. Doty, K.P. Johnston, B.A. Korgel: J. Am. Chem. Soc. **123**, 7797 (2001)
33. P.S. Shah, J.D. Holmes, R.C. Doty, K.P. Johnston, B.A. Korgel: J. Am. Chem. Soc. **122**, 4245 (2000)
34. M.P. Pileni, T. Zemb, C. Petit: Chem. Phys. Lett. **118**, 414 (1985)
35. D.W. Matson, J.L. Fulton, R.D. Smith: Mater. Lett. **6**, 1 (1987)
36. J.P. Caston, C.B. Roberts: J. Phys. Chem. B **104**, 1217 (2000)

Part IV

Fabrication of Nanostructured Bulk Materials and Nanoporous Materials

21

Bulk Nanostructured Materials Obtained by Powder Sintering

F. Bernard and J.-C. Nièpce

21.1 Sintering

21.1.1 Definition

Sintering can be defined as the consolidation of a dispersed material under the action of heat, without total melting of the material. If part of the matter reaches its melting point during this densification, we speak of liquid-phase sintering; otherwise, it is called solid-state sintering. Moreover, if the material is chemically synthesised during sintering, this is called reactive sintering; otherwise, it is called non-reactive sintering.

21.1.2 The Physical Phenomena of Sintering

The two competing phenomena during sintering are (see Fig. 21.1):

- densification, which should generally be favoured,
- growth of grains, which should generally be opposed or at least controlled.

21.1.3 Different Sintering Conditions

Natural Sintering

During natural sintering, the material to be sintered is simply heated in a furnace.

Sintering Under Pressure

- *Hot Pressing (HP).* The work piece is placed in a die and simultaneously subjected to a uniaxial pressure during heating.

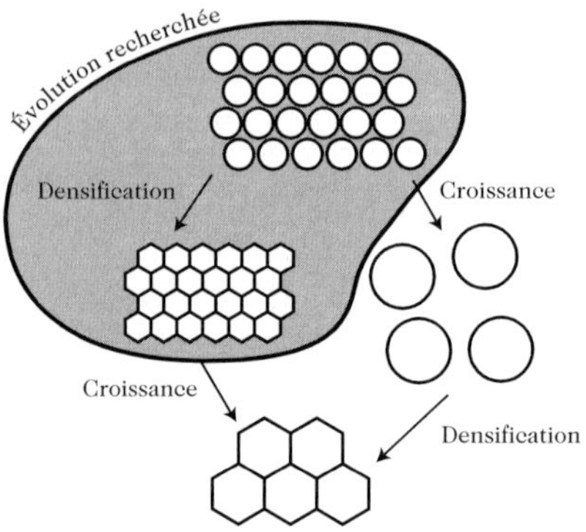

Fig. 21.1. Competing phenomena during sintering

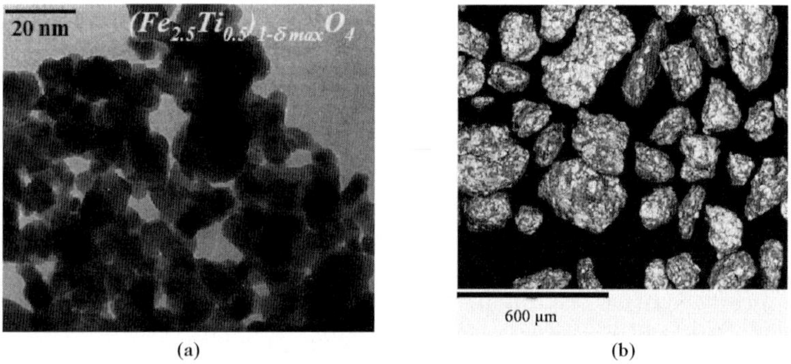

Fig. 21.2. (**a**) SEM image of nanometric titanium ferrite powders prepared by soft chemistry. From [1] © 2001 American Chemical Society. (**b**) SEM image of nanostructured micrometric agglomerates (Fe + Al) prepared by comilling [2]

- *Hot Isostatic Pressing (HIP).* The work piece, usually clad in a deformable (metal or glass) container in which a vacuum has been realised, is placed in a chamber containing a gas under isostatic pressure during heating.

For a given material, these two processes can considerably reduce the sintering temperatures in comparison with natural sintering.

Sintering in Microwaves

The use of microwaves to carry out the sintering stage can provide a good way of heating a powder sample rapidly and uniformly, without inertia or contamination. Heating occurs by dielectric losses.

Sintering Under Electric Discharge

Spark plasma sintering (SPS), or more generally, field activated sintering, consists in placing the work piece in a matrix and subjecting it to a uniaxial

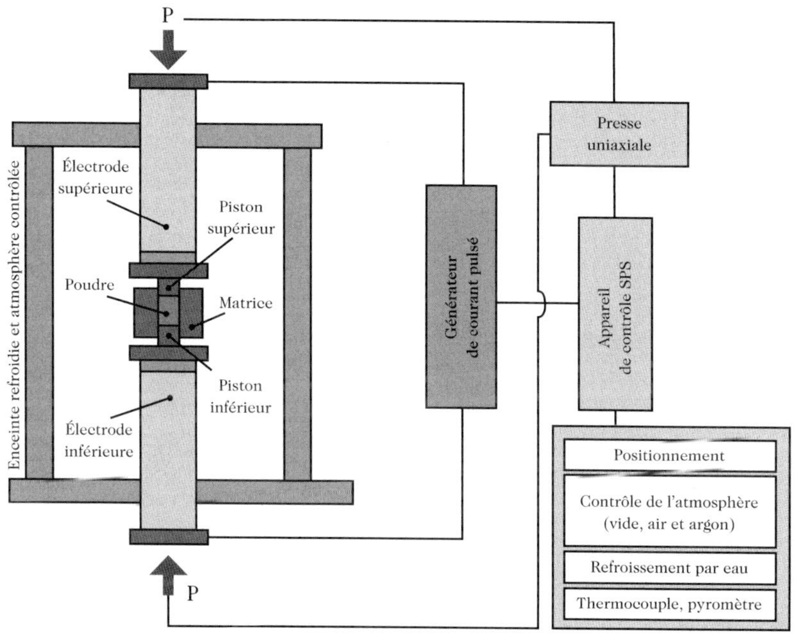

Fig. 21.3. Setup for spark plasma sintering. Courtesy of Sumitomo Coal Mining [3]

pressure coupled with internal 'heating' of the sample by means of a large pulsed current. The SPS technique is described and illustrated below. It provides a way of reducing both the duration and the temperature required for sintering.

21.1.4 Preserving Nanostructure During Sintering

When sintering nanostructured powders, either in the form of individual nanograins (see Fig. 21.2a) or nanostructured micrometric agglomerates (see Fig. 21.2b), it is crucial to restrict grain growth, while promoting the densification process. As a consequence, it is essential to keep sintering times and temperatures as low as possible to preserve, or at least limit the loss of nanostructure.

21.2 Spark Plasma Sintering (SPS)

21.2.1 Basic Principle

The idea behind this process for densifying powdery materials is to combine uniaxial compaction with an intense electric discharge of between 2 000 and 20 000 A in a few ms. Such a process can fully densify a material in a very

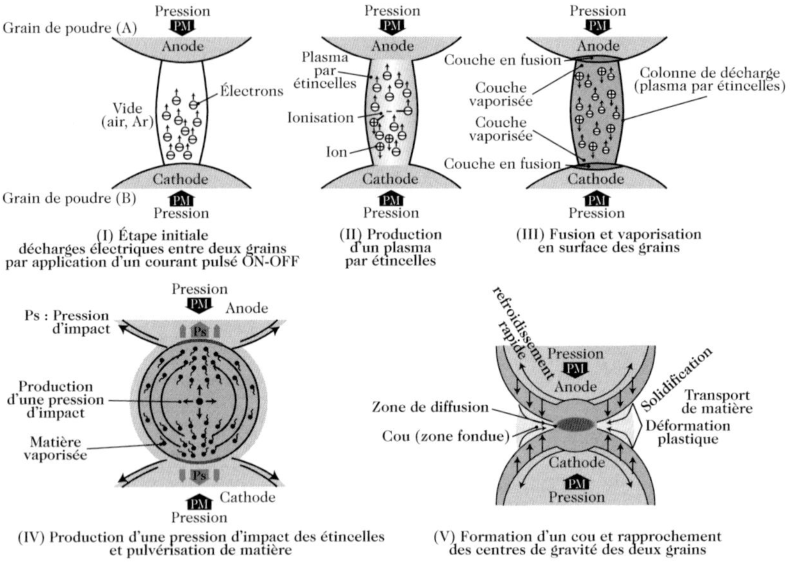

Fig. 21.4. Mechanisms occurring between neighbouring powder particles during an electric pulse generated by the SPS process. Courtesy of M. Tokita, Sumitomo Coal Mining [3]

short processing time compared with other techniques. Figure 21.3 illustrates the setup schematically. The SPS process has the further important advantage that it is easy to implement, since a few grams of powder suffice to obtain, in a few hours, a fairly accurate idea of the required sintering conditions. Indeed, the sintering can be controlled by continuously monitoring the change in sample size through dilatometric measurements.

According to M. Tokita [3], sintering in an SPS machine could be achieved by a spark plasma generated between grains. Repeated application of a low-voltage pulsed current would produce, between neighbouring particles, electrical discharges that would first generate a spark plasma, then an impact pressure due to the sparks and heating by the Joule effect (see Fig. 21.4).

Although they can easily exceed 2 000°C locally, average SPS sintering temperatures across the sample would nevertheless be 200 to 500°C lower than sintering temperatures used in conventional procedures. Moreover, vaporisation, melting and sintering steps can all be carried out in the space of a few minutes (5–20 min), from start to finish, including ramping up the temperature at the beginning and cooling at the end. In addition, surface corrosion is suppressed by activation of the grain surface.

In fact, this sintering method is considered to be very fast, using self-heating inside the powder, rather like self-heating synthesis (SHS) or microwave sintering. Hence, the main difference compared with conventional sintering techniques is that the sample is heated from within rather than

Table 21.1. Materials sintered by SPS [3]

Classification	Materials
Metals	Fe, Cu, Al, Au, Ag, Ni, Cr, Mo, Sn, Ti, W, Be, etc.
Ceramics	Oxides: Al_2O_3, mullite, ZrO_2, MgO, SiO_2, TiO_2, HfO_2 Carbides: SiC, B_4C, TaC, TiC, WC, ZrC, VC Nitrides: Si_3N_4, TaN, TiN, AlN, ZrN, VN Borides: TiB_2, HfB_2, LaB_6, ZrB_2, VB_2 Fluorides: LiF, CaF_2, MgF_2
Cermets	Si_3N_4+Ni, Al_2O_3+Ni, ZrO_2+Ni Al_2O_3+TiC, SUS+ZrO_2, Al_2O_3+SUS SUS+WC/Co, BN+Fe, WC+Co+Fe
Intermetallics	TiAl, FeAl, $MoSi_2$, Si_3Zr_5, NiAl, NbCo, NbAl, $NbAl_3$, $LaBaCuSO_4$, Sm_2Co_{17}
Other materials	Organic materials (polyimide, etc.) and composite materials

from the outside. As a consequence, the significantly reduced sintering times mean that dense bulk materials can be produced within which coarsening due to growth remains almost totally inhibited.

21.2.2 Advantages of the SPS Process

The advantages of SPS lie above all in the low operating cost, shorter sintering times, and lower temperatures than traditional high-pressure sintering processes. Moreover, it is also possible to consolidate and assemble different types of materials such as nanomaterials, composites, and functionally graded materials (FGM) (see Table 21.1).

21.2.3 Illustrations in the Field of Nanomaterials

Non-Reactive Sintering

Nanoceramics such as nano-Si_3N_4 [7], nano-SiC [8], and nano-Al_2O_3 [9] have been sintered by SPS. The microstructure of β-sialon has been controlled by adjusting the electrical stimulation time, as shown in Fig. 21.5.

In the same way, nanometric SiC powders can be consolidated at 1 700°C and 47 MPa, and the resulting microstructure can be modified by adjusting the temperature ramp rate from 100 to 400°C/min, and by altering the period over which electrical pulses are administered from 2 to 5 min [8]. Finally, we mention work by Nygren and coworkers [9] which shows that it is possible to control both the densification and the grain growth at the same time by adjusting the various parameters of the SPS process. In fact, these authors

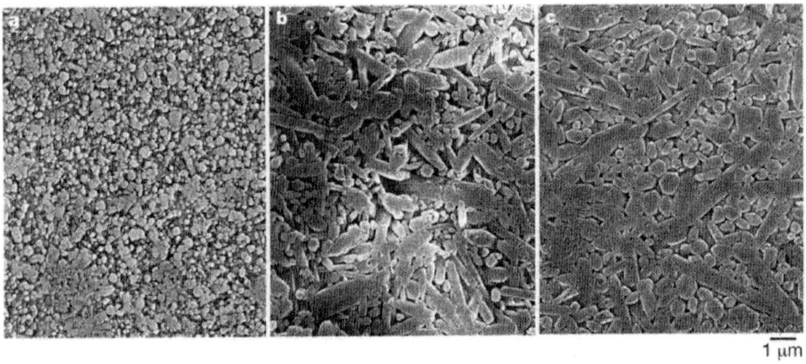

Fig. 21.5. β-sialon produced by SPS. (**a**) Without holding the temperature at 1 600°C. (**b**) Holding the temperature at 1 600°C for 1 min. (**c**) Holding the temperature at 1 600°C for 10 min

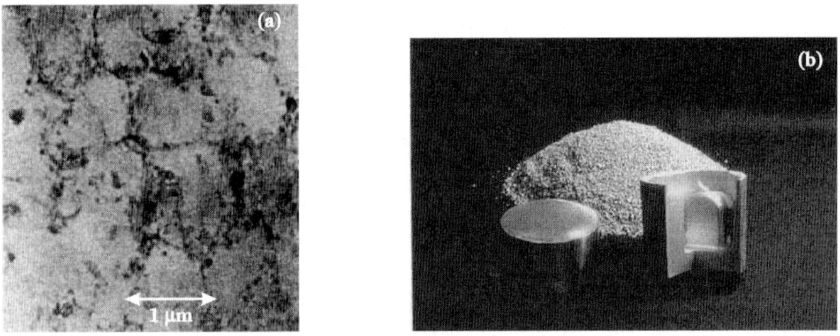

Fig. 21.6. (**a**) TEM image of an Al–17Si alloy synthesised by SPS. (**b**) Description of the different stages leading to the production of a piston made from Al–17Si alloy by cold-forming a nanostructured ingot produced by SPS

have shown that dense, nanostructured aluminas with controlled levels of densification can be produced by the SPS process when the synthesis parameters are optimised.

Sintering of Superplastic Metallic Nano-Alloys

The SPS process has been successfully used to sinter nanometric powders of aluminium–silicon alloys (> 12% at. Si) prepared by atomisation [10]. Preforms obtained from these powders are cylindrical with diameter 60 mm, length 40 mm and density close to 100%. These alloys sintered at low temperature (450–500°C), at low pressure (100–110 MPa), and for less than 20 min (including the whole process of ramping up the temperature, holding it and then cooling), and which possess an extremely fine microstructure (< 1 μm) (see Fig. 21.6a), are then plastically deformed to produce the required 3D object (Fig. 21.6b).

Reactive Sintering

The SPS process has been successfully applied to produce, in one step, dense, nanostructured intermetallic compounds of the Fe/Al system from a mixture of mechanically activated powders [4]. The micrometric aggregates, made from nanocrystals of Fe + Al (Fe–53% at. Al and Fe–60% at. Al) obtained after a short period (4 hr) of high-energy ball milling [5], are introduced into the reaction chamber of the SPS-1050 machine. In fact, this process, which combines uniaxial compaction and intense electrical discharges (5 000 A, 12 ms on, 2 ms off), is able to initiate the exothermic reaction Fe + Al → FeAl to convert all reactants into products in a matter of seconds throughout the sample. The latter is itself subject to a uniaxial load during the consolidation stage. By optimising the process parameters, it has been possible to produce nanostructured (50–100 nm) intermetallic compounds ($Fe_{53}Al_{47}$ and $Fe_{60}Al_{40}$) with relative density close to 99% [6].

References

1. N. Guigue-Millot, Y. Champion, M.J. Hytch, F. Bernard, S. Begin-Colin, P. Perriat: Chemical heterogeneities in nanometric titanomagnetites prepared by soft chemistry: Evidence for Fe-segregation and oxidation kinetics, J. Phys. Chem. B. **105**, No. 29, 7125–7132 (2001)
2. E. Gaffet, F. Bernard, J.C. Niepce, F. Charlot, C. Gras, G. Le Caer, J.L. Guichard, P. Delcroix, A. Mocellin, O. Tillement: Some recent developments in mechanical activation and mechanosynthesis, J. Mat. Chem. **9**, 305–314 (1999)
3. M. Tokita: Trends in advanced SPS systems and FGM technology, J. Soc. Powder Technol. Jpn. **30** (11), 790–804 (1993) and `http://www.scm-sps.com`
4. Z.A. Munir, E. Gaffet, F. Charlot, F. Bernard: One-step synthesis and consolidation of nano-phase materials, US Patent No. 6 200 515 B1, (13 Mars 2001)
5. S. Paris, C. Valot, F. Bernard, E. Gaffet, Z.A. Munir: Investigation of MAFA-PAS processing parameters for producing dense nanostructured materials, J. Mater. Research **18**, 2331–2338 (2003)
6. S. Paris, F. Bernard, E. Gaffet, Z.A. Munir: Spark plasma synthesis from mechanically activated powders: A versatile route for producing dense nanoaluminides, Scripta Mater. **50** (5), 691–696 (2004)
7. Z. Shen, Z. Zhao, H. Peng, M. Nygren: Formation of tough interlocking microstructures in silicon nitride ceramics by dynamic ripening, Nature **417**, 266–269 (2002)
8. Y. Zhou, K. Hirao, M. Toriyama: Very rapid densification of nanometer silicon carbide powder by pulse electric current sintering, J. Am. Ceram. Soc. **83** (3), 654–656 (2000)
9. Z. Shen, M. Johnsson, Z. Zhao, M. Nygren: Spark plasma sintering of alumina, J. Am. Ceram. Soc. **85** (8), 1921–1927 (2002)
10. N. Kuroishi: High-strain-rate superplasticity in sintered pre-forms produced by spark plasma sintering, 67–74 (2001)

22

Self-Assembly of Nanomaterials at Macroscopic Scales

A. Courty

The terms nanomaterial, nanocrystal and nanoparticle all refer to objects with nanometric sizes. The main feature of nanomaterials is that they are made up of a small number of atoms, from a few hundred to a few thousand. Due to their small dimensions, the physical properties of these materials generally differ from those of bulk materials. Over the last few decades, this field of research has developed considerably, both theoretically and experimentally. This development has been particularly pronounced in the three areas where size effects are important, viz., metals (Au, Ag, etc.), semiconductors (CdSe, Ag_2S, etc.), and magnetic materials (CO, Ni, etc.).

Material properties and performance can be improved and optimised by assembling such nanomaterials into 2D or 3D systems. Novel physical properties have been observed in these systems, a result of the interactions between particles. Such collective properties are neither the properties of the individual particles, nor those of the bulk material, and they are favoured by the good organisation of the nanomaterial [1,2]. These nanomaterial lattices are often considered as 'artificial solids' in which the nanomaterials take the place of atoms in 'conventional solids' and their arrangement determines the crystal structure of the superlattice.

In order to manipulate nanomaterials in 2D or 3D lattices, one needs to know their shape, surface properties, charge, polarisability, magnetic dipole moment, and mass. Indeed, these characteristics determine the interactions between the nanometric particles. The nanoparticles must be mobile in order to form ordered arrangements. The idea is thus to achieve self-organisation of the nanomaterial, i.e., to induce a spontaneous large scale organisation of the nanometric objects on a surface without having to intervene at the level of the objects themselves, but simply by providing suitable macroscopic conditions.

The first self-organisation of this kind obtained experimentally over large distances concerned monolayers of spherical Ag_2S nanocrystals (see Fig. 22.1) [3] and spherical CdSe nanocrystals [4] in 1995. Since then, many groups have been able to fabricate 2D and 3D systems for a wide range of materials such as metals [5–10] and oxides [11]. In all these systems, the particles are

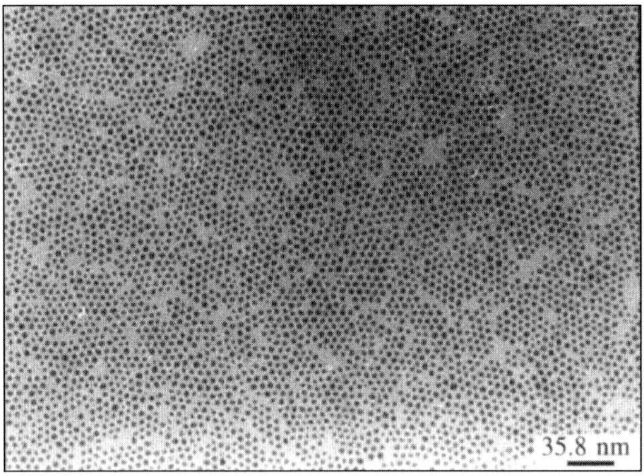

Fig. 22.1. Transmission electron microscope (TEM) image of a monolayer of silver sulfide (Ag_2S) nanoparticles. Taken from [3] © 1995 American Chemical Society

characterised by an inorganic core and are stabilised by an organic surfactant. We shall see later that both the core and the surfactant play key roles in the arrangement of 2D and 3D lattices. Moreover, these superlattices are generally obtained by depositing a solution of nanomaterials on a flat solid substrate and allowing the solvent to evaporate. In this way, the solvent and the substrate supporting the lattice can both have a considerable influence on the final crystal structure.

22.1 Fabrication of Nanomaterials

Historically, the first syntheses of nanoscale metal particles were carried out using aqueous solutions of ionically stabilised colloids [12]. Such particles carry surface charges and are thus characterised by strong interparticle ionic interactions. For particles of micrometric or submicrometric sizes, these interactions can be screened and the colloids then crystallise in the form of stable and ordered aggregates. For metal or semiconductor particles of nanometric sizes, the screening effect is much smaller and these charged particles tend to form unstable, disordered aggregates.

Therefore the main task in producing 2D and 3D lattices of self-organised nanocrystals is to develop methods of synthesis capable of producing particles that are passivated at the surface by organic molecules (carbon chains) and which interact via dispersion forces of van der Waals type. These methods must also be able to produce nanoparticles with well-controlled shapes and sizes. Indeed we shall see that one of the key issues in self-organisation is the reduction of dispersion in the size distribution of the nanocrystals. Only chemical methods have succeeded in achieving such conditions.

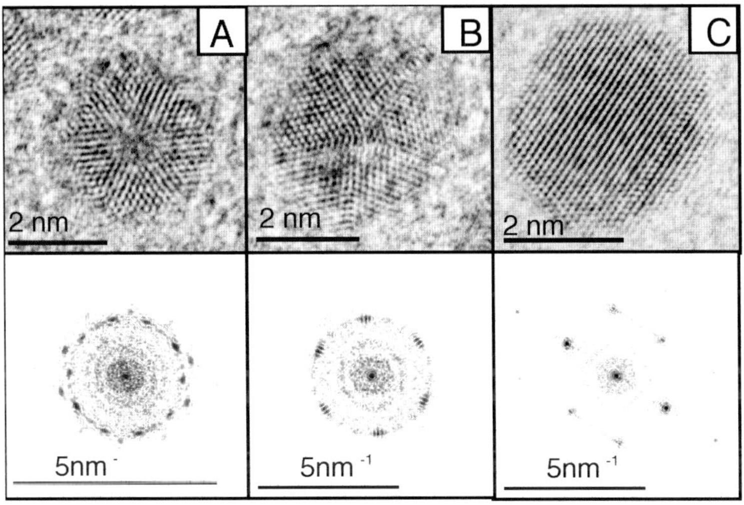

Fig. 22.2. High-resolution transmission electron microscope (TEM) images of silver nanocrystals and their Fourier transforms. Taken from [15] © 2000 American Insitute of Physics. (**A**) Decahedron. (**B**) Icosahedron. (**C**) Cubo-octahedron

Among all these methods, some use micelles as a kind of microreactor [13]. These are water droplets in oil, stabilised by a surfactant whose task is to minimise the interface tension between the two media. These droplets exchange their contents by collisions, before forming independent droplets once again. During these collisions, nanomaterials are formed by chemical reduction of a metal ion or by coprecipitation reactions. This method has the advantage of providing control over the size of the synthesised particles via the size of the micelles. The particles can then by coated with organic molecules, i.e., carbon chains such as alkane thiol, and extracted from the micelle system. This method has been used to fabricate nanocrystals from a wide range of well-crystallised materials with controlled sizes, such as silver, cobalt, copper, and so on [14]. For example, in the case of silver, well-crystallised nanocrystals of diameter 5 nm and low size dispersion (10%) have been observed by transmission electron microscopy (see Fig. 22.2) [15]. They may be either monocrystals with face-centered cubic crystal structure (cubo-octahedra), or polycrystals corresponding to assemblies of tetrahedra with five-fold symmetry (tetrahedra or icosahedra) (see Fig. 22.2).

There are other methods for synthesis using a two-phase water–organic solvent system in which a metal salt, e.g., $AuCl_4^-$, is reduced in the presence of chemical agents which attach themselves onto the particle surfaces (carbon chains) and help to stabilise them. Gold, silver, and cobalt particles have been obtained in this way [16].

Still other methods are based on organometal techniques [17]. Nanocrystals of semiconductors such as CdTe and CdSe have been obtained. The nanocrystals form at high temperatures (> 300°C) in a solvent such as trioctylphosphine oxide (TOPO) which forms covalent bonds at the surface of the nanocrystals. The solvent thus plays the role of a stabilising agent. The preparation is carried out in two steps: a nucleation stage ($T > 300$°C) and a growth stage at lower temperature (around 200°C). This division into two stages has the advantage of producing particles with narrow size distribution for sizes in the range 2–10 nm.

As a last example, it is worth mentioning methods for synthesising nanometric materials using zeolites. These are mesoporous crystalline materials such as GaAs, InP, or AlOSi, within which another material like CdS or ZnS can be incorporated. The pore diameter can be used to control the size of materials formed [18]. The advantage of zeolites is that they provide an ordered system of pores. However, as soon as the volume fraction of inserted materials goes over 20%, the mutual interaction between nanomaterials becomes very strong, making it difficult to understand such structures.

All these chemical methods have thus been used to make nanomaterials with mean diameters in the range 1–10 nm from a wide variety of metals (Ag, Au, Cu, Co, etc.) and semiconductors (CdS, Ag_2S, etc.). However, the inherent problem with all these methods is to reduce the size dispersion of the resulting nanoparticles. In fact, the solubility of the nanomaterials produced, whether they be metallic or semiconducting, depends sensitively on size in solvents like hexane or toluene. This means that precipitation techniques based on the use of a good or bad solvent can be used to separate out different sizes. Many groups have been able to apply this technique to reduce the size distribution of nanomaterials, bearing in mind that this is a key parameter for self-organisation. Examples of good solvent/bad solvent pairs are hexane/pyridine, toluene/methanol, hexane/acetone, toluene/ethanol, and others.

22.2 2D and 3D Nanomaterial Structures

22.2.1 Depositing Nanomaterials on a Solid Substrate

The first examples of self-organisation over large distances were obtained by a simple experimental method which consisted globally in depositing a solution of Ag_2S or CdSe nanocrystals on a flat substrate and letting the solvent evaporate [3]. The nanocrystals then arrange themselves spontaneously into a close-packed hexagonal monolayer (see Fig. 22.1) or in the form of 3D aggregates depending on the amount of matter added (see Fig. 22.3) [19].

The size dispersion was then very low, around 10%. The method was subsequently extended to other metals such as silver and gold [8–11, 20, 21]. This work showed that, in order for nanocrystals to self-organise, they had to have a very low size dispersion (around 13% or less). In further work, the method was extended to nanoparticles of different shapes [22]. The supporting surfaces on which the nanomaterials were deposited were chosen with regard to

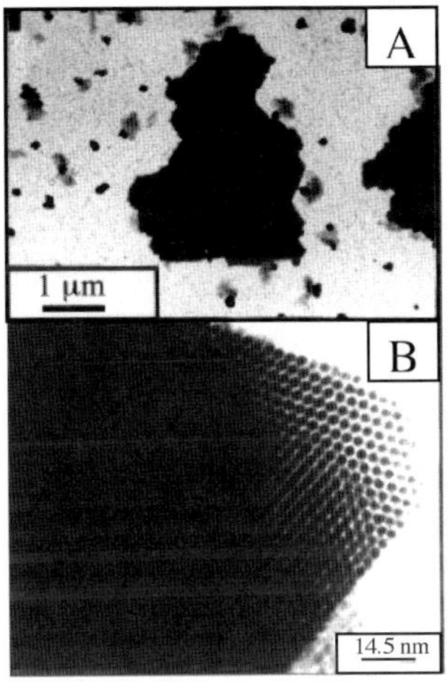

Fig. 22.3. TEM images obtained during deposition of a very concentrated solution of silver sulfide nanocrystals (6×10^{18} particles/ml). (**A**) A large number of 3D Ag_2S nanoparticles. (**B**) High-magnification observation of one crystal [19]

their roughness, which had to be as low as possible in order to avoid defects forming in the organised structure. Conversely, supporting surfaces with very clear physical or chemical textures were also used in order to impose some specific type of organisation on the deposited nanocrystals. This technique often requires more sophisticated preparation of the substrate, but it can yield interesting results including wires [23] and an arbitrary range of different patterns [24].

While 3D lattices can be obtained by depositing droplets of a very concentrated solution, supercrystals with sizes greater than 10 microns can be obtained by quenching the substrate in the solution until all the solvent has evaporated. The advantage with this method of deposition is that the rate of evaporation of the solvent can be adjusted, either through the substrate temperature or through the level of solvent saturation of the atmosphere surrounding the substrate. This method of deposition has been applied to silver nanocrystals (mean diameter 5 nm) synthesised in reverse micelles. When droplets are deposited on a TEM grid, they spontaneously arrange themselves into monolayers or bilayers (see Fig. 22.4A). When the quenching method is used, with a substrate temperature of 35°C during evaporation, TEM observations show silver supercrystals corresponding to stacks of more than 1 000 layers of particles (see Fig. 22.4B) [25]. Moreover, these experiments have demonstrated the effect of the substrate temperature during solvent evaporation [26]. A high temperature favours particle diffusion and interactions. Silver supercrystals are then observed, characterised by shapes with well-defined edges, typical of regular stacks of particle monolayers. At lower temperatures ($T \leq 10°C$), much rougher deposits are observed, containing many voids typical of disorder (see Figs. 22.5A and B).

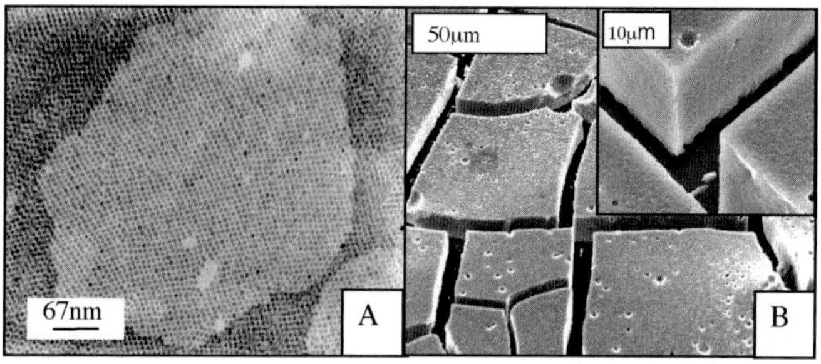

Fig. 22.4. Arrangements of silver nanocrystals obtained (**A**) in 2D by droplet deposition and (**B**) in 3D by quenching the substrate in the solution at 35°C until the solvent evaporates completely [25]

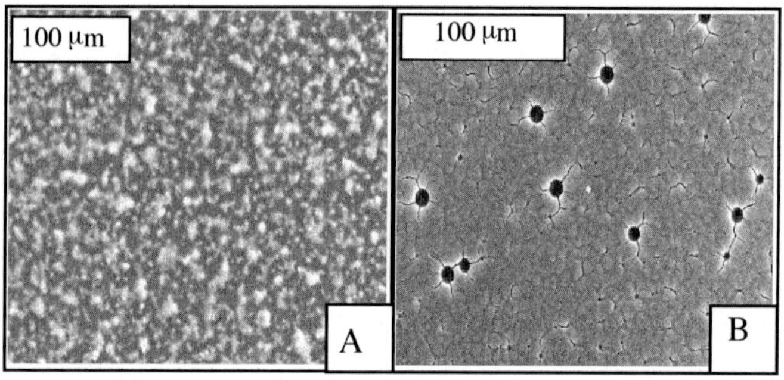

Fig. 22.5. 3D deposits of silver nanocrystals obtained by quenching at substrate temperature (**A**) 0°C and (**B**) 10°C. Taken from [26] © 2001 American Chemical Society

22.2.2 Forces Inducing Self-Organisation

A detailed understanding of the relevant phenomena is required to obtain self-ordered systems. Indeed, even though self-organisation actually occurs without intervention by the experimenter, it is nevertheless induced by forces and can only come about because these forces are already present in the system. For nanoscale objects, the relevant forces are generally short range, such as capillary forces, wetting forces, or van der Waals interactions, which dominate the self-organisation process. Since self-organisation is obtained by depositing a nanocrystal solution on a flat substrate and letting the solvent evaporate, one must first consider particle–particle, particle–solvent, particle–substrate and solvent–substrate interactions in the formation of 2D and 3D lattices. In addition, the solvent in which the nanocrystals are initially immersed will play a role through wetting effects. Indeed, when a droplet of the solution is deposited, it is clear that the way the liquid wets the surface and also its tendency to evaporate will be of considerable importance for the final structure. On smaller scales, the liquid will interact with the substrate and the

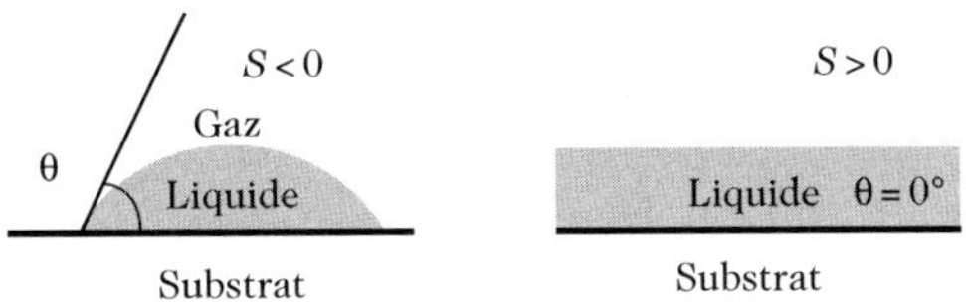

Fig. 22.6. Partial and total wetting

nanocrystals via capillary forces. Finally, one can expect to see other effects becoming significant on even smaller scales, e.g., van der Waals forces, electrostatic interactions if the species are charged, and even steric repulsion due to coatings on the particles. Indeed, molecules coating particles do not serve purely as protective coatings to prevent direct contact between particles, such as might allow them to coalesce; they also play a role in consolidating the lattice.

Wetting and Capillary Forces

Although very different in result, the problems of wetting and capillary forces both arise due to the existence of surface tension. Now all systems tend to minimise their area in order to reduce their interface energy. This explains the formation of spherical droplets and soap bubbles, since the sphere is the shape that encloses a given volume with the smallest possible surface area. Concerning the wetting of a substrate by a liquid, the spreading parameter is defined by [27]

$$S = \gamma_{\rm sg} - (\gamma_{\rm sl} + \gamma) , \tag{22.1}$$

where $\gamma_{\rm sg}$, $\gamma_{\rm sl}$, and γ are the interface tensions of the naked substrate, the liquid-coated substrate, and the liquid with the gas phase, respectively. The wetting state of the supporting substrate on which the nanomaterials are deposited will thus play an important role in their organisation. There are two possible wetting states: total wetting and partial wetting (see Fig. 22.6).

In the case of total wetting ($S > 0$), the solvent will tend to spread suitably across the surface and thereby disperse the particles over a wide region. The resulting deposit will then be very uniform, but it may contain spaces if the amount of particles in the solution is not sufficient. Conversely, partial wetting ($S < 0$) by the solution will cause a droplet to form. When it evaporates, this droplet will therefore tend to get smaller, gradually concentrating its particle content at the center of the droplet. The deposit obtained will now be inhomogeneous and the particles will tend to occur in the form of a large central aggregate rather than in the form of a homogeneous film.

Consider now the particles deposited on a substrate but still surrounded by solvent. When the amount of solvent becomes very small due to evaporation, there comes a moment when a liquid meniscus can form, constituting a

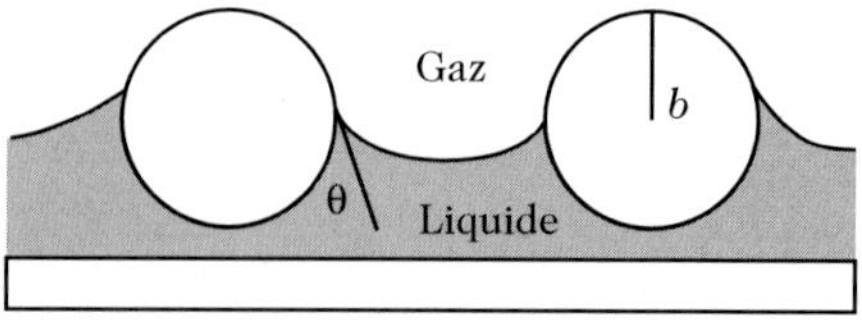

Fig. 22.7. Liquid meniscus between particles

liquid bridge between the particles as shown in Fig. 22.7. Owing to the surface tension of the liquid, this meniscus will exert a capillary force given by [28]

$$F = 4\pi b\gamma \cos\theta \ , \tag{22.2}$$

where b is the particle radius, θ is the angle between the liquid and particle surfaces at the point of contact, and γ is the surface tension. (Examples of the value of γ for several solvents: $9.6 \times 10^{-5}\,\mathrm{J\,m^{-2}K^{-1}}$ for hexane, $1.5 \times 10^{-4}\,\mathrm{J\,m^{-2}K^{-1}}$ for water, and $7.4 \times 10^{-5}\,\mathrm{J\,m^{-2}K^{-1}}$ for decane.) This force tends to bring the particles together and thus favours a close-packed assemblage. If the latter becomes dominant, it can lead to the formation of 3D particle lattices by concentrating particles locally, to the detriment of a 2D film arrangement.

Bearing in mind the two phenomena of wetting and capillary forces, all sorts of situations can be envisaged, from the well-dispersed spreading of the particles to the formation of highly localised aggregates. This shows the importance of being able to make accurate estimates of the orders of magnitude of the two phenomena when trying to predict and explain experimental results.

Dispersion Forces and van der Waals Interactions

At the nanoscale, particles, solvent and substrate interact through dispersion forces. These are due to short-range attractive forces between molecules and are described by fluctuations in their electron density. Indeed, under the effect of thermal vibrations, variations in the electron density of a molecule induce a moment. This moment interacts with the electron cloud of a neighbouring molecule and induces a dipole in it, whose fluctuations will then be synchronised with those of the first molecule. The induced dipoles have the same direction, and a permanent short-range attraction is set up between the two molecules. This interaction, also called the London interaction, is one of the so-called van der Waals interactions. It underlies the cohesion between the molecules and depends on their polarisability, and hence on the number of electrons in each molecule.

The interaction between macroscopic objects is obtained by summing over all pairs of molecules available in the system. This calculation intrinsically takes into account the geometry of the system and the potential between the molecules. The factor containing all terms relating to the polarisability and

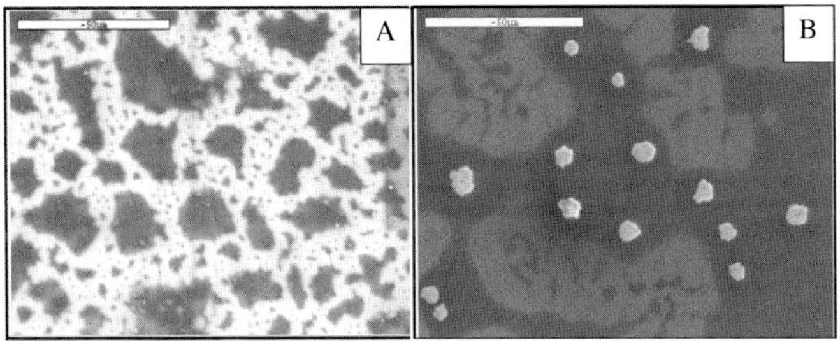

Fig. 22.8. Scanning electron microscope observations of Ag_2S particles deposited (**A**) on molybdenum sulfide, a substrate producing attractive forces, and (**B**) on cleaved graphite, a substrate producing repulsive forces. (**A**) Extensive bright zones correspond to monolayers of nanocrystals with spaces (dark zones corresponding to naked substrate). (**B**) Broad bright zones correspond to monolayers of nanocrystals surrounded by dark zones corresponding to naked substrate. In the dark zones, brighter spots are observed, corresponding to 3D arrangements of nanocrystals. Taken from [31] © 2000 American Chemical Society

the electron density of the molecules is called the Hamaker constant [29]. This constant is used to calculate the interaction energy between particles of a given material:

$$E_{\mathrm{vdW}} = -\frac{A}{12}\left(\frac{4R^2}{C^2-4R^2} + \frac{4R^2}{C^2} + 2\ln\frac{C^2-4R^2}{C^2}\right) , \qquad (22.3)$$

where A is the Hamaker constant and C is the centre-to-centre distance between two particles, e.g., for the attraction between two silver nanocrystals, A has a value close to 1.9 eV [30].

The Hamaker constant is also used to explain the influence of the substrate on the self-organisation of the nanocrystals. The relevance of the chemical characteristics of the supporting surface on which the colloidal solution of nanomaterials is deposited has been demonstrated [31]. Depending on the nature of the substrate, the particle–substrate interactions can be attractive or repulsive. A substrate surface producing attractive interactions with the deposited particles will favour the formation of extended but not very close-packed monolayers, e.g., Ag_2S particles deposited on MoS_2 as shown in Fig. 22.8A, whereas a substrate producing repulsive interactions with the nanomaterials will favour the formation of close-packed monolayers, but also aggregates, e.g., Ag_2S particles deposited on cleaved graphite as shown in Fig. 22.8B.

Another example showing the effect of the substrate on self-organisation is provided by silver nanocrystals deposited on a substrate of cleaved graphite, which produces attractive interactions with the nanocrystals in this case [32]. When the length of the alkyl chain $C_nH_{2n+1}SH$ coating the particles is reduced, i.e., n is reduced from 12 to 8, this favours the formation of less extensive monolayers with a fixed distance between particles (see Fig. 22.9). This

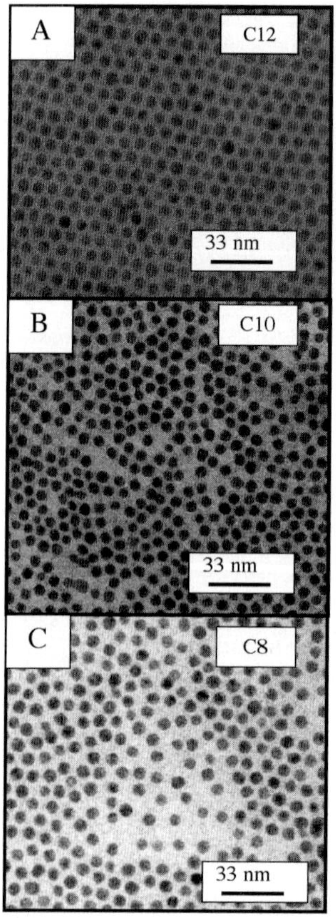

Fig. 22.9. TEM images showing the effect of the length of the alkyl chains coating the silver nanocrystals on their 2D organisation after deposition on cleaved graphite. Nanocrystals coated with (**A**) dodecane thiol $C_{12}H_{25}SH$, (**B**) decane thiol $C_{10}H_{21}SH$, (**C**) octane thiol $C_8H_{17}SH$. Taken from [31] © 2000 American Chemical Society

distance corresponds to the length of a dodecane thiol chain, which is about 2 nm. Therefore this indicates an interpenetration of the chains in the case of nanocrystals coated with dodecane thiol, whereas for nanocrystals coated with shorter chains such as decane thiol or octane thiol, the chains no longer interpenetrate. It seems that for shorter chains the interaction with the substrate is stronger. Since this interaction is attractive, the particles cannot move around easily on the substrate, which would explain a looser distance between particles, i.e., greater than the chain length.

Steric Repulsion

As mentioned earlier, organic molecules (carbon chains like dodecane thiols) passivating the particles play an important role in the construction of 2D and 3D lattices. The steric repulsion potential (E_{steric}) between carbon chains is given by [33]

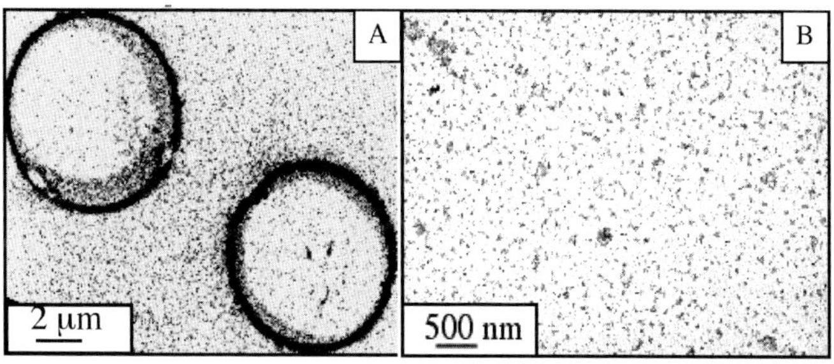

Fig. 22.10. Illustration of Marangoni instabilities. Deposition on a TEM grid with anticapillary tweezers. (**A**) Ferrite nanocrystals dispersed in hexane. (**B**) Ferrite nanocrystals dispersed in decane. Taken from [34] © 2000 American Chemical Society

$$E_{\text{steric}} \approx \frac{100R\delta_{\text{SAM}}^2}{(C-2R)\pi\sigma_{\text{thiol}}} kt \exp\left[-\frac{\pi(C-2R)}{\delta_{\text{SAM}}}\right], \tag{22.4}$$

where R is the particle radius, C is the centre-to-centre distance between two particles, δ_{SAM} is the thickness of the layer of carbon chains, e.g., 18 Å for nanoparticles coated with dodecane thiol molecules, and σ_{thiol} is the diameter of the area occupied by the thiols on the particle surfaces (around 4.3 Å for particles coated with dodecane thiols).

The interaction potential between particles is thus the sum of the van der Waals potential E_{vdW} given in (22.3) and the steric repulsion potential between carbon chains E_{steric} given in (22.4). The particles order themselves like 'soft' spheres with an effective diameter which includes the layer of organic molecules attached to them. If the particles are to self-organise, steric stabilisation provided by these organic molecules must therefore be sufficiently compensated by the interparticle van der Waals attraction.

Bernard–Marangoni Instabilities

We have just seen that the characteristics of nanocrystals, solvent and substrate all play a significant role in the organisation of nanocrystals. But what about the process of solvent evaporation, which depends on the rate of solvent evaporation?

A recent study has shown the influence of solvent evaporation rates on the organisation of nanocrystals [34]. Indeed, it has been shown that a droplet of ferrite particles dispersed in hexane and deposited on a microscope grid held by anticapillary tweezers gives rise to the formation of broad micrometric rings with nanocrystals around the outside, as shown in Fig. 22.10A. If the nanocrystals are now dispersed in decane, which considerably reduces the evaporation rate of the solvent, a random dispersion of particles is observed on the supporting surface, without the formation of rings (see Fig. 22.10B).

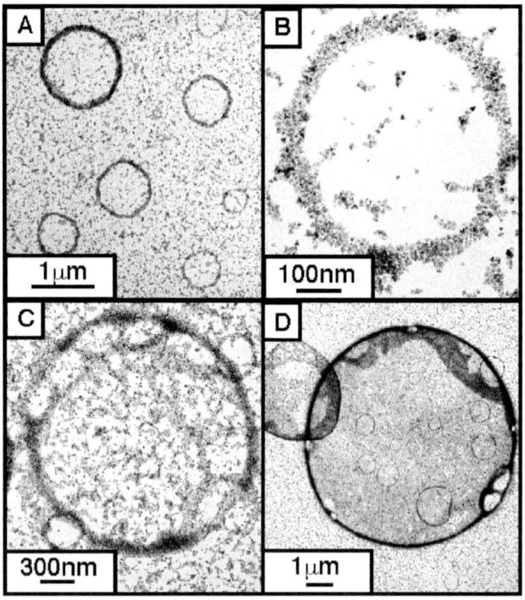

Fig. 22.11. Illustration of Marangoni instabilities for different materials. Formation of rings of nanocrystals obtained by depositing a dilute solution of silver nanocrystals (**A**), copper nanocrystals (**B**), cobalt nanocrystals (**C**), and silver sulfide nanocrystals (**D**). Observations made by TEM. Taken from [34] © 2000 American Chemical Society

This phenomenon has been attributed to Marangoni instabilities. These occur when there is a gradient of surface tension in a liquid medium and they cause periodic convective flow. A high rate of evaporation, as with hexane, favours these instabilities and the formation of rings. This phenomenon has been observed for a wide range of materials such as silver, silver sulfide, and cobalt (see Fig. 22.11).

22.2.3 Crystal Structure of 2D and 3D Nanomaterials

The spontaneous assembly of spheres of uniform size into an ordered lattice is a universal phenomenon observed for objects with diameters ranging over a wide length scale, from atoms (10^{-8} cm), molecules and macromolecules (10^{-6} cm in the case of proteins, polymers and colloids), synthetic spheres and opals (10^{-5} cm), to just about any spherical object (1–10^2 cm in the case of balls, fruit, etc.). The parameters determining the crystal structure of a given substance include not only the intrinsic structure (shape and size) and its hardness, i.e., its compressibility and deformability, but also the type and range of interaction forces between substances. In the limiting case of 'hard spheres', where contact between the spheres is possible, close-packed crystal

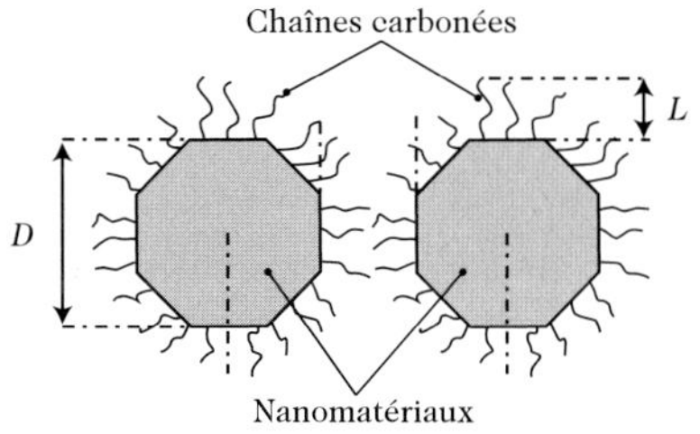

Fig. 22.12. Definition of the determining parameter χ for the crystal structure of a 3D arrangement of nanomaterials coated by carbon chains: $\chi = 2L/D$

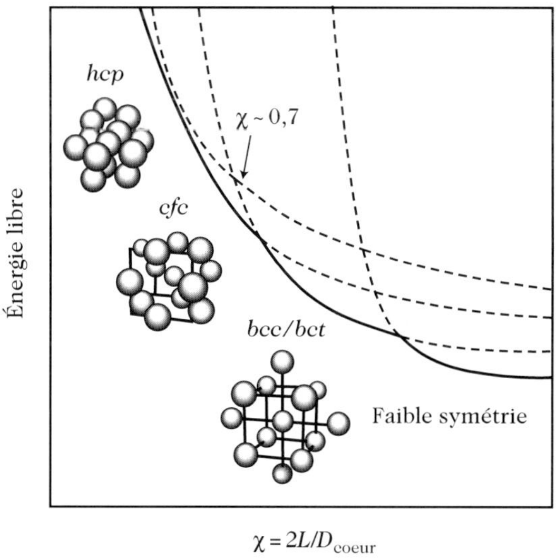

Fig. 22.13. Hypothetical phase diagrams showing stable and metastable phases. *Continuous curves* correspond to the most stable crystal structures. The model structures given are composed of spheres corresponding to a nanomaterial rather than an atom [35]

structures such as face-centered cubic (fcc) or hexagonal close packed (hcp) are optimal. These have the highest packing fraction (74%) and coordination (12 nearest neighbours for each nanoparticle). Metallic or semiconducting nanomaterials coated with organic molecules (carbon chains) can be treated as 'soft spheres', as mentioned in the last section. Since they are characterised by a potential that is attractive at short range, less symmetrical crystal structures, and lower nanoparticle coordination, they may be preferred.

A determining parameter can be defined for the crystal structure of an assembly of nanomaterials coated by carbon chains, viz., $\chi = 2L/D$, where L and D are defined in Fig. 22.12. Depending on the value of χ, different crystal structures have been observed experimentally for a wide range of nanomaterial

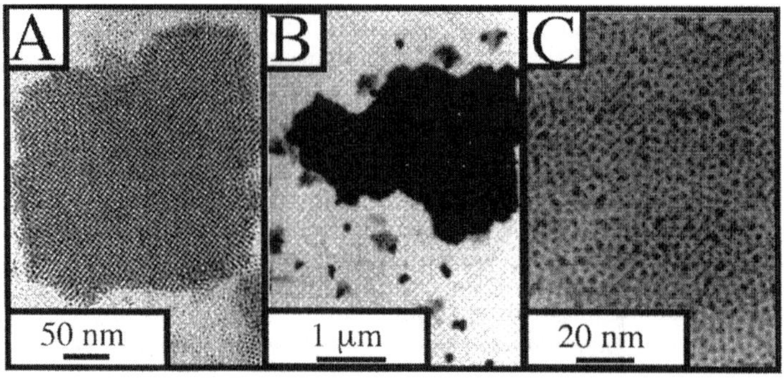

Fig. 22.14. TEM images of Ag_2S nanocrystals coated with different lengths of carbon chains. (**A**) Tetradecane thiol $C_{14}H_{29}SH$. (**B**) and (**C**) Decane thiol $C_{10}H_{21}SH$. (**A**) shows a 3D organisation of nanocrystals while (**B**) and (**C**) show a random 3D arrangement of the latter. Taken from [37] © 1999 American Chemical Society

lattices such as silver, gold, or palladium [35,36], e.g., fcc or hcp structures, and less closely packed structures such as body-centered or simple cubic. All the experimental results obtained show that nanomaterials organise themselves into close-packed (fcc or hcp) 3D lattices for a limiting value of χ of about 0.73. Beyond this value, less closely packed structures are observed viz., body-centered cubic [34,35]. These results point towards the highly schematic phase diagram shown in Fig. 22.13.

In the context of the soft sphere model, it is easy to understand that highly symmetric structures with high levels of coordination (fcc or hcp) will be destabilised when the thickness of the carbon chain monolayer is increased. More 'open' structures, i.e., less closely packed, such as body-centered cubic, will then be favoured. However, some provisos are necessary here. The results do not take into account the conformation of the carbon chain and the presence of defects (if any). Hence it has been observed in the case of Ag_2S particles that alkane thiol ($C_nH_{2n+1}SH$) chains with $n < 8$ will not attach themselves to the particle surfaces owing to the fact that there are too many defects in the chain, and also that for longer chains ($n = 14$) the presence of defects at the end of the chain, although it permits a 2D arrangement, prevents 3D stacks from forming (see Fig. 22.14) [37].

In other work, it has been observed that, when nanomaterials coated with carbon chains self-organise into a 2D hexagonal lattice, the interparticle distance corresponds to one chain length [21, 34]. This suggests an interdigital arrangement of the chains. The ideal situation is therefore to use carbon chains that can pack closely together and thereby consolidate the lattice. One of the parameters determining the crystal structure of 2D and 3D arrangements is also the specific morphology (facets) of the nanomaterial. Indeed, the distribution of carbon chains on the particle surfaces when they are facetted is not uniform. A particle arrangement is favoured in which the crystal faces of the

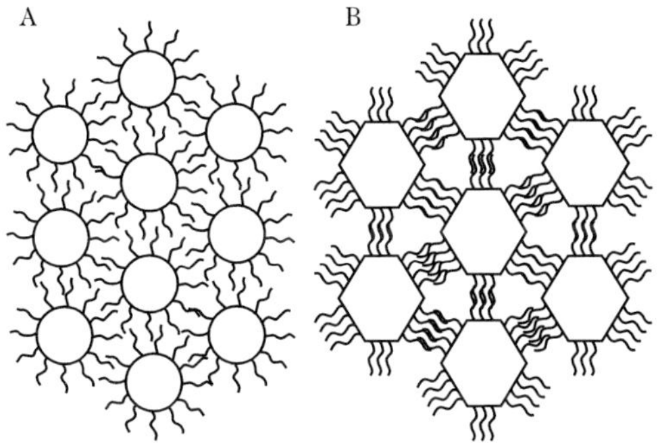

Fig. 22.15. Self-assembly of a lattice of passivated nanocrystals. (**A**) Spherical nanocrystals. (**B**) Facetted nanocrystals [38]

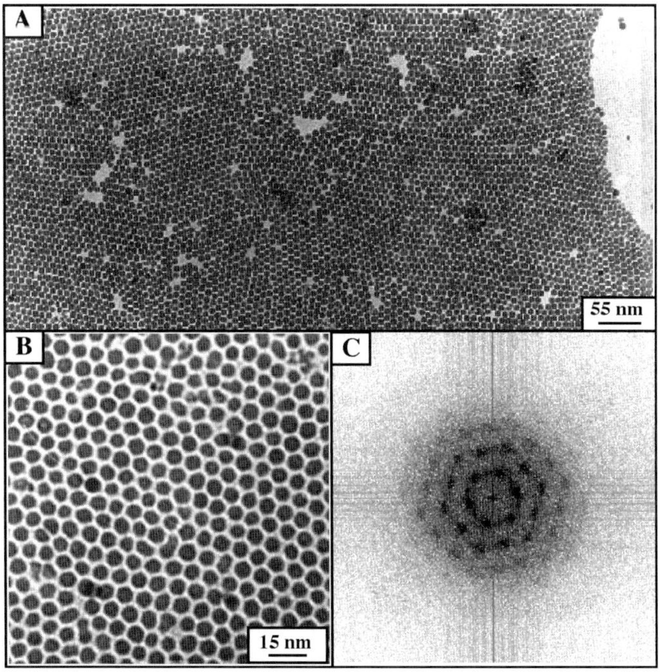

Fig. 22.16. (**A**) Monolayer of cobalt nanocrystals with diameter 7 nm. (**B**) Selected region of the monolayer. (**C**) Fourier transform of the region in (**B**) [39]

particles are parallel to one another (see Fig. 22.15) [38]. This affords optimal packing of the chains.

The transmission electron microscope (TEM) can only be used to characterise nanocrystals and their arrangements when these structures are not too thick, i.e., if they correspond to a stack of only a few particle monolayers. In order to observe thicker stacks of layers, X-ray diffraction is one of the

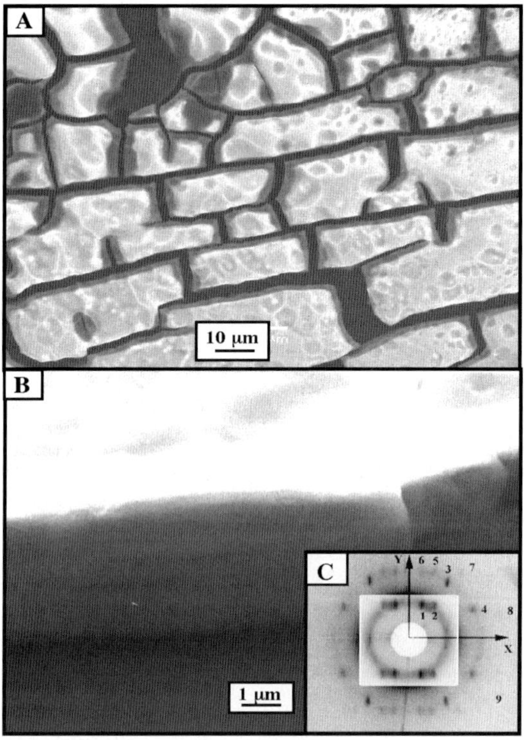

Fig. 22.17. (**A**) and (**B**) Scanning electron microscope (SEM) images of 3D lattices of cobalt nanocrystals. (**C**) X-ray diffraction pattern showing the characteristic points of a face-centered cubic arrangement [39]

best suited techniques. Wide-angle diffraction spectra are directly related to the atomic structure of the nanomaterial, while small-angle spectra are directly associated with the order of the nanomaterial in the 3D lattice. Very recently, 3D arrangements of cobalt nanocrystals with diameter 8 nm and narrow size distributions (around 10%) have been obtained, corresponding to stacks of more than 300 particle layers [39]. These particles, after deposition of a droplet of dilute solution on a cleaved graphite surface, spontaneously organise themselves into a hexagonal lattice, as shown in Fig. 22.16. The Fourier transform in Fig. 22.16C clearly shows the 3 orders of a series of 6 pairs of reflections arranged in a hexagon, characteristic of the hexagonal geometry. If one now uses the technique of deposition by immersing the substrate in a very concentrated colloidal solution (as described previously, which also gave supercrystals of silver nanocrystals), after evaporation of the solvent, scanning electron microscope (SEM) observations reveal the formation of large aggregates of nanomaterials (see Fig. 22.17). The crystal structure of the latter has been characterised by small-angle X-ray diffraction. The diffraction pattern in Fig. 22.17C shows a set of points characteristic of the different planes in an fcc lattice of cobalt nanomaterials.

22.3 Conclusion

In order to manipulate 2D and 3D lattices of nanocrystals, a detailed understanding of the phenomena producing them is clearly a prerequisite. Indeed, although self-organisation happens without direct intervention by the experimenter, it is nevertheless induced by forces such as van der Waals forces between particles, capillary forces due to solvent evaporation processes, steric repulsion between carbon chains coating the nanocrystals, and so on. In addition, we have described the methods generally used to deposit nanocrystals on a substrate: either deposition of droplets of the nanocrystal solution, or quenching of the substrate in this solution until the solvent has completely evaporated. These two methods lead to the formation of 2D and 3D nanocrystal lattices. However, the quenching method can produce 3D lattices corresponding to stacks of several hundred monolayers. Finally, the crystal structure of the 2D and 3D lattices depends on the ratio of the size of the nanocrystals and the length of the carbon chains coating them.

References

1. M.P. Pileni: Appl. Surf. Science **17**, 1 (2001)
2. S.A. Harfenist, Z.L. Wang, M.M. Alvarez, I. Vezmar, R.L. Whetten: J. Phys. Chem. **100**, 13904 (1996)
3. L. Motte, F. Billoudet, M.P. Pileni: J. Phys. Chem. **99**, 16425 (1995)
4. C.B. Murray, C.R. Kagan, M.G. Bawendi: Science, **270**, 1335 (1995)
5. A. Taleb, C. Petit, M.P. Pileni: Chem. Mater. **9**, 950 (1997)
6. S.A. Harfenist, Z.L. Wang, R.L. Whetten, I. Vezmar, M.M. Alvarez: Adv. Mater. **9**, 817 (1997)
7. S.W. Chang, G. Markovitch, J.R. Heath: J. Phys. Chem. B **102**, 6685 (1998)
8. B.A. Korgel, D. Fitzmaurice: Adv. Mater. **10**, 661 (1998)
9. P.C. Ohara, J.R. Heath, W.M. Gelbart: Angew. Chem. Int. Engl. **36**, 1078 (1997)
10. M.J. Hostetler, J.J. Stokes, R.W. Murray: Langmuir **12**, 3604 (1996)
11. J.S. Yin, Z.L. Wang: J. Phys. Chem. B **101**, 8979 (1997)
12. D.A. Handley: In: *Colloidal Gold: Principles, Methods and Applications*, ed. by M.A. Hayat, San Diego Academic (1989) 1–24
13. M.P. Pileni, T. Zemb, C. Petit: Chem. Phys. Lett. **118**, 414 (1985)
14. M.P. Pileni: Langmuir **13**, 3266 (1997)
15. A. Courty, I. Lisiecki, M.P. Pileni: J. Chem. Phys. **116**, 8074 (2002)
16. M. Brust, M. Walker, D. Bethell, D.J. Schiffrin, R. Whyman: J. Chem. Soc. Chem. Commun. 801 (1994)
17. J.E. Bowen Katari, V.L. Colvin, A.P. Alivisatos: J. Phys. Chem. **98**, 4109 (1994)

18. N. Herron, Y. Wang, M. Eddy, G.D. Stucky, D. Cox, K. Molker, T. Bein: J. Am. Chem. Soc. **11**, 530 (1989)
19. L. Motte, F. Billoudet, E. Lacaze, M.P. Pileni: Adv. Mater. **12**, 1018 (1996)
20. C.P. Collier, T. Vossmeyer, J.R. Heath: Annu. Rev. Phys. Chem. **49**, 371 (1998)
21. A. Taleb, C. Petit, M.P. Pileni: J. Chem. Phys. **102**, 2214 (1998)
22. S. Vaucher, J. Fielden, M. Li, A. Dujardin, S. Mann: Nanolett. **2** (3), 225 (2002)
23. D. Wyrwa, N. Beyer, G. Schmid: Nanolett. **2** (4), 419 (2002)
24. Y. Yin, B. Gatles, Y. Xia: J. Am. Chem. Soc. **123**, 8718 (2001)
25. A. Courty, C. Fermon, M.P. Pileni: Adv. Mater. **13**, 254 (2001)
26. A. Courty, O. Araspin, C. Fermon, M.P. Pileni: Langmuir **13**, 254 (2001)
27. P.G. de Gennes: Rev. Mod. Phys. **57**, 827 (1985)
28. A.W. Adamson, A.P. Guest: *Physical Chemistry of Surfaces*, 6th edn., Wiley Interscience (1997)
29. H.C. Hamaker: Physica **4**, 1058 (1937)
30. J.N. Israelachvili: *Intermolecular and Surface Forces*, 2nd edn., Academic Press, New York (1982)
31. L. Motte, E. Lacaze, M. Maillard, M.P. Pileni: Langmuir **16**, 3803 (2000)
32. M.P. Pileni, D. Ingert, I. Lisiecki, Y. Lalatone, A. Courty: Faraday Discussion **125**, 251 (2004)
33. B.A. Korgel, S. Fullman, S. Connolly, D. Fitzmaurice: J. Phys. Chem. B **102**, 8379 (1998)
34. M. Maillard, L. Motte, A.T. Ngo, M.P. Pileni: J. Phys. Chem. B **104**, 11871 (2000)
35. R.L. Whetten, M.N. Shafiyulin, J.T. Khoury, M.M. Alvarez, A. Wilkinson: Acc. Chem. Res. **32**, 397 (1999)
36. P.J. Thomas, G.U. Kulkarini, C.N.R. Rao: J. Phys. Chem. B **104**, 8138 (2000)
37. L. Motte, M.P. Pileni: J. Phys. Chem. B **102**, 4104 (1999)
38. Z.L. Wang: Adv. Mater. **10**, 13 (1998)
39. I. Lisiecki, P.A. Albouy, M.P. Pileni: Adv. Mater. **15**, 712 (2003)

23

Assemblies of Magnetic Nanoparticles

J. Richardi

In the last chapter, we saw how it is possible today to fabricate regular 2D and 3D lattices of magnetic nanoparticles [1–3]. The aim here is to study the specific properties of these systems that can be attributed to the magnetic properties of the particles. Such properties underlie several applications currently under development. We shall then discuss ferrofluids. Magnetic nanoparticles are deposited on a substrate by evaporation of ferrofluids. The properties of these fluids are essential for explaining the structures observed on nanometric and mesoscopic scales in the resulting deposits. Finally, we shall discuss the influence of a magnetic field on the deposition structures.

23.1 Magnetic Properties of Nanoparticle Assemblies

In a magnetic solid, there is competition between short-range exchange interactions and long-range dipole forces, which exist between the spins of the electrons in each atom. This competition favours a parallel orientation of nearby spins and an antiparallel alignment at longer range. For this reason, Weiss domains are observed to form in magnetic solids (see Fig. 23.1A). Within each domain, all atomic dipoles are aligned in the same direction. However, the dipoles in different domains are not in the same direction. But what happens if the solid is reduced to nanometric size? In general, the domains are larger, around a hundred nanometers, than the particles. Hence, particles measuring only around 10 nm are monodomain (see Fig. 23.1B).

For this reason, the dipole moment of a particle is calculated from the magnetisation M_s of the solid at saturation:

$$\mu = \mu_0 M_s V_p \ , \tag{23.1}$$

where M_s is the value of the magnetisation of the solid reached in a strong field if all the domains have the same dipole orientation. The magnetic permeability μ_0 of the vacuum is $4\pi 10^{-7}\,\mathrm{V\,s\,A^{-1}m^{-1}}$ and V_p is the volume of a particle

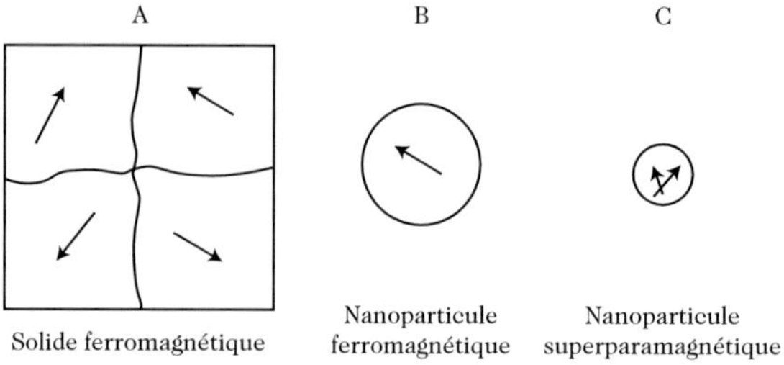

Fig. 23.1. Formation of magnetic domains in solid magnets and magnetic nanoparticles. The orientations of atomic spins in a domain are shown by an *arrow*. For superparamagnetic nanoparticles, the spins are blocked in one direction

calculated from $V_p = \pi/6d^3$, where d is the magnetic diameter, indicating the size of the magnetically active particle. In general, the magnetic diameter is smaller than the geometric size. For cobalt and magnetite (Fe_3O_4) nanoparticles with diameter 10 nm, then since $M_s(\text{Co}) = 14.0 \times 10^5$ A m^{-1}, $M_s(\text{Fe}_3\text{O}_4) = 4.46 \times 10^5$ A m^{-1}, we can calculate the dipole moments 9.2×10^{-25} V s m and 2.9×10^{-25} V s m, respectively, using (23.1).

The fact that nanoparticles comprise single domains permits an application to the magnetic storage of information. Suppose that each magnetic nanoparticle can be addressed as a magnetic bit in which the orientation of the magnetic dipole contains the data. It can be estimated that the data storage density of assemblies of these nanoparticles will be ten times greater than can be achieved by systems based on CoCr alloys, as currently used in industry. However, for such an application, there are several problems to be overcome:

- 2D arrangements of nanoparticles must be highly regular over a scale of a hundred or so nanometers.
- The orientation of the magnetic dipoles must be thermally stable, which is not necessarily the case for magnetic nanoparticles.
- It must be possible to read and write the information on such a length scale.

Regarding the first two problems, progress has been made recently. The last chapter described the necessary conditions for achieving a regular arrangement. In particular, the particles must have a narrow size distribution (around 10%) for regular assembly over a scale of 100 nm. The size distribution depends sensitively on the way the nanoparticles are synthesised. Magnetic particles made from Co, CoO, FePt, Fe_3O_4, Fe_2O_3 and Ni can be synthesised by reducing a metal salt in solution in the presence of a surfactant, or by using micelle systems [1, 4]. Another way of synthesising magnetic nanoparticles is pyrolysis of metal (Co, Fe, Ni, CoPt, etc.) carbonyls in the presence of a surfactant at high temperatures (300°C) [1]. Using a mixture of two surfactants, the diameter and shape of the nanoparticles can be controlled and the size

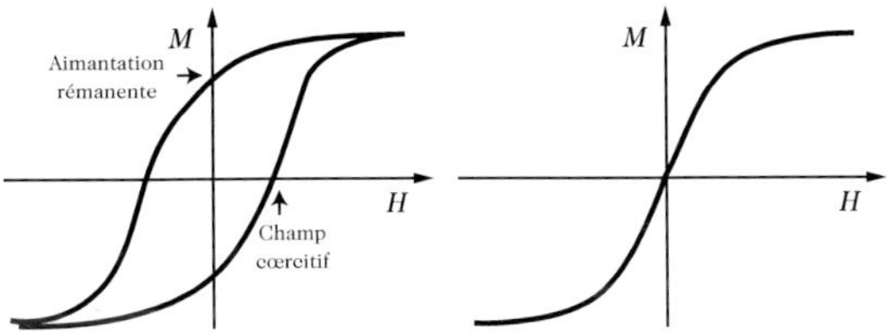

Fig. 23.2. Magnetisation curves of assemblies of (*left*) ferromagnetic and (*right*) superparamagnetic particles. The remanent magnetisation and coercive field disappear for superparamagnetic particles

distribution narrowed [5]. Non-spherical cobalt particles can be produced in this way [5].

A narrow size distribution is obtained by applying the method of size-selective precipitation. For magnetic nanoparticles, a magnetic field can also be used to precipitate out the larger nanoparticles.

The second problem concerns thermal stability of the orientations of the magnetic moments and is related to the small size of the particles. The magnetic behaviour of a magnet is characterised by a hysteresis curve (see Fig. 23.2 left). The curve shows that a minimal field H_c, called the coercive field, is required to change the orientation of the magnetisation. The zero-field magnetisation, called the remanent magnetisation, is not itself zero. The magnetic dipoles of the atoms are coupled by the exchange energy. Reducing the size of the magnetic nanocrystals decreases the number of coupled dipoles resisting spontaneous orientation. The energy required to reorient the magnetic moments of a particle is given by

$$\Delta E = KV_p , \tag{23.2}$$

where K is the anisotropy constant of the material, which depends on the crystal structure, the shape and the surface area of the particles. If the volume V_p of the particles is reduced or if the temperature is increased, ΔE is close to the thermal energy $k_B T$ (4.1×10^{-21} J at 25°C) and the magnetic moment is no longer fixed. For this reason, the hysteresis disappears (see Fig. 23.2 right). The particles are then said to be superparamagnetic, rather than ferromagnetic. All nanoparticles are ferromagnetic at low temperatures. Increasing the temperature reduces the hysteresis and in the end, at the blocking temperature, the particles become superparamagnetic. For example, cobalt has a rather low anisotropy constant $K(\mathrm{Co}) = 4.1 \times 10^5\ \mathrm{J\,m^{-3}}$. For this reason, cobalt nanoparticles, usually 6 nm in size, are superparamagnetic at room temperature. Hence, assemblies of these nanocrystals cannot be used to store data, because the oriented magnetic moments will very quickly become disoriented due to thermal motions.

One way of making ferromagnetic nanoparticle assemblies is to choose a material with very high anisotropy constant, e.g., platinum iron (FePt) with

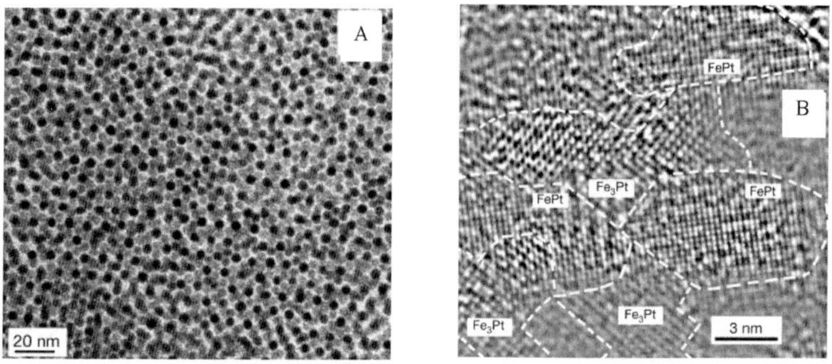

Fig. 23.3. Transmission electron microscope images showing binary assemblies of nanoparticles. (**A**) Fe_3O_4 (particle diameter 4 nm):$Fe_{58}Pt_{42}$ (4 nm) before annealing. Black particles contain platinum. (**B**) The same assembly after annealing, at higher magnification. Taken from [9] with the kind permission of Macmillan Publishers Ltd © 2002

a face-centered tetragonal (fct) crystal structure and $K = 7 \times 10^6\,\mathrm{J\,m^{-3}}$. To this end, 3D assemblies of FePt nanocrystals have been produced [6]. However, the nanocrystals have face-centered cubic (fcc) structure with a low anisotropy constant. But by annealing at 560°C, the crystal structure is transformed into a face-centered tetragonal structure. The assembly of FePt nanocrystals is ferromagnetic at room temperature. The first write and read tests demonstrate the ability of these systems to store data. However, today's devices cannot control the orientation of the moment of each nanoparticle, because the size of one bit is of the order of 50 nm. In particular, the thickness of the assembly film must be reduced from about ten layers to a single layer. This must be achieved without losing the ferromagnetic property. Moreover, particle coalescence during annealing must be avoided.

A great deal of research has been carried out into the effect on magnetic properties of assembling nanoparticles [7]. The blocking temperature T_B can rise if the dipole coupling between nanoparticles increases. Hence, for Ni nanocrystals, T_B is observed to increase by several tens of K when the nanoparticles are organised into a 3D lattice [1]. The blocking temperature T_B also increases if the distance between the particles is reduced by annealing and partial desorption of the surfactant.

Self-assemblies of magnetic nanoparticles have a multitude of specific properties that open the way to potential applications. The effect of a magnetic field on electrical conductivity in such systems has been much studied [8]. Assemblies of magnetic nanoparticles also provide a new way of preparing exchange-coupled magnets. These magnets are nanostructured materials composed of magnetically soft and hard phases that interact by magnetic exchange coupling. (Hard magnetic materials are characterised by high values of the remanent magnetisation and coercive field as compared with soft materials.)

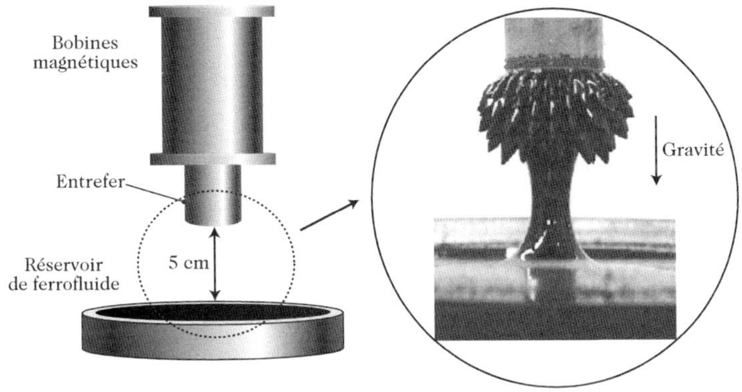

Fig. 23.4. The magnetic force produced by the magnetic coils is strong enough to levitate the ferrofluid. Note also that the ferrofluid close to the magnet is covered with spikes. Taken from [12]

Magnetic materials characterised by exchange coupling are of great interest for applications to permanent magnets, because they have a high magnetic energy. To make these, Zeng et al. [9] mixed dispersions of Fe_3O_4 and FePt nanoparticles and thereby obtained, after deposition, assemblies in which the two types of nanoparticle are randomly distributed (see Fig. 23.3A). By annealing at 650°C, the Fe_3O_4 is reduced to Fe. The crystal structure of the FePt changes from fcc to fct, which is magnetically hard. During annealing, the surfactant desorbs and the nanoparticles coalesce. This allows diffusion between Fe and FePt particles. Hence the iron becomes Fe_3Pt, which is a magnetically soft material (see Fig. 23.3B). Finally, a nanocomposite of FePt and Fe_3Pt is obtained, characterised by exchange coupling. This material has a very high remanent magnetisation and coercive field, giving a high value for the magnetic energy. Magnets with a huge range of applications are thereby produced.

23.2 Structure of Magnetic Nanoparticle Assemblies Deposited Without Field

A very important question concerns the effect of magnetic interactions between particles on the way they arrange themselves. To understand this better, we introduce the idea of a ferrofluid [10, 11]. Magnetic nanoparticles in a solution constitute a ferrofluid, i.e., a magnetic liquid. Deposits of magnetic nanoparticles are very often obtained by evaporation of a ferrofluid. For this reason, structures observed after deposition often correspond to those already existing in the ferrofluid. The fundamentally important point about ferrofluids is that they exhibit a unique association of liquid properties arising from the solvent and magnetic properties arising from the nanoparticles. Ferrofluids

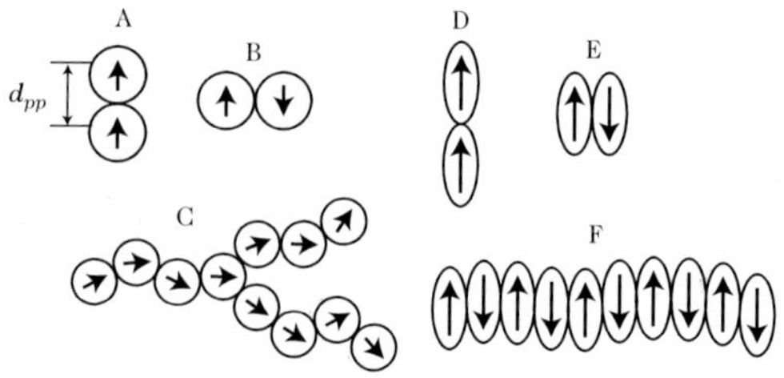

Fig. 23.5. Configurations of high-energy magnetic particles. For spherical particles: (**A**) head-to-tail, (**B**) side-by-side, (**C**) dipolar chain. For long particles: (**D**) head-to-tail, (**E**) side-by-side, (**F**) dipolar band

have a whole range of unique properties. For example, under the effect of an external magnetic field (in the neighbourhood of a magnet), ferrofluids change shape, adopting all sorts of configurations (see Fig. 23.4). The possibility of manipulating these liquids by means of a magnetic field suggests a large number of applications. For example, ferrofluids can be found in magnetically suspended airtight joints, used in hard disks and clean rooms.

For a long time, the structures adopted by nanoparticles in these liquids were not well understood, but thirty years ago, de Gennes and Pincus predicted that magnetic nanoparticles would form flexible chains. They can be viewed as 'living polymers', because the 'bonds' between the particles can be broken by thermal vibrations. The formation of particle chains can be explained by considering the potential between two dipoles $\boldsymbol{\mu}_i$ and $\boldsymbol{\mu}_j$, viz.,

$$u_{\mathrm{dd}} = \frac{1}{4\pi\mu_0}\left[\frac{\boldsymbol{\mu}_i\cdot\boldsymbol{\mu}_j}{r_{ij}^3} - \frac{3(\boldsymbol{\mu}_i\cdot\boldsymbol{r}_{ij})(\boldsymbol{\mu}_j\cdot\boldsymbol{r}_{ij})}{r_{ij}^5}\right], \tag{23.3}$$

where $\boldsymbol{r}_{ij}$ is the separation vector between the two nanoparticles.

This potential has two minima corresponding to head-to-tail and side-by-side configurations, where the dipoles have parallel and antiparallel orientations, respectively, as shown in Figs. 23.5A and B. For spherical particles, the head-to-tail configuration is the most stable. Hence, each particle can attract two particles with these two poles and chains can form, as shown in Fig. 23.5C.

To observe these chains, the dipole attraction must be significantly greater than the thermal energy. The ratio of the dipole energy to the thermal energy is given by

$$\lambda = \frac{\mu^2}{4\pi\mu_0 d_{\mathrm{pp}}^3 kT}, \tag{23.4}$$

where d_{pp} is the distance between the centres of the two nanoparticles at contact (see Fig. 23.5A). Simulations show that the ratio λ must be greater than 4.0 in order to observe chain formation. The material composing the nanoparticles, in the ferrofluids, is often a ferrite such as magnetite Fe_3O_4, or

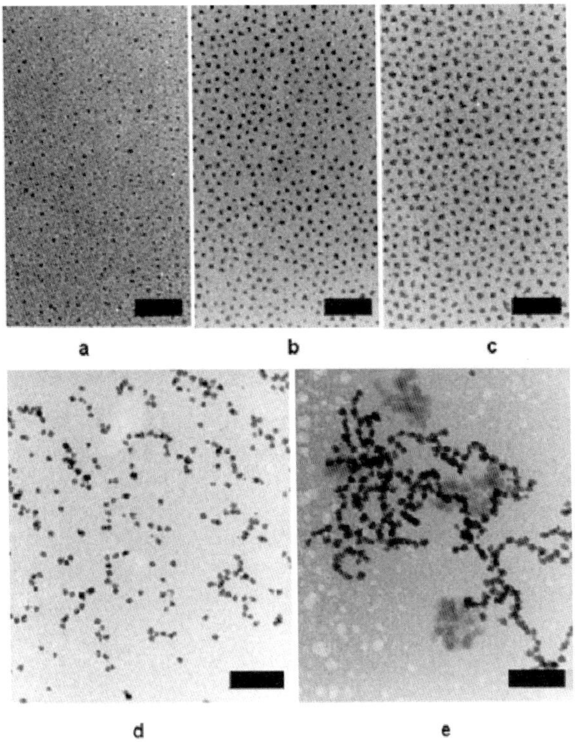

Fig. 23.6. Transmission electron microscope images of vitrified films of iron nanoparticle dispersions for the following magnetic particle diameters: (**a**) 2.6 nm, (**b**) 8.2 nm, (**c**) 10.2 nm, (**d**) 12 nm, (**e**) about 16 nm. *White objects* in (**e**) are artefacts. *Scale bars* 100 nm. Chains are observed to form from a diameter of 12 nm. Taken from [13]

maghemite γ-Fe_2O_3, with $M_s(Fe_3O_4) = 4.46 \times 10^5\,A\,m^{-1}$ and $M_s(\gamma\text{-}Fe_2O_3) = 3.73 \times 10^5\,A\,m^{-1}$. For this type of material and a typical contact distance of 10 nm, we find $\lambda = 1.3$, using (23.1) and (23.4). Chain formation should not therefore be observed for these nanoparticles. This is in good agreement with experiment [13]. However, chain formation is observed for iron nanoparticles, because their magnetisation M_s at saturation is high ($1.5 \times 10^6\,A\,m^{-1}$) and hence their dipole moments are large. Figure 23.6 shows transmission electron microscope (TEM) images of vitrified films of a ferrofluid containing iron nanoparticles. From 12 nm, chain formation is observed (Fig. 23.6d) [13]. The chains do not attract one another to form larger aggregates, in good agreement with simulations, which do not predict condensation in this type of system [14].

Similar structures are observed even after total evaporation of the solvent. For deposits of cobalt nanoparticles with a high value of M_s, viz., $M_s(Co) = 14.0 \times 10^5\,A\,m^{-1}$, chains are observed to form for a particle diameter of 16 nm (see Fig. 23.7) [5].

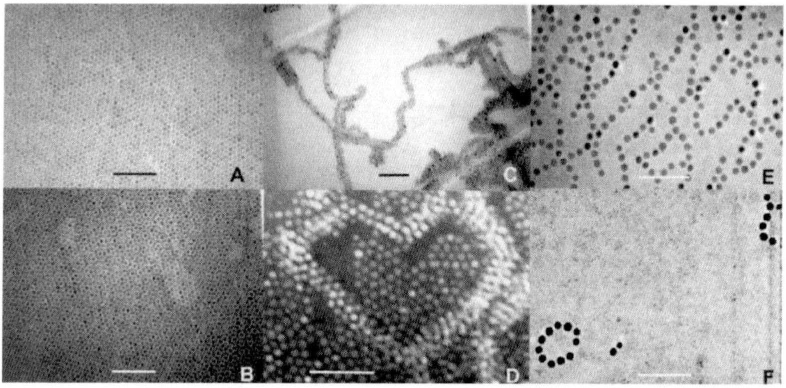

Fig. 23.7. Transmission electron microscope images of a monolayer (**A**), (**C**), (**E**), and a bilayer (**B**), (**D**), (**F**), of cobalt nanocrystals. Particle diameters: (**A**) and (**B**) 10 nm, (**C**) and (**D**) 12 nm, (**E**) and (**F**) 16 nm. *Scale bars* 100 nm. Taken from [5] © 2001 AAAS

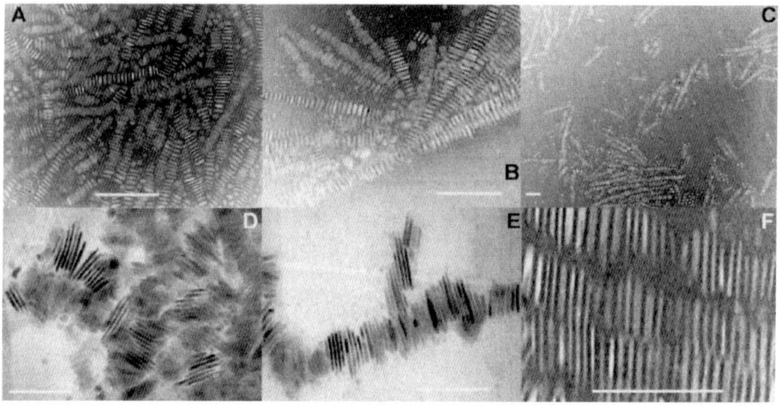

Fig. 23.8. Transmission electron microscope images of non-spherical cobalt particles. Due to dipole interactions, bands of particles are observed to form. *Scale bars* 100 nm. Taken from [5] © 2001 AAAS

Up to now we have only been concerned with spherical magnetic particles. How do the structures change for elongated or disk-shaped particles? We explained earlier that the dipole potential in (23.3) has two minima, corresponding to head-to-tail and side-by-side configurations (see Fig. 23.5). In general, for cigar-shaped magnetic particles, the dipole is in the direction of the longer axis. In the case of cigar-shaped or disk-shaped particles, the distance between the particles is smaller for the side-by-side configuration (see Figs. 23.5D and E). For this reason, the energy of this configuration increases compared with the head-to-tail configuration and it becomes the most stable

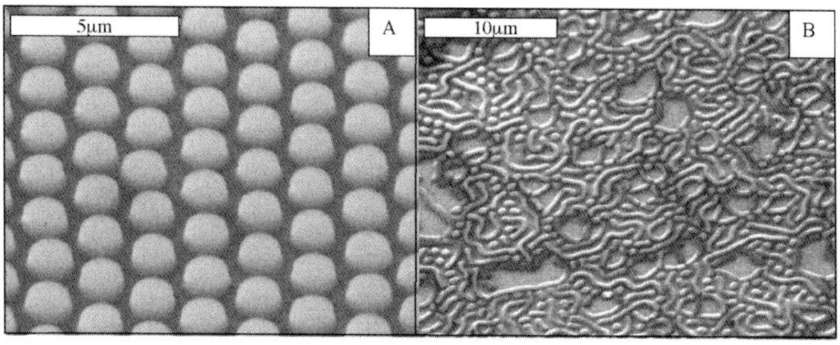

Fig. 23.9. SEM images of two magnetic mesostructures formed from cobalt nanocrystals by evaporating a ferrofluid under a magnetic field normal to the substrate. (**A**) Weak field. (**B**) Strong field. Taken from [15]

configurations. Bands are therefore predicted to form (see Fig. 23.5F), as can be seen experimentally for non-spherical cobalt particles (sec Fig. 23.8).

23.3 Structure of Magnetic Nanoparticle Assemblies Deposited with Field

Since the particles are magnetic, a magnetic field is expected to influence the structure of deposits. A field can be applied perpendicular or parallel to the substrate. We consider first the case of a perpendicular field.

23.3.1 Perpendicular Field

A solution of magnetic cobalt nanoparticles is evaporated while applying a magnetic field perpendicular to the substrate. If the substrate is examined by scanning electron microscope (SEM) after evaporation of the solvent, mesostructures made up of cobalt nanocrystals are observed (see Fig. 23.9) [15]. The strength of the applied field plays an important role in determining this mesostructure. For weak fields, a hexagonal superlattice of micrometric dots forms, as shown in Fig. 23.9A, whereas for strong fields, labyrinths are obtained, as shown in Fig. 23.9B. Similar structures are observed when a magnetic fluid is confined between two glass plates and subjected to a magnetic field normal to the thin film of fluid (Fig. 23.10) [9,16].

These structures are of fundamental importance because they arise in many systems. They are observed in magnetic thin films, superconductors in magnetic fields, and chemical reaction systems (see Fig. 23.10) [16]. These structures are produced by competitition between a short-range attractive force and a long-range repulsive force. In the case of ferrofluid films, the short-range attractive forces are due to the interface tension between the magnetic fluid and its surroundings. The interface energy is given by

$$F_{\mathrm{i}} = \sigma A \,, \tag{23.5}$$

Fig. 23.10. Systems in which labyrinthine structures arise due to competition between a short-range attractive force and a long-range repulsive force. Taken from [16] © 1995 AAAS

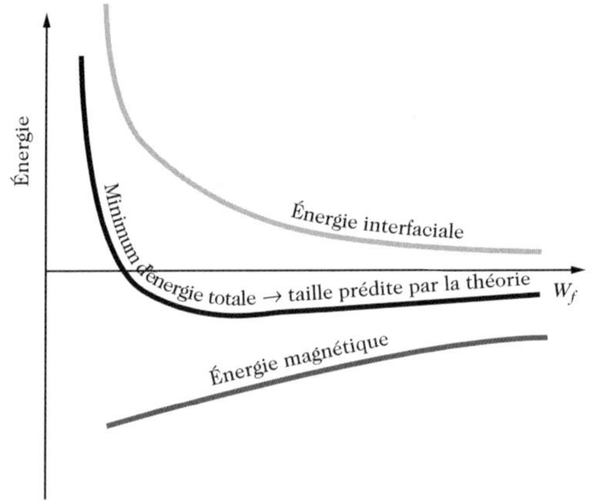

Fig. 23.11. Evolution of interface, magnetic and total energy with the width w_f of the labyrinth fingers

where σ is the interface tension and A is the area of the ferrofluid. F_i increases with A when the structures form, and the width w_f of the labyrinth fingers decreases (see Fig. 23.11).

The long-range repulsive force is related to the magnetic energy:

$$F_\mathrm{m} = -\frac{\mu_0}{2}\langle M\rangle H_0 L A' \,, \tag{23.6}$$

where LA' is the volume of ferrofluid, with L the height of the structures and A' the area of substrate covered by the ferrofluid. $\langle M\rangle$ is the average magnetisation in the structure.

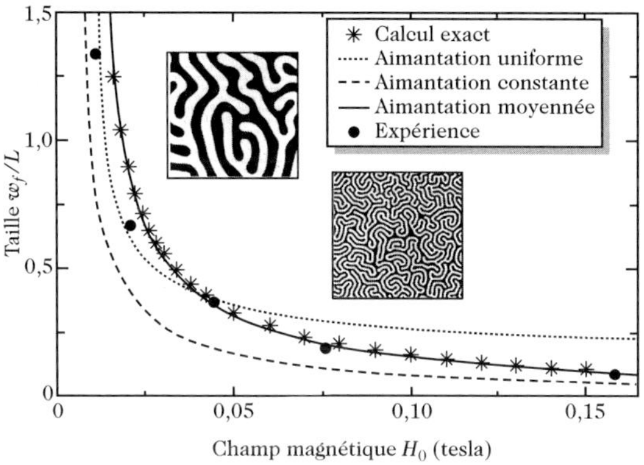

Fig. 23.12. Dependence of the thickness of labyrinth bands on the magnetic field. Theoretical predictions based on certain approximations (uniform, constant or av eraged magnetisation) and an exact calculation are compared with experimental results [17]

The magnetisation M is due to the total field, which is smaller than the applied field H_0 because of the demagnetising field H_d produced by the magnetisation of the ferrofluid:

$$\boldsymbol{M} = \chi(\boldsymbol{H}_0 + \boldsymbol{H}_\mathrm{d}) , \tag{23.7}$$

where χ is the magnetic susceptibility. When structures form, the demagnetising field is reduced. For this reason, the magnetisation increases and the magnetic energy becomes more negative (see Fig. 23.11). The sum of F_i and F_m gives the total energy

$$f_\mathrm{tot} = \frac{F_\mathrm{i} + F_\mathrm{m}}{A'} = -\frac{\mu_0}{2} H_0 L + \sigma \frac{A}{A'} . \tag{23.8}$$

When structures form, F_i is reduced, but F_m increases.

Figure 23.11 shows that the total energy has a minimum, which corresponds to the sizes of structures predicted by the theory. One can thus explain the formation of structures and predict their sizes (see Fig. 23.12). The theoretical predictions are very sensitive to the kind of approximations sometimes used to calculate structure sizes. The exact calculation is in good agreement with experiment [17].

The structures shown in Fig. 23.9 have micrometric sizes. Those observed for the ferrofluids in Fig. 23.10 measure several millimetres. This kind of size reduction can be explained by the lower interface tension in the micrometric system. To obtain structures with the same aspect ratio but on a scale a thousand times smaller, σ must change like the height L of the mesostructures according to (23.8). Hence the reduction in the height of the mesostructures by a factor of 1 000 can be explained by a similar reduction in the interface

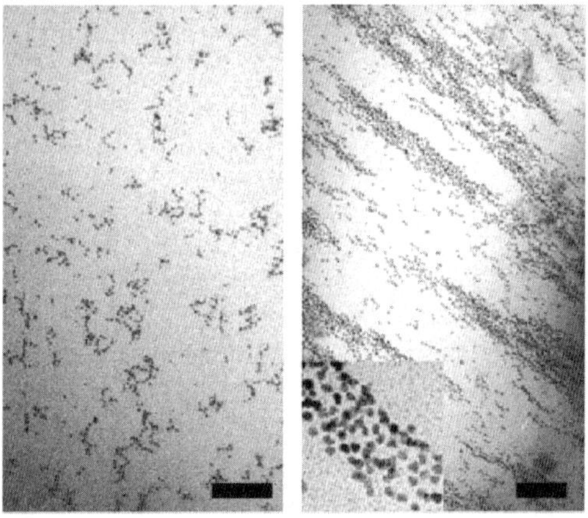

(a) (b)

Fig. 23.13. Transmission electron microscope image of a vitrified ferrofluid film containing iron particles (diameter 12 nm). (**a**) Without field. (**b**) In a strong field. The field induces parallel arrangements of magnetic nanoparticles in the direction of the field (bottom right to top left). *Scale bar* 200 nm. Taken from [13]

tension. As can be seen in Figs. 23.9A and B, experimental observations reveal hexagonal and labyrinthine structures. Theory shows that these structures have very similar energies, despite the radically different morphologies [17]. This explains why the two structures are indeed observed experimentally. However, the formation of these hexagonal and labyrinthine structures is very complex and the underlying dynamics of the process has been the subject of much research [18].

23.3.2 Parallel Field

When a magnetic field is applied parallel to the substrate, cigar-shaped aggregates are produced with the long axis in the direction of the field. Such aggregates have been found for vitrified ferrofluid films (see Fig. 23.13b) [13] and for deposited magnetic nanoparticles (see Figs. 23.14b and c) [19, 20].

The formation of these structures cannot be explained by the theory used for the perpendicular case because the demagnetising field is zero now. Hence the magnetisation and magnetic energy do not change, according to this theory, when the structure forms For particles with a large magnetic moment, structure formation is explained by an orientation of the chains already present in zero field. In contrast with the zero field case, the chains align themselves with the field. The chains attract one another to form bigger aggregates. There are several hypotheses regarding the origin of this attraction between chains, but the details remain rather poorly understood. It is highly surprising to observe chains even for deposition in a field of weakly magnetic nanoparticles

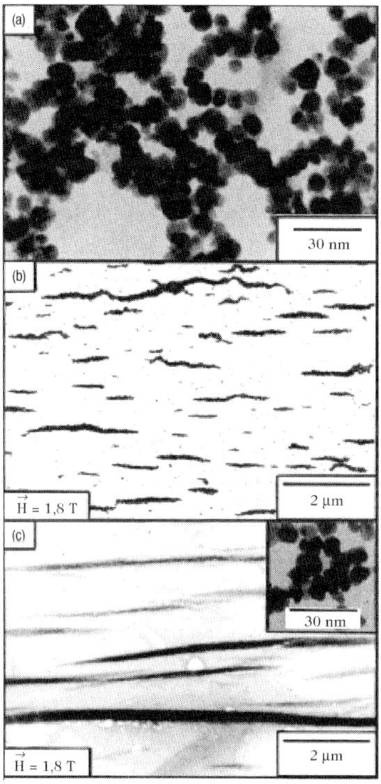

Fig. 23.14. Transmission electron microscope images of deposited cobalt ferrite nanoparticles. (**a**) Without field. (**b**) With field from right to left when only a few particles have been deposited. (**c**) With field from right to left when more particles have been deposited. Taken from [19,20] © 2001 American Chemical Society

for which chain formation is not observed when there is no field. The origin of these structures is probably related to other types of interaction.

To sum up, despite the simple types of interaction existing between magnetic nanoparticles, they are capable of assembling into an astonishing range of different structures. It will be essential to understand and control such structures if they are to be used in applications as different as magnetic data storage or high-performance magnets.

References

1. C.B. Murray, S. Sun, H. Doyle, T. Betley: Monodisperse 3D transition-metal (Co, Ni, Fe) nanoparticles and their assembly into nanocrystal superlattices, MRS Bulletin 985 (2001)
2. C. Petit, A. Taleb, M.P. Pileni: Self-organization of magnetic nanosized cobalt particles, Adv. Mater. **10**, 259 (1998)
3. I. Lisiecki, M.P. Pileni: Face-centred cubic 'supracrystals' of cobalt nanocrystals, Adv. Mater. **15**, 712 (2003)
4. M.P. Pileni: Magnetic fluids: Fabrication, magnetic properties, and organization of nanocrystals, Adv. Funct. Mat. **11**, 323 (2001)
5. V.F. Puntes, K.M. Krishnan, A.P. Alivisatos: Colloidal nanocrystal shape and size control: The case of cobalt, Science **291**, 2115 (2001)

6. S. Sun, C.B. Murray, D. Weller, L. Folks, A. Moser: Monodisperse FePt nanoparticles and ferromagnetic FePt nanocrystal superlattices, Science **287**, 1989 (2000)
7. V. Russier, C. Petit, J. Legrand, M.P. Pileni: Hysteresis of cobalt nanoparticles organized in a two-dimensional network: Dipolar interaction effects, Appl. Surf. Sci. **164**, 193 (2000)
8. C.T. Black, C.B. Murray, R.L. Sandstrom, S. Sun: Spin-dependent tunneling in self-assembled cobalt-nanocrystal superlattices, Science **290**, 1131 (2000)
9. H. Zeng, J. Li, J.P. Liu, Z.L. Wang, S. Sun: Exchange-coupled nanocomposite magnets by nanoparticle self-assembly, Nature **420**, 395 (2002)
10. R.F. Rosensweig: *Ferrohydrodynamics*, Cambridge University Press, Cambridge (1985)
11. J.-C. Bacri, R. Perzynski, D. Salin: Les liquides magnètiques, La Recherche **192**, 1152 (1987)
12. S. Odenbach: Magnetic fluids – Suspensions of magnetic dipoles and their control, J. Phys.: Cond. Matt. **15**, 1497 (2003)
13. K. Butter, P.H.H. Bomans, P.M. Frederik, G.J. Vroege, A.P. Philipse: Direct observation of dipolar chains in iron ferrofluids by cryogenic electron microscopy, Nature Materials **2**, 88 (2003)
14. P.I.C. Teixeira, J.M. Tavares, M.M. Telo da Gamma: The effect of dipolar forces on the structure and thermodynamics of classical fluids, J. Phys. Cond. Matt. **12**, R411 (2000)
15. J. Legrand, A.T. Ngo, C. Petit, M.P. Pileni: Domain shapes and superlattices made of cobalt nanocrystals, Adv. Mater. **13**, 58 (2001)
16. M. Seul, D. Andelman: Domain shapes and patterns: The phenomenology of modulated phases, Science **267**, 476 (1995)
17. J. Richardi, D. Ingert, M.P. Pileni: Theoretical study of field-induced pattern formation in magnetic liquids, Phys. Rev. E **66**, 046306 (2002)
18. A.J. Dickstein, S. Erramilli, R.E. Goldstein, D.P. Jackson, S.A. Langer: Labyrinthine pattern formation in magnetic fluids, Science **261**, 1012 (1993)
19. A.T. Ngo, M.P. Pileni: Nanoparticles of cobalt ferrite: Influence of the applied field on the organization of the nanocrystals on a substrate and on their magnetic properties, Adv. Mater. **12**, 276 (2000)
20. A.T. Ngo, M.P. Pileni: Assemblies of ferrite nanocrystals: Partial orientation of the easy magnetic axes, J. Phys. Chem. **105**, 53 (2001)

24

Nanostructured Coatings

J.-P. Rivière

In many branches of technology where surfaces are playing a growing role, the use of coatings is often the only way to provide surfaces with specific functional properties. For example, the austenitic stainless steels or titanium alloys exhibit poor resistance to wear and low hardness values, which limits the field of applications. The idea then is to develop new solutions which would improve the mechanical performance and durability of objects used in contact and subjected to mechanical forces in hostile gaseous or liquid environments. Hard coatings are generally much sought after to enhance the resistance to wear and corrosion. They are of particular importance because they constitute a class of protective coatings which is already widely used on an industrial scale to improve the hardness and lifetime of cutting tools.

With new vapour deposition processes, treatments and coatings can be combined to produce films with composition gradients or periodic multilayers, for example, associating a hard layer with a more elastic layer, whereupon the structure and properties of coatings can be controlled on a nanometric scale. Films with very high hardness (40–80 GPa) and resistance to wear and corrosion can thereby be fabricated.

There are several different architectures for nanostructured coatings, but the two most important are multilayers made from periodic stacks of bilayers A/B with nanometric thicknesses and nanocomposites comprising two phases, one nanocrystalline and the other usually amorphous. This is shown schematically in Fig. 24.1.

The very high hardness values of this new class of materials have led to a proposed classification: hard for hardnesses in the range 20–40 GPa, superhard for values in the range 40–80 GPa, and ultrahard for values in excess of 80 GPa. These new mechanical properties (see Chap. 8) are mainly due to a size reduction effect in the distribution of phases, i.e., the size of the crystallites or the thickness of the films.

Before describing the vapour deposition techniques used to make these coatings, let us first spell out the main ideas and methodology, giving a few

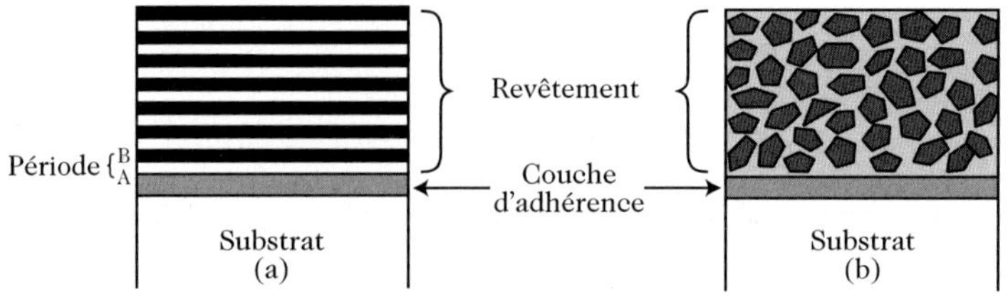

Fig. 24.1. Schematic representation of the structure of the two main types of nanostructured coatings. (**a**) Multilayer with nanometric period A/B. (**b**) Nanocomposite coating formed of small crystallised grains dispersed in an amorphous matrix. For each type of coating, an intermediate adherence layer is often required

examples of coatings in each of the two categories, multilayers and nanocomposites.

24.1 Methodology for Making Superhard Nanostructured Coatings

The hardness of a material reflects its resistance to plastic deformation, which is controlled by the formation and propagation of dislocations under the effects of an applied stress (see Chap. 8). Until the discovery of the new superhard nanostructured materials, the main method for obtaining hard materials was to increase resistance to the motion of dislocations. This can be done by starting with a material like TiN which is already hard and adding a ternary element such as C, Al, or Zr [1–3]. The other solution for obtaining high resistance to plastic deformation is simply to prevent dislocations from forming in the first place, which is done by reducing film thickness or grain size to a value below about 10 nm [4]. This is the first condition to be satisfied when making superhard nanostructured coatings. There are then other more specific criteria for each type of coating (multilayer or nanocomposite), to be specified later. In addition, if these coatings are to display good mechanical performance and durability, a high level of thermal stability is required for most applications. This is why at least one of the phases is composed of a transition metal nitride.

24.1.1 Multilayers with Nanometric Period

These multilayers are characterised by very high interface area/volume ratios. For example, considering a volume of area 1 cm^2 and thickness 1 μm, a multilayer of period $\Lambda = 2$ nm yields an interface area of the order of 1 000 cm^2. In the same volume, a microcrystalline coating with grain size 1 μm only yields an interface area of the order of 10^{-3}–10^{-4} cm^2. It is easy to see why interface effects dominate the structure and properties of nanostructured coatings. The

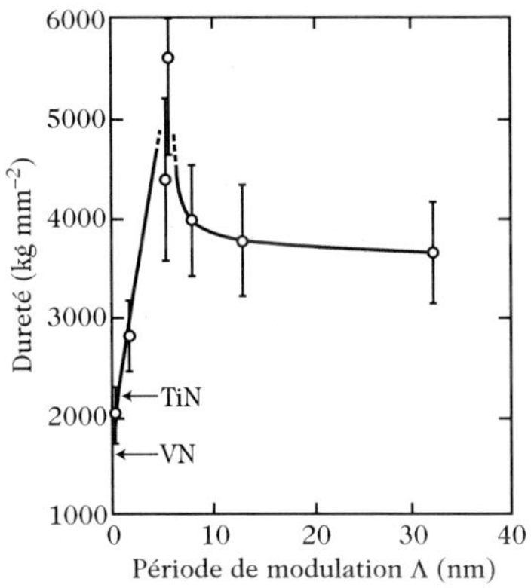

Fig. 24.2. Hardness as a function of the period Λ for TiN/VN multilayers epitaxied in the [100] direction. The TiN and VN layers have the same thickness [5]

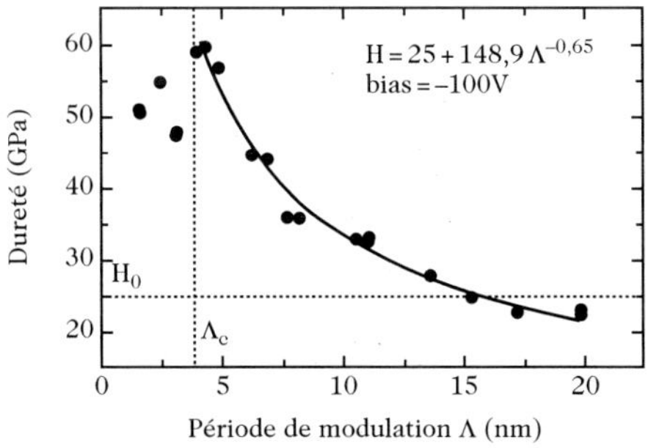

Fig. 24.3. Hardness as a function of the modulation Λ for $Ti_{0.4}Al_{0.6}N$/Mo multilayers comprising 250 bilayers, fabricated by magnetron sputtering with substrate bias $-100\,V$. For $\Lambda_c \geq 4\,nm$, the *continuous curve* represents the fitting to a Hall–Petch law [9]: $H = H_0 + k\Lambda^{-p}$, where k is a constant reflecting the effect of interface hardening and p is an exponent in the range 0.3–0.7

most important parameter controlling the increase in hardness of the stack in an A/B/A/B... multilayer is the period Λ of the bilayer A/B. The main systems studied are either nitride/nitride or metal/nitride multilayers.

The first experimental results showing the effects on the hardness of reducing the bilayer thickness were obtained for the TiN/VN system made up of epitaxied layers of material of the same structure but with different shear moduli [5]. Figure 24.2 shows that, for a bilayer thickness of about 5 nm, the hardness goes through a maximum with value about 2.5 times greater than would be calculated using a simple mixing law. When the TiN and VN layers are not epitaxied but have polycrystalline form, a very similar increase in the hardness is obtained. This is very important for practical applications.

This increased hardness effect has been observed for many other nitride/nitride multilayer systems [6, 7], but also for nitride/metal systems,

Table 24.1. Comparison between properties of multilayer and monolithic coatings [12]. The coefficient of friction μ is defined by $\mu = F_T/F_N$, where F_N is the normal applied force and F_T is the tangential force at contact. The coefficient of wear k is defined by $k = V/F_N S$, where V is the volume of matter removed, F_N is the normal force, and S is the distance moved by one surface relative to the other

Coating	Thickness [μm]	Period [nm]	Hardness [GPa]	Coefficient of friction μ (against Al_2O_3)	Coefficient of wear k [m^3/Nm]
TiAlCrN/TiAlYN	3–4	1.7	36	0.65	2.5×10^{-16}
TiAlN/VN	3–4	3.5	56	0.4	1.3×10^{-17}
CrN/NbN	3–4	3.5	48	0.65	2.5×10^{-15}
Cr/C	1.6	1.5	27	0.17	1.5×10^{-17}
TiN	3–4		28	0.55	7.1×10^{-15}
CrN	3–4		25	0.5	9.0×10^{-15}

where the materials have different crystal structures [8, 9]. Figure 24.3 shows an example of the dependence of the hardness on the modulation period for TiAlN/Mo multilayers.

The metal/nitride systems are particularly useful in many applications, because they combine the elastic properties of the metal with the hardness of the nitride. They thus represent a promising class of coatings for improving surface mechanical properties. Moreover, an important particularity of these systems is their remarkable thermal stability when the metal and nitride constituents are immiscible, as happens for TiAlN/Mo.

Different types of multilayer coatings have been developed [10–12] and can be produced on an industrial scale by magnetron sputtering processes (see Sect. 24.2.3). Table 24.1 gives the main characteristics of several multilayer coatings used in specific applications for high-speed dry cutting or machining tools. These characteristics are compared with those of standard TiN or CrN monolithic coatings.

The first two coatings, TiAlCrN/TiAlYN and TiAlN/VN, were developed for high-speed dry machining (high temperature), and the first in particular for high-temperature machining of titanium alloys. CrN/NbN is a protective coating against corrosion and Cr/C is a coating with very low coefficients of friction and wear.

24.1.2 Nanocomposites

In multilayers, the size reduction effect only operates in one direction, but the concept of confinement can be generalised to three dimensions by synthesising a coating made from small crystallites with sizes below 10 nm, surrounded by a second phase (nanocrystalline or amorphous), forming an intergranular phase with thickness of the order of or less than the nanometer [13, 14]. This type of nanocomposite coating offers far more possibilities for controlling the structure (size, density, and composition of the different phases), and hence the mechanical properties, than multilayers. Moreover, the processes for making them are less sophisticated to implement, an important feature in industrial applications where the parts to be made may have complex geometric shapes.

Simple rules and constitutive laws underlie the fabrication of novel nanocomposite coatings [15]:

- It is hard for dislocations to form in a crystal of nanometric size.
- Sources of dislocations can only be effective if the grain size is less than 10 nm.
- Even if they do manage to form, dislocations are unable to cross an amorphous intergranular region.
- It is hard for a crack to propagate if there is a well-defined interface with very high cohesive energy between the nanograins and the intergranular phase (amorphous or nanocrystalline).

The basic idea is thus to make a material structure such that the usual fracture mechanisms, viz., formation and multiplication of dislocations in a crystal and formation and growth of cracks in an amorphous material, cannot take place.

The proposed methodology [4, 13, 15] for synthesising these coatings can be summed up in a few rules:

- Choose a ternary (or quaternary) system which exhibits, from a thermodynamic standpoint, a strong tendency to segregate into two phases, e.g., the TiSiN system, where the high chemical affinity between Ti and N will lead to segregation of TiN in the form of nanograins surrounded by a second, amorphous phase of Si_xN_y.
- Make sure that the phases formed in this way are totally immiscible in the solid state and exhibit a certain chemical affinity, so that the cohesive energy of the nanocrystal/intergranular phase interface is very high.
- Select in preference an amorphous intergranular phase as thin as possible to minimise the formation and propagation of cracks.
- Carry out the synthesis at a moderate temperature and/or in conditions where grain growth is limited so that they remain smaller than 2–4 nm in diameter.
- Prepare using a technique which produces, at a fairly low temperature, sufficient mobility and chemical activation of the components to favour segregation.

Coatings formed from nanocrystals of a transition metal nitride, written symbolically nc-MN (where MN = TiN, ZrN, NbN, etc.), and an amorphous intergranular silicon nitride phase a-Si_3N_4 have extremely interesting properties. In the formula, 'nc' specifies the nanocrystalline nature of the transition metal nitride while 'a' specifies the amorphous nature of the other, intergranular phase. Among these coatings, the most studied is the system nc-TiN/a-Si_3N_4, because it satisfies all the rules in the above list. It has been synthesised by

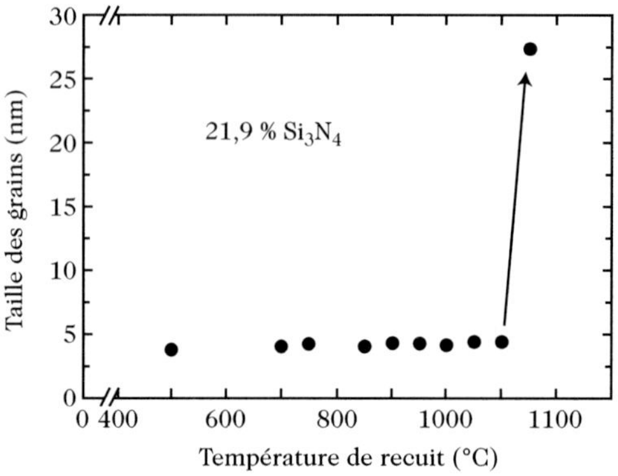

Fig. 24.4. Evolution of the grain size of a nanocomposite nc-TiN/a-Si_3N_4 coating (hardness 48 GPa) when annealed in a hydrogen atmosphere ($p_{H_2} = 10^3$ Pa) [16]

several vapour deposition techniques and we now possess a very complete set of data regarding the relations between conditions of synthesis, structure, and properties [16–20]. A very important feature of this system is its remarkable thermal stability, up to 1 100°C. Indeed, it can be seen from Fig. 24.4 that the grain size of an nc-TiN/a-Si_3N_4 coating (hardness 48 GPa) is not altered by annealing up to temperatures of 1 100°C [16].

The very thin layer of Si_3N_4 (~ 0.5 nm) surrounding the TiN nanograins acts as a very effective diffusion barrier, preventing the TiN grains from growing. The desired properties – hardness and resistance to wear and corrosion – are thereby conserved up to very high temperatures [21].

Comparing hardness measurements made on nanocomposite nc-TiN/a-Si_3N_4 coatings synthesised using very different methods, as shown in Fig. 24.5, it is observed that the hardness goes through a maximum between 40 and 60 GPa for an amorphous phase fraction between 15 and 25 at.%, which corresponds to a grain size in the range 5–10 nm. Under these conditions, the average thickness of the amorphous intergranular phase a-Si_3N_4 is around 0.5 nm, which corresponds to 4–5 interatomic separations in Si_3N_4 [22, 23].

In a nanocrystalline material, plastic deformation arises through two different mechanisms: those acting in the grains and those acting in the grain boundaries. In order to visualise this, it is interesting to examine the general dependence of the hardness of a polycrystalline material on the size of the grains in it [24], as shown in Fig. 24.6. Note that there is a maximum hardness which reflects the important role played by the grain boundaries once their size goes below about 10 nm.

As the grain size is reduced from 100 μm to about 10 nm, the hardness increases according to the Hall–Petch law (see Chap. 8), because plastic deformation is governed by the formation and motion of dislocations within grains. However, for sizes below 10 nm, this type of mechanism becomes less and less relevant and plastic deformation mechanisms in the grain boundaries begin to

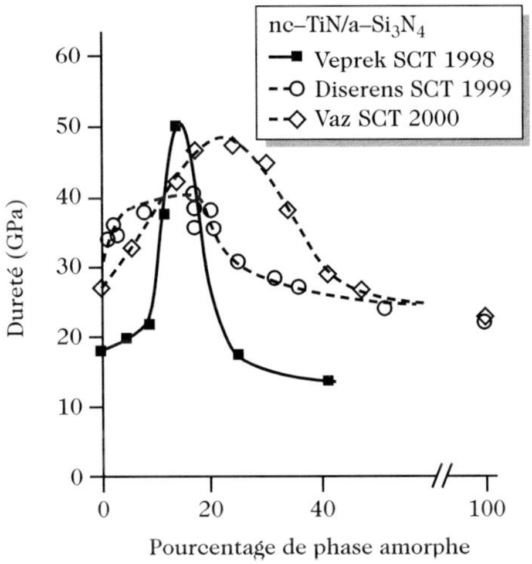

Fig. 24.5. Hardness of nanocomposite nc-TiN/a-Si_3N_4 coatings synthesised using different vapour deposition techniques [PVD (Diserens and Vaz) and CVD (Veprek)], compared for different percentages of amorphous phase [17]. See Sect. 24.2 for details

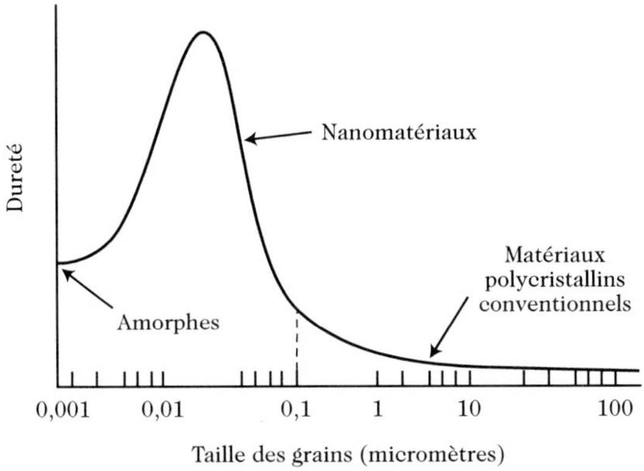

Fig. 24.6. Schematic dependence of the hardness of a polycrystalline material on grain size [24]

dominate. Simulations of the deformation of a nanocrystalline material [25,26] show that softening occurs, i.e., a reduction in hardness, for very small grain sizes (a few nanometers). Plastic deformation then arises mainly due to a large number of small slipping events in the grain boundaries, where the existence of a great many defects allows the rapid diffusion of atoms and voids. This softening effect in nanocrystalline materials is also called the reverse Hall–Petch effect. It is this type of slipping mechanism in grain boundaries that limits the hardness of a single-phase nanocrystalline material. However, in a nanocomposite material comprising several phases, the hardness can be increased by

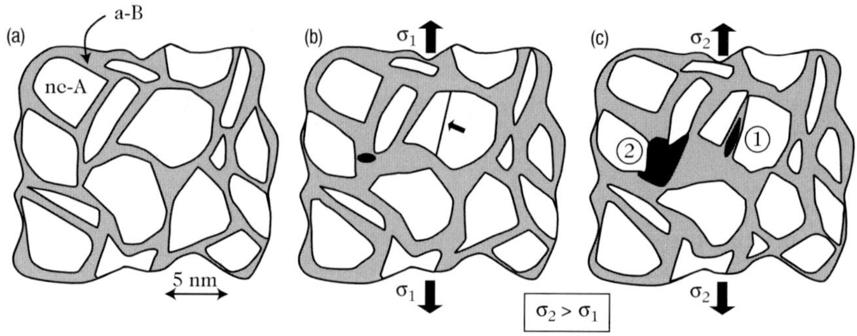

Fig. 24.7. Schematic representation of plastic deformation mechanisms in a nanocrystalline nc-A/a-B coating. (**a**) Initial state before applying the stress. (**b**) When a stress σ_1 is applied, a stacking defect forms in a grain (*small arrow*) and a pore forms in a-B. (**c**) Increasing the stress causes a crack to form and propagate along the stacking defect in grain 1 and likewise for another crack on the pore in a-B [27]

reducing or suppressing slipping in the intergranular phase, through an appropriate choice of microstructure and properties.

Consider what happens when a stress is applied that leads to the rupture of a nanocrystalline coating composed of nc-A nanograins surrounded by an amorphous intergranular phase a-B. The mechanisms are shown schematically in Fig. 24.7 [27].

When a stress is applied, even though the mechanism is unlikely, a dislocation may form at a grain boundary and cross grain 1 (Fig. 24.7b), leaving behind it the stacking defect shown by the arrow in the figure. If the deformation is too great, these stacking defects may cause cracks to form, and these will propagate if the stress increases in grain 1 (Fig. 24.7c). Deformation of the more ductile intergranular a-B phase may generate cracks by coalescence of small cavities or pores. While these cracks remain blocked between the nanograins, there will be no propagation, but when the resistance of the nc-A/a-B interface is no longer sufficient to withstand the stress at the head of the crack, there will be propagation, causing the coating to break.

24.2 Methods of Synthesis

24.2.1 General Principles

The synthesis of nanostructured coatings, either multilayer or nanocomposite, must be carried out at as low a temperature as possible ($< 500°$C) in order to hinder grain growth in the case of nanocomposites and interdiffusion phenomena in the case of multilayers, as well as limiting structural modifications and size variations in the substrates to be coated, e.g., steels or titanium alloys.

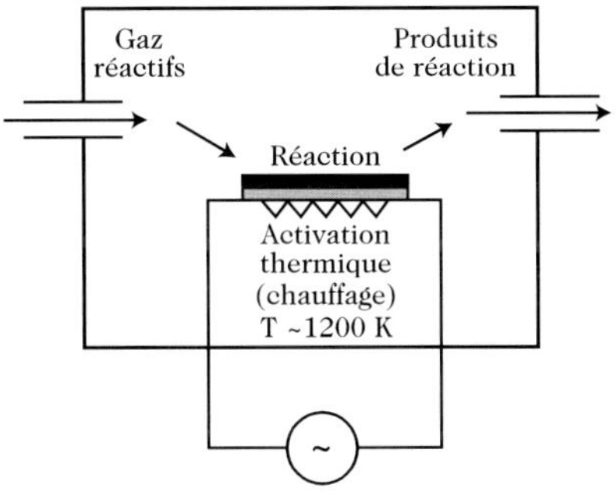

Fig. 24.8. Thermal CVD reactor in which chemical reactions between the reactive gases are activated solely by heating

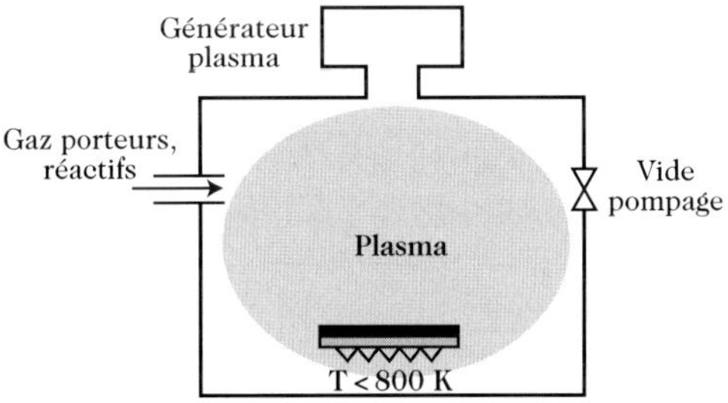

Fig. 24.9. Plasma activated chemical vapour deposition (PACVD) reactor in which production of a plasma activates the chemical reactions between the reactive gases and reduces the deposition temperature

In order to obtain an adequate atomic mobility and/or reactivity of the various species at low temperature, a vapour deposition technique is combined with the use of high-energy ions in the form of a plasma or ion beam. There are two classes of vapour deposition technique, distinguished by the types of phenomenon used to obtain the vapour phase:

- chemical vapour deposition (CVD),
- physical vapour deposition (PVD).

Detailed descriptions of the different vapour deposition techniques can be found in the specialised literature [28]. The CVD process consists in growing a coating from gas precursors comprising the elements of the coating that are made to react on the substrate (see Fig. 24.8). However, the chemical reactions are activated in different ways, either by heat energy (in hot-walled reactors or in cold-walled reactors where only the substrate is heated), or with the help of a plasma (Fig. 24.9).

When the reactor and conditions of deposition are optimised, CVD processes can yield high growth rates (1–5 μm/hr) and uniform growth on substrates with a range of complex geometries. However, there are two disadvantages:

Table 24.2. Main characteristics of different vapour deposition techniques

Technique	Pressure during deposition [Pa]	Speed [μm/hr]	Advantages (+) and disadvantages (−)
CVD and PACVD	$\sim 10^2$	∼ 1–5	Uniform deposition on complex items (+) High deposition rate (+) High temperature in CVD (−) Toxic or corrosive gas precursors (−)
Cathodic sputtering	∼ 1–10	∼ 0.1	Simple to use (+) High deposition pressure (−) Low deposition rate (−)
Magnetron sputtering	$\sim 10^{-1}$–10^{-2}	∼ 1–10	High deposition rate (+) Low pressure (+) Controlled (high or low) temperature (+) High quality of coatings (+)
Direct ion beam sputtering	$\sim 10^{-2}$	0.2–1	Low pressure (+) Controlled (high or low) temperature (+) High quality of coatings (+) Low deposition rate (−)
Cathodic arc deposition	< 10	∼ 1–10	Simplicity of process (+) High deposition rate (+) Industrial process (+) Low quality of coatings (−)

the use of toxic or otherwise hazardous gases raises a number of safety and environmental problems; and the temperature is very high in the thermally activated CVD process ($T > 1\,200$ K) (see Table 24.2). In more recent CVD techniques, where reactions are plasma assisted (plasma activated CVD), the substrate temperature can be reduced to below 800 K, which is a low enough temperature to fabricate nanocomposite coatings.

It is not so easy to achieve multilayers of nanometric thickness with clean interfaces using PACVD, because the flow of atomic vapours must then be modulated. In this case, it is mainly PVD techniques that are used. Indeed, these do not require toxic or hazardous gases and the coatings can be made at very low temperature because the atomic vapour is produced by a purely physical phenomenon, viz., vaporisation of a target by heating, electric arc, or ion sputtering of a target (usually Ar^+) at a sufficient energy (greater than about 30 eV for a metal).

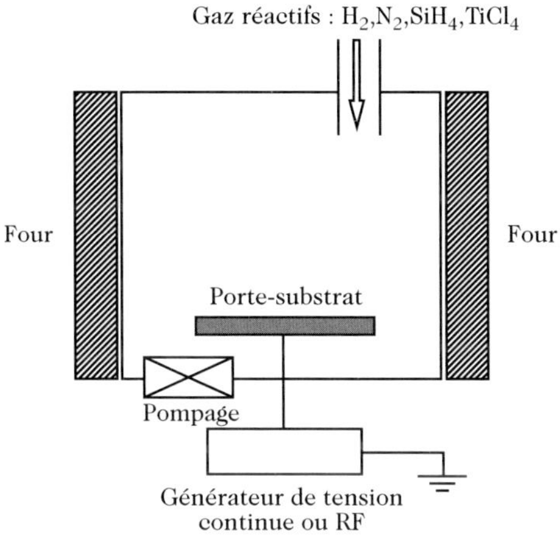

Fig. 24.10. PACVD reactor used to synthesise nanocomposite nc-TiN/a-Si_3N_4 coatings. The plasma is produced by applying a DC or RF (radio frequency) voltage to the substrates [16]

24.2.2 Plasma-Activated Chemical Vapour Deposition (PACVD)

Plasma assisted CVD provides a way of significantly reducing the substrate temperature while yielding deposition rates of several microns per hour thanks to the enhanced reactivity of the precursors and the possibility of accelerating active species towards the substrate in ionic form. The first superhard nanocomposite nc-TiN/a-Si_3N_4 coatings were produced using this technique [4, 29]. The coating forms as a result of chemical reactions obtained by circulating a mixture of reactive gases in the reactor: H_2+ N_2+ $TiCl_4$+ SiH_4. The plasma is produced using an electrode connected to a high-frequency (20–30 MHz) voltage generator, the substrate is heated to 550°C, and for optimal operating conditions (pressure 10^2 Pa, RF power 100 W, gas flow rates[1] of 20 sccm for H_2, 1 sccm for N_2, and 0.1 sccm for $TiCl_4$ and SiH_4), the growth rate is around 3 μm/hr, so that industrial applications can be envisaged.

The problems related to the use of chlorides should be borne in mind. There are two kinds: chemical corrosion of the metal walls of the reactor by the chlorine, and the incorporation of chloride in the coatings, which can lead to premature corrosion of the interfaces. In order to reduce the presence of chlorine in the films and to be able to process items with complex shapes like cutting tools, a reactor with hot walls (an oven) can be used, producing the plasma at higher pressure (5×10^2 Pa) by means of a luminescent discharge by applying a direct voltage of around 700 V to the substrate [16] (Fig. 24.10).

Other types of nanocomposite such as nc-W_2N/a-Si_3N_4 or nc-VN/a-Si_3N_4 have been made using PACVD, producing a plasma by means of a luminescent discharge obtained by applying either a direct voltage or a very high frequency

[1]The unit sccm is the standard cubic centimeter per minute, or cm^3/min.

alternating voltage. Very high hardness values, in the range 50–70 GPa, are obtained for grain sizes of 3–4 nm.

It is also possible to avoid using metal chlorides like $TiCl_4$, and hence avoid chlorine-related problems, by combining PACVD with a PVD technique to obtain a metal vapour. One solution which has been used to synthesise nanocomposite nc-TiN/BN coatings consists in vaporising the metal (Ti) by producing an intense electric arc between an anode and a cathode made from the metal to be evaporated, while boron and nitrogen are introduced into the reactor in gas form, viz., N_2+ B_2H_6 or $B_3N_3H_6$ [16].

24.2.3 Physical Vapour Deposition by Sputtering and Cathodic Arc

Magnetron sputtering is a process that has been considerably developed because it provides an extremely efficient and highly flexible way of building coating architectures with varying degrees of complexity. Deposition rates are high (up to a few μm/min), and by giving the substrate a negative bias, the film can be bombarded during growth, thereby densifying it and increasing its adherence with respect to the substrate.

We must begin by describing cathodic sputtering, before discussing the magnetron process, which consists in superposing a magnetic field on the electric field that produces the discharge, and hence the plasma, in order to increase the sputtering yield.

When a solid is bombarded by ions with high enough energy (> 30 eV), atoms at the surface can be ejected. This surface erosion phenomenon, leading to the formation of an atomic vapour, results from elastic collisions with energy and momentum transfer between the incident ions and the target atoms. We shall not discuss here the ion–matter interaction mechanisms, which can be found in the specialised literature [30]. We shall just mention that a sputtering yield Y can be defined, expressing the number of target atoms ejected per incident ion. It can be calculated from the parameters characterising the incident ions, e.g., mass, energy, angle of incidence, and so on, and the properties of the target, e.g., mass, binding energy, etc. In practice, the value of Y is obtained using a simulation code such as SRIM [31].

The simplest sputtering process is cathodic sputtering, shown schematically in Fig. 24.11. A plasma is produced in a chamber into which an inert gas is introduced (usually argon) at low pressure (1.5–15 Pa) by applying a potential difference of kV order between the substrate holder (anode) and the target (cathode) comprising the material to be sputtered.

The luminescent discharge which forms between the cathode and the anode contains positive ions Ar^+ which are accelerated by the electric field applied at the cathode and cross the chamber to bombard it. This leads to ejection of atoms from the material making up the target, atoms which then condense on the substrate. However, only a very small fraction of the gas atoms are ionised ($\sim 0.01\%$). The ion flux bombarding the target is therefore rather small and

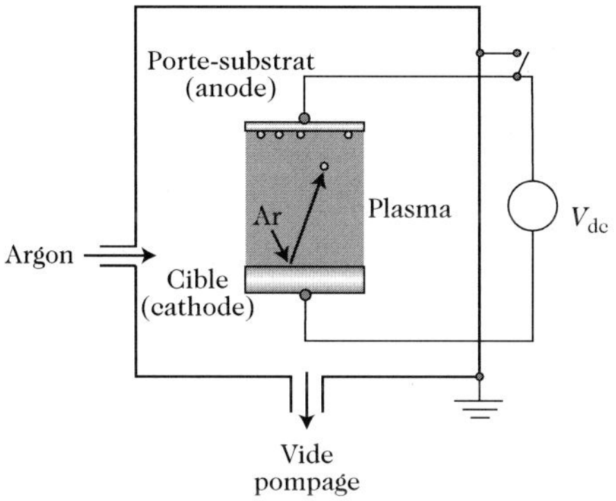

Fig. 24.11. DC cathodic sputtering. The luminescent discharge between the anode and the cathode ionises the argon atoms which then bombard the target and eject surface atoms

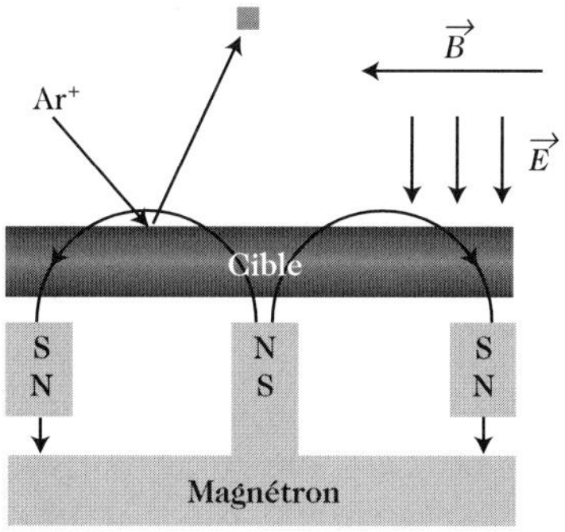

Fig. 24.12. Plane magnetron target showing the magnetic field lines parallel to the target surface. ***E*** is the electric field which produces the discharge and the Ar^+ plasma, while ***B*** is the magnetic field produced by the magnets of the magnetron. The *grey square* represents the sputtered atom

deposition rates at the substrate likewise ($\sim 0.1\,\mu$m/hr). Note also that for DC discharge, when the process is called DC sputtering, the cathode and anode must be metal conductors, limiting the applications. In order to apply to dielectric (insulating) targets, a high-frequency (usually 13.6 MHz) alternating voltage must be used. This is called radio frequency (RF) sputtering.

Magnetron sputtering constitutes an important step forward in the development of PVD processes. A special magnetic device is associated with the cathode to confine electrons close to the target surface and increase the plasma density, and hence the sputtering rate.

The high ion density means that the discharge can be maintained at lower pressure (up to 10^{-2} Pa). Another important parameter is the ion energy, and hence the energy of the atoms deposited on the substrate, which is higher in this process. This plays a key role in the growth of the films, which are

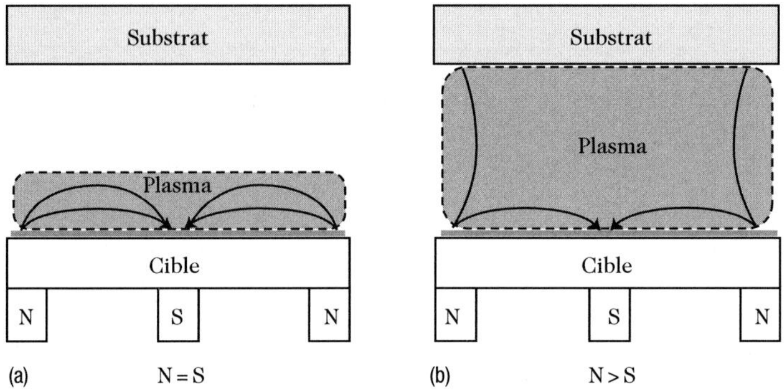

Fig. 24.13. Schematic view of the plasma zone (**a**) for a standard magnetron device and (**b**) for an unbalanced magnetron

denser and adhere better. The magnetron setup is shown schematically for plane geometry in Fig. 24.12.

The cathode, or magnetron target, is equipped with permanent magnets arranged behind the target and which produce a strong magnetic field with closed field lines. The secondary electrons emitted by the cathode under the effect of ion bombardment of the target surface are trapped and follow trajectories that spiral along the field lines. There is increased probability that these electrons will ionise argon atoms.

When the aim is to form metal nitrides or oxides by reactive magnetron sputtering, a reactive gas must be added to the main gas (argon), but RF mode must be used and the pressure of the reactive gas must be controlled. There are many variants on the basic magnetron sputtering process, designed to improve deposition rates and film quality. Indeed, in standard magnetron sputtering systems, the ion current density decreases rapidly when the target–substrate distance is increased. Unbalanced magnetrons are used to produce TiN coatings or coatings of analogous compounds on the industrial scale [32]. The basic idea is shown schematically in Fig. 24.13.

In order to synthesise multilayers of nanometric period and final thickness a few microns, the flux of vapour of the various constituents must be very stable over a time lapse that is long enough to ensure regularity of the stacking process. The most widely used technique consists in placing two or more unbalanced magnetrons opposite one another in closed-field configuration, i.e., so that the magnetic poles facing one another are of opposite kind, as shown schematically in Figs. 24.14 and 24.15. The substrates are placed on a rotating table to deposit alternately the elements of cathodes 1 and 2, or their nitrides in the reactive process.

Another setup that is also widely used is based on the association of four unbalanced magnetrons in closed-field configuration, whereupon the plasma can be trapped in the central region of the deposition chamber in the vicinity of the substrates, as shown in Fig. 24.16. With this type of setup, the ion

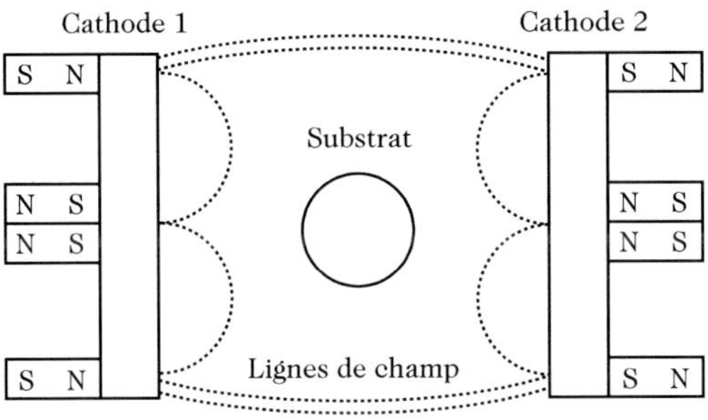

Fig. 24.14. Two unbalanced magnetrons set up opposite one another in closed magnetic field configuration

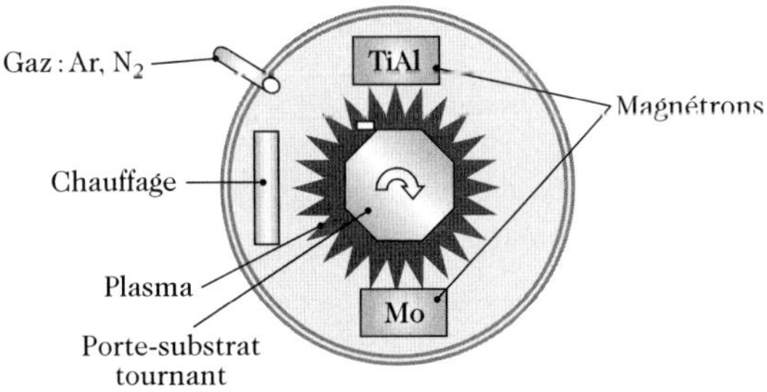

Fig. 24.15. Closed-field unbalanced magnetron setup for synthesising TiAlN/Mo multilayers [33]

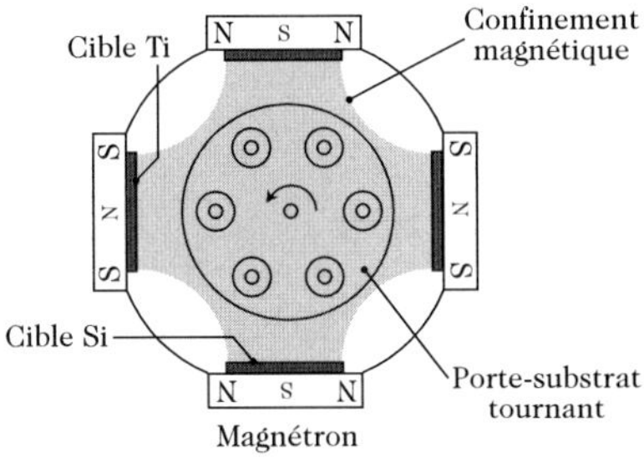

Fig. 24.16. Cross-section of a sputtering setup with four unbalanced magnetrons in closed-field configuration

current density is increased, and with it the deposition rate, so that industrial scale production of multilayers or nanocomposite coatings can be envisaged [11, 12, 34].

Nanocomposite coatings such as nc-TiN/a-Si_3N_4 can also be synthesised using very simple reactive RF magnetron sputtering setups with two Ti and Si targets placed in a horizontal plane, as shown schematically in Fig. 24.17. The substrates are fixed in a holder rotating at 4 rev/min in a plane parallel to the targets (at a separation of around 60 mm). The starting pressure is 10^{-4} Pa

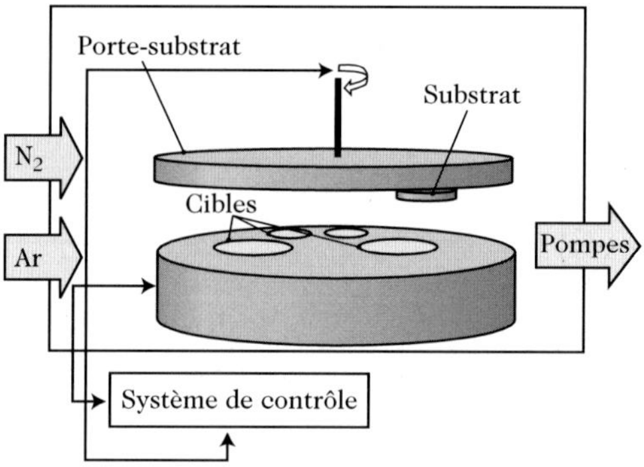

Fig. 24.17. Reactive RF magnetron setup for synthesising nanocomposite ncTiN/a-Si_3N_4 coatings

and climbs to around 4×10^{-1} Pa during deposition. Substrates are heated to 300°C and can be biased with a negative voltage to favour film bombardment during growth, thereby increasing the mobility of surface atoms and ensuring better phase segregation [35].

Hybrid techniques combining cathodic arc evaporation and reactive magnetron sputtering have also been devised recently to synthesise nanocomposite nc-TiN/a-Si_3N_4 coatings [20]. The cathodic arc evaporation technique is highly robust and widely used in industry due to the fact that very high deposition rates can be achieved. The idea is to produce, at rather low pressures, an arc between the metal to be evaporated, e.g., Ti, which constitutes the cathode, and an anode. The cathode material evaporates as a consequence of the very sharp increase in its temperature, accompanied by local melting. The evaporated atoms are ionised by collisions with plasma electrons and accelerated out of the cathode towards the substrate. The main disadvantage with this process is the presence of microparticles (1–100 μm) in the form of liquid droplets in the beam, which leads to films of lower quality. However, there are other setups that can eliminate or significantly reduce this problem. Cathodic arc evaporation is used industrially to produce hard zirconium, molybdenum or titanium carbide or nitride coatings. These coatings are strongly adherent, a consequence of film bombardment during growth, obtained by the negative bias (0.2–2 kV) of the substrate. The setup for cathodic arc deposition is shown schematically in Fig. 24.18.

24.2.4 PVD by Ion Beam Sputtering

In the PVD techniques just described, the target is sputtered by plasma ions whose energy is not accurately determined. It is possible to sputter a target using a beam of ions with controlled flux and energy. This is called ion beam sputtering. Monokinetic ion sources are used (usually producing Ar^+), with which the energy can be varied over the range 0.1–1.5 keV. Current densities

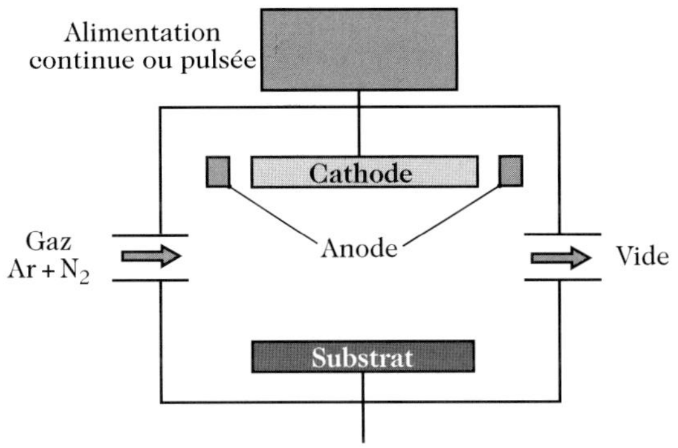

Fig. 24.18. Cathodic arc evaporation, in which the cathode comprises the material to be evaporated

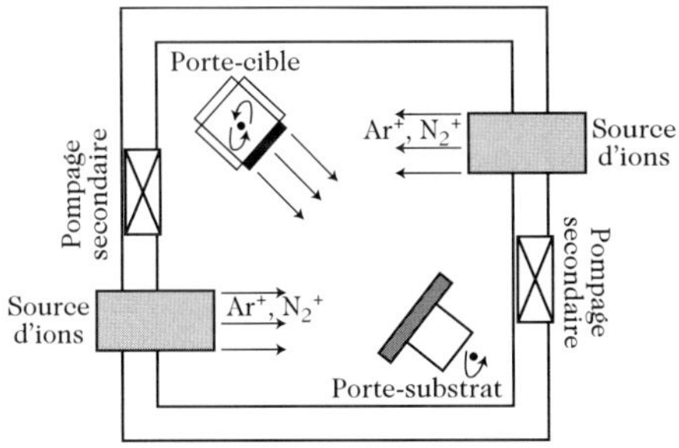

Fig. 24.19. Setup for ion beam sputtering with reactive assistance or otherwise

are high ($\sim 1\,\mathrm{mA/cm^2}$) and the beam can have a broad cross-sectional area (typical diameter $\sim$ 10 cm). Another advantage with this technique is that deposits can be made in higher vacuum ($\sim 10^{-1}$–10^{-2} Pa) than they can with plasma sputtering.

These sources generally work by introducing a gas (argon) into the chamber in which the ions are produced by a discharge. A system of permanent magnets ensures confinement and improves the ionisation yield, while negatively biased grids are used to extract the beam. After its extraction from the source, the ion beam is neutralised to prevent it from broadening under the effect of its own electric field. This is done by introducing electrons into the beam, either by heating a filament (thermoelectronic effect) in Kaufman sources, or using an electron gun for RF ion sources. A second source can be used to bombard the deposit during growth (ion beam assisted deposition or IBAD), either with Ar^+ ions (non-reactive assistance), or with reactive species such as N_2^+ or O_2^+ to synthesise nitrides and oxides. The setup is shown schematically in Fig. 24.19.

Using this technique, multilayers with nanometric period can be synthesised using targets made from different materials [36], since an obturator can protect the substrates while the target is being changed. The whole setup can be computer run and the deposition parameters (current and voltage, gas flow rates, etc.) precisely controlled, to guarantee beam stability and hence good reproducibility of the synthesised layers, which are themselves of high quality. The main disadvantage of this type of technique compared with the previous ones is the low deposition rate, which is only of the order of 0.2–1 μm/hr. Coatings made in this way are therefore rather thin and tend to be reserved for optical or magnetic applications [37,38]. However, ion beam sputtering has the advantage of producing very dense layers with 2D rather than columnar growth. This happens due to the high energy of the sputtered atoms, which is two orders of magnitude higher than can be achieved with thermal evaporation ($\sim$ 0.1–0.5 eV). Calculating the energy distribution of sputtered atoms from an Mo target, for example, by incident Ar^+ ions with energy 1.2 keV, it is found that most of the atoms have energies in the range 10–20 eV, while some have energies as high as 100–150 eV. These high energy values mean that the atoms have high mobility, even at low temperatures, and they also induce a self-bombardment of the growing film, which favours rearrangements at the surface of the film, and even in the bulk. These experimental results have been confirmed by molecular dynamics simulations, which demonstrate the effect of the energy of the atoms on the growth mode.

References

1. D. McIntyre, J.E. Greene, G. Hakansson, J.E. Sundgren, W.D. Munzz: J. Appl. Phys. **67** (3), 1542 (1990)
2. D.M. Teter: MRS Bull. **23** (1), 22 (1998)
3. H. Holleck: J. Vac. Sci. Technol. A **4** (6), 2661 (1996)
4. S. Veprek, S. Reiprich: Thin Solid Films **268**, 64 (1995)
5. U. Helmersson, S. Todporova, S.A. Barnett, J.E. Sundgren, L.C. Markert, J.E. Green: J. Appl. Phys. **62**, 481 (1987)
6. P. Yaskar, S.A. Barnett, J. Rechner, W.D. Sproul: J. Vac. Sci. Technol. A **16** (5), 2913 (1998)
7. Y.H. Chen, K.W. Lee, W.A. Chiou, Y.W. Chung, L.M. Keer: Surf. Coat. Technol. **146–147**, 209 (2001)
8. A. Madan, Y.Y. Wang, S.A. Barnett, C. Engstrom, H. Ljungcrantz, L. Hultman, M. Grimditsch: J. Appl. Phys. **84**, 309 (1996)
9. C.J. Tavares, L. Rebouta, E. Ribeiro, J.P. Riviere, J. Pacaud, M.F. Denanot: Surf. Coat. Technol. **174–175**, 273 (2003)
10. W.D. Munz, D.B. Lewis, P.Eh. Hovsepian, C. Schonjahn, A. Ehiasarian, I.J. Smith: Surf. Eng. **17** (1), 15 (2001)
11. L.A. Donohue, W.D. Munz, D.B. Lewis, J. Cawley, T. Hurksmans, T. Trinh, I. Petrov, J.E. Green: Surf. Coat. Technol. **93**, 69 (1997)
12. W.D. Munz: MRS Bull. **28** (3), 173 (2003)
13. S. Zhang, D. Sun, Y. Fu, H. Du: Surf. Coat.Technol. **167**, 113 (2003)
14. J. Musil: Surf. Coat. Technol. **125**, 322 (2000)
15. S. Veprek: Surf. Coat.Technol. **97**, 15 (1997)

16. S. Veprek, P. Nesladek, A. Niederhofer, F. Glatz, M. Jilek, M. Sima: Surf. Coat. Technol. **108–109**, 138 (1998)
17. J. Patscheider, T. Zehnder, M. Diserens: Surf. Coat. Technol. **147–148**, 201 (2001)
18. F. Vaz, L. Rebouta, P. Goudeau, J. Pacaud, H. Garem, J.P. Riviere, A. Cavaleiro, E. Alves: Surf. Coat. Technol. **133–134**, 307 (2000)
19. X. Sun, J.S. Reid, E. Kolawa, M.A. Nicolet: J. Appl. Phys. **81** (2), 656 (1997)
20. P.J. Martin, A. Bendavid: Surf. Coat. Technol. **163–164**, 245 (2003)
21. M. Diserens, J. Patscheider, F. Levy: Surf. Coat. Technol. **120–121**, 158 (1999)
22. A. Niederhofer, T. Bolom, P. Nesladek, K. Moto, C. Eggs, D.S. Patil, S. Veprek: Surf. Coat. Technol. **146–147**, 183 (2001)
23. R. Chaim: J. Mater. Res. **12** (7) 1828 (1997)
24. J. Schiotz: *Proceedings of the 22nd RISO International Symposium on Materials Science*, Roskilde, Denmark (2001) p. 127
25. J. Schiotz, T. Vegge, F.D. DiTolla: Phys. Rev. B **60** (17), 11971 (1999)
26. J. Schiotz, F.D. DiTolla, K.W. Jacobsen: Nature **391**, 561 (1998)
27. C. Mitterer, P.H. Mayrhofer, M. Beschliesser, P. Losbichler, P. Warbichler, F. Hofer, P.N. Gibson, W. Gissler, H. Hruby, J. Musil, J. Vlcek: Surf. Coat. Technol. **120–121**, 405 (1999)
28. A. Galerie (Ed.): *Traitements de Surface en Phase Vapeur, Mécanique et Ingénierie des Matériaux*, Editions Hermes Sciences, Paris (2002)
29. S. Li, Y. Schi, H. Peng: Plasma Chem. Plasma Process **21** (3), 287 (1992)
30. P. Sigmund: In: *Sputtering by Particle Bombardment I*, Topics in Applied Physics, Vol. 47, ed. by R. Behrisch, Springer-Verlag, Berlin (1981) p. 9
31. J.F. Ziegler, J.P. Biersack, V. Littmark: *The Stopping and Range of Ions in Solids*, Pergamon Press (1986)
32. S.M. Rossnagel: In: *Handbook of Plasma Processing Technology*, ed. by S.M. Rossnagel, J.J. Cuomo, W.D. Westwood, Noyes Pub. Park Ridge NJ (1990) p. 160
33. C.J. Tavares, L. Rebouta, E. Alves, A. Cavaleiro, P. Goudeau, J.P. Riviere, A. Declemy: Thin Solid Films **377–378**, 425 (2000)
34. J. Patscheider: MRS Bull. **28** (3), 180 (2003)
35. F. Vaz, L. Rebouta, P. Goudeau, T. Girardeau, J. Pacaud, J.P. Riviere, A. Traverse: Surf. Coat. Technol. **146–147**, 274 (2001)
36. G. Abadias, C. Jaouen, F. Martin, J. Pacaud, Ph. Djemia, F. Ganot: Phys. Rev. B **65**, 1 (2002)
37. T. Girardeau, S. Camelio, A. Traverse, F. Lignou, J. Allain, A. Naudon, Ph. Guérin: J. Appl. Phys. **90** (4), 1788 (2001)
38. D. Babonneau, J. Briatico, F. Petroff, T. Cabioch, A. Naudon: J. Appl. Phys. **87** (7), 3432 (2000)

25

Dispersion in Solids

D. Babonneau

To make good use of the physical properties of nanoparticles (see Part II of this book), most applications require them to be incorporated within solid materials, such as glasses, ceramics, oxides, etc. The inclusions obtained in this way are thereby protected from chemical (e.g., oxidation) or mechanical (e.g., friction or wear) alteration due to the environment. They are also separated from one another from an electrical or magnetic point of view, which limits this kind of interaction between the particles. The dispersion of particles within solid materials can be achieved in the ordinary way by exploiting precipitation effects in the solid state (e.g., alloys hardened by dispersions of nanometric precipitates, Guinier–Preston zones), but also by a great many less direct methods. Which method is finally chosen depends on several factors such as the kind of matrix and size, shape and spatial organisation of the particles, but also on the volume of material to be produced, the cost, and the difficulty involved in the process.

In this chapter, we shall discuss some of the methods commonly used today, spelling out their specific features. Quite generally, these fall into three categories: chemical methods, physical methods, and mechanical methods. The latter were treated in Chap. 19 and will not be discussed further here. In addition, there are two possible ways of dispersing the nanometric particles. One is the bottom-up approach, which means building together atoms and molecules to constitute a larger structure and incorporating them – either simultaneously or a posteriori – within the host matrix. The other is the top-down approach currently used in microelectronics, which consists in refining the microstructure of some bulk material using more and more sophisticated tools. The chemical and physical methods to be discussed in this chapter use a bottom-up approach, in contrast to mechanical processes, which generally use a top-down approach, and yield nanomaterials in the form of powders or bulk materials with varying degrees of densification.

Fig. 25.1. The Lycurgus cup at the British Museum in London dates from the Roman empire (fourth century A.D.). It appears green in reflected light and red in transmitted light, owing to the presence of nanometric inclusions of gold and silver in the glass. From [4]. See also the colour plate

25.1 Chemical Methods

Chemical methods were the first used to obtain nanometric dispersions in matrices and they are the most highly developed today for synthesising self-organised nanomaterials. In particular, they provide the greater part of the nanomaterials used in optics, where metal inclusions are embedded in bulk glasses or ceramics of varying thicknesses. There are two main types of conventional chemical method here: doped glasses (Sect. 25.1.1) and sol–gels (Sect. 25.1.2).

25.1.1 Synthesis of Doped Glasses

It has been known since ancient times how to synthesise glasses doped by metal powders for decorative applications (see Fig. 25.1). The method is still used today, particularly for synthesising nanomaterials with special optical properties, because it is very easy to implement [1]. Such glasses are usually produced by adding a metal oxide powder and a reducing agent when the glass melts. Sudden cooling of the mixture then yields a perfectly uniform, supersaturated glass, in which the metal remains in the ionised or oxidised state and the reducing agent in the state of lowest valence. A heat treatment of the glass then leads to reduction of the metal ions by the reducing agent and causes the formation of metal nanoparticles in the glass [2]. This nucleation phase can also be activated and controlled by X rays or UV irradiation [3]. At room temperature and below, diffusion processes are negligible and the nanoparticles are completely stabilised within the glassy matrix. One of the particularities of this means of synthesis is that bulk doped glasses can be made, with a narrow size distribution of nanometric inclusions in most cases. However, when low chemical reactivity and low supersaturation concentrations are required, this method can only yield low metal concentrations (of the order of 10^{-6} to 10^{-3} in volume) and it is essentially limited to the noble metals (Cu, Ag, Au).

25.1.2 Sol–Gel Method

New means of synthesis by the name of sol–gel processes were developed during the twentieth century, both in academic research centres and in industry, producing glasses by straightforward polymerisation of molecular precursors in solution, without passing through the molten state [5]. Concerning ceramic materials, the sol–gel approach also provides an alternative to the usual way which proceeds by synthesising powders. Another recent application to the fabrication of hybrid organic–inorganic nanomaterials for optics has only been in existence for twenty years or so. However, progress has been remarkable and sol–gel techniques combined with chemical nanotechnologies have taken their place as a fundamental approach to the development of new materials, with industrial applications in fields as far apart as microelectronics, optoelectronics, data storage, catalysis, the car industry, the life sciences, and many others.

Sol–gel synthesis is carried out in a liquid medium. It uses hydrolysis and condensation of molecular precursors diluted in a solvent (the sol, a solution of solid particles in a liquid). These two chemical reactions lead to the formation of a mesoporous solid structure swollen with liquid (the gel), whose porosity can be controlled on the nanoscale. If an organic surfactant is added to the sol as structuring agent (eliminated at the end of the process by calcination), it is even possible to obtain superlattices of monodispersed pores with sizes that can be adjusted between 1.5 and 10 nm, exhibiting 2D or 3D periodicity [6,7]. Although this is a delicate process for making bulk glasses, the viscosity of the gel is well suited to the fabrication of mesoporous coatings with thicknesses of several hundred nanometers on substrates of different shapes and kinds (e.g., glass, ceramic, metal, polymer) using dip coating, spin coating, or spraying techniques. Such coatings are usually dried at room temperature, leading to a transparent dry gel or xerogel, whose densification at moderate temperatures produces a glass with similar characteristics to a conventional glass.

Synthesising a Nanograin System by the Sol–Gel Method: Ni–SiO_2

The sol–gel technique is based on densification of a precursor previously prepared in the liquid phase (the sol), then subjected to heat treatment at moderate temperature. Thin films of oxide or a mixture of oxides can be obtained for a wide variety of substrates. The film composition is fixed by the dosing of constituents in the sol. The typical thickness of the film is about 100 nm if the dip coating method is used, while this value can be increased by repeating the coating plus heat treatment cycle, to make a multilayer film. In the example illustrated below (Fig. 25.2) [9], starting with solutions of tetraethyl orthosilicate (TEOS) and nickel acetate, composite layers of nickel oxide ($\approx 10\%$) and silicon dioxide ($\approx 90\%$) are deposited on a silicon substrate. In a second stage, the nickel oxide is selectively reduced by annealing in hydrogen at 600°C to yield nanoprecipitates of metallic nickel incorporated in an SiO_2 matrix.

The commonest way to incorporate nanoparticles in a material made by a sol–gel process is to disperse ions in the initial solution in the form of salts.

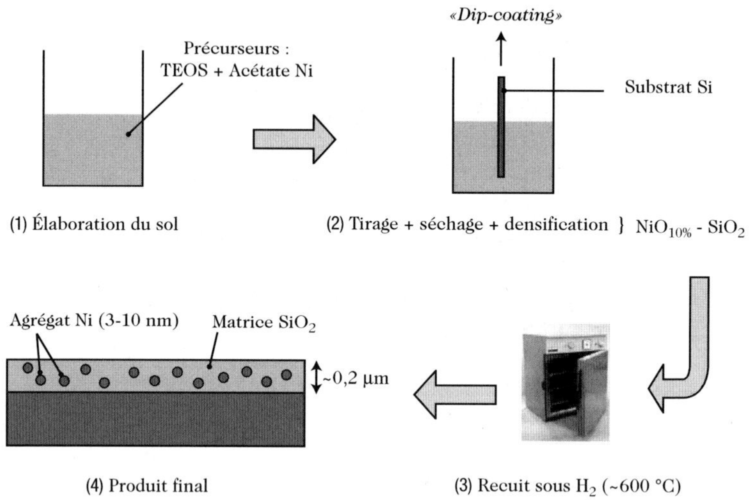

Fig. 25.2. Synthesis of a metallic nickel nanoprecipitate in an SiO_2 matrix by the sol–gel approach

Nanoparticles then form by reduction of these ions and the diffusion of the atoms thereby obtained, which may be activated in various different ways: high temperature heat treatment in hydrogen [8] or low fluence irradiation by heavy ions accelerated to electron stopping regimes (see Sect. 25.2.1) followed by low temperature annealing [10] are two examples among others. From a general point of view, this method usually leads to high dispersion in the size distribution and a random spatial distribution of the inclusions. Another process involves producing the bulk materials by sintering and densification of powders previously treated by immersion in a precursor sol for the dispersed phase to be introduced. This method provides a way of dispersing nanocrystals in intergrain positions, e.g., zirconia-toughened mullite nanocomposites [11], which are mechanically strengthened by the metastable zirconia, with the possibility of colouring the dense ceramic in the bulk. It can also be used to introduce nanocrystals simultaneously into intergrain and intragrain positions, the nanoparticles being coherent with the oxide matrix, e.g., spinel/zirconia nanocomposite, conferring superplasticity properties on the dense, finely grained material (see Fig. 25.3).

Structured mesoporous films synthesised by the sol–gel process are interesting for another reason: they can be functionalised, either during synthesis or afterwards by grafting hydrophobic or hydrophilic functions or specific ligands, or by inclusion of clusters. Indeed, the pores can be used as 'templates' for growing metal or semiconductor nanoparticles in a way that provides many advantages: the possibility of controlling the size, the shape, and the arrangement of the particles, and also of optimising the volume fraction of nanopar-

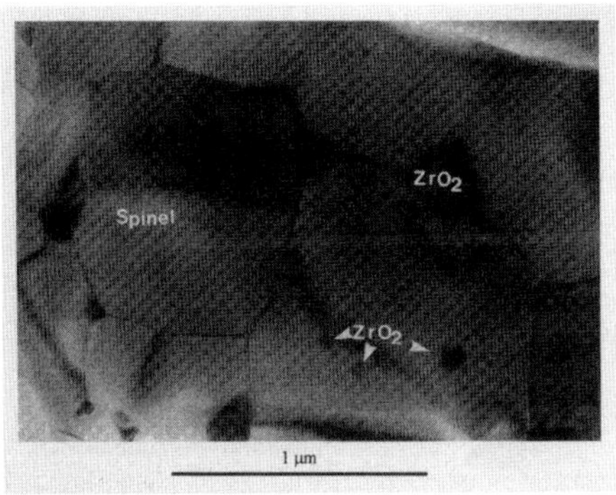

Fig. 25.3. TEM image of a bulk $MgAl_2O_4$–ZrO_2 nanomaterial synthesised by sintering powders obtained by the sol–gel process. Zirconia crystals appear not only in intergrain positions, but also in intragrain positions. From [12]

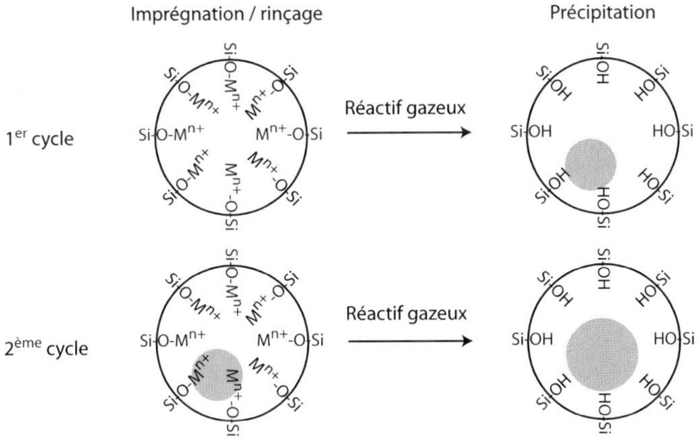

Fig. 25.4. Schematic representation of the general process for growing nanoparticles in mesoporous silica films. From [16]

ticles, while conserving the good mechanical behaviour provided by the host matrix.

A first solution for templating nanoparticles in a nanostructured film is to grow a deposit by physical means at the film surface [13–15]. The atoms thereby deposited can diffuse into the mesoporous structure to fill the pores, although this method only achieves low penetration into the mesoporous material. Another strategy is to grow the nanoparticles chemically in two stages: first adsorption of metal cations at the pore surface by impregnation in solution, followed by precipitation of the particles by reaction with a gas (see Fig. 25.4). A homogeneous, 3D periodic lattice of CdS nanoparticles can be synthesised in this way (see Fig. 25.5), and the method has been extended to sulfides such as ZnS, PbS, and Ag_2S, with potential applications where

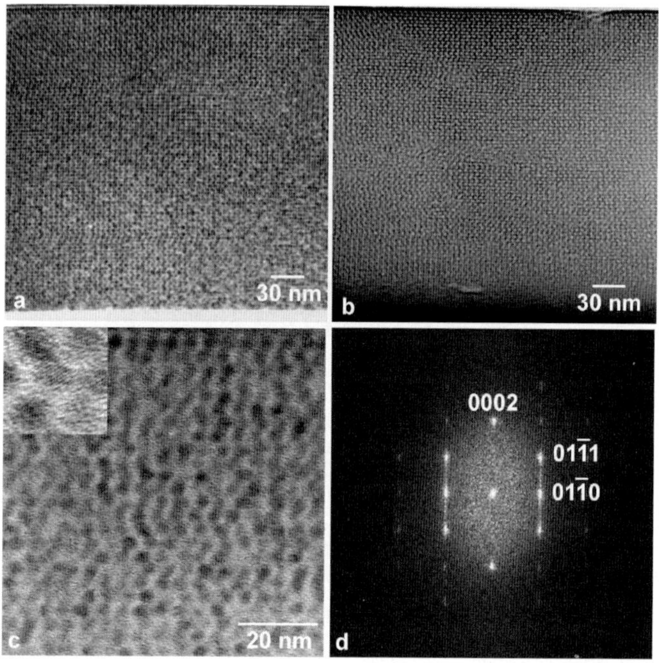

Fig. 25.5. (**a**) TEM image showing a cross-section of a mesoporous silica film with 3D hexagonal structure, saturated with CdS. Note the uniformity of the impregnation and the organisation of the nanoparticle lattice. (**b**) TEM image showing a cross-section of a non-impregnated calcinated mesoporous film. (**c**) High-resolution TEM image of CdS nanoparticles (cross-section). (**d**) Power spectrum of the image in (**a**), corresponding to the axis of the $[2\,\bar{1}\,\bar{1}\,0]$ zone of the 3D hexagonal structure. From [16]

optical or photocatalytic properties are required [16]. A final example of an original use for these pore superlattices provided by structured mesoporous silicas is the production by pyrolysis of arrays of carbon cylinders which, after dissolving the silica, themselves form a carbon lattice able to receive platinum nanoparticles in a very highly dispersed state, for use in catalysis [17].

25.2 Physical Methods

The physical and chemical methods differ in several ways. In particular, the former allow a wider variety of host matrices and control over the microstructure of the inclusions, i.e., their size, shape, composition, and local environment, independently of the host matrix. However, they usually require a larger and more expensive experimental setup, and they generally only allow the synthesis of materials in the form of thin films, whose thickness depends on the technique and the desired application.

Fig. 25.6. Morphology of the facets in a twinned copper precipitate, obtained by implanting the Cu in AlN. From [26] with the kind permission of Elsevier

25.2.1 Ion Implantation

Ion implantation involves bombarding the surface of a material with ions of selected mass and energy. When these ions penetrate into the material, they gradually dissipate their kinetic energy through elastic collisions with nuclei in the target (nuclear stopping) and inelastic interactions with the electron cloud of atoms in the target (electronic stopping). The relative importance of each of these energy loss mechanisms, and hence the final distribution of the ions implanted in the bulk of the material, depend on the energy of the incident ions (from a few tens of keV to a few thousand keV), the nature of the ions (almost the whole of the periodic table from hydrogen to bismuth) and the nature of the substrate (all solid materials can be envisaged). It is therefore a non-equilibrium method that is particularly useful for forcing immiscible elements, whatever they may be, to mix together. The nucleation of ions in nanometric inclusions (metal clusters, carbon nanostructures, bubbles, voids, etc.) can occur by diffusion, either during the implantation itself, or during a post-implantation annealing phase in the case of poorly miscible elements. However, there are several disadvantages with this method, such as damage to the matrix when the ions pass through it and non-uniformity over different depths in the material (with regard to metal concentration but also particle size distribution). Interesting results have nevertheless been obtained in various fields and this physical method is relatively attractive for synthesising nanometric inclusions in materials. We shall now illustrate this potential with three examples.

Ion implantation is one of the oldest physical methods for obtaining clusters of noble or magnetic metals in various crystalline or amorphous, semiconducting, metallic or insulating matrices (see Figs. 25.6 and 25.7). It is thus possible to produce films of clusters buried in a region close to the surface (usually 100 nm), whose physical properties can be adjusted in relation to the kind of matrix and the nature of the implanted ions, as well as the implantation parameters. As an example, ion implantation of gold [18], copper

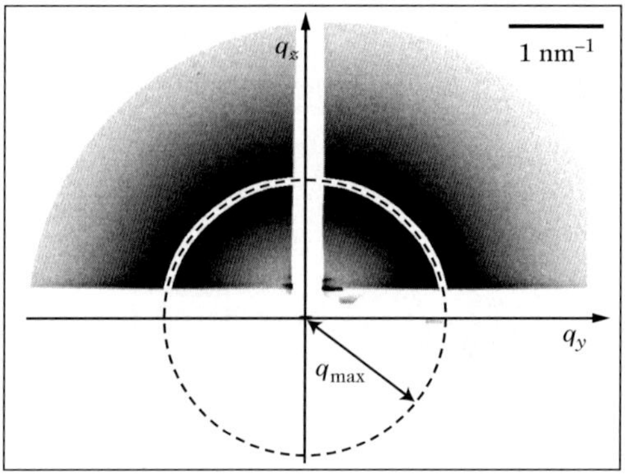

Fig. 25.7. GISAXS image (see text) of nanometric iron inclusions obtained by implantation of $4.8 \times 10^{16}\,\mathrm{cm}^{-2}$ Fe^+ ions in a carbon film. The circular form of the small-angle scattering ring shows that the nanoparticles are spherical and distributed isotropically throughout the matrix. From [27]

or silver [19] in silica glasses has been used to obtain third order nonlinear susceptibilities $|\chi^{(3)}|$ of around 10^{-7} esu, which is a value about 10^8 times higher than can be obtained with non-doped glasses, for applications in opto-electronics. In addition, in the field of magnetism, it is worth mentioning the observation of giant magnetoresistance effects (used today in read heads) in nanomaterials synthesised by implanting cobalt in copper [20].

Characterisation of Nanomaterials by Small-Angle X-Ray Scattering (SAXS)

When a nanostructured material containing a second phase in the form of dispersed nanoparticles is illuminated by an X-ray beam, the immediate angular neigbourhood of the transmitted beam is generally the site of scattering called small-angle X-ray scattering (SAXS). From the profile of the measured scattering, information can be drawn concerning the size, shape, quantity, and spatial arrangement of the nanoparticles. Small-angle X-ray scattering has thus become a favoured technique, along with transmission electron microscopy, for studying heterogeneous media, whether they be liquid, solid, amorphous, or crystalline [21].

Up until the 1980s, small-angle X-ray scattering was only used in transmission mode, the X-ray beam being directed along the normal to the sample surface. More recently, the technique has been adapted to the study of heterogeneities at the surface or in the surface region of a thick sample, where the scattering is difficult to measure in transmission owing to the significant attenuation as the beam crosses the underlying material [22, 23]. By working at grazing incidence (grazing incidence SAXS or GISAXS), the path travelled at the surface of the material increases considerably and it becomes possible to study supported nanoparticles [24] or thin nanopatterned films [25].

Ion implantation has also proved to be an excellent way of dispersing nanostructures of a refractory material such as carbon in metallic matrices. Indeed,

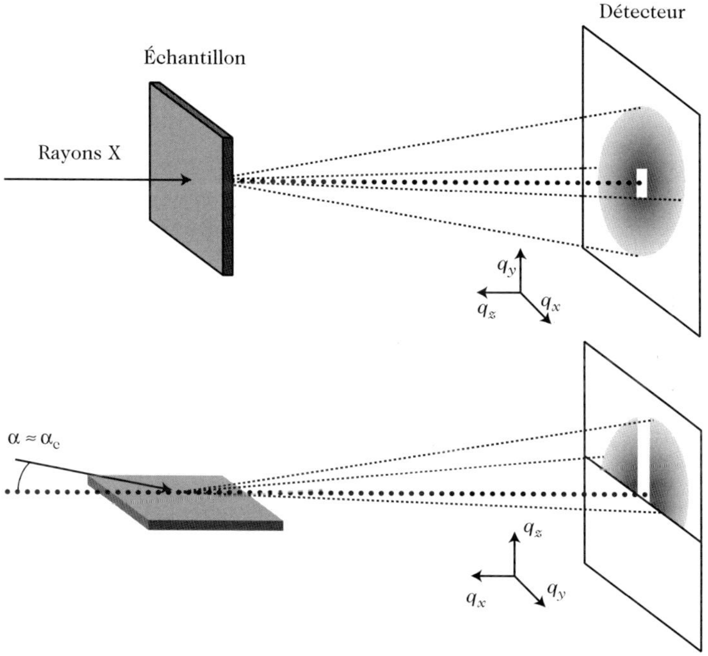

Fig. 25.8. *Upper*: Transmission mode SAXS. Scattering around the transmitted beam is due to the presence of nanoscale heterogeneities in the electron density within the sample. It contains information concerning the morphology and organisation of the nanostructures present throughout the bulk of the material. *Lower*: Grazing incidence SAXS (GISAXS). The angle of incidence is close to the critical angle of total reflection for the sample, in such a way that it is the refracted beam propagating at the surface of the material that plays the role of probe beam for analysing nanostructures at or just below the surface

it has been shown that high temperature ($> 500°$C) implantation of C^+ ions in copper or silver substrates could yield large quantities of structures composed of an arrangement of concentric carbon nanospheres, called carbon onions (see Fig. 25.9). During implantation, the substrate temperature, the fluence (in ions/cm^2), and the flux (in ions/cm^{2}s) are parameters that can be used to control the size distribution of these onions. Moreover, full evaporation of the metal by heat treatment in a vacuum after the implantation yields powders made almost exclusively from carbon onions, whereupon it may be possible to use them as a lubricant in nanomotors, for example.

The third example concerns the formation of voids in silicon-based semiconductors (see Fig. 25.10). These voids, produced by implantation of helium at high fluence ($> 10^{16}$ ions/cm^2) and subsequent annealing, have proved themselves to be efficient traps for metal impurities (proximity gettering of metal impurities) in the silicon [29]. As for metal clusters and carbon onions,

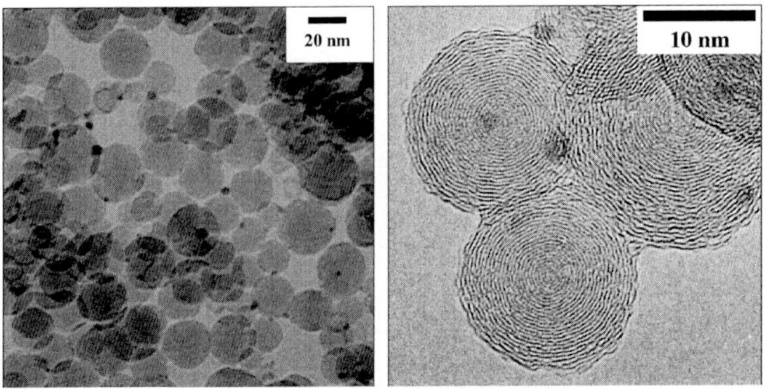

Fig. 25.9. TEM images of carbon onions obtained by C^+ ion implantation in a hot silver substrate. The size distribution of these nanostructures can be controlled by adjusting the substrate temperature during implantation, or indeed the fluence and flux of the implanted ions. From [28]

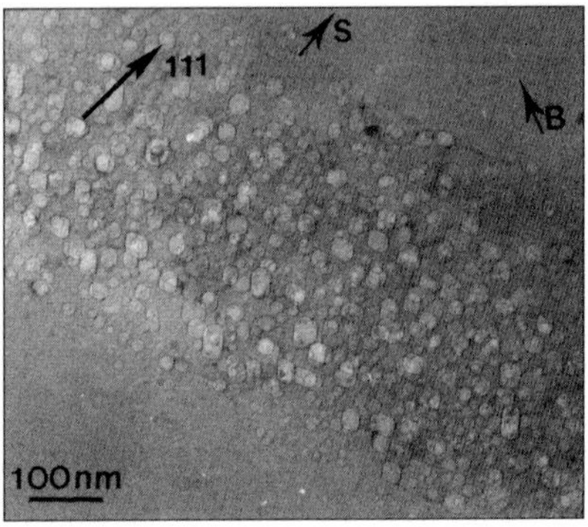

Fig. 25.10. TEM image showing a cross-section of a layer of bubbles obtained by implanting 1×10^{17} He^+/cm^2 at 1.6 MeV in a silicon monocrystal and annealing for 17 hr at 800°C. From [30]

the formation of these voids can be controlled by adjusting the implantation parameters, and one may envisage their incorporation into electronic devices.

Ion Implantation

Figure 25.11 shows a typical ion implantation setup [31]. A gas is ionised in a plasma source. The plasma ions are accelerated to the required energy in an electric field, then manipulated by electrostatic lenses and magnets to be implanted in a target material. Nucleation of the ions implanted in nanometric inclusions in the target can occur by diffusion, either during the implantation itself, or during subsequent annealing.

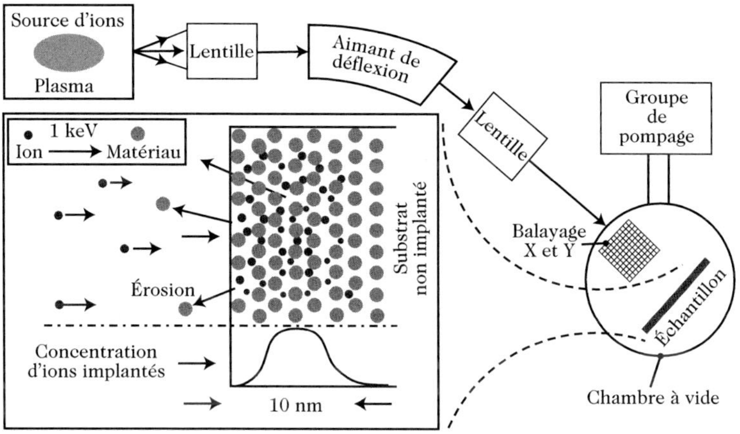

Fig. 25.11. Typical ion implantation setup [31], showing the behaviour of the ions implanted in a material and their effects on the composition and properties of the material

Advantages of Ion Implantation

- Ions of a given isotope can be introduced one by one, thereby keeping control of the concentration.
- The energy of the ions controls the depth and thickness doped by the inclusions.
- Any type of material can be treated in this way, from insulators to semiconductors and conductors.
- It is possible to treat only a part of the material by using a mask.
- Properties and composition of the material are not affected beyond a certain depth.

25.2.2 Vapour Deposition and Sputtering Methods

In many cases today, the synthesis of nanomaterials comprising metal or bimetal clusters in a matrix, in the form of thin films with homogeneous cross-section, is based on the same techniques as those used to make nanostructured coatings (see Chap. 24). These techniques, which use either simultaneous or alternate sputtering – or vapour deposition – of the material comprising the matrix and the material comprising the clusters, were described in the last chapter and will not be discussed further here. We shall simply present some examples to illustrate their use for dispersing nanoparticles in a host matrix.

Codeposition

Codeposition is the simultaneous deposition of atoms of two materials onto a substrate, one material forming clusters and the other the matrix. One advantage over the techniques described above is that the metal concentration

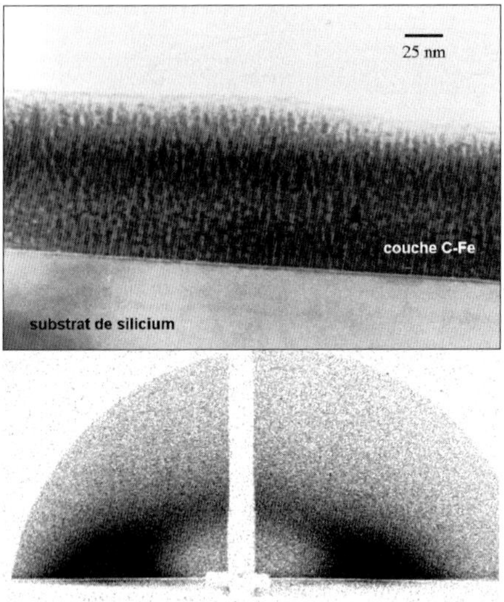

Fig. 25.12. TEM (*top*) and GISAXS (*bottom*) images of a thin film of iron clusters embedded in a carbon matrix, obtained by ion codeposition. The columnar structure of the clusters observed in the TEM image is confirmed by the elliptical shape of the small-angle scattering ring in the GISAXS image. From [34] © 2000 American Institute of Physics

in the matrix can vary from 0 to 100%. On the other hand, the metal concentration and the size of the nanoparticles cannot be controlled independently. When the codeposited elements are poorly miscible, metastable alloys are formed and heat treatment after growth causes segregation and the formation of nanoscale clusters dispersed randomly throughout the matrix. When the elements are immiscible, nucleation and the beginning of cluster growth occurs during deposition by diffusion of the deposited atoms on the instantaneous surface of the sample. The diffusion is then arrested by the coating process as more matter is deposited. This diffusion–aggregation process can then be reactivated by post-deposition heating. In some cases, this kind of growth by surface diffusion can induce column-type growth in the metal clusters (see Fig. 25.12) and the phenomenon can be amplified by carrying out the deposition with reactive assistance [32]. This is an interesting feature when considering the synthesis of clusters with high shape anisotropy, for applications to perpendicular magnetic recording, for example. Note also that, in the field of optics, the codeposition technique allows one to increase the nonlinear susceptibility by an order of magnitude, as compared with ion implantation, when making Au:SiO_2, Au:Al_2O_3, and Au:TiO_2 nanomaterials [33].

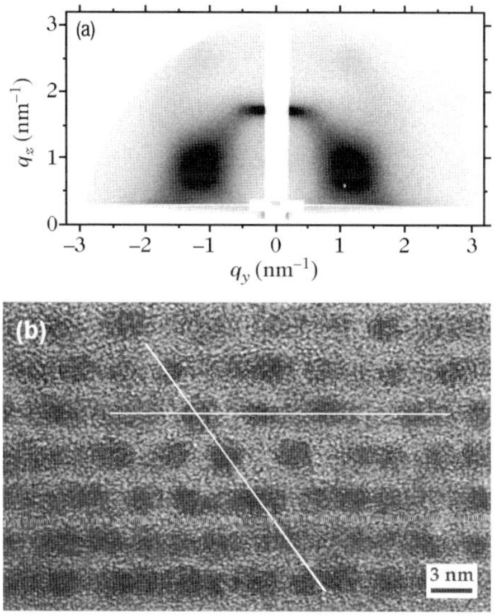

Fig. 25.13. GISAXS image (**a**) and TEM image (**b**) of a discontinuous multilayer (Co 7 Å/Al_2O_3 30 Å) synthesised by alternate sputtering. The self-organised growth revealed by spots on the GISAXS image and also observed by TEM despite projection effects, is attributed to preferential nucleation of cobalt clusters in hollows at the surface of the alumina layers. From [37] © 2000 American Institute of Physics

Alternate Deposition

Alternate deposition is used to obtain metal/insulator multilayers with adjustable nanometric period in which the metallic layers are discontinuous, i.e., take the form of separate 3D islands. The formation of these islands is achieved by the Volmer–Weber mechanism, characteristic of the growth of metallic films on insulators, where there is a large difference in surface energy between the two materials. This method differs from codeposition by the fact that the metal concentration and the nanoparticle size can be controlled independently, the latter generally being determined by the nominal thickness of each metallic nanolayer. Moreover, the vertical separation between two metallic nanolayers can be adjusted very accurately, being determined by the insulator thickness. Alternate sputtering has been used to achieve susceptibilities $|\chi^{(3)}|$ of 3×10^{-6} esu in discontinuous Au/SiO_2 multilayers with 20% volume concentration of gold and average particle size 2 nm [35]. Another novel and interesting feature of these alternating nanoscale deposits is that, depending on the respective thicknesses of metal and insulator, it is possible to induce spontaneous organisation of the clusters from one layer to the next (see Fig. 25.13), so that short-range magnetic interactions can be controlled, for example [36].

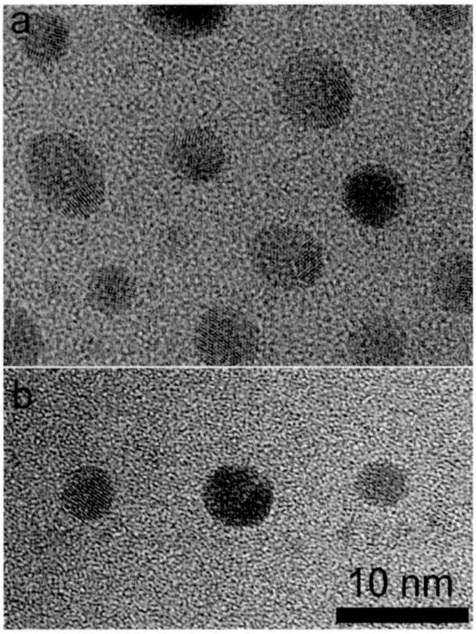

Fig. 25.14. TEM images of silver nanoparticles embedded in Al_2O_3, obtained by pulsed laser deposition. (**a**) Top view. (**b**) Cross-sectional view. Taken from [44]

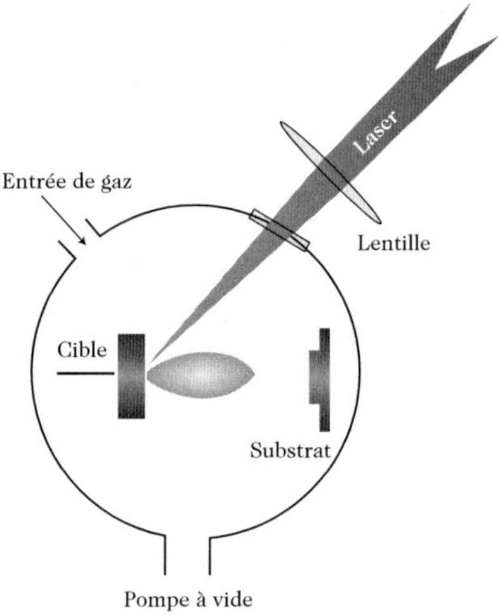

Fig. 25.15. Schematic view of pulsed laser deposition [45]

25.2.3 Pulsed Laser Deposition

Pulsed laser deposition (PLD) is another technique that has been used for several years now to synthesise discontinuous multilayers comprising metal nanoparticles dispersed in oxide matrices. The idea is to focus a pulsed laser beam with high power density (typically of the order of 10–100 MW/cm^2) on the surface of a material target [38]. To synthesise multilayers, the targets are placed successively in the beam and rotated to avoid local damage. The lasers used are usually excimer lasers emitting in the ultraviolet with a pulse width of the order of 20 ns and pulse frequency a few tens of Hz. Under the impact, the target is evaporated locally and the ejected matter, in the form of a plasma, is deposited on a substrate held in place opposite the target. The substrate temperature, the substrate–target distance, and the residual pressure in the chamber, together with the intrinsic characteristics of the laser beam, are parameters than can be exploited to control film growth. In addition, apart from its great flexibility, PLD allows very tight control over the composition of the deposited films. This is due to the fact that the stoichiometry of the target is conserved during transfer of matter onto the substrate. This method has thus proved extremely useful for synthesising Bi, Cu, Ag, Fe, or Co nanoparticles dispersed in a dielectric such as Al_2O_3, whose physical properties (nonlinear optical, magnetic, and magneto-optical properties) have been established in some detail (see Fig. 25.14) [39–43]. Note, however, that the homogeneous surface layer obtained by this method is not generally very wide.

Pulsed Laser Deposition (PLD)

Pulsed laser deposition involves ablation of a target by a laser beam and reception of the ejected matter on a substrate placed opposite the point of impact of the laser beam (see Fig. 25.15). Several features make PLD well-suited to the synthesis of metal clusters in a matrix by alternate deposition of continuous layers of an insulating material and discontinuous nanometric layers of a metal:

- The material is transferred stoichiometrically from the target to the substrate. This facilitates deposition of multi-element materials such as oxides (Al_2O_3, SiO_2, YBCO, etc.).
- There is a wide choice of materials that can be deposited, limited only by absorption at the laser wavelength.

25.2.4 Low Energy Cluster Beam Deposition (LECBD)

In contrast to the methods so far presented, the technique of low energy cluster beam deposition relies on first forming the clusters in free phase and then codepositing them with the matrix on a substrate. One way of producing these free clusters with optimal control over their size, structure, and composition is to use a laser vaporisation source, applying the idea put forward by R.E. Smalley [46] and later developed by P. Milani and W.A. de Heer [47]. A pulsed laser beam is focused on a rod, chosen in accordance with the clusters to be synthesised, and the atomic vapour produced near the surface is quenched by injecting a beam of rare gas. During the quench, the atoms ejected from the target are able to condense to form clusters before actually being deposited. The mixture then undergoes supersonic expansion in a vacuum by passing through a nozzle, and a beam of cold clusters with adjustable average

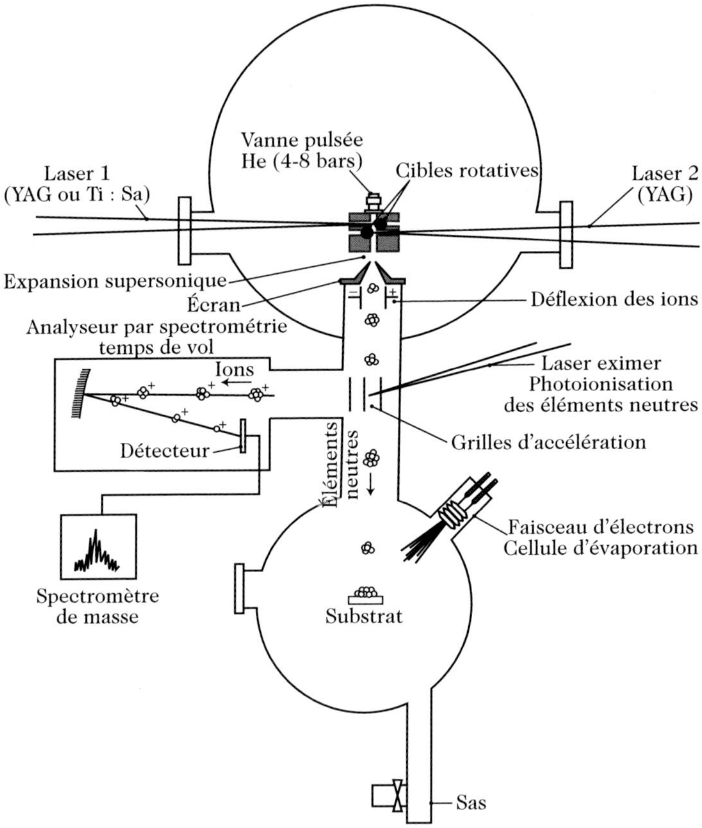

Fig. 25.16. Setup used for LECBD [49]

size is thereby obtained. At the source outlet, a system of deflecting plates extracts ionised species from the beam so that only neutral clusters end up in the deposit. The latter are directed toward the deposition chamber with negligible kinetic energy compared with the cohesive energy per atom and hence do not break up when they collide with the substrate. The matrix in which the clusters are embedded is simultaneously deposited by sputtering or vapour deposition from a second source. One advantage of this method for synthesising nanostructured materials is that the cluster concentration can be adjusted over a wide range of values by controlling the ratio of the speeds of the two beams, and this independently of their average size and their distribution within the matrix. This technique has recently been used to measure the magnetic anisotropy energy of a single cluster made up of 1 000 cobalt atoms in a niobium matrix and to reveal the dominant role of the surface [48].

Low Energy Cluster Beam Deposition (LECBD)

Figure 25.16 shows the experimental setup used to produce free clusters, to study them in the gas phase, and to deposit them to synthesise thin nanostructured films [49]. The clusters are produced in a laser vaporisation source which delivers a collimated beam of particles of different sizes. In a first stage, these clusters are analysed and studied in the gas phase using a mass spectrometer. The size distribution and chemical composition of the clusters are thereby assessed before they are deposited on different types of substrate. Samples made in this way can be analysed using techniques from solid state or surface physics, either in situ or after transfer (in vacuum or in air) to specific installations. To synthesise nanocomposite films in which the clusters are embedded in a matrix, the clusters are deposited simultaneously with the vapour of a material generated by an electron bombardment evaporator in the chamber [50].

References

1. U. Kreibig, M. Vollmer: *Optical Properties of Metal Clusters*, Springer, Berlin (1995)
2. H. Shiomi, K. Umehara: Materials Science Research International **6**, 144–149 (2000)
3. P. Rao, R. Doremus: J. Non-Cryst. Solids **203**, 202–205 (1996)
4. F.E. Wagner, S. Haslbeck, L. Stievano, S. Calogero, Q.A. Pankhurst, K.-P. Martinek: Nature **407**, 691–692 (2000)
5. Revue verre: Vol. 6, No. 5, October 2000
6. Q. Huo, D.I. Margolese, U. Ciesla, P. Feng, T.E. Gier, P. Sieger, R. Leon, P.M. Petroff, F. Schüth, G.D. Stucky: Nature **368**, 317–321 (1994)
7. S. Besson, T. Gacoin, C. Jacquiod, C. Ricolleau, D. Babonneau, J.P. Boilot: J. Mater. Chem. **10**, 1331–1336 (2000)
8. C. Estournès, T. Lutz, J. Happich, T. Quaranta, P. Wissler, J.L. Guille: J. Magn. Magn. Mater. **173**, 83–92 (1997)
9. B. Canut, A. Merlen, V. Teodorescu, C. Ghica, C.S. Sandu, S.M.M. Ramos, C. Bovier, R. Espiau de la Mestre, A. Broniatowxki, H. Bernas: Nucl. Instrum. Meth. B **209**, 335–339 (2001)
10. E. Valentin, H. Bernas, C. Ricolleau, F. Creuzet: Phys. Rev. Lett. **86**, 99–102 (2001)
11. T.M. Kyaw, Y. Okamoto, K. Hayashi: J. Mater. Sci. **32**, 5497–5503 (1997)
12. R. Guinebretière, Z. Oudjedi, B. Soulestin, A. Dauger: J. Mater. Res. **15**, 2482–2487 (2000)
13. Y.S. Tang, S.J. Cai, G.L. Jin, K.L. Wang, H.M. Soyez, B.S. Dunn: Thin Solid Films **321**, 76–80 (1998)
14. C. Renard, C. Ricolleau, E. Fort, S. Besson, T. Gacoin, J.P. Boilot: Appl. Phys. Lett. **80**, 300–302 (2002)
15. Ö. Dag, G.A. Ozin, H. Yang, C. Reber, G. Bussière: Adv. Mater. **11**, 474–480 (1999)
16. S. Besson: Doctoral thesis, Ecole Polytechnique, France (2002)
17. S.H. Joo, S.J. Choi, I. Oh, J. Kwak, Z. Liu, O. Terasaki, R. Ryoo: Nature **412**, 169–172 (2001)

18. K. Fukumi, A. Chayahara, K. Kadono, T. Sakaguchi, Y. Horino, M. Miya, K. Fujii, J. Hayakawa, M. Satou: J. Appl. Phys. **75**, 3075–3080 (1994)
19. K. Uchida, S. Kaneko, S. Omi, C. Hata, H. Tanji, Y. Asahara, J. Ikushima, T. Tokizaki, A. Nakamura: J. Opt. Soc. Am. B **11**, 1236–1243 (1994)
20. H. Errahmani, A. Berrada, S. Colis, G. Schmerber, A. Dinia, D. Muler: Nucl. Instrum. Meth. B **178**, 69–73 (2001)
21. A. Guinier, G. Fournet: *Small-Angle Scattering of X-Rays*, John Wiley, New York (1955)
22. J.R. Levine, J.B. Cohen, Y.W. Chung, P. Georgopoulos: J. Appl. Crystallogr. **22**, 528–532 (1989)
23. A. Naudon, D. Thiaudière: J. Appl. Crystallogr. **30**, 822–827 (1997)
24. G. Renaud et al.: Science **300**, 1416–1419 (2003)
25. D. Babonneau, I.R. Videnovic, M.G. Garnier, P. Oelhafen: Phys. Rev. B **63**, 195401/1–5 (2001)
26. A. Traverse, P. Parent, J. Mimault, S. Hagège, J. Du: Nucl. Instrum. Meth. B **84**, 204–207 (1994)
27. D. Babonneau, A. Naudon, T. Cabioc'h, O. Lyon: J. Appl. Crystallogr. **33**, 437–441 (2000)
28. T. Cabioc'h, J.C. Girard, M. Jaouen, M.F. Denanot, G. Hug: Europhys. Lett. **38**, 471–475 (1997)
29. D.M. Follstaedt, S.M. Myers, G.A. Petersen, J.W. Medernach: J. Elec. Mater. **25**, 157–170 (1996)
30. M.F. Beaufort, E. Oliviero, H. Garem, S. Godet, E. Ntsoensok, C. Blanchard, J.F. Barbot: Phil. Mag. B **80**, 1975–1985 (2000)
31. `www.inrs-ener.uquebec.ca/lafi/`
32. D. Babonneau, M. Jaouen, M.F. Denanot, P. Guérin, F. Petroff: Appl. Phys. Lett. **82**, 3056–3058 (2003)
33. H.B. Liao et al.: Appl. Phys. Lett. **70**, 1–3 (1997); Appl. Phys. B **65**, 673–676 (1997); Appl. Phys. Lett. **72**, 1817–1819 (1998)
34. D. Babonneau, J. Briatico, F. Petroff, T. Cabioc'h, A. Naudon: J. Appl. Phys. **87**, 3432–3443 (2000)
35. S. Debrus, J. Lafait, M. May, N. Pinçon, D. Prot, C. Sella, J. Venturini: J. Appl. Phys. **88**, 4469–4475 (2000)
36. F. Luis, F. Petroff, J.M. Torres, L.M. Garcia, J. Bartolomé, J. Carrey, A. Vaurès: Phys. Rev. Lett. **88**, 217205/1–4 (2002)
37. D. Babonneau, F. Petroff, J.L. Maurice, F. Fettar, A. Vaurès, A. Naudon: Appl. Phys. Lett. **76**, 2892–2894 (2000)
38. K.L. Saenger: Processing of Advanced Materials **2**, 1–24 (1993)
39. R. Serna, J.C.G. de Sande, J.M. Ballesteros, C.N. Afonso: J. Appl. Phys. **84**, 4509–4516 (1998)
40. J.M. Ballesteros, R. Serna, J. Solis, C.N. Afonso, A.K. Petford-Long, D.H. Osborne, R.F. Haglund Jr.: Appl. Phys. Lett. **71**, 2445–2447 (1997)
41. J.C.G. de Sande, R. Serna, J. Gonzalo, C.N. Afonso, D.E. Hole,A. Naudon: J. Appl. Phys. **91**, 1536–1541 (2002)
42. N.M. Dempsey, L. Ranno, D. Givord, J. Gonzalo, R. Serna, G.T. Fei, A.K. Petford-Long, R.C. Doole, D.E. Hole: J. Appl. Phys. **90**, 6268–6274 (2001)
43. V. Dureuil, C. Ricolleau, M. Gandais, C. Grigis, J.P. Lacharme, A. Naudon: J. Cryst. Growth **233**, 737–748 (2001)
44. J.P. Barnes, A.K. Petford-Long, R.C. Doole, R. Serna, J. Gonzalo, A. Suarez-Garcia, C.N. Afonso, D. Hole: Nanotechnology **13**, 465–470 (2002)

45. C.N. Afonso: *Insulating Materials for Optoelectronics*, World Scientific, Singapore (1995)
46. R.E. Smalley: Laser Chem. **2**, 167 (1983)
47. P. Milani, W.A. de Heer: Rev. Sci. Instrum. **61**, 1835–1838 (1990)
48. M. Jamet, W. Wernsdorfer, C. Thirion, D. Mailly, V. Dupuis, P. Mélinon, A. Perez: Phys. Rev. Lett. **86**, 4676–4679 (2001)
49. V. Dupuis, L. Favre, S. Stanescu, J. Tuaillon-Combes, E. Bernstein, A. Perez: *Review of Recent Research Developments in Magnetism and Magnetic Materials*, Transworld Research Network (2003)
50. hplasim2.univ-lyon1.fr/recherche/nano/nano_acti_exp.html

26

Nanoporous Media

J. Patarin, O. Spalla, and F. Di Renzo

26.1 Introduction

This chapter is concerned with ways of synthesising nanoporous solids, and in particular materials with a very narrow pore size distribution. Two main families of solids are relevant here: crystalline microporous solids ($0.3 <$ pore diameter < 1.4 nm) comprising zeolites and related solids (metallophosphates), and ordered mesoporous solids ($2 <$ pore diameter < 30 nm) which, in contrast to the zeolites, possess amorphous inorganic walls (see Sect. 11.4). In both cases, these solids are obtained by introducing inorganic or organic cations, or again molecular assemblages (surfactant micelles) into the reaction medium. These species then act as molecular 'templates' around which the inorganic matter crystallises or polycondenses. After synthesis, these are occluded within its pores and the porous material is obtained when they are extracted or eliminated by calcination. The structure-directing role of these template species will be discussed and illustrated by several examples, and the main mechanisms for forming this kind of nanoporous solid will be presented. X-ray and neutron scattering techniques, which have proved to be particularly well suited for characterising ordered mesoporous solids, will be discussed at the end of the chapter.

26.2 Synthesis of Crystalline Microporous Solids

26.2.1 Methods of Synthesis

Crystalline microporous solids are usually obtained by hydrothermal crystallisation of a gel comprising a liquid and a solid phase. The reaction medium contains:

- sources of framework elements T = Si, Al, P, etc.,
- sources of mineralising agents OH^-, F^-,

Table 26.1. Sources of T elements and mineralising agents

T elements, mineralising agents	Reactants
Si	Colloidal silica, silicon alkoxide [TEOS, $(C_2H_5O)_4Si$], precipitated silica, pyrogenic silica (aerosil), sodium silicate, etc.
Al	Aluminium hydroxide, sodium aluminate, aluminium alkoxide, etc.
P	H_3PO_4, etc.
OH^-	NaOH, KOH, etc.
F^-	NH_4F, NH_4HF_2, HF, etc.

- inorganic cations and/or organic species (neutral molecules or cations),
- a solvent (usually water).

Table 26.1 shows the main sources of T elements and mineralising agents.

After intimate mixing of the different reactants, followed in some cases by an aging period at room temperature (ripening), the reaction mixture is placed in stainless steel autoclaves lined with a protective coating of polytetrafluoroethylene (PTFE, Teflon) to prevent corrosion, and heated.

Zeolites are generally synthesised in a basic medium at temperatures below 200°C and low pressures (autogenous pressure < 20 bar).[1] In the case of related microporous materials (aluminophosphates, etc.), the pH of the reaction medium is general in the range 3–10. The anions OH^- or F^- solubilise the reactive species in the gel (formation of silicates and fluorosilicates) and enable their transfer to the growing crystals. Apart from their mineralising role, the fluoride anions can play a costructuring role by stabilising certain units in the building of the inorganic structure. The time required for this kind of synthesis is extremely variable, but generally somewhere between a few hours and several days.

The product obtained after synthesis, usually called the as-synthesised product, is washed with distilled water, filtered and then dried. This product contains in its pores the inorganic and/or organic cations required for its synthesis (see Sect. 26.2.4) and is not therefore porous. Porosity is reinstated after cation exchange (inorganic cations) and/or calcination (ammonium cations and organic species). Heat treatment is carried out in an oxidising atmosphere (air) at temperatures of the order of 500°C. It is thus essential that these solids should exhibit excellent thermal stability. The latter generally depends on the chemical composition of the inorganic framework. The porosils (crystalline microporous solids with chemical formula SiO_2, see Sect. 11.3) are the most stable materials, with structure remaining intact at temperatures above

[1]The autogenous pressure depends on the temperature and generally follows the power law $[T\,(°C)/100]^4$.

1 000°C. The aluminosilicates and metallophosphates have poorer thermal stability, but above 600°C.

These materials have also been synthesised in non-aqueous media, using ethylene glycol as solvent for example [1, 2], or from dry synthesis gel [3]. However, traces of water are often present in the reactants used. Water may also be produced during the crystallisation process [3, 4]. As an example, we now describe in some detail the synthesis of the LTA zeolite (see Sect. 11.3).

Synthesis of the Linde Type A Zeolite

The Linde type A (LTA) zeolite can be synthesised in many different ways. We describe here the one published in the review *Verified Synthesis of Zeolitic Materials* [5]. In this case, the reaction gel has the following molar composition, expressed in oxide:

$$1.926\ SiO_2 : 3.165\ Na_2O : 1\ Al_2O_3 : 128\ H_2O\,.$$

The reactants used are sodium metasilicate ($Na_2SiO_3{\bullet}5H_2O$, Fisher Scientific), sodium aluminate (NaO_2 : Al_2O_3 : $3H_2O$, Fisher Scientific), caustic soda (NaOH, Fisher Scientific, 99%), and distilled water.

In order to make 10 g of zeolite A (dehydrated), 0.723 g of sodium hydroxide is dissolved in 80 ml of distilled water, and the solution is then divided into two parts and placed in closed polypropylene flasks. In one of the flasks, 8.258 g of sodium aluminate is added and the result is shaken until a clear solution is obtained. In the second flask, 15.48 g sodium metasilicate is dissolved. This solution is then quickly poured into the sodium aluminate solution, whereupon a thick gel is formed. After homogenising in the closed polypropylene flask, the whole ensemble is placed in an oven and heated at 100°C for 3–4 hr in static or stirred conditions. After cooling, the synthesised product (the LTA zeolite) is filtered, then washed with distilled water until the water remains roughly pH neutral (pH = 7–8).

26.2.2 The Crystallisation Process Exemplified by Zeolites

Many studies have been carried out to determine the crystallisation mechanism for these microporous solids. Two hypotheses have been put forward. The first assumes that crystallisation occurs through a gradual reorganisation of the solid phase of the gel, a process in which the liquid phase is not directly involved. This hypothesis has been more or less abandoned at the present time in favour of the second, which assumes that zeolite crystals form in solution or at the gel–solution interface, the seeding and growth of crystals resulting from condensation reactions between certain species present in solution. The gel then only serves as a reservoir for the reactants. Figure 26.1 illustrates this second hypothesis schematically.

It is hard to understand how microporous structures as elaborate as the zeolites could be made in solution from such simple units as TO_4 tetrahedra

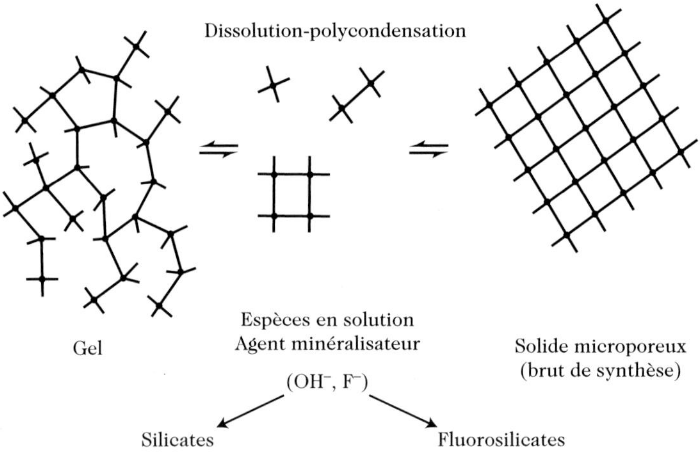

Fig. 26.1. Schematic view of the way crystalline microporous solids may form

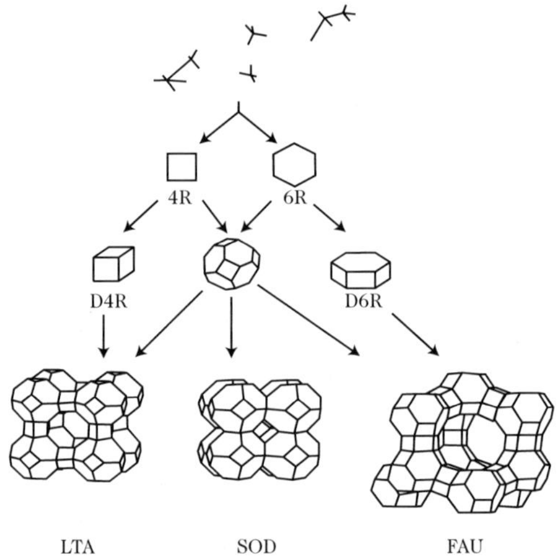

Fig. 26.2. Obtaining the LTA, SOD (sodalite) and FAU (faujasite) structures from 4R and 6R SBUs, with rings of 4 and 6 tetrahedra, respectively. In the structural models, the T elements in the framework are located at the intersection of edges, with oxygen atoms in the middle of the edges

(where T is Si, Al, etc.). It has thus been suggested that more elaborate building blocks may be present in the solution. These building blocks, of which 20 have so far been proposed [6], and which go by the name of secondary building units (SBU), have been determined from structural data. Hence the construction of the FAU (faujasite), LTA (Linde type A) and SOD (sodalite) structures have been based on 4R and 6R SBUs, with rings of 4 and 6 tetrahedra, respectively, as shown schematically in Fig. 26.2.

Although this model seems satisfactory in some ways, since it provides an inorganic prefab for the final construction, it nevertheless remains poorly corroborated owing to the fact that most of the SBUs have never been identified in the solution experimentally. This does not mean that such units do not exist. They may indeed be present in quantities that are too small to be detected as yet. Other models, also based on structural analysis or obtained by molecular modelling, have been proposed to explain the formation of gallophosphates [7] and aluminophosphates [8].

26.2.3 Main Organic Structure-Directing Agents Used to Synthesise Crystalline Microporous Solids

Although zeolites were first synthesised in the presence of inorganic structure-directing agents (alkali and alkaline earth cations), the use of organic species in the reaction medium considerably increased and diversified the number and kinds of inorganic solids that could be made.

The most widely used structure-directing agents are amine derivatives (amines, quaternary ammonium cations, etc.), cyclic or linear ethers, and also coordination compounds (organometallic complexes). The structure-directing agent must remain intact in the reaction medium. It must therefore be chemically and thermally stable. However, decomposition of the organic agent in the reaction medium can also be put to good use to generate the structure-directing species in situ and thereby produce novel materials. This has been illustrated by the synthesis of new microporous aluminophosphates in the presence of alkylformamides. Partial decay of the alkylformamides leads to the corresponding alkylamines, which are then occluded within the pores of the structure [9, 10]. Table 26.2 shows several structure-directing agents used in the synthesis of porosils.

26.2.4 Role of Inorganic Cations and Organic Species

The inorganic cations and/or organic species are occluded within the micropores of the synthesised material. They thus contribute to stabilising the inorganic structure. Stabilising interactions may be Coulomb, hydrogen bond, or van der Waals guest–framework interactions, or they may be guest–guest interactions. This is the case for example with the 18-crown-6/Na^+ complex in the synthesis of the zeolite with structure type EMT, or the paradioxane/Na^+ complex in the synthesis of the zeolite with structure type MAZ.

The various effects of the inorganic cations and organic species can be summed up as follows:

- balance the negative charge of the inorganic framework, e.g., in aluminosilicates,
- fill up the micropores,

Table 26.2. Structure-directing agents used to synthesise porosils (polymorphic varieties of SiO_2)

Organic species	Structure code of porosil
N,N,N-trimethyl-1-adamantammonium cations	AFI
Dibenzyldimethylammonium cations	BEA
Ethylene glycol or trioxane	SOD
Tetrapropylammonium cations	MFI
Di-*n*-propylamine	MTT
Diethylamine	TON
Tetramethylammonium cations	MTN
3,5-dimethyl N,N-diethylpiperidinium cations	MEL
4,4′-trimethylenedipiperidine	MTW
Quinuclidinium cations	AST

- provide structure-directing or template effect by prior organisation of the inorganic species around the organic entity and/or orientation of the crystallisation by adaptation to the shape and symmetry of the organic entity,
- act chemically by modifying the properties of the gel and/or the solution (hydrophilic or hydrophobic character of the inorganic cations and organic species),
- act thermodynamically by stabilising the building units of the inorganic framework.

The specific role of the inorganic cations and organic species is often difficult to ascertain exactly, particularly when the two entities coexist in the final material. This is often the case for zeolites with low Si/Al molar ratios. It is also difficult to establish a clear correlation between the kind of inorganic cation and the resulting microporous structure. However, certain structures can only be obtained in the presence of some specific inorganic cation. Thus the zeolite with structure type FAU can only be obtained in the presence of sodium cations. Likewise, OFF and RHO zeolites require the presence of potassium and caesium cations, respectively [11, 12].

The porosils (microporous silicas) and zeolites rich in silica ($5 < Si/Al < \infty$) are the simplest systems for which a template effect is observed due to the organic species.

26.2.5 Organic Species and the Template Effect

In some cases, there is a good match between the geometries of the organic species and the system of channels in the microporous solid. The organic entity then genuinely plays the role of a template, around which the inorganic framework can build itself. The structuring of water may be an important

Fig. 26.3. TMA^+ cations occluded in the sodalite cage. In this representation, the T elements (T = Si, Al, etc.) making up the cage are located at the intersection of the edges and the oxygens atoms in the middle of the edges

step in explaining this template effect. Moreover, it is worth noting the close analogy between certain clathrates (structures made up of structured water molecules) and certain silicates (clathrasils). This is the case in particular for the tetramethylammonium clathrate $(CH_3)_4NOH{\bullet}5H_2O$, which has the structure of the zeolite sodalite. This zeolite can also be synthesised in the presence of tetramethylammonium cations (TMA^+). The same goes for the structural equivalents zincophosphate (framework Zn,O,P) and aluminophosphate (framework Al,O,P), the cations TMA^+ fitting perfectly into the sodalite cage (see Fig. 26.3).

In addition, a whole new family of hybrid clathrate–silicate materials has been described in the literature [13], strengthening the analogy between these two families of compounds. In the clathrates, water structures itself around large, low charge anions and cations, of hydrophobic nature, and slightly polar molecules like the organic cations [e.g., $(C_3H_7)_4N^+$] in a rejection reaction, isolating them in polyhedral cavities.

The analogies between clathrates and silicates, the existence of hybrid clathrate–silicate compounds, and the structuring of water in the presence of hydrophobic organic species all tend to suggest that the porous inorganic structure results from a kind of water–inorganic species 'replication', followed by condensation of the organic–inorganic species thereby formed. This template effect can be illustrated by the synthesis of silicalite-1 (structure type MFI).

Silicalite-1 with Structure Type MFI

Silicalite-1 is a microporous silica of structure type MFI (see Table 26.1). The crystals, illustrated in Fig. 26.4a, have the form of long parallelepipeds. This structure type is characterised by the presence of two types of interconnected channel, with apertures delimited by 10 T atoms with diameter 0.55 nm (see Fig. 26.4b).

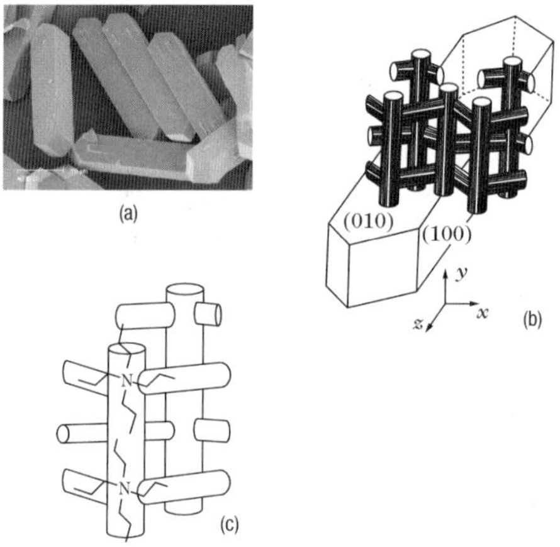

Fig. 26.4. MFI structure type. (**a**) Crystals of silicalite-1. (**b**) Channel system. (**c**) Location of tetrapropylammonium cations at channel intersections

Silicalite-1 can be synthesised in the presence of several organic species. However, a pure form is only obtained when tetrapropylammonium cations are used ($[CH_3(CH_2)_2]_4N^+$, denoted TPA^+). These cations are occluded within the inorganic framework at the intersection of two channel systems, as shown in Fig. 26.4c, with the propyl branches $[CH_3(CH_2)_2]$ pointing into the channels at each intersection. Several studies have revealed the existence of an organic–inorganic entity during the synthesis [14–16], a genuine precursor of the zeolite network. This entity appears to result from the substitution of the hydrophobic hydration layer of TPA^+ cations by silica species. This water–silicate 'replication' might then be thermodynamically favoured, with van der Waals interactions and destructuring of the water molecules contributing respectively to the enthalpy and entropy of the system. The assembly of organic–inorganic entities and the growth of nucleation seeds thus formed would then produce silicalite-1 (see Fig. 26.5).

26.2.6 Porosity of Zeolites and Related Solids

The pore characteristics of zeolitic materials are determined by adsorption of probe molecules, i.e., organic molecules like *n*-hexane or mesitylene, or gases such as nitrogen, argon, and so on. Hence, the adsorption–desorption isotherms of nitrogen at the temperature of liquid nitrogen can be used to evaluate the diameter of the micropores, the internal surface area of the pore

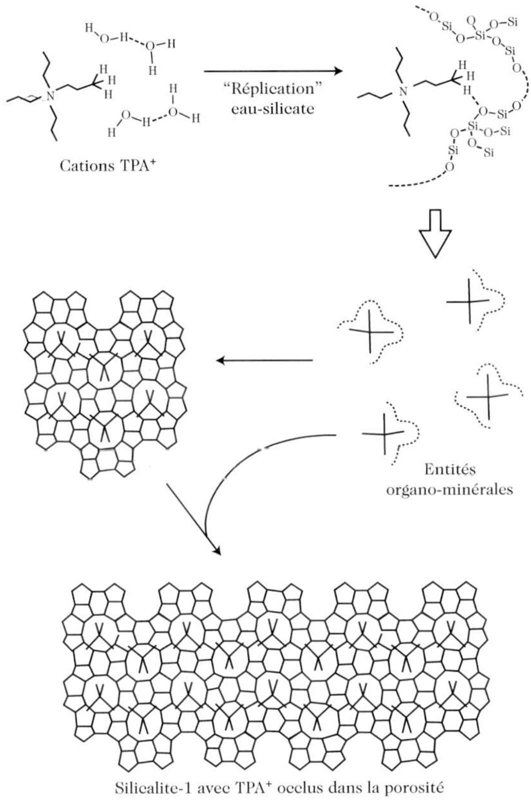

Fig. 26.5. The role of the TPA^+ cation in the silicalite-1 formation mechanism. Adapted from [14]

structure (equivalent[2] specific surface area), and the micropore volume. The average values of these characteristics are as follows:

- diameters of pore apertures 0.28–1.35 nm,
- micropore volume 0.1–0.4 mL/g,
- equivalent specific surface area 300–900 m^2/g.

26.2.7 Applications of Zeolitic Materials

With their perfectly controllable pore structure and pores of molecular dimensions, zeolitic materials are genuine molecular sieves, endowed with shape selectivity properties. The presence of charge compensating cations (alkaline

[2]The equivalent specific surface area for the micropores accounts for the fact that the internal surface area measured by nitrogen adsorption is underestimated. The shape of the nitrogen molecule does not allow it to completely cover the pore area. The same remark can be made for the determination of the micropore volume.

earth, alkali cations, protons, etc.) occluded within the pores of the inorganic framework bestows ion exchange properties and catalytic properties on these materials, which are widely used in industry. In addition, the hydrophobic nature (zeosils, SiO_2) or hydrophilic nature (aluminosilicates) of the inorganic framework, which can be adjusted at will, mean that these solids can be used as specific adsorbents to trap organic molecules or water molecules, and this in either gas or liquid phase.

The three main applications of zeolitic materials are:

- adsorption–drying agents,
- catalysts,
- detergents.

To illustrate the different applications, one can mention adsorption of water molecules (double glazing), gas drying, catalytic cracking (fuel production, increasing the octane number in petrols), and phosphate-free washing powders (fixing of calcium and magnesium). As an indication Figs. 26.6a and b show the consumption in thousands of tonnes and the distribution as a volume percentage for the use of these materials throughout the world in the year 1988 [17]. More recent data [18], relating to the production of synthetic zeolites, shows that the tonnages for the three main applications progressed enormously over the following ten years, since they reached values of 1 050 000, 160 000 and 100 000 for detergents, catalysts, and adsorption–drying agents, respectively, in 1998. Detergents represent almost 70% of the market, the formulation of some washing powders containing up to 40% by weight of zeolite A (LTA structure type).

Many other applications, less important in tonnage and generally using natural zeolites, have also been developed, e.g., for water treatment, processing of nuclear waste, animal feed, or soil improvement. More recently, many patents propose to make use of the anti-bacterial properties of these materials [19, 20] or their ability to delay combustion.

In catalysis, most zeolites are used in the refining and petrochemical industries. In refining, the main applications concern cracking (fluid catalytic cracking or FCC), hydrocracking, C5/C6 isomerisation, and dewaxing. In the petrochemical industry, applications include the transformation of aromatics (isomerisation, alkylation, and dismutation and trans-alkylation reactions).

The use of zeolites in organic fine chemical processes (to synthesise organic molecules with specific properties) is already important and will continue to develop in the years to come. Indeed, heterogeneous catalytic processes, in contrast to homogeneous ones, generally operate on a fixed bed, limit or even completely avoid the problems of corrosion, and allow the catalyst to be recovered.

In this area, among the reactions carried out with zeolitic catalysts, it is worth mentioning the isomerisation of double bonds, skeletal isomerisation, dehydration, dehydrogenation, halogenation, acylation or alkylation of aromatic compounds, selective oxidation or hydrogenation, and many others

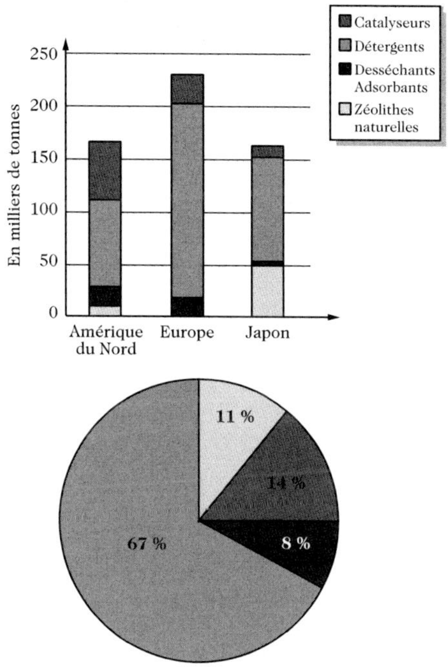

Fig. 26.6. Consumption in thousands of tonnes (**a**) and distribution as a volume percentage of the use of zeolitic materials in the world (**b**) for the year 1988. Adapted from [17]

[21–23]. One of the most striking examples of the last few years has been the industrial development of titanosilicalite TS-1 (MFI zeolite rich in titanium-doped silica) [24] as a phenol hydroxylation catalyst [25].

However, it should be noted that, of the 167 zeolitic structure types so far recorded, only a dozen or so are used in industry or have a significant industrial potential. Table 26.3 lists the main zeolitic materials. Recently, a more detailed review of the catalytic applications of these solids has been published [26].

26.3 Synthesis of Ordered Mesoporous Solids

26.3.1 Methods of Synthesis

Methods for synthesising ordered mesoporous solids are very similar to those described previously for crystalline microporous solids. They differ in the type of organic template introduced into the reaction medium, but also by conditions of synthesis, which are generally milder (temperatures in the range 25–150°C).

The reaction medium contains:

Table 26.3. Zeolitic structures with industrial potential or already in use. Adapted from [18]

Structure type	Application
LTA	Detergency, drying and separation
FAU, MOR	Adsorption and catalysis
LTL	Catalysis: aromatisation
MFI	Adsorption and catalysis
BEA	Catalysis
CHA (silicoaluminophosphate)	Catalysis: conversion of methanol into olefins
FER	Catalysis: skeletal isomerisation of n-butenes
AEL (silicoaluminophosphate), TON	Catalysis: isomerisation of paraffins; reducing the pour point of diesel oils
MTW	Catalysis

- a source for the T element (Si, Al, Ti, etc.),
- an organic structure-directing agent (amphiphilic molecules: amine or quaternary ammonium cation with a long hydrocarbon chain, polyethylene oxide, block copolymers, etc.),
- a solvent (usually water),
- a pH modifier (acid or base).

Time of synthesis is usually shorter too, varying from a few hours to several days. Depending on the temperature, the reaction mixture is placed in polypropylene flasks or in PTFE-lined stainless steel autoclaves. As for the crystalline microporous solids, the product is washed with distilled water, filtered and dried. The organic agent is contained within its pores and can be eliminated by calcination in air at temperatures of the order of 500°C. In these materials, interactions between the organic species and the inorganic walls are generally weaker than those encountered for the crystalline microporous solids and, since the pore diameters are greater, the organic template can in some cases be eliminated by extraction with an organic solvent.

Synthesis of MCM-41 (Mobil Composition of Matter 41)

The MCM-41 phase (see Sect. 11.4) was originally obtained by hydrothermal synthesis (70–150°C) from silicate or aluminosilicate gels, in a basic medium and in the presence of quaternary ammonium cations of the form $C_nH_{2n+1}(CH_3)_3N^+$ (C_nTMA with $8 < n < 18$). The surfactant[3] most commonly used is C_{16}TMA (hexadecyltrimethylammonium cation).

[3]Surfactant micelles are molecular arrangements capable of modifying the surface tension of a solvent, i.e., molecules comprising a polar hydrophilic part and a non-polar hydrophobic part.

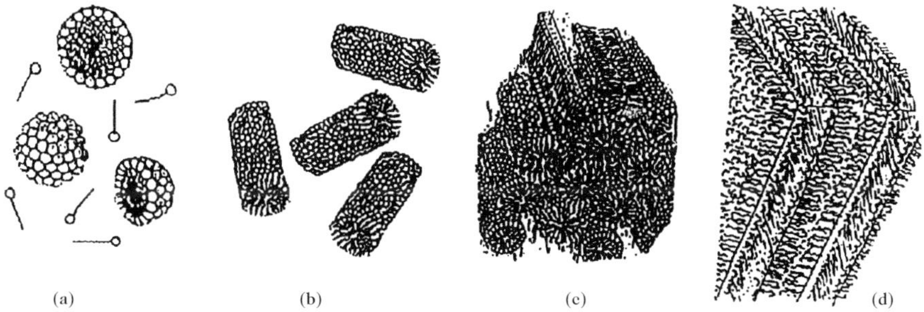

Fig. 26.7. Different micelle arrangements. (**a**) Spherical micelles. (**b**) Cylindrical rods. (**c**) Hexagonal arrangement of cylindrical micelles. (**d**) Lamellar phase

The channel diamter in the MCM-41 phase can be adjusted by changing the length n of the alkyl chain of the surfactant. Organic additives, such as alkanes or mesitylene (1,3,5-trimethylbenzene), can also be added to the reaction medium to increase the diameter of the surfactant micelles. Thus, in the presence of 1,3,5-trimethylbenzene, MCM-41 materials with channel diameters of 10 nm have been synthesised [27].

Since the Mobil patents, silica-based or other MCM-41 type materials, i.e., having an ordered hexagonal arrangement of mesopores, have been obtained with a very wide range of compositions, and at a range of temperatures (20–150°C), in basic, neutral or acidic media, and in the presence of a variety of surfactants.

26.3.2 Definition and Role of the Surfactant

Surfactants are amphiphilic, i.e., they comprise a non-polar hydrophobic or lipophilic part and a polar hydrophilic part. Surfactants are classified according to the nature of their hydrophilic part into cationic, neutral and anionic surfactants.

An aqueous micelle solution can be considered as a kind of two-phase medium. Indeed, above a certain surfactant concentration called the critical micelle concentration (CMC), the surfactant molecules join up to form micelles. As the surfactant concentration increases, various micelle arrangements are observed. Thus in the case of hexadecyltrimethylammonium cations ($C_{16}TMA^+$), as the surfactant concentration in the solution is increased, one observes the formation of spherical micelles, then rods of cylindrical micelles, then hexagonal followed by cubic arrangements of cylindrical micelles, and finally, for high surfactant contents, the formation of a lamellar phase. Several micelle arrangements are shown schematically in Fig. 26.7.

Geometric considerations based on the ratio of the volume V occupied by the hydrocarbon chain of the surfactant molecule and the surface area a_0 of its polar head (Fig. 26.8) can be used to explain and predict these different

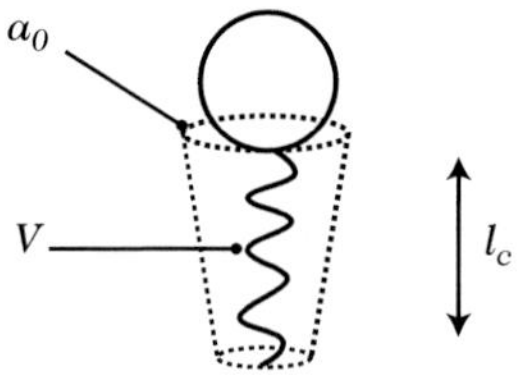

Fig. 26.8. Amphiphilic molecule. a_0 is the optimal effective surface area of the polar head of the surfactant, l_c is the length of its hydrocarbon chain when straightened out, and V is its volume

Table 26.4. Micelle structures corresponding to different values of the stacking factor g, where $g = v/l_c a_0$, with a_0 the effective surface area of the polar head of the surfactant, l_c the length of its hydrocarbon chain when straightened out, and V is its volume

Value of g	Structure
$g < 1/3$	Spherical micelles
$1/3 < g < 1/2$	Cylindrical micelles
$1/2 < g < 1$	Lamellar micelles

arrangements. This ratio g, called the shape factor or stacking parameter, relates the molecular structure of the amphiphilic molecule to the architecture of the aggregates. Table 26.4 specifies several types of arrangement obtained as a function of the value of g.

26.3.3 Mechanisms for the Formation of MCM-41 Phase

These different arrangements underlie the mechanisms put forward by the Mobil research group to explain the formation of the mesoporous silicas with codes MCM-41, MCM-48 and MCM-50. MCM stands for Mobil composition of matter. These are mesoporous silicas patented by Mobil, with hexagonal symmetry (number 41), cubic symmetry (number 48), and lamellar structure (number 50). In the case of MCM-41, a hexagonal arrangement of channels, the authors consider that the surfactant molecules $C_{16}TMA^+$ are ordered in the reaction medium in the form of cylindrical micelles [27]. The quaternary ammonium hydrophilic parts bestow an overall positive charge (S^+) on these micelles. During the reaction, the silicate anions (I^-) arrange themselves around these micelles by virtue of the electrostatic interactions (S^+I^-), the MCM-41 phase resulting from the gradual polycondensation of these inorganic species.

Figure 26.9 illustrates the formation mechanism put forward by these authors. Known as liquid crystal templating (LCT), it has two variants:

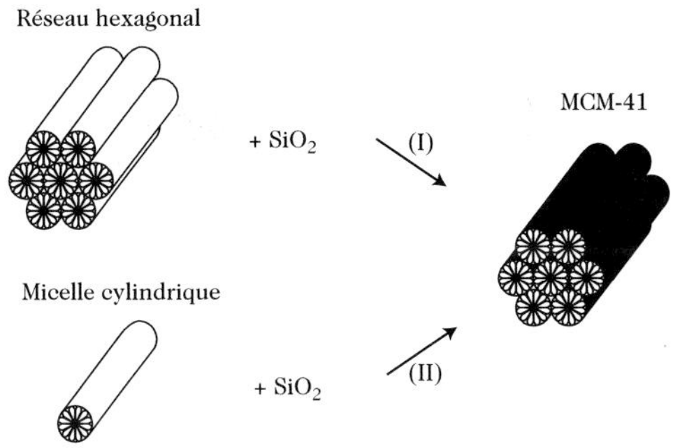

Fig. 26.9. Mechanisms put forward by Beck et al. [27] for the formation of the MCM-41 phase

- The cylindrical surfactant micelles form a hexagonal arrangement, and then the inorganic species polycondense around it to form the walls of the mesoporous solid (see Fig. 26.9, mechanism I).
- The second variant is based on a direct interaction between the inorganic species and the rods of cylindrical micelles, in which the hexagonal array of rods coated in this way leads, after polycondensation of the inorganic species, to the MCM-41 material (see Fig. 26.9, mechanism II).

An important step forward was made in understanding the mechanisms for formation of MCM-41 materials with the work of G.D. Stucky and coworkers at the University of California at Santa Barbara. According to these authors [28–30], the formation of surfactant/silicate structures results from cooperative organisation between organic and inorganic molecular species. Indeed, the MCM-41 phase can be obtained with much lower surfactant concentrations in the reaction medium than those required for the formation of cylindrical micelles, and with silica concentrations for which no precipitation is observed in the absence of any surfactant.

The proposed mechanism is shown in Fig. 26.10. To begin with, multiple bonds form between silica oligomers [double-four ring or D4R units: octamers $(Si_8O_{20})^{8-}$, in particular] and surfactant molecules (monomers). Depending on the surfactant concentration, these molecules are in dynamical equilibrium with the micelle arrangements (spherical or cylindrical micelles). In this process, the micelles serve mainly as a monomer reservoir. The organic–inorganic structures formed in this way initiate the rapid nucleation and precipitation of an ordered network of liquid crystal type, with geometry determined by the charge equilibrium and van der Waals interactions. At this stage, there is little condensation of silica (see Fig. 26.10). The latter is brought about by increasing the duration and/or temperature of the reaction (see Fig. 26.10, stage D). In contrast to the mechanisms proposed at the Mobil research laboratory, this model does not require any prior formation of specific micelle arrangements.

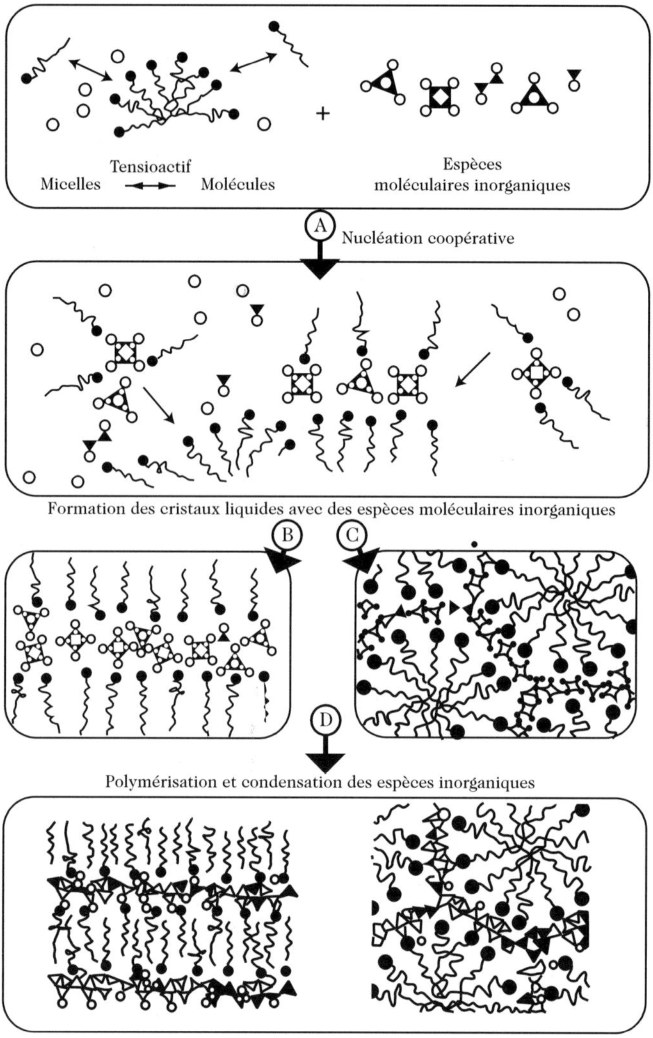

Fig. 26.10. Schematic view of cooperative organisation between surfactant and inorganic species, due to Huo et al. [29]

The degree of polycondensation of the inorganic species depends on parameters like the pH, the duration, and the temperature of the synthesis reaction. A low level of polycondensation (high charge density) would favour a lamellar arrangement of the organic–inorganic compound, while an increase in the degree of polycondensation (lower charger density) would lead to a material with hexagonal symmetry.

Many of the proposed mechanisms assume ion exchange in solution between silicate species and counterions[4] associated with the cationic surfactant (Br^- for hexadecyltrimethylammonium bromide) at the surface of the surfactant micelles. This kind of exchange would then constitute an important stage in the formation of these mesoporous solids. Recently, silicate/counterion exchanges have been studied by Frasch et al. [31,32] using fluorescence properties of probe molecules (pyrene and dipyrenylpropane). This study has given us a better understanding of the modifications induced in the micelles by the inorganic species in precursor solutions of ordered mesoporous silica of MCM-41 type. Fluorescence measurements carried out in a very alkaline medium (pH = 13.6) and before polymerisation of the silica (pH = 11.6) have shown that exchange of counterions located at the micelle surface by silicate anions only occurs at a very low level and that the micelles remain almost spherical. In solution, the ion exchange process cannot therefore represent an important stage in the formation process.

From these results, a new model has been proposed for the formation of ordered mesoporous silicas, in which the key step consists in the formation of silica prepolymers (see Fig. 26.11). The initial aqueous solution (state A) of $C_{16}TMA^+X^-$ (X = Br or Cl) contains spherical micelles in equilibrium with surfactant cations and free counterions. When a very alkaline solution of silicate species is added to the aqueous surfactant solution (stage 1), a very small fraction of the bromide counterions located at the micelle surface are exchanged for OH^- ions or silicate ions (state B). The effect of reducing the pH of the reaction mixture (stage 2) is to initiate polymerisation of the silicate species, forming silica prepolymers with a low degree of polymerisation (state C). These prepolymers can begin to interact with free surfactant ions ($C_{16}TMA^+$). This behaviour agrees with models for the formation of polyelectrolyte/surfactant complexes of opposite charge. After reduction of the pH, the silica prepolymers continue to grow as time goes by (stage 3) and each oligomer is capable of interacting with a larger number of surfactants, producing an extremely efficient cooperative effect. The system then forms hybrid surfactant–silicate micellar aggregates (state D), in agreement with the model put forward by Huo [29] and described above. The surfactant micelles only actually serve as a reservoir of surfactant ions which is gradually consumed. State D corresponds to the system just prior to precipitation of the mesostructured silica/surfactant complex (stage 4). The polymerisation of micelle/silica complexes proceeds during the precipitation–aging stages and the final result is an ordered mesoporous silica.

In fact, the mechanism for formation of these solids depends on the conditions of synthesis and probably comprises several components. However, it seems that the matching of charge densities between organic and inorganic

[4]Counterions are ions of opposite charge. For hexadecyltrimethylammonium bromide, the counterions of the hexadecyltrimethylammonium cations are the bromide anions.

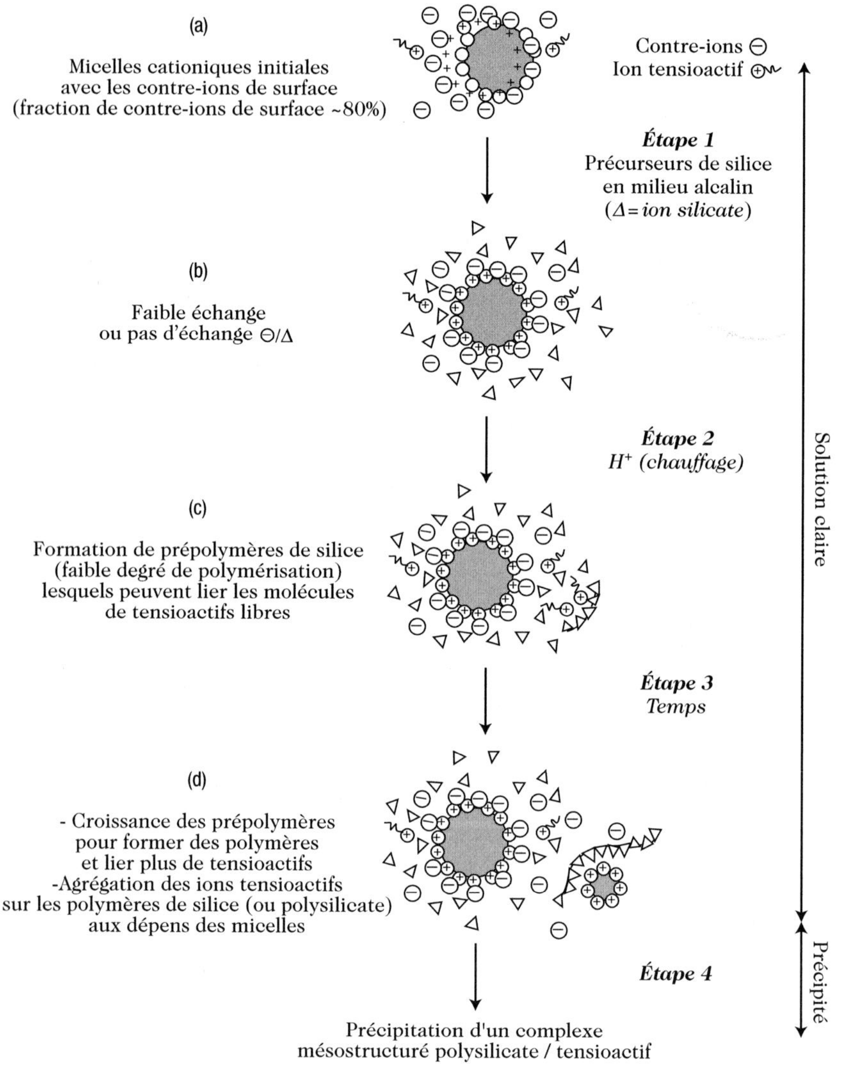

Fig. 26.11. Mechanism for the formation of ordered mesoporous silicas of MCM-41 type as proposed by Frasch et al. [32]

species is a key parameter controlling the formation processes of these phases. From this simple observation, the concepts underlying the formation of mesoporous materials have been generalised and extended to other oxides than silica, and also to sulfides. Apart from the S^+I^- mode (see above), three other electrostatic interaction modes have been proposed [29]. The first of these involves cooperative condensation of cationic inorganic species I^+ with an anionic surfactant S^- (S^-I^+ mode). The two other modes involve conden-

Table 26.5. Examples of mesoporous solids obtained via different interaction modes. MCM-x Mobil Composition of Matter number x, SBA-x Santa Barbara number x, MSU-x Michigan State University number x, HMS hexagonal mesoporous silica

Type of surfactant	Interaction mode	Examples of materials obtained (structure, type)
Cationic (S^+)	S^+I^-	Silica: MCM-41 (hexagonal) MCM-48 (cubic) MCM-50 (lamellar) Tungsten oxide (lamellar, hexagonal) Sb(V) oxide (lamellar, hexagonal, cubic) Tin sulfide (lamellar) Aluminium phosphate (lamellar, hexagonal)
	$S^+X^-I^+$	Silica: SBA-1 (cubic, Pm3a) SBA-2 (3D hexagonal) SBA-3 (hexagonal) Silica (lamellar) Zinc phosphate (lamellar) Zirconium oxide (lamellar, hexagonal)
	$(S^0H^+)(X^-I^+)$	SBA-15 (hexagonal)
	$S^+X^-I^0(X^- = F^-)$	Silica (hexagonal)
Anionic (S^-)	S^-I^+	Oxides of Mg, Al, Ga, Mn, Fe, Co, Ni, Zn (lamellar) Lead oxide (lamellar, hexagonal) Alumina (hexagonal) Tin oxide (hexagonal) Titanium oxide (hexagonal)
	$S^-M^+I^-$	Zinc oxide (lamellar) Alumina (lamellar)
Neutral (S^0)	(S^0I^0)	Silica: HMS (hexagonal) MSU-X (hexagonal) Silica (lamellar, cubic, hexagonal) Oxides of Ti, Al, Zr, Sn (hexagonal)
	S-Me(OEt)	Oxides of Nb, Ta (hexagonal)

sation of charged inorganic species in the presence of surfactant molecules of the same charge, thus requiring the presence of counterions of opposite charge ($S^+X^-I^+$ mode, with $X^- = Cl^-$, Br^-, or $S^-M^+I^-$ mode, with $M^+ = Na^+$, K^+). These counterions are generated by the acid or caustic soda required to acidify or basify the medium.

On top of these mechanisms based on electrostatic interactions, Pinnavaia and coworkers introduced a fifth mode (S^0I^0) using neutral surfactants (S^0). Hence, using primary amines with long carbon chains (S^0), it has been possible to precipitate mesoporous silicas with hexagonal structure (HMS) at neutral pH [33, 34]. Silicas with regular mesoporous structure (MSU-x) have also been synthesised in the presence of neutral surfactants with poly(ethylene oxide) polar head in highly acidic conditions [35, 36]. These reaction conditions lead to solids with thicker inorganic walls and hence to materials that are more stable thermally. The formation mechanism is then based upon hydrogen bonds between the inorganic framework and the surfactant molecules. To these five interaction modes must be added the $S^+X^-I^0$ mode [37], with $X^- = F^-$, and the mode due to Antonelli et al. [38]. In the latter, the surfactant is bound by covalent bonding to the element whose mesoporous oxide one hopes to obtain. Table 26.5 presents examples of mesoporous solids obtained via these various interaction modes. More recently, the silica SBA-15 with hexagonal symmetry has been obtained in an acidic medium in the presence of block copolymers of the form poly(ethylene oxide)–poly(propylene oxide)–poly(ethylene oxide) (PEO–PPO–PEO) as amphiphilic molecule. The best results are obtained when using Pluronic P123: $PEO_{20}PPO_{70}PEO_{20}$ [39]. In a very acidic medium, the silica species are positively charged (I^+) and the oxygens in the ethylene oxide groups are protonated. The interaction mode $(S^0H^+)(X^-I^+)$, close to the one described earlier ($S^+X^-I^+$, with $X^- = Cl^-$, Br^-), has thus been suggested to explain the formation of this material [40].

26.3.4 Characteristics of Mesoporous Silicas Obtained in the Presence of Amphiphilic Molecules

Generally speaking, the most widely used techniques for characterising ordered mesoporous solids are X-ray diffraction (small-angle scattering, see Sect. 26.3.5), transmission electron microscopy (TEM), and gas adsorption volumetry (N_2, Ar).

As an example, an X-ray pattern for the SBA-15 phase, synthesised in the presence of the copolymer $EO_{20}PO_{70}EO_{20}$, is shown in Fig. 26.12a. There are 3 to 5 peaks, characteristic of hexagonal symmetry with lattice spacings d_{100} greater than 11 nm.

The hexagonal mesopore arrangement can be visualised in the TEM image (Fig. 26.12b). In this image, the pore diameter is close to 8–9 nm and the wall thickness is of the order of 2–3 nm. The nitrogen adsorption and desorption isotherms of the product at 77 K after eliminating the block copolymer are shown in Fig. 26.12c. These isotherms have a narrow step at $p/p_0 \sim 0.75$, corresponding to capillary condensation in the mesopores. The higher value of p/p_0 than is observed for MCM-41 materials (where $p/p_0 \sim 0.4$) reflects the much bigger pore diameter.

Average values of the parameters characterising mesoporous silicas synthesised in the presence of amphiphilic molecules are:

- diameters of pore apertures 1.6–30.0 nm,
- micropore volume 0.8–2.5 mL/g,
- equivalent specific surface area 800–1 500 m^2/g,

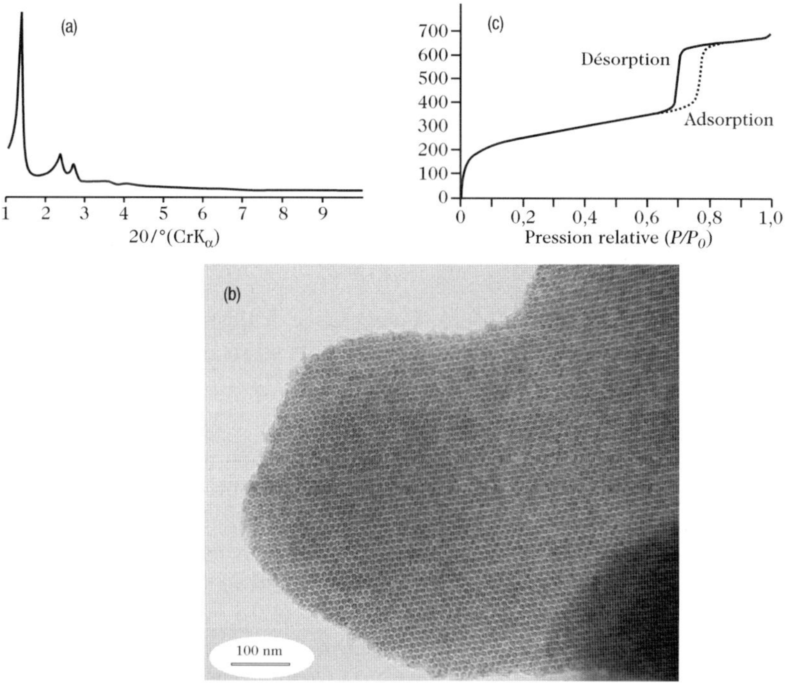

Fig. 26.12. Mesoporous silica SBA-15 synthesised in the presence of the copolymer $EO_{20}PO_{70}EO_{20}$. (**a**) X-ray pattern for the as-synthesised product (CrK_α radiation). The angle 2θ on the horizontal axis is related to the vector $\boldsymbol{q}$ by $q = (4\pi/\lambda)\sin(\theta/2)$. (**b**) Transmission electron microscope image. (**c**) Nitrogen adsorption and desorption isotherms at 77 K

26.3.5 Structural Characterisation of Nanoporous Solids by X-Ray and Neutron Scattering

The macroscopic properties of nanoporous solids are mainly determined by the size and interconnectivity of the pores. It is therefore essential to be able to investigate the structure of these materials on different length scales. X-ray diffraction provides a way of characterising the crystallinity of the inorganic framework, but is not well suited to observation on the scale of one pore (2–50 nm). NMR and other spectroscopic techniques are able to look into the close environment of an atom, but they are unable to characterise on length scales of a few nanometers. So-called semi-local techniques must then be adopted. For example, the nanopore structure can be characterised in particular by electron microscopy, gas adsorption, and small-angle X-ray or neutron scattering. The various adsorption techniques, based on the interaction between the inorganic skeleton and some other matter (a fluid, gas or liquid), are not strictly speaking semi-local techniques, but they can be used to measure certain characteristics of nanoporous media, such as the pore volume and the

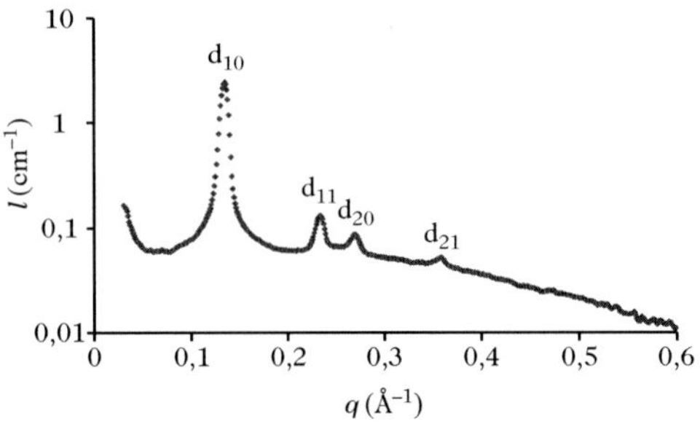

Fig. 26.13. Small-angle X-ray scattering diagram for a hybrid ZrO_2/surfactant material. The position of the Bragg peaks corresponds to an arrangement in hexagonal phase. Taken from [42]

open pore area. However, like electron microscopy, they require a preparatory treatment of the material which can alter its structure.

In contrast to adsorption techniques, small-angle scattering experiments do not require pretreatment of the sample in order to dry it in well-defined conditions. Structural characterisation by adsorption or microscopy can only therefore be used on finalised materials, and cannot be used, for example, to monitor the different stages of the synthesis in situ. Small-angle X-ray and neutron scattering [41] are non-destructive techniques that are particularly well suited to in situ structural observation of materials, and can be used to monitor the various processes involved in the synthesis as they go along. In a scattering experiment, the sample is illuminated by an X-ray, neutron or light beam with wave vector $\boldsymbol{k}_\mathrm{i}$, and the intensity I of the scattered beam produced by elastic interaction (without energy loss) with the matter is measured in a direction making an angle θ (which defines a wave vector $\boldsymbol{k}_\mathrm{s}$) with the incident beam. The scattering is said to be small-angle scattering when the angle θ is no more than a few degrees. One then defines the scattering vector $\boldsymbol{q} = \boldsymbol{k}_\mathrm{s} - \boldsymbol{k}_\mathrm{i}$, which has magnitude

$$|q| = \frac{4\pi}{\lambda} \sin\frac{\theta}{2} ,$$

where λ is the wavelength of the incident beam. The range of scattering vectors $\boldsymbol{q}$ accessible to laboratory equipment and major installations (synchroton, experimental reactors) is 10^{-3}–$6\,\mathrm{nm}^{-1}$ and is perfectly well suited to observing materials in which density fluctuations occur on the nano- and microscales. Finally, note that scattering techniques cannot directly determine the connectivity of the pores. Other techniques are needed here.

Small-Angle Scattering by a Mesoporous Structure

Among the nanoporous solids, the organised mesoporous solids (2–50 nm) have had a special place for 10 years now. The periodic structure of these materials leads to a powder spectrum, in which a series of Bragg peaks can be used to index the structure. The best studied structures are hexagonal or cubic. Figure 26.13 shows a typical example of a scattering diagram for a mesostructured material in which the inorganic matter is ZrO_2. The different Bragg peaks observed here correspond to a honeycomb structure (hexagonal phase). In this precise case, the observation was made before extracting the surfactant (CTABr or hexadecyltrimethylammonium bromide) and the same experiment on the final material showed that removing the CTABr destroyed the hexagonal organisation. This collapse concerns the synthesis chosen as an example here, but fortunately, it is not what usually happens. The synthesis of mesoporous silicas can be very well controlled today.

Note also that there are many mesoporous materials with no periodic structure. One example was discussed in Sect. 11.5. In every case, whether the pores are ordered or not, two essential characteristics of mesoporous solids can be obtained by small-angle scattering: the pore volume and the pore area.

Properties Not Dependent on Organisation

Measuring the Pore Volume

The total pore volume in the sample can be obtained by a simple rule called the invariant rule:

$$\int_0^\infty I(q)q^2\mathrm{d}q = 2\pi^2\phi(1-\phi)(\Delta\rho)^2 \,. \tag{26.1}$$

The integral on the left here does not depend on the nanostructure of the material and is known classically as an invariant. This invariance can be understood as expressing the fact that the total probability for an incident particle or photon to be scattered (integral from zero to infinity) depends only on the amount of matter encountered and not on the manner in which it is organised. On the right-hand side of (26.1), ϕ is the volume fraction of the pores (open and closed) and $\Delta\rho = \rho_{\mathrm{sk}} - \rho_{\mathrm{s}}$ is the difference in the density of the scattering length between the skeleton of the solid and the material contained within the pores, which can be air or a water–surfactant mixture, depending on the stage in the process when the observation is made. Experimentally, it is difficult to obtain a spectrum over a sufficiently broad range of scattering vectors q to be able to determine the invariant. If the material, and hence $\Delta\rho$, are well known, it is then possible to deduce the pore volume. Conversely, if the amount of matter in the sample and hence the value of $1-\phi$ are well known, it is possible to deduce the scattering length density ρ_{sk} of the skeleton. This method has recently been used by F. Né [42] in the case of a mesostructured CTAB–ZrO_2 material. The inorganic framework of this material has lower density than ZrO_2. As a general rule, it is relatively easy to obtain dense walls in the case of SiO_2, but more difficult with more reactive elements such as Ti, Zr or Ce.

Pore Area and Surface State

The total specific surface area S/V (in m^{-1}) of the pores can sometimes be obtained as a simple limit:

$$\lim_{q\to\infty} I(q) = 2\pi(\Delta\rho)^2 \frac{S}{V} \, . \qquad (26.2)$$

G. Porod has established that the limit goes as q^{-4} under the following conditions:

- q greater than the curvature of the interface,
- the transition from medium 1 (the skeleton) to medium 2 (the pore) is short compared with $1/q$ (non-diffuse interface),
- the surface is not rough on a length scale of $1/q$.

Inorganic surfaces are often rough on the atomic scale and there is therefore no Porod regime in this case. The Porod limit, when accessible, refers to the whole interface, without differentiating between occluded pores and open pores. It is therefore essential to compare the value found by small-angle scattering with that found by gas adsorption techniques, which only explore the open pore structure.

An Important Application of Scattering: Observation of Self-Organisation Kinetics

At the present time there is still no general model for the growth of ordered mesoporous solids during the joint precipitation of an organic and an inorganic phase. Small-angle scattering is once again the method of choice for understanding the growth of these materials. The very high flux of X-ray photons produced in very low divergence beams by modern synchrotron sources such as the European Synchroton Radiation Facility (ESRF) in Grenoble, France, can be used to ascertain the formation kinetics of these precipitates, even when the kinetics is very fast (lasting only a few seconds).

The most recent small-angle neutron scattering setups (D22 at the Laue Langevin Institute in Grenoble) can also be used to study growth kinetics. An example of such a neutron scattering study is given in Fig. 26.14. This refers to the precipitation of ZrO_2 in the presence of CTAB [43]. The overall kinetics lasts for 10 s and the time resolution is 200 ms. At the initial time, only the scattering signal due to the micelles is registered. Analysing the way the signal evolves, F. Né et al. [43] showed that, during the precipitation, the micelles serve as a surfactant reservoir for the formation of the mesostructured material, exactly as suggested by the work of Frash et al. [31, 32] described earlier. Joint studies of the kinetics by small-angle X-ray scattering at the ESRF has also shown that the grains formed are initially composed of a rather disordered array of cylinders. Hexagonal order subsequently appears via a first-order transition.

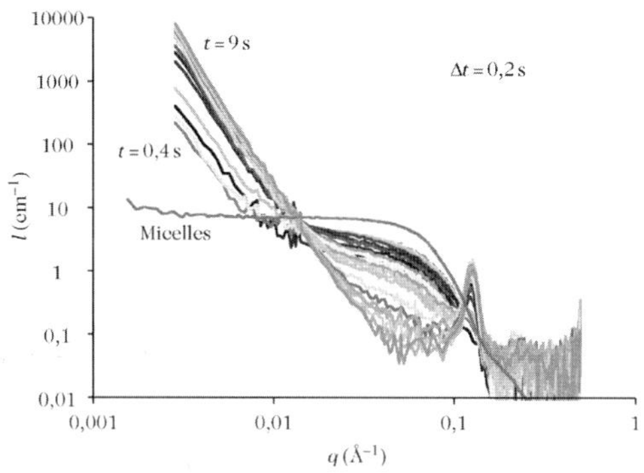

Fig. 26.14. Monitoring the precipitation of a mesostructured ZrO_2 material by small-angle neutron scattering using the D22 at the Laue Langevin Institute in Grenoble, France. The time resolution of 200 ms was sufficient to establish that the micelles play the role of surfactant reservoir. Indeed, the central bump between $q = 0.01$ Å^{-1} and 0.1 Å^{-1} which can be attributed to the micelles steadily diminishes as time goes by, while the peak corresponding to the hexagonal structure appears right from the start. Taken from Né et al. [43]

26.4 Conclusion

There can be no doubt that the technological importance of ordered porous materials will continue to grow over the coming years. Indeed, these materials are good candidates for solving problems of pollution and more general problems regarding environmental protection. By building up an understanding of the mechanisms of synthesis, and in particular the exact role played by organic and inorganic structure-directing species, it should become possible to define ideal conditions for synthesising these materials so as to obtain the desired level of porosity.

References

1. D.M. Bibby, M.P. Dale: Nature **317**, 157 (1985)
2. Q. Huo, R. Xu, S. Li, Z. Ma, J.M. Thomas, R.H. Jones, A.M. Chippindale: Chem. Commun. 875 (1992)
3. R. Althoff, K. Unger, F. Schüth: Microporous Mater. **2**, 557 (1994)
4. D.M. Bibby, N.I. Baxter, D. Grant-Taylor, L.M. Parker: In: M.L. Occelli and H.E Robson (Eds.), *Zeolite Synthesis*, ACS Symposium Series, 398, Washington DC (1989) p. 209
5. H. Robson, K.P. Lillerud: *Verified Synthesis of Zeolitic Materials*, 2nd edn., Synthesis Commission of the International Zeolite Association, Elsevier, Amsterdam (2001) p. 265
6. C. Baerlocher, W.M. Meier, D.H. Olson: *Atlas of Zeolite Framework Types*, 5th edn., Structure Commission of the International Zeolite Association, Elsevier, Amsterdam (2001) p. 302

7. G. Férey: J. Fluorine Chem. **72**, 187 (1995)
8. S. Oliver, A. Kuperman, G.A. Ozin: Angew. Chem. Int. Ed. Engl. **37**, 46 (1998)
9. L. Vidal, C. Marichal, V. Gramlich, J. Patarin, Z. Gabelica: Chem. Mater. **11**, 2728 (1999)
10. L. Vidal, V. Gramlich, J. Patarin, Z. Gabelica: Eur. J. Solid State Inorg. Chem. **35**, 545 (1998)
11. J.M. Bennett, J.A. Gard: Nature **214**, 1005 (1967)
12. H.E. Robson, D.P. Shoemaker, R.A. Ogilvie, P.C. Manor: In: W.M. Meier and J.B. Uytterhoeven (Eds.), *Molecular Sieves*, Third International Conference, Advances in Chemistry, Series 121 (1973) p. 106
13. M. Wiebcke: Chem. Commun. 1507 (1991)
14. S.L. Burkett, M.E. Davis: J. Phys. Chem. **98**, 4647 (1994)
15. P.P.E.A. de Moor, T.P.M. Beelen, B.U. Komanschek, L.W. Beck, P. Wagner, M.E. Davis, R.A. van Santen: Chem. Eur. J. **5**, 2083 (1999)
16. C.E.A. Kirschhock, R. Ravishankar, F. Verspeurt, P.J. Grobet, P.A. Jacobs, J.A. Martens: J. Phys. Chem. B **103**, 4965 (1999)
17. L. Moscou: In: H. van Bekkum et al. (Eds), *Introduction to Zeolite Science and Practice*, Elsevier, Amsterdam (1991) p. 1
18. C. Marcilly: In: A. Galarneau et al. (Eds), *Stud. Surf. Sci. Catal.*, Vol. 135, Elsevier, Amsterdam, 37 (2001)
19. J.P. Guggenbichler, A. Hirsch: PCT Int. Appl., WO 01 09, 229 (2000)
20. M. Liu, X. Li, C. An, P. Li: Faming Zhuanli Shenqing Gongkai Shuomingshu CN 1, **262**, 591 (2000)
21. W.F. Hölderich: Pure & Appl. Chem. **58**, 1383 (1986)
22. W.F. Hölderich, M. Hesse, F. Näumann: Angew. Chem. **27**, 226 (1988)
23. A. Corma, M.J. Climent, H. Garcia, J. Primo: Appl. Catal. **49**, 109 (1989)
24. A. Taramasso, G. Perego, B. Notari: U.S. Pat. 4410501 (1983)
25. A. Esposito, M. Taramasso, C. Neri, F. Buonuomo: U.K. Pat. 2116974 (1985)
26. J.B. Nagy, P. Bodart, I. Hannus, I. Kiricsi: *Synthesis, Characterization and Use of Zeolitic Microporous Materials*, DecaGen Ltd, Hungary (1998) p. 159
27. J.S. Beck, J.C. Vartuli, W.J. Roth, M.E. Leonowicz, C.T. Kresge, K.D. Schmitt, C.T.W. Chu, D.H. Olson, E.W. Sheppard, S.B. McCullen, J.B. Higgins, J.L. Schlenker: J. Am. Chem. Soc. **114**, 10834 (1992)
28. A. Monnier, F. Schüth, Q. Huo, D. Kumar, D.I. Margolese, R.S. Maxwell, G.D. Stucky, M. Krishnamurty, P. Petroff, A. Firouzi, M. Janike, B.F. Chmelka: Science **261**, 1299 (1993)
29. Q. Huo, D.I. Margolese, U. Ciesla, D.G. Demuth, P. Feng, T.E. Gier, P. Sieger, A. Firouzi, B.F. Chmelka, F. Schüth, G.D. Stucky: Chem. Mater. **6**, 1176 (1994)
30. A. Firouzi, D. Kumar, L.M. Bull, T. Besier, P. Sieger, Q. Huo, S.A. Walker, J.A. Zasadzinski, C. Glinka, J. Nicol, D.I. Margolese, G.D. Stucky, B.F. Chmelka: Science **267**, 1138 (1995)
31. J. Frasch, R. Zana, M. Soulard, B. Lebeau, J. Patarin: Langmuir **15**, 2603 (1999)
32. J. Frasch, B. Lebeau, M. Soulard, J. Patarin, R. Zana: Langmuir **16**, 9049 (2000)
33. P.T. Tanev, M. Chibwe, T.J. Pinnavaia: Nature **368**, 321 (1994)
34. P.T. Tanev, T.J. Pinnavaia: Science **267**, 865 (1995)
35. S.A. Bagshaw, E. Prouzet, T.J. Pinnavaia: Science **269**, 1242 (1995)
36. S.A. Bagshaw, T.J. Pinnavaia: Angew. Chem. **35**, 1102 (1996)

37. A.C. Voegtlin, F. Ruch, J.L. Guth, J. Patarin, L. Huve: Microporous Mater. **9**, 95 (1997)
38. D.M. Antonelli, J.Y. Ying: Angew. Chem. Int. Ed. Engl. **35**, 426 (1996)
39. D. Zhao, J. Feng, Q. Huo, N. Melosh, G.H. Fredrickson, B.F. Chmelka, G.D. Stucky: Science **279**, 548 (1998)
40. D. Zhao, Q. Huo, J. Feng, B.F. Chmelka, G.D. Stucky: J. Am. Chem. Soc. **120**, 6024 (1998)
41. P. Lindner, T. Zemb (Eds.): *Neutrons, X-Rays and Light: Scattering Methods Applied to Soft Condensed Matter*, North-Holland, Delta Series, Elsevier (2002)
42. F. Né, Th. Zemb: J. Appl. Cryst. **36**, 1013 (2003)
43. F. Né, F. Testard, Th. Zemb, I. Grillo: Langmuir **19** (20), 8503 (2003)

27

Molecular Imprinting

V. Dufaud and L. Bonneviot

27.1 Introduction

Our senses of smell and taste are able to recognise molecules selectively, to the point where they can even discriminate between different chiral states. This property, called molecular recognition, is essential to all forms of life [1]. It is based on the principle of a specific interaction between a receptor or host and a target molecule, which will be identified among a multitude of others, then selectively adsorbed. If the host is endowed with reactive functions, the attached molecule may be transported or transformed. Enzymes are the archetypal host molecules exploiting the idea of molecular recognition. Their complexation sites comprise a hydrophobic pocket with definite shape within which amino acid residues are located in a precisely defined way. The combined effect of these different characteristics underlies not only the affinity for some specific substrate, but also the transformation of this substrate into the desired product [2]. In fact, the phenomena actually brought into play are much more involved, being made up of an ensemble of physicochemical events that act together in a cooperative way, either simultaneously or sequentially, and in which the molecular processes are difficult to follow in detail.

However, the elegance of naturally occurring systems of molecular recognition has encouraged scientists, perhaps not to try to imitate them, but at least to transpose the main ideas to the laboratory context. Host molecules with low molecular weight called cavitands have in fact been synthesised (crown ethers, cyclodextrins, cryptophanes, cyclophanes, and concave molecules) [3]. It has been shown that their capacity to operate as a specific host relies on the presence of a cavity which has the shape of the molecule to be recognised, and complexation sites (in the form of peptides, amino acids, or sugars) with a definite spatial arrangement that simulates the enzyme concept. However, these hosts are difficult to synthesise, requiring multi-stage synthesis, and are not always well suited to recognising the kind of large molecules commonly found in the chemistry of life.

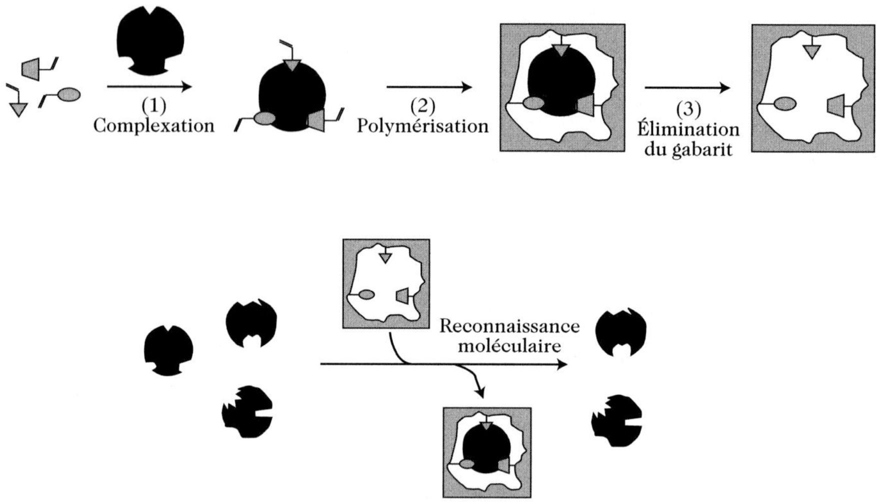

Fig. 27.1. The different stages of molecular imprinting

The most attractive alternative is to synthesise the host by starting with the target molecule as a kind of template. This is the technique known as molecular imprinting [4–6]. The idea is to create a complementary image of the target molecule by polymerising functional monomers in its presence. This simple and flexible methodology was discovered by Dickey et al., who described the imprint of the molecular memory by imprinting silica gels with pigments (alkyl orange dyes) as early as 1949 [7]. However, this technique only really began to take off with the use of organic polymers combining a wide variety of functionalities and better retention of the imprinting effect.

There is a wide range of applications for this technique: affinity chromatography, immunoanalysis, biosensors, catalysis, and so on. It is thus important to develop highly selective hosts but at a reasonable cost. Some example applications will be given, after first describing the fundamental aspects, procedures, and methods involved in molecular imprinting.

27.2 Fundamental Considerations

27.2.1 General Principles

The underlying idea of molecular imprinting is to make organic or inorganic synthetic materials containing cavities capable of specific molecular recognition (see Fig. 27.1). To obtain this specificity, i.e., cavities with a definite shape and functional groups placed with some predetermined orientation, the functional groups, which have a polymerisable form, are bonded to a suitable template at several points (Fig. 27.1, stage 1). The relevant interactions between the functional group and the template can be either covalent or non-covalent. The functional monomer is then polymerised under conditions leading to the formation of a highly cross-linked matrix (Fig. 27.1, stage 2). When the template is eliminated, the imprint remains, characterised by a functionalised cavity. The shape of this cavity and the nature and spacing of the functions constitute points of recognition complementary to those of the target molecule (Fig. 27.1, stage 3).

These complementary sites are then able to recognise the target molecule in an efficient and selective way. The cross-linking of the organic and/or inorganic polymer network serves the purpose of maintaining the 3D structure of the matrix and guaranteeing the memory effect of the imprint once the template has been eliminated.

27.2.2 Role of Complexation Sites During the Imprinting Process

The complexation sites will be put to use throughout the imprinting process, but their function will depend on the stage of the process, the required properties being different during polymerisation, elimination of the template, and subsequent molecular recognition of the target molecule.

To begin with, the interaction between the template and the functional groups must be as strong as possible during the polymerisation and cross-linking process, to produce an imprint with precise spatial positioning of the complementary functions, in terms of their distances and orientations, and to obtain a faithful imprint of the overall shape of the molecule. The template must then be eliminated as completely as possible, in relatively mild conditions so as to preserve this spatial arrangement. Finally, the complexation sites of the imprint left intact after eliminating the template must generate an association (often called complexation) with the target substrate that is as fast and reversible as possibile. Highly specific materials will then be obtained if all these criteria are respected. The relevant interactions can be covalent or non-covalent.

Non-Covalent Approach

Recognition points involve weak interactions like hydrogen bonds, electrostatic interactions, hydrophobic interactions, or coordination bonds. This is the simplest method for producing specific polymers. The monomer–template adducts or assemblies are formed in situ by simpling mixing the reactants. After polymerisation, the template is eliminated by extracting it with a suitable solvent (see Fig. 27.2). During the recognition process between the imprinted polymer and the target molecule, an ensemble of similar non-covalent bonds will be recreated. This non-covalent molecular recognition has been able to produce highly selective materials, with excellent performance in terms of separation, in some cases as good as can be obtained with much stronger covalent interactions. These results were only obtained after rigorous development and optimisation of the method, in particular by K. Mosbach and coworkers [8–10].

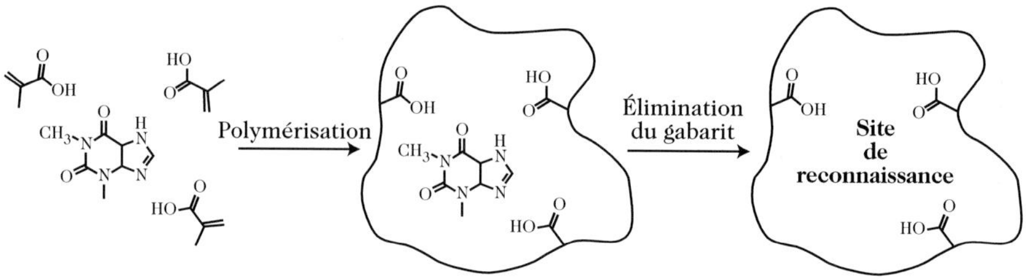

Fig. 27.2. Molecular imprinting by the non-covalent approach. From [10]

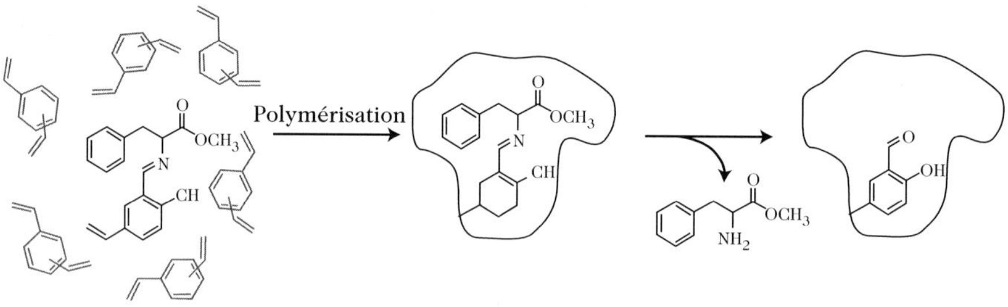

Fig. 27.3. Molecular imprinting by the covalent approach. From [13]

Covalent Approach

Polymerisable derivatives of the target molecule are first synthesised by connecting the functional monomer and the template in one or more points by covalent bonds. After polymerisation, the covalent bonds are chemically cleaved, thereby eliminating the template and freeing the functional sites of the imprint thus produced (see Fig. 27.3). During the recognition process between the imprinted polymer and the target molecule, an ensemble of similar covalent bonds will reform. This covalent molecular imprinting has been able to produce highly selective materials. The first work here was mainly due to Wulff and Shea [11–14].

Choice of Approach

The two types of approach, covalent and non-covalent, each have their advantages and disadvantages. The choice of method will depend on a range of different factors, including context and requirements, e.g., the type of target molecule, the time available to prepare the imprinted materials, the level of selectivity, and so on. The two approaches may be compared as follows:

- The covalent approach is often better in terms of imprinting accuracy. Indeed, the functional monomers are very precisely fixed in space during polymerisation (stoichiometric relationship between template and functional monomer). A wide range of operating conditions can be used since

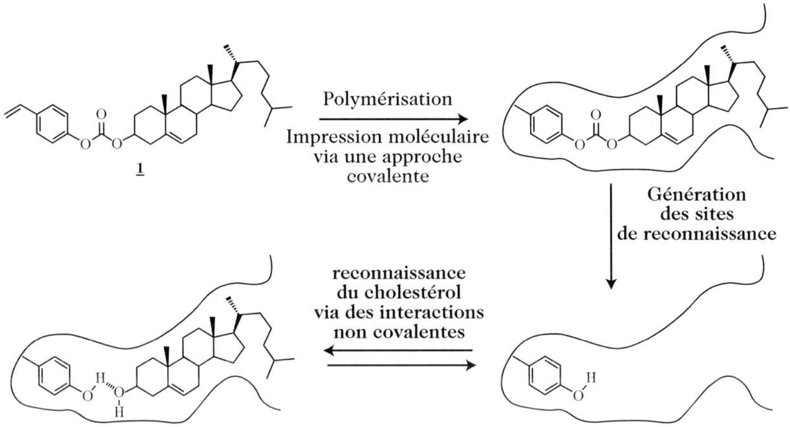

Fig. 27.4. Molecular imprinting of cholesterol using a hybrid covalent and non-covalent approach. Taken from [15]

the covalent bonds are relatively stable (with respect to pH, high temperatures, and polymerisation conditions). It is an attractive method provided that the template can be eliminated easily enough after synthesising the imprint, and provided that the formation of covalent bonds with the target molecule can be sufficiently favoured during molecular recognition. However, the synthesis of specific polymerisable derivatives of the target molecule often proves to be a long and difficult (hence costly) process. Furthermore, the available range of reversible covalent bonds is limited and the rate of complexation (or adsorption)/decomplexation (or desorption) is generally rather slow. Indeed, the covalent bonds must have the time to break and form.

- The non-covalent approach is easy to implement and applies to a wide range of templates. The template is usually eliminated without difficulty and the reversible interactions with the substrate to be recognised are rather fast. An excess of functional monomers, in a ratio of about 4 to 1 in relation to the template, is often needed to ensure saturation of the complexation sites on the template during polymerisation. On the other hand, a considerable fraction of functional groups is incorporated at random, producing non-specific complexation sites in the final polymer. Moreover, the monomer–template adduct thereby formed is labile and non-stoichiometric, and its stability will depend sensitively on the conditions of polymerisation.

Methodologies combining the advantages of the covalent approach (well defined structure and interface between monomer and template) with the advantages of the non-covalent approach (fast adsorption/desorption of the target molecule) were developed several years ago by Whitcombe [15, 16]. These methods can bring improvements and new solutions in certain cases.

The polymer is prepared by a covalent imprinting procedure, but the target molecule is recognised through non-covalent interactions. In the example shown in Fig. 27.4, the carbonate ester 1 functions as a covalent derivative of cholesterol (the target molecule), but at the same time, it is efficiently cleaved with loss of CO_2 to give a phenolic non-covalent recognition site. The recognition of cholesterol by these phenolic sites occurs via hydrogen bonds in non-polar solvents.

27.2.3 Structure and Properties of the Polymer Matrix

In the molecular imprinting process, the structure of the matrix is also crucial in order to conserve an accurate memory of the shape of the cavity and the positions of the functions. It can be organic or inorganic, provided that it possesses the following attributes:

- The structure must be extremely rigid so that the shape of the cavity is retained once the template has been removed.
- All recognition sites must be easily accessible.
- The structure must be mechanically stable so that it can be prepared for applications such as chromatography, membrane sensors, catalysis, etc.
- The structure must be thermally stable to accelerate the recognition process.

In general, the matrix is obtained by polymerisation of basic monomers, cross-linking agents, and functional monomers in the presence of the target molecule. Any type of polymerisation can be used (radical, anionic, cationic, and condensation), the important thing being the stability of the components involved. In particular, it is important to maintain the functional monomer–template interactions, especially in the non-covalent approach. In every case, it is important to maintain structural control on the nanoscale, both in terms of interconnectivity of the pores in the matrix and in terms of uniformity of the complexation sites.

Choice of Matrix

Apart from the possibility of multiple functionalisation, organic polymers by their very nature can procure a certain mechanical flexibility, a parameter required for the complexation/decomplexation kinetics of the target molecule (allowing a reduction in the activation energy of the process). However, there are some disadvantages, such as the heterogeneity of the sites affecting selectivity and a change in the shape of their cavities depending on the polarity of the solvent. On the other hand, inorganic matrices are much more rigid and robust, withstanding more severe reaction conditions and better suited to industrial requirements (high temperatures and pressures, organic solvents and reagents).

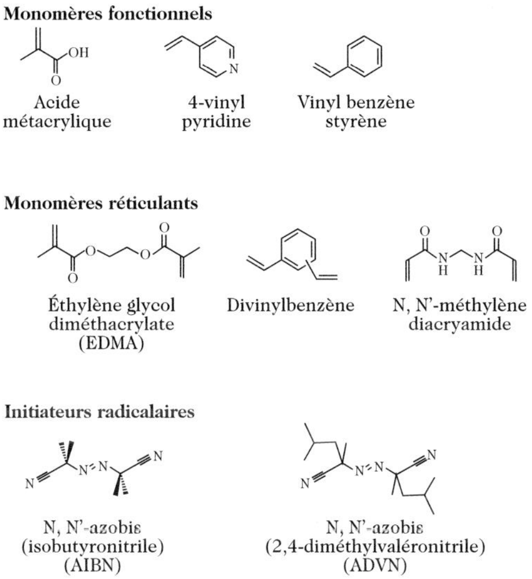

Fig. 27.5. Typical reagents used for molecular imprinting of organic polymers

27.3 Procedures and Methods for Molecular Imprinting

27.3.1 Imprinted Organic Polymers

Experimental Procedures

The synthesis of imprinted organic polymers requires the simultaneous presence of polymerisable functional monomers (associated with the template), cross-linking agents, and suitable solvents for the polymerisation and subsequent extraction of the template (see Fig. 27.5).

Free radical polymerisation is the most commonly used, but this does not mean that other types of polymerisation (e.g., cationic or anionic) are never used. In the first place, the important point is that the structural integrity of the monomer–template assembly should be preserved during the polymerisation process. Secondly, a high level of cross-linking is needed to ensure sufficient rigidity of the polymer network after elimination of the template and to maintain not only the shape of the cavity, but also the spatial arrangement of the complexation points. The solvent also plays a critical part. The presence of a porogenic solvent in the polymerisation mixture will lead to mesoporous and macroporous structures, and this will help in the elimination of the template, as well as improving the accessibility of the recognition sites. In addition, the solvent can play a crucial role in interactions that bind the functional monomer to the template, since it modulates the strength of non-covalent interactions via its polarity. We shall see in the last section of this chapter that it is hard to obtain specific molecular recognition when the solvent used is water.

Table 27.1. Examples of covalent interactions used during the imprinting process. Taken from [5,6]

Polymerisable functional group	Template anchor point	Type of bond
Boric acid 4-vinylphenyl	Diol	Boric acid ester
Amine	Aldehyde	Schiff base
Diol	Ketone	Ketal
Acrylic acid	Amine	Amide
Acrylic acid	Alcohol	Ester
4-vinylphenyl carbonate	Alcohol	Ester
Chelated cobalt	Amino acid	Chelate
Thiol	Thiol	Disulfide

Molecular Imprinting by Covalent Bonding

The choice of covalent bonding to fix the functional monomer to the template is a key parameter in the success of this approach. The criteria that must be satisfied, which seem a priori to be antinomic (see above), are the stability and reversibility of the interactions. Indeed, the covalent bonds must be cleaved after the synthesis in soft conditions that will not damage the imprinted polymer. The available covalent interactions, some of which are listed in Table 27.1, include the boric acid and carbonate esters, acetals, Schiff bases, disulfide bonds, etc.

Molecular Imprinting by Non-Covalent Bonding

In this a priori simpler approach, one only needs to combine the functional monomers with the desired template and copolymerise them in the presence of a cross-linking agent. The monomer/target molecule adducts formed spontaneously in situ by non-covalent interactions are then frozen by polymerisation into a polymer structure. Among the various types of non-covalent interaction available (see Table 27.2), hydrogen bonds remain the most suitable choice for precise recognition, since this type of bond depends sensitively on the distance and orientation of the relevant reactants.

This technique is currently applied to a very broad range of target molecules, in particular pharmaceutical products, herbicides, and biologically active substances, which all have the particularity of possessing polar groups (hydroxyl, carboxyl, amide, amine, etc.), as required in this type of non-covalent interaction.

27.3.2 Imprinted Inorganic Matrices

Experimental Procedures

The sol–gel technique, carried out in mild conditions, is the method of choice for synthesising inorganic matrices. Better still, this technique allows an

Table 27.2. Examples of non-covalent interactions used during the imprinting process. Taken from [5,6]

Polymerisable functional group	Template anchor point
Electrostatic interaction	
Acrylic acid	Amine
Itaconic acid	Amine
Sulfonic acid	Amine
4-vinyl pyridine	Amine
Hydrogen bond	
Acrylic acid	Amine
4-vinyl pyridine	Amine
Hydrophobic bond	
Pigments (intercalation)	Carboxylic acid

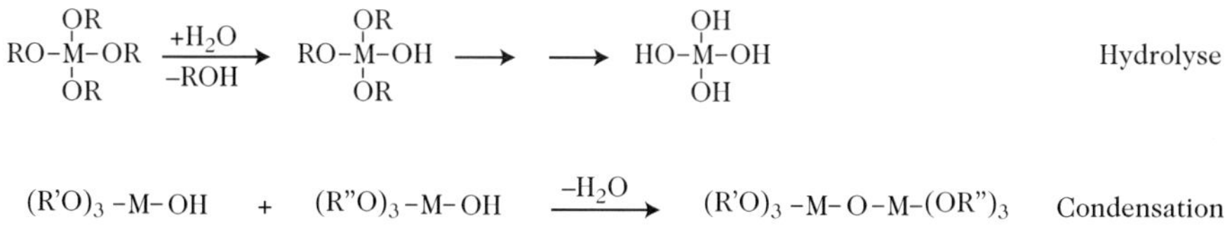

M = Si, Ti, Ge
R = groupement alkyle
R', R" = alkyl, Si ou H

Fig. 27.6. Simplified scheme of sol–gel chemistry

extremely wide range of experimental conditions leading to robust, cross-linked metal gels [17]. In general, condensation of silanes, typically silicon alkoxides, or silicates, involves the use of aqueous and sometimes alcoholic solutions with pH adjusted to catalyse polycondensation (see Fig. 27.6).

In contrast to the synthesis of organic polymers, the presence of a cross-linking agent is not required here, since the monomer is already cross-linked by its very nature, there being four condensation points in tetraethoxysilane, for example. The cross-link density, porosity, and homogeneity of the functional sites depend on the pH of the solution, the degree of ripening, the type of catalysis, and the way the solvent is evaporated, whereupon materials can be tailor-made with specific properties.

Molecular imprinting of an inorganic matrix uses the same type of interaction as those used to produce imprinted organic polymers. For example, to imprint pigments, the non-covalent approach exploits the hydrophobic/hydrophilic character of the matrix. During formation of the silica gel, the template structure will determine the positions of the (hydrophobic) siloxane groups and the (hydrophilic) silanol groups so that they interact in an appropriate and complementary way through hydrogen bonds or van der Waals

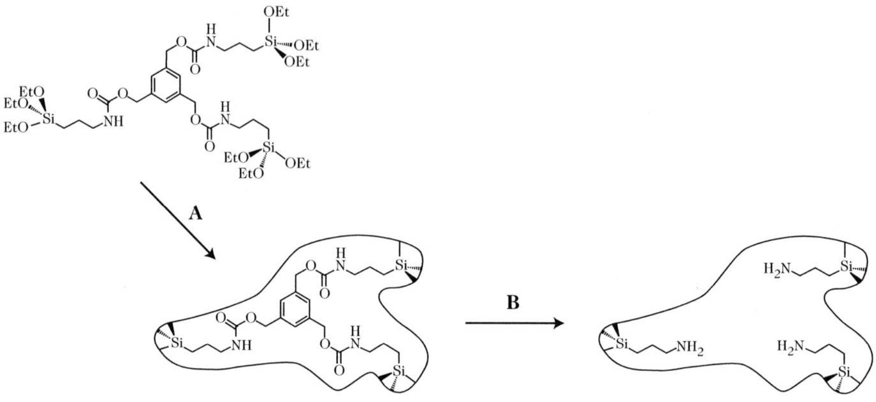

Fig. 27.7. Molecular imprinting in a silica gel. (A) Hydrolysis and sol–gel polycondensation. (B) Elimination of the template. Taken from [18]

interactions with the various sites of the target molecule. Furthermore, by using silanes such as organoalcoxysilanes, viz., $R_xSi(OR)_{4-x}$, as functional monomers during polycondensation or in post-synthesis reaction with surface silanols, it has been possible to imprint gels with metal oxides. The polymerisable double bonds used in the case of organic polymers are then replaced by polycondensable silane groups attached to the complexation sites.

Imprinted metal oxide materials have proved to have highly selective molecular affinities in a wide range of applications in separation and catalysis. Among the many studies reported in the literature, we have chosen three examples which seem to us to be representative. Two of these involve imprinting and/or functionalisation of silica gels in the bulk, while the third concerns surface imprinting of silica.

Covalent Imprinting in Silica Gels

This approach, developed recently by Katz and Davis, uses the molecular imprinting method during the sol–gel synthesis by implementing a specific molecular template containing several precursors of functional groups which remain at the surface of the silica when the spacer between them has been eliminated (see Fig. 27.7) [18]. This generates microporous amorphous silicas with well-defined dimensions (the size of the imprint), containing multiple aminopropyl functions bound covalently to the silica lattice and with a well controlled topology. This can be considered as a genuine example of molecular engineering on the inner surface of the material, combining molecular confinement and topochemistry. The authors of this study showed that these hybrid organic–inorganic materials were capable of accomplishing catalytic reactions with shape selectivity. However, no cooperative effect was demonstrated, probably because the active sites were located too far apart.

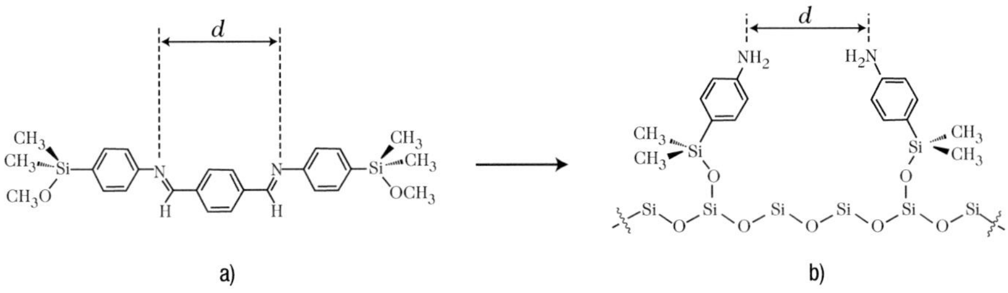

Fig. 27.8. Silica gel synthesised by placing two aromatic amine groups at a controlled distance from one another on the surface. From [12]

Modification of Silica Surfaces

In a conventional molecular imprinting procedure, the selectivity of the imprinted synthetic materials, whether they be organic polymers or silica gels, depends sensitively on the shape of the cavity and the spatial arrangement of the functional groups within it. Wulff and coworkers were nevertheless able to show that silica gels, in which only the surface had been imprinted, were still able to achieve molecular recognition. The observed selectivity is reduced in this case to controlling the distance been two neighbouring complexation sites [12]. This approach uses silane diimine functional monomers (see Fig. 27.8) which are first grafted onto the surface of the silica by means of a condensation reaction between the surface silanols and the alcoxysilane groups of the monomer.

The surface silanols, still accessible, which have not reacted during the grafting process, are then blocked by a silylation agent to prevent non-selective adsorption during the subsequent recognition stage. Finally, the template is cleaved by hydrolysis, which generates host sites comprising two amine groups at a set distance apart, this separation being predetermined by the geometry of the target molecule (see Fig. 27.8b).

Adsorption studies on a series of dialdehydes with various spacings between their aldehyde groups have shown that the dialdehydes used to imprint the silica gel surface were preferentially complexed. Although the reported effects are not very pronounced, this study was the first to mention molecular-level control of the spatial distribution of complexation sites on a silica surface.

Fixing Specific Ligands by Speciation of Transition Metals

Recognition can be applied to the differentiation of metal ions according to their complexation affinities with specific ligands. Much research has been done to fix ligands in a silica gel with a view to heterogenising catalytic species containing transition metals [19]. The same can also be exploited for solid/liquid extraction when treating effluents in the context of environmental protection. For example, by fixing thioalkyl alcoxysilane, preferential retention

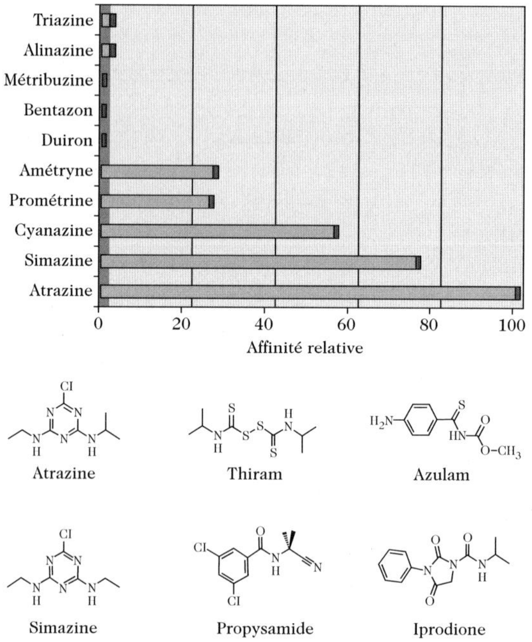

Fig. 27.9. Relative affinity and structures of some molecules occurring in commercialised herbicides, including the target triazines. From [19]

of heavy metals like Hg, Cd, Pb, etc., becomes possible [20, 21]. However, speciation of metals can be improved if the silica can be polycondensed in the presence of an ion of a specific metal serving as a template for the metal ion to be recognised. One then goes from the complexing material to the imprinted complexing material (see below).

27.4 Applications

Among the many applications listed in the bibliography, let us focus on four examples related to several major socioeconomic issues, viz., health (medicines), food, and treatment of pollution, to illustrate the way in which molecularly imprinted materials can be put to use. The four examples are:

- affinity chromatography for analysis of herbicides,
- improving the yield from a key stage in the production of a sweetener,
- chiral separation of amino acids,
- specific elimination of metals in a highly radioactive effluent.

27.4.1 Separating a Mixture of Herbicides

Studies concerning the impact of herbicides on the environment have shown that it will be important to develop sensitive and reliable techniques for separating complex, structurally similar compounds. In order to treat complex samples, a separation by affinity chromatography is often required before the chromatographic analysis itself. Molecularly imprinted materials have been tested on herbicides derived from 1,3,5-triazine, an aromatic heterocycle with three nitrogen atoms [22]. These can form hydrogen bonds with carboxyl groups, which explains the choice of a matrix formed by polymerisation of methacrylic acid (MAA) as functional monomer. Chromatographic tests carried out on polymers imprinted from atrazine show that the capacity factor for the target molecule is 15.5, as compared with 0.55 for the non-imprinted polymer. The relative affinity is represented in Fig. 27.9 by the separation factor expressed as a percentage. In fact, several of the compounds in this family of herbicides have a particular affinity for these imprinted polymers, notably simazine, used on Japanese golf courses. This feature is attributed to the spatial arrangement of the carboxyl groups, which allows multiple hydrogen bonds to form with the 1,3,5-triazine nucleus constituting the very heart of the recognition mechanism. Tests using eluants with variable polarity, such as acetonitrile chloroform, confirm this interpretation. Note that the higher affinity for atrazine clearly shows that there has also been an imprint of the molecule in the polymer. However, the sites exhibit a certain heterogeneity as in polyclonal antibodies. Moreover, these same authors have investigated the effect of the force due to matrix acidity by using another functional monomer, trifluoromethacrylic acid (TFMAA) plus acid, either pure or mixed with MAA. They have thus shown that pure MAA is a better choice for the triazines, whereas TFMAA would be better suited for recognising nicotine [23].

27.4.2 Synthesis of α-Aspartame

The synthesis of α-aspartame involves condensation of an amino diacid, L-aspargic acid (L-AspOH), with L-phenylalanine in its methyl ester form (L-phe-OMe). The amine function of the aspargic acid is protected by benzyl chloroformate (CbzCl) forming Cbz-L-Asp-OH. It is the carboxylic function of the protected amine at α that must react. To avoid peptide condensation at β, a specific enzyme known as thermolysine is used. Unfortunately, this reaction is not favoured thermodynamically. In the presence of the enzyme, it is difficult to displace the reaction by eliminating the water formed. An alternative is therefore to displace the reaction by selectively trapping the condensate, Cbz-α-aspartame. This was done by Ye et al. using an ethylene glycol dimethacrylate polymer imprinted by α-aspartame [24]. After 48 hr, the enzyme reaction occuring at 37°C in the presence of the imprinted polymer achieves a yield greater than 60%, compared with only 20% when the imprinted polymer is absent. This study clearly demonstrates the potential of molecularly imprinted materials for implementing thermodynamically unfavourable enzyme reactions.

27.4.3 Chiral Separation of Amino Acids by Ligand Exchange at a Metal Site

Metal sites are well known for the important part they play in living beings, e.g., for fixing small molecules (oxygen transfer, breathing, etc.), or for catalysing specific reactions (metalloenzymes, oxidases, hydrogenases, etc.). They can also be used as anchor points for molecular recognition. Vidyasankar et al. first applied this idea to separate amino acid analogs of phenylalanine [25]. The molecular imprint is then made from a functional monomer containing a non-chiral complex of copper (II) and its chiral adduct, phenylalanine (the target). The polymer matrix is obtained from a mixture of ethylene glycol methacrylate and the functional monomer (95:5 mol %). The latter is a copper complex, namely, copper (II) [N(4-vinylbenzyl)amine] diacetate or [Cu(VBIDA)], in which one recognises the polymerisable 4-vinylbenzene group, and the chelation principle with the tertiary amine combined with two acetate groups. To make a chromatographic column, the imprinted polymer is deposited on a silica surface (10-μm particles and average pore size 100 Å). The separation coefficient measured for resolution of the L form with respect to the D form is 1.65 (equivalent to a ΔG of 72 J mol^{-1}). The column also separates chiral forms of tyrosine ($\alpha = 1.54$) with similar structure, but its enantioselective structure is much reduced in a racemic mixture of tryptophan or with regard to other aliphatic amino acids. As the imprinted polymer contains no chiral component other than the target molecule which has been removed, selectivity is a combined effect of the constraint due to anchoring by chelation to the metal and also of the shape of the imprint of the target molecule. The authors' discussion concerning the number of recognition points, which is at least three for chiral recognition, remains open. However, they suggest a reasonable recognition scheme in which only the molecule with the right chirality can simultaneously provide a chelating bond at two coordination sites of the copper and also occupy the confined space of the imprint in the polymer (see Fig. 27.10).

27.4.4 Specific Elimination of Lanthanides and Actinides in a Highly Radioactive Effluent

Moreau and coworkers have recently observed that hybrid silicas synthesised according to the scheme outlined below are particularly useful for selectively separating lanthanide and actinide ions in a highly acidic medium (see Fig. 27.11) [26].

In this process, the polycondensable function is $-Si(OEt)_3$ and the complexing function is obtained from the 1,3-propanediethylamide group. The imprint is then made by synthesising the material in the presence of inorganic

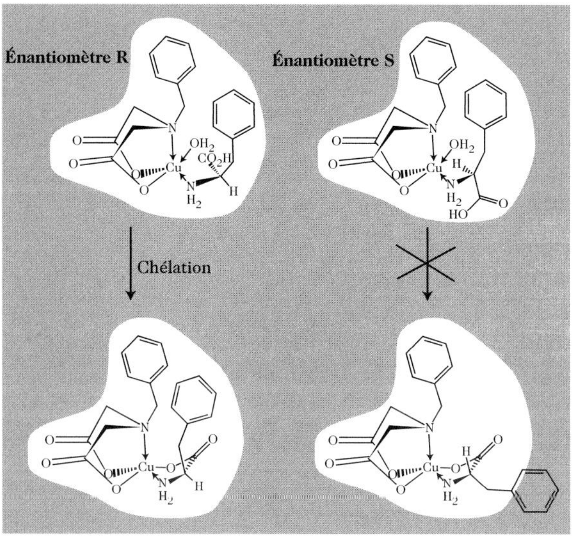

Fig. 27.10. Effect of the imprint on chelation of the D form of phenylalanine (S configuration) in favour of the L form (R configuration). From [23]

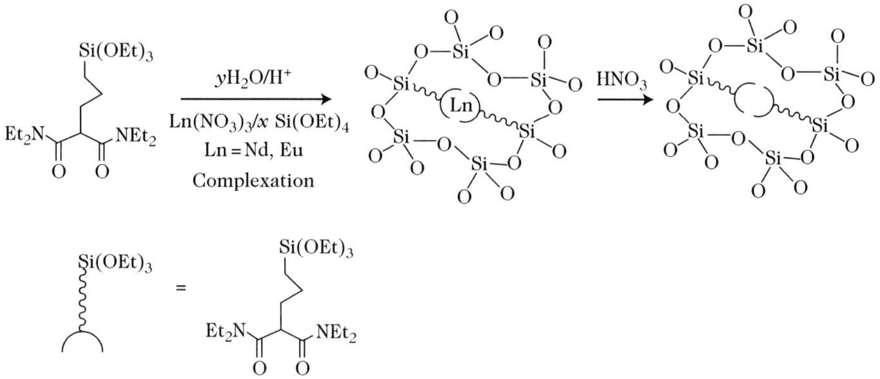

Fig. 27.11. Synthesis of hybrid silicas using Ln^{3+} as inorganic template. From [26]

ions, so that the coordination sphere can be prearranged around an Nd^{3+} or Eu^{3+} ion acting as template. Once the 'template' ion has been eliminated, the material thereby produced is structured locally and exhibits increased extractive properties in terms of the distribution and selectivity factors. For example, when a material imprinted in the presence of lanthanide ions (Eu or Nd) is used to extract Am^{3+}, one finds values of K_d (the distribution coefficient between the extractive solid phase and the effluent liquid phase) four times higher than when the material is synthesised in the absence of the 'template'. On the other hand, the results for extraction of Pu^{4+} remain basically unchanged. This reveals a significant ordering effect within the solid matrix when one uses Eu^{3+} or Nd^{3+} which have coordination chemistry similar to Am^{3+} but different to Pu^{4+}.

27.5 Recent Challenges and Progress

The imprinting of both organic or inorganic synthetic materials has now reached a high level of sophistication, not only in terms of our understanding of the relevant processes, but also with regard to the properties of the resulting materials. However, molecular imprinting is still being developed in more and more different directions, some of which were outlined in the last section, and a certain number of issues remain to be solved which are very likely to require new strategies and new concepts. There follows a brief view of some of the main challenges currently under investigation.

Molecular Imprinting in Water

The reaction medium plays a determining role in the imprinting and recognition stages since it has a significant influence on intermolecular interactions. In many cases, the polymer, synthesised in a non-polar solvent, e.g., toluene, is used in a polar solvent, e.g., water. It may then lose some of its selectivity and its affinity for the target molecule.

Recently, a large number of studies have been carried out to develop methods in which molecular imprinting by the template can be achieved in water. The main problem lies in the fact that water is itself an acceptor and a donor of hydrogen, so that, for this reason, it is able to form hydrogen bonds with the target molecule, thereby entering into competition with the functional monomers.

Several solutions have been put forward recently in the literature [27], exploiting hydrophobic interactions to prearrange the functional monomer–template adduct in water. Polymerisable derivatives of cyclodextrins (non-polar cavities able to form inclusion complexes with non-polar guests in water) have thus been synthesised. During the imprinting process, the latter position themselves in a complementary way to the hydrophobic groups of the target molecule, forming an ordered assembly.

Cooperative Molecular Recognition

Some applications such as immunological tests and sensors require the development of new imprinted materials that are more sensitive and more selective with regard to the target molecule. In general, when polymers are synthesised by molecular imprinting, this is done starting from a single type of functional monomer. But the affinity of a host for a target molecule will be improved if the template can be simultaneously complexed at several points by monomers of different types. This approach has recently been illustrated in the molecular imprinting of cinchonidine, which was polymerised in the presence of two types of monomer, a vinyl derivative of a zinc porphyrin and methacrylic acid [28]. The results show that imprinting efficiency is enhanced when the two types of monomer are combined, suggesting a synergy or cooperation effect in the molecular recognition process.

Producing Analogs of Catalytic Antibodies

The immune system is a biological system producing hosts called antibodies which have the property of blocking foreign molecular intruders. If the foreign agent is a molecule mimicking the transition state of some particular reaction, the created antibody will 'recognise' this transition state whose free energy is reduced compared with the reactants. As a consequence, the reaction will be accelerated. In this case, the target molecule is called a hapten and the host is called a catalytic antibody. Recently, this idea has been transposed to the design of catalytic materials by molecular imprinting.

For example, the crossed aldol condensation of acetophenone with benzaldehyde has been carried out using an artificial catalytic antibody obtained by imprinting the transition state. The hapten was in this case a complex of cobalt (II) dibenzoylmethane bis(4-vinylpyridine) and the matrix a cross-linked polystyrene [29]. The results of the catalysis have shown that the polymer imprinted in this way is not only active and selective in the aldol condensation of acetophenone and of benzaldehyde, but is also capable of increasing the reaction rate by a factor of 8 compared with the reaction in solution. This study represents the first example of C–C bond formation which uses catalysts synthesised by the molecular imprinting technique.

Acknowledgements

The authors would like to thank Gerald P. Niccolai for his invaluable help in designing and producing the figures, and Joël Moreau for his free supply of information concerning imprinted complexing materials.

References

1. J.-M. Lehn: *Supramolecular Chemistry*, VCH Weinheim (1995); T.A. Gilbertson, S. Damak, R. Margolese: Curr. Opin. Neurobiol. **10**, 519–527 (2000); D. Lessing, J.R. Carlson: Curr. Opin. Neurobiol. **9**, 766–771 (1999)
2. A. Fersht: *Enzyme Structure and Mechanism*, New York (1985); T.E. Creighton: *Proteins: Structure and Molecular Properties*, New York (1993)
3. D.J. Cram: Angew. Chem. Int. Ed. Engl. **27**, 1009 (1988); J.-M. Lehn: Angew. Chem. Int. Ed. Engl. **27**, 89 (1988); C.J. Perdersen: Angew. Chem. Int. Ed. Engl. **27**, 1021 (1988)

4. R.A. Bartschard, M. Maeda: *Molecular and Ionic Recognition with Imprinted Polymers*, ACS series, Vol. 703, Washington DC (1997); B. Sellergren: *Molecularly Imprinted Polymers: Man-Made Mimics of Antibodies and Their Applications in Analytical Chemistry*, Vol. 23, Elsevier, Amsterdam (2001); M.E Davis, A. Katz, W.R Ahmad: Chem. Mater. **8**, 1820–1839 (1996)
5. G. Wulff: Angew. Chem. Int. Ed. Engl. **34**, 1812–1832 (1995)
6. M. Komiyama, T. Takeuchi, T. Mukawa, H Asanuma: *Molecular Imprinting: From Fundamentals to Applications*, Wiley-VCH, Weinheim (2003)
7. F.H. Dickey: Proc. Natl. Acad. Sci. U.S.A. **35**, 227–229 (1949)
8. R. Arshady, K. Mosbach: Makromol. Chem. **182**, 687–692 (1981)
9. B. Sellergren, M. Lepistö, K. Mosbach: J. Am. Chem. Soc. **110**, 5853–5860 (1988)
10. G. Vlatakis, L.I. Andersson, R. Muller, K. Mosbach: Nature **361**, 645–647 (1993)
11. G. Wulff, R. Grobe-Einsler, A. Sarhan: Makromol. Chem. **178**, 2817–2824 (1977)
12. G. Wulff, B. Heide, G. Helfmeier: J. Am. Chem. Soc. **108**, 1089–1091 (1986)
13. G. Wulff, J. Vietmeier: Makromol. Chem. **190**, 1717–1726 (1989)
14. K.J. Shea, D.Y. Sasaki: J. Am. Chem. Soc. **113**, 4109–4120 (1991); K.J. Shea, D.Y. Sasaki: J. Am. Chem. Soc. **111**, 3442–3444 (1989); K.J. Shea, T.K. Doughertly: J. Am. Chem. Soc. **108**, 1091–1093 (1986)
15. M.J. Whitcombe, M.E. Rodriguez, P. Villar, E.N. Vulfson: J. Am. Chem. Soc. **117**, 7105–7111 (1995)
16. J.U. Klein, M.J. Whitcombe, F. Mulholland, E.N. Vulfson: Angew. Chem. Int. Ed. Engl. **38**, 2057–2060 (1999)
17. C.J. Brinker, K.D. Keefer, D.W. Schaeffer, C.S. Ashley: J. Non-Cryst. Solids **48**, 47 (1982); C.J. Brinker, G.W. Scherer: *Sol–Gel Science*, Academic Press, San Diego (1990); R.K. Iler: *The Chemistry of Silica*, Wiley, New York (1979)
18. A. Katz, M.E. Davis: Nature **403**, 286–289 (2000)
19. U. Schubert: New J. Chem. **18**, 1049–1058 (1994)
20. A. Bibby, L. Mercier: Chem. Mater. **14**, 1591–1597 (2002)
21. J. Brown, L. Mercier, T.J. Pinnavaia: J. Chem. Soc., Chem. Commun. 69–70 (1999)
22. J. Matsui, M. Okada, T. Takeuchi: Anal. Commun. **34**, 85–87 (1997)
23. J. Matsui, O. Doblhoff-Dier, T. Takeuchi: Anal. Chim. Acta. **343**, 1–3 (1997)
24. L. Ye, O. Ramström, R.-J. Ansell, M.-O. Månsson, K. Mosbach: Biotechnol. Bioeng. **11**, 98–102 (1999)
25. S. Vidyasankar, M. Ru, F.H. Arnold: J. Chromatogr. A **775**, 51–63 (1997)
26. S. Bourg, J.C. Brondic, O. Conocar, J.J.E. Moreau, D. Meyer, M. Wong Chi Man: Chem. Mater. **13**, 491–499 (2001)
27. T. Akiyama, T. Hishiya, H. Asanuma, M. Komiyama: J. Inclu. Phenom. Macrocyclic Chemistry **41**, 149–153 (2001); H. Asanuma, T. Akiyama, K. Kajiya, T. Hishiya, M. Komiyama: Anal. Chim. Acta. **435**, 25–33 (2001)
28. T. Takeuchu, T. Mukawa, J. Matsui, M. Higashi, K.D. Shimizu: Anal. Chem. **73**, 3869–3874 (2001)
29. J. Matsui, I.A. Nicholls, I. Karube, K. Mosbach: J. Org. Chem. **61**, 5414–5417 (1996)

Part V

Applications of Nanomaterials

28

Electronics and Electromagnetism

J.-C. Nièpce and D. Givord

28.1 Multilayer Ceramic Capacitors

Multilayer ceramic capacitors (MLCC) are the capacitors most commonly used in electronic circuits (television, radio, telephone, automobile, aeronautics, space, etc.). The main advantages are low cost, small size, a good level of chemical inertness, due to the fact that they are made from chemically very stable oxide ceramics, and hence good stability in time.

Most of the dielectrics in these capacitors are essentially based on barium titanate, which implies a very high dielectric constant and excellent electrical insulating properties. For more than twenty years now, it has been understood that ceramic dielectrics made with grains measuring a hundred or so nanometers, one of the earliest types of nanomaterial, would be able to meet the requirements of the evolving electronics industry.

28.1.1 What Is a Multilayer Ceramic Capacitor?

In the field of electronic or electrotechnical components, a capacitor is a system of two conductors, generally placed close together with opposing faces separated by an insulator, e.g., two parallel metal plates separated by a certain thickness of air, or two aluminium foil sheets stuck on either side of a sheet of paraffin paper. The two conductors are the electrodes of the capacitor, while the insulator is the dielectric.

The capacitance C of the capacitor is proportional to the effective surface area S of the plates, inversely proportional to the thickness e of the dielectric, and proportional to a characteristic quantity associated with the dielectric material called its permittivity ε (see Fig. 28.1a). The permittivity ε is often replaced by the product $\varepsilon_0\varepsilon_\mathrm{r}$, where ε_0 is characteristic of the physical units and ε_r is a dimensionless quantity, hence independent of the units, characteristic of the dielectric material and called the relative permittivity, or simply the dielectric constant of the material. Hence,

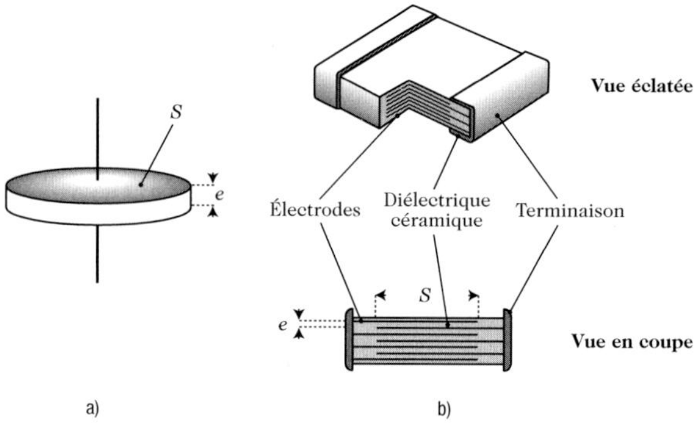

Fig. 28.1. Ceramic capacitors. (**a**) Disk. (**b**) Multilayer

$$C = \frac{\varepsilon S}{e} . \tag{28.1}$$

If the dielectric is a ceramic material, one then has a ceramic capacitor. These materials generally yield high values of ε. To increase the capacitance without making the surface area too big, efforts have been made to reduce the dielectric thickness, moving from a disk technology to a multilayer technology. This leads to multilayer ceramic capacitors (MLCC), comprising n layers of thickness e. This is equivalent to n identical capacitors of thickness e mounted in parallel (see Fig. 28.1b):

$$C = n\frac{\varepsilon S}{e} . \tag{28.2}$$

In practice the dielectric must fulfill at least two functions: it must 'store' electrical charge, an aptitude characterised by its permittivity ε, but it must also be a good electrical insulator, to ensure that the charges cannot transfer from one electrode to the other, an aptitude characterised by its resistivity ρ.

Various technologies are used to make MLCCs, but the process always ends with a stage in which the ceramic and metal layers are cosintered. This restricts the choice of material, but it also imposes an economic constraint. In particular, since the dielectrics are usually oxides, cosintering must be carried out in air. This requires the use of precious metals, usually silver–palladium alloys, to make the internal electrodes.

Cosintering Ceramic and Metal Layers

Before explaining what cosintering involves, it will be useful to recall the basic idea of sintering. The usual definition is: consolidation of a granular agglomerate under the action of heat, with or without melting of one or more components. Put in a more accessible way, sintering is an operation that consists in transforming a workpiece (referred to as the green body in ceramics) made by moulding a dried slurry, itself comprising an assembly of grains that are not permanently fused together (with or without a binding agent), to form the sintered or fired body, made up of grains of the same matter but welded together under the effects of heating. A familiar illustration of sintering occurs when the potter fires a ceramic artefact.

In the case of multilayer ceramic capacitors (MLCC) (see Fig. 28.1b), the green body comprises alternating layers of dielectric and metal. Within this green body, both the dielectric and the metal are in the form of grains, in general a few micrometers in diameter, bound together by organic binders. To obtain the final MLCC, the metal and the dielectric must be sintered simultaneously. This is what is referred to as cosintering. Of course, it requires the metal and dielectric to sinter under the same conditions of temperature and atmosphere.

28.1.2 Market Requirements

Computers and telephones provide a good illustration of the general trend in electronics, namely a steady reduction in price and a spectacular increase in miniaturisation, whilst improving performance and reliability. Among the components of electronic circuits, MLCCs have made a significant contribution to this general trend. The aim in this section will be to show how it has been possible to meet the requirements of this evolution.

Price Reductions

To give an idea of the order of magnitude, one can say that over the last fifteen years (from 1987 to 2001), the average price of an MLCC has been divided by ten. There is no doubt that this has been achieved in part through improved productivity, but technological progress largely supported by the scientific community has also played an important role. This is particularly true of MLCCs with nickel or copper internal electrodes, but also of very fine-grained dielectrics.

Miniaturisation of MLCCs

MLCC manufacturers face two complementary geometrical constraints: they must reduce component volumes to reduce their mass, and they must reduce plate areas to reduce circuit dimensions, whilst maintaining the value of the capacitance. These two constraints have been largely dictated by recent developments in cellular telephones.

Of course everyone is aware of the spectacular successes of miniaturisation in electronics. Even though not all this achievement can be attributed to MLCCs, these have nevertheless made a major contribution, since the capacitance per unit volume has been multiplied by a factor of a thousand in just thirty years.

28.1.3 Constraints Laid Down by these Requirements

Price Reductions

In the production cost of an MLCC, precious metals count for 30–80% depending on the sintering temperature, the format of the capacitor, and the number of internal electrodes. It is not surprising therefore that efforts have concentrated on the question of the precious metals. Various approaches to the problem have been explored, following two main lines of thought: reduce the amount of precious metal used, or replace the precious metal by a less expensive one.

In order to implement the second idea, the technology had to be changed. Either the metal/ceramic cosintering had to be abandoned and one had the so-called injected electrode technology [10], or the cosintering was retained but preventing the oxidation of the metal, by carrying out the cosintering with a low partial pressure of oxygen. It is the second of these choices that has proved itself on the industrial scale and today more than a quarter of all MLCCs are made with non-noble metal internal electrodes, usually nickel, using this process.

In order to reduce the amounts of precious metals, various approaches have also been explored:

- Reduce the thickness of the internal electrodes. These are usually about 1 μm thick and it is impossible to reduce them further without risk of breaking the continuity of the metal film, but metal powders with very fine grains, of the order of a hundred nanometers, are required to obtain an excellent electrode geometry.
- Reduce the palladium content in Ag–Pd alloys by reducing sintering temperatures. To this end, fluxes are introduced at great risk of modifying the behaviour of the dielectric compositions. Another approach involves significantly reducing grain diameters in the precursor powders. Indeed, using powders with grain diameters of a hundred nanometers, the sintering temperature can be reduced by 100–200°C compared with powders with micrometric grains, and hence the palladium content can be reduced by 30–40% (see Fig. 28.2). For the first time, it appeared helpful to use a nanomaterial as dielectric in an MLCC, and more generally, in the synthesis of a ceramic, instead of the same material that would conventionally be composed of micrometric grains, the advantage being that sintering temperatures could be significantly reduced.
- Reduce the number of internal electrodes. This has been the most fruitful line of attack, keeping the Ag–Pd alloys and hence making only minor changes to the technology. To do this, according to (28.2), in order to reduce n while keeping the area S constant, the ratio ε/e must be increased. Dielectrics must therefore be made with higher relative permittivities while the active layers of the dielectric must be made thinner. We recover the same double constraint when considering miniaturisation.

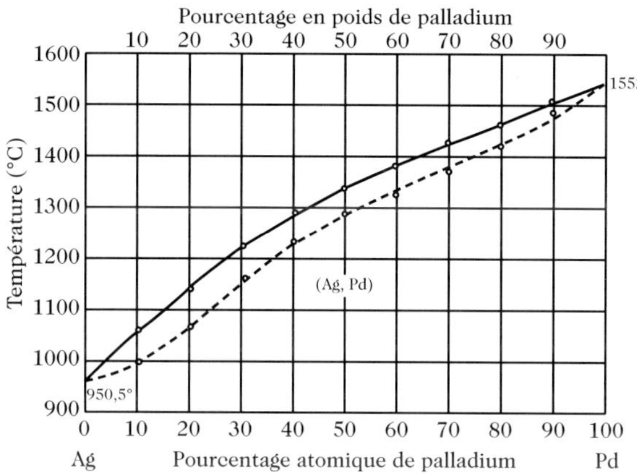

Fig. 28.2. Silver–palladium phase diagram

Miniaturisation of MLCCs

- Reduction of the volume V requires an increase in the capacitance per unit volume C/V. Since $V = nSe$, (28.2) implies that C/V is increased by increasing ε/e^2.
- According to (28.2) and because n must also be reduced (see above), reduction of the plate area S requires an increase in the ratio ε/e.

Therefore, in order to satisfy the requirements of miniaturisation, one must attempt to make dielectrics with higher relative permittivities while reducing the thickness of the active layers of the dielectric. These are the same requirements as those imposed by a reduction in the number of internal electrodes.

28.1.4 $BaTiO_3$ Ceramic Dielectrics with Nanograins: The Favoured Solution

It has already been mentioned above that using nanometric powder grains of the dielectric material provides a way of significantly lowering the palladium content in the alloy used to make the internal electrodes. One appeals here to the influence of grain size on a physical property due to a so-called surface effect. Indeed, as shown in the earlier chapters of this book, it is a quite general fact that the grain size can modify the thermodynamic characteristics of solids through this surface effect.

In addition, it has just been shown that dielectric layers need to be made as thin as possible. Only technological progress has made it possible to achieve some of these ideas. By switching from the disk technology to the multilayer technology, but using conventional precursor powders, the minimal thickness of dielectric layers was reduced from a few tenths of a millimeter to a few tens of micrometers. The capacitance per unit volume was then multiplied by a factor of more than a hundred.

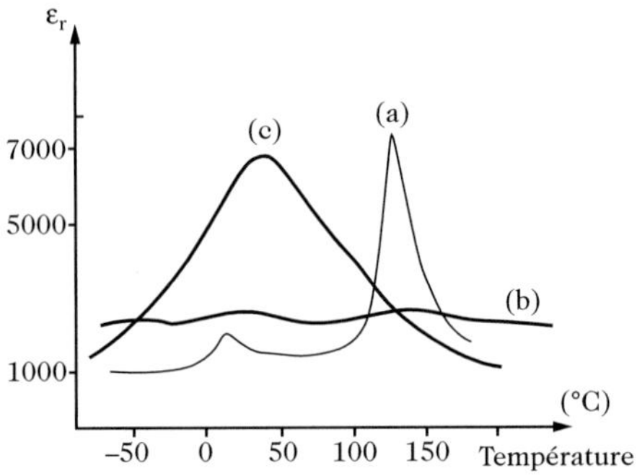

Fig. 28.3. Curves $\varepsilon_r = f(T)$ for a pure $BaTiO_3$ ceramic (**a**) and for X7R (**b**) and Z5U (**c**) ceramics

To further decrease thicknesses, chemists had to develop new ways of synthesising powders. Indeed, to obtain less and less thick dielectric layers in which a significant number of grains make up each layer, ever finer powders had to be used. To make disk capacitors, grains measuring between a few micrometers and several tens of micrometers were perfectly acceptable, but to generate today's multilayer capacitors, grains with sizes in the range 100–500 nm are needed. To achieve this, methods of powder synthesis had to be reconsidered. Whereas all powders used to be synthesised by dry processes, i.e., high temperature reactions between solid precursor oxides, the general trend has been toward synthesis in wet conditions, providing better control over the grain size distribution. Indeed, methods for synthesising powders in liquid media more closely resemble organic synthesis and can be carried out at lower temperatures. This kind of process can be described as soft chemistry. By determining the reaction mechanisms, it has been possible to control nucleation and growth processes, essential if one is to constrain the grain size distribution. The liquid medium is either aqueous or organic, and reaction conditions are more and more often hydrothermal or supercritical. A second need for nanograin dielectrics has thus been imposed on manufacturers. Over the last twenty years, the thicknesses of dielectric layers have dropped from a few tens of micrometers to a few micrometers, a further gain by a factor of a hundred in the capacitance per unit volume.

Let us now turn to the physics of the dielectric and show that nanometric grains will also lead to a very significant increase in the dielectric constant. In the so-called type II MLCCs, the basic constituent of the dielectric is very often barium titanate $BaTiO_3$. For a more detailed discussion regarding barium titanate in its nanocrystalline state, the reader is referred to Chap. 2 of the present book.

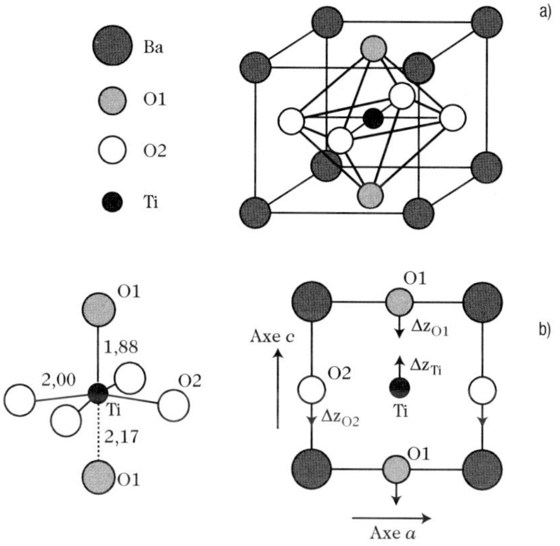

Fig. 28.4. Perovskite crystal structure of $BaTiO_3$. (**a**) Cubic paraelectric form. (**b**) Tetragonal ferroelectric form

Dielectric Characteristics of Barium Titanate

Barium titanate is used in dielectric ceramics because it has a very high dielectric constant during its ferroelectric–paraelectric transition. However, the latter is extremely sensitive to the temperature (see Fig. 28.3). On the other hand, barium titanate cannot be used in the pure state in dielectric ceramics; certain standards must be respected in order to obtain dielectrics for which the variations of ε_r as a function of the applied field are well defined over a given range of temperatures. The X7R and Z5U standards are the most widespread for MLCCs. Figure 28.3 shows typical curves for $\varepsilon_r = f(T)$.

How are the X7R and Z5U dielectrics made? They are generally synthesised from $BaTiO_3$ by carrying out partial substitutions of Ba and Ti by isovalent ions such as Sr or Ca on the Ba site and Zr on the Ti site, in such a way as to produce solid solutions. The Z5U dielectrics are generally composed of grains of the same solid solution, a single phase, with a general formula of type $Ba_{1-x}Ca_xTi_{1-y}Zr_yO_3$, but such that x and y vary from one grain to another. The X7R dielectrics are more sophisticated, being made up of several phases of the above type and with variable composition.

Crystalline Characteristics of Barium Titanate

Like the various solid solutions described above, barium titanate crystallises in the perovskite crystal structure. At room temperature and up to about 120°C, it forms a tetragonal deformation of this standard structure (see Fig. 28.4b).

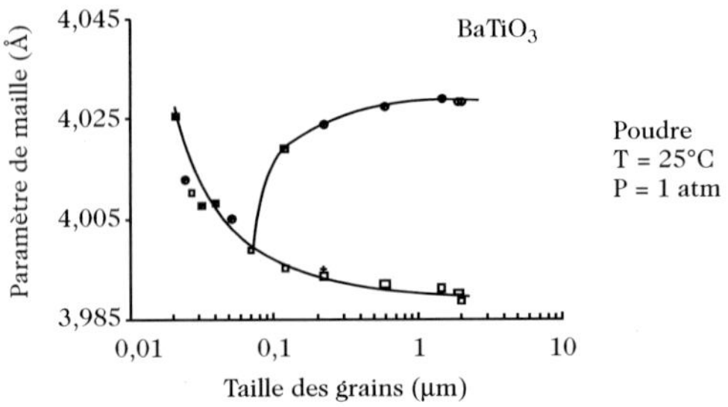

Fig. 28.5. Lattice parameters of $BaTiO_3$ in the powdered state, at 25°C and at standard atmospheric pressure, as a function of grain size [2]

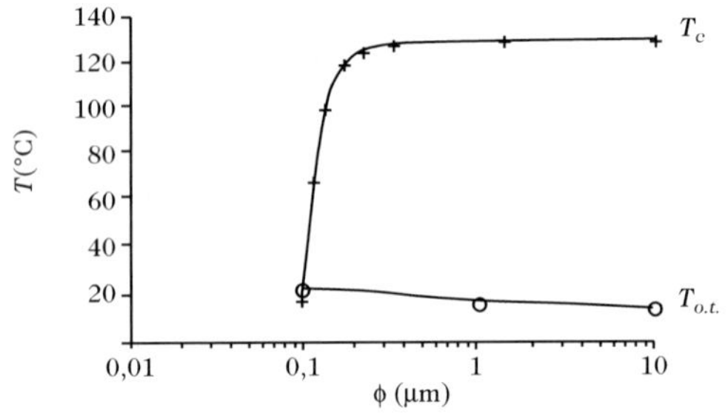

Fig. 28.6. Temperatures T_{ot} and T_c of the orthorhombic–tetragonal and tetragonal–cubic phase transitions, respectively, as a function of grain size for $BaTiO_3$ powders at standard atmospheric pressure [3]

This means that it has a crystal lattice with a dipole moment and the characteristics of a ferroelectric material. Above this temperature, the perovskite lattice is no longer deformed and has cubic symmetry (see Fig. 28.4a). The material is then paraelectric. During the transition from the ferroelectric state to the paraelectric state, the average dielectric constant assumes very high values, as can be seen from Fig. 28.3.

Characteristics of Barium Titanate in the Nanograin State: Influence of Grain Size

As early as the 1950s, Känzig et al. [1] observed that the tetragonality of the $BaTiO_3$ crystal lattice was reduced when the powder grains were made small enough. A more recent investigation [2] with powders covering a broad range of grain size distributions has provided a more detailed picture of this reduction of tetragonality, showing that, if the powder grains are smaller than 100 nm, barium titanate is cubic and paraelectric whatever the temperature (see Fig. 28.5).

Furthermore, it has been shown that the temperature of the phase transition from the ferroelectric to the paraelectric state, at standard atmospheric

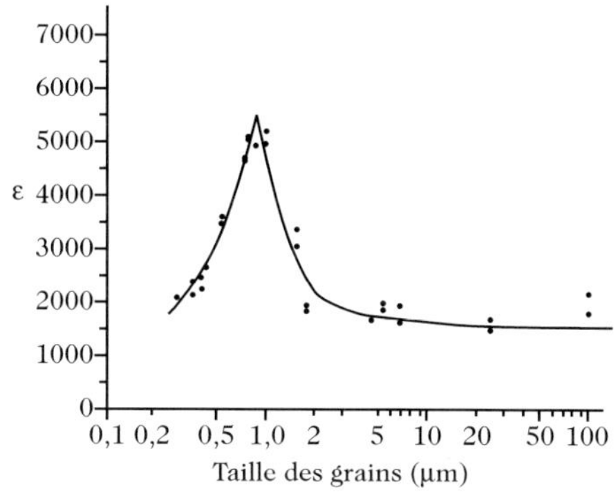

Fig. 28.7. Relative dielectric constant ε_r at 25°C for $BaTiO_3$ in polycrystalline ceramics as a function of the average grain diameter in the ceramic [4, 8]

pressure, falls steadily as the grain size is reduced (see Fig. 28.6) [3]. The transition disappears when the grain size is smaller than 100 nm. However, this phenomenon is much less pronounced in ceramics, i.e., when the solid–gas interface presented by the powder grains is replaced by the solid–solid interface presented by the grains of a ceramic [4]. This phenomenon illustrates the influence of grain size on a physical property through a surface effect [5, 6].

Regarding the dielectric constant of $BaTiO_3$, its dependence on the grain size in this ceramic material has been known for a long time now [7, 8]. As the grain size is reduced, the characteristic curve $\varepsilon_r = f(T)$ for $BaTiO_3$ tends to flatten out. The dielectric constant increases in the neighbourhood of room temperature, whereas it falls during the ferroelectric–paraelectric phase transition near 120°C. However, this trend is not monotonic, as can be seen from Fig. 28.7 [4, 8]: a maximum is obtained for the dielectric constant at room temperature when the grain size is around 700–800 nm and the value of ε_r there is about five times greater than its value for mainly micrometric grains. It is this grain size that is currently sought in $BaTiO_3$ dielectric ceramics.

How can we explain the curve in Fig. 28.7? In order to interpret this curve, we need to know what contributes to the dielectric constant and we need also to explain its dependence on the grain size. A priori, two features of barium titanate which both contribute to the dielectric constant are strongly influenced by the grain size:

- the crystal structure, as shown above,
- the ferroelectric domain microstructure.

Indeed, it has been shown that the ferroelectric domain microstructure in $BaTiO_3$ ceramics is closely related to the grain size (see Fig. 28.8) [9]. Comparing the information obtained from Figs. 28.7 and 28.8, we observe that the maximum value of the dielectric constant coincides with the transition from a microstructure associating domains at 180° and domains at 90°, for grain sizes greater than 0.8 µm, to a microstructure comprising only domains

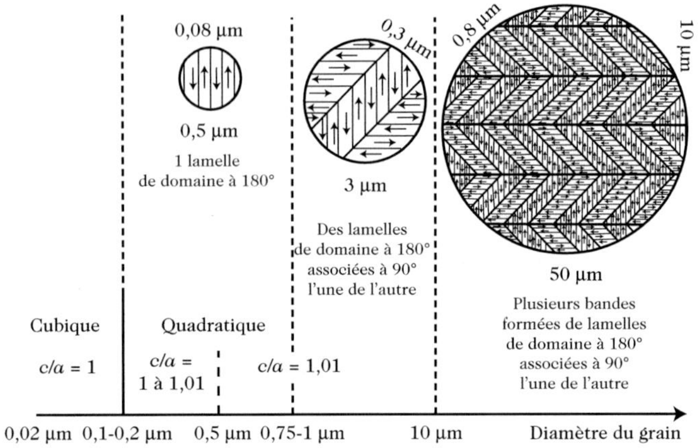

Fig. 28.8. Ferroelectric domain microstructure of $BaTiO_3$ in the polycrystalline ceramic state as a function of the average grain diameter in the ceramic [9]

at 180°, for grain sizes below this value. In this case, the phenomenon illustrates, for the same material, the influence of grain size on a physical property via a confinement effect.

A microstructure with nanometric grains that does not possess domains at 90° also has another advantage: it suppresses almost all dielectric losses of ferroelectric origin.

28.2 Magnetic Recording

The hard disk is an important element in any computer. It stores the information required for the computer to run and the data the user wishes to conserve. It is a complex system using various types of nanostructured magnetic materials.

28.2.1 General Operation

Figure 28.9 shows a photograph of a hard disk with its drive system and the read and write head. The material in which the recorded data is held is deposited at the surface of the disk itself. One bit of information is represented by the orientation of the magnetisation in a given region of the material. Two orientations are possible, with which the states 0 and 1 are associated.

The write head is in fact a tiny electromagnet. When a current goes through the coil, a magnetic field is set up in the iron core. When applied to the region corresponding to a given bit, this field can write the desired information at that site. The head is held at the end of an arm rotating around an axis. It can be positioned at the desired distance from the disk center by

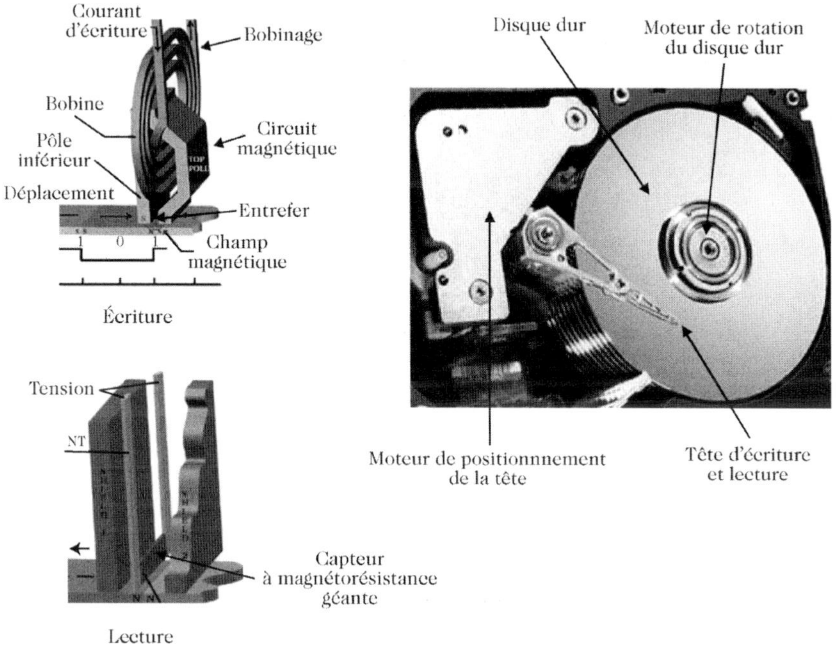

Fig. 28.9. General components of a hard disk

means of its positioning motor (voice coil motor VCM). The disk, driven by its own motor, is always rotating, with a typical angular speed of 100 rev/s. An electric current applied to the coils at a given instant of time and for a lapse of time of a few nanoseconds can generate the write magnetic field in the region of the relevant bit which happens to be passing by the head at that moment.

The read head comprises a giant magnetoresistance (GMR) sensor. The resistance of such a sensor depends on its magnetic state, itself determined by the magnetic field to which it is subjected. One bit of information can be considered as a tiny magnet, source of a magnetic field in its own neighbourhood. When a bit passes close to the read head, the magnetic field generated modifies the magnetic state of the sensor and, through this modification, its electrical resistance. The state of the bit is then translated into an electrical signal.

28.2.2 Recording Materials. Longitudinal and Perpendicular Recording

The smaller the bit size, the higher the recording density. In today's commercially available systems, the two possible orientations of the magnetisation lie in the plane of the disk and recording is said to be longitudinal. In the

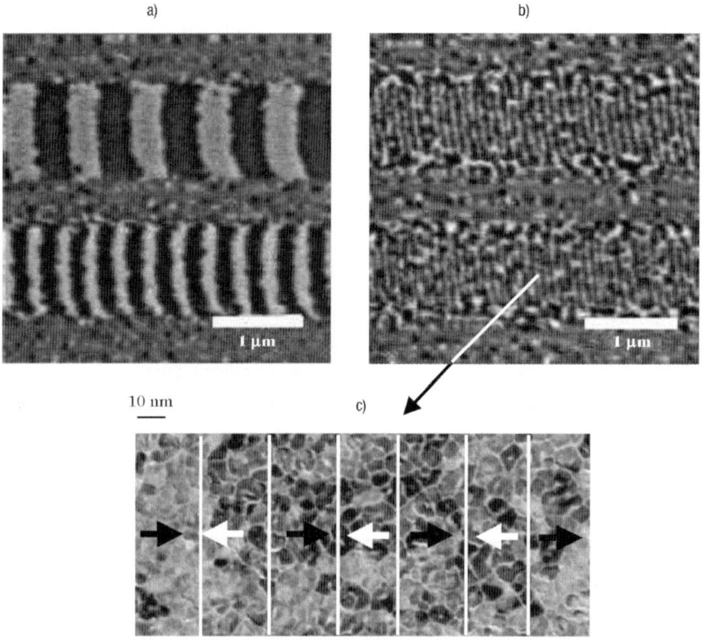

Fig. 28.10. Recording data in a magnetic material

photograph of Fig. 28.10a, bits have width 330 nm (upper part of photo) and 160 nm (lower part of photo). In the photograph of Fig. 28.10b, this width is about 50 nm. In the latter case, the recording density is 3×10^9 bits/cm^2 (or 20 Gbits/in^2). Information is stored in a metal film made from an FeCoCrNi alloy of thickness around 200 nm. Inside a homogeneous magnetic material, it is difficult to define the orientation of the magnetisation within a region independently of the orientation in other regions. The recording material is thus made up of magnetic grains (light or dark grey in Fig. 28.10c), separated from one another by a non-magnetic phase (in white in Fig. 28.10c).

Using current fabrication processes, the crystallographic orientation of each grain cannot be perfectly controlled. The write magnetic field cannot in fact define the orientation of the magnetisation along the easy axis of magnetisation of each grain within a single bit. The fewer grains there are within the bit, the greater will be the fluctuations in the magnetisation direction from one bit to the next. These fluctuations produce an interference signal called read noise, which perturbs the reading of the recorded signal. In order to reduce read noise, a given bit usually comprises several hundred grains. At the highest recording densities achieved today, i.e., 6 Gbits/cm^2, these grains have diameters of the order of 8 nm. However, this reduction in grain size, required to increase the recording density, must already face up to a problem due to another physical phenomenon. Indeed, below a certain critical diameter, the orientation of the magnetisation in a particle will fluctuate spontaneously under the effect of thermal activation (see Sect. 5.5.2). With the alloys used today, we are already close to the limit called the superparamagnetic limit. By imposing a magnetisation direction perpendicular to the plane of the disk, it is possible, by creating rod-shaped particles oriented perpendicularly to the surface, to increase the recording density without reducing the volume of the particles. It is estimated that the perpendicular recording mode will be able to achieve the ultimate density of the order of 20 Gbits/cm^2.

In order to increase the recording density still further, up to values greater than 100 Gbits/cm^2, novel materials will be required, made from quantum bits. In these materials, a single magnetisation direction can be defined. One bit can then comprise a single magnetic grain without being the source of a prohibitive level of read noise.

28.2.3 Write Heads

The size of the electromagnet producing the write magnetic field decreases in proportion to the size of the recording bits. In today's hard disks, the electromagnet is a microsystem fabricated by deposition and lithographic techniques. The electric current in the coils is of the order of 20–50 mA, a value limited by heating of the coils, which must not be above about 30 K. To maximise the write field, the material making up the magnetic circuit must have two properties: high susceptibility in a low field, ensuring that it will be magnetised under the effect of the excitation field created by the coils, and a high magnetisation, since the generated field is proportional to it. The part of the magnetic circuit closest to the disk comprises a stack of FeNi and FeCoTaN alloys. The spontaneous magnetisation $\mu_0 M_s$ of the second of these alloys reaches 2.4 T (the magnetisation of pure iron being 2.2 T). The alloys contains grains with diameters of about 20 nm. Because the grains are nanometric, the exchange coupling between grains can produce the high susceptibility required (see Sect. 5.2).

28.2.4 Read Heads

The read head is a sensor whose core contains a stack of two soft ferromagnetic layers such as FeNi, separated by a non-magnetic layer of Cu or Ag. One of the ferromagnetic layers is coupled with an antiferromagnetic layer of FeMn or MnPt. Its magnetisation is frozen in a given direction by the exchange bias mechanism (see Sect. 5.4). The magnetisation of the other ferromagnetic layer is easily rotated under the effect of an applied field and is said to be free. This whole arrangement is called a spin valve (see Fig. 28.11).

The ferromagnetic and non-magnetic layers typically have thicknesses of about 10 nm. The electrical resistance of this arrangement depends on the respective orientations in the two ferromagnetic layers via the so-called giant

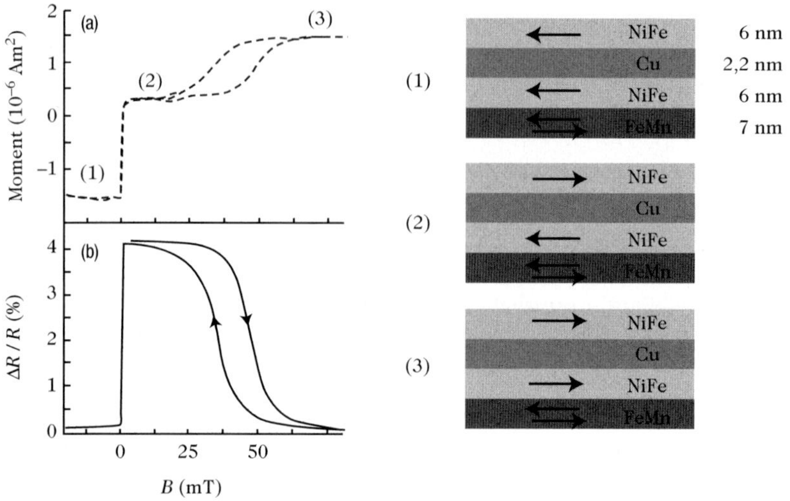

Fig. 28.11. Spin valve structure used in giant magnetoresistance read heads. Note that under normal operating conditions of the sensor, the field at the sensor is less than 10 mT. The magnetisation of the lower NiFe layer shown schematically in (3) is not reversed

magnetoresistance mechanism, described in Chap. 14 of [11]. Giant magnetoresistance is typical of materials with nanometric dimensions, because the thicknesses of the layers must be less than the mean free path of electrons.

The magnetic leakage field produced by the data bit is approximately perpendicular to the initial direction of magnetisation of the free ferromagnetic layer. The resulting rotation of the magnetisation is directly related to the sign of the field generated by the data bit. It is deduced from the measured value of the magnetic resistance. This measurements provides access to the recorded data.

28.2.5 Disk Drive Motor

The motor causing the hard disk to rotate includes magnets made from an alloy of neodymium, iron, and boron, with chemical composition $Nd_2Fe_{14}B$. This material is currently used to make most high performance magnets. In the present case, the magnet is obtained by binding alloy flakes about 30 μm thick within a polymer. This is a reasonably cheap way of making high quality magnets. The diameter of the $Nd_2Fe_{14}B$ grains within the flakes is of the order of 20 nm. At these nanometric sizes, remanence enhancement occurs, as explained in Sect. 5.2, improving the magnetic properties by about 20%.

References

1. M. Anliker, H.R. Brugger, W. Känzig: Helv. Phys. Acta. **27**, 99–124 (1954)
2. G. Caboche, F. Chaput, J.P. Boilot, J.C. Nièpce: Silicates Industriels **58**, No. 5–6, 103–107 (1993)
3. K. Uchino, E. Sadanaga, T. Hirose: J. Am. Ceram. Soc. **72** (8), 1555–1558 (1989)
4. N. Bernaben, A. Leriche, B. Thierry, J.C. Nièpce, R. Waser: Fourth Euro Ceramics **5**, 203–210 (1995)
5. P. Perriat, J.C. Nièpce, G. Caboche: J. Therm. Anal. **41**, 635–649 (1994)
6. P. Perriat, J.C. Nièpce: High Temp. Chem. Proc. **3**, 585–600 (1994)
7. K. Kinoshita, A. Yamaji: J. Appl. Phys. **47**, 371–373 (1976)
8. G. Arlt, D. Hennings, G. De With: J. Appl. Phys. **58**, 1619–1625 (1985)
9. C. Valot-Odot: Doctoral thesis, University of Dijon, France (1996)
10. J.-C. Nièpce, J.-M. Haussonne (Eds.): *$BaTiO_3$: Matériau de base pour les Condensateurs Céramiques*, Vols. I and II, Editions Septima, Paris (1994) (see Vol. II pp. IV85–IV91)
11. C. Dupas, P. Houdy, M. Lahmani (Eds.): *Nanoscience: Nanotechnologies and Nanophysics*, Springer-Verlag, Berlin, Heidelberg (1996)

29

Optics

P. Maestro, M. Chagny, P.-P. Jobert, H. Van Damme, and S. Berthier

29.1 Cosmetics

29.1.1 Introduction

Inorganic nanoparticles provide an excellent example of the way size reduction can lead to novel or improved applications. Catalysis (supported precious metals), magnetic recording (nanoparticles of metal, iron oxide, or chromium oxide) and photography (silver halides), have all been transformed over the past ten years by the use of nanoparticles. New applications have come to light recently, for both complex fluid media and bulk materials, usually through a combination of properties resulting from size reduction, leading to an improved overall performance for the user.

The availability of silica, titanium oxide or rare earth nanoparticles, synthesised on an industrial scale for fifteen years now, and the possibility of designing them for use in a variety of different contexts has led to the development of much more effective and pleasant-to-use sun creams (see below), high performance tyres (see Chap. 30), and diesel vehicles less harmful to the environment (see Chap. 31).

29.1.2 Nano-Titanium Oxides in Cosmetics: Solar Skin Protection

Protection against the sun's UV radiation by means of particles involves two complementary mechanisms: intrinsic absorption due to the electronic properties of the compound, and light scattering and reflection phenomena in which the size and shape of the particles play an important role. Anti-UV inorganic compounds make use of both mechanisms, giving them a considerable lead in this field.

On the other hand, the criteria used to select or eliminate sun creams are becoming tougher and tougher. Of course, the level of UV filtering must be as high as possible, and the absorption spectrum as broad as possible, while at the same time satisfying 'cosmetic' requirements, such as transparency to visible light, so that skin protection is not accompanied by a whitening effect due to the high opacifying power of the inorganic pigments used (TiO_2, ZnO, etc.); development of stable dispersions that can pass through a nozzle without blocking it, in order to design spray systems; and skin protection with as high an index as possible without compromising the aesthetic aspect. Nanometric inorganic solar filters provide a way of satisfying this quest for better overall performance.

Titanium oxide is one of the best industrial white pigments due to its high refractive index and the significant resulting opacifying power. Its perfect transmission of the visible part of the spectrum comes with a very high absorption of UV radiation. This is due to a gap of the order of 3.4 eV, cutting off UV wavelengths below 365 nm, and it is an excellent anti-UV candidate.

Industrial production of titanium oxide nanoparticles was much developed through the 1990s, following the same trend toward wet synthesis as used for silica, but accompanied by a perfectly controlled calcination stage to stabilise one or other of the desired crystal structures (anatase or rutile depending on the application). By adjusting the various parameters controlling the synthesis, a wide variety of average crystallite diameters can be obtained, from 20 to 100 nm. It is also known how to coat these particles with a few atomic layers of silica and alumina. These developments have provided an answer to the needs expressed by the user, but not yet satisfied, as described above.

For the purposes of a solar filter, the following application has been developed:

- Particles of size 60 nm, the best compromise for ensuring optimal UV protection, as determined theoretically by the methods of physical optics applied to concentrated dispersed media, but also good dispersibility and a high level of transparency, in particular, no tendency to reagglomerate, when the cream is applied.
- Spherical particles, ideal for spraying, easy to apply with high covering power, and monodispersity on the nanoscale.
- Silica–alumina surface treatment on the nanoscale (see Fig. 29.1), to control the isoelectric point and hence ensure stability of the dispersions without the need for a surfactant, since it is essential to guarantee high colloidal stability of the product, not only in the pure state, but also in the final preparation.

Today, other possible solutions present themselves:

- Nano-zinc oxides (ZnO) are good candidates, because ZnO has a slightly lower gap than TiO_2 (at 3.3 eV rather than 3.4 eV) and, all other things being equal, is better placed to block out UV radiation than titanium oxide.
- Nano-cerium oxides may eventually prove to be a good compromise, and not only in cosmetics, because it is known how to stabilise them in aqueous or organic dispersions, as can be found in diesel fuel additives.

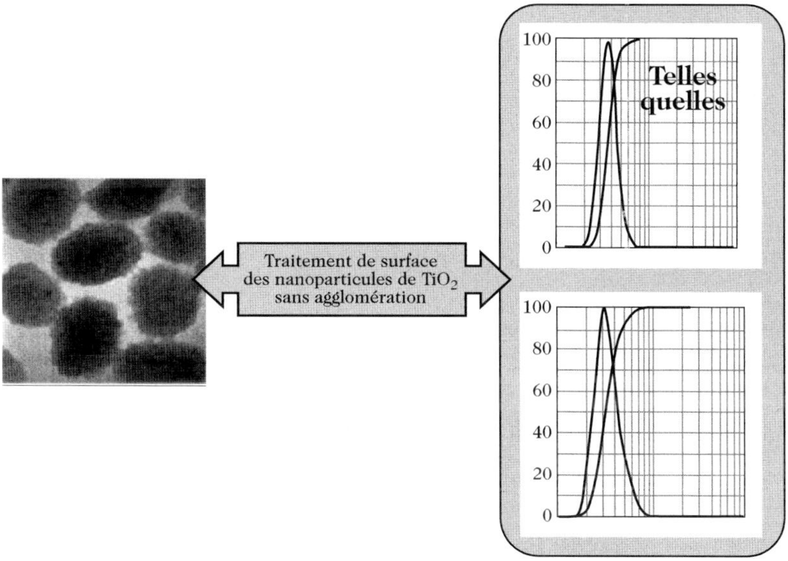

Fig. 29.1. Controlling the surface chemistry of TiO_2 nanoparticles

29.1.3 Conclusion

Although much is said about nanotechnologies today in a context where genuine industrial developments are still hard to find, the many applications of inorganic nanoparticles provide a striking example of the way in which controlled size reduction can be put to use, both in the production process and in the application, to achieve an interesting combination of properties for the final user. Going beyond the applications already mentioned, others are currently under research and development in the field of coatings (optical and mechanical properties), polyolefin or polyamide reinforcement, functional textiles, and so on. To stir the imagination, Fig. 29.2 shows how the use of nanophosphors (dispersions of nanoparticles in a liquid) (see Sect. 29.2) may lead to strong luminescence effects in otherwise transparent systems.

29.2 Nanophosphors

29.2.1 Introduction

Interest in luminescent materials or phosphors, known for decades now, has been revived recently, on both a technological and an industrial scale, with the fabrication of luminescent powders in submicron and nanometer forms. The surprising physical phenomena they give rise to should delight a new generation of scientists, while technologists face the rheological problems involved in their synthesis, and industry discovers previously unheard-of levels

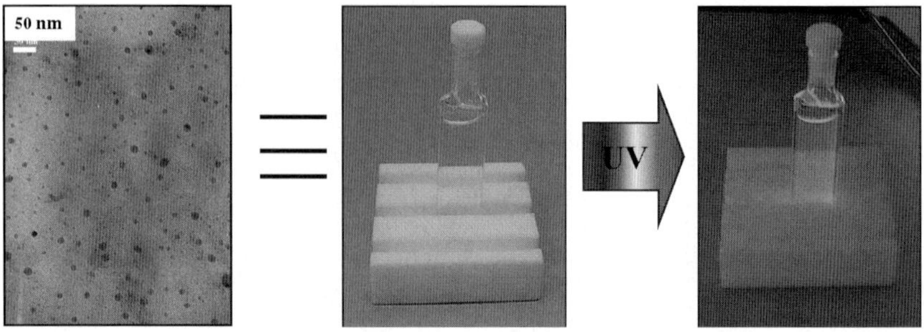

Fig. 29.2. A hidden functionality: transparent phosphors. Nanometric suspension of $LaPO_4$:Eu. See also the colour plate

Fig. 29.3. Phosphors re-emit in the visible after UV excitation. See also the colour plate

of performance, with a renewal of previous applications, reduced production costs, and new uses that could not have been imagined a few years ago.

29.2.2 Phosphors: General Considerations

These luminescent pigments are materials with the property that they re-emit light in the near UV, the visible or the infrared following excitation by UV, visible, or IR light, electron bombardment, or X-ray exposure (see Fig. 29.3). They are energy converters with efficiency expressed in lumen/watt and characterised optically in terms of the following five points: the fluorescence yield, the absorption spectrum, the emission spectrum, the colour point, and the remanence.

Fluorescence Yield

The fluorescence yield refers to the cathodoluminescence in the case of electron excitation and photoluminescence in other cases. These yields can be expressed in $\text{lumen/watt}_{\text{incident}}$, but also in $\text{watt}_{\text{radiated}}/\text{watt}_{\text{incident}}$ using as conversion coefficient the light equivalent ($\text{lumen/watt}_{\text{radiated}}$).

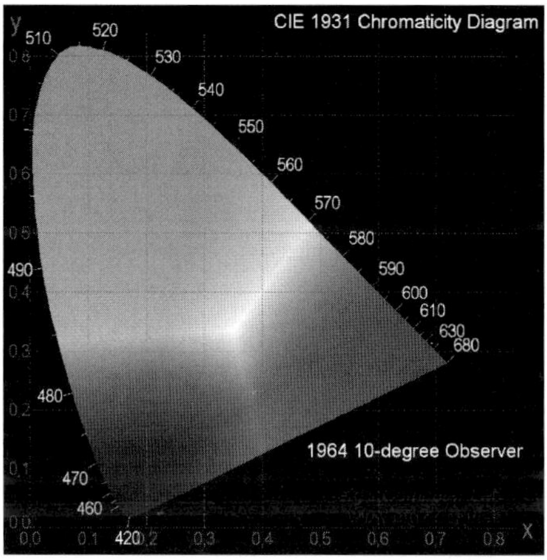

Fig. 29.4. Colour point represented in the chromaticity diagram. See also the colour plate

Emission and Absorption Spectra

Emission and absorption spectra are intrinsic characteristics of a material. Light emission is usually measured with a grating spectrometer or prism and is expressed in watt/steradian/nanometer/m^2.

Colour Point

The colour point describes the colour of a sample in an unambiguous way in a 3D space, usually projected down to two dimensions as in the example of Fig. 29.4. To do this, the emission spectrum is multiplied successively by three spectral sensitivity curves $X(\lambda)$, $Y(\lambda)$, and $Z(\lambda)$, before integrating over the whole spectrum to obtain the quantities X, Y, and Z. The functions $X(\lambda)$, $Y(\lambda)$, and $Z(\lambda)$ are defined in a standard laid down by the International Commission on Illumination (CIE). To project into a 2D space, these results are normalised to give

$$x = \frac{X}{X+Y+Z}\,, \qquad y = \frac{Y}{X+Y+Z}\,, \qquad z = 1-(x+y)\,.$$

The pair (x, y) calculated here characterises the colour of the sample. The point can then be positioned in the classical colour triangle, where perfect white has coordinates $(0.33, 0.33)$. Other representations can be made along similar lines, making use of linear or nonlinear combinations of the above quantities to be able to represent certain specific properties of colour, such as the colour difference.

Remanence

The remanence characterises the continuation of a signal, in this case a light signal, after the source that gave rise to it has been removed. There are various ways of quantifying this temporal quantity (τ_{10}, τ_{90}, $\tau_{10\text{–}90}$). It is usually measured by means of photomultipliers or photodiodes, paying particular attention to the intrinsic characteristics of the measurement chain which may introduce perturbations in the case of ultrashort remanence.

29.2.3 Operating Principle

Two mechanisms underlie inorganic phosphorescence:

- In semiconductors, a dopant creates an energy level in the band gap. During de-excitation, the energy is released, partly in the form of photons and partly in the form of phonons, via the level created by the dopant. This happens for example with manganese-doped zinc silicate, called P1 in the classification by the Joint Electron Device Engineering Council (JEDEC), the cathode ray tube specialists. The degeneracy of the level leads to a broad spectrum.
- In a crystal, insertion of a rare earth can produce transitions within its energy levels. This happens with europium-doped yttrium oxide, which has a spectrum of narrow lines (a few nm at half-maximum) around 613 nm.

In all these cases, the characteristics of the light emission are affected by the crystal lattice and hence by the environment of the dopant. The main factors influencing light-emitting properties are the concentration and degree of oxidation of the dopant and the local stoichiometry (on the scale of the crystal lattice). Some codopants can also modulate these properties, especially the remanence.

Emission by a phosphor is an intrinsic characteristic of matter and so can provide qualitative information about the internal structure, crystallinity, purity, and stoichiometry of a compound.

29.2.4 Industrial Applications

Phosphors are already present in many different fields of activity and arouse even more interest for revisited or innovative applications, especially due to the reduction in their size.

Standard Fields of Use

Naturally, the most obvious applications relate to lighting, with fluorescent tubes emitting white light. A large market is also to be found in TV screens, whether based on cathode ray tubes (zinc sulfide for blue and green, and yttrium oxide for red) or flat plasma screens [oxides and silicates are essential to avoid degradation by very short wavelength (174 nm) UV discharges]. Of course, there is also a difference in the way they are excited. Other applications are worth mentioning, such as RX scintillators for medical use, and anti-forgery tags used for example on the latest generation of bank notes.

Reducing the size of this type of material down to nanometric dimensions opens the way, as for more conventional ceramics (silica or titanium oxides), to novel applications, not only due to the size factor, but also because physico-chemical properties are modified on this length scale. Indeed, as can be seen from developments which involve of some nanostructured material insertion in a matrix, it turns out that macroscopic properties such as hardness, ductility, and surface reactivity, evolve with the size of the elementary particles making up the matter (see Chap. 8). Such changes may constitute a constraint in some respects, but most of the time they offer scope for further development. As usual, and this is the skill of the engineer, compromises must be found to exploit these materials.

One rather positive feature of all this is that our understanding of the way such materials behave requires us to rethink the laws of macroscopic physics governing the kind of matter we usually have to deal with. Nanometric phosphors are a case in point. Indeed, their light-emitting properties are affected by the size factor. In the first place, the fluorescence yield decreases due to an unfavourable interaction between radiation and matter when the grain size becomes much smaller than the exciting light wavelength. However, some surprising behaviour is also observed, bringing in other mechanisms. For example, a grain measuring a few nanometers can be made to fluoresce by exciting at a wavelength of a few hundred nanometers! These ultrafine grain sizes bring with them undeniable advantages in the context of industrial developments.

Applications of Nanophosphors (Nanometric Phosphors)

Ink Jet Compatibility

Today, with these tiny grain sizes, a powdered solid material can be deposited through an aperture without the risk of an aggregate (always possible in a suspension) blocking the nozzle. Compatibility for ink jet applications has already been demonstrated. Many fields of applications are interested in the flexibility represented by such a simple technique, piloted through a computer interface:

- Deposition of nanophosphors at the bottom of the constitutive barriers of plasma screens. Today, this technique is considered by Japanese industry to be standard technology.
- There are also applications in security tagging and markers for product traceability, not to mention less technical areas such as promotional and/or recreational labels on all kinds of materials from glasses and metals to plastics. In this case, it is also the inorganic nature of these ceramic nanoparticles which brings a plus, since they are more stable with regard to external

constraints like high temperatures, corrosive environments, or natural radiation.

Apart from being a simple and flexible technique allowing a high rate of operation, the ink jet and the use of nanopowders that follows from it also present an economic advantage since, for the same given function, the amounts of matter used can be significantly reduced.

Novel Fillers

Nanometric phosphors are also stirring up a lot of interest as fillers in varnishes for cosmetics and other types of make-up product, but also as fillers in technical fibres (light-emitting security cloths). In these applications, the main advantage is the fact that the filler is not visible, because powders with grain size smaller than the visible wavelengths (< 400 nm) no longer interact with visible light (by scattering), allowing it to pass through unhindered and preserving the transparency of the medium.

Tracers and Tags

Nanometric phosphors have great potential in biology and environmental applications. It is once again the size factor which opens the way to improvements:

- In biology, by fixing these tiny particles to a protein or some other active principle, it offers a way of tracking them on their journey through the blood capillaries and even through cell pores.
- In environmental studies, they can be used as non-radioactive and non-polluting markers, stable in any medium, for tracking the flow of a liquid or gas, e.g., to study industrial effluents by monitoring interstitial water or absorption by vegetation, or detecting the presence of a gas.

29.2.5 Conclusion

There can be no doubt that the nanophosphors will lead to new developments in products exploiting light-emission phenomena. The processes developed at DGTec (France) provide a flexible way of producing these materials for a broad range of applications, giving Europe an actor on a stage generally dominated by the Far East.

29.3 Surface Nanoengineering

29.3.1 What Is the Surface Area of a Town?

The surface area of mainland France is around 550 000 km^2. Paris, including the Boulogne woods and Vincennes, covers only about a hundred km^2. But what is the true rather than projected surface area of this town? A moment's thought is enough to realise that the true surface area, measured by the interface with the atmosphere, is immeasurably greater than the ground area. Let us begin by including the area of the rooftops and streets; then add the area of all the walls of the buildings, and the windows, followed by the areas of all the partition walls and floors inside the offices and flats, then the areas of the doors and furniture, then all the objects inside the cupboards, and all the ornaments on the shelves, then the areas of all the cars and their contents, and the endless underground channels and conduits, the road signs, the billboards, and so on and so forth. This calculation has never actually been carried out,[1] but there is a good chance that the difference between the ground area and the true area of the interface with the atmosphere would have to be counted in powers of ten.

All these surfaces get dirty and age, and in the case of glass surfaces, they can lose their transparency in the rain or under the effects of condensation. Given the huge areas involved here, it is not surprising that we spend correspondingly large amounts of time and money to keep them clean and, in the case of glass surfaces, transparent. We have invented windscreen wipers and we spend our time cleaning windows, dusting shelves, repainting woodwork and metal surfaces, rendering walls, and renewing road surfaces. This is not just a question of aesthetics. For the surfaces inside our living areas, it is also a question of hygiene. Avoiding the proliferation of bacteria in places with high population density is a major issue, especially in hospitals.

So what have nanomaterials got to do with all this? Well, it may be that they could provide the solution! Micro- and nanostructuring surfaces is an effective way to control their wettability. And surfaces can be kept clean under the action of solar or artificial light by depositing thin films of nanostructured semiconductors on them. What is more, the cleaning action extends to the surrounding atmosphere, which can be decontaminated in the process! And all this is already commercially available.

29.3.2 Superhydrophobic Surfaces

Water on a glass surface can reduce visibility in two situations, but in both cases, it is the partial wetting of the surface that causes all the problems (see Chap. 12). In the rain, that is, under the impact of water droplets, it is the way the droplets hang on along their triple line that leads to difficulties. The windscreen wiper is a solution, but it can only be used in certain circumstances. One possible line of attack is to make the surface so hydrophobic that the angle of contact will be almost 180°. The droplets then become practically spherical and the slightest slope or the slightest gust of wind will remove them.

[1] And indeed it would have no meaning unless one first defined a suitable resolution or, what amounts to the same thing, the basic length standard to be used.

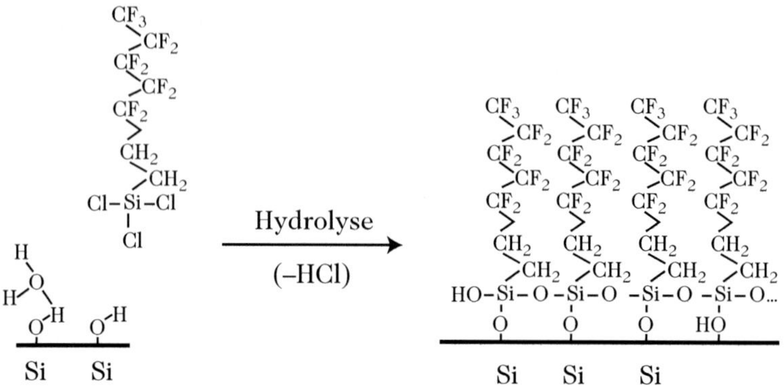

Fig. 29.5. Hydrophobisation of a glass or silica surface by grafting chlorofluorosilanes onto the surface hydroxyl groups of the solid

When there is no rain but the air is damp, it is the formation of condensation that becomes a problem. It is once again the partial wetting that creates this difficulty, because this is what causes nucleation and growth of discrete water droplets that then make the surface scatter light. To prevent the formation of condensation, the solution now is to increase, not the hydrophobicity, but the hydrophilicity of the surface, in such a way that condensation will form a continuous thin film that does not distort visual images, rather than light-scattering droplets.

But how can this be done? Let us begin by describing the hydrophobisation process. It has long been known how to treat glass surfaces. These techniques were first developed to protect glass fibres and make them compatible with polymer matrices in which they served as strengtheners. Since then, these techniques have been applied to flat glass surfaces like windows and hollow glass objects like bottles. In particular, it is known how to make the surface hydroxyl groups of the glass (Si–OH) react in such a way as to graft hydrocarbon chains by covalent bonding (see Fig. 29.5). The surface energy of the solid is considerably reduced and, according to the laws of wetting, the angle of contact is increased for liquids with relatively high surface energy (or surface tension), as predicted by Young's relation:

$$\cos\theta = \frac{\gamma_{SV} - \gamma_{SL}}{\gamma_{LV}} .$$

When fluorinated chains are grafted onto the surface, making it comparable with a Teflon surface, this logic is pushed to its limit. Very high contact angles are then obtained, of the order of 150°.

To go further, another logic must be adopted; one which living beings have been perfecting since the beginning of time. This is surface texturing, by means of hairs, nodules, and spicules of every imaginable kind. One only has to observe the way water droplets roll off the leaves of many plants like the lotus

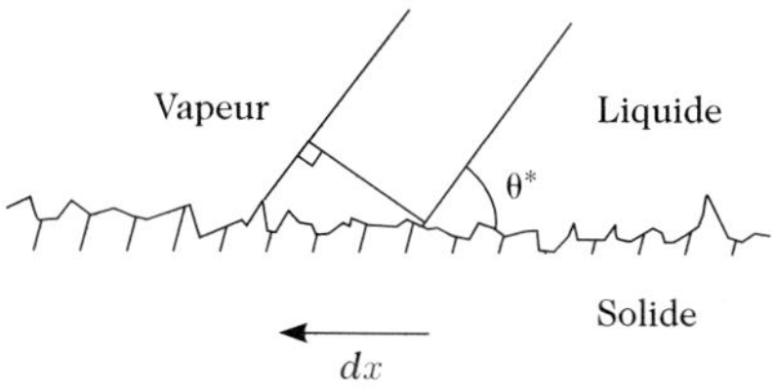

Fig. 29.6. Cross-sectional view of the way the contact line moves forward when a droplet is slightly compressed on a rough solid surface

or the water lily, or off the feathers on the proverbial duck's back, perfectly unable to take hold along the way, to realise that a particularly effective solution has already been found by nature. The main point is to have a rough surface. The effect on the contact angle is not trivial but a simple argument can explain the expected modifications. (For a more detailed discussion, the reader is referred to [1].)

We are concerned with the apparent angle of contact θ^*, assuming that the actual, local angle always obeys Young's relation. Let r by the roughness of the surface, defined as the ratio of the actual path travelled from one point to another to the ideal straight line path. Consider a slight squashing of a droplet, assumed to be large compared with the amplitude of the roughness, causing the contact line to move forward by $\mathrm{d}x$ (see Fig. 29.6). The area of the solid–vapour interface decreases, while that of the solid–liquid interface increases, as does the area of the liquid–vapour interface. The change in the total surface energy is then

$$\mathrm{d}E = r(\gamma_{\mathrm{SL}} - \gamma_{\mathrm{SV}})\mathrm{d}x + \gamma_{\mathrm{LV}}\mathrm{d}x \cos\theta^* \ .$$

Equilibrium corresponds to the energy minimum, i.e., $\mathrm{d}E = 0$. We then have Wenzel's relation:

$$\cos\theta^* = r\cos\theta \ .$$

This relation shows that, whatever the actual equilibrium angle on the microscopic scale, surface roughness acts as an amplifier. Indeed, if $\cos\theta$ is positive, $\cos\theta^*$ will be even more positive, and if $\cos\theta$ is negative, $\cos\theta^*$ will be even more negative. A surface coated with fluorinated chains will thus be made more hydrophobic.

For physical reasons, Wenzel's relation does not apply for all values of the roughness and equilibrium angle. Indeed, since $\cos\theta^*$ cannot be greater than 1 or less than -1, there are pairs of critical values. For example, for $\theta = 120°$, the surface will in theory become totally non-wetting for $r \geq 2$.

In practice, the Wenzel relation suffers from certain shortcomings. It remains valid for low surface roughness but in fact the final result obtained for very rough surfaces, taking into account all the necessary corrections, is

Fig. 29.7. *Upper*: SEM image of a nanopatterned glass surface © Hervé Arribart, Saint-Gobain Recherche. *Lower*: Water droplet of millimetric dimensions placed on a surface made superhydrophobic by a combination of a texturing treatment and a chemical hydrophobisation treatment. Photo courtesy of José Bico

not so very different from what is predicted by Wenzel's relation, at least for hydrophobic surfaces. In this latter case, the main phenomenon neglected by Wenzel is that, if the amplitude of the roughness is large, a non-wetting liquid will not necessarily follow all the anfractuosities of the surface (see Fig. 29.7). Air pockets will form and the droplet may in fact be resting more on these than on the solid surface. The true contact with the solid is then greatly reduced. The condition for air to be trapped is that, in these anfractuosities, a horizontal liquid–air (+ water vapour) interface should be able to form in such a way as to cancel the Laplace pressure, while respecting Young's relation locally. These conditions are easily satisfied on rough surfaces, whether they are relatively regular (see Fig. 29.7) or highly disordered. Such surfaces, referred to as superhydrophobic, have been made, with contact angles greater than 170° (see Fig. 29.7).

29.3.3 Self-Cleaning and Superhydrophilic Surfaces

Water is a nuisance on a glass surface, but grease deposits are even worse. Generally, deposits of dirt or bacterial biofilms are almost unavoidable, whatever the surface. On porous surfaces such as concrete or stone surfaces, one also finds lichens and other plant forms. How could these be removed without scouring, washing, or resurfacing? Photocatalysis on a nanostructured semiconductor currently looks like the most promising process.

The idea of photocatalysis is to catalyse a chemical reaction by means of electronically excited species generated by absorption of photons, usually UV or visible photons. The photocatalyst can be a molecule or a solid. If it is

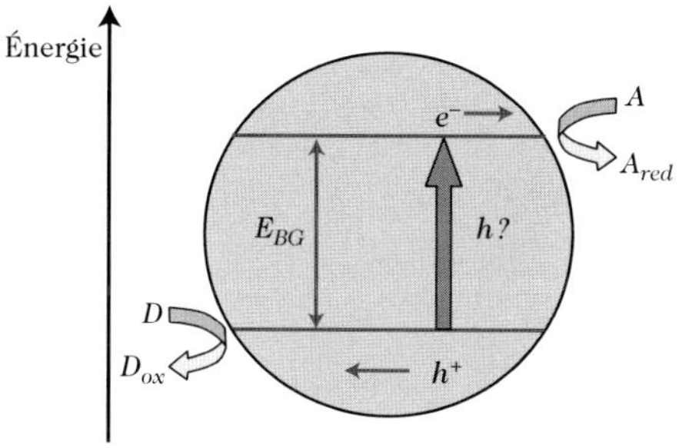

Fig. 29.8. Photocatalysis on a semiconducting particle, with photoinduced formation of an electron–hole pair, followed by two redox reactions

a solid, it will be a semiconductor with a band gap between 3.5 and 2 eV, to make best use of sunlight. This corresponds to a spectral band from the near UV to the near IR. Absorption of a photon of energy at least equal to the width of the band gap for the given material will cause an electron–hole pair e^-–e^+ to form (see Chap. 7). In chemical terms, this corresponds to formation of a reducing species (the electron in the conduction band) and an oxidising species (hole in the valence band). Insofar as direct recombination can be avoided by the fact that charge carriers in the space charge region of the material are kept separate (band bending), these species will be able to react with any adsorbed species, one as an acceptor and the other as a donor. The condition is that the redox potential of the electron must be more reducing than the redox potential of the acceptor and that the redox potential of the hole must be more oxidising than the redox potential of the donor (see Fig. 29.8).[2]

The most widely used and best known photocatalyst is without doubt titanium dioxide TiO_2 in the anatase form. TiO_2 has long been used in many contexts as a white pigment. However, for this purpose it is used in the rutile form, which scatters light better. With a band gap of 3.0 eV, which corresponds to a wavelength of 400 nm, anatase still absorbs a significant fraction of the sun's light. The primary processes of the photocatalytic action can be expressed as follows:

[2]Photocatalysis on a semiconductor is the meeting point between two scientific communities that otherwise have little to say to one another: solid-state physicists and chemists. This leads to some language problems! Solid-state physicists call the chemical potential of the electrons the Fermi level, while chemists speak of the redox potential. In reality, these are the same thing. Another difficulty lies in the reference energy levels. Solid-state physicists situate their energy levels relative to that of the electron in vacuum, and chemists relative to the standard potential of the calomel electrode or the hydrogen electrode.

Fig. 29.9. Film of anatase nanoparticles on a glass substrate. The thickness of the layer is about 200 nm and the grain size is between ten and a hundred nanometers. Courtesy of Saint-Gobain Recherche

$$TiO_2 \xrightarrow{h\nu} e^- + h^+ ,$$

$$\left[Ti^{4+}\text{–}OH^-\right] + h^+ \longrightarrow \left[Ti^{4+}\right] + HO\bullet ,$$

$$\left[Ti^{4+}\right] + e^- \longrightarrow \left[Ti^{3+}\right] .$$

These species can then react with adsorbed water or with organic substances. Even saturated hydrocarbons like methane and ethane are totally oxidised into CO_2 and H_2O.

Anatase is remarkably stable under illumination and easy to obtain in nanostructured form, with a high specific surface area. It is not therefore surprising that in 2003, some 800 patents were filed. Japan alone represents almost 90% of this activity. Nanostructured but nevertheless cohesive coatings of anatase nanoparticles have been deposited on a wide range of different surfaces: flat glass (especially the self-cleaning and superhydrophilic glasses Bioclean of SGG and Sunclean of PPG) (see Fig. 29.9), various polymer materials (but a layer of silica must first be deposited to prevent the polymer itself from decomposing!), and ceramics.

An a priori unexpected property of nanostructured anatase films is their superhydrophilicity, making them very effective anti-condensation coatings. The property is not unique to TiO_2. On a freshly prepared surface, the angle of contact of a water droplet is several tens of degrees. It is only under illumination that it gradually decreases to zero after a few hours. The mechanism is still poorly understood, but it seems likely that the hydrophilic sites are oxygen vacancies in the vicinity of Ti^{3+} ions. In 2002, 6% of the vehicles produced by Toyota were already equipped with outside rearview mirrors coated with TiO_2. All the lighting in road tunnels is equipped with the same type of

Fig. 29.10. The church of *Dives in Misericordia*, in Rome, built with a photocatalytic cement (Italcementi). Architect: Richard Meier

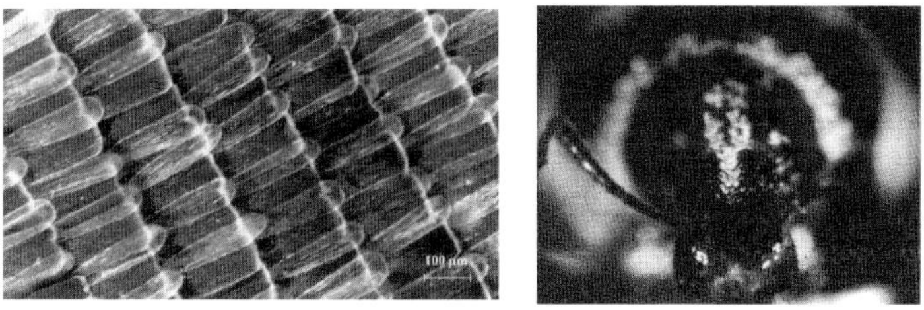

Fig. 29.11. *Left*: A scale from the butterfly *Morpho menelaus* [6]. *Right*: An Australian beetle [7]. No, it is not enamelled! See also the colour plate

glass surface. (The useful light is then the very small fraction of UV generated by the light bulbs.)

Photocatalysis can also be used as a means of synthesising metal nanoparticles. On contact with a solution of a soluble silver salt, the TiO_2 particles are coated with nanometric clusters of metallic silver due to the reducing action of the electrons generated in the conduction band. The method has been applied to silicone elastomers containing TiO_2 particles dispersed in the bulk. The TiO_2 grains are first dispersed in the liquid silicone resin, which is then polymerised. The photoinduced deposition stage is then carried out. The silicone objects produced in this way are remarkable bactericides due to the action of the silver clusters. Many silicone artefacts used in hospitals, e.g., catheters, could benefit from this treatment.

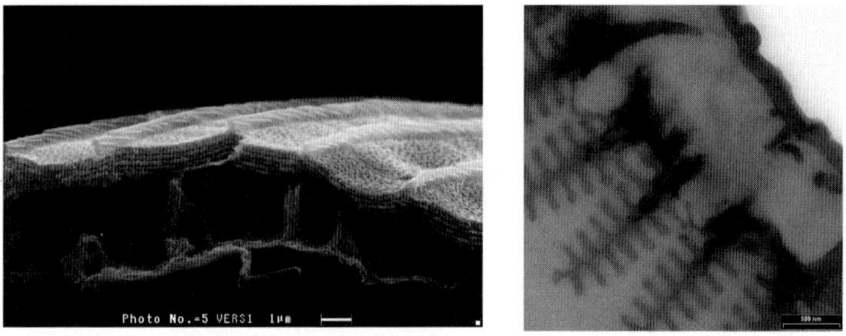

Fig. 29.12. *Left*: Cross-section of the upper membrane of a butterfly wing scale, from *Procris statice* [6]. About a dozen alternating layers of air and chitin are visible. *Right*: Cross-section of the upper part of the wing scales of *Morpho menelaus* [6]. The stacking structure is preserved, but in an alternating form perpendicular to a branch

29.3.4 When Concrete Cleans the Air We Breathe

It is a short step from depositing TiO_2 nanoparticles on glass, ceramic, or polymer surfaces to incorporating them into cements and concretes used to render walls, and this step has recently been taken. The first, expected result is a self-cleaning surface. The *Cité de la Musique* in Chambéry (France) and the church of *Dives in Misericordia* (see Fig. 29.10), recently inaugurated in Rome (December 2003), were built using a cement of this type (Bianco TX Millenium cement made by Italcementi). The striking brightness of these buildings comes from the use of a white cement base containing no ferrous phases, since it is these that give ordinary cement its grey colouration.

The second, less expected result is that these concretes and cements turn out to be very efficient atmospheric decontaminators, especially for the elimination of nitrogen oxides NO_x. Following the first tests, carried out in a 1.5 l chamber, and then a 450 l chamber, experiments were then done on ever larger scales. Several thousand square meters of city streets have been coated with photocatalytic cement. In traffic conditions of more than a thousand vehicles per hour, the observed reduction in the NO_x content of the air compared with an untreated surface was 50% in the summer. A remarkable point is that the cement–TiO_2 nanocomposite turns out to be even more effective than TiO_2 alone, probably due to the adsorbing action of the cement, with its pore structure covering many length scales down to subnanometer.

In conclusion, the prospects offered by surface nanoengineering are impressive. The general application of photocatalytic cements to the scale of whole towns, and why not the whole road network, could radically change our approach to atmospheric pollution and the aesthetic appearance of our towns. Superhydrophilic and self-cleaning glass surfaces will bring with them a considerable improvement. Bactericide treatment, where useful, will make an important contribution to public and private hygiene, without the need for chemical compounds which may themselves be harmful.

29.4 Photonic Crystals

29.4.1 The Colourful World of Birds and Insects

How do butterflies display such beautiful colours? What produces the iridescent colour of a peacock's feathers or the scales of a silver fish? Why does the wing case of some beetles reflect the light in such a metallic way, whatever the angle we view them from (see Fig. 29.11)? Does the secret lie in the ability of these organisms to synthesise exceptional dyes? To concentrate then reduce trace metals? To cover themselves with inorganic pigments? None of these suggestions holds the truth. It is in nature's architecture that we find the key to this enigma. While every living organism is a wonder of organisation, crystalline order is rarely master there. However, in some cases, nature has shown that she can build periodic structures. When such periodicity is of the order of a few tens of nanometers or a few hundred nanometers, which is also the periodicity of the refractive index, the structure will interfere with visible radiation and modify the reflected, transmitted, or absorbed light.

The order generating these light effects is not always three-dimensional. It can be one- or two-dimensional. Multilayers (1D) are commonplace in insects and also in certain types of leaf. They then constitute genuine interference mirrors or filters, producing the characteristic colours of the best reflectors in nature: butterflies, birds, and fish. The periodicity is produced by alternating layers with very different refractive index, generally an organic layer (chitin) and air or a physiological liquid (see Fig. 29.12). When several reflecting filters are brought together, either in an ordered or a disordered way, they can produce a broad band reflector and generate a silver or gold luster, observed in certain insects (beetles) or fish. In butterflies of the *Morpho* genus the layers are stacked up in a laterally discontinuous way, in the form of columns (see Fig. 29.12). The colours are quite extraordinary.

Two-dimensional periodic structures are less common. They generally comprise ordered forests of cylindrical nano-objects that may be filled, in which case they are separated by air or a low-index liquid, or hollow, in the form of pits carved out from a solid matrix (see Fig. 29.13). Cats' eyes, some birds' feathers, and certain butterfly wing scales have structures of the first type, while the hairs on some marine worms have structures of the second type.

It is in the 3D periodic structures that the analogy with crystals made up of atoms or molecules becomes clearest (see Fig. 29.13). The opalescent colours of the wing scales of *Eustales adamantinus* are an example (see Fig. 29.14). And this is not just a chance result. It has long been known that opal is a colloidal crystal, formed by closely packing colloidal beads of silica (see Fig. 29.14). The wing case of the beetle in Fig. 29.11 owes its metallic green coloration to a structure of this type. Like the 2D structures, the 3D periodic structures also exist in negative form, with cavities in a solid matrix. They are then called inverse opals.

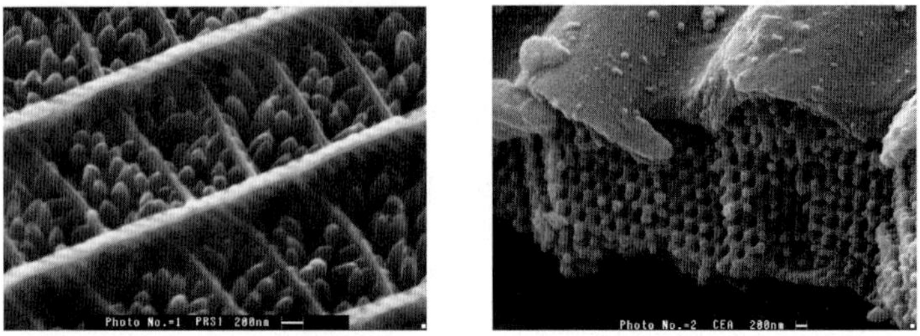

Fig. 29.13. Two- and three-dimensional periodic structures observed in butterfly wing scales. *Left*: A scale from the butterfly *Pieris brasicae*, with a structure composed of rows of boxes containing groups of grains stood on end. The periodicity is rather poorly defined. Scattering dominates over diffraction effects. *Right*: 3D structure of scales from *Eustales adamantinus* [6], regular enough to produce magnificent opalescence effects

Fig. 29.14. Opalescent colours and periodic structure (SEM image) of wing scales from *Eustales adamantinus* [6]. See also the colour plate

29.4.2 Photonic Crystals and Photonic Band Gaps

When these periodic structures extend into the bulk, they are referred to as photonic crystals. (If the periodicity is confined to a plane, the usual term from the study of diffraction is a grating.) The reason for this terminology is that the light interacts with these structures in the same way that electrons interact with a crystal made up of atoms, but replacing the periodicity of the Coulomb potential by the periodicity of the dielectric constant. In both cases, the result is a distribution of energy levels with allowed and forbidden bands, there being no accessible state in the forbidden bands.

A simple visual comparison of Maxwell's equations with the Schrödinger equation shows intuitively why electromagnetic waves and electron wave functions should behave similarly in these contexts. For example, for a harmonic electromagnetic wave in a non-absorbing dielectric medium, i.e., without dielectric losses, the equation governing the electric field is

$$\nabla^2 \boldsymbol{E}(\boldsymbol{r}) - \varepsilon_{\mathrm{r}}(\boldsymbol{r}) \frac{\omega^2}{c^2} \boldsymbol{E}(\boldsymbol{r}) = 0 ,$$

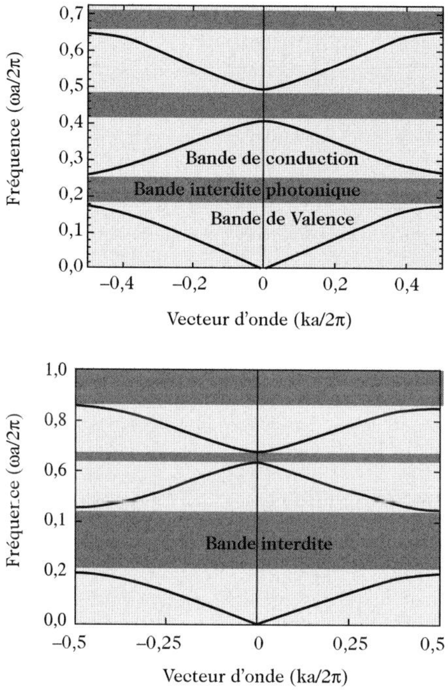

Fig. 29.15. *Upper*: Band structure of a 1D crystal in which the dielectric permittivity varies sinusoidally. The amplitude of the modulation is by a factor of 5 [8]. *Lower*: Band structure of a 1D crystal comprising alternating layers of permittivity $\varepsilon_1 = 13$ and $\varepsilon_2 = 1$, and thickness $d_1 = 0.2a$ and $d_2 = 0.2a$ [9]

where $\boldsymbol{E}(\boldsymbol{r})$ and $\varepsilon_{\mathrm{r}}(\boldsymbol{r})$ are the electric field and the dielectric constant at $\boldsymbol{r}$, respectively, while ω is the angular frequency of the wave and c is the speed of light. This equation should be compared with

$$\nabla^2\psi(\boldsymbol{r}) + \frac{2m}{\hbar^2}\big[E - V(\boldsymbol{r})\big]\psi(\boldsymbol{r}) = 0 \ ,$$

for the wave function of an electron of mass m and energy E in a potential $V(\boldsymbol{r})$. Given the similarity between the two equations, it seems intuitively reasonable to think that a periodic variation in the dielectric constant should lead, for photons, to analogous effects to those produced for electrons by the periodic variation of the electrical potential in which they move in a crystalline solid. One therefore expects to find frequency (or energy) ranges for which light will be able to propagate (allowed bands), and others for which it will not be able to propagate (forbidden bands). If, as we have assumed, the medium is not absorbent, i.e., not coloured in the usual sense of the term, it will behave as a perfect mirror.

In contrast to electrons, which interact strongly with one another, photons interact very little. This means that Maxwell's equations can be solved exactly for a given periodic structure of $\varepsilon_{\mathrm{r}}(\boldsymbol{r})$. The result is expressed in the form of a dispersion relation, relating the frequency or energy of the light to the wavelength λ or the wave vector k.

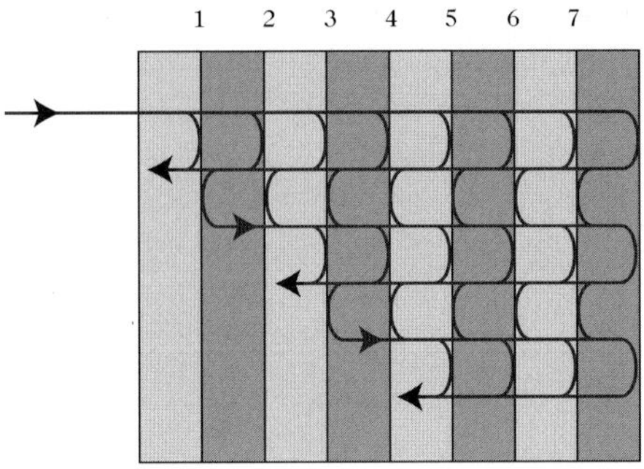

Fig. 29.16. Multiple interference in a Bragg reflector formed by alternating high-index and low-index layers [8]

Bulk photonic structures with 1D, 2D, or 3D periodicity have been obtained in the laboratory along the same lines as those found in nature. The simplest, in the form of 1D alternating stacks of thin dielectric films, with period $\lambda/2$, are referred to as Bragg mirrors. They are excellent mirrors for a given wavelength, but their reflective efficiency falls off as one moves away from normal incidence. It is thus impossible to make a reflecting surface that works for all angles using a 1D photonic structure.

As an illustration, Fig. 29.15 represents the band structure, i.e., the dispersion relations, for a hypothetical crystal in which the dielectric permittivity varies sinusoidally in a single direction, and for a more realistic 1D crystal made from alternating layers of high-index dielectrics and air. In the first case, the allowed bands appear very clearly, separated by a single type of forbidden band. In the second case, there are two types of forbidden band, broad bands of odd order and narrow bands of even order. This can be understood by considering the multiple interference occurring within the crystal (see Fig. 29.16). A wave propagating in the medium undergoes reflection at each interface, with or without phase change, depending on whether it goes from a low-index to a high-index medium, or vice versa. If the crystal has been built in such a way that its periodicity (two layers) is equal to $\lambda/2$, the wave reflected at the interface of order 1 will be in phase with the reflected waves from all the other interfaces of odd order. All these instances of constructive interference lead to total reflection, which amounts to saying that the crystal is a perfect reflector, or again, that it possesses a band gap. When each layer has the same optical thickness $\lambda/4$, the reflected waves are all in phase, whatever the interface on which they reflect. This is the case leading to the broadest odd order band gaps and the disappearance of the even order band gaps.

The next level of complexity is the 2D photonic crystal, formed for example by regularly spaced rods or, alternatively, regularly spaced pits in a matrix, as seen in the natural examples mentioned earlier. These structures are still relatively easy to make. More difficult are genuine 3D photonic crystals, the

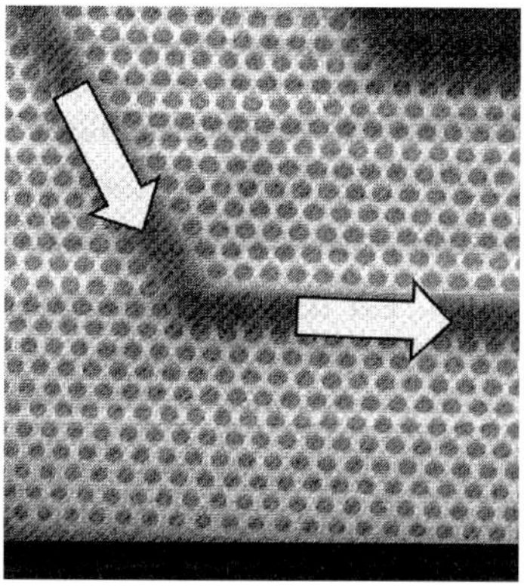

Fig. 29.17. Bent light guide in a 2D photonic crystal composed of rods [8]

only ones that could claim to have an omnidirectional photonic band gap. The most perfect yet achieved have been made using microfabrication techniques developed in microelectronics. However, as we shall see below, self-assembly techniques developed directly within the science of nanomaterials itself should soon provide some interesting alternatives.

29.4.3 Guides and Cavities

Apart from colour effects, photonic crystals also have an enormous potential for the fabrication of components and optical circuits, or more generally, for optical data processing. This can be illustrated for two types of component: wave guides and resonant cavities made in a 2D crystal composed of rods. What happens when a row of rods is removed?

In the terminology of crystallography, an extended defect is created in the periodic structure of the crystal. In optical terms, new dispersion curves are introduced. Some are situated in the band gaps. They correspond to guided modes. The channel produced in the photonic crystal is a light guide with perfectly reflecting walls, because photons propagating in the channel cannot propagate into the crystal (see Fig. 29.17). The guiding principle is thus very different from the one operating in optical fibres, which is based on total internal reflection in going from a high-index material to a low-index material. This allows more complex paths, with sudden changes of direction (see Fig. 29.17), something that is impossible to execute with an optical fibre because the angle of total reflection is then exceeded.

A resonant cavity can also be made with the same 2D array of rods just considered, simply by removing a rod or by replacing one rod by something different, with bigger or smaller diameter than the other rods. The allowed frequency introduced in this way into the band gap depends on the diameter of the rod, which means that the resonance frequency can be controlled. The defect acts both as a trap and as an amplifier. Illuminated from the outside by light at the frequency corresponding to the new allowed frequency, the crystal

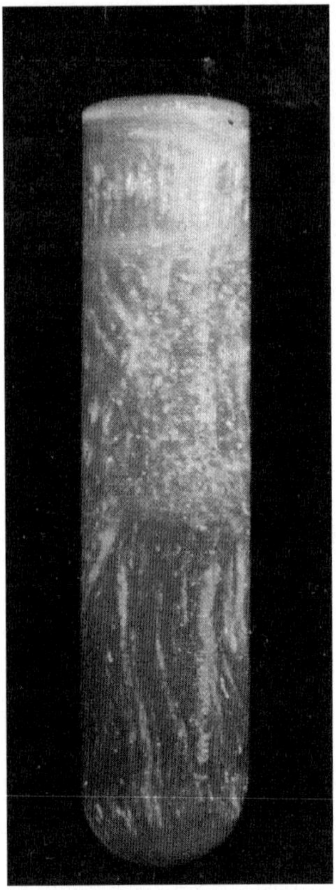

Fig. 29.18. Optical diffraction effects obtained by illuminating a suspension of silica beads of radius 130 nm with white light. The silica beads are coated with a layer of C18 aliphatic chains and the solvent is cyclohexane. Under these conditions, the beads behave as hard spheres. If the tube had not been stirred, it would have had a uniform colour [10]. See also the colour plate

focuses all this light in the cavity and then sends it back out. The cavity acts as a highly efficient relay for the selected wavelength, and also as a filter, because neighbouring wavelengths are still unable to propagate. If an emitter is placed in the cavity, the accumulation of photons in the cavity can lead to cascade effects and hence to a particularly efficient, miniaturised laser.

29.4.4 From Colloidal Crystals to Photonic Crystals

Well before the term 'photonic crystal' appeared on the scene, physical chemists already knew how to make 3D crystals with periodic structures in the range we are concerned with here. As we have seen, natural opal is a crystal of this type, made up of silica beads formed in supersaturated solutions, then lightly fused together. Analogous structures, usually with a close-packed hexagonal or face-centered cubic structure, are relatively easily synthesised in the laboratory. The conditions required for the appearance of long-range order are the same as at the atomic or molecular scales. In an atomic crystal, e.g., a rare gas or metal, or in a molecular crystal, e.g., a sugar crystal, all the atoms or all the molecules are of exactly the same size. Furthermore, if the crystal

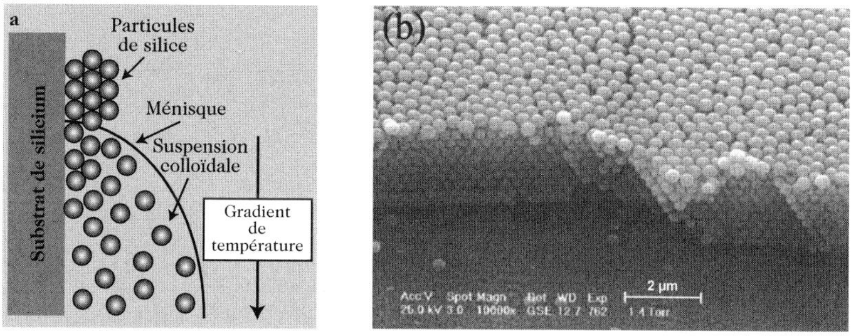

Fig. 29.19. Method for fabricating a monocrystalline film of silica beads (*left*) and the result obtained (*right*). Taken from [11] © 2003 American Institute of Physics

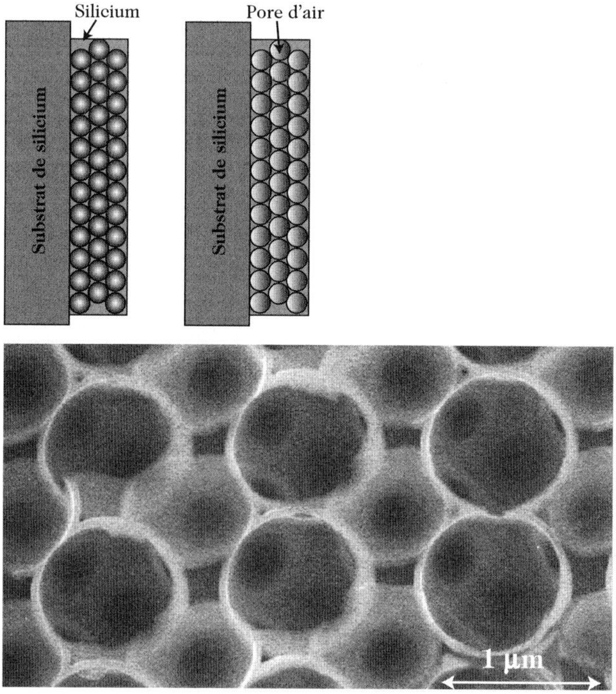

Fig. 29.20. *Upper*: Illustration of the impregnation of a colloidal silica crystal to make an inverse opal with high index contrast. *Lower*: [110] face of a silicon inverse opal [12]

is made from what is known as condensed matter, this is undoubtedly due to attractive forces. On the other hand, in contrast to what is often thought, if it is ordered, this is essentially due to repulsive forces (the impenetrability of the component objects). We then speak of excluded volume effects.

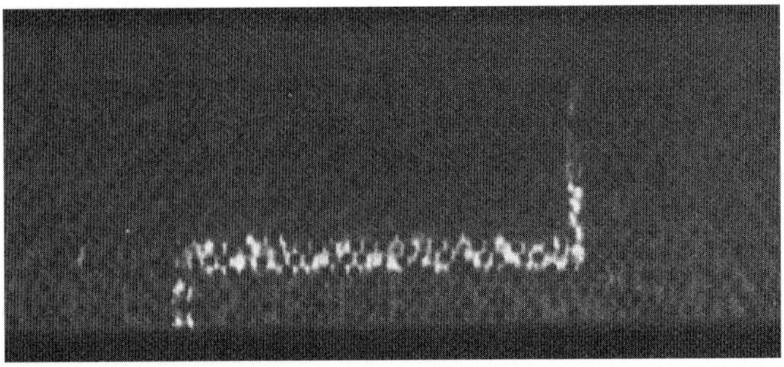

Fig. 29.21. Waveguide made in a photonic crystal built by colloidal moulding [13]

Fig. 29.22. Identity card protected from forgery by depositing alternating layers of inks with very different refractive indexes, using the OVI technique [14]

The same goes for colloidal particles. The size distribution of the beads must be very narrow. The two families of materials for which the synthesis of monodisperse spheres is easiest are the silicas and the polymers. Order appears when the particles come into contact (see Fig. 29.18), but it can also appear in a dilute suspension, in the presence of long-range repulsive forces. The latter arise in particular when the particles are charged and a double electric layer forms within the suspension. The objects arranging themselves in an ordered way are then no longer the particles themselves, but the virtual spheres formed by the particles surrounded by their double layer. Their effective radius is the radius of the actual particles plus the Debye length. Since this length depends on the ionic force (the salinity) of the medium, colloidal crystals can thus be created with controlled lattice parameter by 'dressing' the particles with a double layer of the desired thickness. Optical diffraction effects can then arise, even with particles that are much smaller than the visible wavelengths.

The structures we have been discussing here are fluids. The particles are immersed in a solvent and in general do not form a macroscopic monocrystal, but rather microcrystalline domains. To obtain a solid and useful photonic material, one must make a monocrystal, eliminate the solvent and consolidate the whole assembly in some way. Since the particles come into contact when the solvent is eliminated, they must necessarily have sizes in the visible range of wavelengths, since it is in this domain that effects are sought. Arranging the beads into a monocrystalline form in a given volume and eliminating the solvent without introducing too many stresses, which could alter the original ordering, are much more delicate operations than one might think. The process is easier to implement in the form of a deposit that is not too thick. Some authors have managed to do this in a single operation, by extracting a silicon wafer from a colloidal suspension in a controlled thermal gradient (see Fig. 29.19). Capillary effects and the right ratio beween extraction and evaporation rates leave a thick monocrystalline film adhered to the silicon substrate.

Such a monocrystal, consolidated by moderate sintering, can produce beautiful diffraction effects, but it is not yet a true photonic band gap material. To achieve this, the refractive index contrast must be very high. Between air and silica, the contrast is too low. The index contrast is close to 1.5. Other materials like GaP, Si, Ge or conducting polymers have much higher indices, greater than 3, but they cannot be synthesised in the form of monodisperse beads. The solution here is to use the monocrystal of silica beads as a mould and to impregnate it with one of these other materials by chemical vapour deposition (CVD) or electrodeposition (see Fig. 29.20).[3] It then remains only to eliminate the silica by dissolving it in HF. Finally, an inverse opal is obtained, made from spheres of air surrounded by a film of the impregnation material and possessing the desired photonic band structure (see Fig. 29.20).

The structures obtained in this way are still somewhat imperfect compared with what can be made using lithographic techniques, but the two approaches both deserve to be followed up, the colloidal method offering the advantage of being simpler to implement. In any case, this method has already given rise to operational devices (see Fig. 29.21).

Apart from their use in optical data processing, 1D, 2D, or 3D photonic crystals have many other applications or potential applications in the world of fibres, paints, and cosmetics. A particularly interesting application is protection against forgery. It is much more difficult to copy subtle optical effects obtained by finely tuned control of a nanopattern than it is to reproduce a colour or even a complex mixture of colours obtained using pigments and dyes. Most recent techniques for protecting documents, bank notes, or credit cards are based on the presence of logos comprising a wealth of optical effects. In the case of bank notes and credit cards, current techniques are generally based on the use of holograms. However, the holographic structure is not very thick and can be rather easily detached and forged with a minimum of technical skills. A step toward greater complexity can be taken by adding optical effects by successive deposition of ink layers with different refractive indexes, as in the optically variable ink (OVI) technique, for example (see Fig. 29.22). The

[3]The idea is analogous to that described elsewhere in this book for synthesising porous materials from a mould made from an ordered molecular phase (see Chap. 26). However, the method is different. In the case of porous materials, it is a sol–gel technique.

forgers will certainly have met their match when they have to make micro- and nanostructures with the level of complexity in butterfly scales or beetle wing cases!

References

Surface Nanoengineering

Wetting

1. P.-G. de Gennes, F. Brochart-Wyart, D. Quéré: *Gouttes, bulles, perles et ondes*, 2nd edn., Belin, Paris (2005) with CDROM

Photocatalysis

2. N. Serpone, E. Pellizzetti: *Photocatalysis: Fundamentals and Applications*, John Wiley, New York (1989)
3. M. Kaneko, I. Okura: *Photocatalysis – Science and Technology*, Springer, Berlin (2002)
4. A. Fujishima, K. Hashimoto, T. Watanabe: *TiO_2 Photocatalysis: Fundamentals and Applications*, BKC, Tokyo (1999)

Photocatalytic Cements

5. L. Cassar: Materials Research Society (MRS) Bulletin (May 2004)

Photonic Crystals

6. S. Berthier: *La couleur des papillons, ou l'impérative beauté*, Springer-Verlag France, Paris (2000)
7. P. Vukusic: (School of Physics, University of Exeter, UK, `P.Vukusic@ex.ac.uk`) Natural photonics, Physics World **17**, No 2, pp. 35–39, Institute of Physics, Bristol (Feb 2004)
8. J.-M. Lourtioz, H. Benisty, V. Berger, J.-M. Gérard, D. Maystre, A. Tchelnokov: *Les cristaux photoniques*, Hermès Science et Lavoisier, Paris (2003)
9. J.D. Joannopoulos, R.D. Meade, J.N. Winn: *Photonic Crystals*, Princeton University Press (1995). Figure and citation in [8]
10. A. van Blaaderen: Manipulating electrons and photons with colloid materials, MRS Bulletin **23**, No 10, 39–43, Materials Research Society, Warrendale, PA, USA (Oct 1998)
11. Y. Zhao, K. Wostyn, G. de Schaetzen, K. Clays, L. Hellemans, A. Persoons, M. Szekeres, R.A. Schoonheydt: Appl. Phys. Lett. **82**, 3764 (2003)
12. `http://www.icmm.csic.es/cefe/`
13. P.V. Braun, R.W. Zehner, C.A. White, M.K. Weldon, C. Kloc, S.S. Patel, P. Wiltzius: Epitaxial growth of high dielectric contrast three-dimensional photonic crystals, Adv. Mat. **13**, 721–724 (2002)
14. Optically Variable Ink (OVI), developed by SICPA, Switzerland (OVD Kinegram Technology AG, Switzerland)

General References

15. S. Berthier: *Iridescences, les couleurs physiques des insectes*, Springer-Verlag France, Paris (2003)
16. J.-M. Lourtioz, H. Benisty, V. Berger, J.-M. Gérard, D. Maystre, A. Tchelnokov: *Les cristaux photoniques*, Hermès Science et Lavoisier, Paris (2003)
17. P. Callet: *Couleur-lumière, couleur-matière*, Diderot Editeur, Paris (1998)
18. L. Zupirolli, M.-N. Bussac: *Traité des couleurs*, Presses polytechniques et universitaires romandes, Lausanne (2003)

30

Mechanics

P. Maestro, E. Gaffet, G. Le Caër, A. Mocellin, E. Reynaud, T. Rouxel, M. Soulard, J. Patarin, L. Thilly, and F. Lecouturier

30.1 Silica Precipitates for High-Performance Tyres

Inorganic reinforcements are used in rubber, and in particular in tyre treads for light vehicles, in order to improve the compromise between three key features of tyres: road holding performance or road adherence, especially when the road is wet or snow-covered (road safety), roll resistance (petrol consumption), and resistance to wear (lifetime of the tyre). Over the last ten years, highly dispersible silicas (HDS) developed by Rhodia have been more and more widely used as a substitute for the traditionally used carbon black. The advantage with HDS materials is that they improve road holding and reduce roll resistance, while maintaining the same level of resistance to wear.

30.1.1 Fabrication of Silica Precipitates

Silica precipitates are obtained in an aqueous medium by neutralising a solution of sodium silicate with an acid solution, usually sulfuric acid. This chemical reaction, called the precipitation stage, is carried out in a stirred reactor. This produces an amorphous hydrated silica. The precipitation stage is critical because it determines the intrinsic dispersibility and the final characteristics of the silica. It is also important to note that, because the silica is 100% amorphous, it is completely harmless, as confirmed by all epidemiological and toxicological studies carried out to date.

The silica is then washed, dried, and if necessary, further prepared in whatever way necessary for the final application, in particular to avoid dust emission during handling. At the end of this industrial process, the precipitated silicas may thus occur in any of a number of different forms: powders, microbeads, granulates, and so on. These different forms are obtained by controlled agglomeration after the precipitation stage of aggregates measuring a few tens of nanometers.

After incorporation in the elastomer matrix, it is in fact these aggregates which, once dispersed in the matrix, play the role of a nanometric strengthening agent in the rubber–inorganic filler nanocomposite systems thereby obtained. It is important to use aggregates that can percolate in space, to achieve significant reinforcement while using as little material as possible. If they were not well dispersed, these aggregates would group together into agglomerates with sizes up to several micrometers, and these would then behave more like defects. The consequences would be particularly noticeable when the nanocomposite is subject to extreme conditions.

The aggregates of several tens of nanometers are themselves composed of spherical elementary particles measuring a few nanometers. It is the size of these particles which schematically conditions the matrix–filler contact surface, an essential point for several properties of the nanocomposites.

Chemically, the silica surface mainly comprises silanols (surface OH), which are generally a source of interactions with the different ingredients used to make the nanocomposites. However, in the case of elastomers, the silanols do not provide strong enough interactions to prevent decohesion at the filler–matrix interface when the material is subjected to high stresses or strains. In order to develop a stronger matrix–filler interaction, suitable coupling agents must be used, usually silanes, reacting with the silica via the silanols and with the elastomer via appropriate functions.

30.1.2 Tyres and Other Applications

Michelin have used the highly dispersible silicas (HDS) synthesised by Rhodia to develop a new generation of tyres, the green tyre, with significantly improved performance both in terms of roll resistance – and hence petrol consumption – and in terms of road holding (adherence) on wet or icy surfaces, as illustrated in Fig. 30.1.

Apart from applications to tyres, nanoparticles of precipitated silicas are being more and more widely used in elastomers (silicones and rubbers), to increase hardness, resistance to rupture, and more generally durability of composites. In addition, the dynamical properties of filled elastomers are greatly modified by the presence of these nanometric fillers.

The introduction of inorganic nanoparticles, and in particular silicas, soon proved to be a general way of modifying polymers, giving rise by judicious choice of components and method of synthesis to materials with improved mechanical properties. Many applications are currently under development, which should lead to new, high performance plastics, with applications both in bulk materials and in threads and fibres.

Depending on the type of matrix, e.g., thermoplastic or elastomer, the effects on performance may be different. In thermoplastics, they generally lead to increased rigidity or modulus, exploited to improve the behaviour of the polymer at high temperature and increase its hardness. Under certain conditions, it is even possible to improve the compromise between modulus and shock resistance.

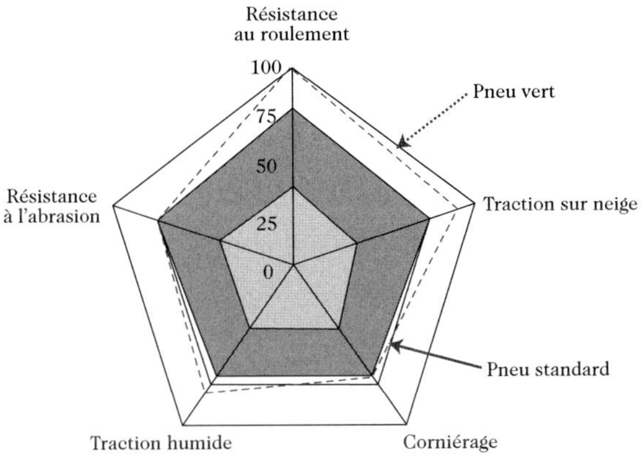

Fig. 30.1. Improved performance of tyres made with dispersible silicas

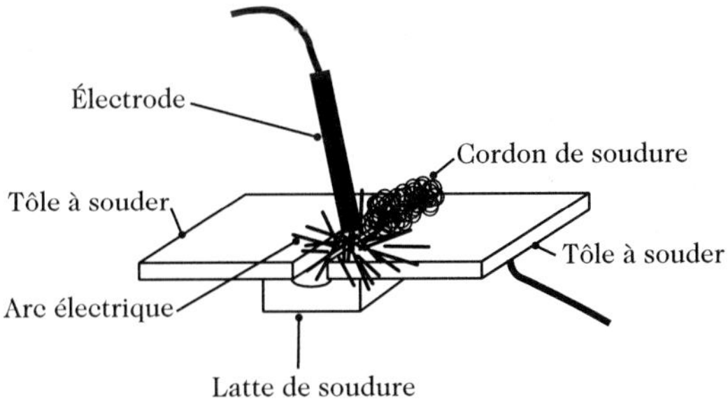

Fig. 30.2. Arc welding two metal sheets

But in every case, the compromise sought between performance requirements will depend on the quality of the dispersed state of the nanoparticles within the composite. The problem is therefore to control not only the synthesis of the nanoparticles, but also the way they are implemented in the material. It is by taking up this double challenge that the nanoparticle industry may be able to assist in the development of new nanocomposites.

30.2 Ceramic–Metal Composite Welding Supports

Arc welding of two steel sheets involves holding them together with an intermediate filler metal, usually steel, to be melted by means of an electric arc set up between the filler material and the workpiece to be welded. The welding supports, bands with different cross-sections depending on the shape of the objects to be welded and the joints to be made, are placed between these objects to retain the molten metal (see Fig. 30.2).

The material used to make the supports must have a high melting temperature, good mechanical strength, good resistance to corrosion by the liquid alloy to prevent the filler metal from sticking to the support, high thermal conductivity, low hygroscopicity, and a low production cost. These conditions have led to the choice of ceramic materials, like inorganic oxides or mixtures of oxides, e.g., cordierite $Mg_2Al_4Si_9O_{18}$. However, the insulating nature of these materials does have some disadvantages. Indeed, it is difficult to hold the workpieces a constant distance apart when they are very big, as happens for example in ship building. Any variation leads to instability in the electric arc, which is harmful to the quality of the weld and makes it difficult to automate certain welding operations. An idea for getting round this problem is simply to disperse a metal conducting phase with a sufficient concentration within the ceramic matrix to produce a conducting composite ceramic–metal material known as a cermet. This dispersion can also be beneficial for the mechanical behaviour [1].

Beyond the special case of the welding supports there is therefore the more general problem of how to synthesise cermets with improved properties, such as mechanical characteristics or absorption of electromagnetic radiation, e.g., nanocermets made from a dispersion of nanometric particles of noble metals in alumina Al_2O_3, silica SiO_2, etc. In order to match and wed the ceramic and the metal, one must of course exploit an intimate alliance of complementary qualities, without on the other hand attempting the impossible union of the rabbit and the carp "whose scales look like enamels" [2]. However, a good many methods for synthesising cermets are more reminiscent of the second than the first. In fact, the difficulties in synthesising cermets, such as alumina–metal cermets, arise from the poor wettability of the ceramic by liquid metals, leading to the formation of pores as the metal solidifies. Furthermore, the poor interfacial cohesion between metal particles and ceramic matrix prevents them from deforming plastically. The resulting down-grading of mechanical properties has been a major hindrance for applications of such cermets. Could there be some other cheap form of genetic therapy that might make this marriage a success? In fact there is at least one method with a nanocrystalline intermediate stage that cuts out the passage via a liquid state. This method, reactive mechanical alloying, is able to synthesise a good many metastable materials that are difficult to prepare by other channels. We shall describe it after a brief review of ceramics and their fragility.

30.2.1 Ceramics

Ceramics form a vast family of materials in which the constituent atoms form ionocovalent bonds between them, e.g., by the combination of metallic with non-metallic elements in oxides such as alumina (Al_2O_3), titanium dioxide (TiO_2), and zirconia (ZrO_2), or in borides, carbides, nitrides, and so on. Hence, alumina, an oxide that can crystallise in several forms including corundum, has a considerable practical importance that can be explained by its very high hardness, its thermal stability, its high melting temperature (2 050°C), its chemical inertness, and its thermal and electrical insulation properties.

Table 30.1. Reaction by high-energy ball milling a powder mixture of aluminium and an oxide

Initial mixture	Reactive mechanical alloying	Final milled mixture
$(3/y)M_xO_y + 2Al$	$\longrightarrow$	$Al_2O_3 + 3(x/y)M$ (1)
Micrometric particles	$\longrightarrow$	Micrometric particles
Initial phases separated at micrometric grain sizes	$\longrightarrow$	Final phases ($\neq$ initial phases) mixed at nanometric grain sizes

Fragility and ductility are not intrinsic properties of materials, but refer to a type of behaviour or a state for a given combination of temperature and load rate [3]. The observation that plastic can "make seals just as well as jewels" by a brief transformation [4] applies equally to ceramics by virtue of their superplastic behaviour in certain ranges of temperature and load rate. However, many commonplace ceramics are fragile when used at room temperature. There are several reasons for this fragility: imperfections within or at the surface, such as pores, cracks, etc., and especially those of micrometric dimensions, and defects such as grain boundaries, dislocations, voids, interstitials, and impurity atoms. These features make up the microstructure, a term also applicable to nanostructured materials, defined by the number, nature, structure, and geometry (size and shape) of the phases and their arrangement with interfaces, the interphases they form between them, and their lattice defects. Mechanical properties can be improved not only by better control of the synthesis, sometimes at a cost, by process engineering methods, but also by controlling the propagation of cracks through an optimisation of the microstructure. The mechanical performance of fragile ceramics can thus be enhanced by incorporating various reinforcements, such as ductile metals like iron, nickel, copper, or various alloys able to attract and arrest cracks.

30.2.2 Reactive Mechanical Alloying and High-Energy Ball Milling

High-energy ball milling of powder mixtures (see Chap. 19) can cause, among other things, certain solid-state reactions in unusual conditions, e.g., aluminothermal reactions [5]. Reactive mechanical alloying is thus one way of synthesising ceramic–metal composite powders, e.g., alumina-(Fe, Cr, Ni, and so on) for a wide range of different compositions. The idea, somewhat simplified, is to carry out the reaction described in Table 30.1 by high-energy ball milling a powder mixture of aluminium and an oxide M_xO_y such as V_2O_5, Cr_2O_3, Fe_2O_3, Fe_3O_4, NiO, etc. An alloy can be obtained under the same conditions by choosing at the outset to mix aluminium and several oxides in proportions determined by the required composition.

Table 30.2. Direct milling of a mixture of a ceramic powder and a metal powder

Initial mixture	High-energy ball milling	Final milled mixture
$Al_2O_3 + M$	⟶	$Al_2O_3 + M$ (2)
Micrometric particles	⟶	Micrometric particles
Initial phases separated at micrometric grain sizes	⟶	Final phases (≡ initial phases) mixed at nanometric grain sizes

A container, containing balls and the powder mixture of the initial reactants in an argon atmosphere, is vigorously stirred at room temperature to give the balls speeds of a few m/s and trap tiny associations of powder in the impacts between balls or between a ball and the container wall (see Chap. 19). The powder particles are thus fractured and the different fragments fused together again during each impact, these transformations being repeated at a frequency of a few hundred Hz (see Fig. 19.2). The initial phases, now mixed together down to the nanometric scale by these chaotic transformations, react together as in the above reaction, in a way which may be anywhere between gradual and explosive depending on the system. The stoichiometric conditions of the reaction determine the volume fraction of the metal phase dispersed in the alumina. This fraction, usually too high for the intended application, can be reduced simply by adding alumina to the initial mixture. Diluting in this way then slows down the oxidation and reduction reactions induced by the milling process, allowing tighter control over the final products and avoiding potentially damaging explosive reactions in some cases.

The ceramic phase (alumina) and metallic phase (metal or alloy) produced by these mechanically stimulated reactions are made up of crystallites (or grains), with sizes between 5 and 20 nm, distributed uniformly throughout the different particles in the milled powder, which all have the same overall chemical composition and usually micrometric dimensions. Each powder particle is therefore a microcermet. The distributions and morphology of the phases can be controlled by the parameters fixed for the mechanical alloying process, i.e., initial reactants, impact strength, milling time, the type of milling tool used, and so on.

Another method is to directly mill a mixture of a ceramic powder (alumina or another oxide) and metal powders to arrive at composite grains. Schematically this is be shown in Table 30.2
Minority phases often form, for example at the oxide–metal interfaces, like hercynite $FeAl_2O_4$ for $M = Fe$. Milling sometimes induces a structural transformation of the initial phases. The method to choose depends on the type of cermet to be synthesised and the constraints imposed by the intended use. It remains only to consolidate these powders in the form of dense samples, optimised for the specific application, e.g., for welding supports. The final microstructure of the resulting cermets generally depends on the chosen method

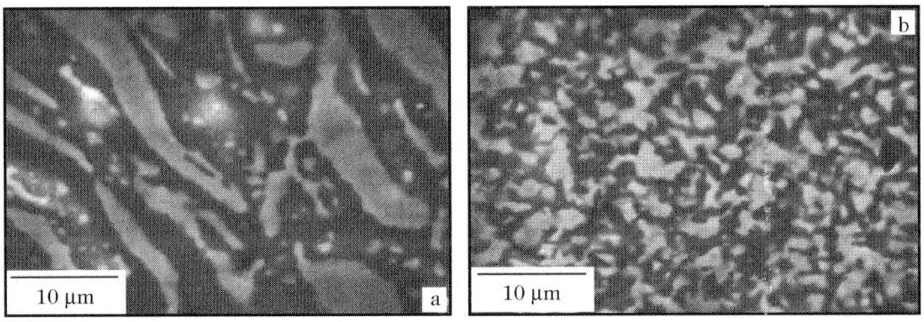

Fig. 30.3. SEM images of consolidated Al_2O_3–Cr 36% composites produced by (**a**) direct milling for 2 h and (**b**) reactive milling for 4 h [6]

of synthesis: reactive mechanical alloying (1) or high-energy ball milling (2) (see Fig. 30.3). Activation of milled powders (see Chap. 19) provides a way of densifying these nanostructured materials at lower temperatures than the melting temperatures of the alloys, at the same time reducing the risk of high porosity.

Some nanostructured particles are sometimes weakly bound together to form large agglomerates that must be broken up by low-energy milling in order to homogenise the particle sizes before the powder is densified. Forming and consolidation can be achieved by conventional means, e.g., dry processing, by simple hot uniaxial pressing, or wet processing, with an extra filtering–pressing of an aqueous dispersion before hot pressing, or natural sintering. An example of synthesis by hot pressing involves heating a sample up to 1 000 K at a rate of 10 K/min, then heating it up to 1 700 K at a rate of 30 K/min under a constant pressure of 30 MPa, holding it at this temperature for 30 min, and finally, bringing it back down to room temperature [6]. The resulting cermet materials, which are no longer nanostructured, are dense, reaching at least 95% of the theoretical density, and homogeneous, with microstructure controlled by the milling conditions used. Whatever the system, e.g., Al_2O_3–Fe, –Cr, –Nb, –V, etc., metal phases have branching morphology emerging from the initial nanometric state with good cohesion at the metal–alumina interfaces [6]. The way the microstructure changes in going from the powder when it has completely reacted to the bulk material has been studied by Guichard [6].

30.2.3 Improving Properties

The mechanical properties of these cermets, such as the hardness, elastic moduli, bending strength, and tensile strength, evolve progressively as a function of the metal content. The specific topology of the arrangement of metal zones reinforces the matrix. Compared with values measured for alumina alone, the tensile strength goes from 4 to 8 $MPa\,m^{1/2}$, the bending strength increases from 350 to 550 MPa, and the hardness decreases from 21 to 12.3 GPa for alumina–chromium composites with a metal content varying from 0 to

36% [6]. This reinforcement depends on the content of the second phase and also on the type of metal used.

Mechanical alloying, which is well suited to industrial scale production, combines technical simplicity with physical complexity in terms of the phenomena occurring during the milling process. Many cermets can be synthesised by this method in two stages, viz., milling followed by densification, producing a wide range of combinations of ceramics with metals, alloys, or intermetallic compounds, with the possibility of adapting the composition of the welding supports to the specific application. If the volume fraction is large enough for the conducting phase to percolate, a fraction which depends on the morphology of the dispersed particles but is of the order of 25–30% vol., the ceramic–metal composite becomes conducting. The dispersion of metal particles increases not only the electrical conductivity, but also the thermal conductivity (product of the thermal diffusivity, the density, and the specific heat capacity at constant pressure and at the relevant temperature), and at the same time reduces the risk of rupture due to thermal shock suffered by the welding supports during the weld process. For example, the thermal diffusivity of a cermet Al_2O_3–26% vol. Cr at 773 K ($\approx 3.8 \times 10^{-6}\,m^2/s$) [1] is about one and a half times that of alumina ($\approx 2.5 \times 10^{-6}\,m^2/s$) and four times that of cordierite ($\approx 1 \times 10^{-6}\,m^2/s$) at the same temperature. The thermal conductivities of alumina and cordierite are ≈ 11 W/mK and ~ 3 W/mK, respectively, at 773 K, while the thermal conductivity of the composite can be estimated to be about twice that of alumina. By passing through the nanocrystalline state and the resulting microstructures after densification, thermal and electrical properties of ceramic–metal composites can be modified simultaneously, while at the same time improving their mechanical properties.

30.3 Reinforced Amorphous Matrices

This section is concerned with macroscopic properties, mainly mechanical, of composite materials in which the main element, the binder or matrix, is a polymer or a glass to which inorganic nanoparticles have been added as a reinforcement. But why associate these two families of matrix materials in the same section? In the science of materials, it is usual to group them together under the same heading, that of disordered materials. Indeed, these families exhibit similar morphology (organisation of atoms or molecules) and macroscopic behaviour in response to thermal or mechanical stimuli, even though the characteristic temperatures at which these types of behaviour are observed are very different for the two families, i.e., around room temperature for the polymers, and much higher temperatures for glasses.

30.3.1 Not All Materials Are Ordered

Some materials have long-range ordered structure, being made up of regularly spaced atoms, with the same appearance as viewed from any point. These are

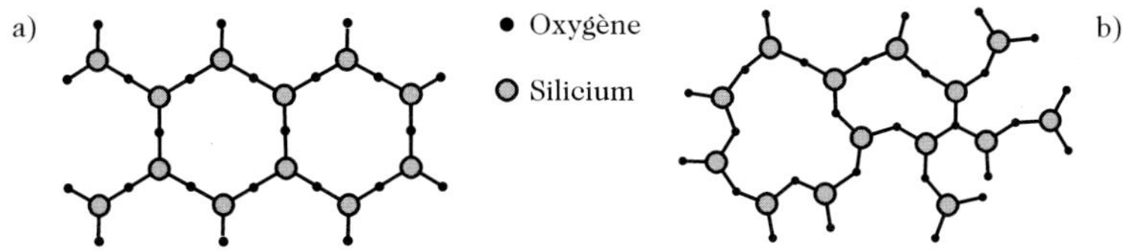

Fig. 30.4. Two-dimensional schematic representation of silica in (**a**) the crystal state and (**b**) the amorphous state. The fourth silicon bond is to be imagined perpendicular to the figure

crystalline materials, among which one finds most inorganic materials (see Fig. 30.4a).

At the other end of the spectrum, some materials have no tendency to order on the atomic scale. For example, Fig. 30.4b shows the structure of glass silica. Starting from a silicon atom, it is hard to predict with certainty where its neighbours will be. At best, given a particular atom, one might have an idea of the probability of the presence of other atoms in its vicinity. This type of material, which includes glasses and polymers, is said to be disordered or amorphous. Note, however, that some polymers display a certain level of atomic organisation. These are called semi-crystalline polymers.

30.3.2 Incorporating Nanoparticles into Amorphous Matrices

A Little History

Additives such as talc were very soon incorporated into synthesised polymer materials as a way of reducing production costs [7]. It was then observed that these additives modified, even improved, certain properties of the material. Among other things, one selection criterion for a material is the temperature at which distortion occurs under a load, i.e., the temperature at which a slab of the material will bend under a given weight. It was observed that this temperature would increase for polymer slabs that had been modified by including particles. So from simple additives, the idea of reinforcement was born.

During the 1960s, development focussed on technical composite materials, expensive to develop but with the kind of properties that enable them to withstand harsh environments, e.g., corrosive atmospheres, space applications, etc. These are composites reinforced by continuous carbon fibres, glass fibres, and Kevlar (a polymer fibre), in which the polymer is just the binder in a very strong structure.

Between the simple additive and the technical composite, there was still a place for purely practical materials, with improved properties and reasonable cost. Since nanocomposites satisfy these criteria, the scientific and technological communities have been interested in them since the beginning of the 1990s.

Table 30.3. Comparison between the properties of Nylon and its composite counterpart (5% clay) [8]

Property	Nylon	Nanocomposite
Tensile strength [MPa]	69	107
Shock resistance [MPa]	2.3	2.8
Thermal expansion [/°C]	13×10^{-5}	6.3×10^{-5}
Heat distortion temperature under load [°C]	65	145
Water uptake [% weight]	0.87	0.51

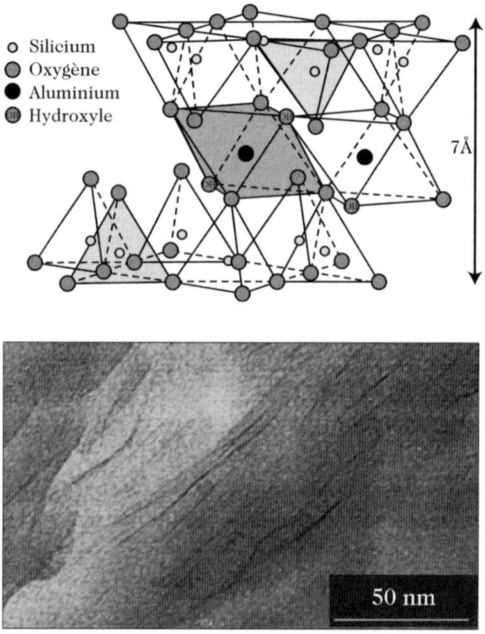

Fig. 30.5. Atomic structure of a clay platelet (*upper*) and transmission electron microscope observation of a Nylon–clay composite (*lower*)

With regard to polymers, as early as 1992, the Toyota research institute was already publising very promising results on a material composed of Nylon (polyamide) and clay (see Table 30.3) [8]. Clay has the form of platelets (see Fig. 30.5). Given the dimensions (thickness 7×10^{-10} m, minimal length 2×10^{-7} m), it is indeed a nanoparticle.

If we compare the properties of the polymer alone with those of the nanocomposite, the results of mechanical tests show that adding nanoparticles to the polymer make it stiffer, i.e., for a given stress, the nanocomposite deforms less than the pure polymer. Concerning the thermal expansion, i.e., the expansion of the material under the effects of heat, it is found that the dimensions of the object are more stable in the presence of nanoparticles. Finally, any polymer is sensitive to water, a characteristic measured by the water uptake by the dry material in an aqueous medium. Incorporating nanoparticles in the plastic material restricts the diffusion of water molecules.

Since these pioneering results, many studies on other systems have confirmed all these advantages and revealed many others. The field of nanocomposites has literally exploded. Among new strengtheners already tested, one could mention carbon nanotubes, alumina nanofibres, POSS (polyhedral oligomeric silsesquioxane, a sort of atomic cage made from silicon), and cellulose whiskers (another polymer). Considerable progress has been made in developing nanocomposites for almost all the polymer families, from the thermoplastics (Plexiglass, Nylon, and so on) to the thermosetting polymers (polyurethanes, epoxy resins) and the elastomers. A lot of work has also been carried out on glass matrices, even though the literature is somewhat less abundant in this domain.

Some Particular Effects

The next section is far from exhaustive, with such rapidly evolving research in the field of disordered materials and nanoparticles. The aim is simply to illustrate through a selection of examples the possibilities currently offered by combining the specific properties of the different phases of the system.

A classic way of testing a material is to place a sample of specific geometry and dimensions (a test piece) into a mechanical testing machine to carry out a tensile test. The test piece is held by two clamps, one fixed and the other mobile. The applied force is then recorded as a function of the extension it produces, and this until the structure is destroyed. The test presented in Fig. 30.6 compares a TPO polymer (a thermoplastic olefin, mixture of polyethylene and polypropylene) with a nanocomposite made by associating TPO with a clay. Clearly, the nanocomposite appears to withstand the mechanical load better than its polymer relative. It exhibits better mechanical properties than the polymer alone, and in particular, destruction occurs later. Such mechanical characteristics are a case much sought after which cannot be generalised by intuition, but which has been confirmed for other pairs of matrix and reinforcement.

By virtue of their improved mechanical properties, these nanocomposite materials have already established themselves in commonplace applications, e.g., a step for getting into a car, commercialised by General Motors [10].

Apart from structural applications, some polymers now play a key role in food packaging, but also packaging of medicines and in vivo uses. Indeed, these particular polymers are biodegradable, although limited by a lack of mechanical strength and functional stability. Such drawbacks can be remedied to some extent by associating nanoscopic strengtheners with these matrices. For example, adding clay to one of these polymers, namely polylactic acid (PLA), leads to improved mechanical behaviour without hindering biodegradability [9]. This confirms the possibility of combining the specific action of

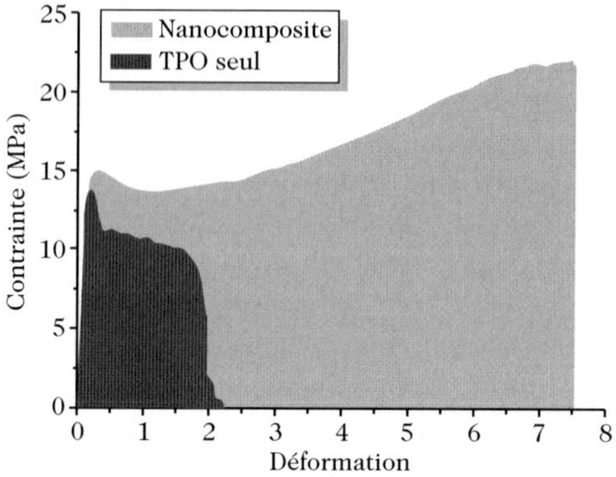

Fig. 30.6. Stress–strain curve for TPO (thermoplastic olefin) and nanocomposite test pieces [9]

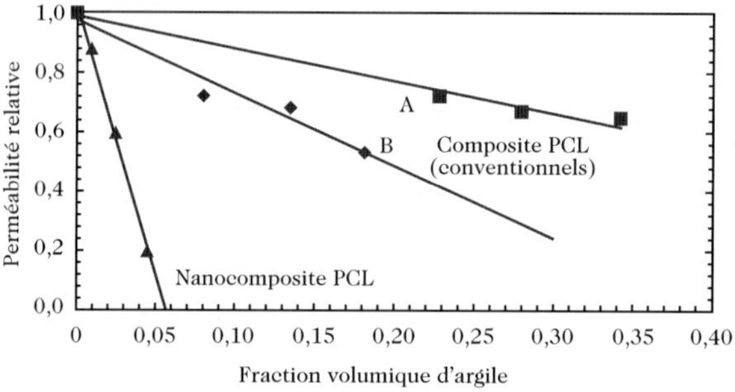

Fig. 30.7. Relative permeability of PCL nanocomposites and conventional PCL composites as a function of the volume fraction of clay [11]

each phase present within the composite material without compromising or deteriorating the resulting structures.

A further illustration of this synergy in nanocomposites concerns the permeability of polymers used in the food industry. These polymers are poorly suited to block the diffusion of gas or liquid molecules. The example shown in Fig. 30.7 concerns polycaprolactone (PCL). The permeability of a nanocomposite is compared with that of conventional composites strengthened by bigger particles, i.e., of micrometric order [11]. Note the restriction of molecular diffusion within the nanocomposite.

This is why nanocomposites are commonly used today for drink bottles (see Fig. 30.8), their properties extending the shelf life of such products. It may even be possible to make beer bottles with these materials, provided the manufacturers can persuade the consumer to exchange the traditional glass bottle for a lighter product.

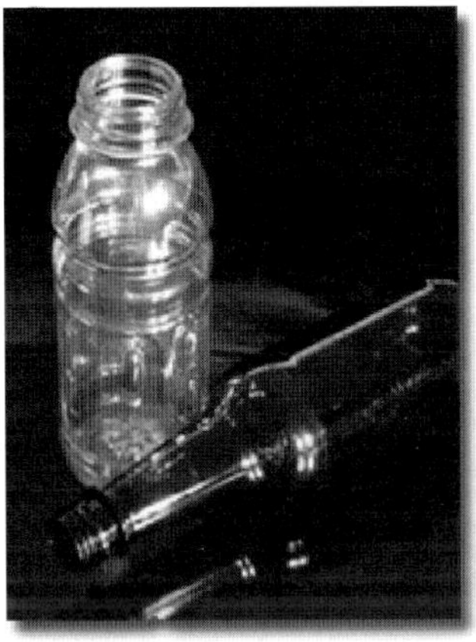

Fig. 30.8. Examples of nanocomposite packaging [12]

30.3.3 Prospects

The field of nanocomposites has developed significantly but in an unequal way depending on the specific area. Of course, combinations of matrix and strengthener have been produced and characterised, but there is still no theory to generalise and predict the observed behaviour. Research is just beginning to identify the key parameters of these materials and their role in the macroscopic behaviour. In this section, we shall discuss three such parameters: the quality with which the strengtheners are distributed within the matrix, the full contact surface between strengthener and matrix, and the local influence of the strengthener surface on the properties of the matrix.

Dispersion of Particles Within a Matrix

The first consequences of the quality of the strengthener particle dispersion in the matrix can be observed in the permeability and transparency of the material. Indeed, with regard to permeability, the smaller and more widely separated the inorganic (non-traversable) particles are, the longer will be the path of the diffusing atom or molecule in order to circumnavigate them. Moreover, small particles are 'invisible' to light from the sun, being too small to modify the path of light rays, whence they will not affect the optical qualities of the matrix in the visible part of the spectrum.

The quality of the dispersion is closely linked to the production process used to make the composite. As an illustration, consider the TEM images in Fig. 30.9, which show the same material, viz., nanometric silica particles embedded in Plexiglass: the composite on the left was obtained by extrusion (working together a mixture of polymer melt and particles), whereas the composite on the right was obtained by stirring and drying a liquid mixture of particles and polymer dissolved in a solvent. The forces involved in working the components together in the latter case are much weaker than those applied

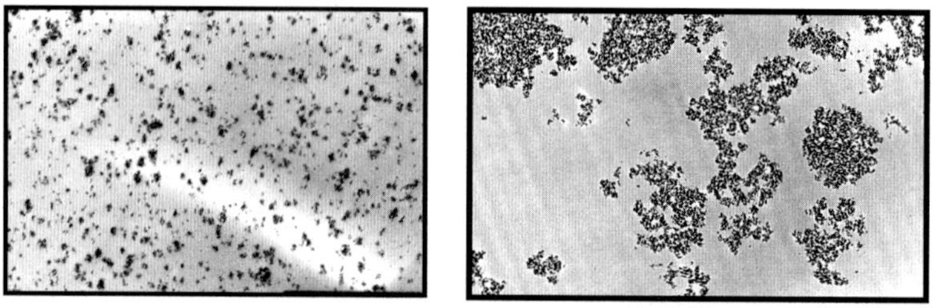

Fig. 30.9. Transmission electron microscope images of two Plexiglass–silica composites. Scale: 6×10^{-6} m by 3.5×10^{-6} m

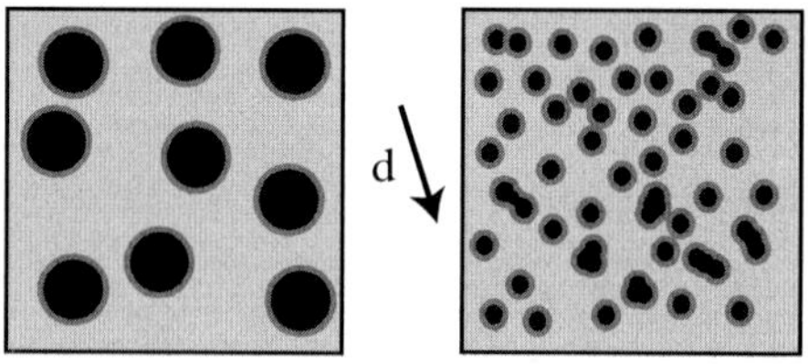

Fig. 30.10. Schematic illustration of the influence of particle size on the interface between strengthener and matrix

during extrusion. They have not been sufficient here to break up the aggregates formed spontaneously by the particles, and the resulting composite has poor dispersion.

The smaller and more extended a particle is, the more its constituent atoms are likely to find themselves at the surface of the particle rather than in its middle (see Fig. 30.10). Considering the areas involved here, the dispersion of the particles will also depend on the chemical compatibility between strengthener and matrix, and some production strategies aim to modify the nanoparticle surfaces in such a way as to encourage bonds at the boundary between strengthener and matrix, so that they concentrate at the interface between the particle and the binding material.

To the Furthest Frontier ...

The idea of interface compatibility has been pushed to the point of creating the matrix by starting from the strengthener, so as to have perfect cohesion between particles and matrix, since they are chemically bound. This is the case for example with nanocomposites of Nylon 6 generated in situ, in which the polymer chains grow from certain sites on the clay sheets. The resulting mechanical properties are better than in other composites, especially under large strains, a consequence of the strong binding between strengthener and matrix. This is in fact a virtuous circle: the chemical compatibility will make it easier to isolate the particles from one another, since they are surrounded by matrix material, and thereby facilitate dispersion.

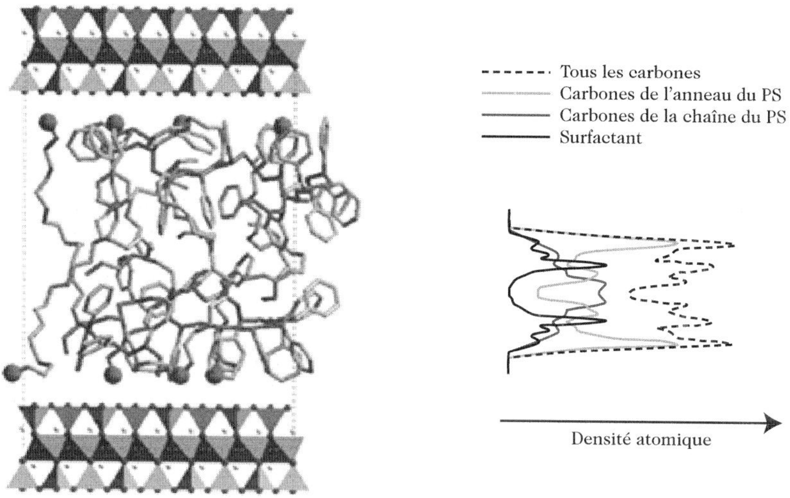

Fig. 30.11. Simulation of the arrangement of polymer molecules (polystyrene) between two clay platelets, at equilibrium, and distribution of carbon atoms in the different units [13]. See also the colour plate

. . . and Beyond

The presence of the inorganic strengthener can affect the matrix over different length scales. One then has a matrix volume called the interphase around each particle, in which its behaviour is modified by the proximity of the rigid surface. The smaller the particles, the more significant this volume will be with regard to the macroscopic behaviour of the material.

The interphase is nevertheless very difficult to observe directly by experimental means. A theoretical approach has thus been developed using molecular dynamics simulation, which studies among other things the geometrical distribution of the various elements making up the nanocomposite (as illustrated in Fig. 30.11). This has considerably improved our understanding of the role played by the interphase, although it is essential to compare the theory with experimental observations.

30.3.4 The Long Road

The field of disordered matrix nanocomposites is in a state of effervescence. The scientific effort is tremendous, but somewhat uneven. The technological effort has followed suit since the first materials have already been commercialised. However, it remains to disentangle the many physical mechanisms called into play!

30.4 Nanoporous Solids as Molecular Springs, Shock Absorbers and Bumpers

30.4.1 Introduction

Heterogeneous systems comprising a nanoporous solid and a non-wetting liquid (put another way, systems made from a liquid and a lyophobic porous solid), thanks to their highly developed interface, can lead to extremely interesting applications. The oldest and best known of these is porosimetry by mercury intrusion [14], used to determine the size distribution of pores in porous and in particular macroporous solids. In the USSR, from 1970, the properties of these systems were put to use in the field of energy storage and dissipation [15]. By virtue of their behaviour as springs, shock absorbers, or bumpers, these systems have been used to build several highly original and successful prototypes, e.g., for opening solar panels on satellites. The work on the intrusion of non-wetting liquids was concerned with liquid metals with very high surface energy. However, it is only very recently that systems made from water and particularly hydrophobic nanoporous matrices have begun to receive some attention. Although they perform less well energetically than systems made from metals, these systems using water have a very promising future that may well lead to a wide range of technological developments and some very exciting applications.

30.4.2 Basic Idea

These applications are based on the following underlying phenomenon: to spread a non-wetting liquid droplet on a solid surface, a certain pressure must be exerted (see Fig. 30.12a). Likewise, to make this liquid penetrate a porous material, the pressure that must be applied is at least equal to the capillary pressure P_c. It will increase if the pores are made smaller or if the liquid is less wetting (see Fig. 30.12b). It obeys the Laplace–Washburn equation, viz.,

$$P_{\mathrm{c}} = -\frac{2\gamma_{\mathrm{LV}}\cos\theta}{r} , \qquad (30.1)$$

where γ_{LV} is the liquid–vapour interface energy,[1] r is the pore radius (assuming the pores to be cylindrical), and θ is the contact angle between the liquid and the lyophobic solid ($\theta \gg 90°$).

During this forced penetration or intrusion into the porous matrix, the bulk liquid is transformed into a multitude of molecular assemblages which soon develop a large interfacial area ($\Delta A > 0$, defined by the variation of the contact area within the pores). During this process, the mechanical energy expended by the pressure forces is thus converted into interface energy. On

[1] One can identify the interface energy γ_{LV} and the surface tension σ of the liquid when the liquid is in the presence of its vapour or a neutral gas.

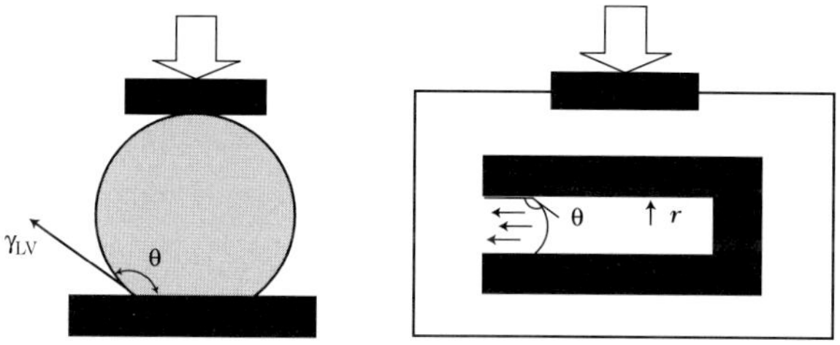

Fig. 30.12. (**a**) Non-wetting liquid in contact with a surface. (**b**) Non-wetting liquid in contact with a porous solid

the microscopic scale, this process results from the breaking of intermolecular bonds within the liquid (corresponding to the energy γ_{LV}) to create new bonds with the solid, these involving an adhesive energy. However, this adhesive energy must be small compared with the interface energy γ_{LV}, in order to obtain a useful non-wetting heterogeneous system. From a thermodynamic standpoint, the increased solid–liquid contact interface is accompanied by an increased free enthalpy $\Delta G > 0$, given by

$$\Delta G = -\gamma_{\mathrm{LV}} \cos\theta \Delta A \,. \tag{30.2}$$

When the pressure is released, the spontaneous evolution of the system, with $\Delta G < 0$, will expel the liquid (extrusion), reducing the solid–liquid interface ($\Delta A < 0$). Although the forces involved here are rather weak (usually van der Waals forces), the fact that they act on a very large surface area (several hundred square metres per gram of solid) means that the process can store and restore a considerable amount of energy in a small volume.

30.4.3 Pressure–Volume Diagram

The behaviour of a heterogeneous system formed from a liquid and a lyophobic porous solid, subjected to increasing pressures, can be divided into three parts labelled A, B, and C in Fig. 30.13. When the system is first put under pressure (A), a very slight decrease in volume is observed because the two quasi-condensed components have high rigidities. When the pressure reaches a certain value P_{i} corresponding to (30.1), there is intrusion of the liquid into the porous matrix, accompanied by a significant decrease in volume. In the diagram $P = f(V)$, this region (B) is represented by an almost horizontal plateau if the pores have constant radius. Finally, when all the accessible pore volume has been filled by the liquid, assumed to be in excess, the system resumes a very slightly compressible profile (C).

When the pressure is relaxed, there are three possible situations:

1. If for some value of the pressure P_{e} close to P_{i} there is spontaneous extrusion of the liquid, accompanied by a significant expansion of the volume,

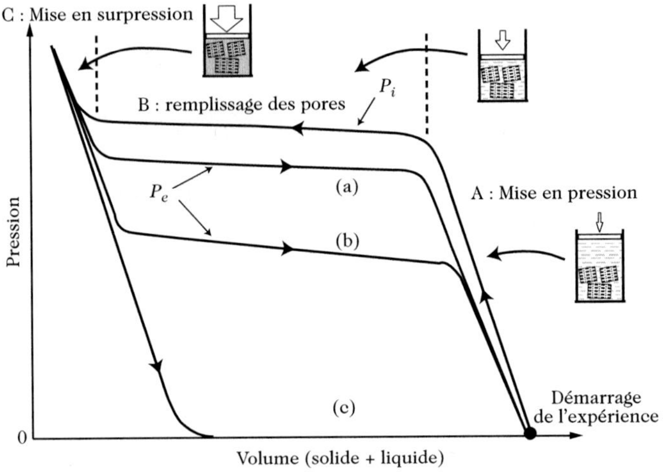

Fig. 30.13. Pressure–volume diagram

the phenomenon then becomes almost reversible and without hysteresis. Such a system, able to accumulate and restore energy, is called a molecular spring. In this case, the phenomenon can be repeated identically for an almost infinite number of intrusion–extrusion cycles.

2. If the liquid comes out of the porous matrix at a considerably lower pressure $P_e \ll P_i$, the phenomenon is still reversible, but there is significant hysteresis. Only a part of the energy is restored in the form of mechanical energy. The system behaves as a shock absorber and resumes its initial behaviour in a new compression cycle.
3. Finally, if under relaxation the phenomenon is not reversible, i.e., the liquid remains held within the pores, the system is able to completely absorb the mechanical energy and behaves like a bumper.

30.4.4 Stored Energy and Restored Energy

Under compression, the mechanical energy received by the system is equal to the total work done:

$$W_{1,2} = \int_{V_1}^{V_2} -P\mathrm{d}V \ . \tag{30.3}$$

In a P–V diagram, this mechanical energy corresponds to the area under the curve, as shown in Fig. 30.14 (left). When the behaviour of the systems is divided into three parts as above, during the first stage (A) and the last stage (C), the energy stored by the system is rather low because the volume only changes very slightly [16]. However, during the intrusion phase (B), the work done is considerable, because it is associated with a large reduction in volume.

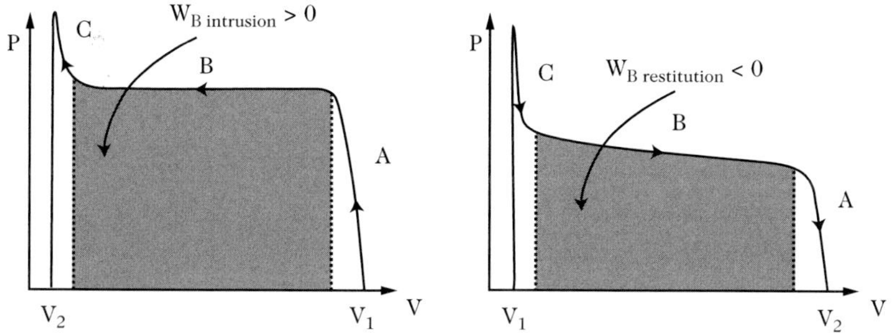

Fig. 30.14. *Left*: Energy stored under compression. *Right*: Energy restored under relaxation

The work done on the system during compression is positive because in (30.3) $\mathrm{d}V < 0$.

During the spontaneous relaxation, the area under the P–V curve represents the work done by the system (see Fig. 30.14 right). In this case, the work done on the system is negative, because in (30.3) $\mathrm{d}V > 0$. If the return path is identical to the outward path, the energy restored is equal to the energy stored during compression. But if there is a difference, the energy lost can be measured on the P–V diagram of the cycle by the difference between the areas under the two curves. Very often, the different values of the stored, restored, or dissipated energies are simply referred to the unit mass of solid, expressing the energy performance e. With W defined by (30.3), we have

$$e = \left| \frac{W}{m} \right| . \tag{30.4}$$

Some very efficient systems can reach a value of 50 J/g. For a reversible system, the energy is restored at constant pressure under relaxation, and hence at constant force, whereas a conventional spring stores and restores the mechanical energy in a way that depends on its elongation.

30.4.5 Causes of Irreversibility

While complete reversibility of the system is easy to understand, the existence of hysteresis is a little more delicate to explain. It is mainly due to two factors: surface heterogeneity and the topology of the pore network. The macroscopic quantities determining the intrusion and extrusion pressures in (30.1) (surface energy and contact angle) are subject to variation on the microscopic scale, owing to impurities and surface roughness, and this will not be without consequence for the question of reversibility. Moreover, the contact angle defined earlier corresponds to a static state. When the liquid is made to move over the surface, one observes that there is an advancing angle θ_{A} and a retraction angle θ_{R}, both of which are different from the static angle θ_{S} (see Fig. 30.15

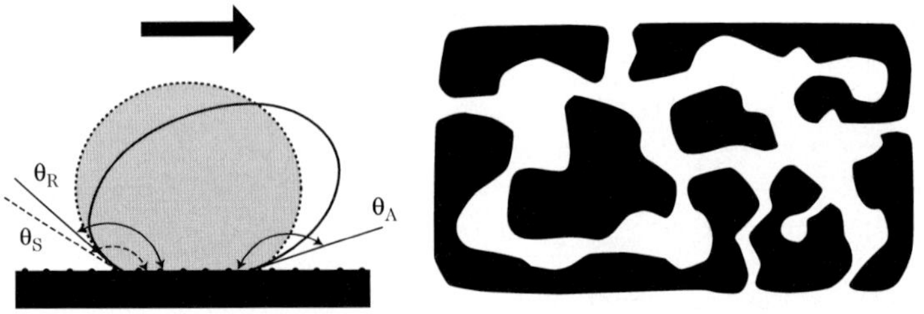

Fig. 30.15. *Left*: Motion of a non-wetting liquid across a rough surface. *Right*: Topology of a disordered pore network

left). In fact, it is observed that $\theta_A > \theta_S > \theta_R$. Now, since θ_A is the relevant angle for intrusion and θ_R is the relevant angle for extrusion, equation (30.1) implies that $P_i > P_c > P_e$.

The path followed by the liquid in the matrix of a disordered amorphous material may involve a rather complex succession of channels, sometimes narrow and sometimes broad, bearing in mind that all the pores are more or less interconnected (see Fig. 30.15 right). Under these conditions, intrusion occurs essentially at high pressure, because this is imposed by the size of the smaller channels before the larger ones are able to fill up. In contrast, extrusion of the liquid occurs at lower pressures, and some of the liquid may be isolated and trapped in the entangled network. For a crystallised or ordered porous material, with a much higher degree of order, this geometric heterogeneity is then limited to the intergranular porosity.

30.4.6 Behaviour of the Solid and Liquid

In order to study such systems, one requires a good understanding of the characteristics of the materials used: pore volume, specific surface area, pore shape, pore dimensions (diameter and length), pore size distribution, and the way the pores are put together to form a network, i.e., the tortuosity, branching, and degree of connectivity of the network.

These thermodynamic systems were originally made from amorphous mesoporous or macroporous silicas and non-wetting liquids such as mercury and low melting point metals or alloys. Indeed, these solid–liquid pairs correspond to good non-wetting conditions because the surface tension γ_{LV} of the liquid is very high and the surface energy of the solid in contact with the vapour is very low. But for large scale applications, these liquids have certain major drawbacks, namely, their toxicity, density and high cost. Among compounds that are liquid at room temperature, water has a high surface tension and in addition possesses certain clear advantages, in particular, the fact that it is non-polluting and has very low cost. Moreover, the very small size of the water molecule, which can be treated as a sphere of diameter 2.8 Å, means that it can enter very small micropores. However, if we decide to use water,

Table 30.4. Characteristics of several hydrophobic solid–water systems [17, 18]

Parameter	Grafted silica gel	Grafted MCM-41	Silicalite-1	β zeolite
Pore volume [cm^3/g]	0.36[a]	0.27[a]	0.19[a]	0.24[a]
Specific area [m^2/g]	300[b]	450[b]	395[b]	560[b]
Pore diameter [nm]	5.7[c]	2.4[c]	0.55[d]	0.7–0.6[d]
Graft density [nm^{-2}]	1.92	1.36	–	–
Intrusion pressure [MPa]	15–45	40–65	99	57
Extrusion pressure [MPa]	0.1–10	15–45	96.5	–[e]
Intrusion pore volume [cm^3/g]	0.19	0.17	0.18	0.25
Stored energy [J/g]	5.7	9.6	18.1	14.5
Restored energy [J/g]	0.6	5.1	17.7	0.2
Dissipated energy [J/g]	5.1	4.5	0.4	14.3

[a] Determined by nitrogen adsorption.
[b] Calculated using the BET (Brunauer, Emmett and Teller) model.
[c] Calculated using the Broekhoff–De Boer model.
[d] Calculated from the structure.
[e] Non-reversible system, no extrusion.

a very hydrophobic porous material must be chosen for the solid phase. Zeolites are then interesting materials to test. These are crystalline microporous solids with a wide range of structures and chemical compositions, discussed in Sect. 11.3. So also are the ordered mesoporous silicas, in particular when silanes are grafted onto them. Table 30.4 lists some examples of hydrophobic nanoporous materials that can be used with water: a silica gel functionalised by dimethyloctylsilane chains, a type MCM-41 silica (Mobil Composition of Matter No. 41) with the same chains grafted on, and two pure silica zeolites synthesised in a fluoride medium, viz., silicalite-1 (F^-) and the β zeolite (F^-).

The ordered mesoporous silicas are formed by condensing silica around surfactant micelles which are then eliminated to leave pores (see Sect. 11.4). Depending on the conditions of synthesis, the resulting phases can be lamellar, cubic, or hexagonal. Hence the hexagonal phase MCM-41 has a simple 1D pore network made up of cylindrical channels with well-defined diameter and wall thickness, characterised by a very high internal surface area and pore volume. Unfortunately, the wall made from amorphous silica has many silanol (Si–OH) bonds making the material hydrophilic. To get round this problem, the surface must be functionalised by grafting hydrophobic chains onto the silanol sites. The grafted molecule is often chlorodimethylalkylsilane, which reacts according to

$$\equiv\mathrm{Si{-}OH} + \mathrm{Cl{-}Si(CH_3)_2{-}R} \quad\longrightarrow\quad \equiv\mathrm{Si{-}O{-}Si(CH_3)_2{-}R} + \mathrm{HCl}\,.$$

The methyl groups provide good lateral coverage of the surface, while the radical R, made from a much longer chain, generally an octyl, possibly with hydrogen atoms substituted by fluorine, contributes to the hydrophobicity of the surface. In addition, a good graft must be uniform over the whole surface and it must not too greatly reduce the pore volume. The same type of graft can be carried out on silica gels displaying for the main part an intergranular porosity with a rather dispersed pore size distribution. The pore characteristics of the two samples of silica studied have been determined by nitrogen adsorption [17] and the values of the different parameters are collected together in Table 30.4.

The zeolites, crystalline microporous solids, have a structure resulting from an assembly of tetrahedra TO_4 (T = Si, Al, etc.), joined together at their oxygen-bearing corners (see Sect. 11.3). The 3D framework built in this way contains channels and cages delimited by apertures made from cycles of n tetrahedra. The hydrophobicity of these materials is directly related to the Si/Al ratio in the framework, and the pure silica forms (zeosils) are distinctly hydrophobic. In the MFI family (structure type Mobil FIve), the pure silica form, known as silicalite-1, has a structure made from two interconnected channel systems with apertures delimited by 10 tetrahedra of dimensions 0.55 nm (see Fig. 26.4b in Sect. 26.2.5).

β zeolite, like the above, can be made with various chemical compositions and in particular, a pure silica form. It is in fact formed by intergrowing at least two distinct but structurally very similar polytypes, one with tetragonal symmetry and the other with monoclinic symmetry. The framework of these two polytypes consists of a 3D network of channels with apertures comprising 12 tetrahedra and diameters in the range 0.56–0.73 nm.

The physicochemical characteristics of two samples of silicalite-1 (F^-) and β zeolite (F^-) have been determined using a range of techniques: scanning electron microscopy, X-ray diffraction, nitrogen adsorption manometry, water adsorption thermogravimetry, and nuclear magnetic resonance of ^{29}Si with magic angle spinning (MAS) [18]. The latter technique is used to show that the silicalite-1 sample, with its very high-resolution spectrum, contains a minimum of defects and a negligible number of hydroxyl groups. The latter would tend to form hydrogen bonds with water and consequently reduce the hydrophobic character of the material. The pore characteristics of these two zeolitic materials are given in Table 30.4.

The compressibility curves, expressing the change in volume of the different systems as a function of the pressure, have been determined at room temperature. The values of the different parameters, viz., pressure at the beginning and end of intrusion and extrusion, occupied pore volume, and energy performance, are all given in Table 30.4. The results show that water–grafted silica gel and water–β zeolite systems behave as molecular bumpers, in which 90% of the energy is dissipated for the first and more for the second. The water–grafted MCM-41 system behaves as a shock absorber, absorbing almost 50%

of the energy, while the water–silicate-1 system is a genuine molecular spring, restoring about 98% of the energy.

It is observed that, for systems made from zeolitic materials, intrusion occurs at much higher pressures than those for grafted silicas. These differences are explained mainly by the pore dimensions. In the latter systems, rather large pressure differences between the beginning and the end of intrusion are also in part related to dispersion in the pore size distribution.

30.4.7 Practical Applications

Various devices have already been applied or undergone tests to produce large forces over significant displacements [18]. These mainly concern the fields of defence, space, and transport. For example, molecular spring actuators have been used to deploy aircraft landing gear, solar panels on satellites, and ejection of a rocket fairing. They are under study or development in the case of submarine guns and swing door counterpoise systems. In these devices, the system is initially compressed, and when the mechanical constraint is removed, it expands spontaneously by extrusion of the liquid.

Thermal keys can also be made by 'freezing' the stored energy during intrusion by solidifying the confined liquid. The system thus conserves its energy, which can be released at any suitable moment simply by heating up to the melting temperature of the compound filling the porous matrix. The liquified compound is then expelled and displaces the actuator rod.

For systems behaving as shock absorbers or bumpers, various uses have been envisaged or have already been subjected to tests, e.g., various types of suspension for vehicles, car bumpers, vibration absorbers, systems for protecting buildings from seismic vibrations, shock absorbers in accidents or for parachuting heavy equipment, and others. For example, collision tests carried out at 35 km/hr against a wall with a standard car (Lada) equipped with a molecular bumper, allowing a maximum displacement of 20 cm, caused absolutely no damage whatever to the bodywork.

External combustion engines based on a new four-stroke thermodynamic cycle and making use of the intrusion–extrusion phenomenon have also been designed. Moreover, the reverse cycle should provide a way of designing new fridges in which the refrigerating fluid – a liquid fluorocarbon or ammonia – would be replaced by a lyophobic heterogeneous system in the form of a colloidal suspension.

To illustrate these different possibilities, Fig. 30.16 shows the diagram and photograph of an actuator which uses a lyophobic heterogeneous system contained in a 16 cm^3 chamber to produce a displacement of 100 mm with a constant force of 780 N for deploying solar panels.

In the initial state, the piston 2 is in the left-hand position (the spindle 3 is outside the body 1). The pore space of the particles, contained in flexible envelopes 7, is empty. A pressure is then applied on the spindle 3, causing the piston to move within the body 1. When the valve 9 opens, the pressure

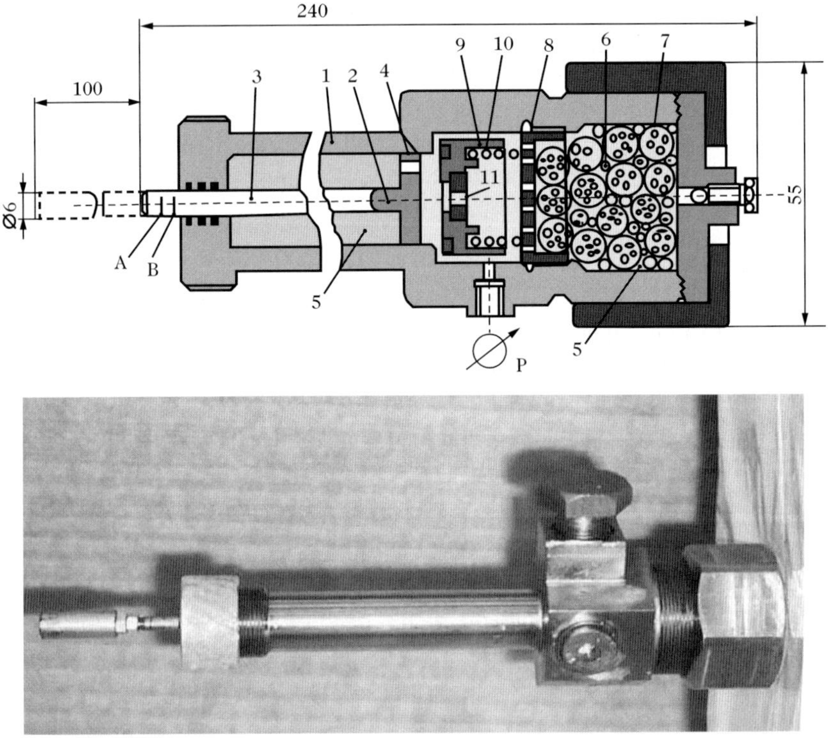

Fig. 30.16. Actuator for opening solar panels [18]. *Upper*: Diagram showing (1) cylindrical body, (2) piston, (3) spindle with markers A and B, (4) opening in the piston, (5) hydraulic oil, (6) porous matrix + non-wetting liquid, (7) flexible envelopes, (8) separation grid, (9) one-way valve, (10) spring, (11) diaphragm. *Lower*: Photograph

is transmitted to the flexible envelopes containing the heterogeneous system, whereupon the liquid penetrates the pore space. The spindle has then moved completely over to the right, to marker A, corresponding to complete filling of the pores.

The spindle is then displaced slightly towards the left by about 3 mm, to bring the pressure down to the spontaneous expulsion value, and held in the position corresponding to marker B. The manometer P measures the load on the device. An electromechanical system holds the spindle in the ‘in’ state. The actuator is ready for the expansion phase. The energy stored at this point is equal to the product of the pore volume and the intrusion pressure. The actuator is triggered in orbit by a command signal which switches off the electromechanical system. Since the actuator has one degree of freedom, its spindle returns to the initial state under the action of the expulsion pressure, moving to the left and doing the work needed to deploy the solar panels. The opening 11 is used to adjust the time required for the operation.

By virtue of their extensive interface, lyophobic heterogeneous systems have the ability to accumulate, dissipate, and restore mechanical energy. The dissipative character is directly related to the form of the hysteresis curve in the pressure–volume diagram.

30.5 High Field Coils

It has been shown that the mechanical properties of materials are strongly influenced by the size parameter, for example, whenever the two characteristic lengths corresponding to the average distance between dislocations and the microstructural dimension come into competition, i.e., whenever the material is nanostructured. This influence can be exploited to develop new nanocomposite materials able to resist extreme and repeated mechanical loads of the kind affecting coils generating very high pulsed magnetic fields.

30.5.1 Specifications for Generating High Pulsed Magnetic Fields

Few research centers around the world can offer the scientific community the possibility of carrying out experiments in strong magnetic fields, although this is essential for developing magnetic materials, semiconductors, and superconductors. Various techniques compete in the race to produce these high magnetic fields: continuous non-destructive magnetic fields can now achieve fields of 45 T (Laboratoire des Champs Magnétiques Intenses in Grenoble, France, and the National High Magnetic Field Laboratory in Florida, USA); and pulsed non-destructive magnetic fields higher than 60 T, developed at the Laboratoire National des Champs Magnétiques Pulsés (LNCMP) in Toulouse, France, for 30 years now, and also at the Los Alamos National Laboratory in the USA. At the present time, the only way to obtain fields above 80 T is to use semi-explosive fields, up to 300 T at the Megagauss Laboratory of the Humboldt University in Berlin, Germany, or explosive fields up to 1 000 T in Tsukuba, Japan, and in Arzamas, Russia, which destroy part or all of the experimental setup.

To overcome this problem, the LNCMP has been developing long-pulse non-destructive pulsed fields for a number of years now. This appears to be the least costly way of achieving fields higher than 60 T [20]. Two features distinguish this development: a new pulse generator comprising 600 capacitors, able to store up to 14 MJ, and high investment in the development of the coils, in particular, in the research and development of materials satisfying the drastic specifications required for generating very high pulsed magnetic fields. These specifications are laid down with respect to the overall objective: production of long-lasting, non-destructive magnetic field pulses.

The width Δt of the magnetic field pulse is defined by

$$\Delta t = T W_{\mathrm{mag}}^{2/3} B_0^{-10/3} [\mathrm{AI}] , \tag{30.5}$$

where T is a geometric factor determined by the structure of the coil and maximal for conventional coils, W_{mag} is the magnetic energy defined by

$$W_{\mathrm{mag}} = \iiint_{\mathrm{volume}} P_{\mathrm{mag}} \mathrm{d}v = \iiint_{\mathrm{volume}} B_0 \frac{1}{2} \mu_0 \mathrm{d}v \ ,$$

P_{mag} is the magnetic pressure, B_0 is the maximum magnetic field at the centre of the coil, and [AI] is the metal action integral given by

$$[\mathrm{AI}] = \int_{63\,\mathrm{K}}^{250\,\mathrm{K}} \frac{c_{\mathrm{v}}}{\rho} \mathrm{d}T \ .$$

For a pulse at 60 T of total duration 150 ms, W_{mag} is of the order of 1 MJ.

For given coil geometry and maximum magnetic field, the three parameters in (30.5) are fixed. In this case, to obtain a long pulse, only the value of the action integral can be varied. This depends on the specific heat capacity c_{v} and the resistivity ρ of the material used to make the coil. We see that the ratio c_{v}/ρ must be a maximum over the temperature range covered during the magnetic field pulse, i.e., from 63 K (nitrogen triple point) to 250 K (coil temperature after firing). In absolute terms, pure copper yields the highest action integral ($[\mathrm{AI}]_{\mathrm{Cu}} = 8.5 \times 10^{16}\,\mathrm{A^2 s\, m^{-4}}$). Note that in practice the only parameter that can actually be adjusted is the resistivity of the conductor, so the conductors used for the coil should be made from copper to provide a minimal resistivity at 63 K. The first condition to be satisfied is therefore

$$\rho_{\mathrm{coil}} \longrightarrow \rho_{\mathrm{Cu}} \qquad \text{(condition 1)} \ .$$

Non-destruction of the coil implies that the stresses developed during the pulse should not go beyond the elastic limit of its constituent material, otherwise irreversible plastic deformation will occur, causing rupture of the conductor and explosion of the coil (and with it the cryostat that contains it). Now during the magnetic field pulse, the Lorentz forces acting on the coil generate stresses with maximal value [21]

$$\sigma_{\mathrm{max}} = 0.5 P_{\mathrm{mag}} \ . \tag{30.6}$$

The condition for the coil not to deform plastically under the effect of these Lorentz forces is thus that its elastic limit must be greater than this maximal stress, viz.,

$$\sigma_{0.2\%}\,(\mathrm{coil}) \geq \sigma_{\mathrm{max}} \qquad \text{(condition 2)} \ .$$

It is interesting to consider several examples, for conventional coils:

$$B_0 = 60\,\mathrm{T} \ , \qquad P_{\mathrm{mag}} = 1\,500\,\mathrm{MPa} \quad \Longrightarrow \quad \sigma_{\mathrm{max}} = 750\,\mathrm{MPa} \ ,$$

$$B_0 = 100\,\mathrm{T} \ , \qquad P_{\mathrm{mag}} = 4\,000\,\mathrm{MPa} \quad \Longrightarrow \quad \sigma_{\mathrm{max}} = 2\,000\,\mathrm{MPa} \ .$$

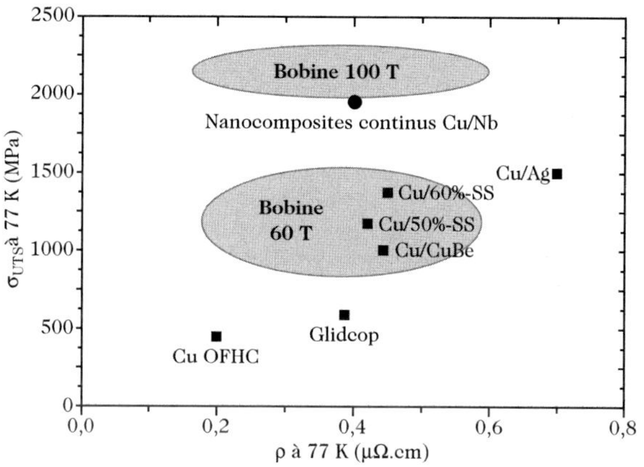

Fig. 30.17. State-of-the-art materials available for making coils for long-pulse magnetic fields: Cu OFHC (oxygen-free high conductivity copper), Glidcop (copper reinforced by Al_2O_3 particles), Cu/CuBe (copper/CuBe alloy macrocomposite), Cu/SS (copper/stainless steel macrocomposite), Cu/Ag (eutectic copper/silver alloy)

Conditions 1 and 2 thus define the specifications that must be met by the material used for the coil. Clearly, and as can be seen from Fig. 30.17, there are few standard materials able to satisfy these requirements for fields above 60 T. It is therefore essential to make specific materials for the design of these coils.

30.5.2 Synthesis of Reinforced Copper Matrix Conductors

According to the above specifications, pure copper can only satisfy the low resistivity condition, since its elastic limit is too low (up to 400 MPa in the cold-drawn state). An obvious thing to do is to try to synthesise a conductor with a reinforced copper matrix that has a high elastic limit. However, the two materials must not be miscible, otherwise there is a risk of downgrading the low resistivity of the copper.

Furthermore, since the conductor must be made in the form of a long wire, it can only be synthesised by cold-drawing, i.e., passing the conductor through a series of calibrated dies of decreasing diameter. The deformation suffered at each pass thus requires among other things that the two component materials must have similar elastic constants, and in particular, similar shear moduli.

Niobium (Nb) presents itself as an obvious choice of strengthener, being totally immiscible with copper up to 900°C, and having a shear modulus (40 GPa) close to that of copper (45 MPa). In addition, due to its body-centered cubic structure, it has the further advantage that its elastic limit increases with decreasing temperature, and in particular under the conditions for use of pulsed field coils.

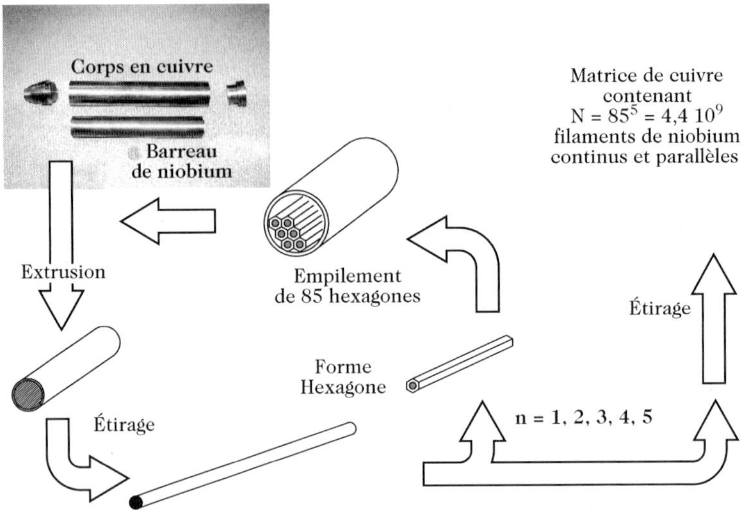

Fig. 30.18. Process for synthesising nanocomposite conductors by repeated extrusion, cold-drawing, and bundling

The way to synthesise Cu/Nb conductors is therefore to reduce dimensions by a series of cold-drawing and bundling phases as shown in Fig. 30.18. The initial billet comprising a niobium rod inserted into a copper matrix is first hot-extruded, then cold-drawn at room temperature until a long wire of hexagonal cross-section is obtained. The latter is then sliced into n segments ($n = 55$ or 85) which are bundled together in a new copper matrix. The billet formed here is subsequently subjected to three (or four) further cycles of hot extrusion, followed by cold-drawing and restacking. The final product is a copper matrix containing $N = n^4$ or $N = n^5$ parallel, continuous niobium filaments, of the same nanometric diameter. In this way conductors containing $N = 55^4 = 9.2$ million, $N = 85^4 = 52.2$ million, and $N = 85^5 = 4.4$ billion filaments have been synthesised [21].

30.5.3 Geometry and Microstructure of Cu/Nb Nanofilamentary Conductors

The main consequence of iterating the cold-drawing and restacking phases is to produce a multiscale structure on 4 (or 5) levels, as shown in Fig. 30.19 for a conductor containing 55^4 filaments. The photograph in Fig. 30.19a, obtained by scanning electron microscopy (SEM), shows an overall view of the cross-section of a Cu/Nb reinforced conductor. In this image, there are 55 grey-shaded hexagons, surrounded by black copper channels. Figure 30.19b shows a magnified view of one of the 55 hexagons. It also contains 55 hexagons separated by copper channels. Figure 30.19c shows a magnified view of one of these hexagons, and it too is composed of 55 hexagons enclosed by copper

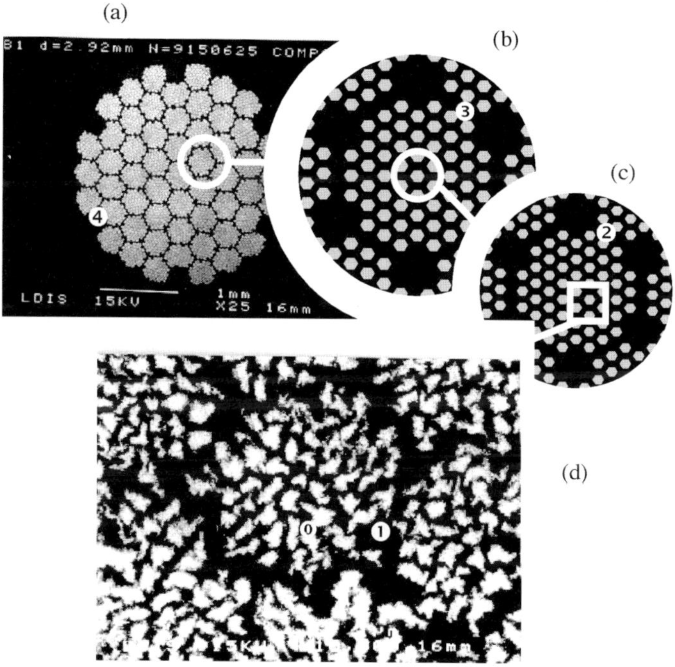

Fig. 30.19. Four-level multiscale structure of Cu/Nb nanocomposites containing 55^4 niobium filaments. (**a**) Full cross-section. (**b**) and (**c**) Successive magnifications. (**d**) Close-up of the filamentary region

channels. The last image (Fig. 30.19d), also obtained by SEM, shows the structure of one of the hexagons in Fig. 30.19c, composed of 55 niobium filaments, separated by interfilamentary copper. There are therefore five levels of copper in the organisation of this structure:

- The finest level, corresponding to the interfilamentary copper of nanometric width, called Cu-0 and indicated by (0) in Fig. 30.19d, separates the niobium filaments.
- The next level up is made of superfilamentary copper, denoted by Cu-1 and indicated by (1) in Fig. 30.19d.
- Levels Cu-2 and Cu-3, indicated by (2) and (3) in Figs. 30.19c and b, respectively.
- The thickest copper, comprising the outer sheath of the conductor, denoted by Cu-4 and indicated by (4) in Fig. 30.19a.

The niobium filaments of nanometric diameter are, for their part, all identical on average. Table 30.5 gives the geometric and mechanical characteristics of various Cu/Nb nanocomposite conductors.

With regard to the microstructure, X-ray diffraction and transmission electron microscope studies of Cu/Nb conductor cross-sections sampled at

Table 30.5. Geometric and mechanical characteristics of various Cu/Nb nanocomposite conductors. d diameter, X volume fraction

$N = 55^4 = 9 \times 10^6$ filaments								
d [mm]	d_{Nb} [nm]	$d_{Cu\text{-}0}$ [nm]	$d_{Cu\text{-}1}$ [nm]	$d_{Cu\text{-}2}$ [nm]	$d_{Cu\text{-}3}$ [µm]	$d_{Cu\text{-}4}$ [µm]	$\sigma_{UTS}(293\,K)$ [MPa]	$\sigma_{UTS}(77\,K)$ [MPa]
0.223	40	7	22	190	1.6	7	1 501	1 951
	X_{Nb}	$X_{Cu\text{-}0}$	$X_{Cu\text{-}1}$	$X_{Cu\text{-}2}$	$X_{Cu\text{-}3}$	$X_{Cu\text{-}4}$		
	0.30	0.102	0.073	0.11	0.16	0.255		
$N = 85^4 = 52.2 \times 10^6$ filaments								
d [mm]	d_{Nb} [nm]	$d_{Cu\text{-}0}$ [nm]	$d_{Cu\text{-}1}$ [nm]	$d_{Cu\text{-}2}$ [µm]	$d_{Cu\text{-}3}$ [µm]	$d_{Cu\text{-}4}$ [µm]	$\sigma_{UTS}(293\,K)$ [MPa]	$\sigma_{UTS}(77\,K)$ [MPa]
2.5	142	43	167	1.8	20.1	226	1 540	1 910
	X_{Nb}	$X_{Cu\text{-}0}$	$X_{Cu\text{-}1}$	$X_{Cu\text{-}2}$	$X_{Cu\text{-}3}$	$X_{Cu\text{-}4}$		
	0.28	0.195	0.090	0.141	0.134	0.16		

different stages of the process shows how the structure of the copper matrix has been affected by the very high strains applied during synthesis: strong [111] texture, nanograin structure and high dislocation density in the wider channels, and monocrystalline interfilamentary copper purged of dislocations. At the same time, the initial polycrystalline structure of the niobium filaments evolves toward a monocrystalline structure of nanometric dimensions, highly textured along a simple [110] crystallographic axis and containing few dislocations. These structural characteristics lead to the conclusion that the niobium filaments are transformed by cold-drawing into nanowhiskers.

Concerning the Cu/Nb interfaces, high-resolution electron microscopy (HREM) has demonstrated the continuity of the (111)Cu and (110)Nb crystallographic planes across the interface (see Fig. 30.20a) and revealed the presence of dislocations at the interface (shown by arrows in Fig. 30.20b). In the copper there is on average an extra (111) half-plane once in every 8 atomic planes, which corresponds to the complete relaxation of the interface for a lattice mismatch of 10.7% [$d_{111}(\text{Cu}) = 2.08$ Å and $d_{110}(\text{Nb}) = 2.33$ Å]. These semi-coherent Cu/Nb interfaces result from the high strains applied to the two materials in purely mechanical initial contact [22].

30.5.4 Physical Properties of Cu/Nb Nanofilamentary Conductors

The ultimate tensile strength (UTS) of continuous Cu/17% vol.Nb and Cu/30% vol.Nb nanocomposites was determined by tensile tests at room temperature and at the temperature of liquid nitrogen. The stress–strain curves of cold-drawn nanocomposites display a very limited plastic regime (plastic strain less than 3%). Figure 30.21 shows the dependence of the UTS at 300 K and 77 K on the diameter d_{Nb} of the whiskers. For whiskers with diameter less than 600 nm, this quantity deviates very rapidly from the values predicted by the rule of mixtures (ROM), since for a Cu/17% vol.Nb conductor, the rule of

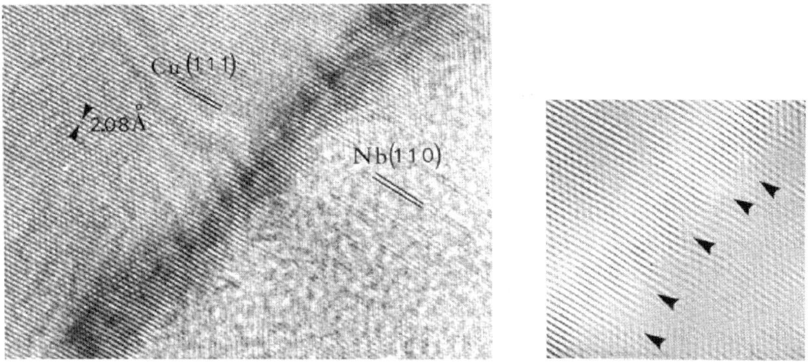

a) b)

Fig. 30.20. (**a**) HREM image of the Cu/Nb interface. (**b**) Image (**a**) processed by filtering to reveal the presence of interfacial dislocations. For the scale: $d_{111}(\mathrm{Cu}) = 2.08\,\text{Å}$ and $d_{110}(\mathrm{Nb}) = 2.33\,\text{Å}$

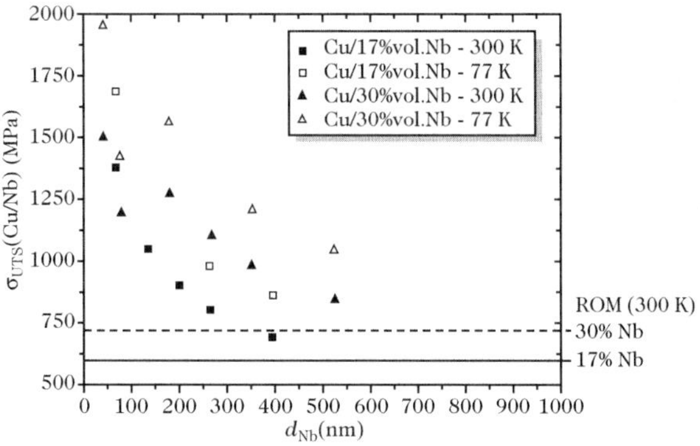

Fig. 30.21. Dependence of the ultimate tensile strength σ_{UTS} of continuous nanocomposites on the niobium whisker diameter d_{Nb} at 77 K and 300 K (strain rate $10^{-4}\,\mathrm{s}^{-1}$)

mixtures predicts a stress of 602 MPa at 300 K, whereas for a Cu/30% vol.Nb conductor, the same rule predicts a stress of 725 MPa at 300 K. In particular, the discrepancy at 300 K is 107% for the Cu/30% vol.Nb conductor containing niobium whiskers of diameter 40 nm. This conductor has a tensile strength of 1 951 MPa at 77 K, which is four times greater than the UTS for pure cold-drawn copper.

Nanofilamentary Cu/Nb materials thus display a quite extraordinary hardening, much greater than predicted by the rule of mixtures for cold-drawn materials, together with an intrinsic hardening due to temperature reduction (Peierls lattice friction mechanism in Nb) [21]. To understand what causes this hardening, tensile tests have been carried out on thin foils of Cu/Nb conductors under the transmission electron microscope.

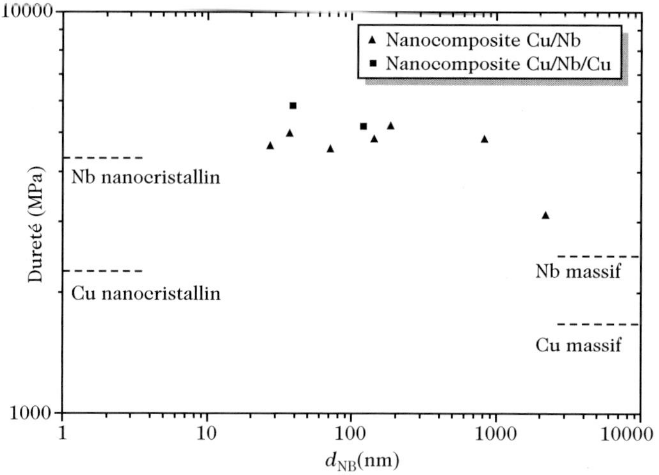

Fig. 30.22. Nanoindentation measurements to determine the hardness of Cu/Nb nanocomposite regions in reinforced conductors as a function of the size d_{Nb} of the niobium whiskers (co-cylindrical Cu/Nb/Cu conductors, niobium whiskers replaced by niobium nanotubes filled with copper)

The observed deformation mechanism consists in the nucleation at Cu/Nb interfaces of single dislocation loops in their slip plane, propagating in the copper until they intercept a neighbouring Cu/Nb interface, where they remain blocked. The niobium whiskers are deformed by a similar mechanism involving single dislocation loops generated at the interfaces. The plasticity of the nanofilamentary conductors due to motion of single dislocations results from size reduction of the microstructure: lack of space prevents dislocations from piling up at the interfaces, so that the Hall–Petch law is no longer valid [23]. The effect of spatial restriction of the interfaces on plasticity mechanisms has been confirmed by a nanoindentation study carried out on reinforced conductors. The local hardness of Cu/Nb nanocomposite regions reaches 5.8 GPa (see Fig. 30.22), i.e., about four times the hardness of bulk copper, and is also greater than that of a nanocrystalline Cu/Nb mixture with the same composition [24].

With regard to electrical properties, the resistivity of reinforced Cu/Nb conductors lies in the range 0.4–0.6 μΩ cm, i.e., between 2 and 3 times the value for pure bulk cold-drawn copper. Concerning the origin of the increased resistivity in Cu/Nb conductors, analysis has shown that it is the nanometric copper channels (with dimension less than 400 nm) which induce the drop in conductivity by diffusion of electrons at the very close Cu/Nb interfaces [21].

30.5.5 Conclusion

The effect of reducing the microstructure to the nanometric scale thus has positive consequences for the mechanical properties (increased elastic limit by confinement of dislocations), but negative consequences for the electrical conductivity (diffusion of electrons at the interfaces in the very narrow channels). A dimensional compromise must therefore be found for nanocomposite reinforced conductors in order to simultaneously optimise mechanical and electrical properties. This compromise should soon be achieved. One aspect is the development of the so-called co-cylindrical Cu/Nb/Cu structure in which the niobium whiskers are replaced by niobium nanotubes filled with copper. Another is to replace the niobium by tantalum (Ta).

References

Ceramic–Metal Composite Welding Supports

1. A. Mocellin, T. Mazet, G. Le Caër: Latte de soudure en matériau composite, Brevet d'invention CNRS, file no. 01.07268, publication no. 02825306, Institut National de la Propriété Industrielle (France)
2. E. Zola: *Le Ventre de Paris* (1873) `http://atilf.inalf.fr`
3. T. Rouxel: Propriétés mécaniques. In: *Propriétés et applications des céramiques*, ed. by P. Boch, Hermes Science, Paris (2001) Chap. 1
4. R. Barthes: *Le Plastique*, Mythologies, Collection Points Seuil, Série Essais, No. 10 (1957) p. 159
5. P. Matteazzi, G. Le Caër: Synthesis of nanocrystalline alumina–metal composites by room-temperature ball-milling of metal oxides and aluminum, Journal of the American Ceramic Society **75**, No. 10 2749–2755 (1992)
6. J.L. Guichard: Doctoral thesis, Institut National Polytechnique de Lorraine, Nancy, France (1998)

Reinforced Amorphous Matrices

7. R. Rothon, M. Hancock: *Particulate-Filled Polymer Composites*, Longman Scientific and Technical, Essex (1995) pp. 1–46
8. A. Usuki, M. Kawasumi, Y. Kojima, A. Okada, T. Kurauchi, O. Kamigaito: J. Mater Res. **8** (5), 1174 (1993)
9. Giannelis Group, Cornell University, USA. Unpublished results (2002)
10. `http://www.plasticstechnology.com`
11. P.B. Messersmith, E.P. Giannelis: J. Poly. Sci. Part A **33**, No. 7, 1047–1057 (2003)
12. `http://www.nanocor.com`
13. D. Schmidt, D. Shah, E.P. Giannelis: Current Opinion in Solid State and Materials Science **6**, 205–212 (2002)

Molecular Springs, Bumpers and Shock Absorbers

14. H.L. Ritter, L.C. Drake: Ind. Eng. Chem. Anal. **17**, 782–786 (1945)
15. V. Eroshenko: C.R. Acad. Sc. Ukraine, Series A **10**, 79–82 (1990)
16. L. Coiffard: Doctoral thesis, Université Blaise Pascal, Clermont-Ferrand, France (2001)
17. T. Martin, B. Lefevre, D. Brunel, A. Galarneau, F. Di Renzo, F. Fajula, P.F. Gobin, J.F. Quinson, G. Vigier: Chem. Comm. 24–25 (2002)
18. V. Eroshenko, R.C. Regis, M. Soulard, J. Patarin: J. Am. Chem. Soc. **123**, 8129–8130 (2001)
19. V. Eroshenko: *Matériaux nanostructurés*, Galerne (2002)

High Field Coils

20. S. Askénazy: Physica B **211**, 56 (1995)
21. F. Dupouy: Doctoral thesis, INSA Toulouse, France (1995); L. Thilly: Doctoral thesis, INSA Toulouse, France (2000)
22. F. Dupouy, E. Snoeck, M.J. Casanove, C. Roucau, J.P. Peyrade, S. Askénazy: Scripta Metall. **34**, 1067 (1996)
23. L. Thilly, M. Véron, O. Ludwig, F. Lecouturier, J.P. Peyrade, S. Askénazy: Phil. Mag. A **82**, 925 (2002)
24. L. Thilly, F. Lecouturier, J. von Stebut: Acta Materialia **50**, 5049 (2002)

31

Biology and the Environment

P. Maestro, P. Couvreur, D. Roux, D. Givord, J.-A. Dalmon, J.-C. Bertolini, and F.J. Cadete Santos Aires

31.1 Inorganic Catalysts for Diesel Engines

Since the 1980s, catalytic antipollution systems for petrol-fuelled vehicles have considerably reduced emissions of CO, NO_x, and unburnt hydrocarbons. They use precious metals as catalyst on a hybrid support made from alumina and a mixture of cerium oxide and zirconium oxide, sometimes doped. Cerium oxide is a key element in these systems, thanks to its triple role as redox agent (maintaining the stoichiometry of the mixture of gases to be processed and hence the efficiency of the precious metal), dispersing and stabilising agent for the nanoparticles of precious metal, and thermal stabiliser for the alumina. However, these systems could only be developed with the help of nanoscale textured powders, able to conserve very high specific surface areas at high temperatures, and hence maintain constant and adequate reactivity over the lifetime of the catalyst (several tens of thousands of kilometers).

More recently, with the benefit of this knowhow in the synthesis of cerium oxides and adapting it to the development of stable dispersions, cerium oxide nanoparticles have found a place in antipollution systems for diesel engines, with the role of consuming the carbon particles resulting from incomplete combustion by diesel engines at relatively low temperatures in the exhaust pipe. Indeed, particle filters developed to trap soots produced in the combustion of diesel fuel are only really useful if they do not need to be replaced or emptied too often.

A major step forward is taken by adding 5-nm nanoparticles of cerium oxide: with a continuous supply of the catalyst in the combustion chamber, the soots are trapped by the cerium oxide nanoparticles and, once in the filter, can be oxidised by the cerium oxide by simply programming a steady increase in temperature in the filter. The efficiency of nanoparticulate cerium oxide means that a few tens of ppm of the additive are enough, both in terms of capturing the soots in the combustion chamber and for the purpose of burning them in the filter. Today, a liter of EOLYS suspension can effectively

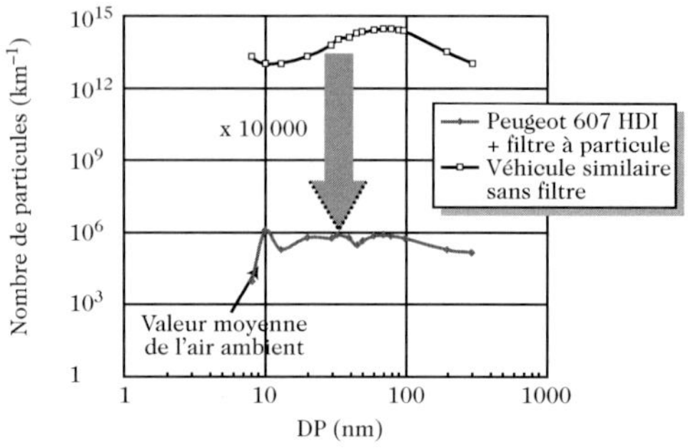

Fig. 31.1. Controlling diesel emissions

limit pollution from a diesel engine for 100 000 kilometers. New generations of additives are currently being developed to work for 200 000 kilometers.

These developments are based on a combination of several areas of knowhow:

- New developments in inorganic synthesis, and the ability using wet synthesis to control particle sizes (in the range 5–100 nm, monodispersed about a given size).
- Techniques from the physical chemistry of dispersed media, which enable one to stabilise suspensions of these nanoparticles not only in either acidic or basic media, but also in aqueous or organic media, thanks to the possibility of carrying out a phase transfer at high concentration – up to 400 g/L – whereas conventional methods for synthesising suspensions of small particles generally involve very dilute solutions. Moreover it is this possibility of stabilising nanoparticles in an organic medium which allows one to make stable and compatible colloidal suspensions without demixing problems in the diesel fuel.
- Control of surface properties by encapsulation of other nanoparticles (as in the case of $Ti0_2$) or grafting and adsorption of organic molecules.
- Synthesis of redispersible powders, improving handling before use.

The additive EOLYS is environmentally friendly, as demonstrated by toxicological studies, and compatible with plastics used in tanks and dosing pumps. In addition, it does not cause wear in the engine, since the nanoparticles are not abrasive in this context. The dispersions are easy to use, in the same manner as an oil, and have no harmful effect on the performance of the engine.

With the EOLYS additive and the particle filter in combination, more than 99% of the particles produced in the diesel engine are eliminated. The oxidising power of the cerium particles, intrinsic to cerium oxide and further increased by the size reduction effect and the increase in specific surface area, considerably reduces the combustion temperature of the soots.

Figure 31.1 compares diesel emissions without a particle filter with emissions from a system equipped with the EOLYS process. The particle content is reduced by a factor of 10 000, and the vehicle emits fewer particles than already exist in the surrounding atmosphere!

31.2 Nanotechnology and New Medicines

31.2.1 Introduction

A major challenge today is to address therapeutic molecules to the sick organ, tissue or cell. Even at the beginning of the last century, the German bacteriologist Paul Ehrlich was already dreaming of a 'magic bullet', able to carry a medicine in a specific way to the relevant site of action. This dream is much closer to reality today, thanks to the development of nanotechnology and the possibility of addressing medicines.

Indeed, many active compounds exhibit physicochemical characteristics (hydrophilicity, molecular weight, etc.) that do nothing to help them get through the biological barriers separating the administration site from the site of action. Other active molecules such as DNA and proteins also encounter enzyme barriers which rapidly cause them to degrade and metabolise. The barriers that need to be crossed are generally highly complex systems involving several elements (epithelium, endothelium, cell membrane) and several components (mechanical and physicochemical barriers, enzyme barriers). In order to obtain therapeutic concentrations at the site of action, one must thus be resigned to a considerable loss of the medicine in other tissues or cells, leading to significant and sometimes even prohibitive levels of toxicity.

For these reasons, drug carriers have been considerably developed over the past few years. With the help of new physicochemical concepts, galenic research has invented submicron administering systems able to protect the active molecule from degradation and also to control the release of the active molecule both in time and in space.

31.2.2 Artificial Carriers: Liposomes and Nanoparticles

Medicine carriers, designed for intravascular injection, are the product of nanotechnology. As discussed above, nanoparticles are generally obtained either by size reduction starting from a larger object, or by assembling smaller, elementary objects, i.e., atoms or molecules. Medicine carriers can be made by either approach, exploiting the fundamental concepts of physical chemistry to build (ordered or disordered) supramolecular entities. In practical terms, the size of carriers intended for injection into the blood stream must be considerably less than the micrometer in order to avoid thromboembolism (blockage of the blood vessel). We shall describe three types of carrier: conventional liposomes and two new types of nanoparticle resulting from recent developments. The first subject is thus a new method for synthesising multilamellar liposomes and the rest concerns polymer nanoparticles with two compartments.

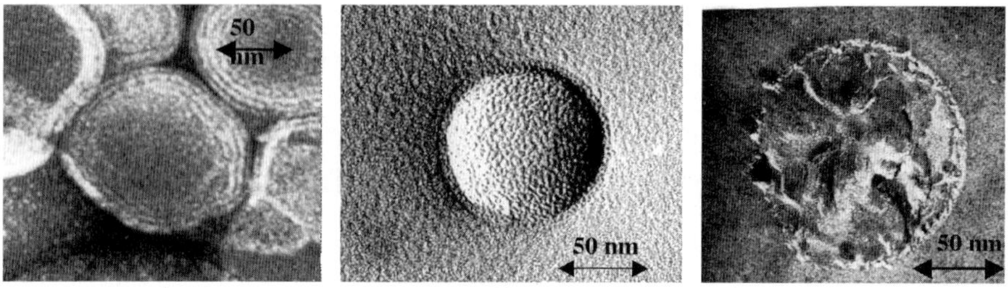

a) b) c)

Fig. 31.2. (**a**) Liposomes are vesicles comprising one or more phospholipid bilayers. (**b**) Nanospheres are particles made from a polymer matrix. (**c**) Nanocapsules comprise an aqueous or oily core, surrounded by a thin polymer membrane

Liposomes are biocompatible and biodegradable vesicular systems comprising a single bilayer (unilamellar liposomes) or several bilayers (multilamellar liposomes) of ordered phospholipids in a lamellar phase, enclosing one or more aqueous compartments (see Fig. 31.2a). Hydrophilic active ingredients can be encapsulated in the aqueous phase, while lipophilic molecules fix themselves in the bilayer(s). A new method for synthesising multilamellar liposomes provides a way of controlling the size of these objects while preserving the multilamellar structure right through to the centre [1].

The nanoparticles are colloidal systems whose structure is generally provided by polymers, preferably biodegradable ones (Fig. 31.2b). They may be of matrix type (nanospheres), in which case the active ingredient can be dispersed or dissolved in the polymer matrix and simply released by diffusion or following the biodegradation of the polymer in the organism. Otherwise they are of reservoir type (nanocapsules), in which case they comprise a central nucleus, generally liquid, surrounded by a thin wall of polymer of thickness no more than a few nanometers.

A new method for synthesising multilamellar liposomes has been discovered recently. The idea is to shear a lamellar phase, leading to a hydrodynamic instability that can be used to obtain a close packing of multilamellar vesicles [1]. When dispersed in water, these vesicles give nanoparticles that are lamellar right through to the centre (see Fig. 31.3). These structures, similar to liposomes, are easier to synthesise and more robust than conventional liposomes [1,4].

Very recently, bicompartmented nanosystems have even been synthesised (see Fig. 31.4). They comprise a lipid compartment, formed mainly by triglycerides and stearylamine stabilised by a phospholipid monolayer, and an aqueous compartment enclosed by a phospholipid bilayer [2]. These open novel prospects for simultaneously carrying hydrophilic and lipophilic molecules and may be used for designing 'Janus' carriers, i.e., systems in which the surface is not homogeneous.

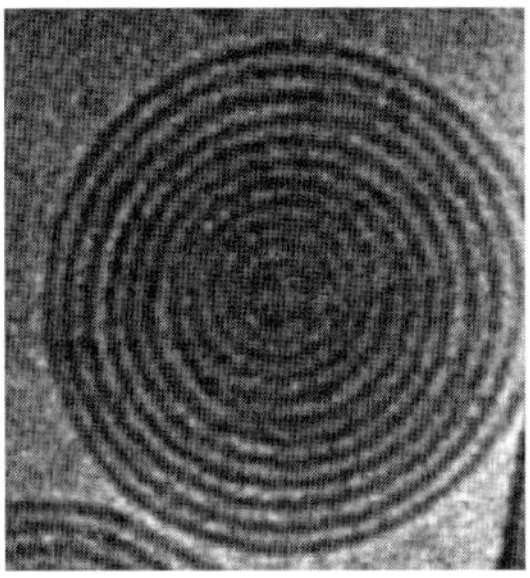

Fig. 31.3. Multilamellar phospholipid vesicles obtained by shearing a lamellar phase and dispersing in water [4]

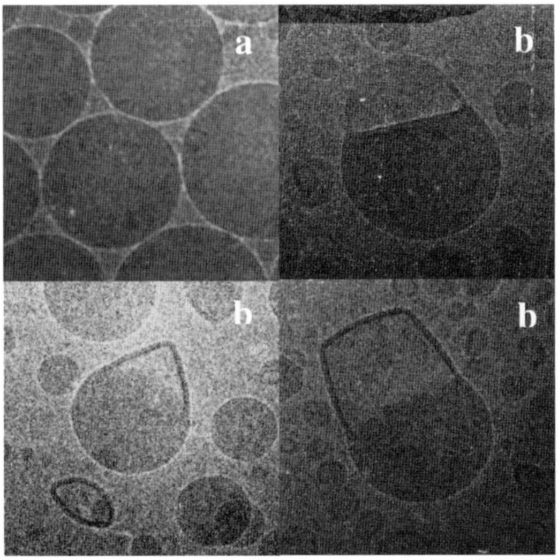

Fig. 31.4. Cryomicroscopic image of a new bicompartmental carrier. The emulsification of triglycerides and phospholipids in an aqueous phase produces a standard emulsion (**a**). Adding a cationic lipid such as stearylamine creates bicompartmental objects (**b**) comprising an oil droplet and an aqueous compartment of circular, rectangular, or triangular shape. Taken from [2]

When administered intravenously, the nanoparticles, like liposomes, interact strongly with plasma proteins owing to their very high specific surface area (several tens of m^2 per g). This surface area favours the creation of strong hydrophobic interactions between the carrier surface and opsonins[1] (IgG, complements, fibronectin, etc.), which are recognised by specific receptors located

[1]Opsonins are plasma proteins with specific receptors on the macrophages. They adsorb onto foreign bodies such as bacteria or particles and thereby facilitate their elimination from the blood compartment. Examples are fibronectin, complements, and Fc fragments of immunoglobulins.

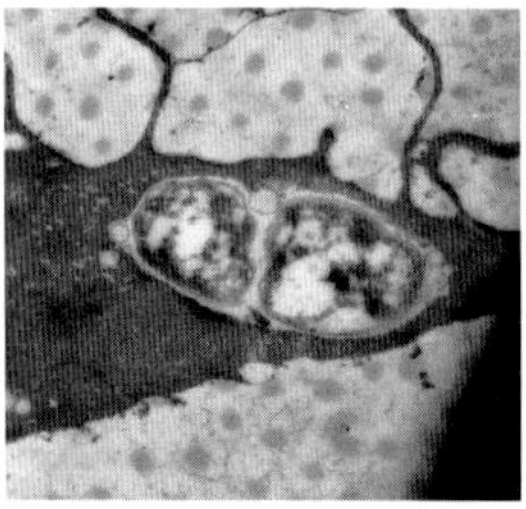

Fig. 31.5. Targeting an intracellular bacterium located in a macrophage endosome. Photo courtesy of H. Alphandary [7]

on macrophages in the mononucleate phagocyte system (MPS) (liver, spleen, bone marrow).

Colloidal carriers coated with opsonins are thus mainly captured by macrophage cells in the liver (Kupffer cells) and macrophages in the outer region of the spleen whose receptors specifically recognise Fc fragments of IgG, the C3bi fragment of the complement, or fibronectin [3].

The field of cancer research has stimulated a great deal of work in this area. For example, the hepatic tropism of first generation carriers has been exploited to treat hepatic metastases. In this case, targeting is indirect since the anticancer medicine is delivered to the Kupffer cells in the liver, which then plays the role of reservoir for the medicine. The active ingredient can then diffuse throughout the tissue and in particular into the neoplastic cells [4]. The hepatosplenic distribution of first generation carriers can also, in some cases, reduce the toxicity of a medicine by diverting it from its toxicological target. For example, the cardiac toxicity of doxorubicine has been reduced by encapsulating in liposomes or nanoparticles. In the same way, the renal toxicity of amphothericine B is significantly diminished after administering to patients in the form of liposomes. Some products resulting from these ideas have recently been put on the market, e.g., Ambisome and Daunosome.

Promising work has also been done on the treatment of intracellular infections. Indeed, when they occur in the blood compartment, bacteria are also opsonised and captured by MPS macrophages. In some situations (immunodepression, opportunistic disease, etc.), the lyosomes of these macrophages provide sanctuaries for the multiplication of intracellular bacteria (see Fig. 31.5). Many antibiotics are rather ineffective against these intracellular germs because they have difficulty diffusing into the cell, where they cannot reach the infected intracellular compartments (endosomes and lyosomes). The tissular (liver, spleen), cellular (macrophages) and subcellular (endosomes/lysosomes) location of first generation carriers therefore makes them a first rate shuttle service for transporting antibiotics into the infected cells (see Fig. 31.5) [5]. Spectacular results have been obtained in experimental infection models, with both liposomes and nanoparticles. Hence, in an experimental infection model for *Salmonella typhimurium*, it has been shown that an antibiotic carried by biodegradable nanoparticles of polyalkylcyanoacrylate was 100 times more effective than when administered in a conventional pharmaceutical form [6].

A third field of application for nanocarriers that has been very widely investigated is genetic therapy. Although new medicine based on this technology has yet appeared on the market, these techniques will probably lead to treatments offering powerful therapeutic advantages. Indeed, it is now possible to envisage direct use of the DNA molecule as a medicine, either to treat genetic disorder, cancer, or cardiovascular disease, or even to design third generation vaccines. DNA is a molecule that cannot naturally enter inside cells. If it is to reach its target, which is the cell nucleus, it must be integrated into a carrier. Viruses are natural carriers of DNA and biologists have learnt how to make viruses containing therapeutic DNA. However, for both biological reasons (immune reactions, the possibility of crossing with viruses) and industrial reasons (it is difficult to control the industrial production of viruses), it will be necessary to develop artificial viruses composed of phospholipids, DNA, and possibly purified proteins. Since the 1980s, it has been known that the association of cationic phospholipids and DNA in vitro provides a way of getting DNA inside cells which can then express the gene [8]. In vivo essays have been disappointing, however, largely because of the cationic nature of the molecules used. Neutral carriers are currently under development [4].

31.2.3 Conclusion

Research carried out at the interface between physics, chemistry, and biology have led to the design of submicroscopic galenic systems able to transport biologically active molecules right up to their tissular, cellular, or even molecular target. Although somewhat futuristic, this kind of progress has already gone beyond the stage of a simple laboratory curiosity, since it has already given rise to new commercially produced medicines for treating cancers and certain infectious diseases.

31.3 Magnetic Nanoparticles and Biomedical Applications

Magnetic nanoparticles have a wide range of applications in the fields of biology and medecine. The first of these applications are those chosen by nature herself (Sects. 31.3.1 and 31.3.2). Several living organisms use nanoparticles as a means of orientation in the Earth's magnetic field. Separation techniques (Sect. 31.3.3) make use of the property of particles to move toward regions where the field is stronger. In magnetic resonance imaging (Sect. 31.3.4), the conditions under which a proton resonates are affected by the magnetic field of a nanoparticle in its environment. The image contrast is then increased. Magnetic hyperthermia is a technique in which the heating of nanoparticles subjected to an alternating magnetic field is used to kill cancer cells (Sect. 31.3.5).

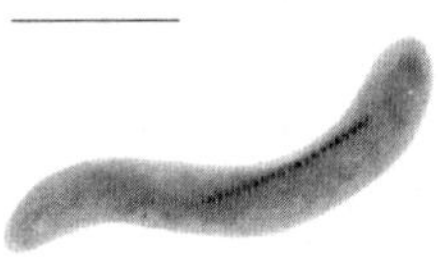

Fig. 31.6. Chains of magnetic particles (*dark*) inside a magnetotactic bacterium. *Scale bar*: 1 μm

31.3.1 Magnetotactic Bacteria

Inside the so-called magnetotactic bacteria, chains of magnetic nanoparticles are observed, e.g., magnetite Fe_3O_4 (see Fig. 31.6). The particles have diameters smaller than the critical radius below which the single-domain magnetisation state is the most stable (see Chap. 5). They thus constitute tiny permanent magnets, sensitive to the Earth's magnetic field along which the bacteria then orient themselves. The main component of the Earth's field is horizontal, directed toward the poles, but there is also a vertical component, ten times weaker. Moving by means of their flagella along the appropriate orientation of the field lines, the bacteria plunge down into deep waters, less rich in oxygen, which provide a more favorable environment for their particular lifestyle. In the southern hemisphere, the orientation of the vertical component of the Earth's field has the opposite sign to its orientation in the northern hemisphere. The magnetisation of the nanoparticles is also reversed, and this remarkable orientation mechanism still works.

The nanoparticles produced by magnetotactic bacteria are of very regular shape and size. This means that they can actually be used to produce nanoparticles. Thanks to the presence of a protective membrane, the particles form homogeneous suspensions. A novel potential application has been envisaged in which the bacteria would produce nanoparticles on the surface of which a protein with specific function would be directly grafted.

31.3.2 Homing Pigeons

It is generally assumed that the orientation mechanism used by homing pigeons involves two main processes: a magnetic compass, called the inclination compass, and a navigation map which is a mental representation of previously overflown locations. A team of German and British scientists have recently put forward a convincing analysis of the magnetic orientation mechanism [11]. This is based on the observation that the pigeons' beaks contain magnetic nanoparticles with diameters in the range 1–5 nm, forming clusters of micronic size (see Fig. 31.7). The nanoparticles within these clusters are superparamagnetic. Even in a weak field such as the Earth's magnetic field, they acquire a substantial magnetisation.

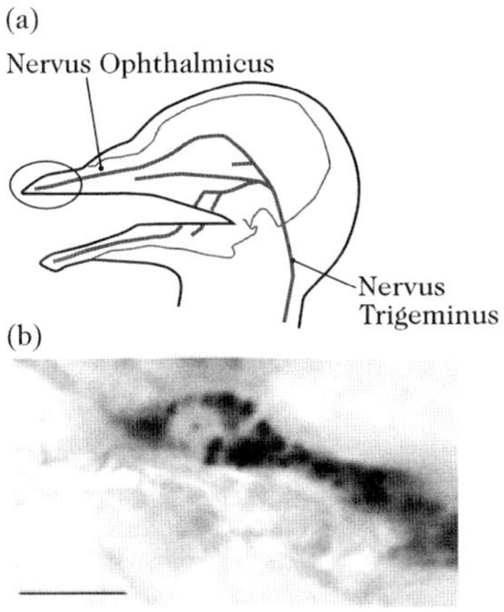

Fig. 31.7. Clusters of magnetic nanoparticles observed in the beak of a pigeon [*dark regions* in (**b**), where the *scale bar* represents 10 μm] and region in which these clusters are distributed [*ellipse* in (**a**)]. Taken from [4]

Under the effect of interactions between the particles resulting from this magnetisation, the rows of particles directed along the field tend to shorten, while those perpendicular to the field tend to grow slightly longer. A signal is then generated at neighbouring nerve ends and transmitted to the brain. This signal does not depend on the sign of the Earth's magnetic field, since the attractive/repulsive force between the particles is specified by the strength of the field and is independent of its sign. It does not procure all the information required for the purposes of orientation. This is why a navigation map is required in addition to the magnetic compass.

31.3.3 Magnetic Separation

The association of particles with an element one hopes to detect in a solution is a well-established analytical process. For example, antibodies can be grafted onto micrometric latex particles. The particles are immersed in the solution to be analysed. If specific microbes for this antibody are present in the solution, they attach themselves to the particles via the antibody, and thereby agglutinate on the particles. This phenomenon is then the signature that the relevant microbe is present.

When the sought microbe is only present at a very low concentration, an immunomagnetic separation technique is used to concentrate them (see Fig. 31.8). In this case, magnetic particles are used instead of latex beads and the microbes attach themselves to them. The particles are then confined by passing them through a column subjected to a magnetic field. The resulting increase in concentration makes the microbes easier to detect if they are in fact present.

Nanoparticles offer a number of advantages over particles of micrometric dimensions: they are biocompatibile, non-toxic, and have less tendency to sediment. The disadvantage is that they are more difficult to concentrate int a magnetic field gradient owing to the fact that the force acting on a particle is

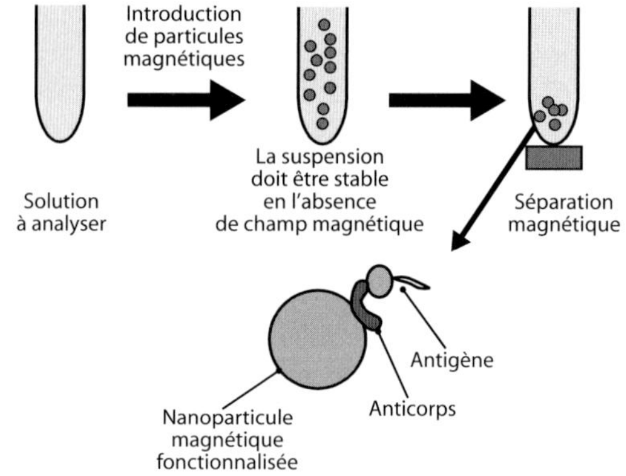

Fig. 31.8. Magnetic separation of different substances

proportional to its magnetic moment. The possibility of using such nanoparticles has been demonstrated in several recent applications, using configurations able to produce high magnetic field gradients.

31.3.4 Magnetic Nanoparticles as MRI Contrast Agents

Magnetic resonance imaging (MRI) uses the phenomenon of nuclear magnetic resonance, observed when an atomic nucleus carries a nuclear magnetic moment μ_{N}. In a magnetic induction field B, applied in the z direction, the nuclear moment precesses around the field direction at a frequency ν with angular frequency $\omega = \gamma B$, where γ is the gyromagnetic ratio. Let us assume schematically that the nuclear spin is $S = 1/2$. There are then two possible states, viz., $S_z = +1/2$ and $S_z = -1/2$. For an assembly of atoms, the populations of each state at temperature T are defined by Boltzmann statistics. Subjecting the system to electromagnetic radiation of energy $E = h\omega = h\gamma B$, transitions can be induced between the two states, imposing a non-equilibrium population distribution on them. When the radiation is switched off, the return to equilibrium occurs in a characteristic time T_1, representative of the interaction between spin and lattice. Moreover, if we assume that at time $t = 0$ the nuclear moment lies along a well-defined direction perpendicular to the z axis, then as soon as $t > 0$, the moment will precess around the field at the frequency specified above, and at the same time it will tend to align itself along the z direction. A phase difference will build up gradually due to the fact that the field is not exactly the same at each atom, so that there is a distribution of Larmor frequencies. The corresponding characteristic time is called the relaxation time T_2.

The substances making up the human body, and in particular fats and water, contain a very high proportion of hydrogen, and medical magnetic resonance imaging makes use of the nuclear spins of protons. The patient is exposed to a macroscopic magnetic induction field B, between 0.5 T (permanent magnet systems) and 5 T (superconducting coils). A known magnetic field gradient is superposed on the constant magnetic field of the main source. The result is that, at each point, the magnetic field has a specific known value. The observed value of the resonance frequency, directly related to the strength of the applied field, is then used to locate the proton that is the source of the observed signal. From one constitutive element of the body to another, the environment of the atoms is not the same, whereupon the relaxation time T_1 is not the same either. One can then deduce to which tissue, muscle, or fat the resonating proton belongs. In particular, magnetic resonance can locate cancerous tumours, which have different characteristic time T_1 to the healthy surrounding tissue.

In some cases, the image contrast is inadequate to be able to extract the required information. For example, it is difficult to detect a tumour in the liver. In this case, contrast agents can be injected. Through their specific chemical or magnetic properties, these modify the relaxation times T_1 or T_2. Magnetic nanoparticles can be used as contrast agents, injected intravenously. For the liver, the healthy regions absorb the nanoparticles and then appear darker on the resonance image because the relaxation time, proportional to the magnetic field seen by the protons, is increased. In contrast, the nanoparticles do not enter the tumours, which thus appear as lighter regions.

31.3.5 Magnetic Nanoparticles and Treatment of Tumours

Over the two last decades, considerable progress has been made in understanding the genetic and molecular mechanisms responsible for the development of cancers. However, the benefits from these studies remain disappointing for patients. In particular, some cancers resist conventional treatment.

The hyperthermia technique consists in heating a tumour up locally, to a temperature of the order of 42–46°C, thereby making it more sensitive to radiotherapy or chemotherapy. One way of carrying out hyperthermia uses superparamagnetic single-domain nanoparticles (see Fig. 31.9). The nanoparticles are grafted onto a ligand, of folic acid in the scheme illustrated in Fig. 31.9. It plays the role of a vector which recognises a tumour and attaches itself to it. When an alternating magnetic field is applied, reversal of the nanoparticle magnetisation, in phase with the field reversal, brings about heating in the nanoparticle. This heating spreads out around it, in the region where the tumour is located.

For this application, the nanometric size of the particles, of the order of 5 nm, reduces sedimentation due to the effects of gravity and facilitates the movement of the nanoparticles from the veins into the tissues. Moreover, the superparamagnetic nature of the particles, due to their nanometric dimensions, prevents them from agglomerating, this property resulting from the dipole interactions between magnetised ferromagnetic particles.

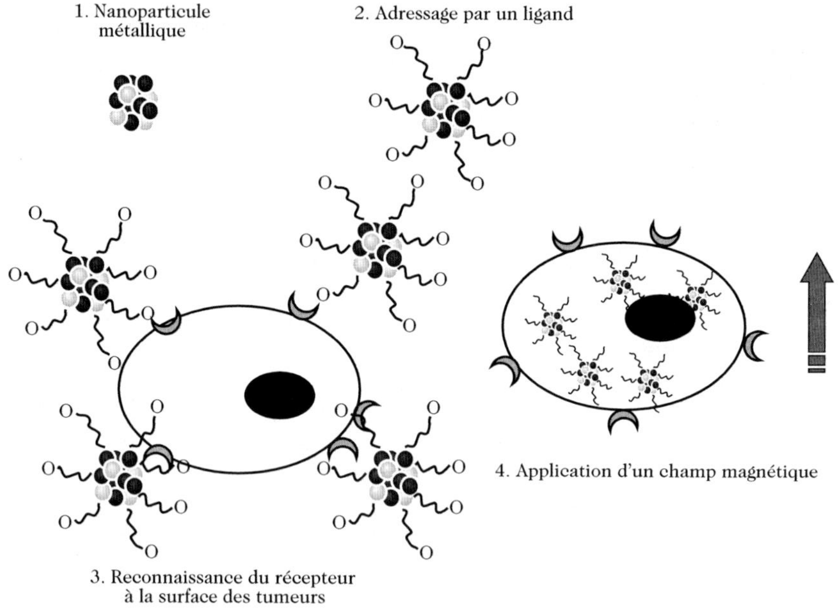

Fig. 31.9. Selective heating of tumorous cells targeted by magnetic nanoparticles

31.4 Zeolitic Membranes for Separation Processes and Catalytic Reactors

31.4.1 Introduction

In this section, it is the pores, i.e., empty spaces within a solid matrix, that have nanometric dimensions. The area/volume ratio of these pores is high and any molecules located within these spaces will be influenced by the pore walls. If we imagine a barrier containing such pores and separating two media, the exchanges between these two media will occur via the pores and can therefore be controlled by specific interactions with the pore walls. This description corresponds to what happens in a porous membrane, used in separation applications.

According to the directives of the International Union of Pure and Applied Chemistry (IUPAC), porous membranes are classified by the pore diameter D_p as follows:

- macroporous membranes $D_p > 50$ nm,
- mesoporous membranes $2 < D_p < 50$ nm,
- microporous membranes $D_p < 2$ nm,
- ultramicroporous membranes $D_p < 0.7$ nm.

It is in the latter class, with dimensions close to molecular sizes, that one can observe selective transport effects by molecular sieving or by specific interaction with the pore walls.

31.4.2 Microporous Membranes

The material making up the membrane can be either organic (a polymer) or inorganic (usually ceramic). Although organic membranes are usually chosen, ceramic membranes with their much higher thermal resistance have attracted a great deal of interest and have potential for many applications. In the following, we shall be concerned with this type of material.

Micropores can be made, as in sol–gel processes, by bringing together solid particles of well-controlled dimensions. It is then the interparticle spaces that constitute the pores. However, size and shape variations in the particle population will lead to a size distribution in the pores, and this will be unfavourable as far as transport selectivity is concerned. A better approach is to synthesise membranes in which the pores are no longer produced by textural effects, but are due to the very structure of the material used, in which case the pores have perfectly well-defined size. This explains the recent development of zeolitic membranes. Indeed, depending on the type of zeolite, one can select the pore size (in the range 0.3–1.2 nm), and also the possibilities for substitution and exchange, which will modify the adsorption properties, and hence also the transport properties.

31.4.3 Zeolitic Membranes: Synthesis and Characterisation

Most ways of synthesising zeolitic membranes are based on the formation of a thin, continuous film of zeolitic material on a macroporous ceramic support. This support, generally based on alpha alumina, provides the mechanical rigidity of the zeolitic film, whilst contributing negligibly to the resistance to matter transfer. The zeolitic film itself is obtained either by direct hydrothermal synthesis or by growth from seeds previously deposited on the macroporous support.

It should be emphasised that the presence of defects in the film (in principle, any pore going right through with diameter greater than the zeolitic pore) will lead to a rapid reduction in selectivity during separation applications, the lesser resistance to transfer through defects favouring non-separative transport. It has been shown that, if the cumulative surface area of defects reaches $1/10^6$ of the membrane area, the membrane will lose all ability to carry out separation. Consequently, the formation of high quality membranes over areas of the order of several square meters is rather difficult. Moreover, during applications, these films may be subject to heat cycles causing long-range tension which, added to effects due to differences in expansion coefficient between the film and the support, can produce fractures in the separating layer.

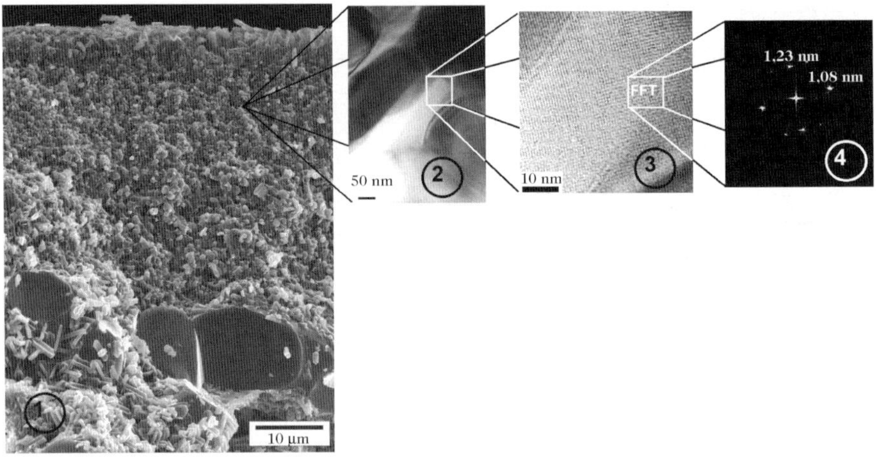

Fig. 31.10. Electron microscope study of a zeolite–alumina composite membrane. Courtesy of M. Aouine, IRC-CNRS, France. (1) SEM image showing the composite structure in cross-section. The tubular support is made from different layers of alpha alumina with variable grain size. (2) TEM image showing the continuum between the zeolite (*light area*) and the alumina (*dark area*). (3) TEM image showing the crystal lattice of the zeolite. (4) Fast Fourier transform (FFT) showing that the material present in the pores has the structure of the Mobil Five zeolite (MFI)

Another method of synthesis, called pore obstruction, developed at the Institut de Recherches sur la Catalyse, in Lyons, France, involves growing zeolite crystals in the macropores of a ceramic support in such a way as to form a composite support–zeolite continuum. The maximal defect size is determined by the support pores and the presence of isolated zeolite crystals in the support matrix limits tension. Characterisation by transmission electron microscope shows that the expected composite material does indeed form (see Fig. 31.10).

31.4.4 Application to Gas Separation

Transmembrane transfer or permeation has been studied for different gases and mixtures of gases. Figure 31.11a shows the permeance (flux divided by the pressure difference across the membrane) of different gases as a function of the temperature. The data obtained show that transfer across this material does not occur by the Knudsen diffusion mechanism which characterises transport in mesopores. In particular, the maximum observed for n-butane and the differences observed for n- and isobutane isomers reveal the significant influence of the diffusivity and adsorption parameters, typical of microporous media. A model involving these parameters provides a perfect fit for the permeance curves shown in Fig. 31.11a.

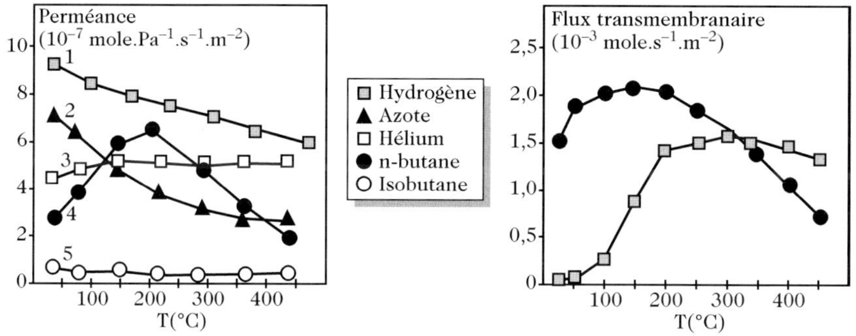

Fig. 31.11. Permeance of pure gases (*left*) and separation of a hydrogen/butane mixture (*right*) as a function of temperature

Owing to the different interactions between the solid (the wall of the zeolite pore) and the components of a mixture, the behaviour of pure gases is not always sufficient to predict the separating ability of the system. Indeed, although hydrogen diffuses more quickly than butane in the pure gas state (see Fig. 31.11a), the separation of a H_2/nC_4H_{10} mixture (see Fig. 31.11b) shows that at room temperature, only the butane crosses the membrane. This is due to the higher adsorption of butane in the zeolite, and the fact that, once adsorbed, the hydrocarbon molecule (0.43 nm) no longer allows the hydrogen molecule (0.29 nm) to pass through the MFI zeolite pore (0.55 nm). There is here a direct effect of the nanometric dimensions of the material. When the temperature is raised, adsorption of n-butane is reduced, and with it its blocking effect, thus allowing hydrogen to get through the membrane. The composition of the cross-membrane flow evolves and, at high temperature, it is the hydrogen that dominates the permeate (see Fig. 31.11b).

In separation experiments, apart from adsorption effects, differences in diffusivity may play a role. For example, for a mixture of n- and isobutane isomers, which have very similar adsorption characteristics, it is the much higher diffusivity of n-butane (10^3 times higher) which favours selective transport of the linear species.

Finally, when one component in the mixture has size greater than the dimensions of the zeolite pore, the molecular sieve effect can yield a very high separation selectivity. The performance obtained is directly related to the presence of defects and this type of test can be used to assess the quality of the material.

31.4.5 Application to a Catalytic Reactor

These selective membranes can be used with a catalyst in a catalytic membrane reactor (CMR). One of the best known applications of the CMR involves thermodynamically controlled reactions. Indeed, in an equilibrium reaction,

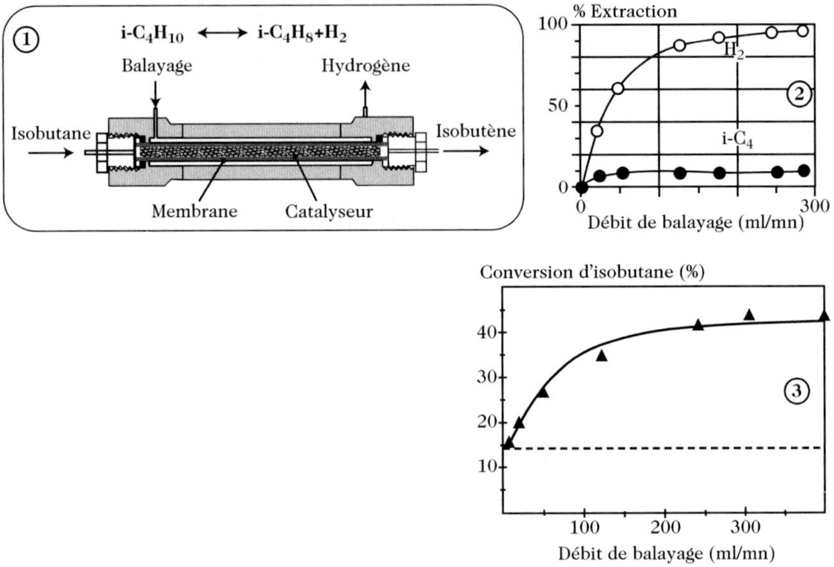

Fig. 31.12. Dehydrogenation of isobutane in a catalytic membrane reactor (CMR). (1) Reaction and diagram of the CMR. The platinum-based catalyst is placed on a fixed bed inside the membrane tube. A purge gas carries off the species extracted by the membrane. (2) Percentage extraction of hydrogen (produced by the reaction) and isobutane (introduced reactant) as a function of the stream rate. (3) CMR performance. When there is no gas feed, the permeation driving force is zero and we recover the conversion rate for the conventional reactor determined by thermodynamics (*dashed line*). When the gas flows, the performance of the CMR may be enhanced by a factor of three over the conventional reactor

extraction of a product shifts the equilibrium in a direction that favours conversion of the reactants. Figure 31.12 illustrates the CMR setup with a zeolite membrane for the dehydrogenation of isobutane. Under the reaction conditions (450°C), little isobutane is adsorbed and it is the reaction product H_2 which has the greatest permeation rate. Shifting the equilibrium by selective extraction of hydrogen allows the CMR to perform significantly better than a conventional reactor.

31.5 Metal Nanoparticles and Catalysis

Noble metals such as Pd, Pt, and Rh are widely used as catalysts. However, these metals are very expensive and it is of interest to obtain extremely active catalysts with a fairly high ratio of reactive surface area to volume in order to economise on the use of such materials. Nanoparticles satisfy these requirements. In order to put them to use, active metal nanoparticles must be supported on a powdered material, itself in the form of beads, platelets, etc., or spread on a physical support of a kind determined by the relevant process.

As an example, consider the catalytic combustion of methane and its application in radiant panels, catalytic burners emitting infrared energy. There is currently a great increase in the use of methane. It represents today some 14.5% of our total energy consumption (25% of fossil fuels). It is one of the largest sources of energy on the planet and is less polluting than carbon or petroleum in the sense that it contains few sulfur compounds and produces less CO_2 for an equivalent amount of energy released.

The point about catalytic combustion as compared with conventional combustion with a flame is the elimination of undesirable reactions emitting pollutants such as the NO_x, which are harmful to the environment. Indeed, NO_x is only produced at temperatures above 1 300–1 500°C, temperatures at which the probability of nitrogen dissociation becomes signficant. Such temperatures are exceeded in certain regions of a flame but not in a catalytic burner (without flame). It is in fact the energy released by the reaction which serves as a source of heat and radiation:

$$CH_4 + 2O_2 \longrightarrow CO_2 + 2H_2O + \text{energy (890 kJ/mole)}.$$

The total energy supplied thus depends a priori on the number of active sites and their reactivity. Furthermore, the activity must be sufficient to convert all the methane. The energy supplied is then controlled by the inflow of the methane/air mixture, heating the support structure to a given temperature.

In practice, the temperature is limited to about 1 000°C, above which the stability of the catalysts is generally rather poor. However, such a temperature is rather high, and the active phase (Pd or Pt for the catalytic combustion of methane) must be deposited on thermostable supports with good heat transfer properties to avoid excess local heating, which may sinter the active phase, and to obtain the required temperature uniformity over the whole surface of the radiant panel.

In the following, we discuss the characteristics and performance with respect to total oxidation of methane for a Pd catalyst supported on Si_3N_4 (a thermoconducting material able to resist high temperatures) and spread on a monolithic cordierite support.

31.5.1 Synthesis and Characterisation of Pd/Si_3N_4 Catalysts

The Si_3N_4 powder used has a specific surface area of 6–7 m^2/g. The grains are mainly crystallised (87%) in the α phase. The metal particles are obtained by impregnating the support with a predetermined amount of a palladium precursor, palladium acetylacetonate $Pd(C_5H_7O_2)_2$, dissolved in toluene. The impregnated powder is then dried and subjected to a series of heat treatments in order to produce the final catalyst. A first annealing in argon at 500°C decomposes the precursor. The particles form without excessive growth.

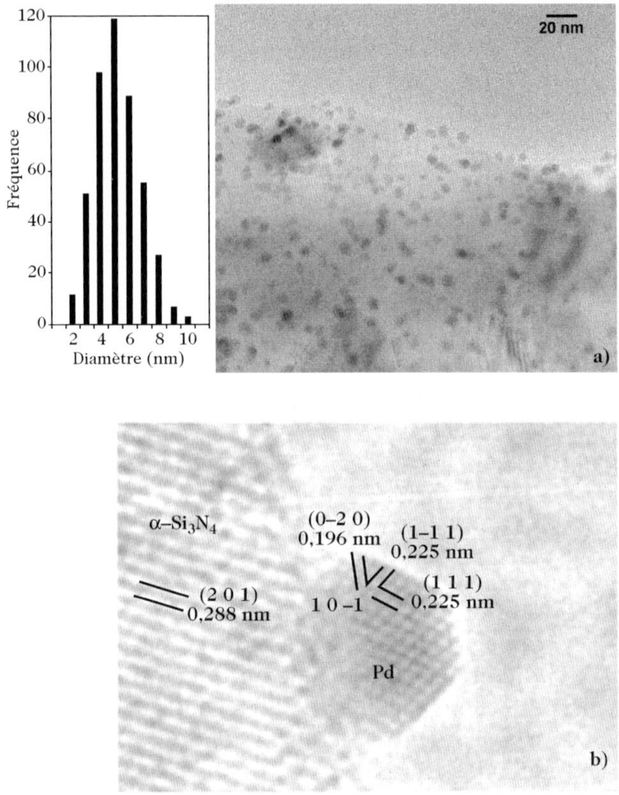

Fig. 31.13. (**a**) Size histogram and TEM image of the catalyst Pd/Si_3N_4. (**b**) High-resolution TEM image of a Pd particle supported on a grain of $–Si_3N_4$. (201) planes are visible on the support and the (111), (1 −1 1), and (0 −2 0) atomic planes visible on the particle allow determination of its orientation as [10 −1]

This stage is followed by calcination at 350°C in oxygen to eliminate carbon species around the particles. This leads to the formation of palladium oxide particles. To obtain metal particles, they have to be reduced in hydrogen at 500°C. Chemical analysis of the resulting catalyst determines the exact metal content to be 1.35 wt.% in the case below. Observation of the final catalyst by transmission electron microscope (TEM) shows that the particles are distributed uniformly on the support (see the TEM image in Fig. 31.13a) and have average size 5.2 nm (see the histogram in Fig. 31.13a). The structure determined by high-resolution TEM in Fig. 31.13b confirms that we do indeed obtain metal particles with the fcc structure of Pd and, for the main part, with the truncated cubo-octahedral shape.

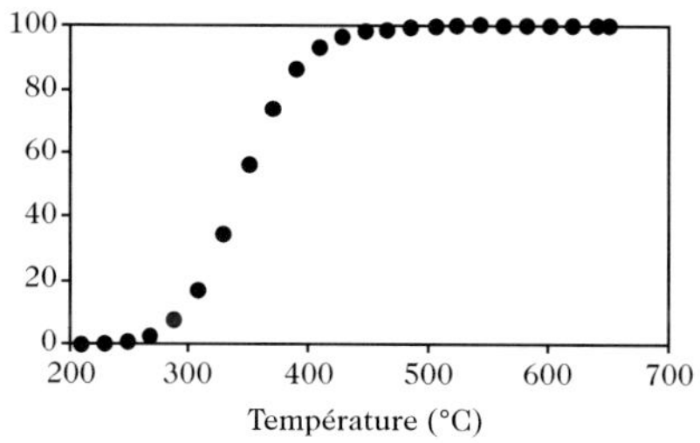

Fig. 31.14. Conversion of methane as a function of temperature in the total oxidation reaction with Pd catalyst (1.35 wt.%)/Si_3N_4

31.5.2 Total Oxidation of Methane: Implementation in the Laboratory

Having determined the physicochemical characteristics of the catalyst, viz., morphology, structure, metal content, etc., its performance as a catalyst, that is, its ability to bring about total oxidation of methane in the present case, must be tested. The reactor used in the laboratory test of the catalyst is a continuous tubular reactor containing 200 mg of catalyst. The gas mixture (N_2 78%, O_2 19.5%, CH_4 2.5%) flows through the reactor and hence the catalyst at a flow rate of 100 ml/min while the temperature of the reactor is increased in a controlled manner (typically at 1°C/min) by means of a cylindrical ceramic oven. The consumption of O_2 and CH_4, and also the formation of CO_2 (the water formed condenses at the reactor outlet), are monitored by mass spectrometry during the temperature ramp. To stabilise the catalyst, the temperature is held at 650°C for three hours. The catalytic test then consists in measuring the conversion of methane as a function of the temperature up to 650°C. The activity curve (conversion of CH_4 as a function of the temperature, as shown in Fig. 31.14) shows that the reaction ignites at around 260°C, to reach 50% conversion at 340°C, and then 100% from 485°C. This kind of activity curve can serve as a reference to monitor the aging of a catalyst which has been used for long periods at high temperatures (800–1 000°C), by repeating this catalytic test under the same conditions.

Under our reaction conditions ($T \leq 650$°C and 40 mbar oxygen), the catalyst particles are oxidised during the whole reaction cycle and it is thus palladium oxide that constitutes the active phase.

31.5.3 Application to Radiant Panels (Infrared Energy Emission)

The transition from a laboratory catalyst to an industrial product requires, among other things, a preparation adapted to the specific application. In the case of radiant panels like the Catherm panels developed by SUNKISS, where methane combustion is used to obtain infrared radiation emission, a suitable preparation involves coating ceramic panels with the catalyst. In

Fig. 31.15. Cordierite panels 150 mm × 150 mm × 8 mm with 0.8-mm channels. *Insert*: Close-up of channels

fact, cordierite panels with dimensions 150 mm × 150 mm × 8 mm and 0.8-mm channels (400 mesh), like those shown in Fig. 31.15, must be coated with a thin film of catalyst (about 50 μm thick) which must adhere perfectly to the cordierite surface, while being mechanically robust and uniformly distributed over the surface of the channels. There are two ways of coating the surface:

- coating with a powder support (Si_3N_4) then impregnating with a Pd precursor,
- direct coating by the Pd/Si_3N_4 catalyst.

For the panels to operate, an ignition stage is required to get the reaction going, whereupon it maintains itself by means of the energy it produces. The air/methane mixture goes through the channels where the methane is converted in the presence of the catalyst. The large number of channels and the uniformity of the catalyst distribution in the channels leads to high uniformity in the infrared emission.

The infrared radiation can be used in different applications such as drying paints or resins, or thermoforming objects. The necessary condition is that the irradiated products should absorb infrared radiation in the same wavelength range as the emission. The fact that there is no flame makes such processes feasible in sensitive environments, e.g., repairs to fuel tanks.

Catalytic systems using metal nanoparticles supported on tubular monoliths, like the ones just described, have a great many applications, especially in the transformation of pollutants such as volatile organic compounds (VOC) in chimneys and catalytic post-combustion for gas-, diesel-, or petrol-fuelled vehicles.

As an example, the catalytic converters on vehicles with petrol-burning internal combustion engines use so-called three-way catalysts in which three different metals (Pt, Pd, and Rh) are deposited on powder supports (modified by other oxides and promoters) coating structures of cordierite type. These are used to:

- eliminate CO by oxidising it to CO_2 on Pt,
- eliminate NO_x by reducing NO with CO into CO_2 and N_2 on Rh,
- burn unburnt hydrocarbons by oxidising them to CO_2 and H_2O on Pd.

References

Nanotechnologies and New Medicines

1. O. Diat, D. Roux: J. de Phys. II France **3**, 9 (1993)
2. H. Teixeira, C. Dubernet, V. Rosilio, S. Benita, C. Lepault, I. Erk, P. Couvreur: New bicompartmental structures are observed when stearylamine is mixed with triglyceride emulsions, Pharm. Res. **17**, 1329–1332 (2000)
3. R.L. Juliano: Factors affecting the clearance kinetics and tissue distribution of liposomes, microspheres and emulsions, Adv. Drug. Deliv. Rev. **2**, 31–54 (1988)
4. D. Roux, P. Chenevier, T. Pott, L. Navailles, O. Regev, O. Mondain Monval: Current Medical Chemistry **11**, 169–177 (2004)
5. P. Couvreur, P. Tulkens, M. Roland, A. Trouet, P. Speiser: Nanocapsules: A new type of lysosomotropic carrier, FEBS Letters **84**, 323–326 (1977)
6. N. Chiannilkulchai, N. Ammoury, B. Caillou, J.P. Devissaguet, P. Couvreur: Hepatic tissue distribution of doxorubicin-loaded nanoparticles after i.v. administration in reticulosarcoma M5076 metastasis-bearing mice, Cancer Chemother. Pharmacol. **26**, 122–126 (1990)
7. O. Balland, H. Pinto-Alphandary, A Viron, E. Puvion, A. Andremont, P. Couvreur: Intracellular distribution of ampicillin in murine macrophages infected with *Salmonella typhimurium* and treated with (3H)ampicillin-loaded nanoparticles, J. Antimicrob. Chemother. **37**, 105–115 (1995)
8. P.L. Felgner, T.R. Gadek, G.M. Ringold, M. Danielsen: Proc. Natl Acad. Sci. USA **84**, 7413 (1987)

Magnetic Nanoparticles and Biomedical Applications

9. C.R. Martin, D.T. Mitchell: Nanomaterials in analytical chemistry, Analytical Chemistry News and Features 322 (1998)
10. P. Tartaj, M. del Puerto Morales, S. Veintemillas-Verdaguer, T. Gonzalez-Carreño, C.J. Serna: The preparation of magnetic nanoparticles for applications in biomedicine, J. Phys. D: Appl. Phys. **36**, R182 (2003)
11. A.F. Davila, G. Fleissner, M. Winklhofer, N. Petersen: A new model for a magnetoreceptor in homing pigeons based on interacting clusters of superparamagnetic magnetite, Physics and Chemistry of the Earth **28**, 647 (2003)

Index